Physics and Chemistry of Finite Systems:
From Clusters to Crystals

NATO ASI Series

Advanced Science Institutes Series

A Series presenting the results of activities sponsored by the NATO Science Committee, which aims at the dissemination of advanced scientific and technological knowledge, with a view to strengthening links between scientific communities.

The Series is published by an international board of publishers in conjunction with the NATO Scientific Affairs Division

A Life Sciences	Plenum Publishing Corporation
B Physics	London and New York
C Mathematical	Kluwer Academic Publishers
and Physical Sciences	Dordrecht, Boston and London
D Behavioural and Social Sciences	
E Applied Sciences	
F Computer and Systems Sciences	Springer-Verlag
G Ecological Sciences	Berlin, Heidelberg, New York, London,
H Cell Biology	Paris and Tokyo
I Global Environmental Change	

NATO-PCO-DATA BASE

The electronic index to the NATO ASI Series provides full bibliographical references (with keywords and/or abstracts) to more than 30000 contributions from international scientists published in all sections of the NATO ASI Series.
Access to the NATO-PCO-DATA BASE is possible in two ways:

– via online FILE 128 (NATO-PCO-DATA BASE) hosted by ESRIN, Via Galileo Galilei, I-00044 Frascati, Italy.

– via CD-ROM "NATO-PCO-DATA BASE" with user-friendly retrieval software in English, French and German (© **Springer Science+Business Media Dordrecht** 1989).

The CD-ROM can be ordered through any member of the Board of Publishers or through NATO-PCO, Overijse, Belgium.

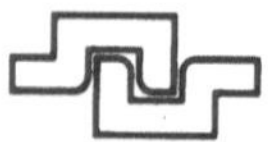

Series C: Mathematical and Physical Sciences - Vol. 374

Physics and Chemistry of Finite Systems: From Clusters to Crystals

Volume II

edited by

P. Jena
S. N. Khanna

and

B. K. Rao

Department of Physics,
Virginia Commonwealth University,
Richmond, VA, U.S.A.

Springer-Science+Business Media, B.V.

Proceedings of the NATO Advanced Research Workshop on
Physics and Chemistry of Finite Systems: From Clusters to Crystals
Richmond, VA, U.S.A.
October 8–12, 1991

Library of Congress Cataloging-in-Publication Data

```
Physics and chemistry of finite systems : from clusters to crystals /
  edited by P. Jena, S.N. Khanna, and B.K. Rao.
      p.   cm. -- (NATO ASI series. Series C, Mathematical and
  physical sciences ; vol. 374)
    "Published in cooperation with NATO Scientific Affairs Division."
    Includes index.
```

ISBN 978-94-017-2647-4   ISBN 978-94-017-2645-0 (eBook)
DOI 10.1007/978-94-017-2645-0

```
    1. Atomic structure--Congresses.   2. Electronic structure-
  -Congresses.  3. Crystals--Electrical properties--Congresses.
  4. Crystals--Optical properties--Congresses.  5. Cluster analysis-
  -Congresses.   I. Jena, P.  II. Khanna, S.N.  III. Rao, B. K.
  IV. Series: NATO ASI series.  Series C, Mathematical and physical
  sciences ; no. 374.
  QC172.P48   1992
  530.4'1--dc20                                          92-16307
```

ISBN 978-94-017-2647-4

Printed on acid-free paper

ELECTRICAL AND OPTICAL PROPERTIES

CLUSTER REACTIONS AND CLUSTER–SUPPORT INTERACTIONS

CLUSTER ASSEMBLIES

MATERIALS INVOLVING CARBON

TRANSPORT PROPERTIES AND ELECTRONIC STRUCTURE OF QUASICRYSTALS

B. D. BIGGS, S. J. POON, and F. S. PIERCE
Department of Physics
University of Virginia
Charlottesville, VA 22901.

ABSTRACT. Transport and specific heat measurements will be discussed on selected icosahedral crystals and their related structural phases to examine the effects of randomness, icosahedral symmetry, and atomic potential on electronic properties. We report the observation of Fermi-surface and Jones-zone boundary interactions and stress their importance in the stability of icosahedral phases. Unusual electronic structure and conductivity behavior in highly ordered icosahedral crystals will be discussed. The effects of structural order on electronic properties and the propensity toward a metal-insulator transition in the AlMnSi, AlCuFe, and AlCuRu systems are ascribed to band-structure effects. Transport in decagonal phases will also be discussed.

1. Introduction

Since the discovery of a metallic phase with icosahedral point group symmetry and no translational symmetry[1], there has been much experimental and theoretical work on the properties of these icosahedral (i-) materials. Studies of these metastable disordered quasicrystals revealed the metallic-glass-like nature of their electron transport properties[2], and also the prominence of the Fermi-surface-Jones-zone (FSJZ) interaction and its importance to the stability of these materials[3]. With the discovery of the stable AlCuFe and AlCuRu i-phases[4] and the confirmation of their long range order[5], it became possible to investigate the effects of quasiperiodicity, randomness, and atomic potential on the electronic properties. Unusual electronic properties are observed, including a low electronic density of states, high resistivities, large Hall coefficients (R_H) which are strongly temperature dependent, and large values of the thermoelectric power S with unusual temperature dependence. In addition, a recent study of the crystalline approximant α-AlMnSi phase reveals electronic properties similar to the stable i-phases.[6] It is found that band structure effects, which are enhanced by the icosahedral point group symmetry, are responsible for the unusual semi-metallic electronic properties seen in these ordered materials. These findings are further confirmed by the suppression of these unusual properties in the amorphous phases and disordered i-phases and in the behavior of the electron transport properties of the i-AlCuRuSi system.

P. Jena et al. (eds.), Physics and Chemistry of Finite Systems: From Clusters to Crystals, Vol. II, 819–828.
© *1992 Kluwer Academic Publishers.*

2. Fermi-surface Jones-zone (FSJZ) Interactions and Their Importance to the Stability of Disordered i-phases

It was suggested[7, 8] that the icosahedral structure might be stabilized through a reduction in the electronic energy when the Fermi-surface interacts with the Jones-zone boundaries. This phase stability criterion is known as the Hume-Rothery rule. In a quasicrystal, there are no Brillouin zones in the conventional sense, but one can construct pseudo-Jones-zones by taking the perpendicular bisecting planes of reciprocal lattice vectors associated with prominent x-ray diffraction peaks. As a result of this FSJZ interaction, a pseudogap is produced in the electronic density of states (DOS), and the positioning of the Fermi level near this minimum of the pseudogap ($2k_F \simeq |\vec{G}|$, where k_F is the Fermi wave vector and $\vec{G}$ is a reciprocal lattice vector) greatly reduces the electronic energy, enhancing the stability of the i-phase. Smith and Ashcroft[9] calculated the DOS of i-Al and found strong Van Hove-like singularities due to the large multiplicity of strong peaks in the structure factor.

Vaks, *et. al.*[8] rephrased this condition in terms of the average valency of the i-phases, since Z is related to $2k_F$. A critical valence Z_c can be calculated for each $\vec{G}$ once the density of electrons is known. Wagner, *et. al.*[3] conducted a study of i-phases in the GaMgZn and AlCuMg systems in an attempt to observe these effects experimentally. The ranges of Z experimentally available in these systems are $\sim$2.17–2.26 and 2.24–2.54 electrons/atom, respectively. The relevant Z_c for these ranges are 2.17 and 2.42, corresponding to the (222100) and (311111)/(222110) reciprocal lattice vectors in the six-dimensional indexing scheme[10]. The Jones-zones corresponding to these vectors are nearly spherical due to the large multiplicity of the $\vec{G}$. Therefore, a spherical Fermi-surface interacting with such a Jones-zone can lead to the disappearance of a significant portion of the free electron Fermi-surface, having dramatic effects on the electronic structure and transport properties.

The results of Wagner, *et. al.*, along with the relevant Jones-zones, are shown in Figure 1. While the phason-induced disorder in these systems will tend to distort the JZ and reduce the effects on the electronic properties, features in the electronic properties do occur close to the predicted values Z_c. These features are explained through the FSJZ interaction as alloy composition (Z) and hence k_F are changed. The effect of the FSJZ interaction in the ordered i-phases with sp-d hybridization will be discussed later.

3. Properties of the Ordered i-crystals

3.1 EXPERIMENTAL RESULTS ON i-PHASES

The structurally ordered, stable i-phases of AlCuFe[11, 12, 13] and AlCuRu[14] were the first i-phases to show barely or semi-metallic behavior, which is not expected in alloys composed of good metals. Because of the high degree of structural order in these materials, meaningful comparisons with theories can be made.

i-Al$_{65}$Cu$_{20}$Ru$_{15}$ has the lowest γ value (0.11 mJ/g-at.K^2) found to date in the i-phases, which is only 10% of the free electron value. The rapid variation of γ with alloy composition ($\sim$0.23 mJ/g-at.K^2 for Al$_{68}$Cu$_{17}$Ru$_{15}$) suggests the presence of distinct features in the DOS in the narrow region of i-phase formation[14].

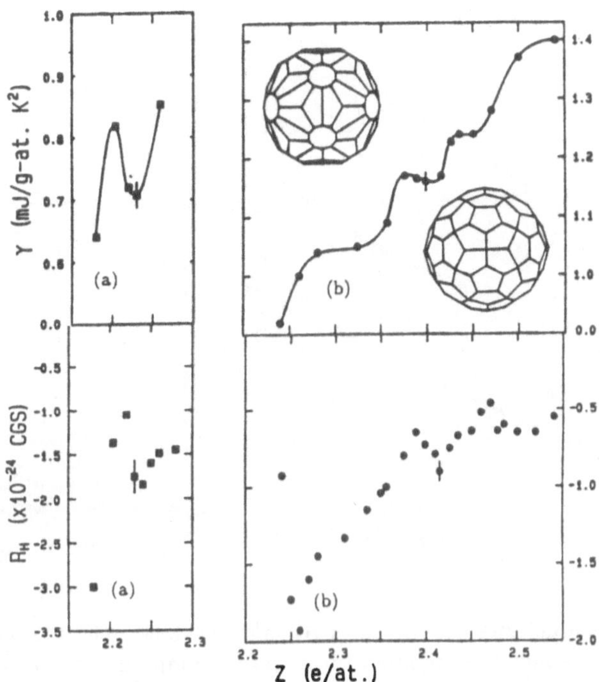

Figure 1: Electronic properties versus the average valence Z of i-alloys of (a) GaMgZn and (b) AlCuMg. Room temperatures values of R_H, and low temperature values of γ are shown, as are the relevant Jones-zone boundaries. S vs. Z shows similar features.

The conductivity data $\sigma(T)$ for i-AlCuRu are shown in Figure 2(a). The $\sigma(300K)$ values are already near the "minimum metallic conductivity" of 200 Ω^{-1}cm^{-1}[15], indicating that these phases are near the metal insulator transition. At low T ($< 5K$), $\sigma(T) \sim \sqrt{T}$, which can be attributed to electron interaction effects, while $\sigma(T) \sim T$ for 10K$<T<$30K as a result of weak localization effects. For higher T, $\sigma(T)$ has a positive curvature which extends to room temperature which is reminiscent of those observed in heavily doped semiconductors[16]. Magnetoresistivity data, which yields positive $\Delta\rho/\rho$ as shown in Figure 2(b), can be successfully analyzed according to the weak localization theory, with reasonable values of the inelastic and spin-orbit scattering times τ_i and τ_{so}[14].

The i-AlCuRu system also shows striking features in $R_H(T)$ and $S(T)$. A large, negative low-T R_H is measured, indicating an effective carrier density 2–3 orders of magnitude lower than the free electron value. In addition, strong temperature dependences of R_H are observed, as shown in Figure 3(a), including a sign change in the x=20 sample. Large values of S as well as unusual behavior in $S(T)$ are also observed, as shown in Figure 3(b), reflecting the change in sign of $R_H(T)$ for the x=20 sample.

i-phase samples of Al$_{63.5}$Cu$_{24.5}$Fe$_{12}$ prepared by annealing melt spun ribbons show similar unusual behavior[13]. $\rho(300K) \sim 2800$ $\mu\Omega$-cm and $\rho(0.5K)/\rho(300K) \sim 1.5$ are observed. By fine tuning the composition, Klein, *et. al.* have found $\rho(4.2K) \sim 10,000$ $\mu\Omega$-cm in

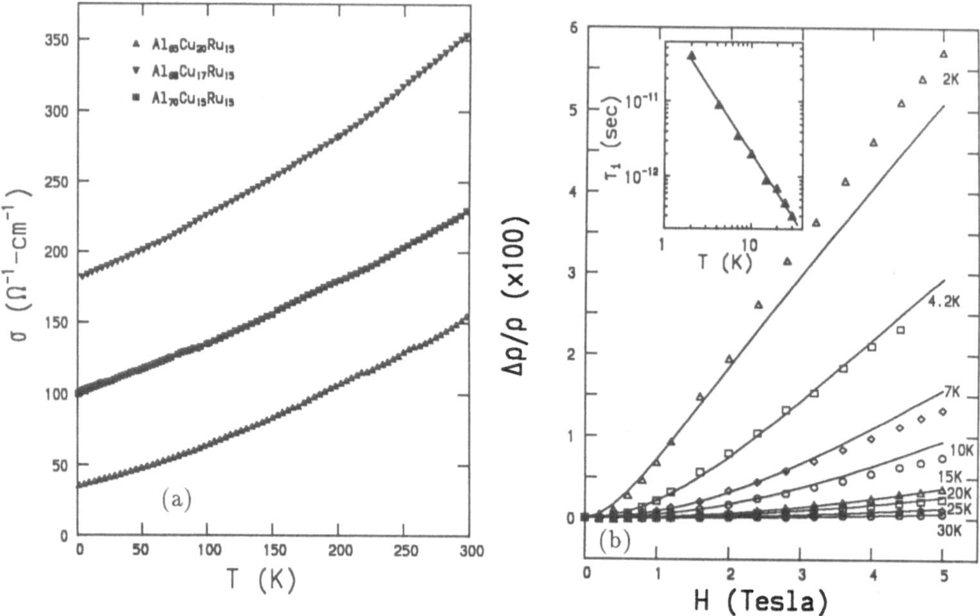

Figure 2: (a) Conductivity vs. T for i-AlCuRu system. (b) Magnetoresistivity for i-Al$_{65}$Cu$_{20}$Ru$_{15}$ at the indicated temperatures. Inset: temperature dependence of the inelastic scattering time τ_i.

i-Al$_{62.5}$Cu$_{25}$Fe$_{12.5}$[12]. $\sigma(T)$ shows behavior very similar to that of i-AlCuRu. Specific heat measurements yield a low $\gamma \sim 0.31$ mJ/g-at.K^2. $R_H(T)$ and $S(T)$ show unusual behavior reminiscent of that seen in i-AlCuRu.

Studies have also been performed on the i-Al$_{65-x}$Cu$_{20}$Ru$_{15}$Si$_x$ system, where $x \leq 1$. The addition of a small amount of Si to this system causes dramatic changes in the electronic properties. Both $\rho(300K)$ and $\rho(0.5)/\rho(300K)$ change dramatically with the small addition of Si. With the addition of as little as 1% Si, $R_H(T)$, while still large and negative, is almost temperature independent, and $S(T)$ has become large and positive with metallic-glass-like temperature dependence. These results are shown in Figure 3. The implication of these results will be discussed later.

3.2 AMORPHOUS PHASES

The amorphous (a-) phases of AlCuRu and AlCuFe show strikingly different behavior from the i-phases. a-Al$_{65}$Cu$_{20}$Ru$_{15}$ sputtered films yield $\rho(300K) \sim$200–300 $\mu\Omega$-cm with $\rho(0.5K)/\rho(300K) \sim 1.14$. $R_H \sim -0.28\times10^{-24}$ cgs independent of temperature, and $S(300K) \sim +3.1\mu$V/K with metallic-glass-like $S(T)$. For a-Al$_{63}$Cu$_{25}$Fe$_{12}$ sputtered films, $\rho(300K) \sim 700$ $\mu\Omega$-cm, $\rho(0.5K)/\rho(300K) \sim 1.08$, $R_H(300K) \sim +1.1\times10^{-24}$ cgs (which shows temperature dependence and field dependence for $T < 200$K due to magnetic effects[13]) and metallic-glass-like $S(T)$ with $S(300K) \sim -2.6\mu$V/K. Therefore, structural

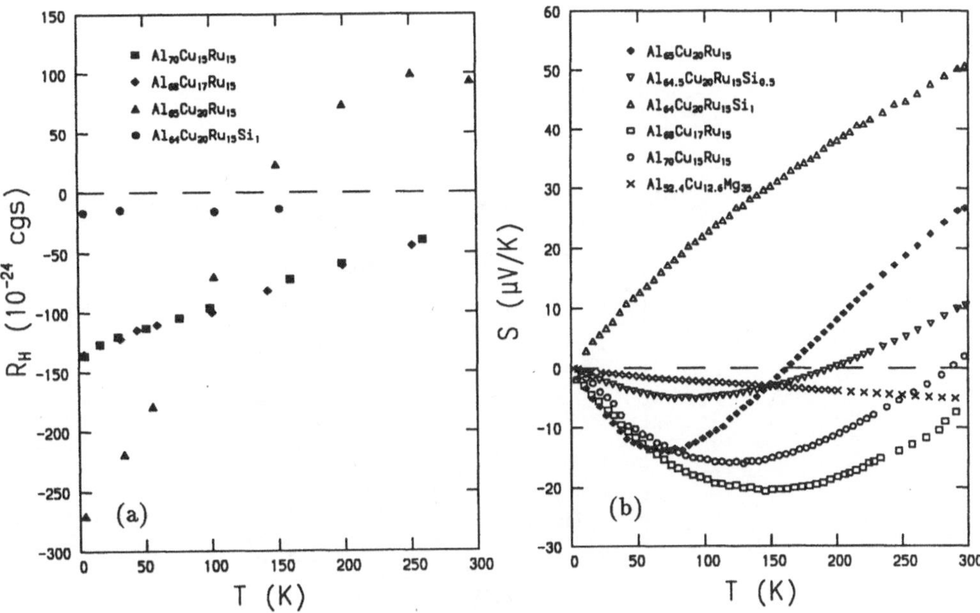

Figure 3: (a) Hall effect and (b) Thermoelectric power of the i-AlCuRu and i-AlCuRuSi systems. Free-electron-like S of i-Al$_{52.4}$Cu$_{12.6}$Mg$_{35}$ is shown for comparison in (b).

disorder restores the metallic-glass-like behavior of these systems.

3.3 PROPERTIES OF α-ALMNSI

In view of the unusual electronic properties observed in the stable i-phases, which are complex structures, the study of low-order crystalline approximant phases would be useful. The α-Al$_{72.5}$Mn$_{17.4}$Si$_{10.1}$ (1/1) crystalline approximant, which is the low-order crystalline analog of the i-Al(TM) (TM = transition metal) class[17], is an excellent choice for study, since band structure calculations for this alloy have been performed[18, 19].

Previous magnetic susceptibility measurements on α-AlMnSi have confirmed that this phase is non-magnetic[20], greatly simplifying the interpretation of the electron transport properties of this material. The annealed as-cast samples yield $\rho(300K) \simeq 3100\ \mu\Omega$-cm and $\rho(0.5K)/\rho(300K) \simeq 2$, as shown in Figure 4(a). In addition, large values of the low temperature Hall coefficient and the thermoelectric power as well as anomalous temperature dependences of R_H and S are apparent, as shown in Figure 4(b). The specific heat yields $\gamma \sim 0.6$ mJ/g-at.K^2, which is significantly lower than the $\gamma \sim 2$ mJ/g-at.K^2 reported previously[21]. All of this semi-metallic behavior is similar to that seen in the i-AlCuRu and i-AlCuFe phases.

824

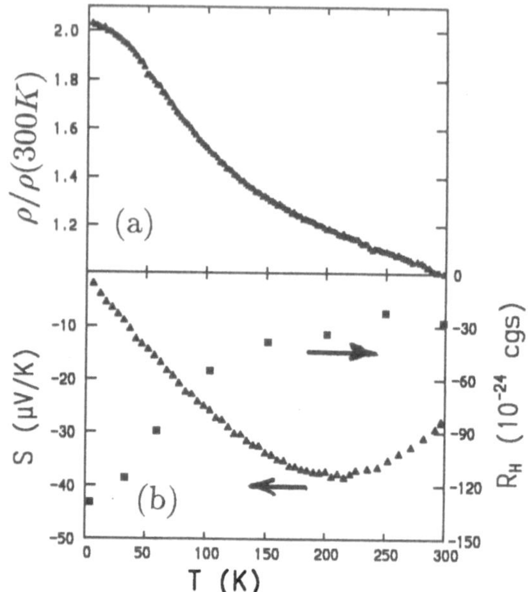

Figure 4: (a) Resistivity data for α-AlMnSi normalized to the 300K value. $\rho(300K) = 3100$ $\mu\Omega$-cm. (b) Hall effect and thermoelectric power data for α-AlMnSi.

3.4 DISCUSSION

3.4.1 FSJZ Effects in Ordered Quasicrystals. The existence of unusual electronic properties in the stable i-phases and in the α-AlMnSi phase raise questions about the origin of this behavior. Since band structure calculations of the α-AlMnSi phase exist[18, 19], this crystalline approximant phase, which contains building blocks of Mackay icosahedra which are also present in i-AlMnSi[17, 22], can shed important new light on the occurrence of these properties in the i-phase. One feature of the band structure calculations is the interplay of the FSJZ interaction and sp-d hybridization in enhancing the stability of these phases. The prominence of the pseudogap, discussed in Section 2, is enhanced in the α- and ordered i-phases due to several factors. Since these crystals are well ordered, the narrow peaks in the structure factor will result in a very pronounced pseudogap. The intensity of the relevant peak in the structure factor, which corresponds to the (611)/(532) peak in the α-phase, and the (221001) peak for the i-phases (or (442002) for FC-type i-crystals), is large which enhances the effect. In the disordered i-phases discussed earlier, the effect of the relevant peak is much more diffuse. In the i-phases, the large multiplicity of the $\vec{G}$ which is due to the unusual structure factor resulting from the icosahedral point group symmetry leads to almost spherical JZ boundaries, which in turn leads to a large fraction of the Fermi surface being involved in the FSJZ interaction, will further increase the prominence of the pseudogap features. Also, strong sp-d hybridization in the α-phase will make the FSJZ interaction more dramatic[18]. Therefore, we can expect the DOS to be complex and to

	α-Al$_{72.5}$Mn$_{17.4}$Si$_{10.1}$	i-Al$_{85-x}$Cu$_x$Ru$_{15}$ (x = 15–20)	i-Al$_{63.5}$Cu$_{24.5}$Fe$_{12}$
σ (Ω^{-1}cm^{-1})	150	100–160	220
γ (mJ/g-at. K^2)	0.6	0.11–0.23	0.31
$-R_H$ (10^{-24} cgs)	127	270–140	110
D (cm^2/s)	0.10	0.3–0.24	0.25
n (10^{20}/cm^3)	5.1	2.4–4.6	5.8
τ/τ_0	$2.5\left(\frac{m_t^*}{m}\right)$	$3.4\text{–}2.8\left(\frac{m_t^*}{m}\right)$	$3.1\left(\frac{m_t^*}{m}\right)$
$|M|^2/|M|_0^2$	$0.8\left(\frac{m}{m_t^*}\right)$	$3\text{–}1.7\left(\frac{m}{m_t^*}\right)$	$1.1\left(\frac{m}{m_t^*}\right)$
ℓ (Å)	$0.22\left(\frac{m_t^*}{m}\right)^{1/2}\ell_0$	$0.5\text{–}0.4\left(\frac{m_t^*}{m}\right)^{1/2}\ell_0$	$0.4\left(\frac{m_t^*}{m}\right)^{1/2}\ell_0$
	$\simeq 1.8\left(\frac{m_t^*}{m}\right)^{1/2}$	$\simeq 4\text{–}3.2\left(\frac{m_t^*}{m}\right)^{1/2}$	$\simeq 3.2\left(\frac{m_t^*}{m}\right)^{1/2}$
$k_F\ell$	$0.26\left(\frac{m_t^*}{m}\right)$	$0.8\text{–}0.6\left(\frac{m_t^*}{m}\right)$	$0.65\left(\frac{m_t^*}{m}\right)$

Table 1: Low temperature electron parameters for the α-AlMnSi, i-AlCuRu, and i-AlCuFe systems. The reference material is i-Al$_{52.4}$Cu$_{12.6}$Mg$_{35}$ with $\gamma \simeq 1.1$ mJ/g-at. K^2, $\ell_0 \simeq 8$ Å, $\sigma \simeq 16,000$ Ω^{-1}cm^{-1}, $k_F\ell \simeq 13$, $D \simeq 5$ cm^2/s, $\tau_0 \simeq 4.5 \times 10^{-16}$ s, $v_F \simeq 1.8 \times 10^8$ cm/s, and $n \simeq 1.3 \times 10^{23}$/cm^3.

deviate substantially from the free electron behavior near E_F.

The results of measurements on the α- and perfect i-phases as well as derived parameters are shown in Table 1. It is clear that a strong FSJZ interaction will lead to γ values much reduced from the free electron value $\gamma \sim 1$ mJ/g-at.K^2 in these alloys, which is observed. In the α-AlMnSi phase, the γ value, while much reduced from the free electron value, is substantially higher than that found in the i-phases. This is in accord with the expected enhancement of the FSJZ interaction in the i-crystal. From the low-T R_H, we can estimate the effective carrier concentration (n_{eff}), which is always an overestimate of n[6]. The effects of the low n_{eff} on σ can be estimates from the expression $\sigma = n_{eff}e^2\tau_{eff}/m_t^*$, where m_t^* is the tangential effective mass. These values can be compared to the reference alloy i-Al$_{52.4}$Cu$_{12.6}$Mn$_{35}$ (chosen because it is in the free electron regime and has a larger-than-interatomic mean free path ℓ of 8 Å), making it clear that a reduced n rather than a reduced τ_{eff}/m_t^* is responsible for the low σ in these materials. It is also clear that τ_{eff}/m_t^* is at least as large as i-AlCuMg, and since m_t^* is expected to be large, based on the highly dispersionless bands in the band structure calculation of α-AlMnSi and the logical inference of dispersionless bands in the i-phases, τ of these barely metallic systems is much larger than that of i-AlCuMg.

3.4.2 Effects of Electronic Structure on Transport. Another feature of the band structure calculation of α-AlMnSi is rapid variation of the DOS on a fine energy scale due to the large number of dispersionless bands. Although the DOS for the i-phases is not known, it is reasonable to conclude that since the i-crystals are ordered, with self-similar atomic arrangements, a rapidly varying DOS is expected. The unusual $S(T)$ and $R_H(T)$ behavior in

826

these materials supports the existence of such fine structures. Since $S(T)$ can be written[15]

$$S(T) = \frac{k_B}{e\sigma} \int \sigma(E) \left(\frac{E - E_F}{k_B T}\right) \left(\frac{\partial f}{\partial E}\right) dE, \tag{1}$$

the observed $S(T)$ can be qualitatively be explained if the Fermi level lies in an asymmetric valley of $\sigma(E)$ which is rapidly varying on a scale of ~ 20 meV. While $R_H(T)$ is much more difficult to interpret, such rapidly varying structures in the DOS can lead to unusual $R_H(T)$[6]. The existence of these structures is also supported by the fact that $S/T \propto (1/\sigma)(d\sigma/dE)|_{E_F}$ is at least an order of magnitude larger in these alloys than in non-magnetic d-band metals[23] and amorphous alloys[24], indicating narrow effective bands or high effective masses.

Further evidence of the existence of this rapidly varying DOS is found in the behavior of the transport properties in the i-AlCuRuSi system. The dramatic changes in $\rho(T)$, $S(T)$, and $R_H(T)$ with a small amount (≤ 1 at.%) of Si as compared with the i-AlCuRu system indicates a substantial change in the DOS near E_F with only a small change in composition.

3.4.3 Disorder Effects. Yet further evidence that band structure effects are responsible for these electronic properties is the effect of structural disorder in these systems. Experimental results for a-AlCuRu presented above indicate that ρ(amorphous) $\sim \rho$(i-phase)/100 and that the transport properties are metallic-glass-like. a-AlCuFe is similarly metallic-glass-like. These results indicate that structural disorder restores the metallic-glass-like properties as expected, since structural disorder will both weaken the FSJZ interaction and wash out any rapidly varying structures in the DOS. A similar conclusion is reached in the disordered i-AlMnSi phases, which show metallic-glass-like electron transport properties, when compared with the α-AlMnSi phase.

3.4.4 The Origin of Electron Localization. In order to investigate the cause of the tendency toward electron localization in these phases, it is useful to list the values of the relevant low-T electronic parameters: the scattering time τ, the Fermi surface averaged quantity $|M|^2$ where $M_{\vec{k}\vec{k}'}$ is the scattering matrix element, the mean free path ℓ $(= (\sigma/e^2)(3m_t^*/nN(E_F))^{1/2})$, and the localization parameter $k_F\ell$ (determined from $k_F\ell = 3m_t^*D/\hbar$). Fermi's golden rule states that $|M|^2 \propto 1/(\tau\gamma)$. The results are shown in Table 1. The results for $|M|^2$ indicate that $|M|^2$ is at most as large as, and depending on the ratio m/m_t^*, where m is the free electron mass, can be smaller than $|M|_0^2$ of i-AlCuMg. This indicates that the ordered systems have low scattering rates, despite the high ρ and the expected strong scattering. The finite lifetime in these structures are due to compositional disorder effects, since the approximant structures contain atomic sites of fractional occupancy. The condition for metallic behavior is $k_F\ell > 1$. It is apparent from Table 1 that if $m_t^*/m > 1$, these systems can be on the metallic side of the metal insulator transition. With the Fermi surface intersecting the JZ boundaries, small pockets of electrons and holes will result in a small effective k_F. The pseudogap will result in a much reduced v_F. These features are consistent with the fact that both $\ell = v_F\tau$ and $k_F\ell$ are reduced from their free electron values as inferred from Table 1, despite an enhanced τ. Therefore, the approach to localization is driven by a band structure mechanism, rather than an enhanced electron scattering rate.

4. Properties of Decagonal Phases

The electronic properties of the decagonal phases, with periodicity along the c-axis and quasiperiodic structure in the a-b plane, have been studied extensively. These samples generally show an anisotropic ρ, with $\rho_{qc}/\rho_c \sim 5$–10[25, 26, 27], where ρ_{qc} is the resistivity in the quasiperiodic plane and ρ_c is measured perpendicular to this plane. Anisotropic R_H and S have also been observed[27]. ρ of $Al_{65}Co_{20}Cu_{15}$ and $Al_{70}Co_{15}Ni_{15}$ has been measured up to 600K, showing a metallic ρ_c and a non-metallic ρ_{qc}[25]. Martin, *et. al.* propose a phonon-assisted tunneling mechanism to explain $\rho_{qc}(T)$. The anisotropy in the thermal conductivities κ_c and κ_{qc} has also been demonstrated[28]. The electronic contribution to κ_{qc} is almost suppressed in contrast to the electronic contribution to κ_c, while the phonon contribution is isotropic, indicating a peculiar scattering mechanism in these quasi-2D materials.

5. Conclusion

We have seen that FSJZ boundary interactions are observed in the disordered *i*-phases, indicating their importance to the stability of the *i*-phase. The unusual electronic behavior in the α-AlMnSi approximant crystal indicates that these properties which also occur in the perfect *i*-phases are due to band structure effects. There appears to be no need to invoke quasiperiodic effects *per se*, at least in the 3D system. However, the band structure effects in the *i*-phases are enhanced by the unique structure factor due to the icosahedral point group symmetry. The effect of structural disorder on the electronic properties supports this conclusion, restoring the metallic-glass-like behavior. In the *i*-phases, the existence of a rapidly varying DOS, as predicted by the band structure calculations for α-AlMnSi, is supported by the changes in the electron transport properties of the *i*-AlCuRu system with small amounts of Si added. In addition, the semi-metallic behavior in these systems is seen to be caused by the band structure mechanisms, rather than enhanced electron scattering.

Acknowledgements — This research is supported by the National Science Foundation Contract No. DMR 90-15538.

References

[1] D. Shechtman, I. Blech, D. Gratias, and J. W. Cahn, Phys. Rev. Lett. **53**, 1951 (1984).

[2] K. M. Wong and S. J. Poon, Phys. Rev. B **34**, 7371 (1986); D. Pavuna, C. Berger, F. Cyrot-Lackmann, P. Germi, and A. Pasturel, Solid State Commun. **59**, 11 (1986); K. Kimura, T. Hashimoto, and S. Takeuchi, J. Phys. Soc. Jpn. **55**, 1810 (1986).

[3] J. L. Wagner, B. D. Biggs, and S. J. Poon, Phys. Rev. Lett. **65**, 203 (1990).

[4] A. Tsai, A. Inoue, and T. Masumoto, Jpn. J. Appl. Phys. **26**, L1505 (1987); A. P. Tsai, A. Inoue, T. Masumoto, and N. Kaoka, Jpn. J. Appl. Phys. **27**, L2252 (1988).

[5] C. A. Guryan, *et. al.*, Phys. Rev. Lett. **62**, 2409 (1989).

[6] B. D. Biggs, F. S. Pierce, and S. J. Poon, to be published.

[7] J. Friedel, Helv. Phys. Acta **61**, 538 (1988).

[8] V. G. Vaks, V. V. Kamyshenko, and G. D. Samolyuk, Phys. Lett. A **132**, 131 (1988).

[9] A. P. Smith and N. W. Ashcroft, Phys. Rev. Lett. **59**, 1365 (1987).

[10] V. Elser, Phys. Rev. Lett. **54**, 1730 (1985).

[11] T. Klein, *et. al.*, Europhys. Lett. **13**, 129 (1990).

[12] T. Klein, C. Berger, D. Mayou, and F. Cyrot-Lackmann, Phys. Rev. Lett. **66**, 2907 (1991).

[13] B. D. Biggs, Y. Li, and S. J. Poon, Phys. Rev. B **43**, 8747 (1991).

[14] B. D. Biggs, S. J. Poon, and N. R. Munirathnam, Phys. Rev. Lett. **65**, 2700 (1990).

[15] N. F. Mott, *Conduction in Non-Crystalline Materials*, Oxford University Press, New York, 1987.

[16] R. F. Milligan, T. F. Rosenbaum, R. N. Bhatt, and G. A. Thomas, in *Electron-Electron Interactions in Disordered Systems*, edited by A. L. Efros and M. Pollak, page 231, North-Holland, Amsterdam, 1985.

[17] V. Elser and C. L. Henley, Phys. Rev. Lett. **55**, 2883 (1985); P. Guyot and M. Audier, Phil. Mag. B **52**, L15 (1985).

[18] T. Fujiwara, Phys. Rev. B **40**, 942 (1989).

[19] T. Fujiwara and T. Yokokawa, in *Quasicrystals*, edited by T. Fujiwara and T. Ogawa, Springer Series in Solid State Sciences, Vol. 93, Heidelberg, 1990, Springer-Verlag, to be published.

[20] J. J. Hauser, H. S. Chen, G. P. Espinosa, and J. V. Waszczak, Phys. Rev. B **34**, 4674 (1986).

[21] M. Maurer, J. Van den Berg, and J. A. Mydosh, Europhys. Lett. **3**, 1103 (1987); K. Wang, P. Garoche, and Y. Calvayrac, J. de Phys. **49(C8)**, 237 (1988).

[22] J. W. Cahn, D. Gratias, and B. Mozer, J. Phys. (France) **49**, 1225 (1988); M. Duneau and C. Oguey, J. Phys. (France) **50**, 135 (1989); M. De Boissieu, C. Janot, and J. M. Dubois, J. Phys.: Condensed Matter **2**, 2499 (1990).

[23] M. V. Vedernikov, Adv. Phys. **18**, 337 (1969).

[24] D. G. Naugle, J. Phys. Chem. Solids **45**, 367 (1984); M. A. Howson and B. L. Gallagher, Phys. Rep. **170**, 265 (1988).

[25] S. Martin, A. F. Hebard, A. R. Kortan, and F. A. Thiel, Phys. Rev. Lett. **67**, 719 (1991).

[26] S. Y. Lin, *et. al.*, Phys. Rev. B **41**, 9625 (1990).

[27] D. L. Zhang, *et. al.*, Phys. Rev. B **41**, 8557 (1990).

[28] D. L. Zhang, *et. al.*, Phys. Rev. Lett. **66**, 2778 (1991).

ELECTRONIC DISTRIBUTIONS OF QUASICRYSTALLINE PHASES

Esther BELIN
Laboratoire de Chimie Physique UA 176
11 rue Pierre et Marie Curie
75231 Paris Cedex 05
France

ABSTRACT. Electronic distributions of various quasicrystalline phases were investigated by Soft X-ray Spectroscopy techniques and compared to those of related phases of different structural states. Complete description of the valence band distributions were obtained. In Al-Mn and Al-Mn-Si alloys, Al 3sp and Mn 3d states are in interaction near and at E_F. In Al-Cu-Fe and Al-Cu-Fe-Cr samples, Fe 3d states interact with low-lying Al 3sp states and Cu 3d states interact mainly with Al 3p middle-band states. It is salient that at E_F a low Al sp DOS and a notable pseudo-gap are observed. Good correlation is found with resistivity values.

1.   Introduction

Quasicrystalline phases are known to exhibit specific crystallographic properties which have led to numerous theoretical as well as experimental investigations. Many studies have also been carried out to ascertain the electronic properties from resistivity, specific heat, and other measurements (see for ex. Klein *et al.* 1990, Berger *et al* 1991) or theoretical investigations of the electronic structure (see for ex. Marcus 1986, Smith and Ashcroft, 1987, Fujiwara and Yokokawa 1991). Recently, calculations in three dimensions of the densities of states (DOS) distributions based on realistic crystalline approximants have become available (Fujiwara 1990). More insight in the understanding of the electronic properties could arise from the determination of how the electronic distributions in these materials are organized .

Several experimental techniques can be applied to obtain the DOS of a material. Soft X-ray Emission (SXES) or photoabsorption (SXAS) spectroscopies offer the possibility to describe separately occupied and (or) unoccupied electronic distributions of a given s,p,d..symmetry around each kind of atoms in the solid; the information collected concern the bulk material. We have used these techniques to achieve partial valence and conduction states distributions in various alloys such as Al-Mn, Al-Cu-Fe, ... according to their structural state and we report the results for the quasicrystals and related crystalline or amorphous counterparts. In this paper, we give a short description of the experimental procedure, then we present and discuss the various results and propose a picture of the valence band states distributions.

P. Jena et al. (eds.), Physics and Chemistry of Finite Systems: From Clusters to Crystals, Vol. II, 829–838.

2.   Experiments

Al-Mn-Si, Al-Cu-Fe and Al-Cu-Fe-Cr alloys were obtained from rapid solidification based techniques (RS); Al-Mn samples were prepared by either RS or ion beam mixing of multilayers (IBMM); all the samples were controlled by X-ray or neutron diffraction or transmission electron microscopy as reported elsewhere (see references in Quasicrystalline Materials 1988, Calvayrac *et al.* 1990, Belin *et al* 1991). Several structural states were achieved : crystalline (C or μC), quasicrystalline (QC: I = icosahedral, D = decagonal) or amorphous (A). The latter for the Al-Mn samples only.

Partial valence band distributions were reached through the different X-ray transitions: Al $L_{2,3}$, Al $K\beta$, Mn $L\alpha$, Cu $L\alpha$ and Fe $L\alpha$ which probe respectively the Al 3s, Al 3p, Mn 3d-4s, Cu 3d-4s and Fe 3d-4s states distributions (valence band --->1s or $2p_{3/2}$ levels). Note that due to transition probabilities, the intensity of the 3d states distribution is much higher than that of the 4s. Let us also mention that no satisfactory Cr $L\alpha$ spectra could be obtained to complete the valence distributions of the Al-Cu-Fe-Cr samples. Al p states of the conduction band were probed from the investigation of Al K spectra (1s ---> conduction band). Details concerning the different instrumental set-ups which we used are reported elsewhere (Traverse *et al* 1988, Belin and Traverse 1991, Belin *et al* 1991).

Complementary photoelectron spectroscopy measurements provide the binding energies of the Al, Mn, Cu and Fe $2p_{3/2}$ levels involved in the X-ray transitions, as referrred to the Fermi level (E_F). Since the Al 1s level binding energy cannot be obtained directly, we also measured the energy of the Al $K\alpha_{1,2}$ emission line: $1s<----2p_{3/2}$. Combining these measurements, we locate the Fermi level on the X-ray transition energy scale of the various emission and photoabsorption spectra within $\pm$ 0.2 eV for Al and Mn and $\pm$ 0.3 eV for Cu and Fe respectively.

3.   Results

The experimental curves are adjusted in the binding energy scale except for Al_2O_3 curves that are given in the X-ray transition energy scale.This allows to identify the ranges where oxide might be involved in the spectra of the various samples when its contribution cannot be removed from the raw experimental data.Whereas no shift is observed in the limits of the experimental precision for Al and Mn inner levels in the Al-Mn and Al-Mn-Si alloys, with respect to the pure metals, a shift of a few tenths of eV is seen for the various elements in Al-Cu-Fe and Al-Cu-Fe-Cr samples. For a given composition of the alloy, no significant difference is found from one sample to another whatever the structural state is.

Our SXES curves are normalized at the same height between the maximum and the bottom of the curves in ranges where the variation of intensity is negligible. The Al photoabsorption curves are normalized between the bottom before the jump and the intensity at about E_F - 5 eV, indeed, in this energy range, no oxide is involved in the spectra.

Figure 1-A shows the Al 3p curves for Al(1), C-$Al_{86}Mn_{14}$(2), I-$Al_{79}Mn_{21}$(3), D-$Al_{78}Mn_{22}$(4) and A-$Al_{79}Mn_{21}$(5). In the alloys with respect to the pure metal, the full width at half-maximum (FWHM) is reduced by about 1 eV; the emission edges retain the arctg shape but the slope is less and less abrupt with increasing Mn concentration in the sample. The maximum A of the distribution, at about E_F - 2 eV, is split into two parts in the C and D samples whereas it is rounded in the I and A ones. At half maximum intensity, Al edge crosses the Fermi level, this is not the case for the alloys, the distance δ to E_F being for example 0.35 eV in I-Al$_{79}$Mn$_{21}$ and 0.4 eV in A-Al$_{79}$Mn$_{21}$. At E_F, the intensity varies from one sample to another: we measured 50 and 42% for C- and I-Al$_{86}$Mn$_{14}$, 46 and 42% for C- and I-Al$_{84}$Mn$_{16}$, 32 for D-Al$_{80}$Mn$_{20}$, 40, 34 and 28% for C-, I- and A-Al$_{79}$Mn$_{21}$ and 28% for D-Al$_{78}$Mn$_{22}$.

The maximum of the Al 3p distribution in C- Al-Cu-Fe phases is split into two parts as shown figure 1-B for Al$_{55}$Cu$_{33}$Fe$_{12}$(2) a second peak, denoted B, is observed at E_F+ 5.3 eV; this broadens the entire distribution with respect to pure Al so the FWHM is about 1 eV wider. In I-Al$_{63}$Cu$_{25}$Fe$_{12}$(3), peak A is unsplit and peak B is slightly less contrasted.The distance δ to E_F is 0.45 eV for the C-phase and 0.7 eV for the I- one.

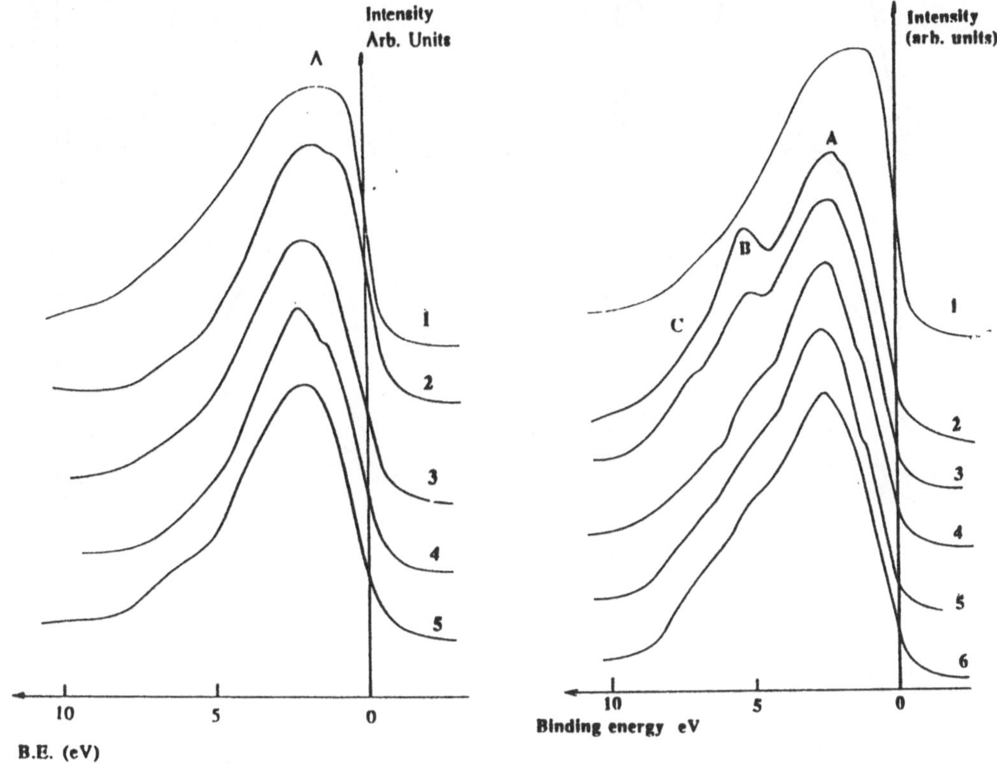

Fig.1A: Al 3p distributions in Al and Al-Mn alloys

Fig.1B: Al 3p distributions in Al and Al-Cu-Fe and Al-Cu-Fe-Cr alloys

In the Al$_{65,5}$Cu$_{18,5}$Fe$_8$Cr$_8$ alloys, peak B is seen as a shoulder. The maximum A retain the rounded shape in all phases but for μC- and D- samples, the edge is splitted into two parts of approximately the same slope. The distance δ is 0.8 eV for the μC-phase, 0.9 for the D-one and 0.85 for I- Al$_{65,5}$Cu$_{18,5}$Fe$_8$Cr$_8$. All the curves Al-Cu-Fe and Al-Cu-Fe-Cr exhibit a faint shoulder C at about E_F+ 7.

Figure 2-A display the Al 3s distribution curves in Al and the Al-Mn and Al-Mn-Si alloys namely: Al (1), C- and I- $Al_{73}Mn_{21}Si_6$ (3 & 4),C- and I-$Al_{86}Mn_{14}$(5 & 6), D-$Al_{80}Mn_{20}$(7) and D-$Al_{78}Mn_{22}$(8). They are consistent with previous data from Bruhwiller *et al.* 1987 and Ederer *et al.* 1988. Figure 2-B refers to samples C-$Al_{55}Cu_{33}Fe_{12}$(3), I-$Al_{62}Cu_{25,5}Fe_{12,5}$(4), μC-, D- and I-$Al_{65,5}Cu_{18,5}Fe_8Cr_8$ respectively (5),(6) and (7); curves of Al(1) and Al_2O_3 (2) are also shown. From E_F to over about 4.5 eV, the shapes of the different curves are little affected by oxidation.

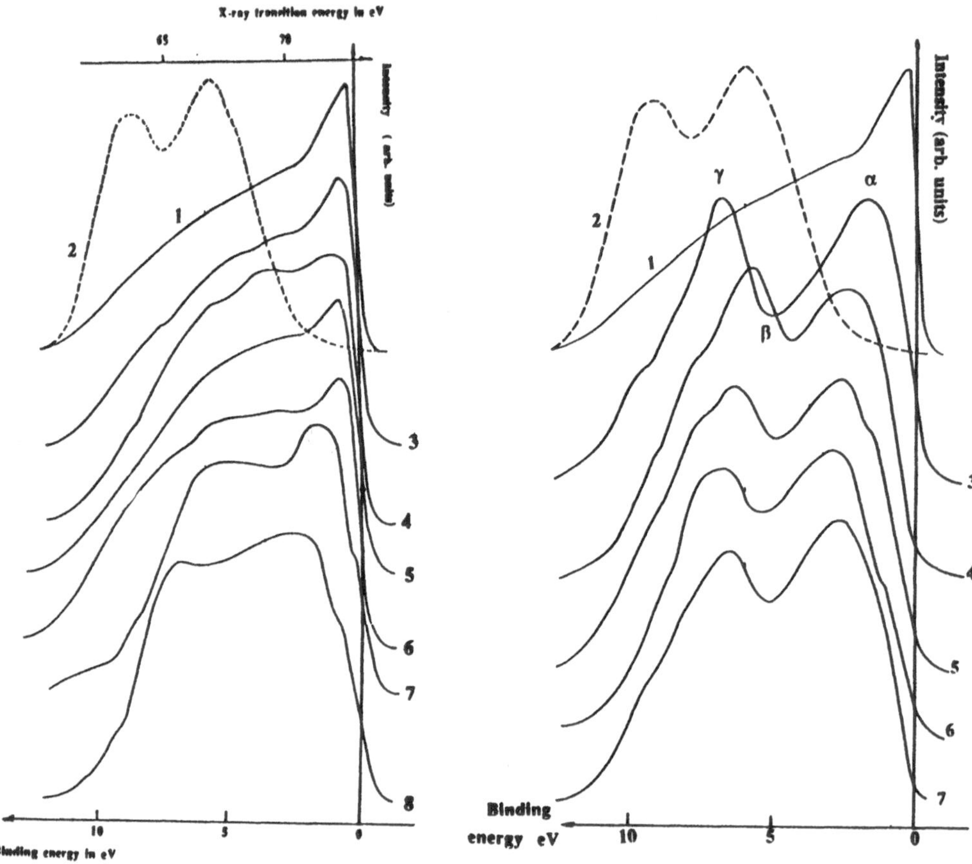

Fig. 2A: Al 3s distributions in Al, Al_2O_3 and Al-Mn alloys

Fig.2B: Al 3s distributions in Al, Al_2O_3, Al-Cu-Fe and Al-Cu-Fe-Cr alloys

The curves of C-Al-Mn-Si and Al-Mn phases show a prominent peak near E_F like in pure Al, this peak progressively vanishes in the QC alloys with increasing Mn content. Similarly, the edges which are abrupt for the crystals tend to progressively become less steep. Note that for D-alloys, the emission edges are two-stepped. The intensities at E_F, within the experimental precision (± 1) are consistent with the values fround for the Al 3p distributions.

The Mn Lα, Cu Lα and Fe Lα curves of the alloys are very similar to those of the pure metals but their FWHM are slightly narrower and tend to decrease a little from C, to QC phases. In the Al-Mn and Al-Mn-Si alloys, with increasing Mn concentration, the maximum of Lα tend to be a little closer E_F and the intensity at the Fermi level to slightly increase. With respect to the pure metals, in the Al-Cu-Fe and Al-Cu-Fe-Cr samples, the maximum of Fe Lα bands are closer E_F by about 1 eV while the Cu Lα ones are pulled down by about the same quantity towards high binding energies.

The Al K photoabsorption curves are displayed in figure 3 for Al(1), C-,I- and A-$Al_{86}Mn_{14}$(2,3 and 5), D-$Al_{80}Mn_{20}$(4), I-$Al_{63}Cu_{25}Fe_{12}$(6) and Al_2O_3 (7). This shows that the shapes of the curves are significant in the energy range E_F-5 eV, therefore, the bumps noted F on curves 3 and 4 cannot be totally accounted for by the oxide contribution. The edges of all the curvres are less abrupt than in pure metal. The intensities at E_F are decreasing from C- to QC alloys and in the Al-Mn samples with increasing Mn concentration in the alloy.

4.   Discussion

Let us stress that, both SXES and SXAS experiments on Al-Mn alloys reveal no differences according to the RS or IBMM preparation mode.The experimental curves given in the binding energy scale, allow adjustment of the different curves together and thus direct comparison of the electronic distributions becomes possible.

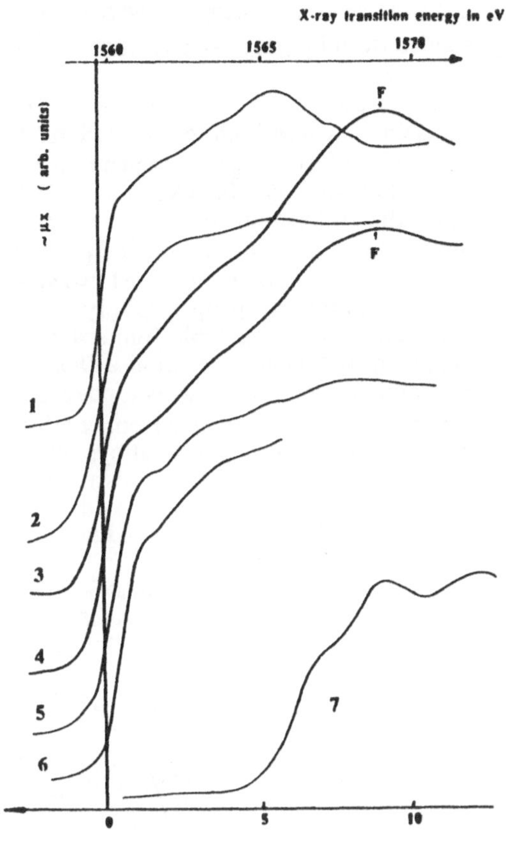

Fig.3: Al p distributions in Al, Al_2O_3, Al-Mn and Al-Cu-Fe alloys

4.1. Al-Mn AND Al-Mn-Si ALLOYS

Figure 4 shows the different partial distribution curves obtained for C- and I-$Al_{86}Mn_{14}$ and D-$Al_{80}Mn_{20}$ adjusted as mentionned above. Similar figures are achieved from the other Al-Mn and Al-Mn-Si alloys. With increasing energies from E_F, the valence band is organized as follows: Al 3p states are hybridized with Al 3s and are in interaction with Mn 3d-4s states, then they are Al 3sp and practically Al 3s-like in the high-lying part of the band.

For a given structural state, the Al 3sp intensity decreases at E_F when Mn is increased in the sample (Belin and Traverse 1991), all the same, for a given Mn concentration, the Al 3sp DOS at E_F decreases when going from C- to I-, D- and A-alloys. This lowering of the Al

834

3sp DOS at E_F is consistent with resistivity measurements (see references in Belin and Traverse 1991) which show that ρ increases whatever the strucrural state is with increasing Mn concentration in the alloy and that also, for a given concentration, ρ increases from C- to I- and from D- to A- phases. At the same time, as the sp DOS at E_F decrease, the edges of the Al 3p and 3s distributions are progressively repelled from E_F, as a consequence a gap opens in the Al p states distributions at half maximum intensity.

This is confirmed in figure 5 for I-$Al_{86}Mn_{14}$ where both Al 3p and Al p curves are adjusted together: a gap of total width of about 0.55 ± 0.1 eV. In a parallel way, the maximun of the Mn Lα distribution tends to approach E_F; as a result, the interaction between Al 3sp and Mn 3d states is progressively less complete (see for example figure 4-3 as compared to 4-2) and the Mn d-s DOS at E_F is quite noticeable. However, no direct information can be deduced concerning magnetic properties of these alloys. Note that, enhancement of magnetism with increasing Mn content has been evidenced by Gozlan et al 1991, which is in line with the present observations.

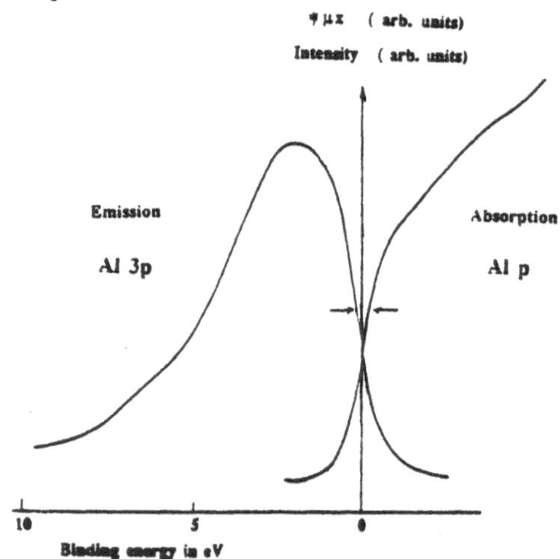

Fig.4: Valence band distributions in Al-Mn alloys

Fig. 5: Filled Al 3p and empty Al p distributions in I- $Al_{86}Mn_{14}$

With increasing Mn in the alloys and when going from C- to QC phases, the prominent peak near E_F in the Al $L_{2,3}$ curves tend to vanish and the edge to be repelled from E_F. The interaction with d and sp electrons at and in the vicinity of E_F becomes thus less and less pronounced and the valence electrons acquire progressively a more and more localized character. Then, the free-electron model is lsuggested to be ess and less adapted to the interpretation of the electronic properties. Notice that the photoabsorption experiments show a modification in the p states distribution at about E_F-7 eV in QC phases with respect to their C- or A- counterparts which reveal changes in hybridization of the empty p states.

Let us compare the experimental results with theoretical predictions. Several authors suggested that a virtual bound state must exist in the QC alloys, as an example, from Mayou *et al.* 1988, E_F must be at the maximum of the Mn 3d DOS distribution. This is in contradiction with our measurements which show that ,although in a region of high DOS, the Fermi level is not located at the top of the Mn Lα curve. Recently, a calculation of the valence band states distributions has been performed by Fujiwara 1989,1990 for crystalline $\alpha Al_{114}Mn_{24}$ which is an approximant of the icosahedral QC phase. From the partial calculated DOS, it comes that (i) a pseudogap is located in the middle of the Mn 3d band, (ii) the Fermi level is lying in the middle of the pseudogap, (iii) Al 3p states of noticeable intensity at E_F have the maximum intensity at E_F+ 2 eV, (iiii) high lying states are Al 3s-like and of very weak intesity at E_F . No virtual bound states arise from this model. Our experimental results for the valence band agree in part with these predictions. Refering to the empty states distributions, Fujiwara's calculations which concern a very short energy range (about 5 eV) predict the presence of a peak at 1.2 eV and then a monotonically increase of intensity. This peak could correspond to the limit of the Al p absorption jump.

Note that, to interprete disordered transition metal based systems, Mayou *et al.* 1986 and Cyrot-Lackmann *et al.* 1988 proposed a simple model, in a Bethe lattice approximation including sd hybridization effects and accounting for chemical environment; they predict in particular the formation of a pseudogap in $Al_{100-x}Mn_x$ for x about 20%. This model also is in good agreement with our results.

No information concerning magnetism of the Al-Mn QC alloys could be derived from the study of the Mn 3d states distribution.

4.2. Al-Cu-Fe AND Al-Cu-Fe-Cr ALLOYS

The Al 3p and Al 3s dsitributions for C-$Al_{46}Cu_{36}Fe_{18}$(1) and I-$Al_{63}Cu_{25}Fe_{12}$(2) are presented figure 6. As the two edges are superposed, the corresponding states are completely sp hybridized. They are still 3 sp-like over E_F and E_F + about 3 eV, then isolated 3p states are found (maximum B of the Al 3p curve coincides with minimum β of the Al 3s one; in curve (2), oxide contribution showed by vertical bars affects the Al 3s intensity in this energy range), beyond, the states are 3s slightly 3p hybridized (coincidence of structures C and γ of the 3p and 3s distributions).

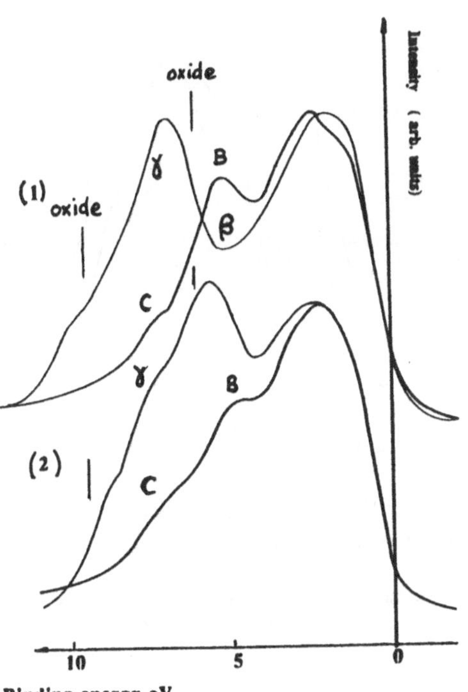

Fig.6: Al 3p and 3s distributions in Al-Cu-Fe alloys

The same is true for the Al-Cu-Fe-Cr alloys despite the Al 3p isolated states are less contrasted.

Adjusting the Lα curves to the Al 3p and 3s ones, we obtain figures like to figures 7-1 and 7-2 respectively for I-Al$_{63}$Cu$_{25}$Fe$_{12}$ and D-Al$_{65.5}$Cu$_{18.5}$Fe$_8$Cr$_8$. The valence band is organized as follows : (i) Al 3sp states are in interaction with Fe 3d ones over about E$_F$ + 2.5 eV, (ii) Cu 3d states are present at about E$_F$+4 eV while isolated Al 3p states are observed around E$_F$+5 eV, (iii) high-lying Al 3s states slightly 3p hybridized are found at about E$_F$+7.5 eV.

Mori *et al.* 1991 have studied the valence band distributions of I-Al-Cu-Fe by photoelectron spectroscopy. This technique concerns the surface of the sample and provides the total valence band distribution . The authors observed a quite intense structure at E$_F$+4 eV which they attibute to Cu d states. Matsuo *et al.* 1989 studied magnetic properties of the same alloys, they concluded that virtual high lying Fe 3d states must be located very low from the Fermi level. This last conclusions is not in agreement with our results.

The Al 3s distributions in the alloys display a two wide peaked shape (the intensity of the high energy peak might be enhanced due to oxidation). The interactions which we observe in the valence band of the various alloys could be understood by assuming a charge transfer from Al to Cu and Fe orbitals. This could give to Al a somewhat ionic character with, as a consequence, the related electronic distributions in the alloys acquire a character somewhat less extended than that of pure Al; thus, the free-electron model should not be involved to interprete the electronic properties of the Al-Cu-Fe and Al-Cu-Fe-Cr alloys.

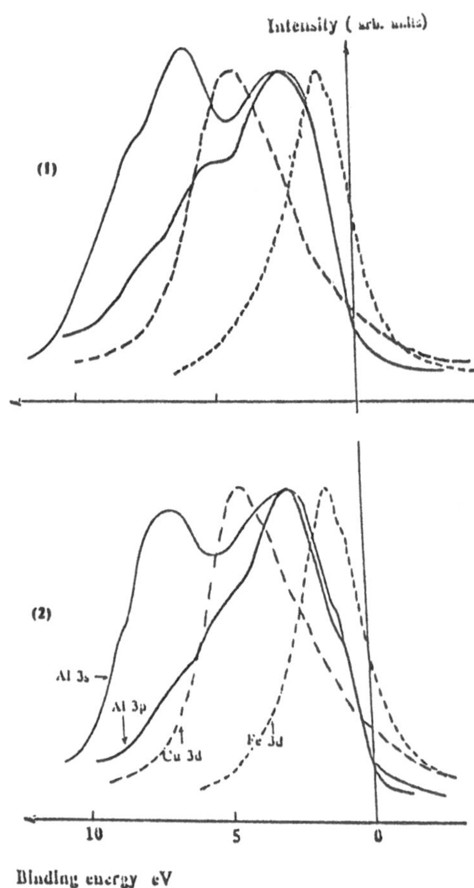

Fig.7: Valence band distributions in
Al-Cu-Fe-Cr (1) and Al-Cu-Fe (2) alloys

This is supported by the fact that a pseudo-gap is present at E$_F$ (see values of δ as given above) and that very low Al 3 sp DOS are measured at E$_F$: 21% and 12% for the C-Al-Cu-Fe phases and I-Al$_{63}$Cu$_{25}$Fe$_{12}$, 13%, 12% and 11% respectively for μC-, D- and I-Al$_{65.5}$Cu$_{18.5}$Fe$_8$Cr samples respectively.

These observations are consistent with low total DOS measured for I- Al-Cu-Fe alloys from specific heat measurements (Berger *et al.* 1991) and with very high resistivity values

(> 1000 $\mu\Omega$cm) usually reported for these samples. It is to mention that from Fujiwara and Yokokawa 1991's calculation, a low DOS at E_F is found in $Al_{13}Fe_4$ due to the existence of a pseudo-gap dividing the transition metal into two occupied and unoccupied subbands and due to E_F lying in the pseudo-gap. Unexpected high resistivity values, have already been observed from alloys other than the Al-Mn or Al-Cu-Fe. For example, for Ti based alloys, Prekul and Sasovskaya 1979 interpreted the electronic properies by the existence of a pseudo-gap in the electronic structure

Figure 8 shows the adjustment of the AlKβ and K curves for I- $Al_{63}Cu_{25}Fe_{12}$. This allows to measure the pseudo-gap value which is found to be 1.6 ± 0.1 eV wide.

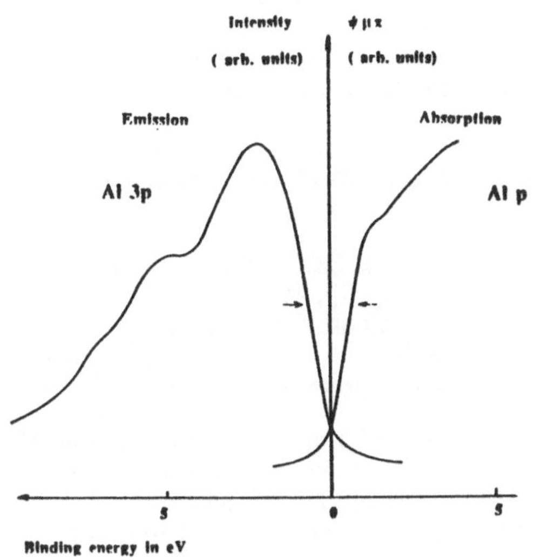

Fig.8: Filled Al3p and empty Al p states in I- $Al_{55}Cu_{33}Fe_{12}$

5. Conclusion

From Soft X-ray spectroscopy measurements, s, p and d states distributions in Al-Mn, Al-Mn-Si, Al-Cu-Fe and Al-Cu-Fe-Cr alloys were determined. The positionning of the Fermi level make possible to describe the valence band distributions and evidence the different interactions which take place i.e.: Al 3sp-Mn 3d near E_F in Al-Mn and Al-Mn-Si samples, Al3sp-Fe3d near E_F and Al3p-Cu3d in the middle of the band in the Al-Cu-Fe and Al-Cu-Fe-Cr alloys . The presence of a pseudo-gap at E_F is observed and the Fermi level is found to lie in it; low Al 3 sp DOS at the Fermi level are measured. The free-electron model is suggested not to be appropriate to the description of the electronic properties of the Al-Mn QC and all the Al-Cu-Fe and Al-Cu-Fe-Cr alloys.

The most important features of the DOS are observed to be the same in the samples whatever the structural state is; they seem to be governed by local order, i.e. an order resulting from the chemical bonds (cf.Marcus 1986). However, fine structure of the sub-bands is related to modifications in the structural state (ex. splitting of peaks or edges...). The general trend which stands out from our experiments is that the more characteristic features: low Al DOS at E_F and pseudo-gap opening, are more marked for the QC phases than for the C copunterparts and, as a consequence, the pseudo-gap is wider in the QC than in related crystals.

6. References.

Belin,E. and Traverse,A. (1991) "Al 3p electronic distributions in Al-Mn crystalline, quasicrystalline and amorphous alloys" J. of Phys. Cond. Mat. 3, 2157-2166.

Belin,E. Kojnok,J. Sadoc,A. Traverse,A. Harmelin,M. Berger,C. and Dubois,J.M. (1991) "Electronic distributions of Al-Mn and Al-Mn-Si alloys" to be publised in J.

of Phys. Cond. Mat.

Berger,C. Gozlan,A. Lasjaunias,J.C. Fourcaudot,G. and Cyrot-Lackmann,F. (1991) "Electronic properties of quasicrystals" Physica Scipta 35, 90-94.

Bruhwiler,P.A. Shen,Y. Schnatterly,S.E. and Poon,S.J. (1987) " Comparison of the icosahedral and α structures of AlMnSi with the use of soft X-ray emission spectroscopy" Phys. Rev. B 36, 7347-7352.

Calvayrac,Y Quivy,A. Bessiere,M. Lefebvre,S. Cornier-Quiquandon,M. and Gratias,D. (1990) " Icosahedral AlCuFe alloys: towards ideal quasicrystals" J. de Physique
(France) 51, 417-431.

Cyrot-Lackmann,F. Mayou,D. and Nguyen Manh-D (1988)"Electronic structure of amorphous transition metal alloys" Mat. Sci. and Eng. 99, 245-251.

Ederer,D.L. Schaefer,R. Tsang,K.L. Zhang,C.H. Calcott,T.A. and Arakawa,E.T. (1988) "Electronic structure of the icosahedral and other phases of aluminium-manganese alloys studied by soft X-ray emission spectroscopy" Phys. Rev. B 37, 8594-8597.

Fujiwara,T. (1989) " Electronic structure in the Al-Mn crystalline analog of quasicrystals" Phys. Rev. B 40, 927-946. - (1990)"Electronic structure in three dimensional quasicrystals" J. of Non Cryst.   Solids 117 & 118, 844-847.

Fujiwara,T. and Yokokawa,T. (1991)"Universal pseudo-gap at the Fermi energy in quasicrystals" Phys. Rev. Lett. 66, 333-336.

Gozlan,A. Berger,C. Fourcaudot,G.Omari,R. Lasjaunias,J.C.and Préjean,J.J. (1991) " Anomalous hall effect related to the magnetization inpure decagonal Al-Mn phases" Phys. Rev. B 44,575-583.

Klein T. Gozlan,A. Berger,C. Cyrot-LackmannF. Calvayrac, Y. and Quivy,A. (1990)"Anomalous transport properties in pure AlCuFe phases of high structural quality"
"Europhys. Letters 13,129-134

Marcus, M.A. (1986) " Comparison of electronic properties of quasiperiodic and periodic lattices in two and three dimensions" Phys. Rev. B 34,5981-5983

Matsuo, S. Nakano, H. Ishimasa, T. and Fukano, Y. (1989)" Magnetic properties and the electronic structure of a stable Al-Cu-Fe icosahedral phase " J. of Phys. Cond. Mat. 1, 6893-6899.

Mayou, D. Nguyen Manh-D, Pasturel, A. and Cyrot-Lackmann, F. (1986) " Model for hybridization effect in disordered systems " Phys. Rev. B 33,3384-3391.

Mayou, D. Maret, M. and Pasturel A. (1988) " Electronic structure and stability of structural. different Al_4Mn alloys" in Quasicrystalline Materials"Janot,C.and Dubois,J.M. Eds. World Scientific Publications, Singapore, pp.409-416.

Mori, M. Matsuo, S. Ishimasa, T. Matsuura, T . Kamiya, K. Inokuchi, H. and Matsukawa, T. (1991)"Photoemission study of an Al-Cu-Fe icosahedral phase"J.of Phys. Cond. Mat. 3, 767-771.

Prekul, A.F. and Sasovskaya, I.I. (1979)"Pecularities of infrared absorption in Ti-based alloys " Sol. State Com. 30, 91-93.

"Quasicrystalline Materials" (1988) Janot,C. and Dubois,J.M. Eds. World Scientific Publications, Singapore.

Smith, A.P. and Ashcroft, N.W. (1987) " Pseudopotentials in quasicrystals" Phys. Rev. Let. 59, 1365-1368.

Traverse, A. Dumoulin, L. Belin, E. and Sénémaud, C. (1988) " Densities of states in quasicristalline alloys. First results" in "Quasicrystalline Materials" Janot, C. and Dubois,J.M. Eds. World Scientific Publications, Singapore, pp.399-408.

METAL CLUSTER MOLECULES: FROM MOLECULE TO METAL

L.J. de Jongh, J. Baak, H.B. Brom, D. van der Putten,
J.M. van Ruitenbeek and R.C. Thiel
Kamerlingh Onnes Laboratory, Leiden University
P.O. Box 9506 – 2300 RA LEIDEN – The Netherlands

1 Introduction

Metal cluster compounds (MCC-'s) present an interesting class of cluster solids which is very suitable to study the transition to "metallic behavior" of a metal cluster as its size increases [1]. An MCC consists of identical macromolecules (the metal cluster molecules), which can be ionic or neutral and which are composed of a metal core (cluster) containing a given number (n) of metal atoms. The core is surrounded by a ligand "shell" formed by ligand atoms (Cl, I, O, ...) or ligand molecules (CO, PPh_3, ...) which are chemically bonded to metal atoms at the surface of the metal cores. Since chemical compounds are involved, a sample of a given MCC contains only one particular type of macromolecule (provided that it is pure, of course), and thus presents a macroscopically large collection of identical metal clusters, mutually separated by the ligand shells (plus the counterions in case of the ionic forms). Thus the ligand shells provide a highly effective means of "chemical stabilization" of the small metal particles (analogous to metal colloids). In going from one compound to the other, the type of metal atom in the clusters or the size n of the clusters can be varied. At present a few hundreds of MCC-'s are known already, most of them with cluster sizes ranging up to $n = 20 - 30$, and with many of the transition metal elements (Fe, Co, Ni, Mo, Ru, Rh, Pd, Ag, Os, Ir, Pt, Au, ...). For some elements even giant MCC-'s are known, the metal core size becoming as large as $n = 309$ for Pt and $n = 561$ for Pd. Clearly, such sizes are already closely approaching those of the smallest metal particles in metal colloids ($n \simeq 10^3 - 10^4$ atoms). Not surprisingly therefore, these giant MCC-'s offer very promising possibilities for the study of the physical and chemical properties of catalysts. In particular the comparison of the properties of the ligated and the bare metal clusters (obtained by removing the ligand shells) are of considerable importance in this regard.

Since several reviews on the chemical and structural aspects of the MCC-'s exist already in the literature [2,3,4], we shall here only give a few examples to illustrate the above points. In fig.1 three examples of the smaller MCC-'s are shown, two of which belong to the large class of metal carbonyl clusters. All three are ionic molecules, and thus form crystalline ionic cluster solids in combination with suitable counterions. The regular packing of the

P. Jena et al. (eds.), Physics and Chemistry of Finite Systems: From Clusters to Crystals, Vol. II, 839–851.

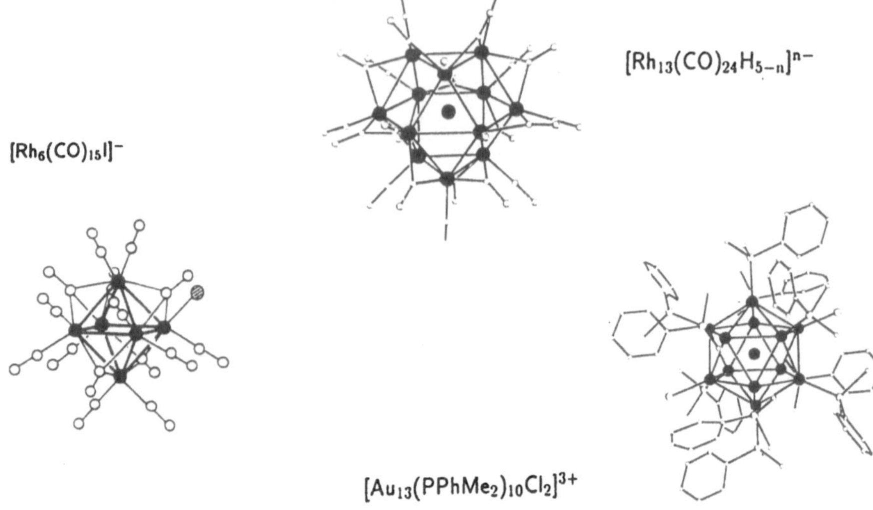

Fig.1: Molecular structure of $[Rh_6(CO)_{14}(\eta^3\text{-}C_3H_5)]^-$, $[Rh_{13}(CO)_{24}H_{5-n}]^{n-}$ and $[Au_{13}(PPhMe_2)_{10}Cl_2]^{3+}$.

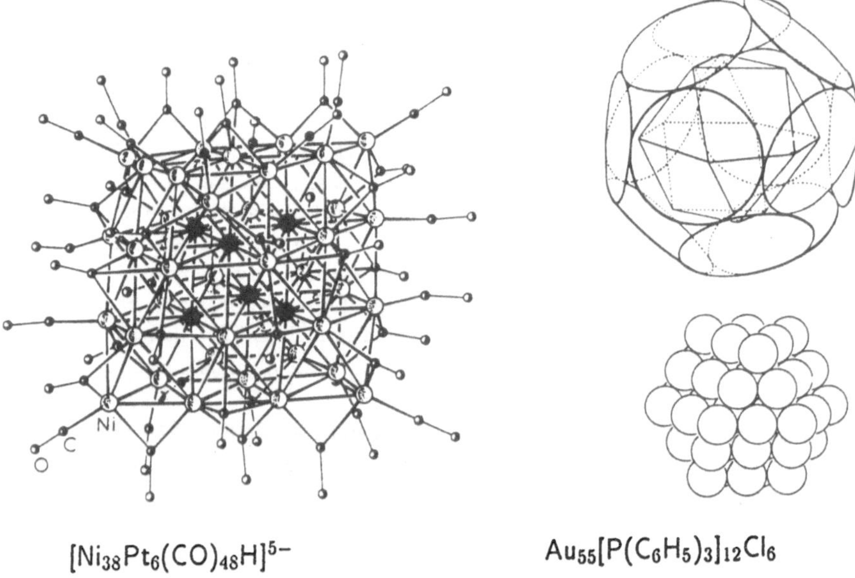

Fig.2: Structures of the giant metal-cluster molecules $[Ni_{38}Pt_6(CO)_{48}H]^{5-}$ and $M_{55}L_{12}Cl_x$, where M = Au, Pt, Rh, Ru or Co. For the latter the way in which the umbrella-shaped organic ligand molecules envelop the 55 atom metal core (below) is sketched schematically (up).

large molecules is of course reminiscent of other cluster solids like the Chevrel phases, solid boron and, last but not least, the recently discovered fullerenes.

As examples of large MCC-'s we show in fig.3 one of the many large (ionic) metal carbonyl clusters obtained by the group in Milano [2] and in addition the structure of the neutral cluster molecules $M_{55}L_{12}Cl_x$ synthesized by Schmid and coworkers in Essen [3]. In these "Schmid-clusters", the metal atom can be either Au, Pt, Ru, Rh or Co. Depending on the metal atom, x is either 6 or 20 and the ligand L is e.g. PPh_3, PMe_3, $P(t\text{-}Bu)_3$, $As(t\text{-}Bu)_3$, or $P(p\text{-}tolyl)_3$. In addition, for M=Au a *water-soluble* MCC has been obtained by exchange of PPh_3 in $Au_{55}(PPh_3)_{12}Cl_6$ by the sulfonated derivative, yielding $Au_{55}L_{12}Cl_6$ with $L=PPh_2C_6H_4SO_3Na \cdot 2H_2O$. So far crystalline samples of the Schmid-clusters could not be obtained, the materials being either in the form of a random packing of neutral molecules (like in a glass) or of extremely small crystallites ($< 0.1 - 0.01$ μm). The same holds for the giant cluster molecules $Pd_{570\pm30}L_{60\pm3}(OAC)_{180\pm10}O_{190\pm10}$ described by Moiseev and coworkers [5] and $Pt_{309}Phen^*_{36}O_{30\pm10}$ [6] and $Pd_{561}Phen_{36}O_{190-200}$ [3] synthesized by Schmid and coworkers. In what follows we shall often abbreviate the MCC-'s by giving the type and number of metal atoms as M_n, e.g. Pd_{561} for the above $Pd_{561}Phen_{36}O_{190-200}$. We point out that the M_{55} clusters and the Pt_{309} and Pd_{561} clusters represent the two-shell, four-shell and five-shell members, respectively, of the series of n-shell magic atom-number clusters obtained by surrounding a given atom progressively with shells of atoms of its kind. This yields the magic atom-number clusters M_{13} (realized in Rh_{13}, cf. fig.1), M_{55}, M_{147}, M_{309}, M_{561}, etc. ..., for both cuboctahedral and icosahedral packing. The cuboctahedral clusters are shown in fig.3. Although Moiseev and coworkers report [5] their giant Pd-clusters to be icosahedral, the metal cores of the Schmid clusters M_{55}, Pt_{309} and Pd_{561} are cuboctahedral, as evidenced by powder X-ray diffraction, high-resolution electron microscopy and, more indirectly by physical experiments like EXAFS and Mössbauer spectroscopy.

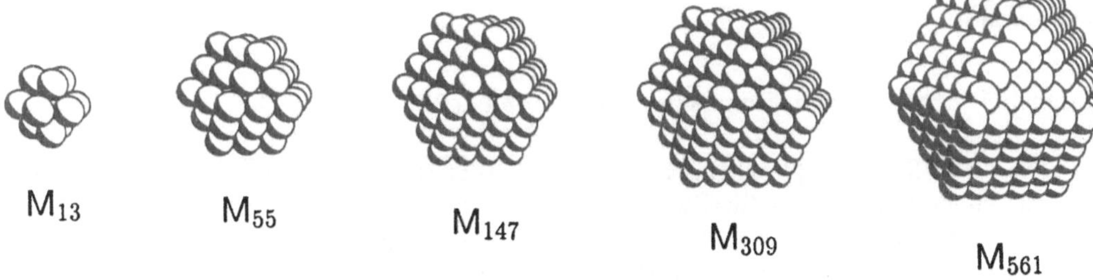

Fig.3: Magic-number clusters M_n obtained by surrounding a given atom by successive shells of atoms (the illustration is for cuboctahedral packing).

2 Transition from molecular to metallic behaviour

Although the ligand shells are essential in preventing the coalescence of the metal particles forming the cores of the metal cluster molecules, it is obvious that their influence on the chemical and physical properties of the cores should be considerable, in particular for the smaller MCC-'s. The ligands are chemically bonded to "surface" metal atoms in the outer shell of the metal clusters, and accordingly there will be charge transfer between the ligand shells and the metal cores, the nature and extent of the charge transfer being dependent on the type of chemical bond formed (covalent, ionic, etc. ...). In the course of the investigations which we shall describe below, we have become convinced that the influence of the ligand shell on the possible "metallic properties" of the metal cores is, in all probability, mainly restricted to the "surface" shell of metal atoms. As it turns out, this is in keeping with the recent developments in surface chemistry [7], where it is found that for the majority of the investigated examples of chemical bonds formed by molecules chemisorbed onto (macroscopic) metal surfaces, the major transfer processes are with the atom (or atoms) to which the ligand molecule is coordinated, the disturbances caused by the chemisorption decaying quite rapidly when moving radially away from the metal atom(s) involved in the bond. In fact it is well-known that the chemical bonds between ligands and surface metal atoms in the MCC-'s are very similar to those found in surface chemistry, where the same molecules are bonded to a macroscopic metal surface.

In what follows we shall briefly summarize a number of different physical measurements on transition metal MCC-'s of increasingly larger cluster sizes, which were meant to study the evolution towards metallic behavior, and which can be interpreted in a consistent way in terms of such a localized scenario for the ligand-metal interaction. The picture which emerges is that what we shall call the "inner-core" or "volume" metal atoms in the metal clusters in many respects behave already quite similar to the corresponding bulk metals, apparently because these inner- core atoms find themselves in an environment that is almost identical to the bulk. As we shall see, this holds even for surprisingly small metal cluster sizes, depending of course on the "metallic property" considered and on the temperature range involved. We should add that similar conclusions have been reached in quantum-theoretical studies of bare and ligated metal clusters based on molecular-orbital type of approaches.

Example 1: Structural and electronic properties of $Au_{55}(PPh_3)_{12}Cl_6$

The EXAFS studies [8] on this material show that the Au_{55} metal cores have (almost perfect) cuboctahedral structure, with a single metal-metal distance of 2.803 Å , i.e. only slightly smaller than the bulk gold value of 2.878 Å . The comparison of the X-ray absorption edge with bulk data shows that the cluster has most of the features of the bulk, be they attenuated and broadened due to the small cluster size. The main edge is at the same position as in the bulk, which would suggest that no formal charge resides on the metal core, just as in the bulk [8].

Photoelectron spectra [9] on this material also show striking similarities with bulk behavior. The X_α valence band spectrum shows a $5d_{5/2}$-$5d_{3/2}$ valence level spin-orbit splitting that is equal to the bulk Au value of 2.3 eV within the experimental error. Also the width of the 5d band is comparable to the bulk. All these features are indicative of a considerable d-d overlap between the Au atoms forming the metal cluster cores.

Lastly, the ^{197}Au Mössbauer spectra of the Au$_{55}$ cluster could be interpreted [10] in terms of four contributions from different Au sites, namely the 13 Au atoms which form the inner core of the Au$_{55}$ cuboctahedron, the 24 uncoordinated peripheral atoms, the 12 peripheral atoms coordinated to the PPh$_3$ groups, and the 6 peripheral atoms coordinated to the Cl atoms. Thus one can distinguish three different types of "surface" sites in addition to the "volume" site. Indeed, the Mössbauer parameters of the "volume" sites are very close to bulk values, the quadrupole splitting being zero due to the high symmetry, and the isomer shift being nearly equal to the bulk value. On the other hand the Mössbauer parameters found for the coordinated surface sites lie in the range of parameters appropriate for non-conducting Au compounds.

Example 2: Magnetic properties and electronic structure of large Ni carbonyl clusters

Quite recently [11], the electronic structure of two high-nuclearity carbonylated Ni clusters, [Ni$_{32}$C$_6$(CO)$_{32}$]$^{n-}$ and [Ni$_{44}$(CO)$_{48}$]$^{n-}$, with n=0-6, has been investigated by means of the linear combination of Gaussian-type orbitals (LCGTO) local density functional (LDF) method. All-electron, spin-polarized calculations were performed, in order to determine the magnetic nature of the ground states of both the bare and the carbonylated Ni clusters, and thereby study the effect of the ligand coordination.

In agreement with earlier work on smaller Ni clusters [12], it was found that the *bare* Ni clusters are in high-spin ground states. The average magnetic moment per Ni atom is found to be about $0.7 - 0.8$ μ_B for *both* surface and volume Ni atoms. The fact that this value is somewhat higher than that of the bulk (0.60 μ_B) is attributed to the larger Ni-Ni distances used in the calculations (as appropriate for the experimental examples [2] of such clusters). This agrees with the notion that the electronic configuration for Ni in small, bare Ni clusters is close to $3d^9 4s^1$, the metal-metal bond being due primarily to the 4s conduction band electrons. This would leave one hole in the fairly localized 3d shell of each Ni-atom, however, as in the above example of the Au$_{55}$ cluster, there is additional substantial d-d overlap between nearest-neighboring metal atoms, which reduces the average number of unpaired electrons per atom. It is quite interesting to note that the bonding of the Ni atoms in the bare Ni clusters already resembles that of bulk Ni to such an extent as to lead to a ferromagnetic spin-ordering, with an average moment per atom comparable to the bulk.

The addition of the CO shell to the Ni metal cores is found to have dramatic effects on the magnetism. The magnetic moments of the surface Ni atoms are completely quenched, those of the volume atoms are reduced but still have a sizable value of 0.5 μ_B per atom. These results are in keeping with previous calculations on smaller clusters [12], from which it could be concluded that carbonylation of Ni clusters so small as to have only surface atoms leads to a complete suppression of the magnetic moment of the cluster. The quenching of the Ni-moments due to interaction with CO molecules is an effect which also occurs when CO is chemisorbed on Ni-metal surfaces. The electronic mechanism behind this effect is ascribed to a transfer of electrons from the 4s orbitals into the 3d shell of the Ni atoms.

Such transfer is induced by the combination of: (i) the repulsive interaction of the 4s-derived metal MO with the 5σ MO of the CO (destabilizing the 4s orbitals well above the cluster HOMO level, so that the electrons are excited into the 3d band), and: (ii) the π back-donation from d_π metal orbitals to the $2\pi^*$ MO of CO, which leaves the Ni atoms slightly positive and causes a further electron transfer from 4s orbitals into the 3d shells.

The net result of this ligand-induced transfer mechanism, which can be seen as the molecular analogue of the high-spin/low-spin transition for a transition metal ion in a crystal field of varying strength, can be interpreted as a change in the electronic structure of the surface atoms into a formal $3d^{10}$-like atomic configuration (diamagnetic). The volume Ni atoms, on the other hand, retain most of their original $3d^9 4s^1$ character, so that they still show magnetic behavior similar to that of the bulk metal.

These calculations provide a satisfactory explanation for the magnetic properties of a series of high-nuclearity Ni clusters studied in our laboratory [13]. In fact, the above mentioned theoretically considered Ni clusters served as model systems for these real Ni carbonyl clusters. From the experimental results for the temperature dependence of the magnetic susceptibility and of the high-field magnetization, it followed that the total magnetic moment per cluster was only of the order of 4 to 9 μ_B, even though the clusters contained 34 to 38 Ni atoms. The strong reduction of the cluster magnetic moment was at the time tentatively ascribed to the effect of the ligands, an interpretation that now appears to be fully corroborated by the LDF calculations [11].

Example 3: Metallic Knight shift in the ^{195}Pt-NMR line in Pt$_{309}$Phen*$_{36}$O$_{30}$

At this conference ^{195}Pt-NMR lineshape and spin-lattice relaxation measurements taken at 77 K have been reported for Pt$_{309}$Phen*$_{36}$O$_{30}$ by Van der Putten et al. [14]. Also at this meeting, Van der Klink has presented an overview of ^{195}Pt-NMR data on Pt-catalysts [15]. As explained in his review, two features of the NMR signal are of interest in determining whether the Pt nuclear spins are in a "metallic" or in an "insulating" environment, namely the position of the resonance line and the nuclear-spin lattice relaxation time T_1. In diamagnetic, nonconducting Pt compounds the NMR frequency, $\omega_0 = 2\pi\nu_o$, in an applied magnetic field is given by $\omega_0 = \gamma(1 + K_c)B_0$, where γ is the magnetogyric ratio for the nuclear spin, and $-K_c$ is the so-called chemical shift which arises from the orbital contributions to the field-induced magnetization. Chemical shifts are usually small (0.1-1%), and the ^{195}Pt-NMR frequencies for insulating Pt compounds lie indeed in a narrow range of $B_0/\nu_0 = 1.08 - 1.10$ (kG/MHz). For metals, on the other hand, an additional shift (the Knight shift) is introduced by the Pauli-spin susceptibility of the conduction electrons, which shift can be considerably larger than the chemical shift. For Pt metal it amounts to -3.4%, the resonance occurring at 1.138 kG/MHz.

The second and perhaps even more sensitive criterion for metallic behavior as probed by NMR is the magnitude and temperature dependence of the nuclear-spin lattice relaxation time T_1. In insulators, with no electronic magnetic moments, the T_1 is very long, and becomes essentially infinite at liquid ^{4}He temperatures, the reason being that there is no direct energetic contact between nuclear spins and phonons. For metals the conduction electron spins provide an efficient relaxation channel and consequently T_1 is relatively short, even at 4.2 K ($\simeq 0.01 - 0.1$ s). The temperature dependence of T_1 is given by the Korringa relation $\hbar/T_1 \propto K^2 k_B T$, where K is the Knight shift.

For small Pt particles in catalysts the observed NMR lines tend to be very broad [15], which is attributed to the fact that Pt nuclei of the surface atoms of the (bare) Pt particles do not exhibit observable Knight shifts, while Pt nuclei of (volume) atoms located more towards the center of the particles will have a more bulk-like Knight shift. In the paper of Van der Klink [15] a "shell-model" is discussed to treat this effect quantitatively. In that analysis, a local density of states (LDOS) is proposed of the 5d electrons which decreases exponentially towards the surface of the clusters, whereas the LDOS of the 6s electron would remain almost constant throughout the cluster. Since the 5d-LDOS and 6s-LDOS contribute to the Knight-shift with opposite sign, an effective cancellation occurs for the surface atoms, which nevertheless should still be considered as metallic judging from the short T_1 which is found also for the surface atoms of these *bare* Pt clusters.

The ^{195}Pt-NMR signal of Pt$_{309}$Phen*$_{36}$O$_{30}$ measured at 77 K by Van der Putten [14] is also quite broad. On the basis of the observed field-dependence of the T_1, the signal could be analysed in terms of two contributions, one which peaks at $B_0/\nu_0 = 1.096$ kG/MHz, and is characterized by very long T_1 ($> 10^2$ s), and a second centered around $B_0/\nu_0 = 1.110$ kG/MHz (i.e. with a K one third of the bulk Knight shift) which has metallic-like values of T_1. It seems a logical conclusion to assign the first peak to the 162 Pt atoms that are in the surface layer of the Pt$_{309}$ clusters of the metal cluster molecules. Since these atoms are chemically coordinated by the ligand shell surrounding the cluster, one may expect these surface atoms to be "nonmetallic", i.e. with zero Knight-shift *and* nonmetallic (very long) T_1. The second peak is then attributed to the 147 volume atoms forming the inner core of the Pt$_{309}$ clusters. To these (Pt$_{147}$) *inner* cores, similar arguments as proposed by Van der Klink for *bare* Pt clusters should apply, that is one should expect the 5d-LDOS to decrease exponentially towards the surface of these Pt$_{147}$ inner cores. Since the surface shell of an 147 atom cluster contains $147 - 55 = 92$ atoms (cf. fig.3) it is obvious that the average K value for the Pt$_{147}$ inner core should be substantially reduced with respect to the bulk Pt value, in particular if one assumes the 92 surface-shell atoms to have near-zero K, as deduced for the Pt catalysts. Although it is difficult to give a true quantitative analysis of the data at this moment, the fact that the observed Knight shift of the metallic peak is only one third of the bulk value seems to be plausibly explained by such an argument.

Example 4: Magnetic susceptibility and linear electronic term in the specific heat of Pd$_{561}$Phen$_{36}$O$_{200}$

Another hallmark of "metallic" behavior is the appearance of an electronic contribution to the specific heat in the form of a linear term: $C_{el} = \gamma T$. The coefficient γ is equal to $\gamma = \frac{1}{3}\pi^2 D(E_F) k_B^2$ and thus directly proportional to the density of states around the Fermi energy. Like the metallic Knight-shift, the observation of such a linear specific heat term requires $D(E_F)$ to be at least quasi-continuous, i.e. there should be no gaps around E_F larger than the relevant thermal energy $k_B T$. Since the electronic specific heat contribution only becomes observable at *low* temperatures (below 1 K), due to the predominance of the lattice vibrational contributions at higher temperatures, it represents a sensitive test indeed. In fact in previous specific heat data on Au$_{55}$(PPh$_3$)$_{12}$Cl$_6$ in our laboratory [10] such a linear term was not observed down to 0.5 K. Extending these measurements down to lower temperatures, Goll et al. [16] found no evidence for the linear term down to 60 mK.

An important conclusion that could be drawn from the analysis of the specific heat of $Au_{55}(PPh_3)_{12}Cl_6$, however, was that the data below 50 K can be fully accounted for by the sum of two vibrational contributions, namely the vibrations of the atoms inside each of the metal clusters plus the vibrations of the cluster as a whole. The first becomes negligible below a few K due to finite-size effects on the phonon spectrum of such small particles, whereas the second essentially yields a cubic law ($\propto T^3$) for temperatures below a few K. Contributions of the ligands only become important at much higher temperatures (> 50 K), due to the relatively high intramolecular vibrational energies of the ligand molecules.

At this conference Baak et al. [17] have now reported recent specific heat experiments on the much larger cluster molecule $Pd_{561}Phen_{36}O_{200}$, which do show the appearance of a linear term below 1 K, down to the lowest temperature of 0.2 K reached in these first experiments. As was the case for the Au_{55} cluster, the low- temperature phonon specific heat of Pd_{561} can be analysed in terms of the above mentioned two vibrational contributions.

The value of γ for bulk Pd is known to be one of the largest measured for metallic elements, i.e. $\gamma = 9.42$ mJ mol^{-1}K^{-2}, which can be related to the high value of $D(E_F)$ for bulk Pd. For the Pd_{561} cluster Baak et al. [17] find a γ of about one third of the bulk value. It is interesting to compare this result with the previous analysis of the magnetic susceptibility for this cluster compound by Van Ruitenbeek et al. [18]. The paramagnetic spin susceptibility of bulk Pd is known to be very much enhanced with respect to the free-electron Pauli-spin susceptibility due to exchange-correlation effects. If $\chi_0 = \mu_B^2 D(E_F)$ denotes the unenhanced spin susceptibility, then the enhanced susceptibility is given by $\chi_e = \chi_0/(1 - \alpha\chi_0)$, where α is the (Stoner) enhancement factor (the exchange correlation integral.) The susceptibility of Pd_{561} was found [18] to be about 1/3 of the bulk Pd susceptibility at low temperatures (below 150 K). The appearance of the same ratio (roughly) as in the electronic specific heat values is quite suggestive, since both physical quantities are directly related to the (effective) density of states near E_F.

Guided by the previously discussed examples we would thus explain the factor of 1/3 by a very strong reduction of the 4d-LDOS for the surface atoms, due to the ligand coordination shell, combined with a somewhat lower reduction for the remaining volume atoms. It should be noted here that, even for as large a cluster as Pd_{561}, there still is a large fraction $561 - 309 = 252$ of atoms at the surface.

3 Conclusions from these examples

A first general conclusion from the above examples seems to be that the experimental results on MCC-'s of different cluster size all seem to agree with the idea that the influence of the ligand shell on the electronic properties of the metal cluster cores is mainly restricted to the ligated surface atoms, thereby confirming the predictions from the LDF calculations on large bare and carbonylated Ni clusters. We recall that the underlying principle is a ligand-induced charge transfer from the s-band into the d-shells of the surface atoms. Since Ni is at the end of the 3d series this will result in a dramatic lowering of the LDOS: the surface metal atom 3d-shells become filled and the Fermi-energy is shifted *above* the d-band, so that the configuration of these surface atoms will become (grosso modo) similar as that of Cu, for which the $D(E_F)$ value is known to be an order of magnitude smaller than for Ni. As an illustration of this idea, we compare in figs.4 and 5 the DOS calculated [11] for the bare and carbonylated Ni_{44} cluster with the known band structures of bulk Ni and Cu metal. One may observe that the d-LDOS of the surface atoms of the $[Ni_{44}(CO)_{48}]$ cluster qualitatively resembles that of bulk Cu (with a somewhat reduced DOS at the top of the d-band).

Since Ni, Pd and Pt are in the same row of the periodic table, as are Cu, Ag and Au, one may extrapolate the result for Ni, in the sense that the effect of the ligand coordination on the surface atoms would be the same for Pd and Pt clusters as for the Ni clusters, namely in that by the ligand-induced s→d charge transfer the d-shells of the surface atoms become completely filled, leading to a strong reduction in the d-LDOS. For the d-metal cluster compounds, therefore, "d-metallic" behaviour for sufficiently large clusters is only to be expected for the inner core (volume) atoms, i.e. particles of size Pt_{147} and Pd_{309} for $Pt_{309}Phen^*_{36}O_{30}$ and $Pd_{561}Pen_{36}O_{200}$, respectively. Since these Pt_{147} and Pd_{309} particles are still quite small, their $D(E_F)$ values will be less than the corresponding bulk values, and we recall the result claimed by Van der Klink [15] that the d-LDOS decreases towards the surface also for *bare* Pt metal particles, reaching a value at the surface about 50% below that in the interior. So, besides the almost complete suppression of the d-LDOS of the ligand-coordinated surface metal atoms, also the d-LDOS of the volume-atoms will be affected. It is of interest to see if we can arrive at the above- encountered factor 1/3 by such arguments. The Pd_{561} cluster is built up in the sequence 1+12+42+92+162+252. According to the above reasoning, the d-LDOS of the outer 252 atoms should be completely quenched by the ligands, whereas Van der Klink predicts the next shell of 162 atoms to be reduced by 50%, and the next-next shell of 92 by 10%. Thus, to compare with bulk Pd we should effectively count 1+12+42+83+81=219 atoms instead of 561. The ratio 219/561=0.31 is as close to 1/3 as one could hope for with such a rule-of-thumb argument (for instance we have neglected the s-LDOS since it is so much smaller; however, it should evidently also be taken into account in a final analysis, as well as size effects on the enhancement factor α).

The ligand-induced s-d transfer also provides a simple explanation for the measured isomer shift (IS) value of the 13 volume Au atoms which form the inner core of the Au_{55} cluster. Although the IS value of the Au_{13} inner core is close to the bulk Au value of -1.2 mm/s, it is slightly more negative, i.e. -1.4 mm/s [10]. This does not seem too disquieting were it not for the fact that the EXAFS measurements have clearly shown the Au-Au distance in the Au_{55} cluster to be reduced by about 3.5% relative to the bulk, and correcting for this would change the bulk-value to -0.5 mm/s, leaving the much larger difference of -0.9 mm/s. Now, the main contributions to the Au IS are expected to come from the 5d and 6s electrons, namely +8.0 mm/s per 6s electron and -1.6 mm/s per 5d electron. It then follows that an s→d charge transfer of only 10% of the (average) s charge density would be sufficient to account for the observed difference between the IS-value of the Au_{13} inner (volume) atoms and the bulk Au value corrected for contraction. Since the effect of the ligands cannot be expected to be fully limited to the outermost shell of metal atoms of the cluster, this result seems acceptable. We recall that the LDF calculations for the high-nuclearity Ni clusters likewise predict some reduction of the effective moment per atom from 0.7 to $0.5\mu_B$ also for the inner (volume) Ni atoms, which is in keeping with the above reasoning.

848

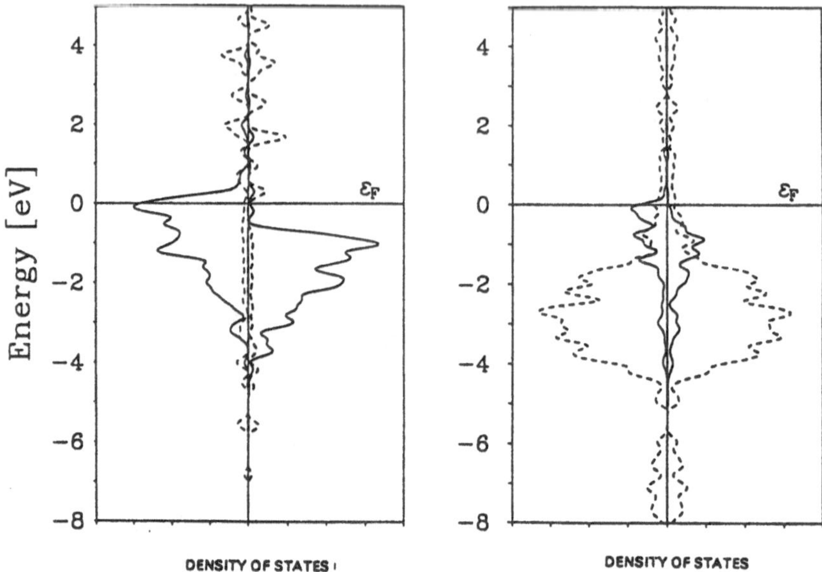

Fig.4: Density of states (in arbitrary units) as calculated in ref.[11] for the bare Ni_{44} cluster (left frame), and for the carbonylated form $[Ni_{44}(CO)_{48}]^{6-}$ (right frame). One-electron energies were obtained from spin-polarised calculations (left parts and right parts corresponding to minority and majority spins, respectively), and by applying a Gaussian broadening of 0.1 eV to each level. In left frame dashed and solid lines are for Ni 4sp and Ni 3d contributions. In right frame dashed and solid lines are for the surface and volume Ni atoms, respectively.

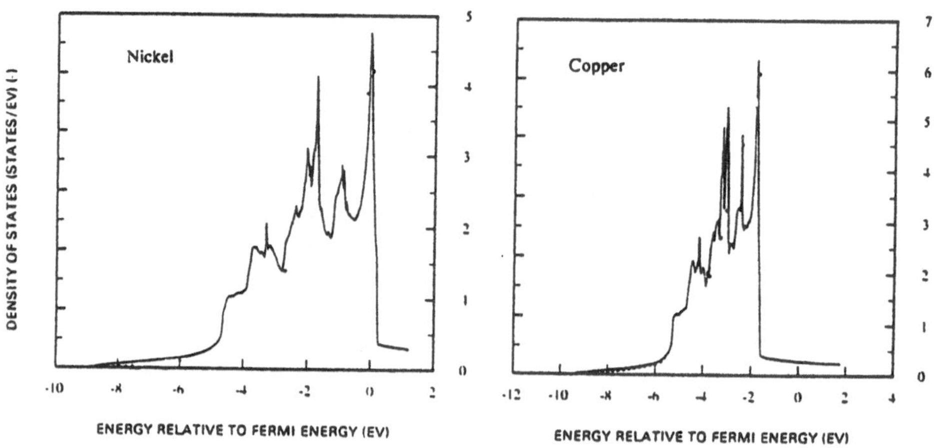

Fig.5: Comparison of (total) DOS for Ni and Cu (from ref.[21]).

The second important conclusion from the experimental studies is that for d-metal clusters of size larger than, say, 10^2 atoms, the energy level spectrum near E_F is quasicontinuous. Such a conclusion cannot be drawn from the Mössbauer data on $Au_{55}(PPh_3)_{12}Cl_6$, nor from the magnetic behavior of the large Ni clusters, since for both these properties a continuous level distribution is not needed. However, in order to display an electronic linear specific heat term, or a metallic Knight shift, such a quasicontinuum indeed is a necessary prerequisite. More precisely, the level spacings should be smaller than the thermal energies involved in the experiment. Now, for large metal clusters the average spacing can be roughly estimated from the quantity $2E_F/N$, where E_F is the energy of the HOMO-level and $\frac{1}{2}N$ is the number of occupied-valence orbitals (N is the total number of valence electrons in the cluster). Here we have ignored degeneracy of the orbital levels, which will be allowed when the symmetry is low, boundaries of the cluster are irregular, spin-orbit coupling is present, etc.. With Fermi energies of the order of 5 to 7 eV, and with 10 valence (s and d) electrons per atom, one then arrives at average level spacings of the order of 20-30 K and 40-50 K for the Pd_{561} and the Pt_{309} compounds, respectively. In the NMR experiment on Pt_{309} the temperature was 77 K, but the electronic specific heat term of Pd_{561} is found down to 0.2 K, i.e. to much smaller thermal energies. A reduction of the average level spacing near E_F could be caused e.g. by many-body effects, or by symmetry effects. However, we would expect that, in order to explain the apparent quasicontinuous level structure, a broadening of the molecular energy-level structure with widths of the order of 10 K (a few meV) still needs to be invoked. Various possible sources come to mind, as there are the finite-lifetime broadening (energy-time uncertainty relation) and the interactions between the valence electrons within the cluster. These are broadening mechanisms for the individual molecules; in addition a small *inter*cluster electron transfer of a few meV would also have the effect of broadening each molecular level into a mini-band. This situation is resemblant of that found for molecular and solid C_{60} [19]. Since C_{60} is such a highly symmetric molecule, and since the number of valence electrons ($60 \times 4 = 240$) is relatively small, even the fairly large intercluster transfer integral of the order of a few 0.1 meV between neighbouring C_{60} clusters does not lead to mutual overlap of the minibands in the condensed solid. For Pt_{309} and Pd_{561} the situation is different in that the number of valence electrons per cluster is very much larger and the symmetry lower, so that the spacings between the molecular energy levels will be indeed two orders of magnitude smaller than for C_{60}, so that even a much smaller intercluster transfer could lead to overlapping, molecular-orbital/derived minibands. An additional broadening effect in the Pt_{309} and Pd_{561} compounds will arise from the noncrystalline packing in the solid. It is well-known that randomness will introduce broadening and may lead to a smearing of the (HOMO-LUMO) derived energy gap above E_F. The result is a quasicontinuous DOS with localized states around E_F. This model has been proposed [20] to explain the observed diffuse hopping-type of conduction found for the Pd_{561} and the Au_{55} compound. In particular, for the Pd_{561} compound no activated conduction behavior was found down to liquid ^{4}He temperatures, indicating that, if an energy gap would exist above E_F, it should be smaller than 0.4 meV. Similar indications for the absence of an energy gap near E_F were found from the frequency dependence of the conductivity in those materials. An important task for the future is, therefore, to try to determine experimentally to what extent the apparent level broadening in the Pt_{309} and Pd_{561} compounds is due to *intra*molecular or to *inter*molecular mechanisms.

This work is part of the research program of the Leiden Materials Science Center, and is supported by the "Stichting voor Fundamenteel Onderzoek der Materie" (FOM), which is sponsored by the "Nederlandse Organisatie voor Wetenschappelijk Onderzoek" (NWO). One of us (D.v.d.P.) acknowledge the financial support by an IOP grant (BP203). Furthermore, we wish to acknowledge the earlier financial support of the Commission of the European Communities under contract ST2J-0084.

References

[1] See e.g. L.J. de Jongh, J.A.O. de Aguiar, H.B. Brom, G. Longoni, J.M. van Ruitenbeek, G. Schmid, H.H.A. Smit, M.P.J. van Staveren and R.C. Thiel, Z. Phys. D12 (1989) 445; L.J. de Jongh, H.B. Brom, J.M. van Ruitenbeek, R.C. Thiel, G. Schmid, G. Longoni, A. Ceriotti, R.E. Benfield and R. Zanoni in: *"Cluster models for surface and bulk phenomena"*, eds. G. Pacchioni and P.G. Bagus, Plenum (1991) (NATO ASI Series B, Physics).

[2] G. Longoni, A. Ceriotti, M. Marchionna and G. Piro, in: *"Surface Organometallic Chemistry: Molecular Approaches to Surface Catalysis"*, eds. J.M. Basset et al., Kluwer (1988) p.157 (and references in this review).

[3] G. Schmid, Polyhedron 7 (1988) 2321; Endeavour, New Series 14 (1990) 172; Aspects of Homogeneous Catalysis 7 (1990) 1, ed. R. Ugo, Kluwer.

[4] "The Chemistry of Metal Cluster Complexes", eds. D.F. Shriver, H.D. Kaesz, R.D. Adams, VCH Publishers, 1990.

[5] M.N. Vargaftik, V.P. Zagorodnikov, I.P. Stolyarov, I.I. Moiseev, V.A. Likholobov, D.I. Kochubey, A.l. Chuvilin, V.I. Zaikovsky, K.I. Zamaraev, and G.I. Timofeeva, A novel giant palladium cluster, J. Chem. Soc. Chem. Commun. 937 (1985).

[6] G. Schmid, B. Morun, and J.-O. Malm, Angew. Chem. Int. Ed. Engl. 28 (1989) 778.

[7] See e.g. the Proceedings of the Erice Workshop on "Cluster models for surface and bulk phenomena", eds. G. Pacchioni and P.S. Bagus, Plenum (1991). (NATO ASI Series B, Physics).

[8] M.C. Fairbanks, R.E. Benfield, R.J. Newport and G. Schmid, Solid State Commun. 74 (1990) 431; M.A. Marcus, M.P. Andrews, J. Zegenhagen, A.S. Bommannavar and P. Montano, Phys. Rev. B42 (1990) 3312.

[9] M. Quinten, I. Sander, P. Steiner, U. Kreibig, K. Fauth and G. Schmid, Z. Phys. D20 (1991) 377; and data by R. Zanoni cited in ref.1 (2nd paper).

[10] H.H.A. Smit, R.C. Thiel, L.J. de Jongh, G. Schmid and N. Klein, Solid State Commun. 65 (1988) 915; H.H.A. Smit, P.R. Nugteren, R.C. Thiel, and L.J. de Jongh, Physica B 153 (1988) 33.

[11] N. Rösch, L. Ackermann, G. Pacchioni and B.I. Dunlap, J. Chem. Phys. 95 (1991) 7004.

[12] F. Raatz and D.R. Salahub, Surface Science 176 (1986) 219; G.F. Holland, D.E. Ellis, and W.C. Trogler, J. Chem. Phys. 83 (1985) 3507; G. Pacchioni and P. Fantucci, Chem. Phys. Lett. 134(1987) 407; G. Pacchioni and N. Rösch, Inorg. Chem. 29 (1990) 2901.

[13] B.J. Pronk, H.B. Brom and L.J. de Jongh, Solid State Commun.59 (1986) 349; L.J. de Jongh, Physica B 155 (1989) 289.

[14] D. van der Putten, H.B. Brom, L.J. de Jongh and G. Schmid, this conference.

[15] J.J. van der Klink, this conference.

[16] G. Goll, H. von Löhneisen, U. Kreibig and G. Schmid, Z. Phys. D 20 (1991) 329.

[17] J. Baak, H.B. Brom and L.J. de Jongh, this conference.

[18] J.M. van Ruitenbeek, M.J.G.M. Jurgens, G. Schmid, D.A. van Leeuwen, H.W. Zandbergen and L.J. de Jongh, Z. Phys. D 19 (1991) 267.

[19] See e.g. S. Saito and A. Sohiyama, Phys. Rev. Lett. 66 (1991) 2637.

[20] M.P.J. van Staveren, H.B. Brom and L.J. de Jongh, Physics Reports, 208 (1991) 1-96; M.P.J. van Staveren, J.T. Moonen, H.B. Brom, L.J. de Jongh and G. Schmid, Z. Phys. D 12 (1989) 461; H.B. Brom, M.P.J. van Staveren and L.J. de Jongh, Z. Phys. D 20 (1991) 281.

[21] V.L. Moruzzi, J.F. Janak and A.R. Williams, *"Calculated Electronic Properties of Metals"*, Pergamon, 1978; D.A. Papaconstantopoulos, *"The Bandstructure of Elemental Solids"*, Plenum, 1986.

COLLECTIVE OPTICAL EXCITATION IN FREE METAL CLUSTERS

C.BRECHIGNAC, Ph.CAHUZAC, M.de FRUTOS,
N.KEBAILI, J.LEYGNIER, J.Ph.ROUX, A.SARFATI.
Laboratoire Aimé Cotton
Campus d'Orsay Bât. 505
91405 Orsay cedex, France

Metal cluster provides an ideal object to study a many body system. In particular it consists of many electrons which strongly interact each other. Two kinds of electrons may be distinguished i.e. : the valence electrons which become more delocalized as the cluster size increases, and the core electrons which may remain localized around each nucleus whatever the cluster size is. The correlations between the electrons, within each variety, induce collective effects revealed by the observation of giant resonances.

Giant resonances were already observed in excitation spectra of nuclei, atoms and molecules [1] and more recently in excitation spectra of clusters [2-6]. This paper will discuss the two types of giant resonances observed in clusters.

The one developped in alkali clusters, in the few eV energy range [2-6], is explained by the collective excitation of the valence "s" electrons in the non coulombic effective potential well, created by the ion cores and the electrons themselves. Such an effect is more pronounced for large sizes involving a large number of electrons. Emphasis will be done on the evolution of the giant resonance versus cluster size and its convergence to the bulk plasmon.

On the other hand the giant resonances observed in the 20 to 120 eV energy range in the absorption spectra of antimony clusters probe the collective excitation of the inner 4d electronic shell of each constituent atom [7]. It can be qualitatively understood as araising from the presence of the short range inner well in the effective potential of excited "d" electrons and is described as an atomic property. Any change in the shape or the position of such a giant resonance must be related to a perturbation of atomic environment.

1.GIANT RESONANCE IN ALKALI CLUSTERS

The giant dipole resonances recently observed in the few eV range in the excitation spectra of alkali clusters are attributed to the collective oscillation of the delocalized valence "s" electrons against the ionic cores and can be viewed as the optical response of a Fermi gas of electrons in a potential well. In that respect they are similar to the giant resonances in nuclei which occur around 20 Mev and are attributed to the collective oscillations of the protons against the neutrons.The principle of the experiment is based on the photoevaporation spectroscopy of mass selected species. Briefly, neutral clusters

P. Jena et al. (eds.), Physics and Chemistry of Finite Systems: From Clusters to Crystals, Vol. II, 853–860.
© 1992 *Kluwer Academic Publishers.*

are generated by adiabatic expansion of the neat metallic vapor. Several nozzles are used in order to obtain clusters X_n in the successive size ranges : n=1 to 50, n=50 to 500 and n=500 to 1500. The clusters are ionized by a pulsed nitrogen laser and accelerated to 5kV typically. Mass spectrometry is achieved by a tandem time-of-flight system. The accelerated cluster ion bunches enter a first field-free tube where they spatially resolve into separated ion packets within a mass resolution larger than 200. After size selection, a given cluster ion bunch entersdecelerating-accelerating region where it interacts with a second pulsed laser. The ion products from photon interaction are mass analysed by a second time-of flight. It is well known that for metal clusters the electronic excitation resulting from visible photon absorption, relaxes very rapidly among the numerous vibrational modes providing unimolecular evaporation. For small sizes several evaporations of neutral atoms occur after a single photon absorption, within the observable time window of the decelerating-accelerating region. For large sizes single photon induced-evaporation cannot be observed within the few microsecond of the residence time in the interacting region. Multistep photon excitation is used in order to heat the parent prior to the first evaporation [8]. With such a procedure the temperature of the clusters increases from 450K to 850K for K_9^+ to K_{900}^+ respectively. Assuming that the only possible channel for excited clusters proceeds via non radiative process, the photoabsorption cross sections are deduced from the dissociation patterns. Figure 1 shows the absorption cross sections measured in the energy range 1 to 4 eV for the potassium cluster ions K_9^+, K_{21}^+, K_{500}^+ and K_{900}^+. For K_{500}^+ and K_{900}^+ the parent masses are selected including more or less few masses. Experimental data of each cluster are fitted by the resonance profile which may attributed to the dipole collective excitation :

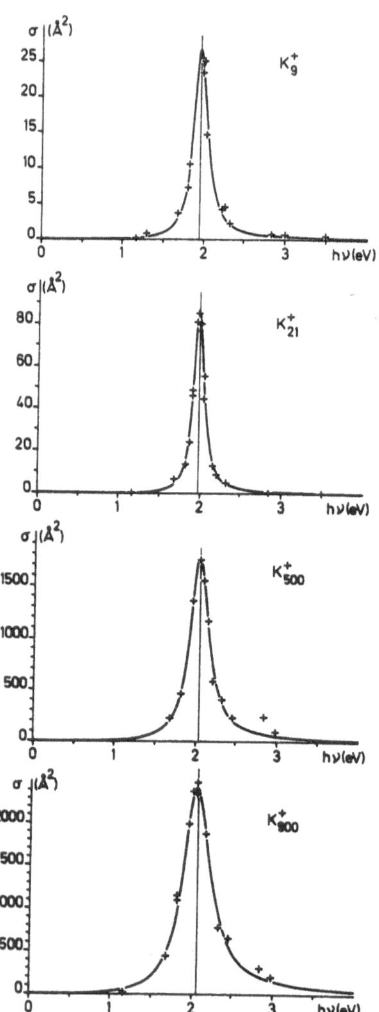

Figure 1 : Measured Photoabsorption cross section for K_9^+, K_{21}^+, K_{500}^+ and K_{900}^+ Their resonance energies are 1.93 eV, 1.98 eV, 2.03 eV and 2.05 eV respectively.

$$\sigma_n(h\nu) = \sigma_n(h\nu_0) \frac{[h\nu\Gamma]^2}{[h^2\nu_0^2 - h^2\nu^2]^2 + [h\nu\Gamma]^2} \tag{1}$$

The fitting parameters are the resonance energy $h\nu_0$, the maximum cross section $\sigma_n(h\nu_0)$ and Γ the width of the resonance. For the small size closed shell clusters K_9^+ and K_{21}^+, having a spherical symmetry, single peak resonance is expected, while for open shell clusters, the resonance splits up into several peaks as we have previously shown for Na_{11}^+. For large sizes the single peak fit indicates that the deformation of large clusters is less pronounced than for small ones in agreement with the Nilson diagram tendency [9]. However for large masses the cluster deformation can contribute to broaden the cross section profile. As cluster size increases the absorption cross section rises to extremely high values as 2000 $A^{\circ 2}$ for K_{900}^+ in agreement with the sum rule. This shows that for such large clusters all the valence electrons are equally involved in the collective excitation.

The blue shift behavior is of fundamental interest to test the validity of the different models. A simple model for the collective resonance is contructed in terms of the Mie solution for oscillating conducting sphere driven by an electromagnetic fiel. The optical response is given by a damped resonance form (1) with a resonance frequency :

$$h\nu_0 = h \left[\frac{n\,e^2}{m\,\alpha} \right]^{1/2} \tag{2}$$

where α is the static polarizability of the sphere.
For a classical monovalent conducting sphere

$$\alpha = [r_s\, n^{1/3} + \delta]^3 \tag{3}$$

where r_s is the Wigner-Seitz radius and δ the spill out of the electron cloud over the ionic sphere. In the large size limit the classical resonance frequency varies as $n^{-1/3}$:

$$h\nu_0 = h \left[\frac{e^2}{m\,r_s^3} \right]^{1/2} \left[1 - \frac{3}{2} \frac{\delta}{r_s} n^{-1/3} \right] \tag{4}$$

With this assumption the blue shift as cluster size increases is due to the spill out of the electrons. It has to be noted that the first factor of the product is the surface plasmon of the bulk having a spherical symetry. By analogy with nuclear physics the many-body formalism known as the RPA (Radom Phase Approximation) has been extended to clusters [10]. We calculated the electromagnetic response of the closed shell potassium cluster ions K_n^+ for n=9,21,41,59,93,139,255,339,441,563 from the available program of Bertsch [10], that we have modified for large masses.

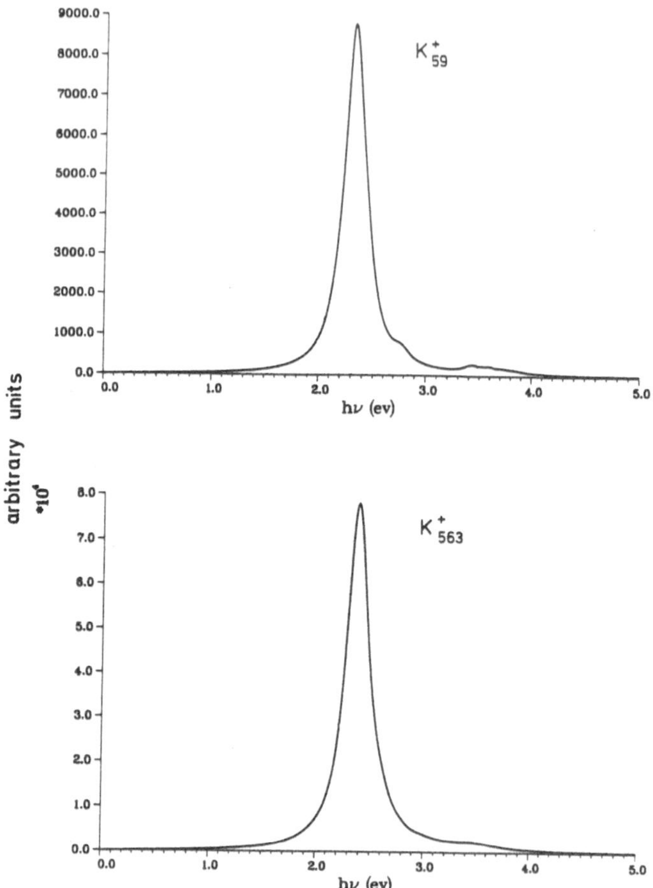

Figure 2: Calculated photoabsorption cross section for the two closed shell K_{59}^+ and K_{561}^+ clusters after modification of Bertsch's programe [10]. The width of the resonance is arbitrarily fixed to 0.2 eV. The Wigner seitz radius is taken as r_s=4.86 a.u.

Two examples of such calculation are given in figure 2. It is clearly seen that the oscillator strength of the collective state is concentrated in a large peak shifted to the blue as cluster size increases. However a fragmentation of the oscillator strength is visible in the blue wing of the main peak which is consistent with experimental data of large masses. On figure 3 are plotted the resonance frequencies of the calculated and the experimental data referred to plasmon bulk value as well as the classical behavior of a conducting sphere. It is clearly seen that experimental points lie below the theoretical ones and slowly converge to the bulk. This discrepancy asks an interesting question whether or not ionic core positions play a role in the electronic collective response. Moreover the experimental procedure implies "hot" clusters which may be taken into account more precisely in the calculation.

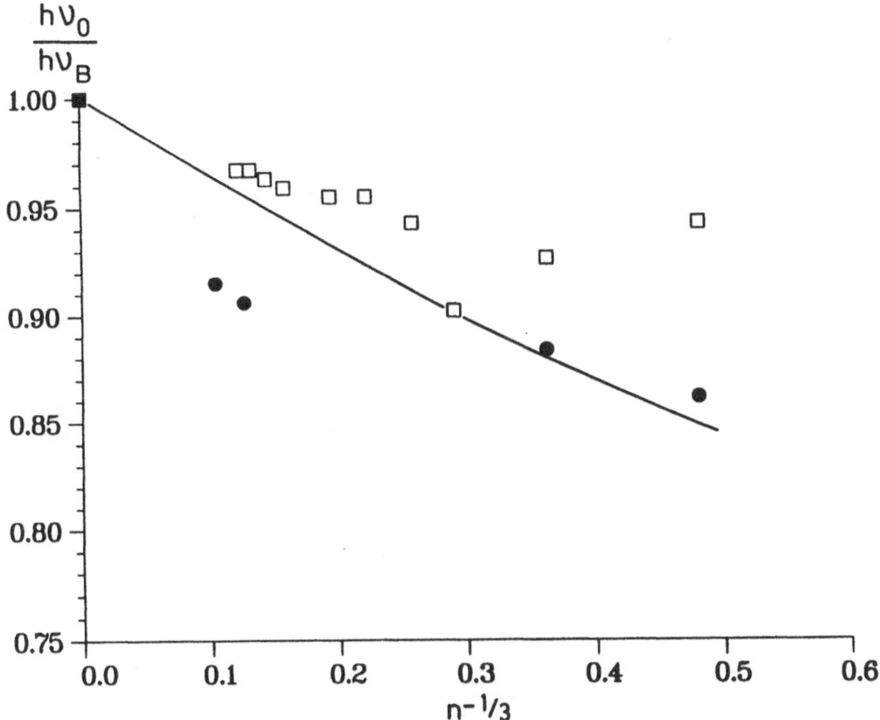

Figure 3 : Resonance frequency referred to the bulk limit versus $n^{-1/3}$. The solid line represents the classical conducting sphere calculation. Squares are the result of LDA/RPA calculation and dots are the experimental values.

2.ATOMIC GIANT RESONANCE IN ANTIMONY CLUSTERS

Many body effect can be observed in free atom when the number of electrons involved in a shell is not too small. For example, this is the case for the elements having a "d" closed shell leading to d $\rightarrow \varepsilon f$ excitation. However cooperative effects are most significant when the inner well of the effective potential of the excited electron becomes more binding. Then it involves resonant ionization of inner-shell electrons for which the escaping electron is temporally trapped by a centrifugal barrier. This appears as a broad resonance in the continuum of ionization. The persistence of this resonance in solid indicates that initial and final atomic state wave functions are only little perturbed by the environment.

We have studied the atomic 4d-εf giant resonance in antimony clusters as a function of cluster size in order to test how the molecular construction could affect the atomic behavior. In fact, besides being important semiconductor doping materials, the group V elements are well suited for study molecular construction. Vapor phases of these elements are mainly composed of tetramer species for P and As; atoms, dimers and tetramers for Sb; atoms and dimers for Bi. It has been shown that antimony clusters generated by gas aggregation technique mainly consist of Sb_{4p} clusters [11].

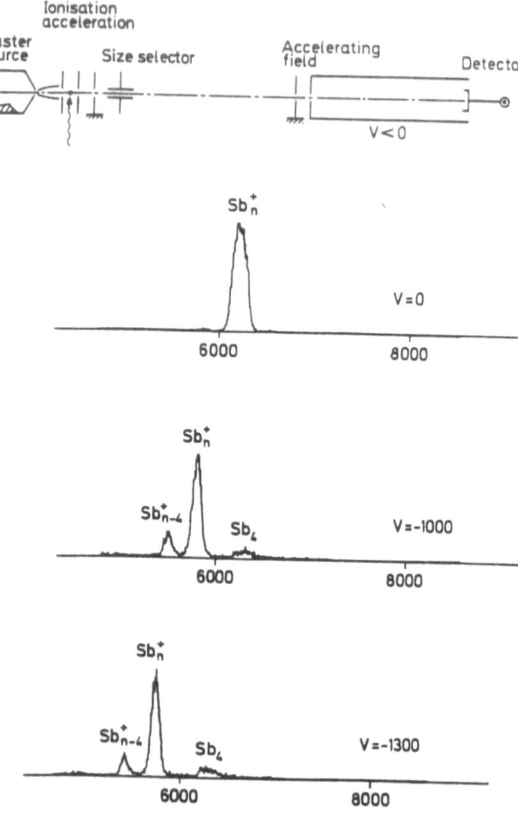

Figure 4 : Unimolecular dissociation of Sb$_n$. Upper diagram represents the experimental set up involving a tandem time of flight. Lower traces are three unimolecular dissociation spectra for three different accelerating voltages. Notice that the neutral ejected tetramer Sb$_4$ remains at the time arrival of the parent.

To go further we performed the unimolecular dissociation of mass selected cluster ions Sb$_n^+$ and put into evidence (figure 4) that for $n \geq 8$ all cluster ions evaporate a tetramer. Only the dimer is observed for Sb$_5^+$ and Sb$_6^+$ For Sb$_7^+$ both channels compete.

The photoionization cross section of neutral antimony cluster beam formed by gas-aggregation technique is achieved in 25 eV to 120 eV range by synchrotron radiation. The monochromatized-undulator radiation exit of the Super Anneau de Collisions d'Orsay (super ACO) storage ring provides more than 10^{14} photons A$^{\circ -1}$ s^{-1} which is focused at right angles to the neutral cluster beam to ionize the clusters. The cluster ions are then mass selected by a commercial quadrupole mass spectrometer modified to reach more than 8000 mass units. Photoion signal is normalized with respect to the incident photon flux. Figure 5 represents the binding-energy diagram of monoelectronic levels in atomic and bulk animony. In this energy range the photoionization involves the inner 4d-np transitions as well as the 4d-εf giant resonance. Concerning the 4d-np excitation, the corresponding energy is larger than the 5p ionization potential, and the ionization is achieved by Auger process.

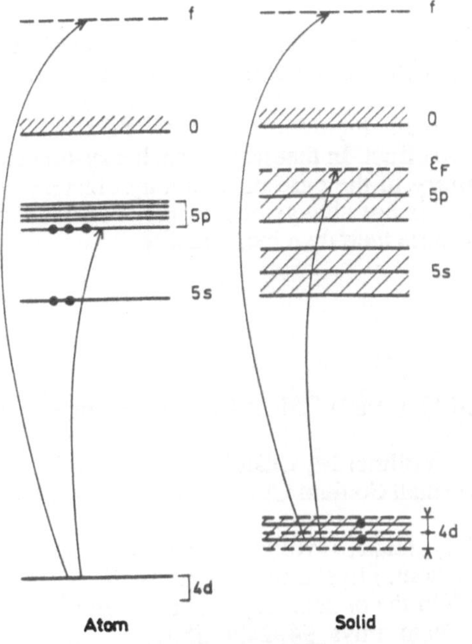

Figure 5 : Binding energy diagram of monoelectronic levels in antimony. Arrows represent the absorption from the 4d level to the 5p and εf levels respectively.

As cluster size increases this transition energy must converge to the 4d binding energy of the bulk referred to the Fermi Level. Concerning the 4d-εf excitation, which could be considered as the coherent excitation of the whole 4d atomic subshell, the ionization provides from the direct coupling between the resonance level and the continuum of ionization. On figure 6 is shown an example of the photoionization cross section of Sb_5 where the atomic behavior is clearly present. We have recently demonstrated [7] that such a behavior depends on the cluster size and on the cluster formation. In particular when the antimony cluster distribution mainly consists of Sb_{4p} species the atomic like ionization spectrum disappears. Two possibilities may be envisaged for such a spectacular behavior either the atomic transitions are forbidden in the tetramer-building block cluster, or the fragmentation, which cannot be ruled out from an experiment, is totally different for Sb_{4p} than for other clusters contrarily to unimolecular dissociation. In any case the cluster ionization by collective atomic excitation is perturbed by cluster environment.

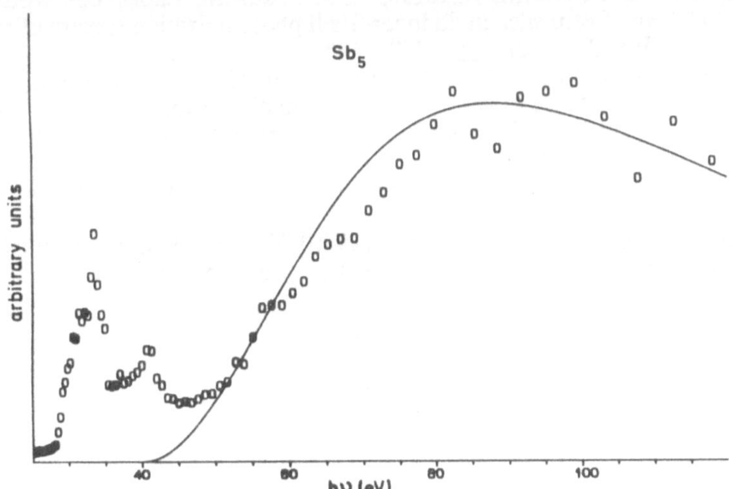

Figure 6 : Photoionization spectra of Sb_5 from ref [7].

3.CONCLUSION

The existence of giant resonances in excitation spectra of clusters and their evolution with cluster size is an interesting tool for studying the bonding of the clusters. In the case of alkalies the collective excitation of the valence "s" electrons is developped as cluster size increases. We have shown that up to one thousand of atoms the correlations between the valence electrons induce collective effect. In that respect such a system can be considered as a super atom. On the contrary in the case of antimony clusters the collective effect which exists in the free atom is not always observable. For peculiar tetramer block architecture constituting atoms loose their own individuality.

REFERENCES

[1] Connerade.J.P, Esteva.J.M, Karnatak.R.C, (1987) "Giant Resonances in atoms, Molecules and Solids" Plenum New york.

[2] de Heer.N, Selby.K, Kresin.V, Masui.J, Vollmer.M, Châtelain.A, Knight.W, (1987) "Collective Dipole Oscillation in small Sodium Clusters" Phys. Rev. Lett. 59, 1805-1808.

[3] Bréchignac.C, Cahuzac.Ph, Carlier.F, Leygnier.J, (1989) "Collective excitation in closed-shell Potassium cluster ions" Chem. Phys. Lett. 164, 433-437.

[4] Wang.C, Pollack.S, Kappes.M, (1991) "On the optical response of Ni and its relation to computational prediction" J.Chem. Phys. 94, 2496-2501.

[5] Fallgren.H, Martin.T.P, (1990) "Photoabsorption of Cs_8 and Cs_{10} clusters" Chem. Phys. Lett. 168, 233-238.

[6] Blanc.J, Broyer.M, Chevaleyre.J, Dugourd.Ph, Kühling.H, Labastie.P, Ulbricht.M, Wolf.J.P, Wöste.L, (1991) "High resolution spectroscopy of small metal clusters" Z. Phys. D 19, 7-12.

[7] Bréchignac.C, Broyer.M, Cahuzac.Ph, de Frutos.M, Labastie.P, Roux.J.Ph, (1991) "Shape Resonance in 4d inner-Shell photoionization spectra of antimony clusters" Phys. Rev. Lett. 67, 1222-1225.

[8] Bréchignac.C, Cahuzac.Ph, Kebaili.N, Leygnier.J, Sarfati.A, Submitted "Collective Resonance in large free potassium cluster ions".

[9] Clemenger.K, (1985) "Ellipsoïdal shell structure in free-electron metal clusters" Phys. Rev. B 32, 1359-1362.

[10] Bertsch (1990) "An RPA program for Jellium spheres" Compt.Physics commun 60, 247-259.

[11] Sattler.K, Mühlbach.J, Rechnagel.E, (1980) "Generation of metal clusters containing from 2 to 500 atoms" Phys. Rev. Lett. 45, 821-824.

A FIRST PRINCIPLES INVESTIGATION OF ALUMINUM CLUSTERS: GEOMETRIES, REACTIVITIES, STABILITIES AND POLARIZABILITIES

Mark R. Pederson
Complex Systems Theory Branch - 4692
Naval Research Laboratory
Washington DC 20375-50000

ABSTRACT. The local-spin density (LSD) formalism has been used to predict equilibrium geometries, cohesive energies and electron affinities for a variety of aluminum clusters (N=6, 8, 13, 23 and 55). For a 13-atom cluster, the icosahedral geometry is energetically more stable than the octahedral geometry. In contrast, for the 55-atom cluster, octahedral symmetry is preferred over icosahedral symmetry. Of the moderate sized clusters that we have studied, we note the 13- and 23- atom clusters exhibit large electron affinities.

1. Introduction

Due to the potential for new technologies and the inherent fundamental interest, there has been a substantial amount of scientific effort aimed at an enhanced understanding of clusters. From the technological point of view, the electronic structure of clusters is expected to be tunable with respect to size and composition. Providing clusters of arbitrary size and composition may be synthesized as stable entities, it may then be possible to tailor make devices with prescribed electronic or optical characteristics. Because of their relative simplicity, many researchers have been concentrating on aluminum clusters.[1-10]

Recent experimental work by Castleman and coworkers have indicated that negatively charged 13- and 23- atom aluminum clusters exhibit anomolously low reactivities.[5] Within the context of the jellium model, several researchers have noted that the low reactivity of these clusters is at least partially explained by the fact that these cluster sizes correspond to closed shell systems consisting of 40- and 70- electrons respectively.[5-6] The response of aluminum clusters to external fields has also been measured recently by de Heer et al[8] and there have been studies within the framework of the jellium model by several researchers. [4,9,10] The results of the experiments suggest that the jellium models provide relatively accurate polarizabilities for aluminum clusters larger than about 40 atoms. However, for smaller clusters, the precise details of the electronic structure appear to be rather important for quantitatively accounting for size-dependent oscillations in the observed polarizabilities. In addition to the dipole-coupled band gap, polarizabilities may also depend rather strongly on the geometries of atoms. Regarding geometries, we note that there has been a great deal of theoretical interest in predicting the transition between octahedral and icosahedral geometries for metallic systems.[1-3] Although very large nobel-gas clusters (up to 3000 atoms) have been observed to be icosahedrally coordinated, the transition for metals has not been experimentally determined.

In this paper, we present results from a variety of simulations on aluminum clusters. The theoretical and computational framework that has been used is briefly discussed in Sec. 2. In Sec. 3, we discuss our calculations on icosahedral and octahedral (13 and 55 atom) aluminum clusters and compare the results to recent work by Yi et al[2] and Cheng et al.[1] To help understand the

P. Jena et al. (eds.), Physics and Chemistry of Finite Systems: From Clusters to Crystals, Vol. II, 861–866.
© 1992 Kluwer Academic Publishers.

high abundances of negatively charged 13- and 23- atom aluminum clusters we compare the electron affinities and electronic structure for these systems to that of several other clusters of aluminum atoms. The negatively charged 13- and 23- atom ground state clusters are indeed closed shell and the electron affinities of these systems are found to be relatively large in comparison to other cluster sizes. Finally, in Sec. 5, we discuss the static polarizabilities for several clusters and compare to the experimental work of de Heer[8] and the existing jellium calculations[4,9-10].

2. Theoretical and Computational Details

The calculations discussed here have employed our all-electron density-functional-based cluster codes which have been discussed in detail in a variety of places recently.[11-14] Here, we briefly note that the total energy of a system consisting of N electrons and M nuclei is parameterized within the local spin density approximation in terms of the electronic spin densities and the nuclear positions. The spin densities are derived from a set of Kohn-Sham orbitals[15] that self-consistently satisfy a Schroedinger equation. In this work, solution of the Schroedinger equation is accomplished by expanding the Kohn-Sham orbitals in terms of a linear combination of Gaussian type functions. The orbital expansion coefficients are adjusted to minimize the total energy. A recent advance that has been important for calculations on metallic clusters is a method that allows for a variational treatment of the occupation numbers. For metallic clusters the occupation numbers associated with states at the Fermi level range between zero and unity and there is no _a priori_ method for choosing the correct value. In Ref. [13], a method that allows one to correctly determine the Fermi-level occupations is discussed. To find the equilibrium geometries, we calculate the Hellmann-Feynman forces[12] and use the conjugate gradient algorithm to adjust the nuclear positions. A discussion of our implementation of these techniques appears in Refs. [14].

3. The Icosahedral-Octahedral Transition for Aluminum Clusters

The transition from icosahedral to octahedral symmetries in metallic clusters appears to be a clear-cut example of a problem that is easier to address from a theoretical point of view than an experimental point of view. To address the question of the icosahedral-octahedral energy difference for the metallic aluminum clusters we have performed quasi-dyanamical simulations on 13- and 55-atom clusters. Due to the fact that these codes have been designed to study isolated clusters, rather than periodic supercells, and also that the codes allow for geometrical optimizations within symmetry constraints, the codes are particularily well suited for addressing this question. We discuss the 13- and 55- atom clusters separately.

3.1 THIRTEEN ATOM ALUMINUM CLUSTERS

For the neutral 13-atom clusters, we have performed complete optimization of the icosahedral and octahedral nuclear and electronic degrees of freedom. The equilibrium geometry for the icosahedral ground state is realized by placing an atom at (0,0,0) and twelve atoms at the icosahedral sites that are generated from (4.347,2.687,0). The equilibrium geometry for the octahedral ground state is obtained by placing an atom at (0,0,0) and twelve atoms at the octahedral sites which are equivalent to (3.615,3.615,0). We find that the neutral icosahedral geometry is more stable by 1.1 eV. For comparison we note that Cheng _et al_[1] and Yi _et al_[2] (both using a density functional approximation but different numerical schemes) also find the neutral icosahedral geometry to be more stable by 1.6 and 0.5 eV respectively.

The Fermi levels for the icosahedral and octahedral ground states are found to be -5.44 and -5.25 eV respectively. For very large metallic systems, the Fermi level corresponds to the electron affinity. Roughly speaking, the electron affinity increases with the magnitude of the Fermi level. This implies that the negatively charged icosahedral cluster is expected to be further stabilized in comparison to the negatively charged octahedral cluster since the Fermi level of the 13-atom icosahedral cluster is lower than that of the octahedral cluster. This is indeed what is found. As shown in Table 1, the

Table 1. Nearest-Neighbor distance (R), cohesive energy per atom (U), Fermi level (E_f), number of electrons/total states at the Fermi level [$N(E_f)$], and the electron affinity (A) for the neutral thirteen-atom icosahedral and octahedral aluminum clusters.

Symmetry	R	U	E_f	$N(E_f)$	A
Octahedral	5.11	3.05	-5.25	5/6	3.13
Icosahedral	5.10	3.14	-5.44	11/12	3.70

electron affinity for the icosahedral cluster (3.70 eV) is 0.57 eV larger than that of the octahedral cluster. Also apparent from Table 1 is the fact that partially occupied states at the Fermi level are expected to lead to slight Jahn-Teller distortions for the unpolarized neutral species. However the addition of one electron leads to a shell closing for both clusters. This point is discussed further in Sec. 4.

3.2. FIFTY-FIVE ATOM CLUSTERS

Complete geometrical optimizations on the spin-unpolarized icosahedral and octahedral 55-atom aluminum clusters have been performed as well. The resulting geometries were optimized until the force on each atom was smaller than 0.13 eV/Bohr. The octahedral ground state consists of atoms at all sites that are equivalent to (0,0,0), (3.677, 3.677, 0), (7.724,0,0), (7.566,3,724,3.724) and (7.342,7.342,0) where all coordinates are in Bohr. The icosahedral ground state geometry is defined by placing atoms on all sites that are equivalent to (0,0,0), (4.309,2.663,0), (0, 0,8.844) and (8.723, 5.428,0). For the spin unpolarized case, the octahedral cluster is found to be 0.65 eV lower than the icosahedral cluster. This result is in accord with two other theoretical calculations that predict an octahedral geometry that is more stable by 0.5 eV[1] and 1.9 eV[2]. In Fig. 1, the electronic structure and occupation numbers of the states near the Fermi level is presented for both equilibrium geometries. As shown in the figure, for the octahedral Fermi level, there is an accidental degeneracy between two 3-fold representations and one 1-fold representation. The seven available electrons prefer to occupy the three-fold representations leaving the 1-fold representation devoid of electrons. In addition to the exact accidental degeneracy at the Fermi level, it is interesting to note that two representations occupied by a total of ten electrons are nearly degenerate lying only 0.3 eV below the Fermi level. Also pictured in Fig. 1, is the character of the states near the Fermi level for the icosahedral geometry. The icosahedral Fermi level is elevated with respect to the octahedral Fermi level and the number of electrons at the Fermi level has decreased to three. In analogy to the thirteen atom case, the cluster that is energetically more stable has a lower Fermi level indicating that the negatively charged fifty-five atom octahedral cluster would be further stabilized over the negatively charged icosahedral cluster.

While the spin unpolarized systems are expected to be slightly unstable due to the partially occupied states at the Fermi level, for the octahedral case, the instability can be quenched by allowing for spin polarization. For the octahedral cluster, it is clear from Fig. 1 that the accidental degeneracies between the two three-fold and the single one-fold representations at the Fermi level allow the spins of the seven electrons to align and form a closed shell system. This calculation has been carried out leading to a spin polarized (S=7/2) system that is 0.16 eV lower than the spin unpolarized system. The resulting eigenvalues shift slightly from that shown in Fig. 1 with the one-fold representation appearing below the three-fold representations. The icosahedral cluster is also susceptible to spin-polarization. However the effects due to spin polarization are expected to be smaller than in the octahedral system since there are only three electrons at the Fermi level.

To summarize the results of this section, we note that the 13-atom icosahedral cluster is definitely more stable than the octahedral cluster but the situation is reversed for the 55-atom case.

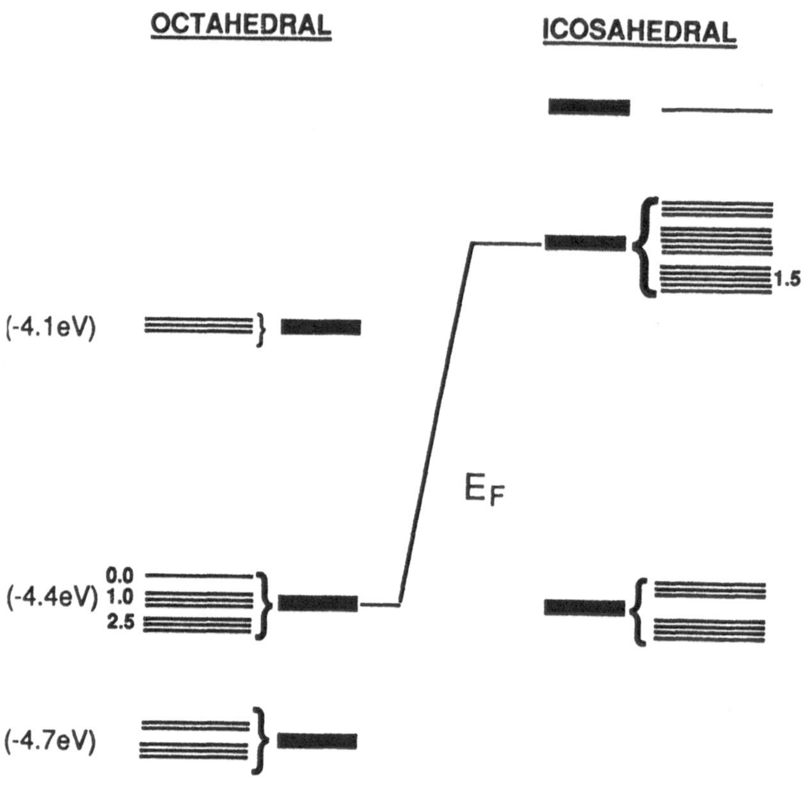

Figure 1. A comparison of the electronic structure for the spin unpolarized fifty-five atom aluminum clusters for octahedral and icosahedral clusters. In the octahedral case there are three accidental dengenacies which yield a total of fourteen degenerate states to be occupied by seven electrons. Accidental degeneracies persist in the icosahedral case also. For the icosahedral symmetry there are a total twenty-four states occupied by three electrons.

4. The Low Reactivity of negatively charge Al_{23} and Al_{13} clusters

Recent work by Leuchtner et al[6] indicates that both Al_{23} and Al_{13} are relatively unreactive in comparison with other aluminum clusters. Commensurate with the observed low reactivity of Al_{23} is a pronounced dip in the polarizability per atom observed by de Heer et al;[8] to the best of our knowledge, such data for Al_{13} is unavailable. In the context of the jellium picture, the low reactivity of Al_{23} and Al_{13} has been explained by the fact that the negatively charged clusters correspond to closed shell systems. While the calculations of this paper confirm that these negatively charged systems are indeed closed shell the relatively large electron affinities for these systems serve to further quantify the low reactivity.

There are several high-symmetry twenty-three atom aluminum clusters that are expected to be quite stable and it is relatively difficult to predict which ones are expected to be the most stable from bonding arguments. To address this question, we have performed optimizations of several different high symmetry clusters. While we have looked at a total of five different structures, in this paper we discuss only one of these structures. A complete description of the other structures will appear elsewhere

One low-energy Al_{23} cluster that has been found is best described as a nineteen atom FCC seed with four additional atoms placed along the (1,1,1) (1,-1,-1) (-1,-1, 1) and (-1,1,-1) directions. The overall symmetry of the cluster is reduced to tetrahedral and the commensurate relaxations of the inner nineteen atoms from their ideal octahedral sites is relatively severe. However, since clusters are often obtained by sputtering off of aluminum surfaces, the existence of 19-atom FCC seeds is quite possible. For the ground state of this cluster we find a binding energy per atom of 3.27 eV per atom. In Table 2, the binding energy per atom is exhibitted as a function of the number of atoms and symmetry for several different cluster sizes (N=6, 8, 13, 23 and 55). Since the cohesive energy per atom is not observed to be anomolously large for Al_{23} and Al_{13} it is indeed unlikely that the high abundance of these structures can be explained in terms of cohesive energies. However, the electron affinities of icosahedral thirteen atom and the octahedral twenty-three atom cluster are quite a bit larger than those associated with the smaller clusters. The results of Table 2 suggest two effects that would lead to the observation of a high abundance of negatively charged thirteen and twenty-three atom clusters. First given a gas with a uniform size distribution of neutral clusters and a significantly smaller number of free electrons, the clusters with the largest electron affinity would preferentially soak up the extra electrons leading to a higher abundance of negative charge states of the clusters with large electron affinities. Second, the reactivity of a negatively charged cluster is expected to decrease as the (neutral) electron affinity increases. Since electron affinities tend to be large for systems that need one electron to close a shell, the latter point is related to the arguments for an abundance of 13- and 23- atom clusters that are based on shell closing.

Table 2. Cohesive energies per atom (U), electron affinities (A) and polarizabilities per atom (P) as a function of size and symmetry. In this table the Gap refers to the smallest energy difference between occupied and unoccupied states that are dipole coupled. The labels O_h, Y and T_d refer to octahedral, icosahedral and tetrahedral symmetry respectively.

Cluster	Symmetry	U(eV)	A(eV)	Gap (eV)	P (A^3)
Al_6	O_h	-2.45	-2.80	1.66	6.3
Al_8	O_h	-2.48	-1.82	0.14	7.4
Al_{13}	O_h	-3.05	-3.13	1.66	6.0
Al_{13}	Y	-3.14	-3.70		
Al_{23}	T_d	-3.27	-3.60		
Al_{55}	O_h	-3.47			
Al_{55}	Y	-3.46			

5. Static Polarizabilities of Aluminum Atoms

As a first step toward a first-principles understanding of electronic polarizabilities of clusters we have placed several different octahedral aluminum clusters in DC electric fields and calculated their response. This is accomplished by adding a term to the external potential of the form $\underline{E}.\underline{r}$. For the octahedral clusters discussed here that are symmetric under inversion, all odd derivatives vanish. Further, the invariance of the clusters under rotations by 180° about one of the principal axes guarantees that the mixed second derivatives vanish as well. Hence, for the systems discussed here, the polarizability is simply the second derivative the total energy with respect to the applied field. In Table 2 the polarizabilities for Al_n (n=6, 8 and 13) are presented along with the dipole coupled band gap. For n=6 and 13, the dipole coupled band gap is coincidentally the same and the polarizabilities are similar as well. For n=8, the dipole coupled band gap is very small (0.15 eV) and the polarizability per atom is observed to increase.

In recent experimental work De Heer $\underline{et\ al}$[8] measured electronic polarizabilities as a function of cluster size. They found that for cluster sizes larger than 40 atoms, polarizabilities from jellium calculations[4,9,10] compared favorably with experiment. However, for smaller clusters, the jellium calculations tended to overestimate the experimental measurements and did not reproduce the oscillations with cluster size. While there are no experimental measurements below n=17, the polarizabilities presented in Table 2 are smaller than those taken from jellium calculations and do indeed exhibit oscillations as a function of cluster size.

6. References

[1] H. P. Cheng, R. S. Berry, and R. L. Whetten, Phys. Rev. B 43, 10647 (1991).
[2] J. Y. Yi, D. J. Oh, J. Bernholc and R. Car, Chem. Phys. Lett. 174, 461 (1990).
[3] L. L. Boyer, M. R. Pederson, K. A. Jackson, and J. Q. Broughton, Proc. 1990 MRS Fall Meeting.
[4] A. Rubio, L. C. Balbas and J. A. Alonso, Solid State Commun. 75, 139 (1990).
[5] A. Rubio, L. C. Balbas and J. A. Alonso, Physica B 168, 1991.
[6] R. E. Leuchtner, A. C. Harms, A. W. Castleman, J. Chem. Phys. (1990); R. E. Leuchtner, A. C. Harms and A. W. Castleman, J. Chem. Phys. 91, 2753 (1989).
[7] K. E. Schriver, J. L. Persson, E. C. Honea and R. L. Whetten, Phys. Rev. Lett. 64, 2539 (1990).
[8] W. A. de Heer, P. Milani and A. Chatelain, Phys. Rev. Lett. 63, 2834 (1989).
[9] Puska et al, Phys. Rev. B 31, 3487 (1985).
[10] Kresin, Phys. Rev. B 39, 3042 (1989).
[11] M. R. Pederson and K. A. Jackson, Phys. Rev. B 41, 7453, (1990).
[12] K. A. Jackson and M. R. Pederson, Phys. Rev. B 42, 3276, (1990).
[13] M. R. Pederson and K. A. Jackson, Phys. Rev. B 43, 7312, (1990).
[14] M. R. Pederson, in Proceedings of the 3rd International Conference on Supercomputing and Second World Supercomputing Exhibition, Edited by L. P. Kartashev and S. I. Kartashev (International Supercomputing Institute, New York, 1988), Vol I, p 179.
[15] P. Hohenberg and W. Kohn, Phys. Rev. 136, B864 (1964); W. Kohn and L. J. Sham, Phys. Rev. 140, A1133 (1965).

OPTICAL PROPERTIES OF MACROSCOPIC MANY-CLUSTER MATTER

U. KREIBIG
I. Physical Institute of the RWTH
Sommerfeldstraße 28
D-5100 Aachen / Germany

ABSTRACT. Real cluster samples in nature as well as in technology usually contain large numbers of clusters. Most commonly they are macroscopic in 2 or 3 dimensions. Such systems have been designated CLUSTER MATTER. A classification scheme for various kinds of these substances is presented and used to distinguish between contributions of the individual clusters and the cluster collectives to their optical properties. Noble metal clusters are treated as examples since they allow of producing well defined model samples, and their optical properties are governed by distinct plasmon polariton excitations. The theory of Mie, the generalized Mie theory, effective medium theories and the Oseen theory are introduced to describe the optical properties of the clusters, the cluster aggregates and the samples as a whole.

In the second part optical properties of the cluster collectives due to four different kinds of cluster-interaction effects are illustrated by experimental results: The percolation, the electromagnetic coupling via near field scattering waves, the plasmon band inversion due to multiple extinction and the transition of the far field scattering towards regular geometrical reflection.

1. Introduction

Fundamental cluster research is presently concentrated upon the individual cluster. However, technical and practical applications usually require systems of very many clusters forming samples which are macroscopic in two or three dimensions. Also most of the wide-spread occuring natural cluster systems are of this type. This kind of matter has been designated CLUSTER MATTER. The clusters may be distributed in free space or in gaseous, liquid or solid embedding substances, they may be mobile or fixed at their locations, the sample topology, i.e. the local arrangement of the clusters, may be regular or statistically distributed, the clusters may be separated or touching each other: there is a vast variety of structural and, consequently, of physical properties these materials can have. CLUSTER MATTER is a particular case of inhomogeneous or granular matter, since typical structural elements are large compared to the atomic scale. It may also be regarded to belong to the wide class of mesoscopic systems. (In conformity with present trends, the use of the term "cluster" is not limited to specific size regions. For the following we merely assume the size to be markedly smaller than the wavelength of light.)

In this paper, an attempt to classify the broad variety of sample topologies is followed by a brief review of the optical properties which are determined by the interplay of the properties of the INDIVIDUAL building units, i.e. the clusters, and the COLLECTIVE

P. Jena et al. (eds.), Physics and Chemistry of Finite Systems: From Clusters to Crystals, Vol. II, 867–879.

properties, both of the building blocks and of the extended sample. Two series of extinction spectra in Fig. 1, obtained from many cluster samples, produced with almost identical Au clusters show that the increase of cluster packing density can result in strongly differing optical properties. As will be explained in the following, these differences are due to the influences of different collective effects.

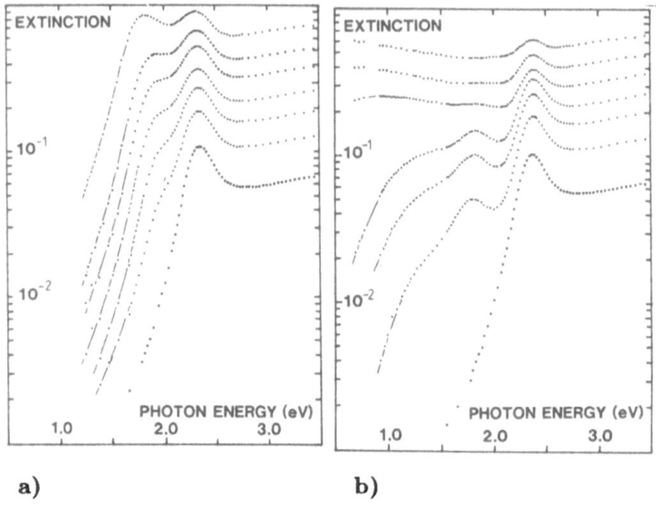

Fig. 1a),b):

Measured extinction spectra of Au-cluster-systems with varied amount of aggregation, increasing from bottom to top.

The first curve is the spectrum of separated clusters.

Cluster-sizes $\sim$17nm. (The spectra are shifted vertically for arbitrary amounts.) (From [5].)

Throughout, we assume the cluster sizes to be beyond the molecule-solid state transition but small compared to the wavelengths of light, and we restrict ourselves to noble metal clusters whereever numerical or experimental results are discussed. These clusters have the advantage to allow for producing fairly well defined cluster matter samples and their optical spectra to be sensitively modulated by distinct plasmon polariton features. The theory of Mie and Debye [1] is introduced for the building units, i.e. the clusters, the generalized Mie theory (GMT) [2] is applied to the description of building blocks (i.e. cluster aggregates) and an effective medium ansatz may, under proper conditions be used for the samples as a whole [3].

Four different kinds of collective cluster-interaction effects which give rise to the macroscopic optical properties of the samples can be distinguished:

The strongest is the direct electronic and lattice coupling due to the coalescence of neighboring clusters, which leads to electrical percolation and, eventually, to "nanostructured" ("nanocrystalline", "nanophase") material.

The second is the electromagnetic coupling via near-field scattered waves which leads to changes of the effective cluster polarization. It is described by the GMT.

The third and fourth collective effect concern the transmitted and scattered far field waves, respectively. It is the multiple extinction effect which always occurs in many-cluster systems and is important if these are optically "thick". Then, e.g. plasmon band inversion in rough analogy to the line inversion in atomic spectroscopy may occur in the optical spectra. In special, if scattering and absorption in the sample differ in their wavelength dependences, which is usual for larger clusters and for samples containing more than one kind of scatterers, the measurable spectral features may be markedly changed.

The fourth collective effect concerns the Mie scattering waves, the angular distribution of which is drastically influenced by interference in samples of proper topologies. Then, the distributed Mie scattering of the individual clusters is transformed into the regular geometric optical beams of quasi-homogeneous matter, as was described by Oseen [4] for the arrangement of atomic or molecular dipoles.

All four effects were experimentally identified and are demonstrated here by selected results. They manifest that optical measurements may, under appropriate conditions concerning the sample structure, yield insight not only into the electronic and geometric structure of the individual clusters but also into the different collective properties which are responsible for the widely varying optical properties of real CLUSTER MATTER samples. On the other hand they point to the difficulties to extract individual cluster properties from experiments performed on CLUSTER MATTER.

2. Classification of cluster matter

Homogeneous matter (i.e. matter which is inhomogeneous on an atomic scale, only) can be classified between single-crystalline and amorphous as limiting cases. Analogously, CLUSTER MATTER spreads this field with the single cluster as building unit. Yet, in reality, the crystalline ordering has only been successfully obtained in very few cases, e.g. by dense-sphere packing. More common, by far, are topologies with particular statistical local distributions, and additional intermediate structural elements consisting of a multitude of clusters each, are often important, as, e.g., in quasi-fractal samples (see Section 5). These elements are called the building blocks.

Tab. 1 gives a systematic compilation of the ample field of various kinds of topologies. The building units, i.e. the clusters, described by the given set of parameters are thought to form more or less densely packed building blocks, i.e. cluster aggregates by coagulation (the clusters remain isolated units, even if the interlayers between neighboring clusters become small) or by (partial) coalescence (neighboring clusters are touching and grain boundaries grow between them due to atomic diffusion) or by precipitation or by Oswald ripening.

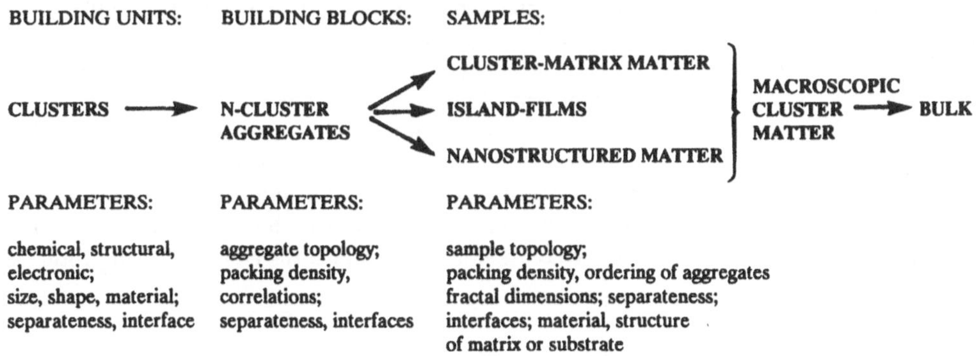

Table 1

These aggregates can be of ordered or statistical shapes and again, a set of parameters, which in given in the Table, determines their physical behavior. They reflect the large scale inhomogeneities typical of various kinds of macroscopic CLUSTER MATTER samples. Only in the case of statistically homogeneous local arrangement of the clusters, these building blocks are absent or, as alternative description, the whole sample forms one building block.

Hence, different classification criteria may be introduced than for homogeneous matter. First, the amounts of interactions among building units of same kind are important, i.e. the relative contributions of individual clusters and the cluster collectives to the resulting physical properties. Second, there are the differences between constituents of different kinds as clusters and the media they are embedded in.

The property "separateness", introduced in Tab. 1 characterizes qualitatively the differences between clusters and their surrounding, i.e. the formation of separating interfaces. It is mainly this property that distinguishes cluster matter from homogeneous material. The differences may be due to discontinuities in the chemical composition and/or physical material properties, like the density, the electrical conductivity or the formation of domains during a phase transition in homogeneous materials as, e.g. the clusters with lattice ordering during the liquid-solid transition.

If the clusters are included in an embedding and stabilizing medium, we have CLUSTER-MATRIX-MATTER. Alike the clusters, these media may be metallic, semiconductors or dielectrics. Usually, they are neccessary for stabilization since the cluster state is thermodynamically unstable due to the energy stored in the surface/interface.

The matrix may, itself, be homogeneous (crystalline, glassy, liquid, gaseous) or inhomogeneous (e.g. zeolithes, aerogels, silver bromide grains etc.). If absent, densely packed clusters directly touch and grow together more or less spontaneously by coalescence. Afterwards, they are separated only by grain boundaries and by cavities. This effect can be increased by applying external pressure.

Eventually, extremely fine-grained polycrystalline substances are produced called "NANOSTRUCTURED" (recent names were also "nanocrystalline" or "nanophase") MATERIAL, large portions of which consist of less ordered grain boundary regions. In general, collective properties are more important in nanostructured material than in cluster-matrix matter. As an example, the below described plasmon polariton bands typical of metallic single clusters, are completely lost in nanostructured samples.

It depends on the regarded physical property, whether the primary building units, the clusters, have to be regarded to be transformed into new larger, irregular units after coalescence or not. E.g. the electrically conducting areas of metallic clusters are increased, while the regions of ordered, lattice like structure remain roughly unchanged.

An intermediate place between cluster matrix matter and nanostructured material is taken by the ISLAND FILMS, which in one direction are stabilized by the additional medium of the substrate while in the two remaining dimensions, the clusters are unprotected and coalescence can occur.

3. The optical properties of cluster matter

3.1. THE CLUSTERS (BUILDINGS UNITS)

Sufficiently beyond the cluster size region of the molecule-solid state transition the dielectric response of matter on incident electromagnetic waves is formulated for spherical symmetry by the electromagnetic theory of Mie. In good metals it is governed, in the NUV, VIS and IR regions, by elementary excitations, i.e. plasmon or phonon polaritons of spherical symmetry.

Usually, embedding media are present or selected which do not contribute to the imaginary part of the polarization, yet in the case of low volume concentrations f of the clusters they determine the real part. This is the case of the original theories of Mie and of Debye. Extensions to absorbing matrices exist.

Fig. 2a shows, as an example, contributions of plasmon modes and of eddy current modes of different spherical multipole orders to the absorption and to the scattering spectrum of 100 nm Al clusters. Fig. 2b gives, also taken as an example, the angular distribution of the near field around a 20 nm Au cluster. These fields are important for the cluster interaction described in the next Section.

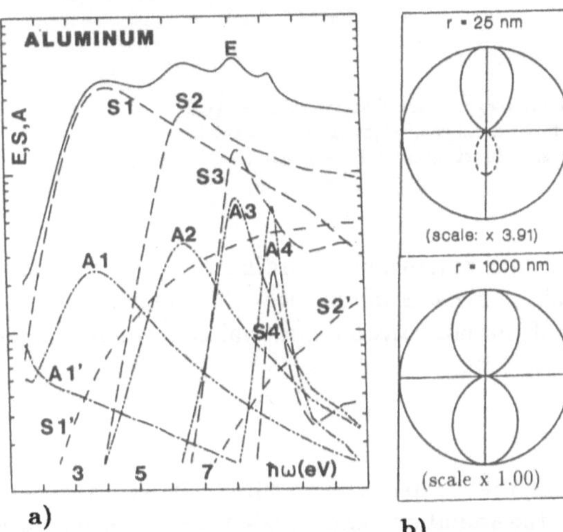

Fig. 2: a) Decomposition of the extinction spectrum (E) of 100 nm Al clusters into plasmon and eddy current contributions to absorption (A) and scattering (S) of different multipolar orders (indices 1 to 4). Primes: eddy current contributions.

b) Scattering characteristics (polar diagrams of the radial component of the Poynting vector) of a 20 nm Au cluster in the near-field (distance r = 25 nm) and the far-field (r = 1000 nm) regions. The dashed curve indicates Poynting vectors directing towards the cluster. (Calculations from Mie's theory [10].)

3.2. THE MANY-CLUSTER AGGREGATES (BUILDING BLOCKS)

The electromagnetic near-fields of different multipolar orders around each excited cluster give rise to electromagnetic coupling of all neighbors in densely packed cluster matrix matter samples. Depending on the ratios R/λ and D_{ij}/λ (R: cluster radius, D_{ij}: cluster distances in the aggregates) the retardation of these fields can play an important role. To obtain the optical properties of the aggregates, these near-fields may be summed up for each cluster to give additional contributions to the effective incident field. Including this correction, which, however, requires extensive theoretical and numerical work, the dielectric response of the whole aggregate can be determined following Mie's ansatz. Recently, the thus resulting "generalized Mie theory" (GMT) has been successfully advanced to high numerical precision [5]. As an example, Fig. 3 shows the optical extinction spectra of various Ag cluster aggregates, including $\ell = 1$ and 2 multipoles and retardation effects without any approximations. The main presupposition is that there are spacings between all clusters, though they may be small, to keep the single clusters isolated, individual units with polarizability unchanged compared to the single Mie cluster. Hence, coalescence aggregates cannot be described by GMT, which is, thus restricted to cluster-matrix matter (including the particular case of vacuum the clusters are embedded in). Since contributions of the

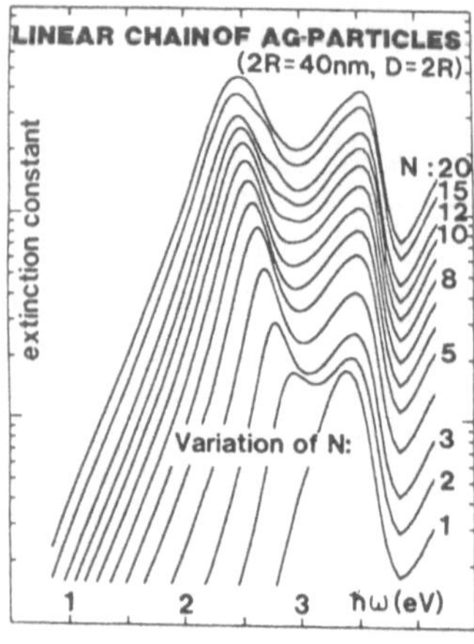

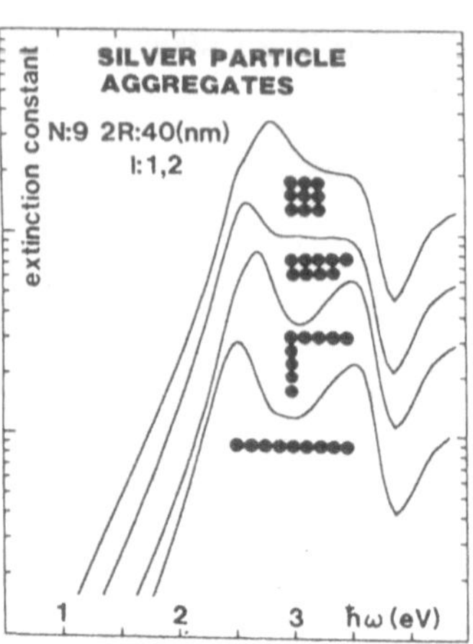

Fig. 3: Optical extinction spectra of Ag cluster coagulation aggregates of various shapes. (Unpublished GMT calculations by M. Quinten). N: number of clusters; ℓ : multipolar order; D: center-to-center distances. The dielectric function of bulk Ag is used.

single clusters are computed and are added up numerically according to their individual local positions, the GMT is limited to small aggregates of less, say 10^2 clusters, and the optical response of extended aggregates or of the macroscopic material as a whole cannot be evaluated this way.

3.3. THE MACROSCOPIC EFFECTIVE MEDIUM

In principle, there is no upper limitation for the cluster sizes in GMT. If, yet, they are very small compared to the wavelength of the radiation, and quasi-static conditions are fulfilled for volumes containing many clusters, then effective medium theories, usually applied to samples with separated single clusters can be introduced to yield an expression for the effective dielectric function of the macroscopic material consisting of large numbers of aggregates [3,6]. Fig. 4 shows as, an example, the extinction of small Ag cluster aggregates (linear triplets), statistically distributed with varying factor in the macroscopic sample. Suitably averaged aggregate polarizabilities which may be computed from GMT or from simplyfied versions of it, have been inserted into the effective medium ansatz of Lorenz-Lorentz-M.Garnett. The influences of the aggregate-aggregate interaction are obvious as they shift and raise the plasmon bands with increasing aggregate filling factor. These interactions are also electromagnetic ones in the effective medium theories though introduced in a more indirect and cumulative way than in the GMT. The advantage of this type of theories is the yield of macroscopic optical constants, the disadvantage is that single cluster properties and details of the sample topology are replaced by rough statistical distributions, and that it is limited to clusters as small that scattering can be neglected. A detailed description of electromagnetic cluster interactions is going to appear.

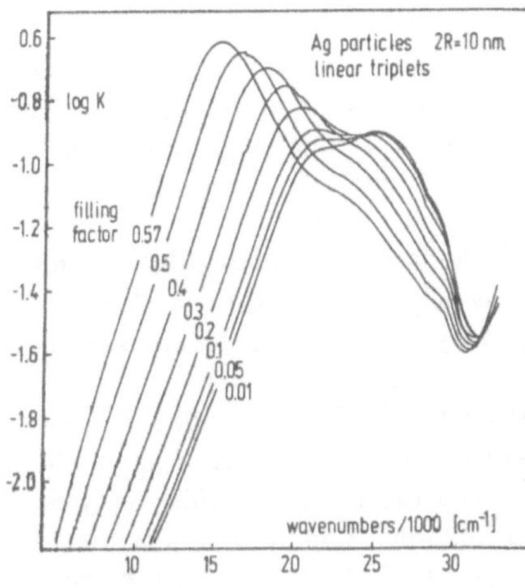

Fig. 4: Calculated absorption spectra of Effective Media consisting of linear Ag cluster triplets, randomly distributed with varying volume filling. The clusters (diameters 10 nm) are almost touching. (From [3]).

In contrast to Fig. 3 size effects are included in the dielectric function of the cluster material which partly smear out the double peak structures.

4. Discussion

Comparing Figs. 2, 3 and 4 it is obvious that the optical properties, starting from the individual cluster spectra change strongly when collective effects occur. The splitting of the cluster plasmon band into mainly two shifted components which is the consequence of the electromagnetic coupling effects, explains the double peak structure and the broadening of the experimental spectra of Fig. 1 a and the increase with increased packing density and, thus, coupling strengths. So, regarding the changes in the optical spectra, i.e. the increase of collective effects we can follow the development from the isolated clusters towards the bulk material. The electromagnetic interaction theories, however, cannot explain the flat and extremely broad measured spectral features of Fig. 1 b at high packing densities, even if effective medium effects are included at their maximum. This discrepancy will find its explanation in the next Section.

Systematic investigations of aggregate topologies [5] have shown that the splitting of the cluster plasmon peak is a direct measure of the degree of anisometry of the cluster aggregates contained in the sample. Hence, the double peak structure of Fig. 1a points to a considerable amount of chainlike cluster aggregates which are typical of diffusion limited aggregation in colloidal systems. This result was confirmed by TEM analysis of the samples under consideration.

But even at high packing densities of the aggregates, selective bands are left in the spectra, though they are strongly shifted, as in Fig. 4. As a consequence, even at dense packing the optical properties of cluster matter strongly differ from the behavior of the homogeneous bulk metal, as long as the cluster packing is restricted to coagulation and the clusters survive as individuals.

The absorption spectrum of bulk Au would, instead, be almost flat in the spectral region of Fig. 1 [8], and, thus, resemble the broadest spectrum of Fig. 1 b. In the following Section this similarity will be explained as due to the occurrence of coalescence of clusters which, by creating larger metallic units means an additional step towards the bulk crystalline matter.

5. Coalescence aggregation

In the following, an example is given to show that it is possible to follow the transition from the cluster matrix matter to nanostructured material by the according optical spectra. This transition was induced by admitting, subsequently, coalescence among neighboring clusters to form larger metallic units in a cluster matrix matter sample where the clusters were originally separated by narrow interlayers.

The samples used were aqueous colloidal systems, each cluster surrounded by a stabilizing dielectric interlayer. Such samples proved to be especially well suited to produce series of samples of varying cluster packing density by simply varying the mean thickness of these interlayers.

By chemical means or by moderate heat treatment coalescence can be induced in these samples. E.g., precursor states of percolation in quasi fractal cluster matrix systems have been, thus, studied earlier [9]. Here we will only present the results of this analysis to demonstrate the direct interaction effect described in Section I and to demonstrate that this particular step of cluster matter towards the homogeneous bulk material can be clearly observed in the spectra of optical extinction.

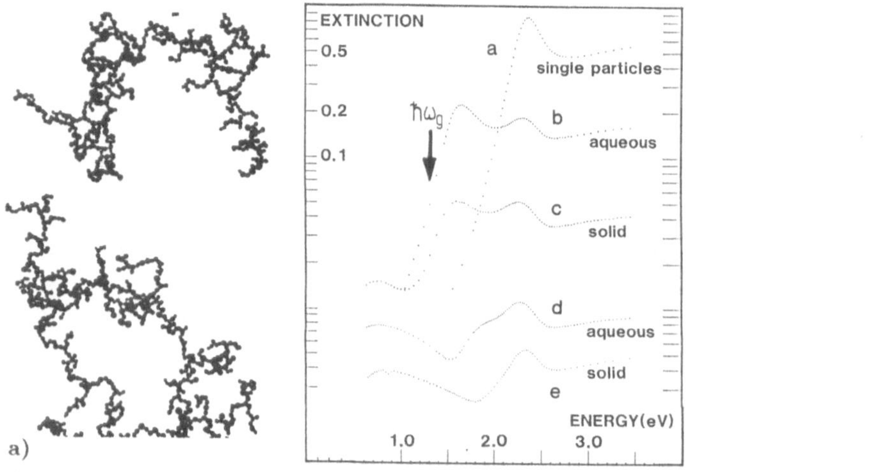

Fig. 5:   a) Quasi-fractal coagulation aggregates of 38 nm Au-clusters.
b) Optical extinction spectra measured before (curves b,c) and after (curves d,e) partial coalescence. The embedding media were H_2O and solid gelatin, respectively.
c) TEM micrograph of resulting aggregates.

Fig. 5 a shows a section of an original sample with well separated, yet closely connected clusters forming quasi-fractal coagulation aggregates. The corresponding optical extinction spectrum in Fig. 6 comprises the above mentioned typical two-peak structure which is due to electromagnetic coupling in anisometric cluster aggregates. As shown earlier [9] the amount of peak splitting depends on the aggregate topology and takes a finite asymptotic maximum for linear chains irrespective of their lengths. This limit is marked with $\hbar\omega_g$ in Fig. 5 b and is almost reached by the investigated sample.

After moderate coalescence between neighboring clusters, i.e. removing of the separating interlayers and formation of extended grain boundaries, which can be identified in Fig. 5 c, the optical spectra are changed drastically as demonstrated in Fig. 5 b. Now the low frequency peak (which is due to electromagnetic excitation with electrical vectors along the chain axis) has almost vanished, and, instead, a broad peak clearly beyond the limit $\hbar\omega_g$ has grown up.

This peak is assigned to the new larger and irregularly shaped metallic units i.e. percolation precursors formed from the clusters by coalescence. It should be pointed out that now, collective elementary excitations of the conduction electrons only take place in these new units as a whole, since charge conservation is now restricted to them. As shown in [9] the position of this new peak points to a limited correlation length of these percolation paths of about 10 cluster diameters, a result which was confirmed by TEM analysis.

As a consequence, a detailed analysis of optical spectra give insight into the formation of grain boundaries between metallic clusters, i.e. the elementary processes leading to percolation and to "nanostructured" material. The larger the metallic units become, the more shifts the absorption into the IR region. When it reaches zero frequency, the percolation structures are macroscopic and the sample acts similar to the bulk metal.

6. The transition of Mie scattering to regular reflection

This collective interaction effect does not concern the optical response of the clusters but the scattered and transmitted far field waves. They stem from all clusters in the sample, which in the most simple approach can be regarded as point dipoles in analogy to the atomic dipoles of quasi-homogeneous matter. Due to their phase relations to the incident wave these partial waves (or: Huyghens elementary waves) have finite coherency and interfere with each other. It depends on the topology to which extend the resulting macroscopic far fields, which are the only accessible to usual optical spectroscopy, are influenced. Three extremal cases are

the samples with filling factors as low that the degree of coherency goes to zero,

the samples with inhomogeneous statistical distributions of cluster positions where interference does not produce constructive or destructive interference of the observable wave,

the samples with regular crystal-like or dense cluster arrangements, where interference produces the regular beams of geometrical optics, as known from atomic matter, while waves in all other directions are cancelled.

For the atomic matter case, the problem has been intensively studied by Oseen [4]. For the example of plane incident waves, the resulting waves can be divided into a remaining scattered wave with irregular angular distribution and the two regular geometric optical plane waves, namely the reflected one and the transmitted one. The relative intensities of the latter two waves being proportional to the average filling factor f, the remaining scattering wave (i.e. the incoherent part) is determined by the local variations of the cluster packing measured by $\delta f = < (f_{local} - f)^2 >^{1/2}$, where f_{local} is an average over small volumes containing only few clusters. Obviously, δf goes to zero for regular sample topologies.

This effect does not only occur in atomic matter but also in proper samples of cluster matrix matter as has been shown recently [10].

Again, we restrict ourselves to demonstrate essential experimental results. In Fig. 6 a the optical extinction of 20 nm Au clusters is shown, both for a highly diluted and for the two dimensional, extraordinarily homogeneous sample depicted in the inset. The clusters are densely packed with separating interlayers which in this case consist of phosphine molecules bound at the cluster surface [11]. As discussed in Section 3 the plasmon polariton structure of the individual clusters is changed into a broad, shifted band.

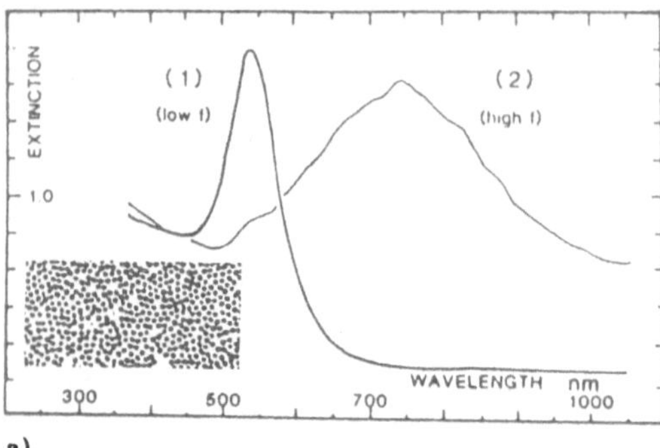

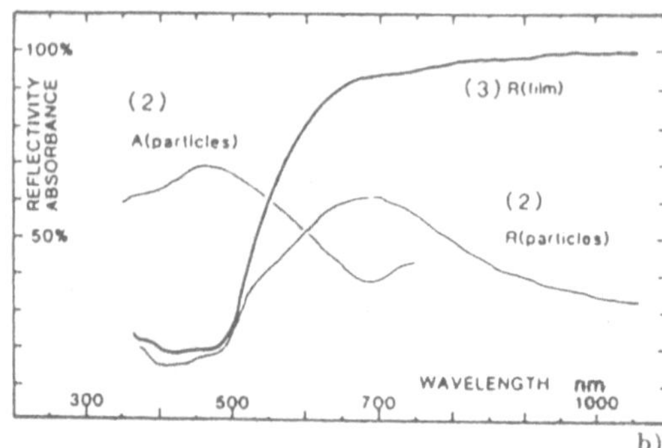

Fig. 6:   Measured optical spectra of Au cluster matrix matter and of Au film.
a) Extinction of separated clusters (curve (1)) and densely packed clusters (curve (2)).
Part of the sample is shown in the inset.
b) Pure absorption from photothermal measurements (curve (2)A) and reflection (curve
(2)R) of the sample shown left, and, for comparison, the reflection of a plane Au film
(curve (3)R). Cluster size: 20 nm.

In Fig. 6 b the true absorption and the reflection at about 40° are plotted and, for
comparison, the according reflection spectrum of a thick, continuous Au film. Even in this
sample we find again, that the spectral features of the cluster sample strongly differ from
the bulk in showing a peak structure reminding to the surface plasmon of the conduction
electrons of the individual cluster. In particular, the high IR reflectivity of Au which is
caused by the conduction electrons in the extended metal is still missing. The reason is that
due to the separating interlayers the charge conservation and, hence, the spatial extension
of elementary electronic excitations are limited to the single clusters.

Series of similar samples, yet with varying filling factor f were investigated in search for
the "Oseen-effect" in cluster matter. These samples were supported by quartz-substrates.
The polarized beam of the 514.5-nm line of an Ar-Laser was used as incident light, and
the Brewster angle of the quartz substrate was selected to suppress additional reflection.
The angular distribution of the backward directed light was then recorded, which, as was
controlled, was due to the Au clusters, with all other contributions being neglectibly small.
Fig. 7 shows several results.

Obviously there are strong variations of the width and the hight of the "reflected"
beam. The curves are normalized to equal including area in order to compensate the
strong differences due to the different f's. This area proves to be roughly linear in f. The
peak maximum which we ascribe to the regular reflected beam, decreases with decreasing
filling factor while the width   changes in a non-monotonous way. The increase of width is
attributed to (incoherent) scattering. In fact, at the highest value of f the width shrinks
to the width of the apparatus profile of the detector system and equals thus the angular
distribution of light reflected at the compact Au foil.

Following the theory of Oseen we can explain the non-monotonous broadening when f
is decreased: Then, as show TEM micrographs, the local arrangement of clusters in the
sample becomes strongly inhomogeneous due to the existence of extended dense cluster
aggregates surrounded by arrays with clearly lower local values for f_{local}. Hence, δf is large
and this causes broad angular distributions. At the lowest f of Fig. 7 (f = 0.09), these
extended aggregates have vanished, leaving a quite homogeneous arrangement of, mainly,

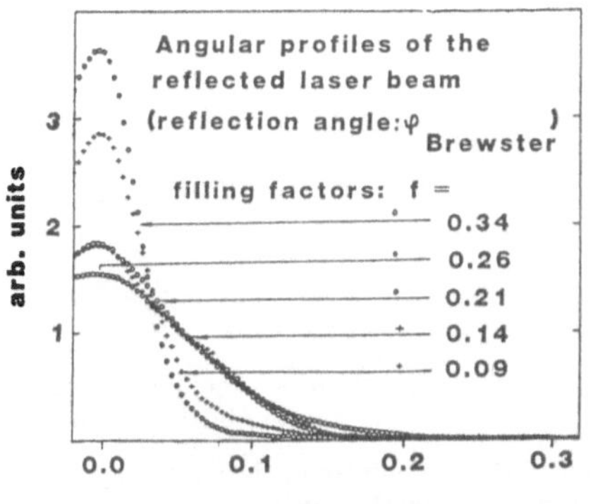
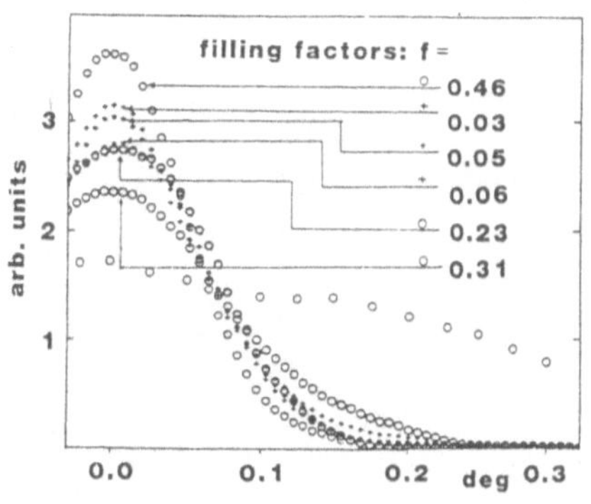

Fig. 7: Angular profiles of a laser beam "reflected" at cluster matrix matter samples of varying volume filling. The reflection angle is the Brewster angle of the quartz substrate. The two figures show results from two different sample series. The broadest curve in the right-hand figure is due to a diluted liquid sample of filling factor $f \sim 10^{-4}$. It was not normalized as described in the text for the other spectra.

single clusters. Obviously, the filling factor is still high enough to ensure sufficient coherency of the scattered light and, so, the width again shrinks almost to the value of the direct "reflected" beam. It is shown that for further diminished f the width increases again toward the Mie scattering characteristic provided the space coherency of the incident light is no longer high enough to include sufficient numbers of clusters in the coherency volume.

7. Multiple extinction

To illustrate the fourth of the mentioned collective contributions to the optical properties of cluster matter, an experiment concerning multiple extinction is discussed in the following [12].

Extinction being due to, both, absorption and scattering as shown in Fig. 2, multiple effects include higher order scattering processes and scattering followed by absorption. Absorption, obviously, is always of first order. Multiple effects increase with the optical density i.e., the filling factor f and the thickness of the samples, but as will be shown, resulting changes in the spectral features of the scattered light may also be independent of f and, hence occur at arbitrarily diluted systems.

The samples in consideration were three-component systems, consisting of Au-clusters ($2\overline{R} = 6nm$) of filling factor f_1 and of Latex clusters ($2\overline{R} = 46nm$) of filling factor f_2, both embedded statistically in aqueous solution.

The Au clusters show absorption with a sharp increase at the plasmon polariton frequency, but they are too small to scatter considerably. The Latex clusters, in contrast, do not absorb in theVIS but strongly scatter close to Rayleigh's λ^{-4}-law, i.e. without spectral selectivities. The resulting scattering spectrum, measured integrally via Ulbricht-sphere is shown in Fig. 8: The monotonous Latex spectrum is interupted by a marked decrease around the Au plasmon peak. This effect can be explained assuming second order extinction processes, at least and, probably, also higher order processes.

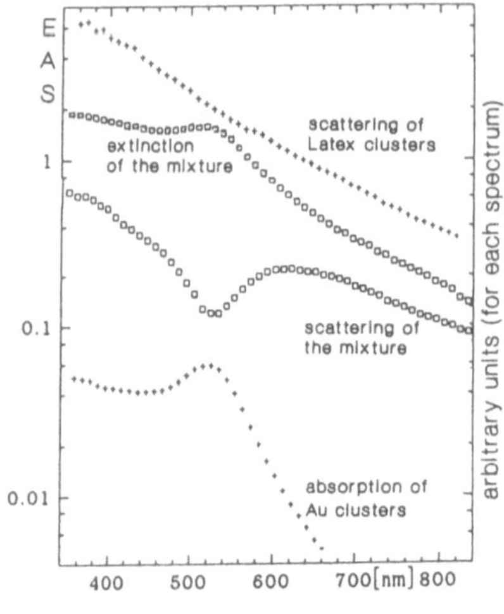

Fig. 8: Measured optical spectra of an aqueous system containing Au clusters (6 nm) and Latex clusters (46 nm).

Due to higher order extinction processes the scattering of the Latex is strongly influenced by the purely absorptive Au clusters.

Then, around $\lambda \sim 520nm$, light scattered by the Latex spheres is re-absorbed by Au clusters and is, thus, missing in the measured spectra. The important feature is that this effect depends on the ratio f_1/f_2, rather than on the absolute values of f_j and cannot be removed by moderate diluting the system. As a consequence, if a scattering system contains impurity components with non-negligible extinction - or, vice versa - an extinction system contains spurious scattering components like dust or air bubbles the scattering and extinction spectra can, in general, no longer be used to determine cluster sizes directly. The situation is analogous for systems of metal colloids where absorption and scattering spectra do not coincide on the wavelength scale. This is the case if particles are large or have broad size distributions.

However, by using a detailed model for the higher-order extinction processes, it was possible to separate the concurrent single processes, as will be described in detail elsewhere [12].

8. Résumé and Acknowledgements

Most commonly clusters are met in extended many-cluster systems i.e. in CLUSTER-MATTER. Containing coagulation and coalescence aggregates to various amounts, this kind of matter covers the ample field from the well distributed and separate single cluster to the bulk-like material. The attendant drastic variations of physical properties are due to several cluster interaction effects which bring up specific behaviour of the cluster collective as a whole, in addition to the single cluster properties. The optical properties were drawn upon to demonstrate this transition from the isolated cluster to bulk-like matter. Some theoretical concepts were briefly outlined and were supported by experimental evidence.

The financial support by the Deutsche Forschungsgemeinschaft - Schwerpunkt "Anorganische Cluster" - for part of the presented experimental work is gretefully acknowledged.

REFERENCES

[1]   Mie, G., Ann. Physik 25, 377 (1908)
       Debye, P., Ann. Physik 30, 57 (1909)
[2]   Ausloos, M., Gerardy, J.M., Phys. Rev. B 22, 4950 (1980);
       Phys. Rev. B 25, 4204 (1982); Phys. Rev. B 27, 6446 (1983)
       Quinten, M., Kreibig, U., Schoenauer, D., Genzel, L., Surface Sci. 156, 741 (1985),
       Z. Physik D 12, 521 (1989)
[3]   Kreibig, U., in: Contribution of Cluster Physics to Material Science
       and Technology", Proc. ASI, Agde/France 1982, ed. J. Davenas,
       R. Rabette; Nijhoff, Den Haag 1986
[4]   Oseen, C.W., see: Max Born "Optik", Springer, Heidelberg 1965
[5]   Quinten, M., Surface Sci. 172, 557 (1986)
       Thesis (1989)
[6]   Felderhof, U., Jones, R., Z. Physik B 62, 43 (1985)
[7]   Vollmer, M., Kreibig, U. "Optical Properties of Clusters"
       to appear in "Springer Series on Material Science", Springer, Heidelberg 1992
[8]   e.g. Kreibig, U., Genzel, L., Surface Sci. 156, 678 (1985), Fig. 14
[9]   Schoenauer, D., Quinten, M., Kreibig, U., Z. Physik D 12, 527 (1989)
[10]   Dusemund, B., Hoffmann, A., Salzmann, T., Kreibig, U., Schmid, G.,
       Z. Physik D 20, 305 (1991)
       Dusemund, B., Diploma work (1991), unpublished
[11]   Schmid, G., Structure and Bonding 62 (1985) and ref.s therein.
       We are indebted to Prof. Schmid for providing us with these samples.
[12]   Hoffmann, A., Salzmann, T., Kreibig, U., to be published

DIPOLE RESONANCE SYSTEMATICS IN METAL PARTICLES AND ATOMIC NUCLEI

MUSTANSIR BARMA and R.S. BHALERAO
Tata Institute of Fundamental Research
Homi Bhabha Road, Colaba
Bombay 400 005, India

ABSTRACT. Photon-induced collective dipole oscillations in metal particles and atomic nuclei are examined as a function of the number of constituent fermions A. We find that the shell-structure-linked oscillations in the full width at half maximum (FWHM) of the resonant photoneutron cross section in nuclei, earlier recognized for $A > 63$, in fact hold over the entire periodic table. Similar oscillatory tendencies are seen in data on separated metal clusters also. If the oscillations are discounted, the FWHM for nuclei is seen to decrease with increasing A, consistent with $A^{-1/3}$, a dependence earlier known to hold in metal particles. On rescaling the FWHMs by the respective Fermi energies and the inverse radii by the Fermi wave vectors, the data sets for metal particles and nuclei become comparable in magnitude. We give a schematic theoretical description of the systematics.

1. INTRODUCTION

There are points of strong resemblance between metal particles or clusters and atomic nuclei [1] – both finite Fermi systems. Here we focus on one phenomenon, namely, the response of these systems to electromagnetic radiation. In both systems, the field resonantly excites a collective dipolar mode. In metal particles, the Mie resonance involves displacing the conduction electron cloud with respect to the background of positive ions, with electromagnetic restoring forces [2]. In nuclei, it is the giant dipole resonance (GDR), in which protons are displaced with respect to neutrons and strong interactions provide the restoring force [3]. In both cases, the wavelength of the radiation at resonance far exceeds the size of the system.

An immediate consequence of the different nature of the restoring forces is that the A-dependence of the resonance frequency ω_0 is quite different in the two cases. In metal particles ω_0 is related to the plasma frequency, and thus is roughly A-independent. In nuclei the range of the restoring force is generally smaller than the nuclear radius, and thus ω_0 decreases with A, intermediate between $A^{-1/3}$ and $A^{-1/6}$ [3]. Remarkably, however, as shown below, the resonance *widths* display similar systematics in the two cases. In Section 2, we discuss the available data and these similarities, while in Section 3, we present a schematic theoretical description for the size dependence of the width. A detailed discussion can be found in [4].

P. Jena et al. (eds.), Physics and Chemistry of Finite Systems: From Clusters to Crystals, Vol. II, 881–886.
© 1992 *Kluwer Academic Publishers.*

2. RESONANCE WIDTH SYSTEMATICS

2.1. Oscillations of FWHM

Photoabsorption cross sections have been measured in a variety of metal particles [5] and clusters [6], and also in a large number of nuclei [7]. A simple, single characterization of the resonance spectrum is provided by the full width at half maximum (FWHM). The plots of FWHM vs A for nuclei with $A > 90$ [8a] and $166 > A > 63$ [8b] show that the FWHM exhibits systematic oscillations with local minima near spherical, near-magic nuclei. We wanted to see whether these systematics are also present in lighter nuclei and in metal clusters.

We examined the data for about 120 nuclei ranging from ^3He to ^{239}Pu. It was straightforward to determine the FWHM for those nuclei where the data exhibit a single peak. For cases with two or more (overlapping) peaks, we found the FWHM by drawing a smooth curve with a single maximum through the data points, trying to ensure that the areas under the smooth and experimental curves were nearly equal. Nuclei where the data seem incomplete or where there is too much structure were ignored.

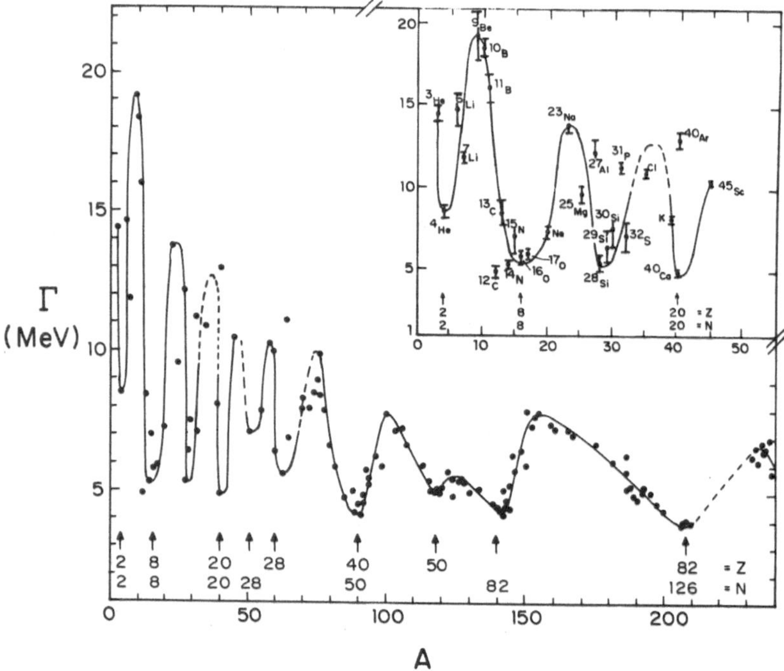

Figure 1. FWHM Γ vs A for nuclei. Note the systematic modulations in the curve, with minima at the proton (Z) or neutron (N) magic numbers. The minimum at $A = 28$ corresponds to ^{28}Si; see [4]. Dashed lines indicate regions of sparse or nonexistent data.

Our results are displayed in Fig. 1. Examination (see the inset) shows that the FWHM Γ continues to oscillate with local minima near the magic numbers, even for light nuclei. A detailed discussion of the curve is given in [4]. The uncertainties in Γ are at most ~ 1 MeV, while the amplitude of oscillations is much larger implying that the oscillations are statistically significant. They are as systematic and pronounced as those for large A.

Optical absorption experiments on metal particles have been performed on two types of samples: (i) free metal clusters in which precise size separation is achieved by mass spectroscopy. (ii) metal clusters embedded in various matrices, such as glass or solid Ar. Particles are isolated from each other, but some spread in size cannot be avoided. Experiments on samples of type (i) have been performed on Na clusters with 2-40 conduction electrons [6]. Full lineshapes are not yet available, making the extraction of the FWHM difficult. But the data indicate that, as with nuclei, there is a strong response over a relatively narrow frequency interval in the case of magic numbers, and over a much broader frequency interval in other cases. In the latter case, the line shows splittings, which can be interpreted in terms of shape deformations [6]. A larger range of sizes ($\sim 10 - 100A$) has been investigated [5] in experiments on samples of type (ii), but no noticeable oscillations in Γ versus radius R have been observed. This is probably because (a) oscillations are averaged out due to the distribution of sizes, and (b) the amplitude of shell-structure-linked oscillations decreases with increasing size, and is small for the above range of sizes.

2.2. Global Trends and Scaling

Experiments on samples of type (ii) have revealed a systematic dependence [5] of the averaged FWHM Γ_{av} on R:

$$\Gamma_{av} = K\,\frac{\hbar v_F}{R} + \Gamma_\infty, \tag{1}$$

where v_F is the Fermi velocity, K a matrix-dependent constant of order unity and Γ_∞ the width in the bulk metal.

A similar $1/R$ dependence has been seen earlier [9] for nuclei in the range $A > 50$. It is interesting to examine whether this trend persists for $A < 50$, and to compare the metal particle and nuclear data. On dividing across by the Fermi energy ϵ_F, we see that Eq. (1) predicts that Γ_{av}/ϵ_F is a linear function of $(k_F R)^{-1}$, where k_F is the Fermi wave vector. Figure 2 shows the scaled data, using $\epsilon_F = 38$ MeV and $k_F = 1.36$ fm^{-1} for nuclei, and $\epsilon_F = 5.49$ eV and $k_F = 1.20A^{-1}$ for Ag particles. We used RMS radii $\tilde{R}$, as these are well determined for nuclei; for Ag particles, we took $\tilde{R}$ to be given by $\sqrt{3/5}$ times the quoted radii. Since we are interested in the overall downward trend seen in Fig. 1, we have replotted points corresponding to singly or doubly magic nuclei from the lower envelope of the curve in Fig. 1; the line marked 'magic' is the best fit line. Thus the data are consistent with a linear dependence on $(k_F \tilde{R})^{-1}$. We also examined the average downward trend of the oscillatory curve in Fig. 1, and fit it to a linear dependence. We conclude that Eq. (1) holds to a good approximation for nuclei also. In particular, it is interesting to see how well the doubly magic nuclei follow a straight line.

884

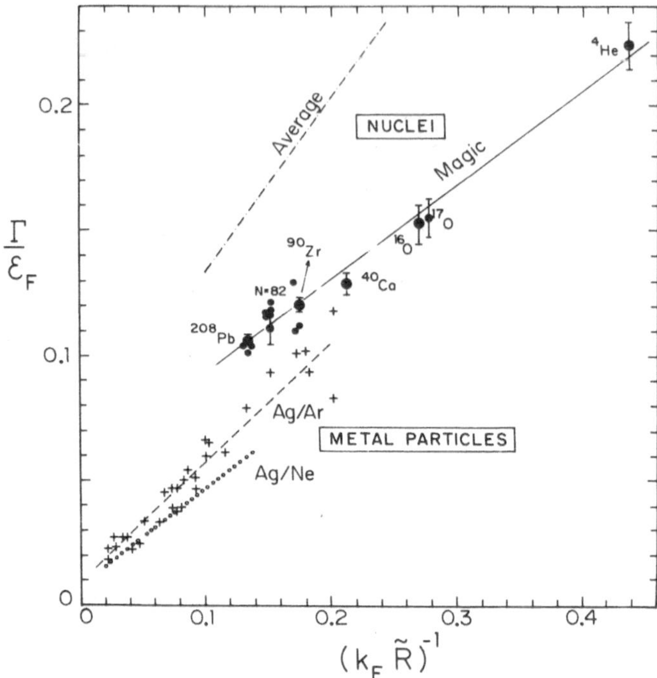

Figure 2. Γ/ϵ_F vs $(k_F\tilde{R})^{-1}$ for metal particles and nuclei. The dashed and dotted lines are best fits for Ag/Ar (+) and Ag/Ne [5]. The nuclei shown here are singly (•) or doubly (⬤) magic nuclei from the lower envelope of the oscillating curve in Fig. 1. The solid straight line is the best fit to this data set. The dot-dashed line indicates the 'average' trend of the oscillatory curve in Fig. 1.

Scaling Γ by ϵ_F and R^{-1} by k_F has allowed us to compare nuclear and metal particle data which differ by 5-6 orders of magnitude in length and energy. Note, however, that the intercepts on the Γ/ϵ_F axis, for the two sets of data in Fig. 2, are quite different. The significance of this will be discussed in the following section.

3. THEORETICAL CONSIDERATIONS

3.1. Global Trends

The FWHM receives two types of contributions: (i) The intrinsic width Γ_i comes from the finite lifetime of the resonance, and is present in all cases. It receives contributions from a variety of sources to be discussed below. It is the only contribution in spherical cases. (ii) In nonspherical cases, static deformations in shape can lead to two distinct resonance frequencies, corresponding to a splitting of the line. In such cases, the FWHM receives additional contributions.

Γ_i can be written [3] as the sum of three terms $\Gamma_i = \Delta\Gamma + \Gamma^\uparrow + \Gamma^\downarrow$, reflecting contributions from distinct physical effects. The fragmentation width $\Delta\Gamma$ arises from the finite system analogue of Landau damping. It occurs due to the scattering of the fermions from the 'wall' or 'surface' of the self-consistent mean field potential. $\Gamma^\uparrow$ is the escape or decay width corresponding to the direct coupling of the $(1p-1h)$ doorway state to the continuum, giving rise to its decay into a free fermion and an $(A-1)$ residue. Finally, the spreading width $\Gamma^\downarrow$ is due to the coupling of the doorway state to more complicated $(2p-2h)$ states of the system, the transition occurring on account of two-body collisions. Macroscopically, it can be modelled by introducing viscous effects.

The R-dependence of $\Delta\Gamma$ and $\Gamma^\uparrow$ may be estimated classically by considering the frequency of collisions with the surface [10]. If individual fermions moving with Fermi velocity are reflected off the surface, then the lifetime of the state is the mean time between surface hits, $R/c_1 v_F$. Here c_1 is a constant of order unity. The width is then the inverse lifetime $c_1 v_F/R$. This agrees with a quantum calculation of $\Delta\Gamma$ within an infinite square well model [11]. The subject of one-body dissipation has been discussed extensively in the nuclear physics literature. For example, Blocki et al. [12] have presented a classical derivation of the socalled wall formula which gives the rate of dissipation of energy due to the one-body mechanism, and Yannouleas [12] has presented a microscopic derivation of this formula starting from the RPA dynamics. The application to the GDR widths was done by Myres et al. [9] as already discussed above.

We generalize this heuristic argument to the present instance, where the well is not infinitely deep. Every time a fermion hits the wall, it has a probability p of escaping, and a corresponding probability $(1-p)$ of being reflected. The former possibility leads to an escape width $\Gamma^\uparrow = pc_1 v_F/R$, while the latter possibility corresponds to the fragmentation width $\Delta\Gamma = (1-p)c_1 v_F/R$. The sum of the one-body rates $\Gamma^\uparrow$ and $\Delta\Gamma$ is then $c_1 v_F/R$.

$\Gamma^\downarrow$ arises from two-particle collisions. We expect it to vary smoothly with energy and size for magic cases, since the collisional mean free path λ shows similar smooth variations. The average time between collisions is $\sim \lambda/v_F$ and hence $\Gamma^\downarrow \sim v_F/\lambda$.

The total intrinsic width is thus expected to be $\Gamma_i = \Gamma^\downarrow_\infty + c_1 v_F/R$. This is in agreement with Eq. (1) and the data presented in Fig. 2.

From Fig. 2, we see that if the straight lines are extrapolated leftward, the resulting intercept on the Γ/ϵ_F axis is much larger for nuclei than for metal particles. This is reasonable, as $k_F\lambda$ is much smaller in nuclear matter than in bulk metals at room temperature, reflecting the greater effect of collisions in the former case.

3.2. Oscillations of FWHM

One may distinguish between two types of effects which give rise to the oscillations of the FWHM. We illustrate these by considering two broad representative regions for nuclei, namely $150 < A < 190$ and $80 < A < 150$.

In the range $150 < A < 190$, Dietrich and Berman [7] have fitted two-component Lorentz curves to the data. This indicates a splitting of the line, due to deformation of the nucleus. We denote the two resonance energies by ω_{01} and ω_{02}, with $\omega_{01} < \omega_{02}$, and the corresponding widths by Γ_1 and Γ_2. For a spheroidal deformation, the A−dependence of ω_0 leads us to expect that ω_{01} and ω_{02} correspond to oscillations along the semimajor and semiminor axes respectively. The R−dependence of Γ_i

then implies that $\Gamma_1 < \Gamma_2$. This is indeed found to be true, for $150 < A < 190$, for the values of the widths tabulated in [7]. In fact, with two exceptions, this is true for all the nuclei listed. The corresponding cross sections σ_1 and σ_2 also generally satisfy $\sigma_1 < \sigma_2$. The increase of the FWHM away from the spherical cases can be ascribed, at least partially, to the fact that deformations produce a splitting of the line, and also cause $\Gamma_2 > \Gamma(> \Gamma_1)$, where Γ would be the width if the nucleus were undeformed. Since deformations of the shape follow shell-structure systematics, with smallest deformations close to the magic numbers, so does the FWHM.

In the region $80 < A < 150$, Dietrich and Berman [7] have fitted one-component Lorentz curves to the data (with two exceptions). They call these nuclei spherical. As is clear from Fig. 1, oscillations of the FWHM versus A in this region are as prominent as those for $150 < A < 190$. Thus, even when the line is unsplit, the width of the best-fit Lorentzian oscillates as a function of A, with minima at the magic numbers. This indicates that the intrinsic width Γ_i can itself show shell-structure linked oscillations. Bergère [8a] has discussed the origin of these oscillations.

In conclusion, the systematics which have been pointed out and discussed in this paper bring the task of a theory into better focus. While a complete theory has not been presented here, we have given a schematic theoretical description which allows one to understand at least the principal trends.

REFERENCES

[1] S. Sugano, in *Microclusters*, edited by S. Sugano, Y. Nishina and S. Ohnishi (Springer, Berlin, 1987).

[2] S.P. Apell, J.Giraldo and S. Lundqvist, Phase Transitions **24-26**, 577 (1990).

[3] See, e.g., A. van der Woude, Prog. Part. Nucl. Phys. **18**, 217 (1987).

[4] M. Barma and R.S. Bhalerao, TIFR preprint, submitted for publication.

[5] U. Kreibig and L. Genzel, Surf. Sci. **156**, 678 (1985); K.P. Charlé, W. Schulze and B. Winter, Z. Phys. D**12**, 471 (1989).

[6] K. Selby, V. Kresin, J. Masui, M. Vollmer, A. Scheidemann and W.D. Knight, Z. Phys. D **19**, 43 (1991).

[7] S.S. Dietrich and B.L. Berman, At. Data Nucl. Data Tables **38**, 199 (1988).

[8] (a) R. Bergère, in *Photonuclear Reactions*, edited by S. Costa and C. Schaerf, Vol. 61 of Lecture Notes in Physics (Springer, Berlin, 1977), Chapter III, Fig. 23a, p. 114; (b) K.A. Snover, Ann. Rev. Nucl. Part. Sci. **36**, 545 (1986).

[9] W.D. Myers, W.J. Swiatecki, T. Kodama, L.J. El-Jaick and E.R. Hilf, Phys. Rev. C **15**, 2032 (1977).

[10] U. Kreibig, J. Phys. F **4**, 999 (1974).

[11] A. Kawabata and R. Kubo, J. Phys. Soc. Jpn. **21**, 1765 (1966); M. Barma and V. Subrahmanyam, J. Phys. Cond. Matter, **1**, 7681 (1989).

[12] J. Blocki, Y. Boneh, J.R. Nix, J. Randrup, M. Robel, A.J. Sierk and W.J. Swiatecki, Ann. Phys. (NY) **113**, 330 (1978); C. Yannouleas, Nucl. Phys. A **439**, 336 (1985).

OBSERVATION OF A SPIN-FORBIDDEN OPTICAL TRANSITION OF Mo₂ IN A MOLYBDENUM CLUSTER BEAM

P.S. BECHTHOLD, H. HANDSCHUH, H. NOACK and W. EBERHARDT
Institut für Festkörperforschung
Forschungszentrum Jülich
D-5170 Jülich, Federal Republic of Germany

ABSTRACT. Dye laser induced weak optical transitions of jet-cooled Mo_2 isotopomers are observed in the 518 to 521 nm spectral range by resonant two photon ionization (R2PI) spectroscopy. ArF (6.42 eV) or KrF (5.0 eV) photons are used for the ionization. For each isotopic combination a narrow line ($\Delta\lambda \approx 0.02 nm FWHM$) is observed which shows an isotope shift of 8 cm^{-1}/amu. The lifetime of the resonantly excited intermediate state is $4.8 \pm .1\mu s$ indicating a nominally spin-forbidden transition. Using an intracavity etalon for high resolution spectroscopy ($\Delta\tilde{\nu} = 0.04 cm^{-1}$) the line could be identified as the band head of a R-branch of the $a^3\Sigma_u^+ \leftarrow X^1\Sigma_g^+$ intercombination transition whereby a higher vibrational level of the triplet state is occupied. Tentative assignments of vibronic levels are given.

1. INTRODUCTION

The geometry and electronic structure of small transition-metal-particles is of great importance for the understanding of the fundamental mechanisms of heterogeneous catalysis and surface chemistry. Therefore transition metal molecules and clusters have attracted much attention in recent years. Their geometries and electronic structures can be derived from spectroscopic properties [1]. The information however, often suffers from large uncertainties. This is particularly the case when intercombination bands are considered. This report deals with the $a^3\Sigma_u^+ \leftarrow X^1\Sigma_g^+$ intercombination transition of Mo_2 studied by means of resonant 2-photon-ionization spectroscopy. We demonstrate that this transition shows an unusual isotope selectivity which can be used for isotope enrichment by laser irradiation.

2. EXPERIMENT

Gas phase molybdenum dimers were first produced by pyrolytic [2] and photolytic [3] decomposition of gaseous molybdenum hexacarbonyls, $Mo(CO)_6$. In this work Mo_2 molecules are produced in a molybdenum cluster beam. The metal particles are generated in a laser vaporization cluster source [4] and formed into a molecular beam,

P. Jena et al. (eds.), Physics and Chemistry of Finite Systems: From Clusters to Crystals, Vol. II, 887–892.

which is skimmed and after propagation through a differential pumping stage enters a grid-free reflectron time-of-flight mass spectrometer[5]. For vaporization a KrF laser is focused on a continuously rotating and propagating molybdenum target rod at a pulse energy of $\approx 180 mJ/pulse$ and a repetition rate of 30 Hz. Helium carrier gas is supplied through a pulsed piezoelectric valve synchronously driven at a backing pressure of 10 atm. In the mass spectrometer the particles are photoionized by ArF or KrF radiation. The beam of a dye laser counter propagates the cluster beam to excite the particles prior to ionization. The ions are detected by dual channel plates, and the time-of-flight mass spectra are signal averaged over several hundred laser pulses in a digital storage oscilloscope. The dye laser can be suitably advanced with respect to the ionization laser to measure excited state life times.

3. RESULTS AND DISCUSSION

Natural molybdenum consists of a mixture of 7 stable isotopes with masses in the 92-100 amu mass range where only the masses 93 amu and 99 amu are missing. When forming dimers we find 28 possible isotopic combinations which are distributed between 15 possible masses ranging from 184-200 amu, where now the masses at 185 and 199 amu are absent. From the natural abundance ratios of the isotopes we calculate expected relative abundances of the dimers as shown in Fig. 1. The observed time-of-flight mass spectra nicely fit this distribution within statistical errors when we simply ionize with ArF or KrF radiation. This usually holds true even if we increase the dimer ion signals with a low resolution dye laser ($\Delta \tilde{\nu} = 0.2 cm^{-1}$) by resonant 2-photon-ionization via the well known $A^1\Sigma_u^+ \leftarrow X^1\Sigma_g^+$ -transition [2-4]. In a similar way as described by Hopkins et al. [4] we use the hot band analysis of this transition to characterize the vibrational and rotational level populations of the ground electronic state. We estimate the vibrational temperature of the dimers in the molecular beam as $T_{vib} = 330 \pm 30 K$ and the rotational temperature as $T_{rot} = 165 \pm 10 K$.

In some narrow wavelength ranges we find striking deviations from the expected dimer distribution where the signal of a particular dimer ion is selectively strongly enhanced while the intensity ratios of the other ions still correspond to their natural abundances. The relative enhancement is larger for KrF ionization where we are closer to the ionization limit. An example is shown in Fig. 2 for dimers with mass 188 amu where the excitation wavelength was adjusted to 518.26 nm. When tuning the dye laser we find that a narrow band exists for each of the 15 masses where the respective dimer signal is selectively increased. The energetic separations between these bands being about 8 cm^{-1}/amu. Two corresponding optical spectra for masses 188 and 190 amu are displayed in Fig. 3. For mass 188 amu one sees the narrow new transition at 518.26 nm and on the right the broad rotational R- and P-branch distributions of the $A^1\Sigma_u^+ \leftarrow X^1\Sigma_g^+$ -transition. For mass 190 amu the new transition coincides with the R-branch of the $A \leftarrow X$ transition. Thus, whereas the isotope shift for the $A^1\Sigma_u^+ \leftarrow X^1\Sigma_g^+$ -transition is very small as expected, the new transition shows a dramatic isotope shift. The splitting of the new line for mass 190 is also

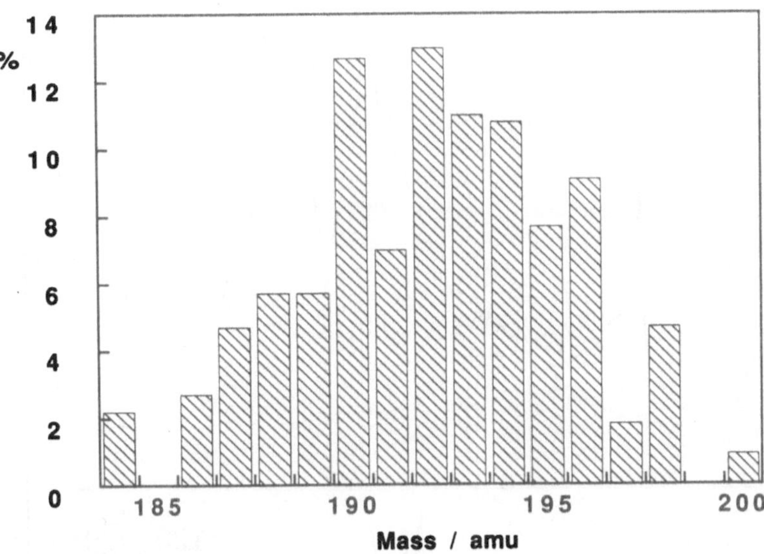

Figure 1. Relative abundances of molybdenum dimers calculated from the natural abundance ratios of the isotopes.

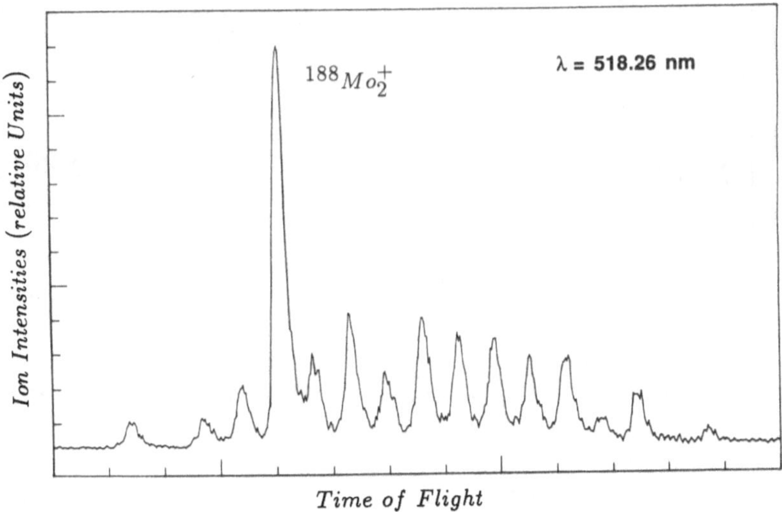

Figure 2. Time-of-flight mass spectrum of molybdenum dimers with a selectively enhanced ion signal at mass 188 due to the resonant excitation at 518.26 nm. The abundance ratios of the other ions correspond to their natural distribution as shown in Fig. 1. Ionization is obtained with KrF radiation (5.0 eV).

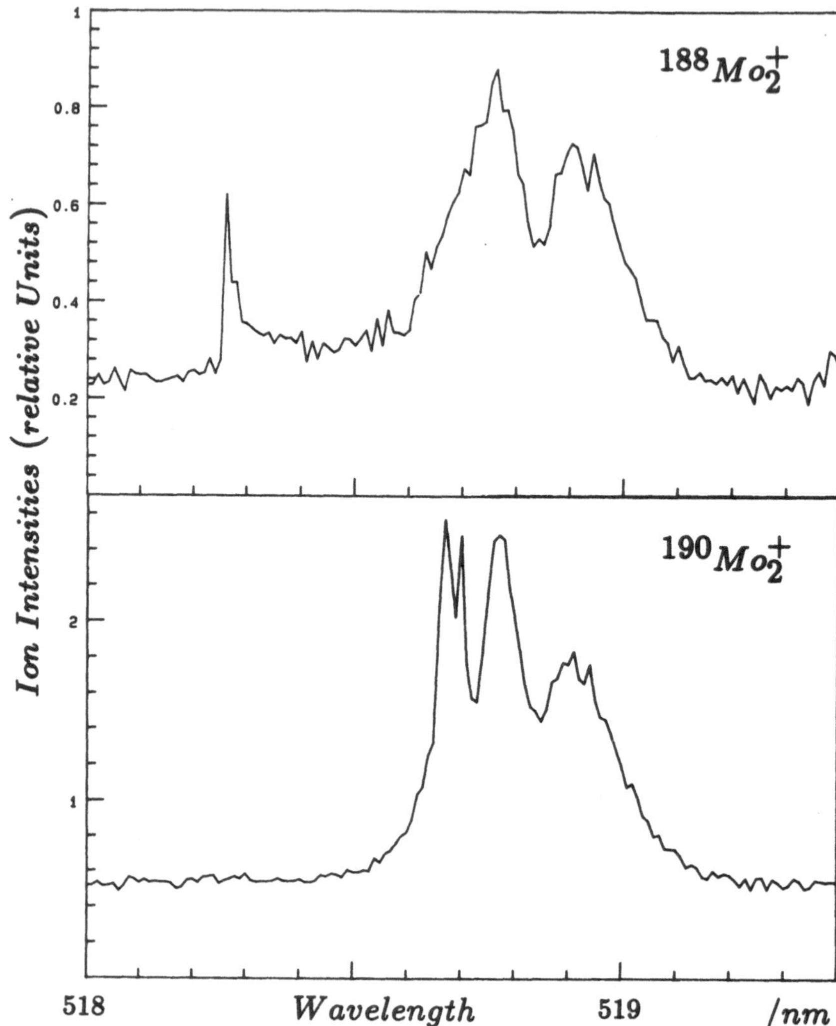

Figure 3. Optical excitation spectra for masses 188 and 190 amu, respectively, as obtained by resonant two-photon-ionization spectroscopy. For mass 188 amu one sees the narrow new transition at 518.26 nm and on the right the broad rotational R- and P-branch distributions of the $A^1\Sigma_u^+ \leftarrow X^1\Sigma_g^+$ -transition. For mass 190 amu the new transition coincides with the R-branch of the $A \leftarrow X$ transition. Here the splitting of the new line is due to the three isotopic combinations (92-98, 94-96, 95-95) and occurs because of the slightly different reduced masses of the dimers formed.

due to the strong isotope effect and occurs because the three isotopic combinations (92-98, 94-96, 95-95) have slightly different reduced masses.

The lifetime of the excited intermediate state was measured to be $4.8 \pm .1 \mu s$ indicating a nominally spin-forbidden transition. Since the lifetime of the $A^1\Sigma_u^+$-state is only 18 ns [4] it is easy to separate the new transition from the $A \leftarrow X$ transition in R2PI-spectra by simply advancing the dye laser by some hundred ns with respect to the ionization laser. With such conditions some additional transitions were detected. The observed new transition energies are summarized and plotted as a function of the dimer masses in Fig. 4. The transitions corresponding to the circles show the strongest intensities. All the other transitions are very weak and sometimes difficult to extract from the background noise. Besides the strong isotope shift one sees three closely spaced lines resembling the (0,0), (1,1), ... vibrational sequence observed for the $A^1\Sigma_u^+ \leftarrow X^1\Sigma_g^+$ transition [4] and one line at a larger energy separation which is assigned to transitions into the next lower vibrational level.

The extreme isotope effects can only be explained if one assumes that the observed electronic transitions are associated with large changes of the vibrational quantum

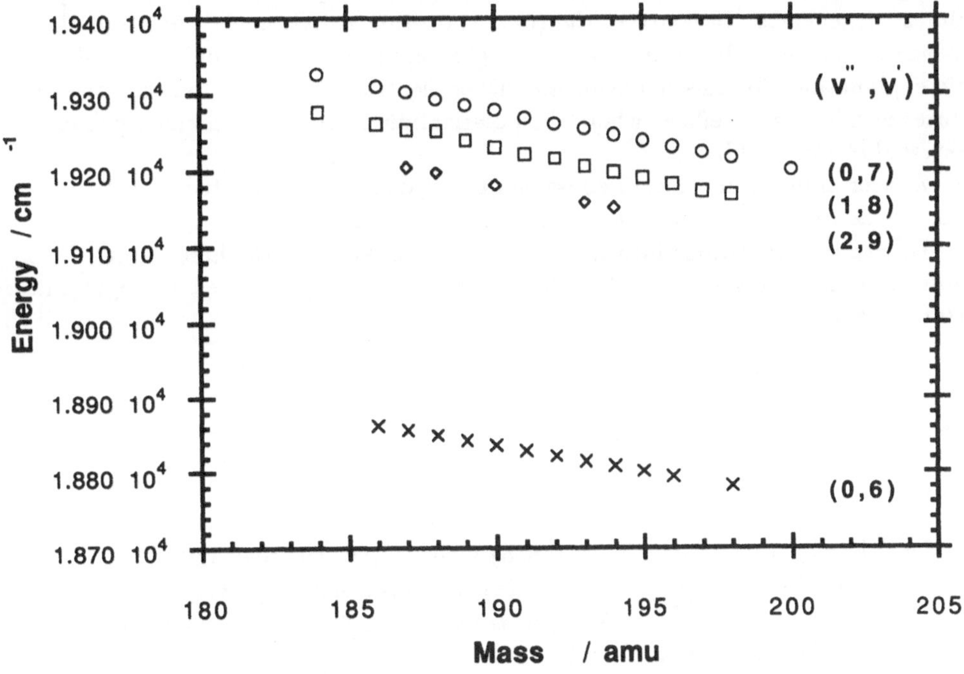

Figure 4. Summary of the observed rovibronic transition energies of the $a^3\Sigma_u^+ \leftarrow X^1\Sigma_g^+$ -transition plotted as a function of the dimer masses. The tentative vibronic assignment is given on the right side (see text for details). The transitions corresponding to the circles show the strongest intensities. All the other transitions are very weak and sometimes difficult to extract from the background noise.

numbers. In such cases the isotope shifts sum up over all intermediate vibrational spacings. Thus a high vibrational quantum state will be occupied in the long living intermediate excited state. Rotationally resolved spectra ($\Delta\tilde{\nu} = 0.04cm^{-1}$) of the upper-line-transitions (circles) of Fig. 4 identify these as band heads of R-branches of rovibronic transitions. These spectra are currently evaluated. A rough tentative assignment of the vibrational sequences can be obtained, however, if we omit the influence of the rotations. In this case we treat the levels as if they were band origins instead of band heads. The effect is a moderate more or less uniform shift of the transition energies which presumably is irrelevant for the vibrational sequence. With these assumptions the transitions of Fig. 4 are tentatively assigned to (0-6),(0-7),(1-8) and (2-9) vibronic sequences, respectively.

A comparison of our results with generalized valence bond calculations for all the possible spin states dissociating to the ground state atoms by Goodgame and Goddard [6] reveals that the observed peaks have to be assigned to the spin-forbidden $a^3\Sigma_u^+ \leftarrow X^1\Sigma_g^+$ transition. The $a^3\Sigma_u^+$-state is a $\delta\delta^*$-state [6]. The observed transitions become allowed due to spin-orbit coupling. We note the unusual intensity of a strong (0-7) transition together with a weak (1-8) transition, which are energetically almost degenerate. In contrast the (0-6) transition is extremely weak and the (0-8) transition was not observed. This is a highly unusual Franck-Condon profile. Possible explanations for this behavior are either electronic coupling between the excited states or a local field effect, where the polarization of the electron cloud enhances the external electric field.

A laser induced fluorescent emission observed for matrixisolated Mo_2 [7] may also trace back to the $a^3\Sigma_u^+$-state as the intermediate state.

Due to the transitions into the higher vibrational levels the here reported isotope shifts are extremely large. This effect principally could be used for light induced isotope separation.

ACKNOWLEDGEMENTS

The technical assistance of F.P. Johnen and A. Hrastnik is greatfully acknowledged.

REFERENCES

1. For a Review see: Morse, M.D. (1986) Chem. Rev. 86, 1049-1109
2. Becker, K.H. and Schürgers, M. (1971), Z. Naturforschung 26A, 2072
3. Efremov, Y.M., Samoilova, A.N., Kozhukhovsky, V.B. and Gurvich, L.V. (1978) J. Molecular Spectrosc. 73, 430
4. Hopkins,J.B., Langridge-Smith, P.R.R., Morse, M.D. and Smalley, R.E. (1983) J. Chem. Phys. 78, 1627
5. Bechthold, P.S., Mihelčić, M. and Wingerath, K., patent pending
6. Goodgame,M.M. and Goddard III, W.A. (1985) Phys. Rev. Lett. 54, 661
7. Pellin, M.J., Foosnaes, T. and Gruen, D.M. (1981) J. Chem. Phys. 74, 5547

N. D. BHASKAR
The Aerospace Corporation
P.O.Box 92957
Los Angeles, CA 90009

ABSTRACT

Experimental investigations of surface plasmon absorption in cesium cluster ions (Cs_n^+ n=4-21) are reported. Photoabsorption in simple metal clusters is believed to be dominated by collective excitation of the delocalized electrons. Our investigations are directed towards obtaining the absorption spectra as a function of cluster size, and for clusters of different metals. The width of the absorption spectrum reflects the damping mechanisms responsible for the decay of the collective excitations. We use a liquid metal ion source (LMIS) for cluster generation. LMIS produces 'hot' clusters. Consequently, a substantial thermal broadening of the resonance spectrum is expected. A mass selected pulsed beam of Cs_n^+ interacts with a collinear counter-propagating pulsed laser beam. We measure the loss in intensity (photodepletion) of the parent clusters in a time-of-flight mass spectrometer. Photoabsorption cross sections for clusters in the range of 4-9 atoms, at 600 nm, are about 1-3 $(\overset{\circ}{A})^2$/atom. The typical absorption width is found to be larger than 100 nm.

INTRODUCTION

Surface plasmon absorption in small clusters of simple metals is currently a subject of intense interest. Many experimental and theoretical investigations have been reported and many are currently in progress. Optical absorption cross sections of free neutral clusters of sodium were first measured by de Heer et al. (Ref. 1, 2). They interpreted the absorption data on the basis of collective excitation of the delocalized electrons, using the electronic shell model. Subsequently, optical absorption spectra of potassium cluster ions (K_9^+ and K_{21}^+) were obtained by Brechignac et al. (Ref. 3). They observed a blue-shift and line narrowing with cluster size. Kresin's calculation of the collective resonance frequencies (Ref. 4) agrees well with the experimentally measured results of Brechignac et al. (Ref. 3). Yannouleas et al. (Ref. 5) and Bertsch et al. (Ref. 6) have studied the broadening mechanisms in small clusters. These investigations clearly show that the contributions to the broadening of the plasma resonance are quite different in the charged and neutral clusters. The potentials of small charged clusters are significantly deeper than the corresponding neutrals. Landau damping is hypothesized to be almost absent in small charged clusters. The resonance damping mechanism in the charged clusters is dominated by the coupling of the plasmon to the thermal fluctuations. A lucid experimental investigation was reported by Dam et al. (Ref. 7) who used a 'hot'

P. Jena et al. (eds.), Physics and Chemistry of Finite Systems: From Clusters to Crystals, Vol. II, 893–898.
© 1992 *Kluwer Academic Publishers.*

source of cluster ions to study the influence of the thermal fluctuations on the width of the resonance.

In our laboratory we are investigating the optical properties of mass selected cluster ions of simple metals (Na, K, Rb and Cs). In the area of surface plasmon absorption, our goal is to systematically study the widths and resonance frequencies as a function of cluster size, and to some extent, as a function of cluster temperature. The influence of cluster structure on the shape of the resonance is also of interest. Using the simple theoretical model (Ref. 2) we hope to extract the static polarizabilities of the cluster ions which has not been measured thus far.

EXPERIMENT

A schematic of our experimental apparatus is shown in Fig. 1. We use a capillary type liquid metal ion source (LMIS) (Ref. 8) for cluster generation. The capillary is filled with the liquid metal of interest. Cluster ions are extracted from the tungsten capillary needle by applying a strong electric field (on the order of volts/angstrom) to the tip. The cluster ion beam is collimated by an electrostatic lens assembly. Using two pairs of deflecting plates (A and B) a mass selected pulsed cluster ion beam is produced. A collinear counter-propagating pulsed tunable dye laser beam of approximately 1 microsecond duration interacts with the pulsed ion beam at the entrance of a 2 meter long time-of-flight drift tube. Very good spatial and temporal overlap of the two beams is essential for obtaining reliable measurements of cross sections. We also use the fixed frequency lines of an argon ion laser for these experiments.

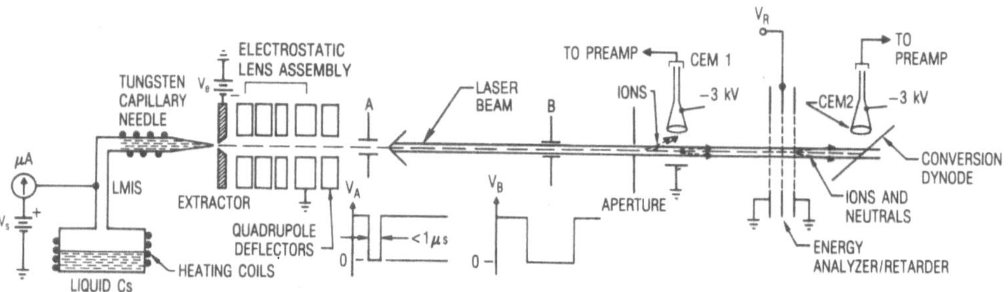

Fig. 1 Experimental apparatus for the study of surface plasmon absorption in Cs_n^+.

RESULTS AND DISCUSSIONS

A typical mass spectrum of the cluster ions is shown in Fig. 2. The shell closing features at n=9, 19, 21 are clearly observed. In addition, the relative abundance of 13-mer is large, reflecting the importance of stable geometric configurations. For details please see Ref. 9, 10 and 11. In Figs. 3 and 4 we show the time-of-flight mass spectra of Cs_n^+ with laser beam off and on, for different cluster sizes. By adjusting the time delay between the pulsed ion beam and the laser pulse, any desired mass range can be made to interact with the laser beam.

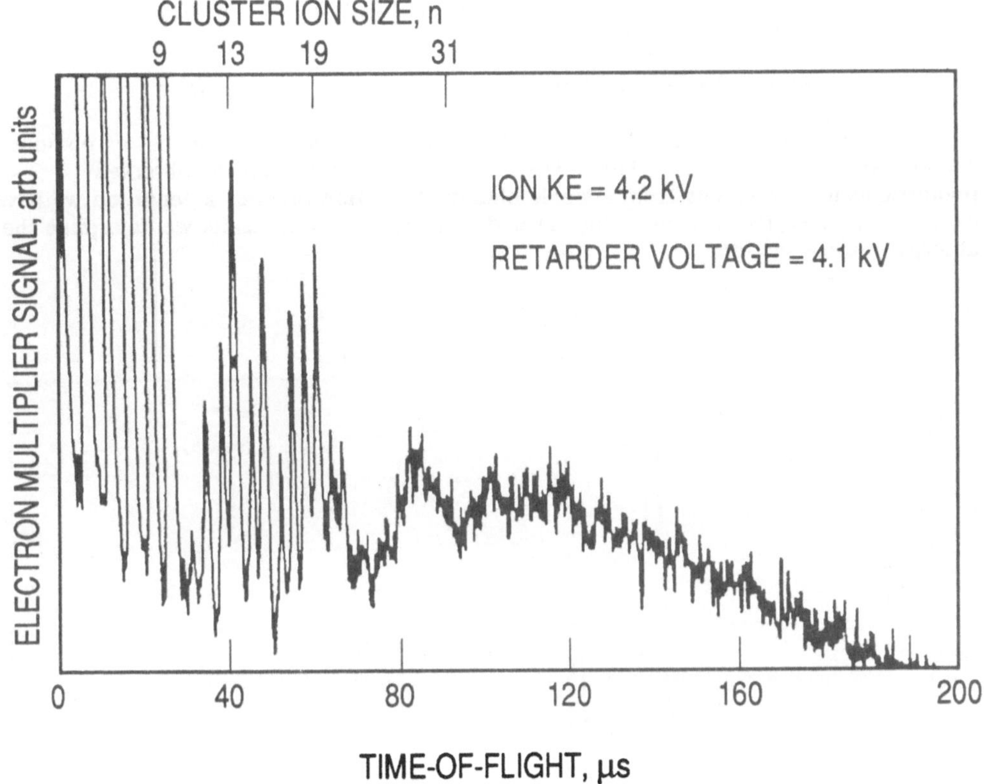

Fig. 2 Mass spectrum of Rb_n^+. Magic numbers corresponding to the shell closing numbers are clearly present. Rb_{13}^+ also shows a maximum in relative abundance.

The sharp reduction in the cluster beam intensity when the laser beam is present arises due to photofragmentation. In the fragmentation process the kinetic energy is partitioned in inverse proportion of the masses of the fragments; though the parent and the fragment ion kinetic energies are different, they travel with essentially the same speed. However, the fragments have a much larger angular divergence than the parents and therefore some of the fragments miss the detector. In our experiments we use the energy retarder to prevent the fragment ions from reaching the detector, without relying on the differential divergence for separation. The electrostatic energy retarder is set to transmit only the parent ions. Although the neutral fragments are not affected by the retarder, they do not reach the off-axis ion detector. This enables us to obtain a reliable measurement of the fractional loss in the parent ion signal due to the photofragmentation process.

For a particular parent cluster ion size n, let N_n and N_n^0 denote the cluster beam signals for laser on and off respectively. Then

$$N_n = N_n^0 \exp(-(\sigma \phi)) \qquad (1)$$

where σ is the single-photon absorption cross section, and ϕ is the number of photons per unit area per laser pulse. This relationship is valid only when the observed photofragmentation results from a single photon absorption process, a condition we have experimentally verified. Combining (1) with our experimental results we determine the absorption cross sections.

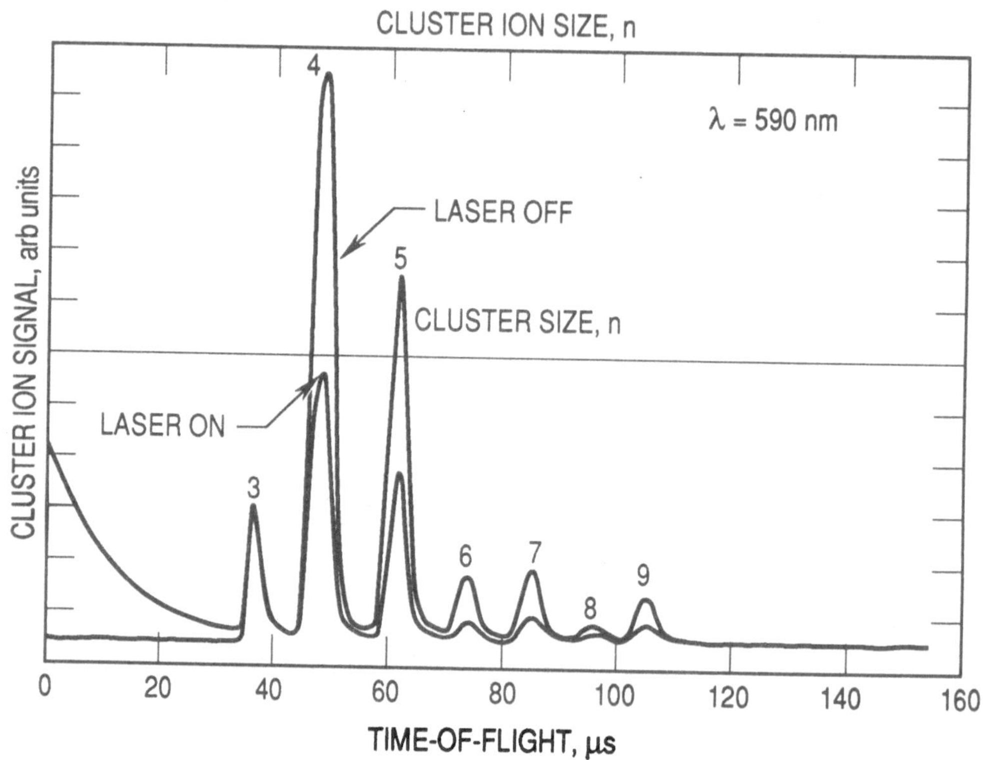

Fig. 3 Mass spectrum of Cs_n^+ with the laser off and on for n=4-9

We have measured the photoabsorption cross sections for various cluster sizes as a function of the laser wavelength. Our preliminary investigations show broad absorption features, with typical spectral widths of about 100 nm or more. Typically the peak absorption cross section is 1-3 $(\text{Å})^2$ per atom. The peak of the absorption cross section for n=4-9 lies in the range of 560-610 nm laser wavelength. In the limited range of laser wavelength that we have scanned thus far we see no evidence of multiple resonance peaks.

The large absorption width is due to the high intrinsic temperature of the cluster ions. Cluster ions extracted from LMIS are known to be 'hot', with internal temperature as large as 650 K (Ref. 7). Clusters produced by this source cool only by evaporation. These clusters are expected to be substantially hotter than those generated by the supersonic sources where collisional cooling is very effective. These large widths are due to thermal shape fluctuations of the cluster core. The absence of multiple resonances, even for 'non-spherical' structure, is perhaps due to the rapid thermal fluctuations, resulting in smeared, effective spherical shapes. Experiments are in progress to obtain detailed absorption spectrum for each cluster size (n=4-21). These results will be presented in subsequent publications.

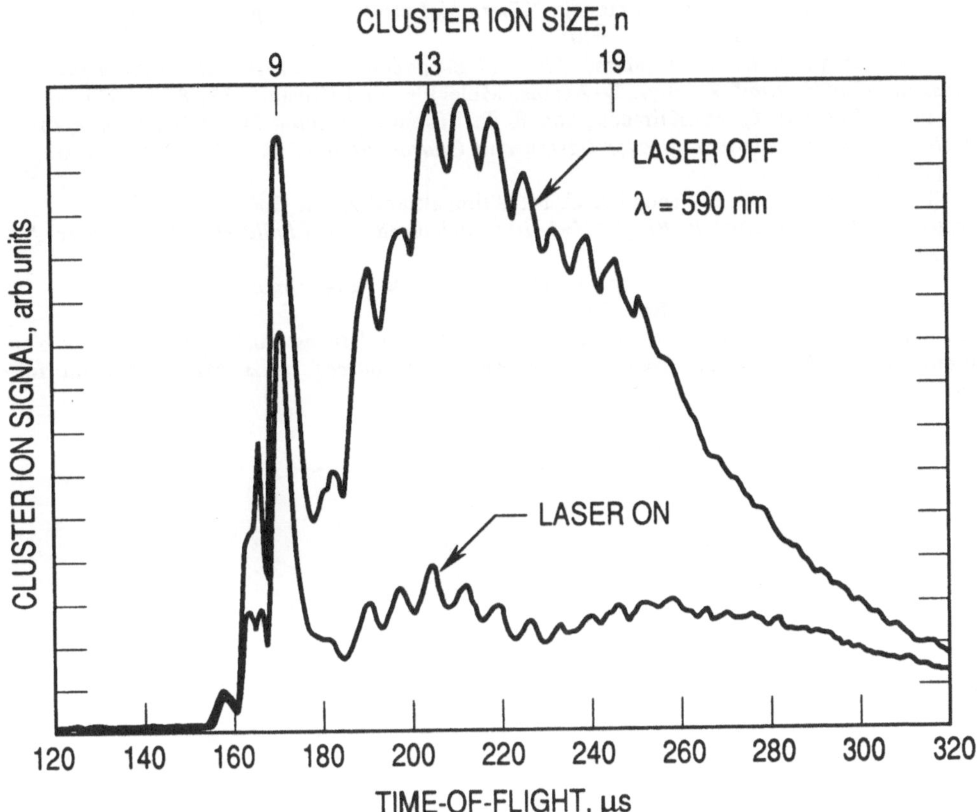

Fig. 4 Mass spectrum of Cs_n^+ with the laser off and on for n=10-21

ACKNOWLEDGEMENT

This work was supported by the Aerospace Sponsored Research Program. I thank Dr. C. Klimcak and Mr. J. Sutter for technical assistance with the laser.

REFERENCES

1.   W. A. de Heer, K. Selby, V. Kresin, J. Masui, M. Vollmer, A. Chatelain, and W. D. Knight, *"Collective Dipole Oscillations in Small Sodium Clusters"*, Phys. Rev. Lett., **59**, 1805 (1987).

2.   K. Selby, M. Vollmer, J. Masui, V. Kresin, W. A. de Heer, and W. D. Knight, *"Surface Plasma Resonances in Free Metal Clusters,"* Phys. Rev. **B40**, 5417 (1989).

3.   C. Brechignac, Ph. Cahuzac, F. Carlier, and J. Leygnier, *"Collective Excitation in Closed-Shell Potassium Cluster Ions,"* Chem. Phys. Lett. **164**, 433 (1989).

4.   V. Kresin, *"Collective Resonance Frequency of Metal Cluster Ions,"* Phys. Rev. **B40**, 12507 (1990).

5.   C. Yannouleas, J. M. Pacheo, and R. A. Broglia, *"Microscopic Structure of the Plasma Resonance in Charged Potassium Microclusters,"* Phys. Rev. **B41**, 6088 (1990).

6.   G. F. Bertsch, and D. Tomanek, *"Thermal Line Broadening in Small Metal Clusters,"* Phys. Rev. **B40**, 2749 (1990).

7.   N. Dam and W. A. Saunders, *"Thermal Broadening of Plasma Absorption in Potassium Cluster Ions"* Z. Phys. D-Atoms, Molecules and Clusters, **19**, 85 (1991).

8.   N. D. Bhaskar, C. M. Klimcak, and R. P. Frueholz, *"Liquid Metal Ion Source for Cluster Ions of Metals and Alloys: Design and Characteristics,"* Rev. Sci. Instrum. **61**, 366 (1990).

9.   N. D. Bhaskar, R. P. Frueholz, C. M. Klimcak and R. A. Cook, " *Evidence of Electronic Shell Structure in Rb_N^+ (N=1-100) Produced in a Liquid Metal Ion Source,"* Phys. Rev. **B36**, 4418 (1987).

10.   N. D. Bhaskar, C. M. Klimcak and R. A. Cook, *"Electronic Shell Structure Effects in Cs_n^+,"* Phys. Rev. **B42**, 9147 (1990).

11.   N. D. Bhaskar and C. M. Klimcak, *"Experimental Investigations of Structure and Stability of Rb_n^+ Extracted from a Liquid Metal Ion Source,"* Int. J. Mass Spectrom. Ion Proc. (in press).

Quantum Chemical Investigation of Absorption Spectra of Small Alkali Metal Clusters; Molecular Dimensionality Transition (2D–3D)

V. Bonačić–Koutecký, J. Pittner, C. Fuchs, P. Fantucci
and J. Koutecký
Institut für Physikalische und Theoretische
Chemie, Freie Universität Berlin
Takustr. 3, D–1000 Berlin 33, Germany

ABSTRACT: The ab initio configuration interaction transition energies and oscillator strengths obtained for the equilibrium geometries of small alkali metal clusters account fully for the spectroscopic patterns obtained by depletion technique and make possible the structural assignments. Specific geometrical and electronic properties responsible for excitations in small s^1 clusters allow for interpretation of theoretical and experimental findings in the framework of the molecular many–electron description. Similarities and differences between the optical response properties of Na_n and Li_n clusters have been found and discussed Molecular dimensionality transition (2D–3D) seems to take place for Li_6 and for Na_7.

I. Introduction

Theoretical investigation of excited states of alkali metal clusters offers a good opportunity to study structural and electronic characteristic features as a function of the cluster size [1]. A comparison of quantum chemical predictions of optically allowed transitions for the stable ground state geometries with the recorded depletion spectra represents a suitable tool for a) the structural assignment and b) interpretation of characteristic patterns which contain relatively small number of intense transitions and large number of weak ones [1]. For this purpose ab–initio type methods which account adequately for electronic correlation effects (such as multireference configuration interaction–MRD–CI) are well suited since they are free from apriori assumptions which can in principle be well established and justified in other fields. Moreover, in addition to the predictive power they can serve as a guidance to find out the essentials which must be included in the simplified approaches for description of ground and excited states of simple metal clusters.

The first photodissociation measurements on clusters larger than tetramers have been carried out for Na_n (n=8,9,10,12,16 and 20) but with low resolution and for selected visible wave lengths [2]. Since they exhibited small number of broad maxima, the classical collective electronic excitations in spherical, spheroidal or elipsoidal metal droplets have been used for their interpretation. Pronounced single

P. Jena et al. (eds.), Physics and Chemistry of Finite Systems: From Clusters to Crystals, Vol. II, 899–906.

intense transition was observed for cationic clusters Na_n^+ and K_n^+ (n=9,21) resembling a "giant resonance" found for nuclei [3]. The photoabsorption spectra of Cs_8 and $Cs_{10}O$ have been interpreted as due to plasmon or plasmon—enhanced excitations [4]. In addition to the classical approach the approximate quantum mechanical treatment in the framework of the time dependent local density approximation (TDLDA) based on jellium model has been applied to the calculations of photoabsorption cross sections [5]. The calculated maxima have been attributed to "surface plasmons" or to volume plasmons at higher energies.

From the high resolution spectra of Na_n (n=3–8,20) [6–8], Li_n (n=2–4,6–8) [9–11] and Li_yNa_x [12] which cover the spectral region up to 3.2 eV, the presence of quantum effects became apparent and needs for quantum molecular interpretation evident [13–18]. It has been shown on a number of examples that the ab—initio MRD—CI transition energies and oscillator strengths calculated for the lowest energy structures determined in the earlier work on the ground state properties [19,20] are in excellent agreement with the recorded absorption spectra [6–12], demonstrating clearly that the structural aspects cannot be ignored. A comparison of the CI predictions with those obtained by the ab initio random phase approximation (RPA) based on the H–F SCF procedure serves to investigate the importance of treating single and double excitation at the same footing which is not the case in the RPA [1,21]. Moreover, the results obtained from the ab initio RPA calculations are useful for comparison with the predictions of the RPA based on jellium model [22], in order to find out the influence of exact potential and the position of the nuclei in within equivalent approximate treatments of the correlation effects.

In this paper the evidence for the transition from two—dimensional to three—dimensional structures of Na_n and Li_n clusters will be shown. The successful structural assignment of topologies determined at T=0 to the recorded depletion spectra suggests that a) the atoms in clusters are not so mobile as assumed although the experiments are not carried out at T=0, and that b) the experimental temperature does not seem to be considerably higher than 300K for small clusters. This is also supported by the quantum molecular dynamic work on Na_n at the T≠0 [23].

II. Quantum Molecular Assignment of the Na_n(n=4–8) and Li_n(n=4,6–8) Clusters.

In alkali metal clusters the bonding is characterized by multicenter delocalized bonds and very small clusters do not assume highly symmetrical shapes due to Jahn—Teller deformation [1]. Tetramers with four valence electrons assume 2D rhombic geometries but the octamers having sufficient number of valence electrons have 3D compact symmetrical T_d geometries built from the tetrahedral subunits [19,20]. This structural information has been confirmed by comparing the ab–initio CI predictions with the depletion spectra. For both Li_4 and Na_4 rhombic tetramers three intense transitions located at $\sim$ 1.7–1.8 eV and in the 2.5–3.0 eV energy

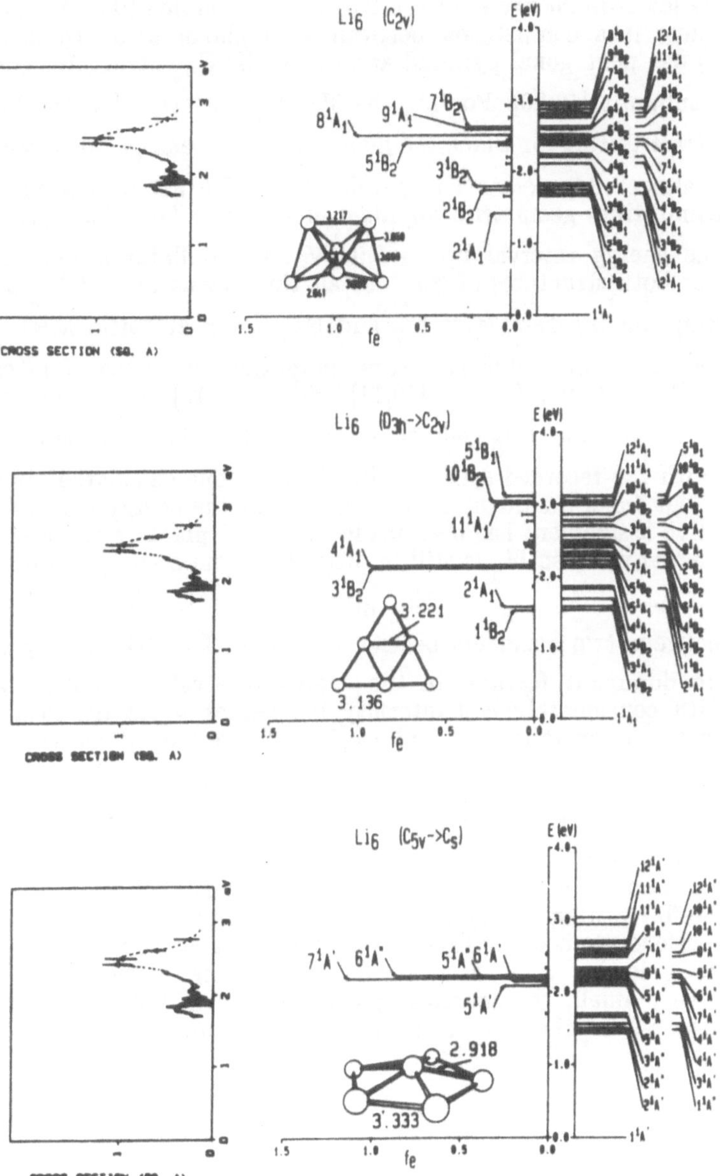

Figure 1: Comparison of photodepletion spectrum and CI predicted optically allowed transitions and oscillator strengths f_e for the three Li_6 structure [11].

interval as well as a larger number of weak bands have been found theoretically [13,15,16,10] and experimentally [6,9]. The 3D Na_8 and Li_8 give rise to one dominant feature located at $\sim$ 2.5 eV and red shifted fine structure [6,14,15,11]. The question can be raised at which cluster size the transition from planar to three dimensional structures occurs. Trapezoidal planar structures of Na_5 and Li_5 are substantially lower in energy than the trigonal bipyramids (0.2 eV). In the case of hexamers there is a competition between a) a planar structure built from the triangles, b) flat pentagonal pyramid and c) the 3D C_{2v} structures containing the tetrahedral subunits [19,20]. For Li_6 the 3D C_{2v} structure has the lowest energy [19b] in contrast to Na_6 for which the planar and flat pentagonal pyramid have the lowest and almost degenerate CI energies [20]. The 3D pentagonal bipyramids represent equilibrium geometries for Li_7 and Na_7 [19a,20]. The optically allowed states in the energy interval up to 3.0 eV and oscillator strengths have been calculated for both structures of Na_5, for all three structures of Li_6 and Na_6 and for the pentagonal bipyramids of Li_7 and Na_7 using ab initio MRD–CI method with all electron and effective core potential corrected for core–valence polarization (ECP–CPP) for Li_n [10,11] and Na_n [18] series, respectively. The predicted spectrum for Na_5 planar structure [18] is in substantially better agreement with the recorded spectrum [8] than the one calculated for the trigonal bipyramid. The broad dominant band is located in the energy interval 2.0–2.2 eV and a tentative assignment has been made to the 2D planar structure. Theoretical and experimental work on Li_5 is still in progress. From a comparison of calculated and recorded spectra for Li_6 and Na_6 given in Figures 1 and 2 a fine structural difference between two hexamers became apparent. The 3D C_{2v} structure of Li_6 gives rise to dominant features in the energy interval 2.5–2.7 eV and to fine structure with considerably less intensity located at $\sim$ 1.8 eV in the complete agreement with the experimental finding [11]. The predicted spectra of other two structures with the dominant features at $\sim$ 2.1 do not correspond to any of the measured features. In contrast, the calculated transition energies and oscillator strengths for the equivalent topologies of Na_6 [18] in particularly for the planar 2D geometry coincide well with the recorded spectrum [8] which is characterized by a dominant transition at $\sim$ 2.1 and the considerably weaker shoulder at $\sim$ 2.8–3.0 eV. Although the flat pentagonal pyramid does not represent a local minimum on the HF energy surface, its contribution to the recorded spectrum cannot be excluded presently for following reasons: the energy difference between this and planar structure is very small, the spectroscopical patterns of both structures are very similar and, moreover, the experiment is not carried out at T=0. However, the 3D

C_{2v} structure of Na_6 with considerably higher energy 0.25 eV gives rise to transition energies and oscillator strengths which do not confine with the experimental finding. Therefore the contribution of the 3D structure to the recorded spectrum of Na_6 can be ruled out. Figures 1 and 2 illustrate clearly that Li_6 seems to be three–dimensional and Na_6 is still planar or close to planarity. It is likely that the sp hybridization is stronger for Li than for Na. The findings of

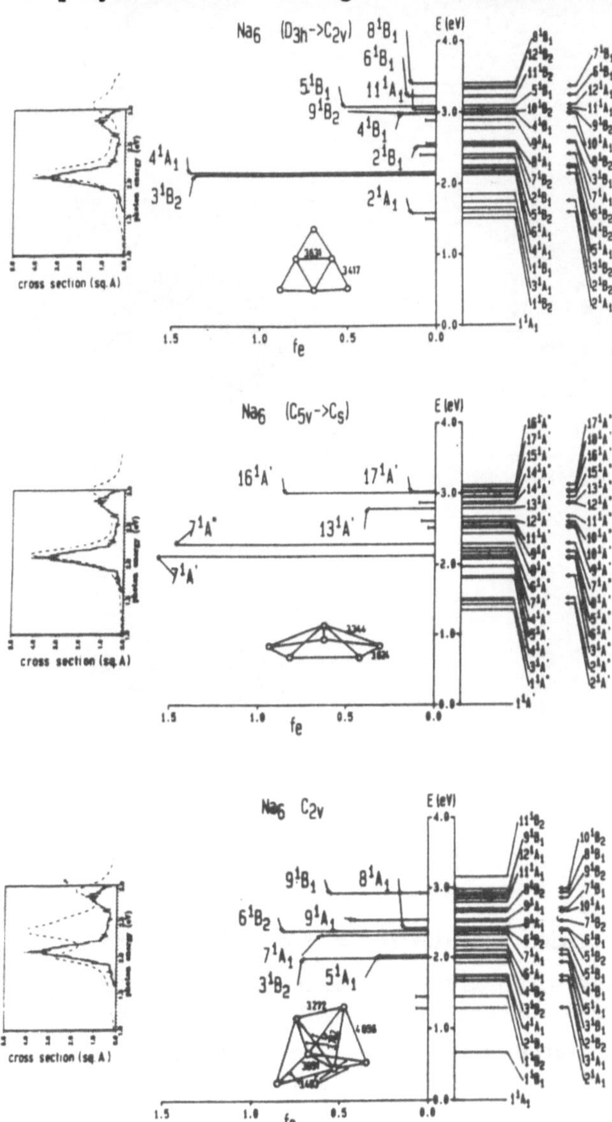

Figure 2: Comparison of photodepletion spectrum [8] and CI predictions for three Na_6 structures [18].

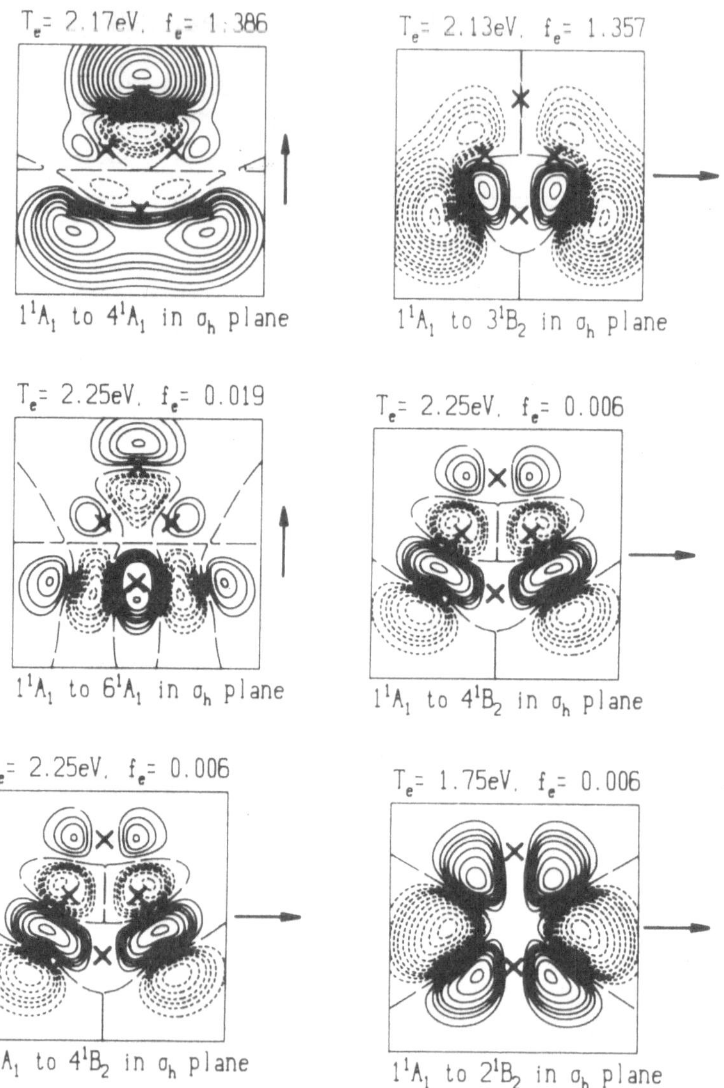

Figure 3: The transition densities $D_{on} = \int \rho_{on}(r)\ \alpha d\beta$ with $\alpha \neq \beta = x,y,z$ of Na_6 for a) the pair of states with dominant intensity and b), c) with negligible intensities. The values of transition energies T_e and oscillator strengths f_e are indicated. Arrows label polarization axis. Dashed and full lines correspond to negative and positive values. x stands for position of nuclei.

Figures 1 and 2 demonstrate that the position of nuclei is important and therefore the structural aspects should not be neglected because of possible experimental conditions (e.g. unknown exact temperature). If the average contribution of all three types of structures, provided they have comparable weights, would be responsible for recorded spectrum, three dominant features located at $\sim$ 2.2, 2.5 and 2.8 eV would be present. Since this is not the case, it seems that the temperature of the experiment is not high enough that the vibrations are able to bring cluster from one class of related geometries to the other one. Pentagonal bipyramids for Li_7 and Na_7 give rise to the spectra with dominant transition at $\sim$ 2.5 eV which is in agreement with the experimental finding. It seems that the intense transition at $\sim$ 2.5 eV is characteristic for the three dimensional structures of Li_6, Li_7, Na_7, Li_8 and Na_8.

III. Discussion

The analysis of the wavefunctions of excited states in terms of single and double excitations shows that the excitation in small alkali metal clusters is characterized by interaction among very few leading excitations as is the case in many molecular excited states. A comparison of the dominant features in the wavefunctions of the states to which transitions with small and large oscillator strengths have been calculated indicates the occurence of interference phenomena, since the same configurations only differing in the amount and the sign of their contributions determine to a large extent the composition of the wavefunctions. This can be illustrated by quantities related to the transition densities between the ground and the excited states. For the dominant transitions to the $4\,^1A_1$ and $3\,^1B_2$ states of Na_6 the compensation of positive and negative values is relatively small as shown in Figure 3 and the resulting large oscillator strength is due to large transition densities located outside the frame of the nuclei. Figure 3 also illustrates examples of interference phenomena for which the compensation of positive and negative transition density regions is almost perfect and the oscillator strength is negligible.

The appearance of relatively small number of intense transitions and a very large number of weak ones in the depletion spectra of small alkali metal clusters is connected with the symmetry of the geometry given by the position of nuclei and with the role of p functions which participate in the delocalized bonding and in polarization connected with the molecular excitations.

The predictive power of quantum chemical methods, although still limited to small clusters with a small number of valence electrons, have been demonstrated by successful structural assignment and interpretation of the optical response properties.

References:
1.   Bonačić–Koutecký, V.; Fantucci, P. and Koutecký, J. (1991), Chem. Reviews, 91, 1035–1108.
2.   a) De Heer, W.A.; Selby, K..; Kresin, V.; Masui, J.; Vollmer, M.; Chatelain, A. and Knight, W.D. (1987), Phys. Rev. Lett., 59, 1805–1808.

    b) Selby, K.; Kresin, V.; Vollmer, M.; de Heer, W.A.; Scheidemann, A. and Knight, W.D. (1991), Phys. Rev. B 43, 4565–4578.

3.   a) Bréchignac, C.; Cahuzac, P.; Carlier, F. and Leygnier, J. (1989), Chem. Phys. Lett., 164, 433–437.

    b) Bréchignac, C.; Cahuzac, P.; Carlier, F.; de Frutos, M. and Leygnier, J. (1991), Z. Phys. D – Atoms, Molecules and Clusters, 19, 1–6.

4.   a) Fallgren, H. and Martin, T.P. (1990), Chem. Phys. Lett., 168, 233–238.

    b) Fallgren, H., Bowen, K.H. and Martin, T.P., (1991), Z. Phys. D – Atoms, Molecules and Clusters, 19, 81–84.

5.   a) W. Ekardt (1984), Phys. Rev. Lett., 52, 1925–1928.

    b) W. Ekardt (1985), Phys. Rev. B 31, 6360–6370.

6.   Wang, C.; Pollack, S.; Cameron, D. and Kappes, M.M. (1990), J. Chem. Phys., 93, 3787–3801.

7.   Pollack, S; Wang, S. and Kappes, M.M. (1991), J. Chem. Phys. 94, 2496–2501

8.   Wang, C.R.Ch.; Pollack, S.; Dahlseid, T. and Kappes, M.M., J. Chem. Phys., in press.

9.   Broyer, M.; Chevaleyre, J.; Dugourd, P.; Wolf, J.–P. and Wöste, (1990) Phys. Rev. A 42, 6954–6957.

10.   Blanc, J.; Bonačić–Koutecký, V.; Broyer, M.; Chevaleyre, J.; Dugourd, P.; Koutecky, J.; Scheuch, C.; Wolf, J.–P. and Wöste, L., J. Chem. Phys., in press

11.   Dugourd, P.; Blanc, J.; Bonačić–Koutecký, V.; Broyer, M.; Chevaleyre, J.; Koutecky, J.; Pittner, J.; Wolf, J.–P. and Wöste, L., Phys. Rev. Lett., in press.

12.   Pollack, S.; Wang, C.R.Ch.; Dahlseid, T. and Kappes, M.M., J. Chem. Phys., in press.

13.   Bonačić–Koutecký, V.; Fantucci, P. and Koutecký, J. (1990), Chem. Phys. Lett. 166, 32–38.

14.   Bonačić–Koutecký, V.; Kappes, M.M.; Fantucci, P. and Koutecký, J. (1990), Chem. Phys. Lett. 170, 26–34.

15.   Bonačić–Koutecky, V.; Fantucci, P. and Koutecký, J. (1990), J. Chem. Phys. 93, 3902–3825.

16.   Bonačić–Koutecký, V.; Fantucci, P. and Koutecký, J. (1988), Chem. Phys. Lett. 146, 518–523.

17.   Bonačić–Koutecký, V.; Gaus, J.; Guest, M.F. and Koutecký, J., J. Chem. Phys., in press.

18.   Bonačić–Koutecký, V.; Pittner, J.; Scheuch, C.; Guest, M.F. and Koutecky, J., J. Chem. Phys., in press.

19.   a) Boustani, I.; Pewestorf, W.; Fantucci, P.; Bonačić–Koutecký, V. and Koutecký, J. (1987), Phys. Rev. B 35, 9437–9450.

    b) Koutecky, J.; Boustani, I. and Bonacic–Koutecky, V. (1990), Int. J. Quant. Chem. 38, 149–161.

20.   Bonačić–Koutecký, V.; Fantucci, P. and Koutecký, J. (1988), Phys. Rev. B 37, 4369–4374.

21.   Gatti, C.; Polezzo, S. and Fantucci, P. (1990), Chem. Phys. Lett., 175, 645–654.

22.   Yannouleas, C.; Broglia, R.A.; Brack, M. and Bortignon, P.F. (1989) Phys. Rev. Lett., 63, 255–258; Yannouleas, C.; Pacheco, J.M.; and Broglia, R.A. (1990) Phys. Rev. B 41, 6088–6091; Reinhardt, P.–G.; Brack, M. and Genzken, O.(1990) Phys. Rev. B 41, 5568–5582; Yannouleas, C. Broglia, R.A.,(1991) to be published.

23.   Röthlisberger, U. and Andreoni, W. (1991) J. Chem. Phys. 94, 8129–8151.

RAMAN STUDIES OF NANOCRYSTALLINE AlO(OH) PREPARED BY SOL-GEL TECHNIQUES

C. J. DOSS[*], A. G. KALLIANOS[+], A. L. RITTER[*], and R. ZALLEN[*]

[*]*Department of Physics, Virginia Tech, Blacksburg, VA 24061 USA*
[+]*Philip Morris Research Center, Richmond, VA 23261 USA*

Careful Raman-scattering studies were carried out on AlO(OH) gels prepared by the Yoldas technique, using the position and lineshape of the characteristic boehmite band near 360 cm^{-1} as a marker of the structural progress of the sol-gel reaction. At short reaction times, the band is shifted and asymmetrically broadened with respect to the bulk-crystal line; with increasing time, it sharpens and approaches the crystal-band position. We interpret these results in terms of finite-size effects of nanocrystallinity (microcrystal dimensions of order 10 nm) on the Raman lineshape. These experiments demonstrate that Raman scattering is effective for monitoring microcrystal sizes in sol-gel systems, and provides a tool for investigating the kinetics of microcrystallite growth.

1. Introduction

Since the mid seventies, a sol-gel process pioneered by Yoldas [1] has been used for preparing boehmite AlO(OH) gels. While it is generally accepted that the complex gels produced by this technique contain a microcrystalline boehmite component, relatively little has been quantitatively established. In particular, it has not been possible to monitor the evolution of microcrystallite size by x-ray techniques, which reveal only very broad features for the early gels [2].

In this paper, we report results of a study in which we have used Raman scattering to address the issue of structural evolution during the preparation of boehmite gels (also called gelatinous boehmite or sol-gel alumina). We have found that the linewidth and peak position of the band which is the primary Raman signature of crystalline boehmite exhibits small but systematic changes during the course of sol-gel processing. At short times, the line is downshifted and asymmetrically broadened with respect to

P. Jena et al. (eds.), Physics and Chemistry of Finite Systems: From Clusters to Crystals, Vol. II, 907–912.
© 1992 *Kluwer Academic Publishers.*

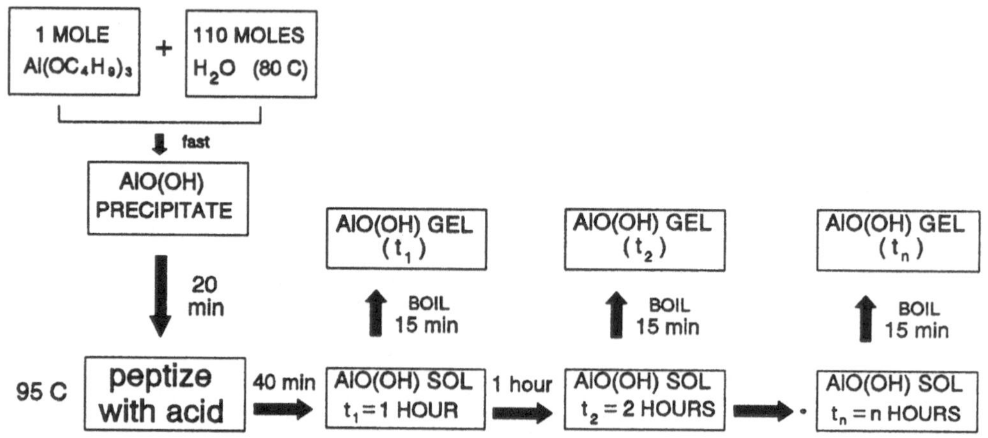

Figure 1. Flow chart of the procedure for preparing wet boehmite gel samples.

the crystal feature; with increasing time it sharpens and shifts toward the crystal position. Reasoning by analogy with similar effects seen in nanocrystalline semiconductors, we interpret these observations in terms of finite-size effects which attenuate with increasing microcrystal size. This Raman technique, which is convenient to carry out on sol-gel samples, thus provides a useful handle on the increase in microcrystal size, for boehmite nanocrystals, during the early stages of sol-gel preparation.

2. Experiment

Figure 1 shows the method used for making the AlO(OH) wet gel samples. This procedure, based on Yoldas' technique, begins by adding doubly distilled aluminum sec-butoxide (ASB) to water at 80° C. The hydrolysis reaction proceeds very fast and produces an amorphous precipitate. The sol is then heated to 95° C which drives off most of the alcohol produced in the hydrolysis step. After 20 minutes, the sol is peptized with HCl or HNO_3 using acid/ASB molar ratios of .07 or .14. At selected times, sol samples were removed from their reaction vessel and boiled until gelation occurred. The wet gels were sealed and stored at room temperature.

Another set of sols, referred to here as "no acid" samples, were produced without the peptization step. For these unpeptized samples, removal from the reaction chamber, at selected times, was followed by centrifuging and then drying at 110° C for at least 24 hours. These samples were dry within 2 hours, halting any further crystal growth.

Raman spectra were obtained at room temperature using a SPEX 1403 Raman spectrometer equipped with a GaAs photomultiplier and photon-counting electronics. All spectra were obtained in the 90° scattering configuration. A CW argon ion laser operating at 514.5 nm was the excitation source. All measurements were performed with an instrumental bandwidth of 3.8 cm^{-1} and a 0.5 cm^{-1} grating step size. Linewidth estimates were corrected for instrumental broadening.

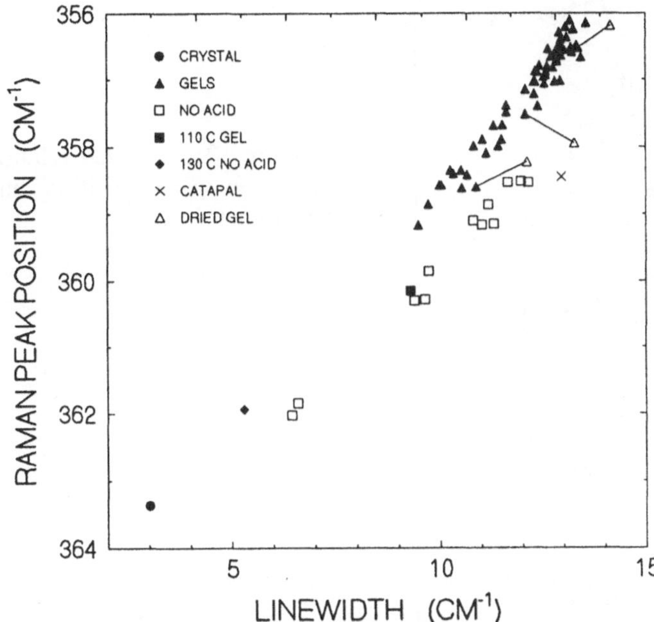

Figure 2. The correlation between peak position and linewidth for the 360 cm^{-1} band observed in the gels. At lower left is the point corresponding to bulk crystalline boehmite.

3. Spectral Evidence for Nanocrystal Size Effects

The dominant Raman-active Al-O stretching vibration in crystalline AlO(OH) occurs near 360 cm^{-1}. The D_{2h}^{17}-symmetry boehmite structure consists of layers with strong ionic/covalent bonding within each layer and weaker hydrogen bonding between layers; two layers intersect each primitive cell [3]. In the 360 cm^{-1} mode, the aluminum atoms within one layer move together as a "rigid sublattice" against the oppositely-moving rigid sublattice of oxygen atoms. The motion in the second layer is similar, but with opposite phase, so that the crystal mode has A_g symmetry and is Raman active. The Davydov partner of this vibration (ie., the mode in which adjacent layers move in phase) is infrared active and Raman-inactive, while for a single layer -- if it occurred "by itself" (for example, as layer fragments in an amorphous phase) -- the corresponding mode is <u>also</u> Raman-inactive. Thus the occurrence of the 360 cm^{-1} band in the Raman spectrum requires the presence of AlO(OH) crystals or microcrystals; it does not merely reflect, say, octahedral coordination about Al or even the presence of individual AlO(OH) layers.

910

Figure 2 represents our central finding. In the many gel samples we prepared, the 360 cm^{-1} band is present but is found to exhibit small but significant shifts in position and changes in lineshape. Figure 2 displays the peak position and the linewidth (full width at half maximum) for each of our gel samples. To provide a standard of comparison, the position/linewidth point corresponding to bulk crystalline boehmite is included at the lower left.

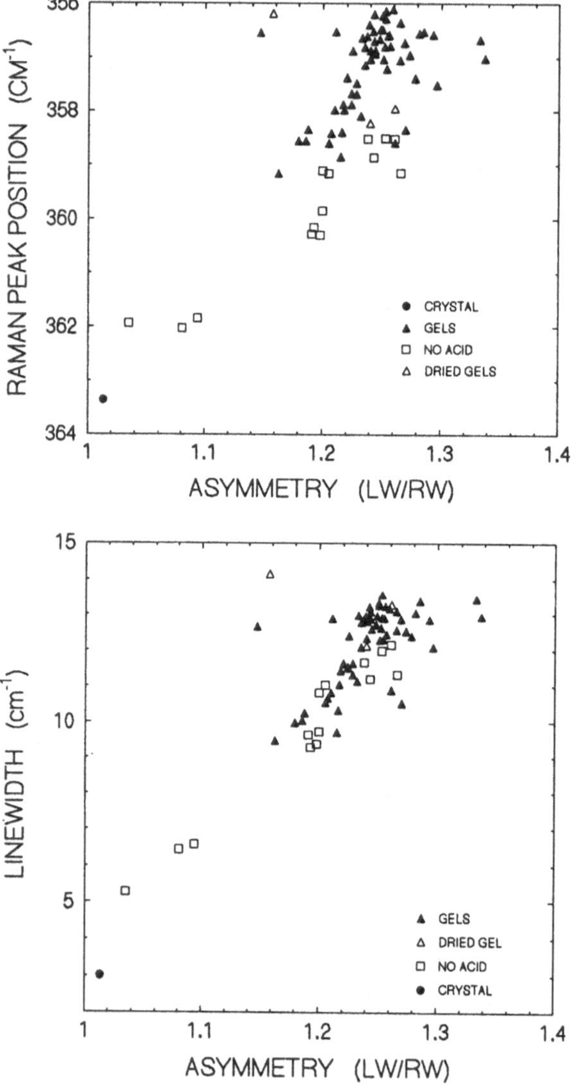

Figure 3. Correlation between peak position and asymmetry.

Figure 4. Correlation between linewidth and asymmetry.

The clear correlation between Raman linewidth and peak position, exhibited in Figure 2, resembles a well-documented effect observed in nanocrystalline semiconductors [4]. We tentatively interpret our results for AlO(OH) in terms of the same general mechanism: the finite-size effect of nanocrystallinity (microcrystal dimensions of order 10 nm) on the Raman lineshape [5]. The gist of the argument is as follows. For a nanocrystal of size L, the strict "infinite-crystal" k-space selection rule is replaced by a relaxed version characterized by a k-space uncertainty of order (1/L). The smaller is L, the larger is the shift and broadening of the Raman band.

The points at the upper right in Fig. 2, for which the boehmite-like band is substantially broadened and shifted relative to the bulk-crystal Raman band, correspond to the earliest times (t_1, t_2, ····) in the preparation procedure of Fig. 1. With increasing reaction time, the linewidth narrows and the peak shifts toward the bulk-crystal position. Our Raman experiments thus monitor the increase of the characteristic boehmite nanocrystal size, as the sol-gel reaction proceeds.

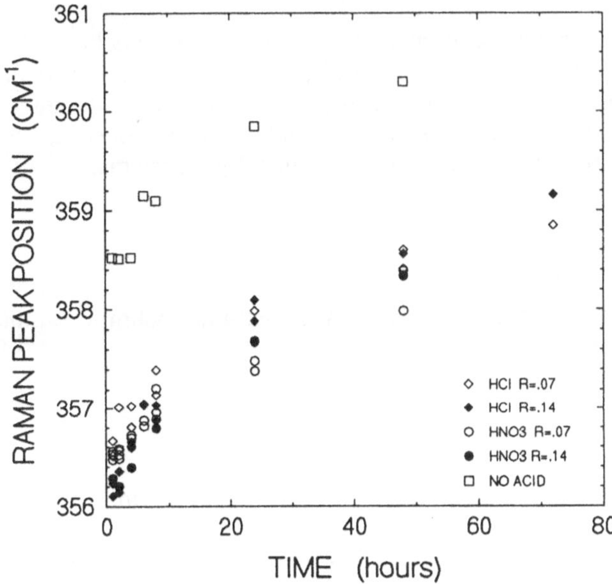

Figure 5. Microcrystallite growth kinetics monitored by the peak position of the boehmite-like Raman band. R is the acid / ASB molar ratio.

The no-acid samples plotted in Fig. 2 tend to lie slightly lower than the gel samples. This is a very small but interesting effect, and appears to be connected with the 110° C drying step used to prepare these samples. The three pairs of points linked by lines in Fig. 2 represent gel samples (solid triangles) later subjected to 110° C drying (open triangles). The drying introduces a small additional linewidth broadening.

Figures 3 and 4 show correlations involving another aspect of the Raman lineshape, the band asymmetry. The asymmetry is defined here as LW/RW, where LW is the left width (half width at half maximum, from the peak position to the low-frequency side) and RW is the right width (similar half width, on the high-frequency side). The asymmetry is small, so that the scatter in these correlations is substantial. Nevertheless, figures 3 and 4 reveal that the degree of asymmetry is correlated with both the shift and the broadening of the band relative to the bulk-crystal values. This behavior is consistent with the finite-size effect interpretation of the Raman changes [4, 5].

In figure 5 we indicate the microcrystal growth kinetics in a plot of Raman peak position versus time. The no-acid samples are offset with respect to the gels, starting off at larger crystal sizes. For the gels, it is noteworthy that neither acid type nor concentration significantly affects the rate of crystal growth.

4. Summary

We have seen systematic changes in the position and lineshape of the boehmite-like Raman band in a variety of gels prepared as indicated in figure 1, as well as in samples prepared without acid. The correlations shown in figures 2, 3, and 4 strongly suggest that we are seeing finite-size effects in boehmite nanocrystals, and demonstrate that Raman scattering is useful for monitoring microcrystal sizes in sol-gel systems.

Acknowledgments

This research was supported by Philip Morris, USA. A crystalline boehmite sample was kindly provided by Karl Wefers.

References

[1] Yoldas, B. E. (1975) "Alumina sol preparation from alkoxides", Amer. Ceram. Soc. Bull. 54, 289-290.

[2] Wefers, K., and Misra, C., (1987) "Oxides and Hydroxides of Aluminum" (Alcoa Laboratories).

[3] Inoue, M., Kondo, Y., and Tomoyuki I. (1988) "An ethylene glycol derivative of boehmite", Inorg. Chem. 27 vol. 2, 215-221.

[4] Tiong, K. K., Amirtharaj, P. M., Pollak, F. H., and Aspnes, D. E. (1984) "Effects of As$^+$ ion implantation on the Raman spectra of GaAs: spatial correlation interpretation", Appl. Phys. Lett. 44, 122-124; Holtz, M., Zallen, R., Brafman, O., and Matteson, S. (1988) "Raman-scattering depth profile of the structure of ion-implanted GaAs", Phys. Rev. B 37, 4609-4617.

[5] Richter, H., Wang, Z. P., and Ley, L. (1981) "The one-phonon Raman spectrum in microcrystalline silicon", Solid State Commun. 39, 625-629.

SURFACE INDUCED IONIZATION OF WATER CLUSTERS.

D.Yu.DUBOV and A.A.VOSTRIKOV
Institute of Thermophysics
Siberian Branch of the USSR Acad.Sci.
Novosibirsk 630090,
USSR

ABSTRACT. The paper presents the results of the molecular beam studies of the collisional ionization of water clusters. Both positive and negative ions were observed in scattered flow when clusters $(H_2O)_{n>300}$ collided with various solid surfaces at the velocity about 1.3 km/s. In measurements of total ion currents the probability of charge removal from the surface was determined in dependence on the incident angle, on the mean cluster size (up to 3 000 molecules), and on the target material. The proposed model includes 1) the formation of an ion pair during ionic dissociation of a vibrationally excited molecule in water cluster colliding with surface; 2) asymmetric neutralization of an ion pair by surface; 3) inertial removal (recoil) of cluster ion from the surface. However some suprising features observed in the measured angular and energetic distributions of emitted charged particles, viz. rainbow spatial pattern of emission and the correlation between mean energy and scattered angle reversed to usual scattering, show that the emission of charged particles may be only partially determined by inertial removal.

1.INTRODUCTION

The electric processes in the lower atmosphere are based on the water droplet and aerosol elecrification, which are so far poorly understood processes. The complicated dependence of electrification pattern on the variety of factors and conditions is one of the major unsolved problems. This fact makes large difficulties for obtaining the reliable data in a macrophysical experiment. The principal possibility of using of the molecular beam method for the collisional ionization study on the elementary event level in the well-controlled conditions was demonstrated by us earlier, when we observed the microscopic analog of this process [1]. The electrification of the surface was initially detected at normal incidence of water cluster beam to the stainless steel target as a current to the target. The absence of ions in the incident beam was checked. The signal was absent in analogous experiments with $(CO_2)_{n<10\ 000}$ and $(N_2O)_{n<10\ 000}$ clusters and was observed in the case of water clusters

P. Jena et al. (eds.), Physics and Chemistry of Finite Systems: From Clusters to Crystals, Vol. II, 913–918.
© 1992 *Kluwer Academic Publishers.*

for all the surfaces investigated.

2. EXPERIMENTAL

The experiments were performed in the molecular beam generator described in details elsewhere [2]. The vapour temperature in the nozzle source was kept constant, so the mean cluster size, $\bar{N}$, in a beam was varied by varying the pressure P_0 in the source. The measurement of $\bar{N}$ was carried out via retarding potential technique [3,4]. A sonic nozzle with the diameter of 1 mm was used. Under these conditions the directed velocity of water clusters in the beam is almost equal to the top velocity of a gasdynamic expansion [5,6], which in this case is about $1.3 \cdot 10^3$ m/s. Hence the kinetic energy of water N-mer was $\epsilon(N) = 0.17 \cdot N$ eV.

The planar targets were placed in the chamber with the pressure $< 2 \cdot 10^{-4}$ Pa and could be rotated so that the incident angle of the beam, θ_i, varied from 0° to 80°. The charged particles emitted from the target were collected by a cylinder enveloping the target [2]. To achieve a complete collecting the charges of one sign and retarding the charges of opposite sign the voltage $U = \pm 300$ V was applied between target and collector.

A spatial pattern of scattered neutrals, J_s, currents of emitted charged particles, $I_e^{+,-}$, and currents of scattered cluster ions, $I_s^{+,-}$, were measured by the intensity detector, channel multiplier, and Faraday cup collector, respectively. In latter case clusters were preliminary ionized by electrons with the energy of 30 or 8 eV. All the detectors could revolve around the target in the principal plane. The angle resolution of detectors was about 4°. The energy analysis of emitted and scattered ions was carried out applying the retarding potential to the grids in front of channeltron or collector entrance.

3. RESULTS

The total currents of emitted charged particles were measured for various targets (gold, Duralumin, steel, germanium, fiberglass laminate) in dependence on incident angle and on mean cluster size in the incident beam. The measured dependences on θ_i shown in Fig.1 are characterized by sharp maxima at angles close to grazing, $\theta_{i,max} = 70^\circ$, (although the currents are detected even at normal incidence). It seems to be connected with the features of neutral cluster scattering, i.e. large probability of reflection from the surface of cluster having a large tangential velocity [7-9]. The sign and value of equilibrium current detected at $U = 0$ V depend on target material, and for the target of germanium – on the angle θ_i. It is also to be noted that the target material effects significantly both on the absolute value of current and especially on the ratio of the currents of opposite signs.

This can be also seen in Fig.2, where the values of currents at $\theta_i = \theta_{i,max}$ are presented versus mean cluster size $\bar{N}$. All the curves of Fig.2 are normalized to the flux J_c of water clusters in the beam, so they are the resulting probabilities φ of the emission of elementary charge at collision of water cluster with $N = \bar{N}$ with surface. The maximum value of φ is seen to be 10^{-4} for positive charge emission at $\bar{N} =

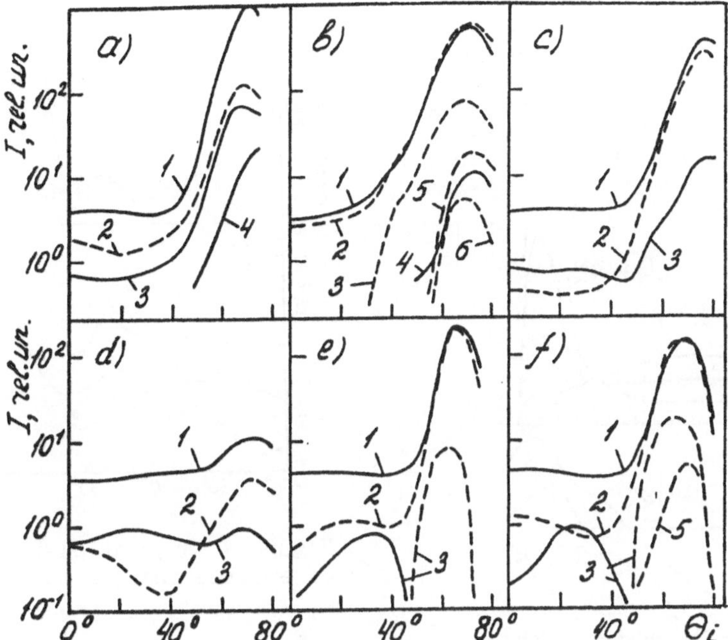

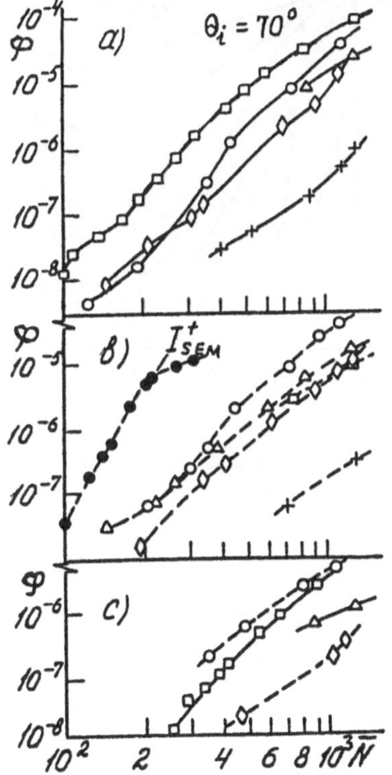

Figure 1. Total currents of emitted ions I^+ (——) and I^- (----) versus angle θ_i. Curves 1-3 – $\bar{N}$ = 1200, 4(a) - 270, 4-6- - 350. Targets: a) Duralumin, b) steel, c) gold, d) fiberglass laminate, e)germanium ⟨110⟩, f) germanium ⟨100⟩. Curves 1,4 – total positive currents; 2,5 – total negative currents; 3,6 – currents at U = 0 V.

Figure 2. Total currents normalized to cluster flux versus size $\bar{N}$. Targets: Duralumin (□), steel (o), gold (△), germanium ⟨100⟩ (◊), fiberglass laminate (+). a – positive currents, b – negative currents, c – currents at U = = 0 V.

= 1500 and Duralumin target. The current to collector should be noted to begin at $\bar{N} > 100$, and in the case when the entrance of multiplier is used as a target – see curve I_{SEM} in Fig.2 – the multiplier operated in usual way detects the signal even at

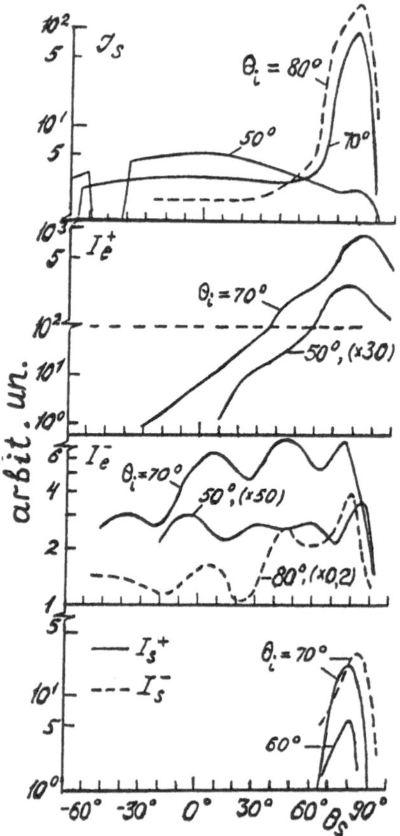

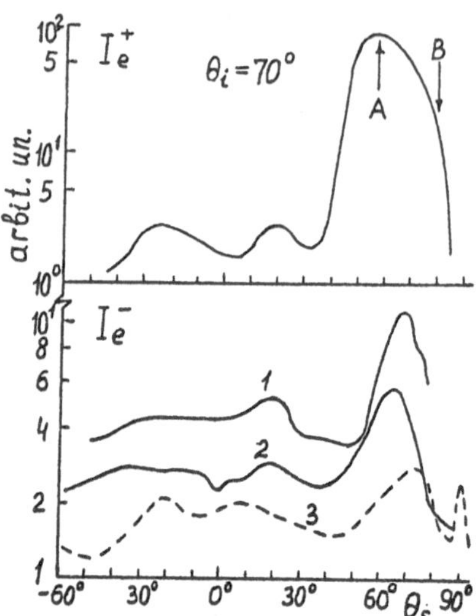

Figure 4. Angular distribution of emitted charged particles.

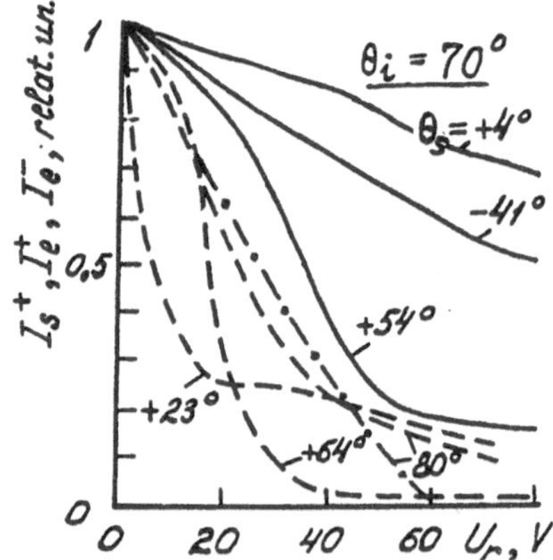

Figure 5. Emitted ion currents versus voltage U_r.

Figure 3. Angular distribution of emitted particles compared with scattered ones.

$\bar{N} > 40$. But taking into account the spread size distribution of water clusters in the beam [6], we estimate the threshold size for ionization in our experiments to be $N_* = 300$ molecules [2].

The angular distributions of emitted and scattered particles are compared in Figs.3 and 4. Here and in Fig.5 are the results for $\bar{N} = 1\ 500$; Duralumin (Figs.3 and 5) and steel (Fig.4) targets. One can see that with the increase of θ_i the neutral J_s and charged I_s^+ - components of beam scattering acquire a sharply pronounced lobe character. As to the dependences $I_e^+(\theta_a)$ and $I_e^-(\theta_a)$ there is no direct relation between the character of cluster scattering and ion emission observed. Moreover the spatial pattern of negative ion current differs qualitatively from the $I_e^+(\theta_a)$. The additional maxima are distinctly seen at the curves $I_e^-(\theta_a)$.

The analogous differences between I_a^+ and I_a^- have been found in the dependences on retarding voltage U_r (Fig.5). It can be seen that with the increase of θ_a the mean kinetic energy of emitted positive charges increases in accordance with the behaviour of scattered neutrals [8] and ions [10]; just as the energy of negative charges emitted to normal and back directions exceeds the energy at large θ_a. The dependences of $I_a^-(\theta_a)$ and $I_a^-(U_r)$ should be noted to be more sensitive to the surface conditions, i.e. target material and temperature, surface cleanness. So Figure 5 shows the transformation of initial dependence $I_a^-(\theta_a)$ (1) after ten minutes bombardment by the prepared cluster ions (2) and after warming-up to 440 K followed by cooling till 320 K (3).

4.DISCUSSION

The formation of ions in water cluster has been related by us with the ionic dissociation (autoprotolysis) of molecules excited by collision:

$$(H_2O)_N + surface \longrightarrow [(H_2O)_i{}^*(H_2O)_j] \longrightarrow [H_3O^+(H_2O)_k \cdot OH^-(H_2O)_l] \qquad (1)$$

In liquid water the activation energy E_a of (1) due to the strong hydratation of ions is only < 0.6 eV and makes possible to stimulate the reaction by the vibrational excitation of H_2O molecule [11]. The estimation of E_a for clusters [12,2] has shown that this value decreases rapidly with the increase of N so that we suppose the similar pathway of ion formation.

The separation of ions may occur in two processes: 1) the asymmetric neutralization of the cluster ions during the interaction with surfaces seems to be the main one. It is pointed out by the observation of current to target and by the large size (energy) of emitted ions. 2) The break-up of cluster into fragments containing ions of different signs may prevail at small incident angles.

The removal of charged particles from the surface could be explained by the inertial mechanism similar to the cluster scattering. But in this way we are in great difficulties to account for our results obtained for I_a^-. We suppose that the processes which may be useful for interpretation of these results are: 1) charge removal with cluster desorption which has the similar kinematics [13] or 2) the break-up of cluster having the crystal structure. The full explanation is planned to be the aim of our further studies.

5.REFERENCES

1. Vostrikov, A.A., Dubov, D.Yu., and Predtechenskiy, M.R. (1987) 'Ionization of water clusters by surface collision', Chem.Phys.Lett. 139, 124-128.

2. Vostrikov, A.A., Dubov, D.Yu., and Gilyova, V.P. (1989) 'Fragmentation of charged clusters during collisions of water clusters with
electrons and surfaces', in E.P.Muntz et al (eds.), Prog.Astron.
Aeron. 117, 335-353.
3. Bauchert, J. and Hagena, O.F. (1965) 'Massenbestimmung ionisierter
Agglomerate in kondensierten Molekularstrahlen nach einer elect-
rischen Gegenfeldmethode', Z.Naturforsch.A 20, 1135-1142.
4. Vostrikov, A.A. and Predtechenskiy, M.R. (1985) 'Interaction of
electrons with CO_2 van der Waals clusters', Sov.Phys.-Tech.Phys.
30, 529-534.
5. Dreyfuss, D. and Wachman, H.Y. (1982) 'Measurement of relative concentrations and velocities of small clusters (n < 40) in expanding water vapor flows', J.Chem.Phys. 76, 2031-2042.
6. Vostrikov, A.A. and Dubov, D.Yu. (1991) 'Cluster generation in a
free jet for molecular beam studies', in A.E.Beylich (ed.), Rare-
fied Gas Dynamics, VCH, Weinheim, pp.1156-1163.
7. Mironov, S.G., Rebrov, A.K, Semyachkin, B.Ye., and Vostrikov, A.A.
(1981) 'Molecular clusters: formation at free expansion and with
vibrational energy pumping, cluster-surface interaction', Surface
Sci. 106, 212-218.
8. Dreyfuss, D. and Wachman, H.Y. (1981) 'Scattering of water cluster
beams from surfaces', in S.S.Fisher (ed.) Prog.Astron.Aeron. 74,
pt.1, 183-187.
9. Holland, R.J., Xu, G.Q., Levkoff, L., Robertson, A.,Jr. and Bernasek, S.L. (1988) 'Experimental studies of the dynamics of nitrogen van der Waals cluster scattering from metal surfaces', J.Chem.
Phys. 88, 7952-7963.
10.Vostrikov, A.A. and Dubov, D.Yu. (1990) 'Angular and energy distribution of charged particles formed during the scattering of neutral water clusters', Sov.Tech.Phys.Lett.(USA) 16, 27-28.
11.Natzle, W.C. and Moore, C.B. (1985) 'Recombination of H^+ and OH^- in
pure liquid water', J.Phys.Chem. 89, 2605-2612.
12.Compton R.N. (1985) 'Negative-ion states', in S.P.McGlynn et al.
(eds), Photophysics and Photochemistry in the Vacuum Ultraviolet,
D.Reidel Publishing Co., 261-295.
13.Cho, C.-C., Polanyi, J.C., and Stanners, C.D. (1988) 'Photoejection
of clusters from HBr adsorbate: $(HBr)_n$, n < 4', J.Phys.Chem. 92,
6859-6861.

Charge Separation Reactions of Doubly Charged Xe Clusters

M. FIEBER, E. HOLUB-KRAPPE, J. LEHMANN, AND T. DREWELLO
Hahn-Meitner-Institut, Berlin, Germany
and
A. DING
Optisches Institut, Technische Universität Berlin, Berlin, Germany

ABSTRACT: An experiment has been designed to investigate the double photoionization of Xe-Clusters. TEPICO and PIPICO techniques have been applied to investigate the Coulomb fragmentation of doubly charged Xe clusters. It is shown that the ionization can be described by an Auger process at higher photon energies and by a secondary ionization process at lower photon energies.

1.   Introduction

The charge separation reactions of doubly charged van-der-Waals clusters into two singly charged fragments, a process commonly referred to as Coulomb explosion, has been a topic of great interest in the last decade. The effect that stable doubly charged van-der-Waals clusters can be observed only above a certain (so-called critical) size has mainly been attributed to charge separations of unstable (lower-sized) clusters. There has been no experimental evidence for these kind of dissociation at all in the case of van-der-Waals clusters. Only very recently three different experimental approaches succeeded in providing the desired information.

First, Lezius and Märk[1] probed the occurrence of charge separation reactions by the detailed analysis of electron impact ionization efficiency curves of Ar_n clusters. Gotts and Stace[2] attributed some broad collision-induced fragment ion signals of mass-selected $(CO_2)_n^{2+}$ clusters to charge separation, and Rühl et al.[3] provided an instructive insight into the actual pathways of the charge separation reactions of doubly charged Ar_n clusters applying photoion-photoion-coincidence techniques (PIPICO).

The present report aims to elucidate further the charge separation behaviour of doubly charged van-der Waals clusters. The PIPICO-method has been applied to record the charge separation reactions of doubly charged Xe_n^{++} clusters generated by supersonic beam expansion. The observed coulomb fragmentation is discussed, together with the corresponding kinetic energy released in these reactions as a function of the energy of the ionizing photon.

919

P. Jena et al. (eds.), Physics and Chemistry of Finite Systems: From Clusters to Crystals, Vol. II, 919–924.
© *1992 Kluwer Academic Publishers.*

2.   Experimental

The experimental set-up has been described previously in detail[4]. Briefly, the Xe_n clusters were prepared by supersonic expansion of pure Xe gas. VUV light was obtained from the Berlin synchrotron facility BESSY, which was monochromatized using a toroidal grating monochromator (TGM7) covering the range between 10 eV and 125 eV. A time-of-fight spectrometer was used to perform both the TEPICO (Threshold Electron Photo Ion COincidence) and PIPICO(Photo Ion Photo Ion COincidence) experiments. In both methods ions are analyzed in terms of the differences in flight time. While TEPICO experiments measure the difference between the arrival time of the electron and the correlated ion, the PIPICO method determines the differences in flight time between different ions generated by the fragmentation of a parent ion.

3.   Results and Discussion

3.1   TEPICO SPECTRA:

TEPICO spectra were produced by forming electron-ion pairs through photoionization and measuring the flight time of the ion in respect to the correlated electron. Additionally the electrons passed e zero kinetic energy analyser. Fig 1. shows an example of such a spectrum generated by the photoionization of Xe-clusters with high energy photons(107.7 eV). The TEPICO spectrum is characterized by the signals for the triply, doubly and singly charged Xe atoms followed by a series of singly charged cluster ions. Besides the signals for the multiple charged ions, all other signals are considerably broader than expected for a simple ionization. It is apparent that the singly ionized cluster ions Xe_n^+ show distinctly broadened peaks. As it is confirmed by the PIPICO experiments this is the result of a violent fragmentation of the doubly ionized clu-

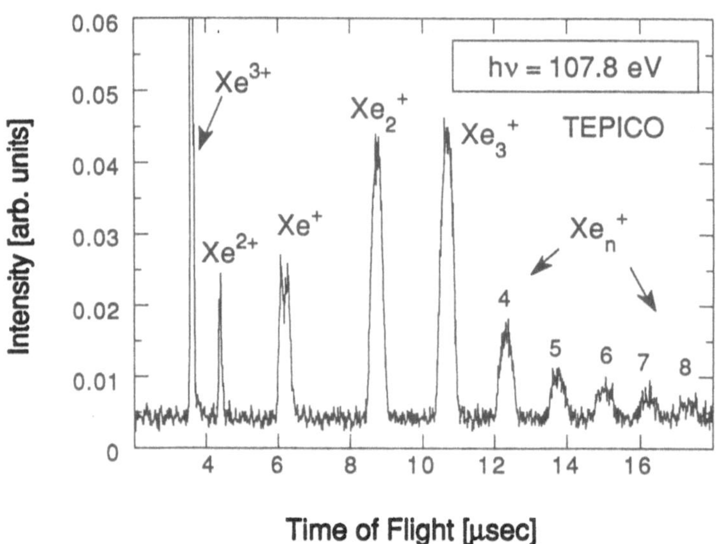

Fig. 1: TEPICO-spectra of Xe clusters taken at 107.8 eV

sters. Considering the energy of the ionizing photon the width of these peaks can be mainly attributed to the enormous energy released in charge separation reactions of unstable multiple charged clusters.

With a given acceleration energy U (usually 960 V) the average kinetic energy released ΔE_{kin} can be obtained from the width of the corresponding mass peak by the equation

$$\Delta E_{kin} = (2.\Delta t/t)^2.eU \qquad [1]$$

$2\Delta t$ is the width of the broadened peak and t the average flight time of the fragment ion. These values are displayed in Fig. 2 as a function of cluster fragment ion size.

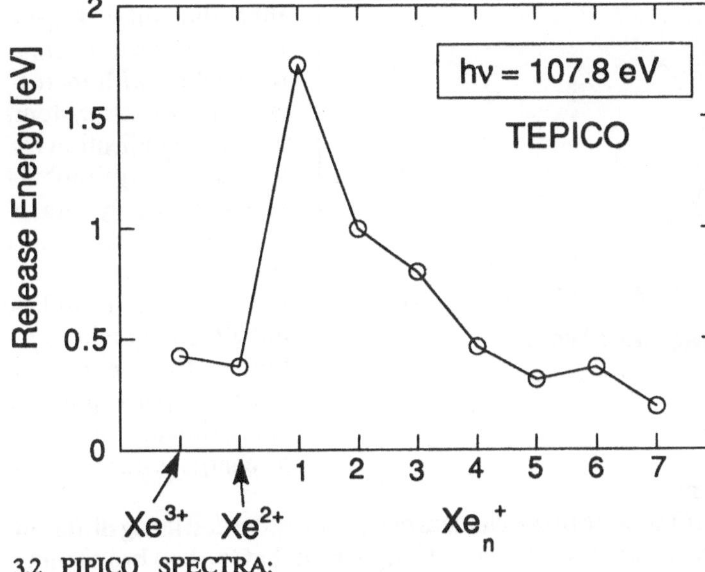

Fig. 2:

Kinetic energy release of Xe_n^+ as a function of cluster size (photon energy: 107.8eV,

stagnation pressure: 7.5 bar)

3.2 PIPICO SPECTRA:

PIPICO spectra were taken using essentially an identical set-up, but different triggering conditions. Doubly charged ions fragment into ion pairs (masses m_1, m_2) which separate quickly because of Coulomb repulsion. The difference in flight time between these ions is greatest if the ions are released parallel to the acceleration axis. As they undergo an additional acceleration by the electric field the lighter of the two fragments (index 1) is the first to arrive at the detector, and is used to trigger the start pulse of the time-of-flight electronics. The method then measures the difference of the flight time of the two ionic fragments. The evaluation uses equation [1] with slight modifications

$$\Delta E_{kin,2} = \{m_2/(m_1+m_2).2\Delta t/t\}^2 \qquad [2]$$

In order to obtain the total kinetic energy released into both fragments ΔE_{kin} has to be corrected for the second fragment ion. One obtains for the total energy

$$\Delta E_{tot} = \Delta E_1 \cdot (m_1 + m_2)/m_1 \qquad m_2 > m_1 \qquad\qquad [3]$$

Fig 3 compares the total fragmentation energies obtained from TEPICO and PIPICO spectra. In the case of the TEPICO-spectra the other fragment ion is not known. Depending on the choice of the second fragment different total energies are obtained.

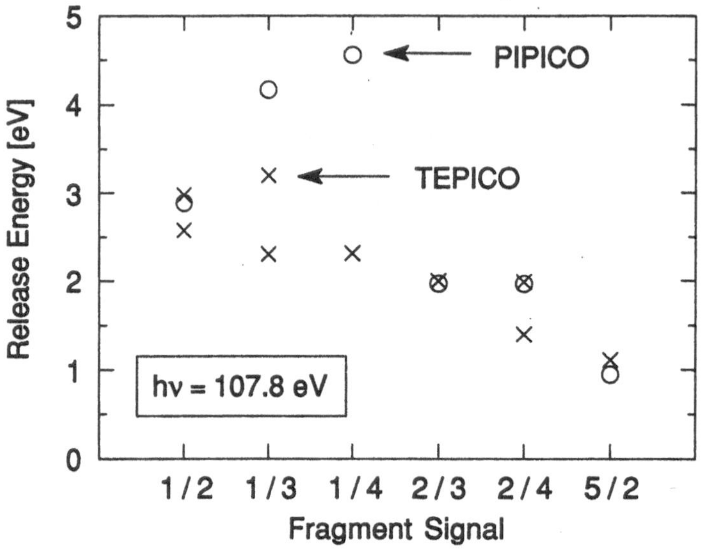

Fig. 3: Total kinetic energy released through fragmentation

PIPICO spectra define both fragment ions uniquely. Except for two points both methods produce very similar results. It should be noted that the total kinetic energy release decreases slightly with increasing size of the system. This is an indication for a certain charge mobility while the system is undergoing fragmentation. From the known mobility of holes in liquid Xe one obtains mobilities for the cluster ion which correspond to approximately 50 ps for a charge transfer between two adjacent Xe atoms.

Both TEPICO and PIPICO spectra of Xe clusters have been recorded using different photon energies. The latter are displayed in Fig 3. It is apparent that the line broadening decreases significantly when the photon energy is lowered. The change in line width seems to coincide with the ionization threshold of the 4d electrons (67.55 eV [5]). This behaviour seems to mirror the effect of several different ionization mechanisms:

If the photon energy is sufficiently high, an inner shell electron can be removed initiating an Auger process and generating a doubly charged atom in the cluster, which immediately decays by charge tranfer of one of the charges onto a neighbouring atom. Below the threshold of this process doubly charged clusters can be generated by secondary ionization, i.e. the electron produced by the photon has an energy high enough to ionize another atom in the cluster. Both processes produce different charge densities in the cluster: while the Auger process produces clusters with charges on adjacent atoms and therefore a strong repulsion, the secondary ionization process generates charges fairly far apart which show only a weak repulsion. It is our conclusion that this is the reason for the change in linewidth in Fig4. It has to be kept in mind, however, that rapid charge migration would smear out the effect described above and make a distinction between these two processes much more difficult.

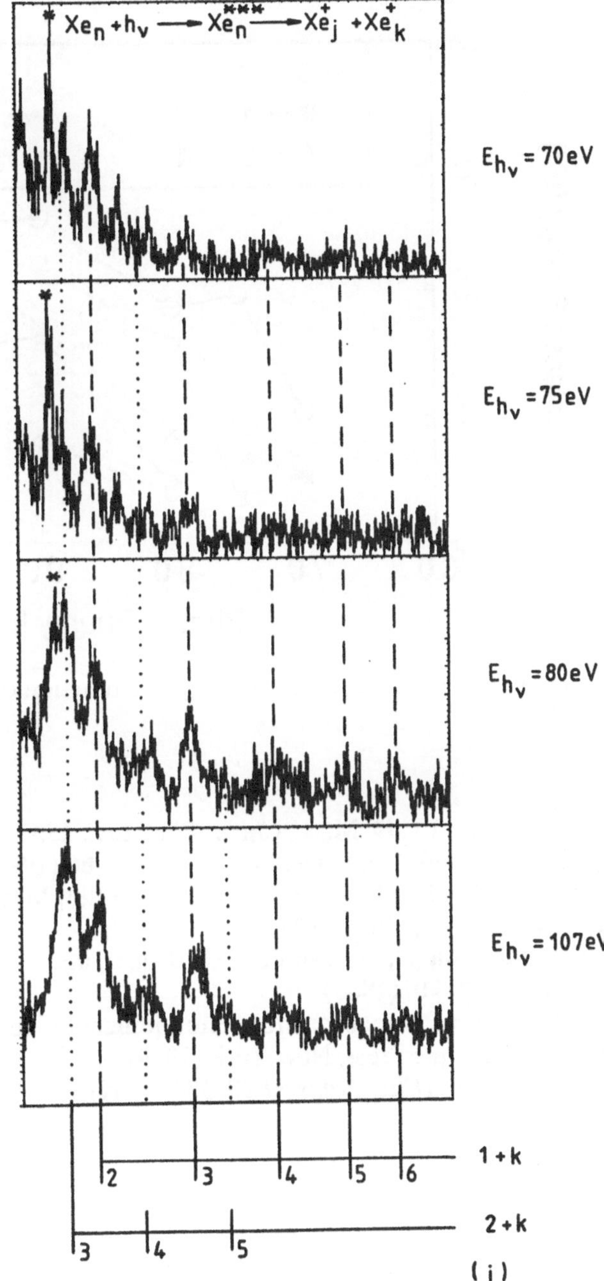

Fig. 4:

PIPICO-spectra of Xe$_n^+$-clusters taken at different photon energies. The stagnation pressure was 7 bar.

Fig. 5 shows the kinetic energy of the detected cluster ion fragment as a function of photon wavelength for the PIPICO experiments. It shows clearly a decrease towards lower photon energies and a steep fall off below 70 eV.

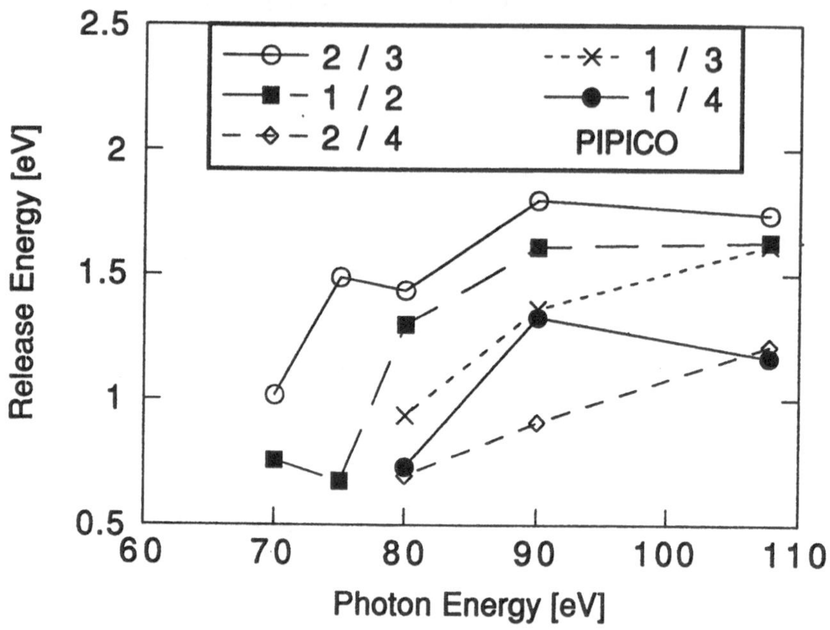

Fig. 5: PIPICO release energy as a function of photon energy

References:

[1]   M. Lezius and T. D. Märk, *Chem. Phys. Lett.* **155** (1989) 496.
[2]   N. G. Gotts and A. J. Stace, *Phys. Rev. Lett.* **66** (1991) 21.
[3]   E. Rühl, C. Schmale, H. W. Jochims, E. Biller, M. Simon and H. Baumgärtel, *J. Chem. Phys.*, in print (1991)
[4]   E. Holub-Krappe, G. Ganteför, G. Bröker, and A. Ding
       Z. Phys. D10, 319 (1988)
[5]   J. Berkowitz, Photoabsorption, Photoionization and Photoelectron Spectroscopy, p 182, Academic Press, New York (1979)
[6]   W.F. Schmidt, *IEEE Trans EI 19*, 389 (1984)

PHOTOIONIZATION OF SOLVATED Cs ATOMS

K. FUKE, F. MISAIZU, K. TSUKAMOTO, and M. SANEKATA

Institute for Molecular Science, Myodaiji, Okazaki 444, Japan

ABSTRACT. Cesium atoms solvated by polar molecules are studied by one-photon ionization and time-of-flight mass spectroscopy. Ionization potentials (IPs) of $Cs(H_2O)_n$ and $Cs(CH_3CN)_n$ are found to be constant for $n \geq 4$ (3.1 eV) and $n \geq 12$ (2.4 eV), respectively, while the IPs for $Cs(NH_3)_n$ and $Cs(CH_3OH)_n$ decrease monotonically with increasing n. $Cs(NH_3)_n$ gives a limit value of 1.4 eV, which coincides with the bulk value. Enhanced stability at n=20 is also observed for the $Cs(H_2O)_n$ clusters. These features are discussed in connection with the structure of the cluster and the stability of the excess electrons in the condensed phase.

1. Introduction

Electrons in fluids and solids play important roles in many aspects of physical and chemical phenomena, and have been the subject of numerous investigations for several decades [1].

Advances in molecular beam technique have opened new approaches to a microscopic investigation of the excess electrons in fluids. Recently, negatively charged water and ammonia clusters, $(H_2O)_n^-$ and $(NH_3)_n^-$, were prepared and its vertical detachment energies (VDE) have been determined [2,3]. On the other hand, Hertel and coworkers prepared the prototypes such as $Na(H_2O)_n$ and $Na(NH_3)_n$ [4,5], which enable us to link the macroscopic with microscopic properties of alkali metal-solvent systems. In analogy of the bulk behavior of an alkali metal atom [1], at sufficiently large n, the valence electron of alkali atom embedded in the clusters is transferred to a solvent molecule and the ground state may have an ion-pair character. Therefore, the photoionization threshold as a function of n is expected to include size-dependent information on the stability of the solvated atom, and also, on the excess electron state in clusters.

In this paper we present results of photoionization-threshold measurements of $Cs(H_2O)_n$, $Cs(NH_3)_n$, $Cs(CH_3OH)_n$ and $Cs(CH_3CN)_n$. These clusters are found to show individual features in the size dependence of the ionization thresholds. These features are discussed in connection with

P. Jena et al. (eds.), Physics and Chemistry of Finite Systems: From Clusters to Crystals, Vol. II, 925–930.
© 1992 *Kluwer Academic Publishers.*

926

the stability of the excess electron in the bulk phase of these solvents.

2. Experimental

The experimental apparatus used in the present study consists of a three-stage differentially evacuated chamber which includes a 'pick-up' cluster source [4] and a reflectron-type time-of-flight (TOF) mass spectrometer. The cluster source is composed of two pulse valves arranged at right angles. Solvent molecular clusters were produced by expansion of 2-atm Ar gas mixed with sample gas from the first pulse valve. Pulsed Cs atom beam was formed by expansion of the neat vapor from the second valve (heated to about 350 °C). Cs atom beam was injected at 10-15 mm downstream from the first nozzle. Solvated Cs clusters produced by subsequent multiple collisions were ionized by crossing an excimer-pumped dye laser(<1 mJ/cm^2) at 25-cm downstream from the first nozzle. The ionization thresholds were determined by scanning the photon energy with an interval of 0.03 eV in the region of 3.65-2.14 eV (340-580 nm).

3. Results

3.1 Cs(H$_2$O)$_n$ CLUSTERS

Typical photoionization mass spectrum of Cs(H$_2$O)$_n$ at 360 nm is shown in fig. 1a. No evaporation of constituent H$_2$O molecules from the cluster is expected after ionization at this photon energy. An anomalously enhanced peak is observed at n = 20, which is the same magic number as that found in the mass spectrum of H$_3$O$^+$(H$_2$O)$_n$ [6]. The ionization thresholds, IP(n), determined for n = 1-21 are plotted in fig. 2a as a function of $(n+1)^{-1/3}$, which is approximately inversely proportional to the cluster radius.

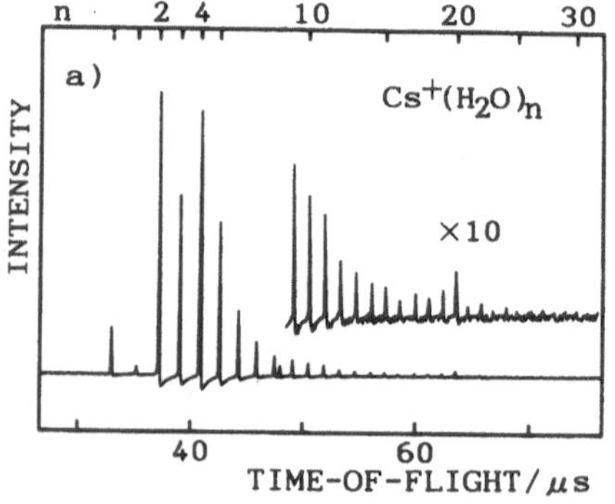

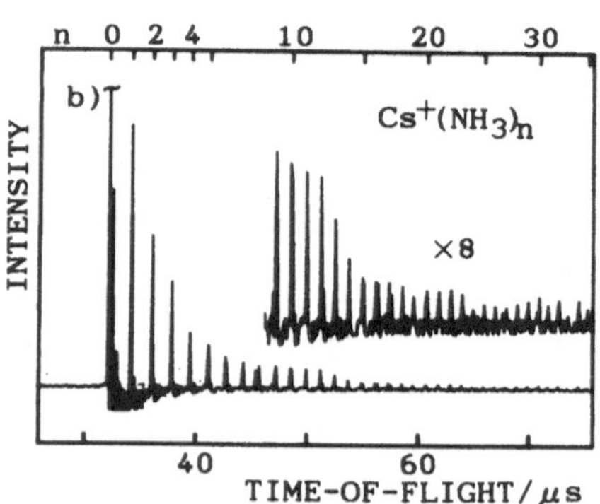

Figure 1. Typical photoionization mass spectra of solvated Cs atom clusters. a) Cs(H$_2$O)$_n$ (n $\leq$ 30). Ionization laser wavelength (λ)is 360 nm. b) Cs(NH$_3$)$_n$ (n $\leq$ 35). λ= 308 nm.

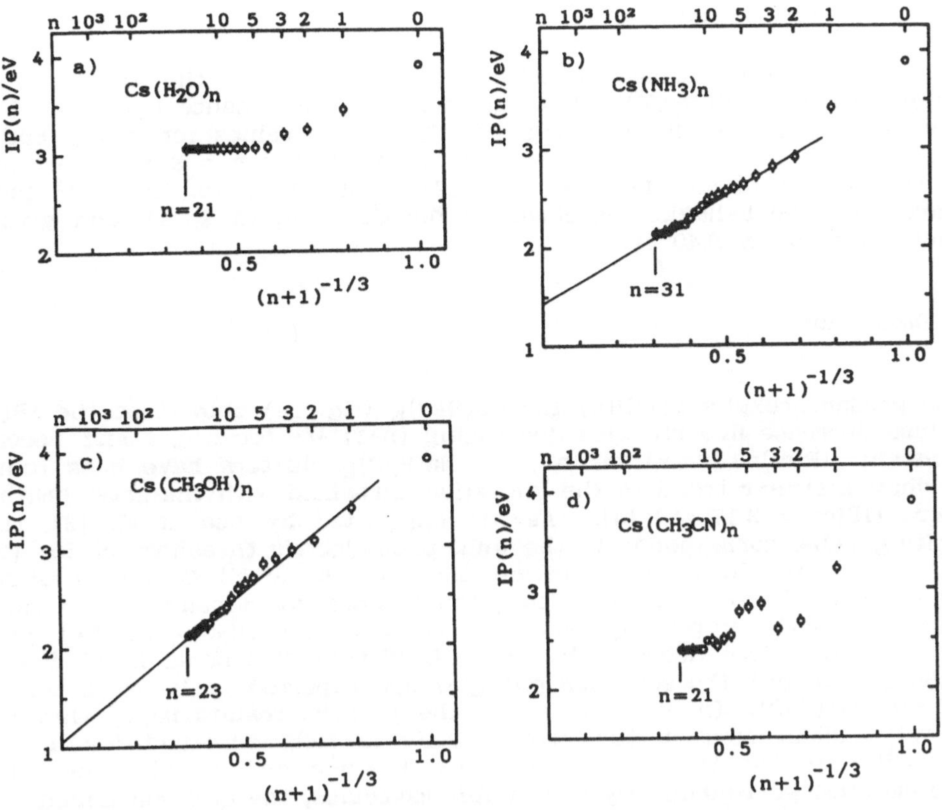

Figure 2. Ionization potentials (IP(n)) of CsM$_n$ clusters plotted vs. $(n+1)^{-1/3}$. Error bars, indicated at each point for $1 \leq n \leq 10$ and at 5 point interval for $n \geq 10$, include the uncertainty in determining IP(n) and the deviation caused by field ionization. a) M=H$_2$O ($n \leq 21$). b) M=NH$_3$ ($n \leq 31$). The result of the least-squares fitting for $n \geq 2$ is also shown. The extrapolated value to $(n+1)^{-1/3} = 0$ ($n \rightarrow \infty$) is 1.41 eV. The correlation coefficient of the fitted line, r, is 0.990. c) M=CH$_3$OH ($n \leq 23$). The extrapolated value is 1.08 eV. r = 0.992. d) M=CH$_3$CN ($n \leq 21$).

3.2 Cs(NH$_3$)$_n$ and Cs(CH$_3$OH)$_n$ CLUSTERS

Fig. 1b shows a typical one-photon ionization mass spectrum of Cs(NH$_3$)$_n$ ($n \leq 35$) at 308 nm. A step in the size distribution at $n \sim 10$ is observed at any laser wavelength between 308 and 480 nm. This step may correspond to the filling of the first coordination shell around the cesium ion according to the theoretical work by Klein and coworkers [7]. As shown in fig. 2b, IP(n) of Cs(NH$_3$)$_n$ decreases almost linearly with $(n+1)^{-1/3}$ for $n \geq 3$. The intercept of the fitted line at $(n+1)^{-1/3} = 0$ ($n \rightarrow \infty$) is 1.4 eV. This value, which agrees also in this case with the results of a photoionization experiment for Na(NH$_3$)$_n$ [5], coincides with the intercept value from the fitted line of VDE for (NH$_3$)$_n^-$ [3]. Cs(CH$_3$OH)$_n$ clusters show a similar trend in IP(n) with a limiting value of 1.1 eV (fig. 2c).

3.3 $Cs(CH_3CN)_n$ CLUSTERS

In contrast to the cases of other solvated clusters, the IP(n) values determined for $Cs(CH_3CN)_n$ show an anomalous dependence upon $(n+1)^{-1/3}$ for $n \leq 11$, as shown in fig. 2d. The IP(n) values for $n = 2$ and 3 are lower than those for $n = 4$-6, and the value for $n = 9$ is lower than those for $n = 10$ and 11. For $n = 12$-21, the IP(n) vs. $(n+1)^{-1/3}$ plots show a plateau behavior, as observed for $Cs(H_2O)_n$ ($n \geq 4$), and give a constant value of 2.40 eV.

4. Discussion

The present results of IP(n) for $Cs(H_2O)_n$ (fig. 2a) show that the IP(n) values decrease linearly with decreasing $(n+1)^{-1/3}$ for $n \leq 4$ and become constant $(IP(\infty)=3.12$ eV) for $n \geq 4$. $Na(H_2O)_n$ clusters have been found to show a similar trend in the ionization threshold with the same limiting value $(IP(\infty) = 3.17$ eV) [4]. As was suggested by Lee et al. [3], this limiting value corresponds to the bulk photoelectric threshold of ice (3.2 eV). These results seem to indicate that, for the alkali atom surrounded by more than four water molecules, the IP does not depend on the metal atom but on the solvent species. In the case of $Cs(H_2O)_n$ ($n \geq 4$) clusters, the IP values decrease by only 0.8 eV from that of an isolated atom (3.89 eV), though the solvation energies are reported to be more than 2 eV for $Cs^+(H_2O)_n$ ($n > 4$)[8]. Thus, the present results imply that the solvated alkali atom in the ground state is fully screened for $n \geq 4$. According to the theoretical study by Kestner and Dhar[9] for the sodium atom surrounded by four water molecules, the electron density of the 3s electron is very diffuse and extends over the region occupied by the solvent molecules. As a result, the sodium atom interacts with H_2O molecules by an ionic type bonding. These predictions are consistent with the above observations.

Figure 3. Schematic diagram of the energies of the lowest covalent (a), ion-pair (b), and the ion-ground (c) states for CsM_n (M: solvent molecules) vs. n. The ion-pair state is expected to correlate with the ground state of the solvated electron plus the solvated Cs^+ ion in bulk fluids. The ground-state electronic character of the neutral cluster changes from covalent to ion-pair type at a size, n^*.

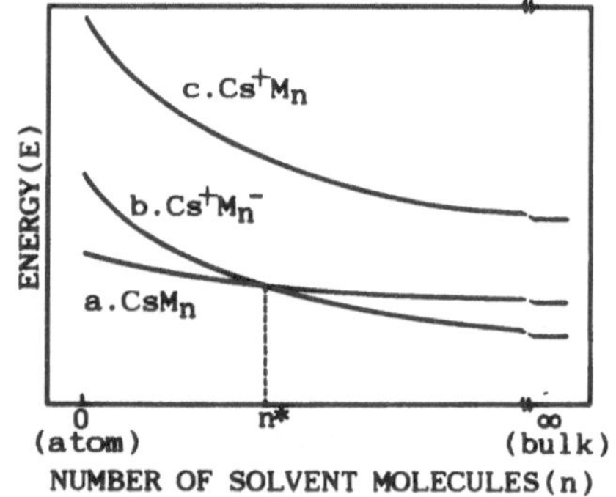

An alkali metal atom has been known to be ionized spontaneously in a polar fluid and be converted to a contact ion-pair or a solvated alkali metal ion plus a solvated electron [1]. This bulk state may correspond to the lowest ion-pair (or a Rydberg-like) state of the clusters. Then, this bulk behavior can be related to the properties of solvated alkali atom clusters as follows. Figure 3 draws schematically the energy levels of the covalent, ion-pair, and ion ground states for a solvated Cs atom as a function of n. The energy levels of the ion ground states decrease monotonically with increasing n, as can be seen from the enthalpies of solvation for the Cs^+ ion [8]. For small clusters, the ground state is of the covalent type and is expected to be stabilized very slowly with increasing n (fig. 3). On the other hand, the ion-pair state is stabilized much more rapidly than the covalent-type state. Since the ion-pair state correlates with the ground state of the solvated electron plus the solvated Cs^+ ion in the bulk, the electronic character of clusters in the ground state is expected to change from the covalent type to the ion-pair type at a certain critical cluster size (n^*). Based on the above theoretical results, the curvature of the E vs. n curve for the ion-pair state (fig. 3) is expected to be similar to that of the ionic state for the $Cs(H_2O)_n$ ($n \geq 4$) clusters; the interaction in both states is mostly electrostatic in nature. Therefore, the observed plateau in the IP(n) vs. $(n+1)^{-1/3}$ plots for $Cs(H_2O)_n$ (fig. 2a) can be understood by the cluster size dependences of the ion-pair and ionic levels, which suggests that $n^*=4$.

The mass spectrum shows the enhanced stability of $Cs(H_2O)_{20}^+$ cluster (fig. 1a). This value of n often appears as the magic number in the mass spectra of the $H_3O^+(H_2O)_n$ clusters [6]. There are several possible explanations for the $Cs(H_2O)_{20}^+$ dominance such as (1) the predominance of the preformed $(H_2O)_n$ clusters and (2) the special stability of $Cs(H_2O)_{20}^+$ or (3) $Cs(H_2O)_{20}$. Since the distribution of the neutral $(H_2O)_n$ clusters has been known to have no anomaly in the size-range discussed here [6] and the enhanced stability at n=20 is observed even at the photon energy close to the ionization threshold (3.2 eV), where the fragmentation due to the excess energy at ionization seems to be negligible, the explanations (1) and (2) can be excluded. On the other hand, the theoretical calculations for $Na(H_2O)_n$ mentioned above predict that the core of the clusters for $n \geq 4$ has an ionic character. As with sodium, Cs is expected to interact with water molecules by an ionic-type bonding. Therefore, it is reasonable to ascribe the observed anomaly to the enhanced stability of the $Cs(H_2O)_{20}$ cluster in the ground state (explanation (3)); this observation seems to support the scheme shown in fig. 3. It is believed from both experimental and theoretical [6] studies that $H_3O^+(H_2O)_{20}$ forms a clathrate structure (deformed pentagonal dodecahedron) with the H_3O^+ ion in the center of the hydrogen-bonding networks of 20 water molecules. Thus, the $Cs(H_2O)_{20}$ cluster is expected to be in the ion-pair state in which the ion core has an ion-clathrate structure and the excess electron is distributed outside the core.

As shown in fig. 2b, the IP values of the $Cs(NH_3)_n$ clusters can be smoothly extrapolated to that of the bulk value. Since the threshold cluster size of the $(NH_3)_n^-$ cluster formation is found to be 35 [2], these results seem to imply that n^* is larger than 35. An alternative explanation is that, for $Cs(NH_3)_n$, the electronic character of the ground state may change succeedingly from the covalent type to the ion-pair type with increasing the cluster size. Since the $Cs(CH_3OH)_n$ clusters show a similar trend in IP(n) (fig. 2c), the analogous arguments seem to be valid.

The intriguing feature of the IP(n) vs. $(n+1)^{-1/3}$ plots of the

$Cs(CH_3CN)_n$ clusters is the anomalous change of IP(n) for $n \leq 11$ (fig. 2d). The rises near the threshold in the ionization efficiency curves for n=4-6 are found to be as steep as that for n=1, while those for n=2, 3 and for $n \geq 7$ are much slower. These results suggest that the differences in the geometry of neutral and ionic clusters for n=4-6 are much closer than those for n=2 and 3. These anomalous changes in the geometry may be due to the large permanent dipole moment of acetonitrile. As in the case of $Cs(H_2O)_n$, the IP(n) vs. $(n+1)^{-1/3}$ plots of $Cs(CH_3CN)_n$ show a plateau for $n \geq 12$ ($n^*=12$), which gives the bulk value of 2.4 eV. According to the ESR studies for γ-irradiated acetonitrile crystal, a dimer radical anion, $(CH_3CN)_2^-$, is found to be stable; electrons are trapped by a deep potential well formed in the dimer and a well-defined anisotropic hyperfine structure is observed in the ESR spectrum [10]. Thus, the behavior of the electrons in the bulk is consistent with the finding that the solvated Cs atom is fully screened by a rather small number of acetonitrile molecules as seen in the IP(n) vs. $(n+1)^{-1/3}$ plots (fig. 2d).

The critical number n^* observed in the present study may have some correlation with the cluster size, n_{th}, at which the negatively charged cluster starts to be observed. Several groups have reported n_{th} obtained by electron attachment to molecular clusters: Stable negative ions are produced from n = 2 for $(H_2O)_n$ [2], 10 for $(CH_3CN)_n$ [11], and n = 35 for $(NH_3)_n$ [2].

Acknowledgement

The authors thank I. V. Hertel for sending a preprint of the work of his group.

References

[1] Jortner, J. and Kestner, N. R. eds., (1973) 'Electrons in Fluids', Springer.

[2] Haberland, H., Ludewigt, C., Schindler, H.-G., and Worsnop, D. R. (1985) 'Clusters of water and ammonia with excess electrons', Surface Sci. 156, 157-164.

[3] Lee, G. H., Arnold, S. T., Eaton, J. G., Sarkas, H. W., Bowen, K. H., Ludewigt, C., and Haberland, H. (1991) 'Negative ion photoelectron spectroscopy of solvated electron cluster anions, $(H_2O)_n^-$ and $(NH_3)_n^-$', Z. Phys. D20, 9-12.

[4] Schulz, C. P., Haugstätter, R., Tittes, H.-U., and Hertel, I. V. (1988) 'Free sodium-water clusters', Z. Phys. D10, 279-290.

[5] Hertel, I. V., Hüglin, C., Nitsch, C., and Schulz, C. P. (1991) 'Photoionization of $Na(NH_3)_n$ and $Na(H_2O)_n$ clusters', Phys. Rev. Lett. 67, 1767.

[6] Nagashima, U., Shinohara, H., Nishi, N., and Tanaka, H. (1986) 'Enhanced stability of ion-clathrate structures for magic number water clusters', J. Chem. Phys. 84, 209-214.

[7] Marchi, M., Sprik, M., and Klein, M. L. (1990) 'Solvation and ionization of alkali metals in liquid ammonia', J. Phys. C2, 5833-5848.

[8] Keesee, R. G. and Castleman, Jr., A. W. (1986) 'Thermochemical data on gas-phase ion-molecule association and clustering reactions', J. Phys. Chem. Ref. Data 15, 1011-1071.

[9] Kestner, N. R. and Dhar, S. (1987) 'Theoretical studies of sodium water clusters', in J. Jortner (ed.), Large Finite Systems, D. Reidel Publishing company, pp. 209-215.

[10] Williams, F. and Sprague, E. D. (1982) 'Novel radical anions and hydrogen atom tunneling in the solid state', Acc. Chem. Res. 15, 408-415.

[11] Kondow, T. (1987) 'Ionization of clusters in collision with high-Rydberg rare gas atoms', J. Phys. Chem. 91, 1307-1316.

LINEAR AND NONLINEAR OPTICAL PROPERTIES OF CuCl-NANOCRYSTALLITES: EVOLUTION FROM CLUSTER-LIKE TO BULK-LIKE BEHAVIOR

S.GAPONENKO, I.GERMANENKO, L.ZIMIN, V.LEBED
and I.MALINOVSKII
Stepanov Institute of Physics, Minsk
SU-220072 USSR
M.VASIL'EV, E.PODOROVA and V.TSEKHOMSKII
Vavilov State Optical Institute, S.-Peterburg
SU-199364 USSR
V.GOLUBKOV
Grebenshchikov Inst. of Silicate Chemistry,
S.-Peterburg SU-199164 USSR

ABSTRACT. CuCl nanocrystallites with size ranging from 25 to 150 Å imbedded in a transparent glass matrix are investigated by the means of pump-and-probe laser spectroscopy as well as by traditional optical techniques. The melting temperature, the photo-induced blue shift of the exciton absorption and the excitation spectrum of excitons were found to be size-dependent. The results are interpreted in terms of the size quantization of excitons, the coherence of excitonic states, the interaction of two excitation in quantum box and $\vec{k}$ -selection rule relaxing for small crystallites.

1. MATERIALS and METHODS

CuCl nanocrystallites (NC's) were formed due to controlled heat treatment of the glass containing Cu and Cl. The NC's have a quasispheric shape because the growth temperature was higher than the CuCl melting temperature T_m. The latter was found to exhibit a pronounced decrease with decreasing NC's radius R. An additional fine annealing at temperature slightly below T_m was performed to obtain more perfect crystal structure resulting in a sharpening of the excitonic lines. The average $\bar{R}$ values were measured by the small-angle X-ray scattering technique.

P. Jena et al. (eds.), Physics and Chemistry of Finite Systems: From Clusters to Crystals, Vol. II, 931–936.
© 1992 *Kluwer Academic Publishers.*

2. RESULTS and DISCUSSION

2.1. Nonlinear Absorption

The absorption and the luminescence spectra exhibit well
pronounced quantum-size effect due to dependence of the exciton energy upon the size of quantum box (QB). The energy
shift of the Z_3- and Z_{12}-excitonic absorption bands versus
$\bar{R}$ (fig.1) is satisfactorily fitted by the simple theory
taking into account the quantization of the exciton motion
in the spheric quantum well with infinite potential walls [1].
Nonlinear absorption (NA) appears mainly as a photo-
induced blue shift of exciton absorption band at the excitation in the band maximum (fig.2). For $R \geqslant 75$ Å the Z_3-line
shifts as a whole (fig.2a). For $R \leq 35$ Å the Z_3-line changes
assymetrically: the long-wave wing which is formed by the
NC's with larger R remains constant, the maximum of the
band and the short-wave edge are strongly shifted to higher
$\hbar\omega$ (fig.2b). This result is interpreted in terms of the
initial inhomogeneous broadening of the excitonic band and
size-dependent nonlinear response. The values of the spectral
shift for $R \leq 35$ Å agree with the energy of interaction of
two excitonlike excitations in the QB [2]. For details see
[1,3].
It has been found that the saturation intensity I_s
decreases sharply with decreasing $\bar{R}$ for $\bar{R} \leq 35$ Å . The
decrease in I_s is attributed to the decrease in the propa-
bility of exciton radiative decay for smaller NC's. The
latter is possible if exciton can be treated as a coherent
excitation over the whole NC. The reduction of I_s agrees as
to the order of the magnitude with the theoretical predictions [2,4,5] and with the direct measurement of the size-dependent exciton lifetime [6].

2.2. Luminescsence

Emission spectra show the excitonic luminescence due to direct annihilation of Z_3-exciton without noticable manifestation of phonon replicas (fig.3). The observed ES are slightly
shifted as compared to the proper absorption spectra due to
the reabsorbtion effect. With decreasing $\bar{R}$ the evolution
from the bulk-like excitation spectrum to the spectrum,
characteristic of NC's was observed. In the former case ($R \geqslant$

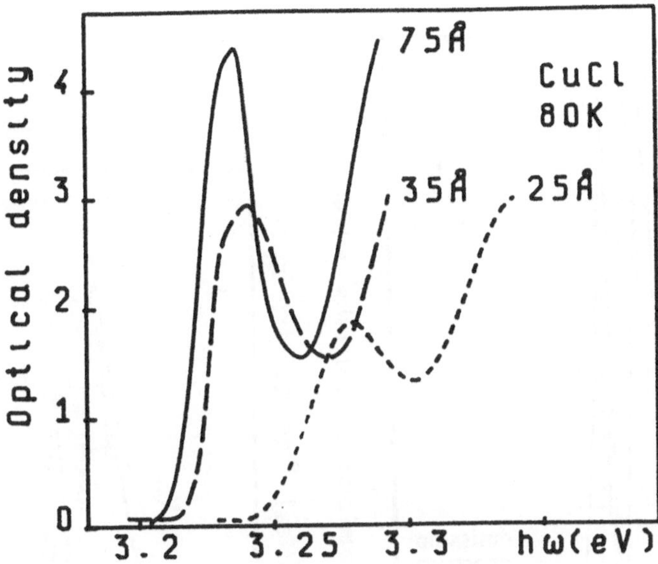

Fig.1   Linear spectra of Z_3-exciton absorption for various $\bar{R}$

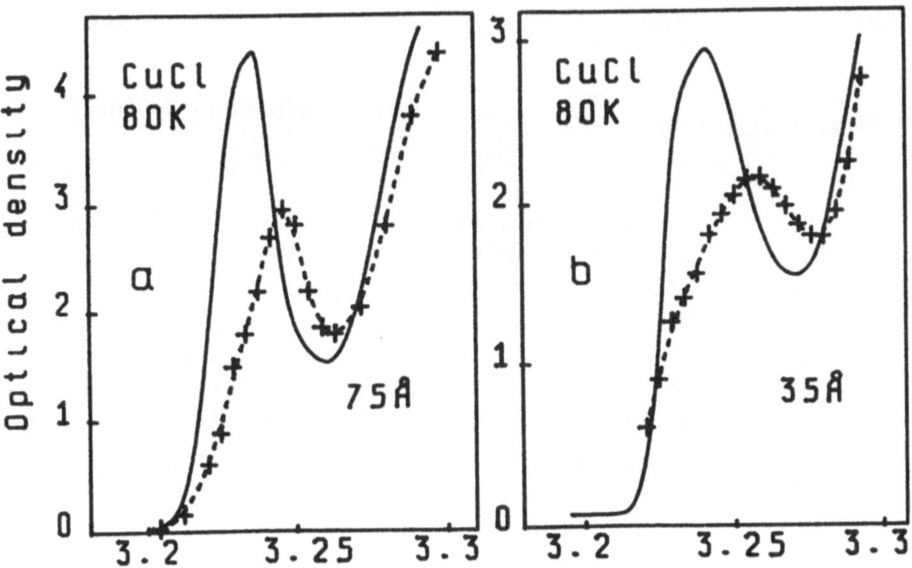

Fig.2   Absorption spectra of unexcited (lines) samples and under ecxitation in the Z_3-band maximum (crosses) for two R values

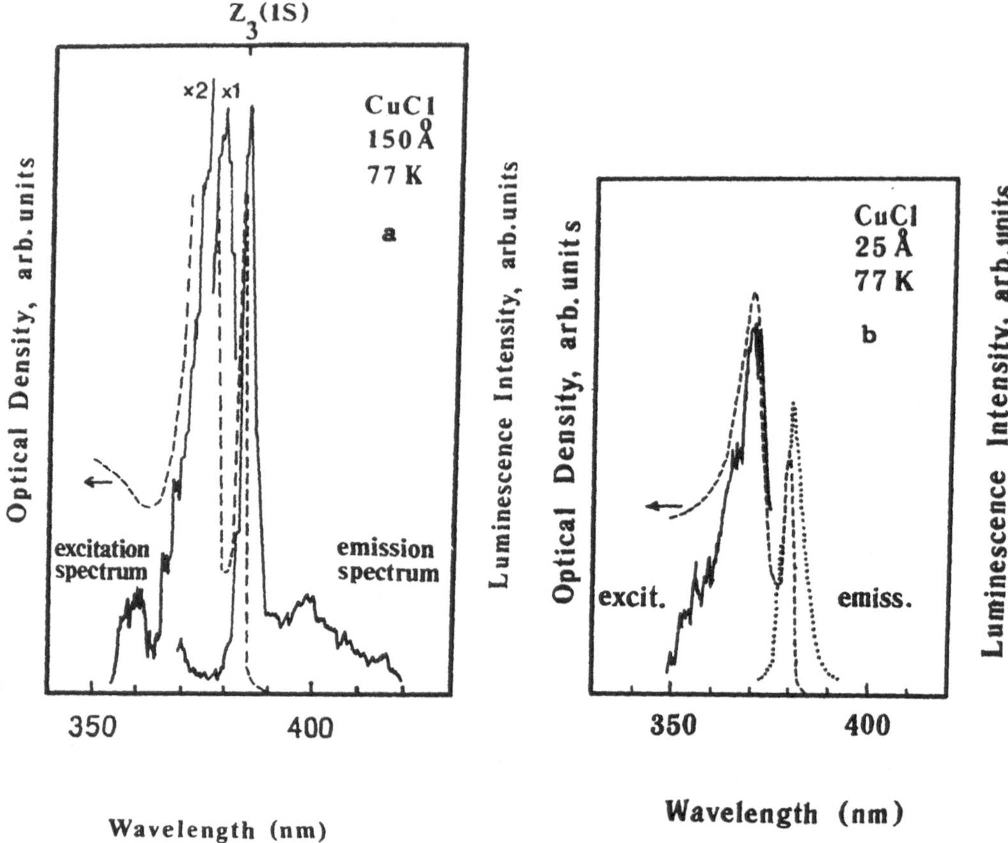

Fig. 3 Absorption, emission and excitation spectra of CuCl-doped glass for two values of mean NC's radius. Emission spectra are registered at excitation wavelength 360 nm. Note the change of scale (x2) at the shortwave wing of the excitation spectrum a). The additional excitation band at $\lambda = 360$ nm is due to the $Z_3(2S) \rightarrow Z_3(1S)$ process. E_g of the bulk CuCl corresponds to $\lambda_g = 362$ nm

100 Å) the excitation maximum is shifted from Z_3-exciton absorption maximum, and the high-energy part shows[3] a step-wise structure correllating with the phonon energy. In the latter case (R<35 Å) the excitation spectrum replicates the absorption one. The shape of excitation spectrum presented at fig. 3a is determined by both energy and momentum conservation. Thus the spectrum does not coincide with the absorption one. The similarity of the excitation and absorption spectra for smaller NC's leads to a conclusion that $\vec{k}$-rule relaxes for NC with a few tens of atoms inall dimensions. The other manifestation of the $\vec{k}$-rule relaxing in NC's is discussed in [7].

For all the samples investigated the interband ($\hbar\omega > E_g$) excitation do not produce neither Z_3-luminescence nor nonlinearity in absorption [1,3]. The absorption by glass is not the reason because the matrix itself is transparent. The processes that may be involved are i)fast e-h-recombination without exciton formation and ii)entry of the 'free' electron into the matrix. Both are closely connected because of the increasing probability of Auger process with decreasing R due to $\vec{k}$-rule violation and strong wavefunction overlapping.The Auger process results in the absence of excitons under excitation by $\hbar\omega > E_g$ and in the entry of electron into the matrix. The latter give rise to additional absorption, i.e. photodarkening. From the other side the short e-h-lifetime may be also due to the capture of electron by the sur - face Cu-atom. Thus appeared $\overset{-}{Cu}$-ions in turn lead to photodarkening according to the existing theory of the photochemisry of Cu-doped glasses.

2.3. Photodarkening

The photodarkening effect, i.e. the long-live photoinduced absorption in the visible range after illumination by the near-UV light, was found for all the samples. With decreasing $\bar{R}$ the effect is weakened, being negligible for $R \leq 20$ Å. The close consideration of luminescent and photochemical properties of CuCl-doped glasses will be the subject of future investigation.

ACKNOWLEDGEMENTS

The helpful discussions with Dr.L.Bányai are strongly acknowledged.

Minsk group wish to dedicate the present paper to the 60th
anniversary of Prof.V.P.Gribkovskii.

References

1. Zimin, L., Gaponenko, S., Malinovskii, I., Germanenko, I.,
   Podorova, E. and Tsekhomskii, V. (1990) 'Copper chloride
   nonlinear optical absorption under quantum confinement',
   J.Mod.Optics. 37, 829-834.

2. Hanamura, E. (1988) 'Very large optical nonlinearity of
   semiconductor microcrystallites', Phys.Rev. B 37, 1273-
   1279.

3. Zimin, L., Gaponenko, S., Lebed, V., Malinovskii, I. and
   Germanenko, I. (1990) 'Nonlinear optical absorption of
   CuCl and CdSSe microcrystallites under quantum confine-
   ment', J.Lumin. 46, 101-107.

4. Takagahara, T. (1987) 'Excitonic optical nonlinearity and
   exciton dynamics in semiconductor quantum dots', Phys.
   Rev. B 36, 9293-9296.

5. Kayanuma, Y. (1988) 'Quantum-size effects of interacting
   electrons and holes in semiconductor microcrystals', Phys.
   Rev. B 38, 9797-9805.

6. Itoh, T., Furumiya, M., Ikehara, T. and Gourdon, C. (1990)
   'Size-dependent radiative decay time of confined excitons
   in CuCl', Sol.St.Comm. 73, 271-274.

7. Uhrig, A., Banyai, L., Gaponenko, S., Wörner, A.,
   Neuroth, N. and Klingshirn, C. (1991) 'Linear and nonlinear optical studies of CdSSe quantum dots', Z.Phys.D 20,
   345-348.

DIPOLE EXCITATIONS OF CLOSED-SHELL ALKALI-METAL CLUSTERS

C. GUET and W. R. JOHNSON*
DRFMC/SPHAT
Centre d' Etudes Nucléaires Grenoble
BP 85X, 38041 Grenoble
France

ABSTRACT Excitation energies and associated oscillator strengths for dipole-excited states of alkali-metal clusters – treated as jellium spheres – are calculated in the random-phase approximation(RPA). Closed-shell systems with 8, 20, 34, 40, 58 and 92 delocalized electrons are considered. The ground state is described in the Hartree-Fock(HF) approximation. The excitation spectrum is determined by solving the RPA equations. Exchange contributions are naturally taken into account. The theoretical oscillator-strength distribution is found to be very different for neutral and charged clusters. Static polarizabilities of the clusters are calculated and compared with experimental values and with other calculations.

1. Introduction

It is now well-established [1] that gross properties of simple metallic clusters with from ten to a few hundred atoms are understood in terms of the quantal arrangement of delocalized electrons, moving in their mutual mean field and a smooth positive background (jellium model). To a first approximation the dynamics associated with the moving conduction electrons govern the stability of the metallic cluster as well as its response to an external electromagnetic field. In that respect the optical response of alkali-metal clusters has been the object of thorough experimental investigation revealing the existence of a giant dipole resonance interpreted as an electronic collective excitation (surface plasmons) [2,3]. A substantial theoretical effort [4-8] has accompanied these experimental findings.In analogy with the infinite case, the finite electron may-body system is usually described in the local density approximation (LDA). This has the advantage of taking the strong screening of the Coulomb interaction into account in a rather simple way. However, it probably fails to describe the outer part of the electronic distribution correctly, since the local exchange potential falls faster than $1/r$. This fact might explain why static polarizabilities in the LDA are lower than the measured ones [9]. Although most of the discrepancy probably originates from the crudeness of the jellium approximation it is worth trying to assess carefully the LDA. In this paper, we calculate the dipole oscillator strength distribution and the static polarizability of closed-shell sodium clusters by using the RPA. As in other jellium models, it is assumed that the positive ions form a homogeneous sharp-edged spherical distribution. Our calculations differ from previous works based on the LDA in that the exchange interaction is taken into account fully. The ground-state is described in the HF approximation, leading to a non-local HF potential. The spectrum of

937

P. Jena et al. (eds.), Physics and Chemistry of Finite Systems: From Clusters to Crystals, Vol. II, 937–942.
© 1992 *Kluwer Academic Publishers.*

physical excited states is then obtained by solving the RPA matrix equation. Oscillators strengths and static polarizabilities are discussed with regard to experimental data and LDA predictions. The major approximation remains the assumption of a uniform charged background. We discuss how a diffuse surface modifies the results.

2. Hartree-Fock Approximation for the Ground State

A jellium cluster consists of free electrons in a homogeneous, positively charged background. In alkali metals the delocalized electrons are the valence electrons, one electron per atom. The hamiltonian describing the electron is written:

$$h_0 = \frac{p^2}{2m} + V_{bkg}(r) + U(r)$$

where $V_{bkg}(r)$ is the positively charged background potential, and $U(r)$ accounts in some approximation for the electron-electron interaction. Since we assume a constant density distribution for the jellium ions, we have:

$$V_{bkg}(r) = \begin{cases} -\dfrac{A}{2R} \left[3 - \left(\dfrac{r}{R}\right)^2 \right] & r \leq R \\[2ex] -\dfrac{A}{r} & r > R \end{cases}$$

where A is the number of ions, and $R = A^{1/3} r_s$. The Wigner-Seitz radius r_s is assigned its bulk value, i.e. 4 a_0 for sodium (a_0 being the Bohr radius). The many-body hamiltonian describing the system made of Z delocalized electrons in the presence of A ions is then:

$$H = H_0 + V,$$

$$H_0 = \sum_{i=1}^{Z} h_0(r_i),$$

$$V = \frac{1}{2} \sum_{ij} \frac{e^2}{r_{ij}} - \sum_i U(r_i).$$

In the present work, the potential U is the non-local Hartree-Fock potential defined as:

$$V_{HF} \, u_a(r) = \sum_b \int \frac{dr'}{|r'-r|} \left[u_b^\dagger(r') u_b(r') \, u_a(r) - u_b^\dagger(r') u_a(r') \, u_b(r) \right]$$

where the u's are the single-particle wave-functions and the summation extends over occupied states. Restricting to closed-shell systems we are led to solve coupled radial HF equations for the occupied orbitals. This is a standard procedure in atomic physics; a numerical code has been written to yield single-particle orbitals u_a and eigenenergies (ε_i) to a very high accuracy (typically one part in 10^8). Total binding energies are defined as:

$$E_b = E_{HF} + E_{bkg},$$

$$E_{HF} = \sum_a \varepsilon_a - \frac{1}{2} \sum_{ab} \left(\langle ab| \frac{1}{|r'-r|} |ab\rangle - \langle ba| \frac{1}{|r'-r|} |ab\rangle \right).$$

It is worth noticing that for a neutral system the direct terms almost cancel the self-energy of the positively-charged background $E_{bkg} = \frac{3}{5}\frac{A^2}{R}$ (as it would do exactly for a uniform system). Therefore, the remaining exchange and kinetic terms account mainly for the total cluster energy at the HF level. The binding energies per atom for closed-shell sodium clusters are reported in the next table together with the predicted HF bulk value, $\varepsilon = 30.07/r_s^2 - 12.46/r_s = -1.236 eV$. One sees that this bulk value is reached approximatively for all systems. It is well-known that the correlation energy (beyond the HF approximation) increases the binding of a uniform electron liquid. Presently, we do not consider this contribution to the ground state and refer to a recent work where LDA predictions are compared to pure HF ones [10].

TABLE 1. Hartree- Fock ground-state energy per particle of Na- clusters ($r_s = 4$ a_0) for different sizes.

A	8	20	34	40	58	92	∞
$\frac{E_b}{A}$ (eV)	-1.280	-1.237	-1.238	-1.200	-1.241	-1.237	-1.236

3. Random Phase Approximation for the Dipole Excitations

Previous microscopic calculations [4-6] of the optical response of small alkali-metal clusters have been carried out within the framework of the RPA which allows one to take into account the strong screening of the electron-electron interaction. In all these calculations it has been found that the average dipole resonance frequencies are systematically blue-shifted with respect to experiment [2,3]. They all start from an effective density dependent interaction, thus avoiding the numerical complications which are brought by the exact treatment of exchange. Here we study the vibration modes when the ground state is a Slater determinant obtained by solving the HF equations discussed above. There are different ways of deriving the RPA equations which may be put into matrix equation:

$$\begin{pmatrix} A & B \\ B^* & A^* \end{pmatrix} \begin{pmatrix} X^k \\ Y^k \end{pmatrix} = \omega_k \begin{pmatrix} X^k \\ -Y^k \end{pmatrix}$$

$$A_{ma,nb} = (\varepsilon_m - \varepsilon_a)\, \delta_{ab}\, \delta_{mn} + < mb \,|V|\, an >,$$

$$B_{ma,nb} = < mn \,|V|\, ab >,$$

$$< ij \,|V|\, kl > = < ij \,|\, \frac{1}{|r'-r|} \,|\, kl > - < ij \,|\, \frac{1}{|r'-r|} \,|\, lk >,$$

where the indices a,b (m,n) refer to the hole (particle) states. The matrix A contains the matrix elements of the Coulomb interaction between particle-hole excitations whereas the matrix B is composed of the matrix elements of that interaction between the ground state and two particle-hole excitations. The positive eigenvalues ω_k are the excitation energies of the system. The corresponding eigenvectors representing the physical states are expressed as linear combinations of forward-going and backward-going amplitudes X_{ma}^k and Y_{ma}^k . A

numerical solution of this eigenvalue problem requires a complete set of single-particle states. In order to discretize the continuum states it is convenient to confine the cluster to a cavity of finite radius. By choosing this cavity radius sufficiently large, the low-lying bound states in the cavity can be brought arbitrarily close to the actual HF bound states. Although the cavity spectrum is discrete, it is infinite. A finite HF pseudospectrum is built by expanding the HF orbitals in terms of a finite number of B-splines. The low-lying states in this pseudospectrum can be adjusted very accurately to the HF states by an appropriate choice of the number and order of the B-splines used. For more details see ref[11].

An angular reduction of the RPA equation is performed with the restriction to dipole excitations. An external electric dipole field gives rise to a set of excitation channels. The radial amplitudes which are associated with these channels are expanded in terms of the pseudostates. In order to reach a very high accuracy 50 B-splines of order 7 have been used. As an example we have 10 channels for Z=40 thus resulting in 466 by 466 matrices A and B.

The dipole transition amplitude from the ground state to the kth excited state is expressed in terms of reduced matrix elements of the dipole operator, d, as:

$$q_k = \sum_{am} (X_{ma}^k - Y_{ma}^k) < u_m \| d \| u_a> ,$$

and the corresponding oscillator strength is defined as:

$$f_k = \frac{4}{3} \omega_k q_k^2 .$$

The f-sumrule , $\Sigma_k f_k = Z$, is satisfied to better than one part in 10^5 providing a numerical test of the RPA calculation. The oscillator strength distributions are shown in figure 1. for three closed-shell systems both neutral or singly charged. To compare the results for different cluster sizes the percentage of the sum rule contributed by each line is plotted. A few observations ought to be made. As already shown by RPA calculations based on LDA,a few states exhaust the f sum rule. These states which lie below the ionisation threshold, around 3.4 eV and 4.6eV for neutral and charged clusters respectively, are bunched and lead to the observed giant dipole resonance.There is a marked difference between neutral and charged clusters: in the later case which is characterized by a relatively higher ionisation potential the dipole oscillation is shared by fewer states (a single peak exhausts 80% of the sum rule in case of Na_{21}^+) but lies at about the same frequency. As the size of the cluster grows the splitting into several states also grows. A weak dependence of the average excitation frequency on the number of atoms is observed. A calculation on Na_{58} not reported here confirms this trend. A straightforward comparison with experimental data on photoabsorption cannot be performed since these electronic vibrations couple to the ionic jellium density fluctuations [12], a fact which the present model does not embrace. However it appears that the observed resonance centroid that has been measured on Na_{20} (~2.4 eV)lies close the theoretical prediction(~2.6 eV). With regard to density dependent RPA calculations a systematic—although small—red-shift is observed but is not sufficient to resolve the discrepancy with experiment. This is consistent with our

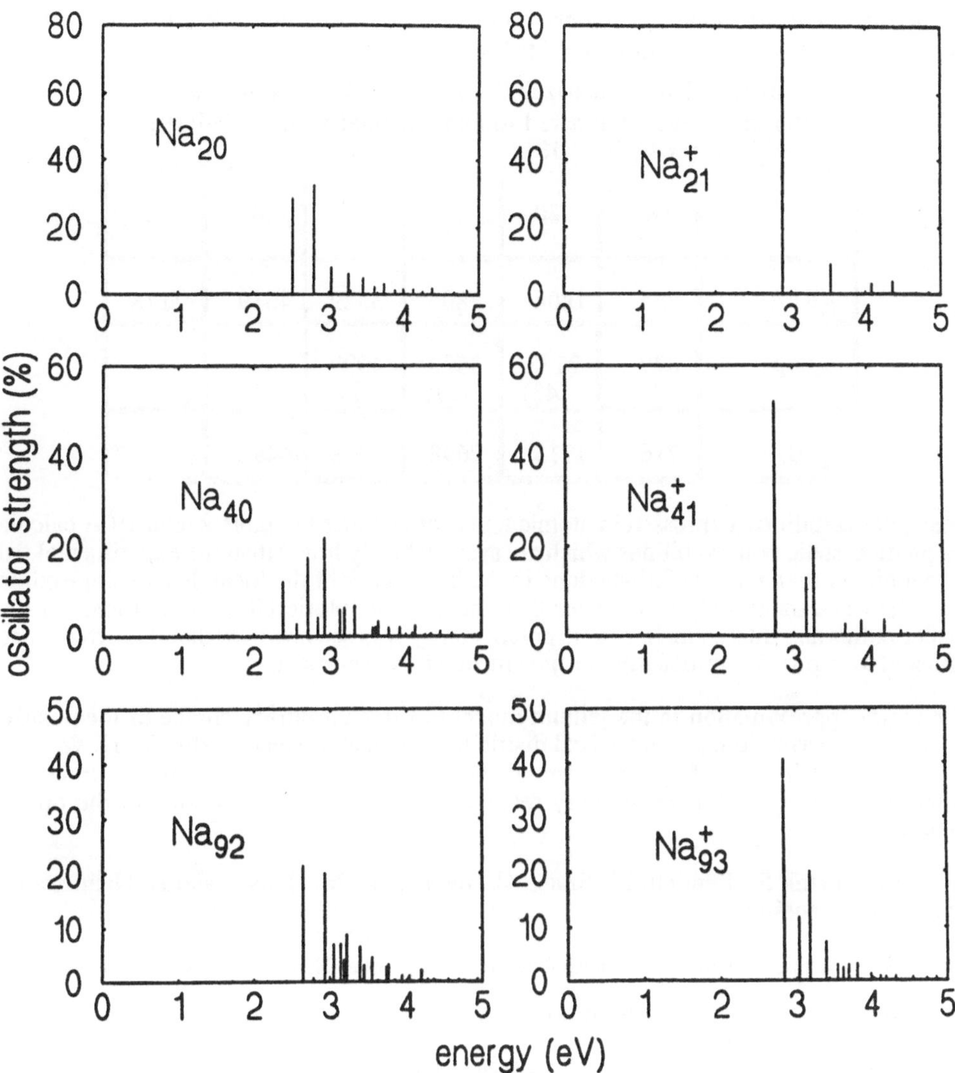

FIGURE 1. Oscillator-strength distribution, expressed as a percentage of the total f-sumrule, as a function of the excitation energy.

own prediction of the static dipole polarizabilities which are immediately derived from the f-distribution through the relation, $\alpha = \Sigma\, f_k/\omega_k^2$.

TABLE 2. Static polarizabilities α^{RPA} of Na-clusters for different sizes A compared to experimental values α^{exp} [9] and to LDA predictions [13]

A	8	20	34	40	58	92
α^{RPA}	755	1808	2806	3529	4619	7178
α^{exp}	879 (17)	2138 (43)	3520 (17)	4090 (77)		
α^{LDA}	716	1715	2698	3328	4492	

These polarizabilities, expressed in atomic units, are reported in table 2. Our RPA calculation predicts static polarizabilities which are systematically lower than the experimental values but higher than the predictions done in the framework of the local density approximation. The LDA mean field falls faster than the 1/r asymptotic HF potential leading to a smaller radius and thus a smaller static polarizability. We see however that this effect is not sufficient to improve significantly the agreement with experiment.

The major approximation in the jellium model lies in an arbitrary choice of the positive background distribution: a sharp-edged distribution with an assigned value for r_S. We have considered a diffuse surface background the shape being optimized to minimize the ground-state energy. Our preliminary results show strong modifications to the optical response.

We wish to thank S. Blundell, M. Brack, C. Bréchignac, M. Hansen and H. Nishioka for stimulating discussions.

* Permanent address: Department of Physics, University of Notre Dame, Notre Dame, Indiana 46556, USA.
1. For a review, see W.A. de Heer, W. D. Knight, M. Y. Chou, and M. L. Cohen, Solid State Physics **40**, (1987).
2. C. Bréchignac, P. Cahuzac, F. Carlier and J. Leygnier, Phys. Rev. Lett. **63**, 1368 (1989).
3. K. Selbt et al., Phys. Rev. **B 43**, 4565 (1991).
4.W. Ekart, Phys. Rev. **B 31**, 6360 (1985).
5. D. E. Beck, Phys. Rev. **B 35**, 7325 (1987).
6. C. Yannouleas, R. A. Broglia, M. Brack, and P. F. Bortignon, Phys. Rev. Lett. **63**, 255 (1989).
7. M. Brack, Phys. Rev. **B 39**, 3533 (1989).
8. G. F. Bertsch, Computer Physics Comm. **60**, 247 (1990).
9. W. D. Knight, K. Clemenger, W. A. de Heer, and W. A. Saunders, Phys. Rev. **B 31**, 445, (1985).
10. M. S. Hansen, H. Nishioka, preprint.
11. W. R. Johnson, S. A. Blundell, and J. Sapirstein, Phys. Rev. **A 37**, 307, (1988).
12. G. F. Bertsch and D. Tomanek, Phys. Rev. **B 40**, 2749 (1989).
13. P.Stampfli and K. H. Bennemann, in Physics and Chemistry of Small Clusters, ed. by P. Jena, B.K. Rao and S. N. Khanna, 473 (1986) and refs therein.

OPTICAL SPECTROSCOPY OF ARGON CLUSTER IONS

H. HABERLAND, B. v. ISSENDORFF, H. KORNMEIER, W. ORLIK, T. KOLAR,
C. LUDEWIGT, T. REINERS, and A. RISCH
Fakultät für Physik, Universität Freiburg, D-7800 Freiburg, GERMANY

Photoabsorption cross sections have been measured for Ar_n^+, $(3 \leq n \leq 60)$, using a new apparatus built for the optical spectroscopy of mass selected cluster ions. The localization of the positive charge in argon cluster ions is discussed.

Neutral rare gas clusters have one of the simplest kind of chemical binding. Every atom is bound to its neighbours by the same kind of weak undirectional van der Waals force. The first electronic excitation is far in the vacuum UV region [1]. An electron is removed from a strongly antibonding orbital if the cluster is ionised. Consequently the bonding becomes much stronger around the charge, and differs in character, direction and strength. The optical absorption shifts to the visible. The absorption is due to the charge localisation on a molecular chromophore. It was originally deduced from energetic considerations and condensed state data [2], that the positive charge in a rare gas cluster ion localizes after about a picosecond on a dimer ion, e.g. Ar_2^+. An argon cluster ion Ar_n^+ would thus have

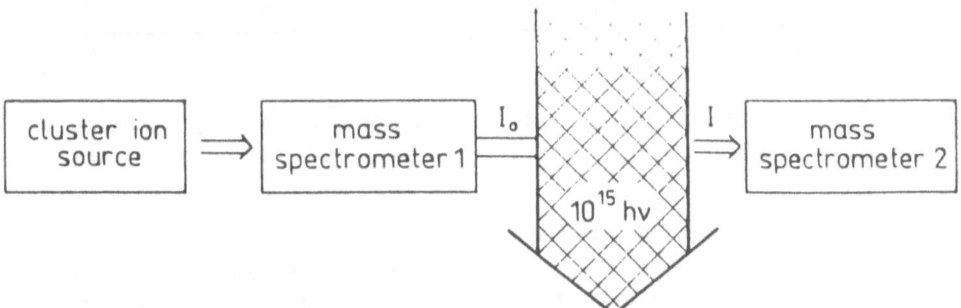

Fig. 1: *Schematic of the experimental setup. Cluster ions of one mass and intensity I_0 are selected in mass spectrometer 1. The photon interaction leads to a decrease of the cluster ion intensity, which is measured in mass spectrometer 2. The photons act effectively as "absorbers" for the extremely tenuous ion beam.*

P. Jena et al. (eds.), Physics and Chemistry of Finite Systems: From Clusters to Crystals, Vol. II, 943–947.

a $(Ar_2^+)Ar_{n-2}$ structure. Later experimental [3, 4, 5, 6, 7] and theoretical [8, 9, 10] results pointed to charge localization on a larger unit: Ar_n^+ with n = 3 or 4. This view is not universally shared: Stace and coworkers [11] deduce from their data that the charge does localize on a dimer ion. However Deluca and Johnson [12] and Bowers et al.[13] favour an $Ar_2^+ - Ar$ double minimum potential, a notion which is contested by Gadea and Malrieu [14]. Carnovale et al. [15] deduce from their photoelectron spectra n = 3, 7, or 13, depending on cluster size.

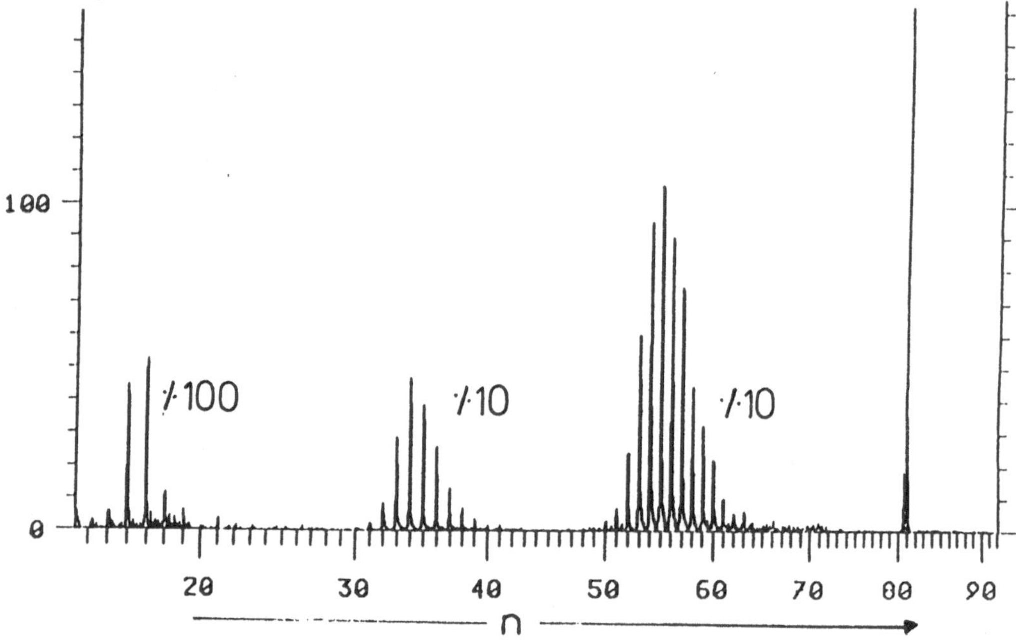

Fig.2: *Photofragmentation of Ar_{81}^+ at 610 nm. Three photons are necessary to fragment n = 81 down to n ≈ 17*

The apparatus has been described in detail elsewhere[16]. The basic principle is shown and explained in fig. 1. A photofragmentation spectrum is shown in fig. 2. The cluster Ar_{81}^+ was selected by the first mass spectrometer, irradiated by 610 nm photons ($h\nu$ = 2.03 eV), and the resulting spectrum of photofragmented ions recorded in mass spectrometer 2. For low photon fluences, only the distribution around n ≈ 55 is observed. On the average 26 Ar atoms are evaporated for a 2 eV atom. This gives immediately an upper limit of 2 eV/26 = 78 meV for the dissociation energy of argon cluster ions in this mass range. This value has to be corrected for the kinetic energy carried away by the evaporating atoms and the different internal energy contents of the clusters at selection and detection time.

If the laser fluence increases, the cluster ion fragments around n = 55 can absorb a second photon from the laser pulse. The time scale for fragmentation is rapid compared to the length of the laser pulse (10 to 15 ns). If these fragments absorb again a photon, one obtains the distribution near n = 17.

Fig. 3 shows all the data measured for argon. Only the data for n = 3 and n = 23 have been known earlier. Fig. 4 shows how the maxima shift with the cluster size. They remain at first constant, shift then in an irregular fashion and remain constant for n $\geq$ 21.

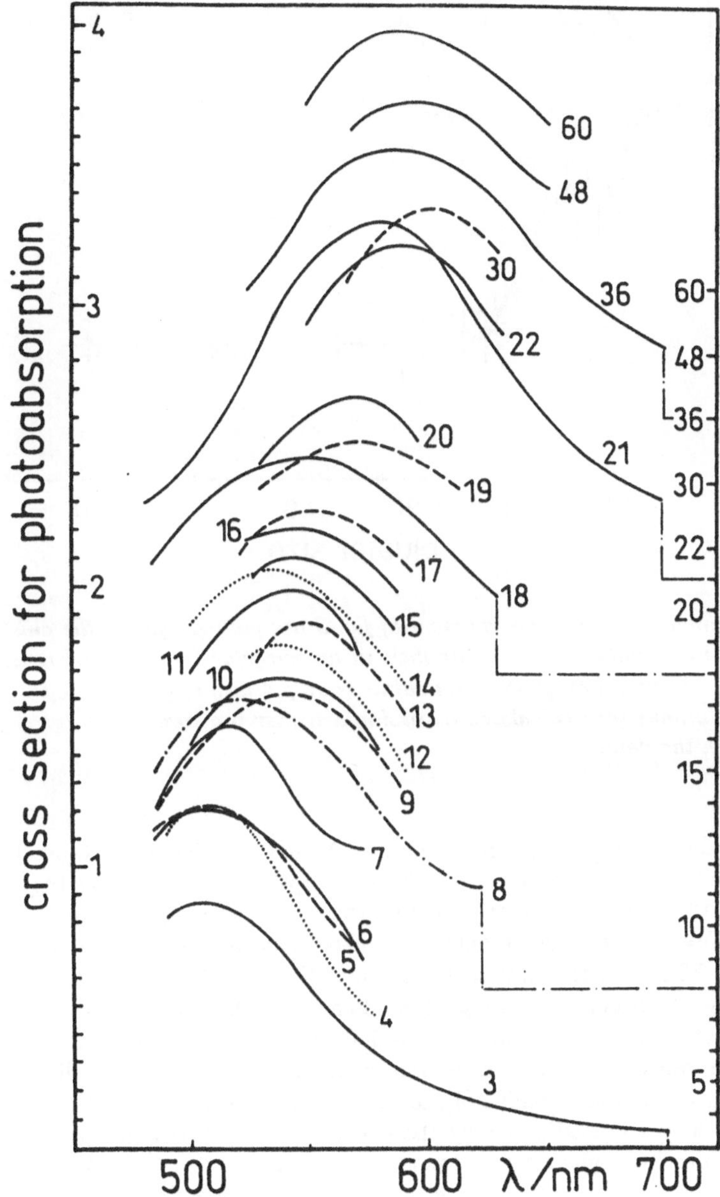

Fig. 3: *Photoabsorption cross sections of* Ar_n^+, *for n between 3 and 60. Only one broad distribution is always observed. The curves are shifted vertically for clarity. The zero of each curve is marked at the right vertical scale.*

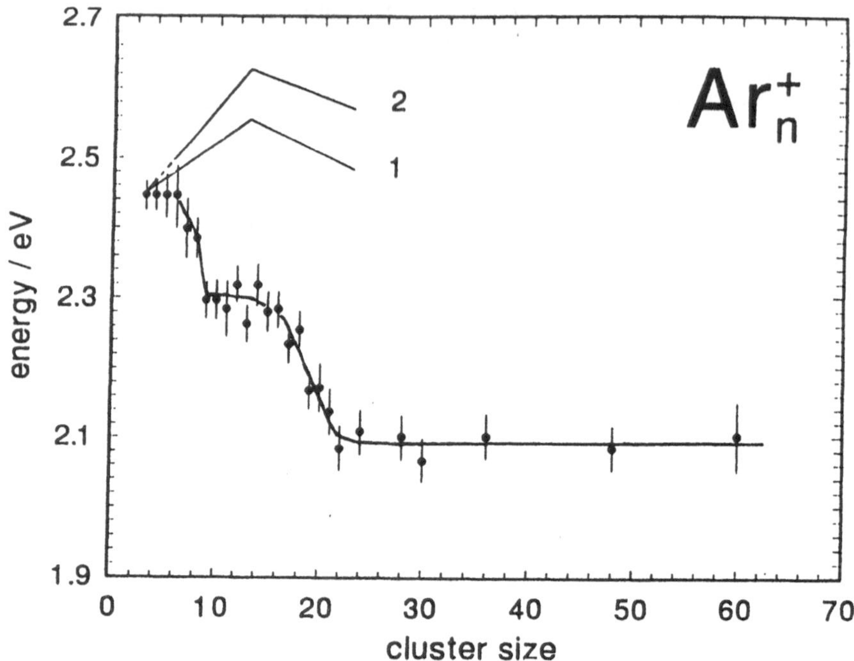

Fig. 4: *The energies of the peak maxima of fig. 3 are plotted against the cluster size. The peak does not move until $n = 6$, shifts then in an irregular fashion, and remains constant above $n = 21$. The line through the data points is only drawn to guide the eye. The two lines marked 1 and 2 have been calculated assuming a trimer ion core. Only for $n \leq 6$ are they compatible with the data.*

The spectrum for Ar_3^+ agrees reasonably with that calculated by Gadea [17]. A rough calculation has been performed how the trimer absorption would shift as a function of cluster size, assuming that the charge distribution on the trimer chromophore does not change with cluster size. Taking into account only the difference in polarization energies, one obtains line 2 of fig. 3. The van der Waals contribution to the optical shift was calculated using the theory by Longuett–Higgins and Plople [18]. The sum of both contributions gives line 1.

No agreement between calculation assuming and experiment is obtained for $n \geq 6$, i.e. the charge does not localize on a trimer ion for the larger clusters. The dimer ion does not absorb strongly in the visible region [4], so that our data do not support the idea of a dimer ion core. The only alternative left by the calculations [8, 9] is a charge localization on a tetramer ion core. No explanation of the structure between $n = 6$ and 21 in fig. 3 has been found so far. A calculation of the optical properties of argon (and other rare gas) cluster ions would be highly desirable to check these conclusions.

This work was supported by the Deutsche Forschungsgemeinschaft through SFB 276–TP C5.

References

[1] T.Möller, Z.Phys.D20, 1 (1991)

[2] H.Haberland, Surf.Sci.156, 305 (1985)

[3] C.R.Albertoni, R.Kuhn, H.W.Sarkas and A.W.Castleman Jr., J.Chem.Phys.87, 5043 (1987)

[4] N.E.Levinger, D.Ray, M.L.Alexander, W.C.Lineberger, J.Chem.Phys. 89, 5654 (1988)

[5] J.T.Snodgrass, C.M.Roehl and M.T.Bowers, Chem.Phys.Lett. 159,10 (1989)

[6] T.Nagata, J.Hirokawa, T.Ikegami, T.Kondow and S.Iwata, Chem.Phys.Lett.171, 433 (1990)

[7] W.Kamke, J.deVries, J.Krauss, E.Kaiser, B.Kamke and I.V.Hertel, Z.Phys.D14, 339 (1989)

[8] P.J.Kuntz and J.Valldorf, Z.Phys.D8, 195 (1988)

[9] M.Amarouche, G.Durand and J.P.Malrieu, J.Chem.Phys.88, 1010 (1988)

[10] H.-U.Böhmer and S.D.Peyerimhoff, Z.Phys.D11,239 (1989)

[11] C.A.Woodward, J.E.Upham, A.J.Stace and N.J.Murrell, J.Chem.Phys.91, 7612 (1989)

[12] M.J.Deluca and M.A.Johnson, Chem.Phys.Lett. 162, 445 (1989)

[13] M.T.Bowers, W.E.Palke, K.Robins, C.Roehl and S.Walsh, Chem.Phys.Lett. 180, 235 (1991)

[14] F.X.Gadea and J.P.Malrieu, submitted to Chem.Phys.Lett.

[15] F.Carnovale, J.B.Peel, and G.R.Rothwell, J.Chem.Phys.95, 1473 (1991)

[16] H.Haberland, H.Kornmeier, C.Ludewigt, A.Risch and M.Schmidt, Rev.Sci.Instrum, accepted

[17] F.X.Gadea and M.Amarouche, Chem.Phys.140,385 (1990), and F.X.Gadea, Z.Phys.D20, 25 (1991)

[18] H.C.Longuet-Higgins and J.A.Pople, J.Chem.Phys.27, 192 (1957)

Observation of Cluster Specific Excitations in Rare Gas Clusters

M. Joppien, J.Wörmer, R.Müller and T. Möller
II. Institut für Exp. Physik, Universität Hamburg D-2000 Hamburg 50
Federal Republic of Germany

ABSTRACT

Absorption spectra of rare gas clusters (N = 50-500 atoms) exhibit extra absorption bands previously not reported in condensed rare gases. They show up if the cluster radius is comparable to the radius of the electronic excitation. From the sharpness of the bands and from the energetic position it is concluded that these excitations have a character in between molecular Rydberg states and excitons of the solid.

INTRODUCTION

The development of the electronic structure as single atoms are put together to form a solid has been subject of a great variety of research in the past decade[1]. In case of non-metallic clusters the evolution of electronically excited energy levels is particularly appealing because a gradual transition from molecular valence and Rydberg states to the exciton bands of the solid is expected. Recently, experiments to investigate the evolution of electronically excited energy levels of rare gas clusters in very large mass range (N = 2 - 10^6) have been started[2,3]. It turned out that the transition to solid state properties, namely, the appearance of excitonic absorption bands takes place in a very large mass range, depending on the size of the electronic excitation. So far, all detected electronic excitations in rare gas clusters correspond either to molecular Rydberg states or to excitons of the solid. Here we report on the first observation of extra absorption bands in medium size Ar_N and Xe_N clusters (N = 50-500) previously not reported in condensed Ar and Xe gas. They show up if the cluster radius is comparable to the radius of the electronic excitation. From the energetic position it is concluded that these excited states are characterized by extended orbitals with a radius somewhat larger than the nearest-neighbour distance.

P. Jena et al. (eds.), Physics and Chemistry of Finite Systems: From Clusters to Crystals, Vol. II, 949–955.

Model calculations in the effective mass aproximation give evidence that the electron has a high probability to stay inside as well as outside the cluster [4]. The narrowing of the absorption bands compared to the Wannier excitons of the solid indicates that these excitations are a new type of electronic excitations in non-metallic material with a character in between molecular Rydberg states and excitons of the solid. In analogy to 'confined' [5] or 'zero dimensional' [6] excitons they might be regarded as **'cluster excitons'** in which the atomic structure and the number of shells of the cluster seem to play a cruical role.

EXPERIMENT

The mesurements were performed at the experimental station Clulu behind the high-intensity beamline Superlumi at HASYLAB (Hamburg). In brief, rare gas clusters are generated in a nozzle expansion of pure gas or a gas mixture ('seeded beam') using large conical nozzles. After passing a skimmer, the cluster beam crosses the beam of monochromatized synchrotron radiation. The size distribution in the cluster beam is analysed with a time-of-flight mass spectrometer in a different set of measurements. Typically, the width (fwhm) corresponds to the average number N of atom per cluster. The fluorescence light is detected undispersed with a closed channel plate detector (CsTe photo-cathode, window cutoff wavelength 112 nm). Fluorescence excitation spectra are recorded by scanning the monochromator.

RESULTS AND DISCUSSION

Figure 1 shows fluorescence excitation spectra of Ar_N, Kr_N and Xe_N clusters containing roughly 100 atoms in an energy range from the lowest excitation up to 2 eV above the ionisation limit. Below the ionisation limit the fluorescence yield corresponds to the absorption coefficient, because dark relaxation channels are of minor importance [2]. Appart from a few sharp 'extra' bands marked with arrows all bands have a 1:1 correspondence in the solid state limit. Consequently, these bands are assigned to the well known bulk and surface excitons of solid Ar, Kr and Xe. The different members of the exciton series are denoted by a main quantum number n. Since the radius of the excitons [8] is proportional to n^2 only excitons with

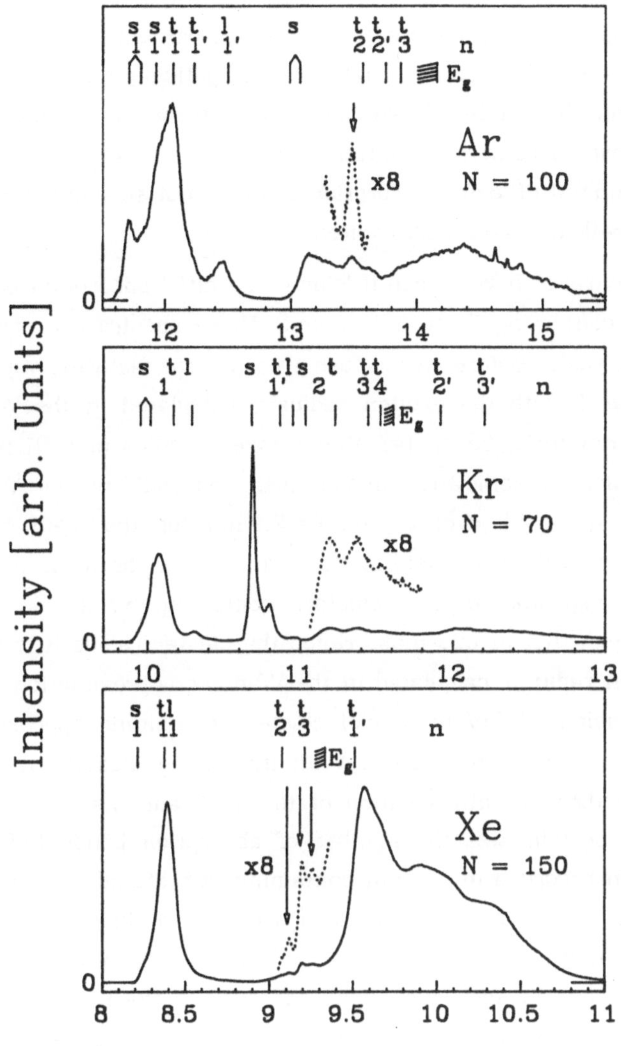

Figure 1: Fluorescence excitation spectra of small Ar, Kr and Xe clusters. The cluster specific excitations ('cluster excitons') are indicated by arrows. In small Kr clusters pronounced 'cluster excitons' are missing. Presumably, they are energetically degenerated with the n = 2 excitons of solid Kr. The average cluster size N [atoms / cluster] and the energetic position of bulk (t,l) and surface (s) excitons of the solid [8] are given in the figure.

small n (n=1 Frenkel-type and n=2 Wannier-type) contribute to the absorption spectra of clusters containing approximately hundred atoms in agreement with previously reported findings [2,3]. The 'extra' absorption bands are energetically close to the n=2 and n=3 Wannier excitons. It will be shown below, that these bands are in fact 'extra' absorption bands which remarkably differ from the n=2 Wannier excitons. In Kr_N clusters, the assignment of 'extra' absorption bands is not straightly foreward, because of their overlap with the n = 2 bulk excitons.

A very rich structure of 'extra' or how we call it 'cluster specific' absorption bands is observed in XeN clusters containing 50 to 500 atoms. These spectra are presented in figure 2 in an enlarged scale. Before we discuss the spectral features, we like to compare the cluster radius R with the exciton radius r calculated in the Wannier approximation. Clusters containing 55 or 147 atoms have a radius of 0.91 nm and 1.26 nm respectively, which is somewhat smaler than the radii of the Wannier excitons (r = 1.28 nm for the n = 2 exciton and r = 2.9 nm for the n = 3 exciton, respectively). Since we deal with free clusters, the formation of the n = 2 and n = 3 excitons therefore is not expected for Xe_N clusters containing 50 to 150 atoms. However, the numbers show, that in this mass range the cluster radius R is exactly in the range of the exciton radius r calculated in the Wannier approximation. In the energy range under discussion (8.8eV to 9.3eV) clusters containing approximately 20 atoms or less are completely transparent. With increasing cluster size sharp absorption bands close to the energetic position of the n = 2 and n = 3 excitons of solid Xe appear. Up to a certain size the number of absorption bands increases. In particular, the spectrum recorded in a beam containing 265 atoms per cluster at the average yields seven absorption bands in this energy range. in clusters containing more than 500 atoms the number of absorption bands decreases. Very large clusters (N = 25000) show only one strong absorption band assigned to the n = 2 exciton of the solid and additionally a weak shoulder at 9.3 eV due to the n = 3 exciton. The width of the n = 2 exciton is nearly identical with the one in solid Xe [8]. It is obvious that the widths of some of the 'extra' absorbtion bands are remarkably smaller. For the interpretation of the extra absorption bands we propose the following picture: These bands are due to bound electron-hole states where the hole is only moving in one of the shells of the cluster. This would explain that the number of absorption bands increases with increasing cluster size, at least for clusters containing less than 500 atoms. Model calculations [4] indicate, that the radial probability of the

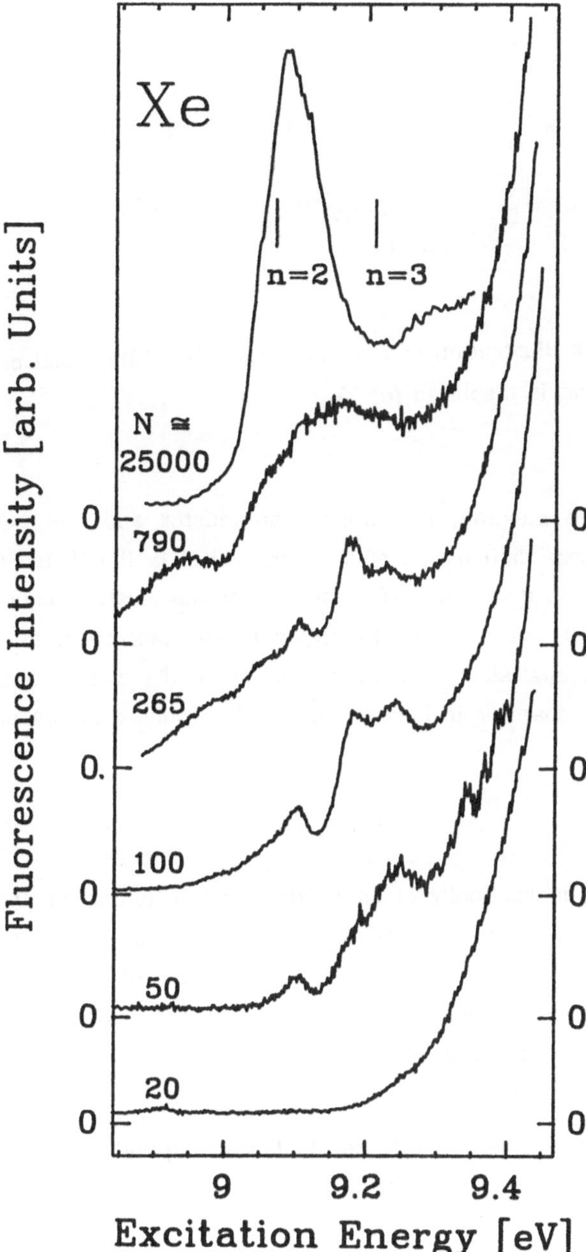

Figure 2 : Fluorescence excitation spectra of Xe clusters in the vicinity of the $n=2$ and $n=3$ excitons of solid Xe. The average cluster size N [atoms /cluster] is given in the figure.

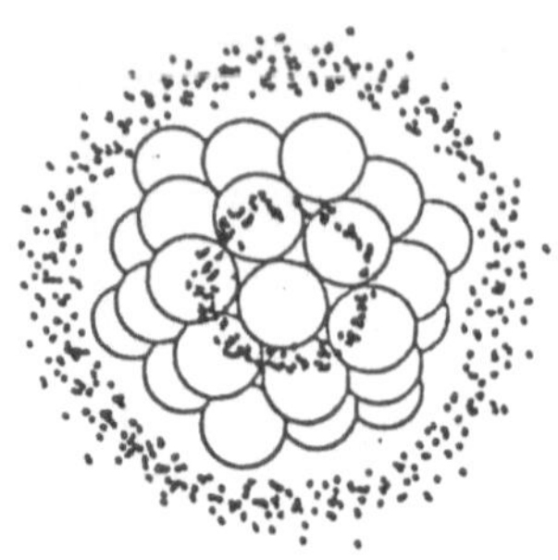

Figure 3 : Schematic illustration of a 'cluster exciton' . The radial probability of an excited electron is displayed for Xe_{55}.

excited electron has two maxima, one inside the cluster and one outside of it. For Xe_{55}, the radial charge distribution of such an orbital is illustrated in figure 3. The large probability to find the excited electron outside the cluster might be responsible for the sharpness of some of the absorption bands. Moreover, the location of the two maxima of the radial electron density relative to the cluster indicates, that these excitations have a charcter in between molecular Rydberg states and excitons of the solid.

CONCLUSION

Extra absorption bands energetically close to the $n = 2$ excitons of the solid are reported for Ar_N and Xe_N clusters containing between 50 and 500 atoms. These bands are assigned to 'cluster excitons' which are a new type of electronic excitation with a character in between molecular Rydberg states and excitons of the solid. The appearance of 'cluster excitons' is a direct consequence of the finite size and the shell structure of the cluster.

Financial support from the Bundesministerium für Forschung und Technologie (BMFT) is kindly acknowledged.

References

1 Small Particles and Inorganic Clusters, Z. Phys. D 19 & 20, (1991)

2 J. Stapelfeldt, J. Wörmer and T. Möller, Phys. Rev. Lett. 62, 98 (1989)

3 J. Wörmer and T. Möller, Z. Phys. D 20, 39 (1991)

4 M. Joppien, J. Stapelfeldt, J. Wörmer and T. Möller, to be published

5 H.M. Schmidt, H. Weller, Chem. Phys. Lett. 129, 615 (1986)

6 L.E. Brus, IEEE Journal QE 22, 1909 (1986)

7 H. Wilcke, W. Böhmer, R. Haensel and N. Schwentner, Nucl. Instr. Meth. 208, 59 (1983)

8 Electronic Excitations in Condensed Rare Gases, N. Schwentner, E.E. Koch and J.Jortner, Springer Tracts in Modern Physics, Vol. 107, Springer, Berlin (1987)

INELASTIC COLLISIONS OF ELECTRONS WITH SMALL METAL CLUSTERS

VITALY V. KRESIN*, ADI SCHEIDEMANN**, and W.D.KNIGHT
Department of Physics
University of California
Berkeley, California 94720

ABSTRACT. We have carried out an investigation of inelastic collisions between electrons (E=0.1-30 eV) and free neutral size-selected sodium clusters. We determined absolute inelastic scattering cross sections as a function of both electron energy and cluster size. Cross sections increase at impact energies below $\cong 0.5$ eV, and are essentially constant at higher energies. The dominant depletion processes are electron attachment and collision-induced fragmentation. The inelastic interaction range is found to considerably exceed the cluster radius. This reflects the influence of the attractive long-range polarization interaction, and establishes a connection with studies of electromagnetic response properties of metal clusters.

1. Introduction

Electron beams are a valuable tool for the study of clusters (see., e.g., [1]). They provide insight into such properties as structures, stabilities, electron affinities, and ionization potentials. All these topics are currently of considerable interest in cluster research. Furthermore, spectroscopic information on small metal clusters is presently limited to dipole laser-induced transitions (e.g., photoionization [2,3] and giant resonances [4]). Electron impact experiments are a unique way to study cluster response properties involving higher-order multipoles.

Size-selected clusters free of external influences can be produced only in molecular beams. The density of clusters in a beam is extremely low, and this makes electron-impact experiments difficult. Thus earlier studies of collisions between electrons and metal clusters have been limited to ionization threshold measurements [5]; absolute cross section measurements have been previously reported only for dimers [6,7].

P. Jena et al. (eds.), Physics and Chemistry of Finite Systems: From Clusters to Crystals, Vol. II, 957–962.
© 1992 *Kluwer Academic Publishers.*

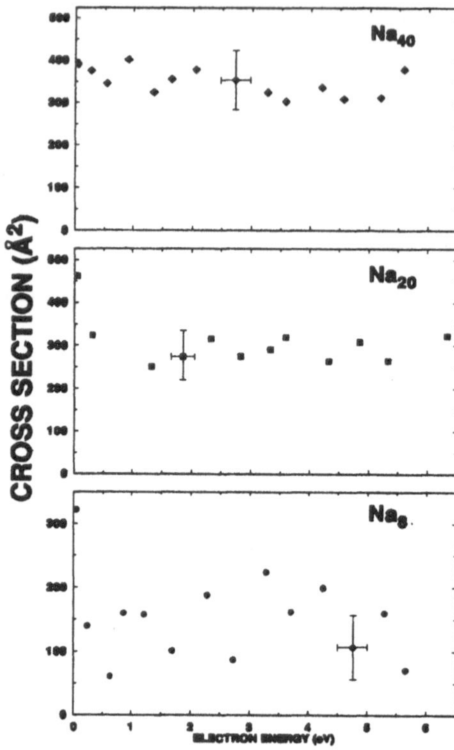

Figure 1. Absolute electron-impact depletion cross sections for three spherical clusters.

The experiment described here [8] is the first measurement of absolute interaction cross sections carried out on free neutral mass selected clusters (Na$_N$, N=8,9,20,40). We have measured absolute total inelastic scattering cross sections as a function of both cluster size and electron energy.

2. Experimental method

The experimental arrangement has been described elsewhere [8]; an outline follows.

The essential feature is that we determine the scattering cross sections by measuring the electron-induced depletion of the cluster beam, rather than by detecting the scattered electrons. This is analogous to the approaches used in laser-induced spectroscopy of giant dipole resonances [4] and in measurements of elastic electron scattering by beams of atoms and alkali dimers [6,9]. It is important to note that in our experiment the kinematic conditions are such that the observed depletion is due to *inelastic* scattering.

Halfway between the source and the detector, the supersonic beam of neutral clusters is crossed by a monoenergetic electron beam. The electron gun is based on the design described in [10] and can produce current densities in excess of 1 mA/cm² with a high ($\cong 0.4$ eV) energy resolution. Following an inelastic collision, a cluster is removed from the highly collimated beam, resulting in a decrease in the counting rate at the detector. By measuring the change in the counting rate for a given cluster size, the absolute collision cross section can be determined.

3. Results

Fig. 1 shows the total inelastic scattering cross sections for spherical closed-shell clusters [2] with N=8, 20, and 40 valence electrons. To within the experimental accuracy, the cross sections for the open-shell spheroidal cluster Na_9 were similar to those of Na_8. The results represent the average of a large number of data points acquired in a series of experimental runs. Error bars indicate the experimental uncertainty in the results ($\pm 50\%$ for Na_8, $\pm 20\%$ for Na_{20} and Na_{40}); the scatter reflects the fact that the attainable depletion ratios are very small (less than 1%).

4. Analysis

In general, several mechanisms can remove a neutral Na_N cluster from our beam following an inelastic collision: electron attachment, electron-induced ionization, or electron-induced fragmentation. Fragmentation may proceed either directly or via an intermediate stage, with sufficient excess energy deposited into the cluster. However, in order for fragmentation to take place, the amount of energy deposited by the incident electron must exceed the cluster binding energy, which is ≈ 0.5-1 eV [11].

Note that the cross section curves display two distinct regimes. Above $\cong 1$ eV, the cross section is essentially constant; below $\cong 0.5$ eV, a rise is visible [12]. Let us discuss these two regions in turn.

4.1 RISE REGION ($E \leq 0.5$ eV)

This regime corresponds to low electron energies, where only s-wave scattering contributes significantly to the collision process. In this case, inelastic scattering is described by the "$1/v$" law [13], according to which the inelastic scattering cross section varies inversely with the electron velocity. This is qualitatively consistent with our observations.

The kinetic energy of these electrons is smaller than that needed for fragmentation or ionization. We conclude that the dominant channel of cluster depletion in the s-wave scattering region must be electron attachment.

4.2 PLATEAU REGION (E≳1 eV)

At higher energies, angular momenta up to $\ell = k\mathcal{A}$ will contribute to the scattering process, where $\mathcal{A}$ is the range of the inelastic electron-cluster interaction and k is the wave number of the incident electron. (This is an example of the multipolar character of electron-cluster interactions). It can be shown [13] that in this case there is an upper bound on the inelastic scattering cross section:

$$\sigma_{in}^{max} = \pi\mathcal{A}^2. \tag{1}$$

The fact that above ≈1 eV the observed depletion cross sections are basically constant, taken together with the very weakly bound structure of sodium clusters, indicate that in the plateau region the cross sections are essentially given by the expression (1).

The validity of Eq. (1) is confirmed by the data shown in Fig. 2. In agreement with the expected constancy of the depletion cross sections in the plateau region, the cross sections of Na_{40} at electron energies of 10, 20 and 30 eV are seen not to differ from those at lower energies.

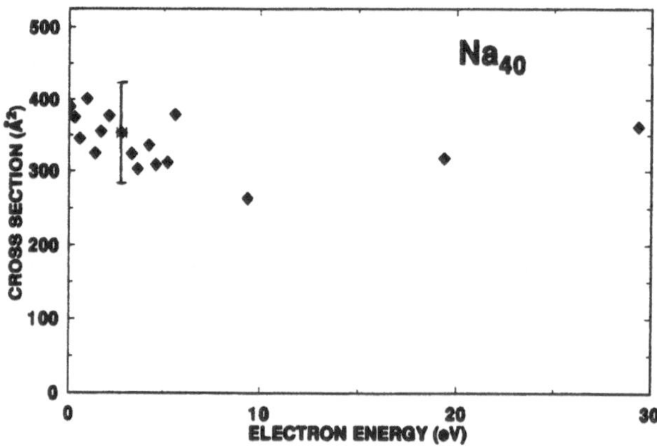

Figure 2. Electron-impact depletion cross sections for Na_{40}.

Electrons in the plateau region carry sufficient kinetic energy (>1 eV) to cause cluster fragmentation. At higher energies, additional channels come into play above their respective thresholds (e.g., excitation of surface plasmons above $\simeq 2.3$eV [4] and ionization above $\simeq 3$eV [3]); however, these do not appear as prominent features on top of the overall cross section data. Thus fragmentation must be the primary mechanism of cluster depletion in this energy range.

4.3 INELASTIC INTERACTION RANGE

The experimental data, together with Eq. (1), allow us to determine the inelastic interaction range, and we find

$$\mathcal{A} \cong (1.4-1.7)R, \tag{2}$$

where R is the radius of the uniform positive jellium sphere commonly used to represent the cluster background [2].

It is interesting to note that $\mathcal{A}$ is greater than both the extent of the spill-out of the cluster's valence electron cloud and the range of the potential which confines this cloud. At distances $r \approx \mathcal{A}$, the valence electron density and the strength of the single-particle potential become extremely small, i.e., less than 1% of their values at $r=R$ [14]. The large value of $\mathcal{A}$ may be due to the influence of the attractive long-range polarization interaction $\sim \alpha/r^4$, where α is the electric polarizability of the cluster [8]. This also explains why for Na_{40} the ratio $\mathcal{A}/R$ appears to be slightly smaller than for the other clusters: it is known that polarizability per atom decreases with cluster size (see, e.g., [2,15]). (Accounting for the polarization potential is important also for the analysis of elastic scattering [16].)

The threshold for s-wave behavior is consistent with the value (2) of the inelastic interaction range: the condition $k\mathcal{A} \cong 1$ is satisfied for electron kinetic energies $\lesssim 0.2$ eV, which agrees with the measured onset of the rise region of the cross section data. Furthermore, as the cluster size increases, so does the interaction radius $\mathcal{A}$, and the threshold for s-wave scattering moves to lower energies.

5. Conclusions

We have carried out the first measurement of absolute inelastic cross sections for collisions between electrons and size-selected free metal clusters. The knowledge of these cross sections is important for studies of interactions between clusters and charged-particle beams. A consistent description of the observed phenomena has been presented. The cross sections exhibit s-wave scattering at low energies, and are essentially constant for energies higher than ≈1 eV. The former regime corresponds to electron attachment, and the latter to collision-induced fragmentation.

The effective range for interaction between electrons and neutral clusters has been determined and found to considerably exceed the cluster dimensions. This effect is due to the long-range attractive polarization potential, and provides a bridge between the studies of electron scattering and of other cluster response properties, such as electric polarizabilities and dipole resonances.

This work was supported by the U.S. National Science Foundation under Grant No. DMR-89-13414; A.S. was supported by the Deutsche Forschungsgemeinschaft.

* Present address: LLNL, Livermore, CA 94550
**Present address: Department of Chemistry, University of Washington, Seattle, WA 98195

6. References

[1] T.D.Märk, Int. J. Mass Spectrom. Ion Proc. 79, 1 (1987).
[2] W.A. de Heer et al., in Solid State Physics, Vol. 40., ed. by H.Ehrenreich and D.Turnbull (Academic,N.Y.,1987).
[3] M.M.Kappes et al., Chem.Phys.Lett. 143, 251 (1988).
[4] K.Selby et al., Phys.Rev.B 43, 4565 (1991).
[5] A.Hoareau, B.Cabaud and P.Melinon, Surf.Sci. 106, 195 (1981); R.E.Walstedt and R.F.Bell, Phys.Rev.A 33, 2830 (1986); and references therein.
[6] T.M.Miller, A.Kasdan, and B.Bederson, Phys.Rev.A 25, 1777 (1982)
[7] K.Franzreb, A.Wucher, and H.Oechsner, Z.Phys. D 19, 77 (1991).
[8] V.V.Kresin, A.Scheidemann, and W.D.Knight, Phys.Rev.A 44, 4106 (1991).
[9] A.Kasdan, T.M.Miller, and B.Bederson, Phys.Rev.A 8, 1562 (1973).
[10] R.E.Collins et al., Rev.Sci.Instrum. 41, 1403 (1970).
[11] J.L.Martins, J.Buttet, and R.Car, Phys.Rev.B 31, 1804 (1985); P.Fantucci, V.Bonačić-Koutecký, and J.Koutecký, Z.Phys.D 12, 307 (1989).
[12] The rise is not resolved for Na_{40} due to the fact that it falls into an energy range where the electron gun current is too small to provide observable depletion.
[13] N.F.Mott and H.S.W.Massey, The Theory of Atomic Collisions, 3rd ed. (Oxford University Press, 1965).
[14] V.V.Kresin, Phys.Rev.B 38, 3741 (1988).
[15] V.V.Kresin, Phys.Rev.B 42, 3247 (1990).
[16] B.Wassermann and W.Ekardt, Z.Phys.D 19, 97 (1991).

A "GIANT RESONANCE" IN IRON: ATOM-CLUSTERS-CRYSTAL

L. I. KURKINA ,O. V. FARBEROVICH
Voronezh State University
University sq.,1 , SU-394693, Voronezh, USSR

V. S. STEPANYUK, A. A. KATSNELSON
Moscow State University
Lenin Hills, SU-117334, Moscow, USSR

A. SZASZ
Eotvos University
Muzeum krt. 6-8, H-1088, Budapest, Hungary

ABSTRACT. Using the time-dependent local spin-density
approximation and the model "atom in jellium" photoabsorption
cross-sections of iron clusters containing from 9 to 127 atoms
have been calculated in the energy region of the giant resonance.
The change of the shape of the giant resonance from a free Fe atom
to iron clusters and bulk metal has been studied. It has been
obtained that the intensity of the resonance is nonmonotonically
decreased with the increase of iron atomic aggregation size.

1. INTRODUCTION

It is known that in photoabsorption spectra of atomic 3d-metals
the strong asymmetric peak near the energy of the 3d- excitation
is dominated. The peak is interpreted as the giant autoionization
resonance [1,2], and is caused by the large overlap (hence, the
strong interaction) between the localized 3p and 3d wave functions
and by the existence of unoccupied 3d states. In this case the 3p
→ 3d excitation which results in the discrete state coupled with
the ionization continua is possible. The interference between the
discrete excitations and the ionization channels brings to
appearance of wide asymmetric resonances in photoabsorption
spectra of 3d-elements.
   In this work the change of the giant resonance shape from a
Fe atom to iron clusters and bulk metal has been studied.

2. MODEL AND METHOD

The calculations of the dipole photoabsorption cross-section

P. Jena et al. (eds.), Physics and Chemistry of Finite Systems: From Clusters to Crystals, Vol. II, 963–967.
© 1992 Kluwer Academic Publishers.

$$\sigma(\omega) = \frac{16\pi^2\omega|e|}{3c} \, \text{Im} \, \sum_{s} \int_{0}^{\infty} \delta n^s(r,\omega) r^3 \, dr$$

for a free atom and small metal clusters of iron were carried out within the time-dependent local spin-density approximation [3,4] using the self-consistent solution of the set of equations

$$\delta n^s(\vec{r},\omega) = \int \chi_o^s(\vec{r},\vec{r}\,',\omega) \, \delta V^s(\vec{r}\,',\omega) \, d\vec{r}\,',$$

$$\delta V^s(\vec{r},\omega) = V_{ext}(\vec{r},\omega) + V_{ind}^s(\vec{r},\omega),$$

$$\hspace{10cm}(1)$$

$$V_{ind}^s(\vec{r},\omega) = 2 \sum_{s'} \int \frac{\delta n^{s'}(\vec{r}\,',\omega)}{|\vec{r} - \vec{r}\,'|} \, d\vec{r}\,' + \sum_{s'} \frac{\partial V_{xc}^s(\vec{r})}{\partial n^{s'}(\vec{r})} \, \delta n^{s'}(\vec{r},\omega),$$

where $V_{ext}(\vec{r},\omega)$ and $V_{ind}(\vec{r},\omega)$ being, respectively, external field of frequency ω and induced field, $\delta n(\vec{r},\omega)$ is the change in the charge density. $\chi_o(\vec{r},\vec{r}\,',\omega)$ is the susceptibility function in the independent particle approximation:

$$\chi_o^s(\vec{r},\vec{r}\,',\omega) = \sum_{i}^{occ} \psi_{is}(\vec{r})\psi_{is}^*(\vec{r}\,') \, G(\vec{r},\vec{r}\,',E_{is}+\omega) +$$

$$+ \sum_{i}^{occ} \psi_{is}^*(\vec{r})\psi_{is}(\vec{r}\,') \, G^+(\vec{r},\vec{r}\,', E_{is}-\omega)$$

where $G(\vec{r},\vec{r}\,',\omega)$ is the Green's function. Energy levels E_{is} and wave functions ψ_{is} of the ground state are obtained as the solution of the Kohn-Sham equations of the local spin-density approach.

In the calculations of the photoabsorption by iron clusters we modeled every atom of a cluster by a Fe atom embedded in a jellium sphere with the radius which equal to the distence between the atom and the cluster surface. Difference between photoabsorption cross-sections of a Fe atom in the jellium sphere and the same jellium sphere (but without the inner atom) was considered as the photoabsorption of the single atom which is "dressed" in the cluster s-d interaction. The total photoabsorption spectrum of the cluster is obtained as the sum of cross-sections of these cluster "quasiatoms".

We considered spherical Fe-clusters with a BCC geometry structure (cluster atoms are round the central atom like coordination spheres of BCC structure with a=2.87 Å).

3. RESULTS

The calculated photoabsorption spectra of a free Fe atom,
spherical iron clusters containing from 9 to 127 atoms, and bulk
BCC-iron in the energy range 50–60 eV are shown in figure 1. As
can be seen in the figure, the theoretical photoabsorption
cross-section for atomic Fe will rightly reproduce the shape of
the experimentally observed spectrum [1] if the calculated curve
is moved as whole to the right by 2 eV. Necessity to move the
theoretical results along an energy scale is bound with the use
of the local density approximation which gives the not quite
accurate energy eigenvalues and self-consistent potential of the
ground state [5]. Since the photoabsorption spectra of iron
clusters are obtained within the local density approach, one can
suppose that the cluster spectra are also displaced in the lower
energy region.
We have analysed the partial contribution of separate
electron excitations to the total photoabsorption cross-section as
well as the change of the resonance position with the decrease of
the induced part of theself-consistent potential (1) (apparently,
as induced contribution to the self-consistent potential is
decreased to zero, the energy of the resonance maximum will be
neared to the energy of the discrete excitation with which the
autoionization resonance is connected [6]). The reseach showed
that the giant resonance in the photoabsorption cross-secton of
atomic iron is due by the interference between the discrete $3p_\downarrow \rightarrow$
$3d_\downarrow$ transition and continua $3d^\uparrow \rightarrow \varepsilon p^\uparrow$ (the peak 1 in figure 1a)
and $3d^\uparrow \rightarrow \varepsilon f^\uparrow$ (the peak 2 in figure 1a) excitations.
Photoabsorption spectra of small iron clusters have been
analysed within the framework of the model of "atom in jellium".
The upper part of the potential well for the system "jellium
sphere with embedded central Fe-atom" is different from the atomic
potential considerably and resembles the spherical potential well.
For this reason in comparison with a Fe atom the system "Fe-atom
in spherical-finite jellium" can have several discrete d-levels.
And what is more, as distinct from a free Fe atom several discrete
d-states of "Fe atom in jellium" can be unoccupied. According to
the selection rule discrete transitions with p-shellson this
d-levels are possible. Energies of discrete $3p \rightarrow nd$ transitions
have near values and lie in the atomic giant resonance region.
Since wave functions of 3p-shell and unoccupied d-shells are
overlap considerably, several discrete $3p \rightarrow nd$ transitions can have
autoionization character. That results in the complication of
structures of photoabsorption spectra for "Fe atom in jellium" in
comparison with the cross-section of a free Fe atom. In addition
the insertion of a Fe atom in jellium and the increase of the
sphere-jellium size cause the change of the energy position and

966

spin splitting of shells of the system. This effect also brings to
the displacement of the resonance maxima in the photoabsorption
cross-sections.

As a result the photoabsorption spectrum of an iron cluster
consists of "quasiatoms" cross-sections with different structures.
Summation of this "quasiatomic" spectra and normalization of the
sum per atom give the cluster resonance profile which is more
spreaded than that of a free Fe atom.

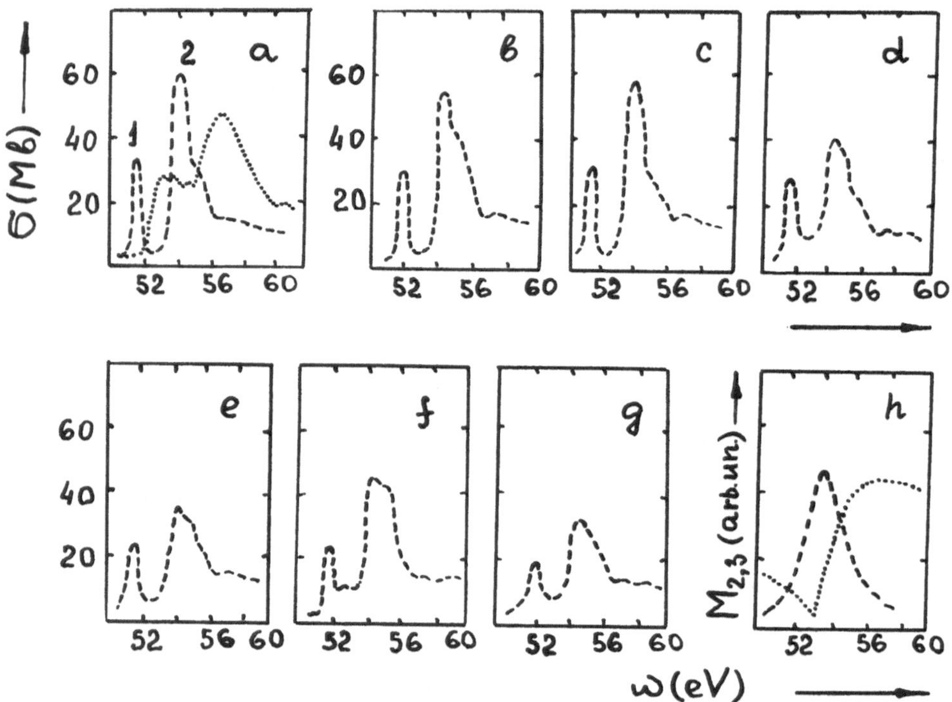

Figure 1. A variation of the giant resonance profile depending on
a size of an iron atomic aggregation: the cross-section of atomic
Fe (a) (broken curve, our calculation; dotted curve, experimental
data [1]); cross-sections of iron clusters containing 9 (b), 17
(c), 28 (d), 66 (e), 84 (f), 127 (g) atoms; $M_{2,3}$-absorption
spectrum of bulk iron (h) (broken curve, our calculation; dotted
curve, experiment [7]); (experimental data are presented in
arbitrary units)

Figure 1 illustrates the change of the shape of the giant resonance in iron with the increase of the atomic aggregation size. As can be seen in the figure, the decrease of resonance intensity with a cluster size is the nonmonotonic process.

Our calculations of bulk iron absorption in the energy region 50-60 eV (by the linear augmented plane wave method) as well as the well-known experimental data [7] have shown that the wide single peak without a fine structure is presented in the absorption spectrum of solid state iron.

4. REFERENCES

[1] Meyer M., Prescher Th., von Raven E., Richter M.,Schmidt E., Sonntag B., and Wetzel H.E. (1986) 'Decay channels of core excitation resonances in 3d and 4f-metal atoms', Z.Phys.D 2, 347-362.

[2] Amusia M.Ya. (1987) Atomic photoeffect, Moscow, Nauka (in Russian).

[3] Zangwill A. and Soven P. (1980) 'Density-functional approach to local-field effects in finite sistems: Photoabsorption in the rare gases', Phys.Rev.A 21, 1561-1572.

[4] Rajagopal A.K. (1978) 'Linear-response functions in spin-density-functional theory', Phys.Rev.B 17, 2980-2988.

[5] March N.H. and Lundqvist S.(eds.) Theory of the inhomogeneous electron gas, New York and London, Plenum Press.

[6] Nuroh K., Stott M.J., and Zaremba E. (1982) 'Calculation of the 4d-subshell photoabsorption spectra of Ba, Ba , and Ba^{++}', Phys.Rev.Lett. 49, 862-866.

[7] Shun-ichi Nakai, Hiroo Nakamori, Akihiro Tomita, Kenjiro Tsutsumi, Hatsuo Nakamura (1974) 'Soft-X-ray $M_{2,3}$-absorption spectra of some transition-metal holides', Phys.Rev.B 9, 1870-1873.

SPECTROSCOPY OF MASS-SELECTED ZIRCONIUM DIMERS IN ARGON

Z. HU, Q. ZHOU, J. R. LOMBARDI AND D. M. LINDSAY

Chemistry Department, Center for Analysis of Structures & Interfaces (CASI)

The City College of New York (CCNY)

New York, NY 10031, USA

ABSTRACT. We report absorption ("scattering depletion") spectra and Raman measurements for Zr_2 in an argon matrix prepared by the mass selected ion deposition technique. The principal dimer absorption bands occur at 388 nm and near 630 nm where we observe a vibrational progression with $\omega_0' = 225$ (15) cm^{-1} for $T_0 = 15,230$ (15) cm^{-1}. Resonance Raman spectra, obtained by exciting into the 630 nm band, give $\omega_e'' = 305.7$ (35) cm^{-1} and $\omega_e x_e'' = 0.5$ (7) cm^{-1}. The Raman data are consistent with either a $^1\Sigma_g^+$ or a $^3\Delta_g$ ground state.

1. Introduction.

A profound understanding of the bonding properties of small, transition metal clusters can best be achieved through a symbiosis of experiment and calculation. [1] On the one hand, however, theoretical progress has been impeded by difficulties in treating the weak, multiple d-d bonding of the transition metal elements, [2] while on the other hand the experimental data base remains sparse and (in some instances) clouded by uncertain spectral assignments. [3]

In this paper, we present absorption and (heretofore unreported) Raman measurements on Zr_2 in an argon matrix. Our samples (prepared by neutralizing a mass-selected beam of dimer ions) are relatively monodispersed. Accordingly, there is little or no interference between the spectroscopic transitions of the dimer and those from atoms or larger clusters. Matrices prepared by the mass-selected ion deposition technique are, however, fairly dilute ($< 10^{16}$ dimers/cm^2) and often highly scattering. To partially overcome these disadvantages, we employ a variation on the usual absorption process, which we term "Scattering Depletion Spectroscopy" (SDS). This involves detecting the light *scattered* at 90° to that incident. Thus, if the sample absorbs at a particular wavelength, the scattered light will be depleted at this wavelength and so also contains the absorption spectrum of

P. Jena et al. (eds.), Physics and Chemistry of Finite Systems: From Clusters to Crystals, Vol. II, 969–976.

the sample. In addition to the results obtained in this work, we have successfully applied the SDS technique to Ta_2, [4] W_2 [4] and Nb_2 [5].

2. Experimental.

The CCNY cluster deposition source has been described in previous publications. [5,6] Briefly, an argon ion beam (typically 15 mA at 25 keV) from a "CORDIS" source [7] sputters cluster ions from a cooled, zirconium target (Goodfellow, 99.8%). Secondary ions are extracted, mass-selected using a Wien filter, bent by 10° (in order to eliminate neutral sputtered products) and then guided to the deposition region by electrostatic lenses. Ion currents, measured at 10 eV on a Faraday plate in the deposition region, were: Zr^+ (120 nA), Zr_2^+ (110 nA) and Zr_3^+ (25 nA).

Zirconium ions were co-deposited (at about 14 K) with Ar and electrons (from a tungsten filament) on a CaF_2 plate mounted on a Displex refrigerator. Matrices were grown at 5 - 10 µ/hour with an Ar:metal dilution ratio of approximately 10^4:1. The deposition region is surrounded by a "Faraday cage" whose potential with respect to the sputtering target controls the kinetic energy (10 eV for the experiments described here) of the deposited ions. The energy distribution of the arriving ions, measured by applying a retarding potential to the Faraday plate, was centered close to the target potential and had a FWHM of about 10 eV. Fragmentation may be estimated by comparing the intensities of atomic excitation features in a dimer deposition with those obtained from depositions of the atom under similar conditions. By this yardstick, we estimate that about 1 - 2% of the zirconium dimers (neutral bond energy, $D_0 = 3.2$ eV) [8] were fragmented, as compared to the similarly determined parameters [5,6] 10 - 15% for V_2 ($D_0 = 2.8$ eV) [9] and 1/2% for Nb_2 ($D_0 = 5.0$ eV) [8].

3. Spectra and Analysis.

Matrix samples were interrogated *in-situ* using both absorption and Raman spectroscopy, as described in detail elsewhere. [5,6] For the former measurements we employed a tungsten lamp, dispersed by a (computer controlled) Spex 1/4 m monochromator (calibrated with a Hg lamp), reflected off a plane mirror and then focused onto the matrix sample. The mirror is mounted on a stepping motor, which allows the focused light to be scanned (in several steps) across the sample. Raman spectra were recorded using a dye laser and

Rhodamine 6G. Typically, 30 mW were focused down to an estimated 50 μ spot. Any fluorescence from the dye was removed by predispersing the laser beam with a grating. Both the Raman *and* (as described in more detail below) the absorption measurements were made by collecting the light scattered at 90° to that incident. The Raman signal was dispersed by a 1/4 m Spex "Doublemate" monochromator (also under computer control) and then detected by a cooled photomultiplier tube (PMT) and amplifier/discriminator using photon counting techniques and an AT386 compatible computer. The absorption signal was recorded by the same detection system but with the PMT mounted directly on the deposition chamber. Raman excitation frequencies were measured by scanning through the Rayleigh line. The "Doublemate" was carefully calibrated by scanning through several Hg lines in second order.

3.1. SCATTERING DEPLETION SPECTROSCOPY

Figure 1 shows a "Scattering Depletion Spectrum" (SDS) of dizirconium in an unannealed argon matrix at 14 K. The spectrum (resolution 0.4 nm) was acquired following a 2 hour deposition of 125 nA of Zr_2^+ (250 nA-h) [10] at a deposition energy of 10 eV. The ordinate in Fig. 1 (the "SDS signal") is the ratio of the "reference intensity" to the "signal intensity". The latter corresponds to light scattered from a region near the center of the 8 mm wide sample, whereas the former pertains to that scattered near an edge. Prior to

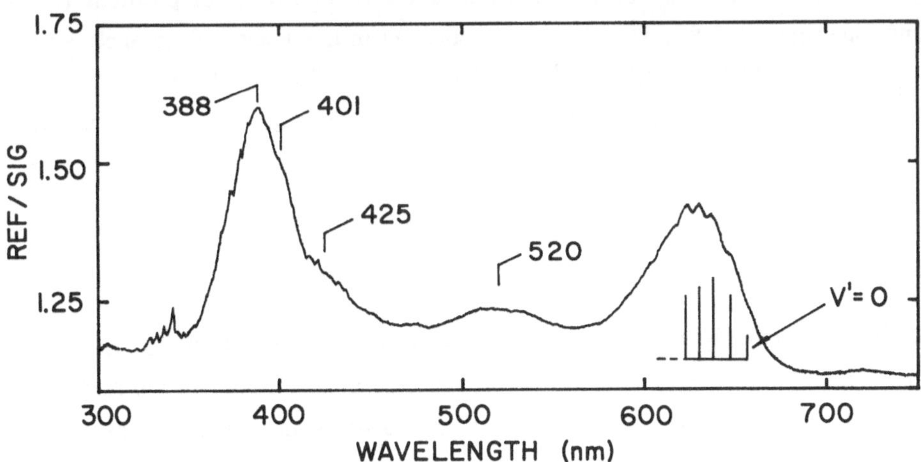

Figure 1. Absorption ("Scattering Depletion") spectrum of Zr_2 in argon.

recording scattering depletion spectra, the optical density profile of the sample is obtained by recording single beam spectra as a function of lateral position on the sample. It is generally found that the sample is non uniform, with about twice as much absorption in the center as compared to near an edge. Thus the reference also contains absorption information but to a lesser degree than the signal.

The absorption spectrum of dizirconium (Fig. 1) consists of an intense band at 388 nm (with shoulders at 401 and 425 nm as indicated), a weaker feature near 520 nm and a structured absorption centered at about 630 nm. The relatively sharp features between 300 and 450 nm arise from absorption by atomic zirconium and have been reported previously. [11,12] As the dimer has no detectable fluorescence, atomic transitions are most easily identified by examining excitation spectra (not shown). Klotzbücher and Ozin identified four Zr dimer bands (390, 422, 585 and 615 nm) in matrices containing a distribution of cluster sizes. [12] While our spectra clearly show absorption features near 390 nm and in the region 600 - 650 nm, we do not observe a strong peak at 422 nm, nor do we see a resolved feature at 585 nm as shown in Fig. 6c of Ref. 12.

Our 630 nm band includes a vibrational progression which ends rather abruptly after the fifth member. The SDS "signal" spectrum for this region was simulated by adjusting the positions and relative intensities of 5 overlapping Gaussian functions assumed to have the same width (250 ± 10 cm^{-1} FWHM). A satisfactory match (illustrated by the stick spectrum in Fig. 1) could be obtained only for a limited range of positions (± 10 cm^{-1}) and intensities (± 5 %). The frequencies obtained in this manner were analyzed to give $\omega_0' = 225 \pm 15$ cm^{-1} with $T_0 = 15,230 \pm 15$ cm^{-1} where the latter frequency [13] corresponds to the (nominal) origin notated $v' = 0$ in Fig. 1. The data fit both a harmonic and an anharmonic potential, but our analysis is not yet capable of distinguishing between these two cases.

3.2. RAMAN SPECTRA

Figure 2 shows a typical Raman spectrum (a different sample from that of Fig. 1) for Zr_2 in an argon matrix. Raman spectra were recorded at eleven wavelengths in the range 600 - 625 nm and were also observed using the 454.5 and 457.9 nm argon ion laser lines, but these latter spectra were relatively weak. Because of a combination of atomic fluorescence and an increasing linewidth (due in part to isotope effects) with increasing Raman shift, no more than four transitions could be assigned. The Raman data (summarized in Table I) were analyzed to give (one standard deviation uncertainty in parentheses) $\omega_e'' = 305.7\ (35)\ cm^{-1}$ with $\omega_e x_e'' = 0.5\ (7)\ cm^{-1}$.

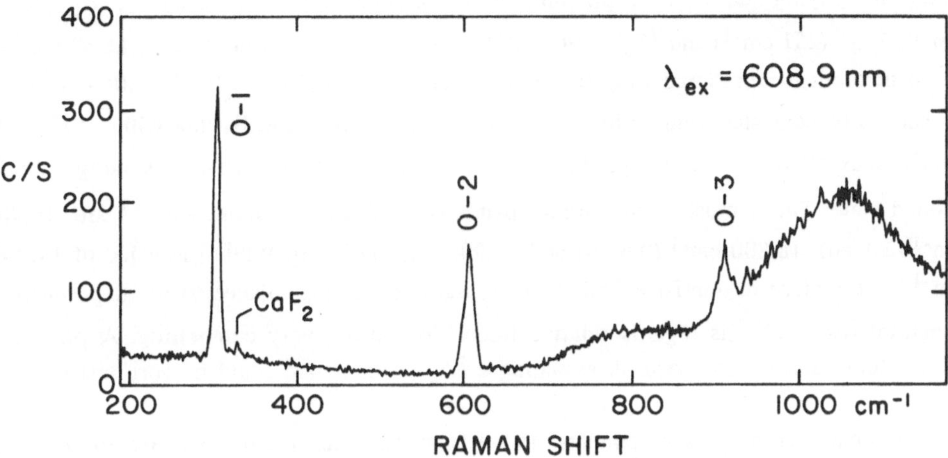

Figure 2. Raman spectrum (Rhodamine 6G) of Zr_2 in argon at 14 K.

Under higher resolution (not shown), the first three Stokes transitions are seen to consist of a principal band (degraded towards the Rayleigh line) plus a *higher* frequency "subcomponent" displaced by $3.3\ (3)\ v''\ cm^{-1}$. The shape of the principal band was well accounted for by simulating overlapping Gaussians corresponding to the 15 possible isotopic combinations [14] spaced (by ca. 1 cm⁻¹ for $v'' = 2$) according to their relative reduced masses. [15] No anti-Stokes Raman transitions could be observed, which implies that we are not observing "hot band" features arising from radiative processes. [15] The relative intensity (uncertainty ± 0.4) of the two Raman components (also determined by simulation) was 1.7:1 for all three Stokes transitions. This intensity ratio was invariant (within error) to annealing, to changes (15 and 25 K) in temperature, to the exciting wavelength and to rotating a polarizer in the path of the scattered light.

974

4. Discussion.

Two theoretical papers on dizirconium have recently been published. [2,16] As the calculations of Ref. 2 include both external correlation and p orbitals (important for determining transition moments), we focus on the conclusions in this paper alone. Four low ($<$ 1500 cm^{-1}) lying states (ω_e in parentheses) are predicted: $^1\Sigma_g^+$ (338 cm^{-1}), $^3\Delta_g$ (305 cm^{-1}), $^3\Sigma_u^+$ (251 cm^{-1}) and $^7\Sigma_u^+$ (196 cm^{-1}). Given the estimated uncertainty ($\pm$ 30 cm^{-1}) [17] in theoretical vibrational frequencies, our Raman measurements ($\omega_e'' = 306$ cm^{-1}) can be said to be consistent with either a $^1\Sigma_g^+$ or a $^3\Delta_g$ ground state, but not with $^3\Sigma_u^+$ and $^7\Sigma_u^+$. Bauschlicher et al. also predicted the dimer absorption spectrum, assuming a $^1\Sigma_g^+$ ground state. The 3 most intense dimer bands (intensities in parentheses) [17] are: 14,900 cm^{-1} (2.1 au), 18,800 cm^{-1} (0.6 au) and 21,900 cm^{-1} (1.2 au). While the value of 14,900 cm^{-1} is very close to our $T_0 = 15,230$ cm^{-1}, the overall resemblance to the experimental spectrum (especially as regards relative intensities) is not very compelling. A prediction of the dimer absorption spectrum assuming a $^3\Delta_g$ ground state could be very instructive.

The occurrence of "subcomponents" in a Raman spectrum (as noted above for Zr$_2$) can be interpreted in two ways. One possibility (as previously invoked for V$_2$) [15] is an interaction within a single matrix site which lifts the spatial and/or spin degeneracy of the dimer. A similar argument here would exclude $^1\Sigma_g^+$ as the ground state, but we feel that other possibilities cannot at present be ruled out. Thus, Zr$_2$ might be trapped in two different matrix sites for which the dimer vibrational frequencies differ by 3 cm^{-1} (1%). For this situation, the resistance of the "subcomponent" structure to (e.g.) annealing is perhaps surprising. Finally, we note one similarity between Zr$_2$ and Nb$_2$.[5] Both dimers have strong absorption bands (and a vibrational progression) around 650 nm. This lowest energy (visible) transition has been interpreted in terms of a $\delta \to \delta^*$ transition. [5,19] While consistent with several ground state configurations for Nb$_2$, [1] such an interpretation would not be possible if $^1\Sigma_g^+$ (configuration: $d\sigma_g^2 d\pi_u^4 s\sigma_g^2$) were the ground state of Zr$_2$.

5. Acknowledgements.
We thank C. W. Bauschlicher for helpful discussions. We also thank Shunguo Shen and Bo Shen for technical support. This work was supported by the National Science Foundation under Cooperative Agreement No. RII-8802964 and Grant No. CHE-9112897 and by The City University of New York PSC-BHE Faculty Research Award Program.

TABLE 1. Raman Frequency Shifts (cm^{-1}) for Zr_2 in an Argon Matrix.

λ_{ex} (nm)	$v'' = 1$	$v'' = 2$	$v'' = 3$	$v'' = 4$
601.9	304.0	607.2	909.0	1209.2
604.8	302.0	605.7	908.0	1213.0
606.0	302.9	605.4	911.3	1201.0
607.5	304.2	606.9	909.4	1211.5
609.6	306.0	608.0	911.7	
609.7	303.9	605.9	909.7	
610.3	303.8	608.4	909.8	
614.0	303.8	608.2	901.2	1213.5
620.4	301.8	601.7		
623.4	302.4	605.4	903.1	
625.6	304.0	605.1		
Mean (σ)	303.5(12)	606.2(19)	908.1(36)	1209.6(51)

6. References.

1. See, for example: *Comparison of ab initio Quantum Chemistry with Experiment*, ed. by R. J. Bartlett (Reidel, Boston, 1985).

2. C. W. Bauschlicher, H. Partridge, S. R. Langhoff and M. Rosi, J. Chem. Phys. **95**, 1057 (1991).

3. For example: W. Harbich, S. Fedrigo, F. Meyer, D. M. Lindsay, J. Lignieres, J. C. Rivoal and D. Kreisle, J. Chem. Phys. **93**, 8535 (1990).

4. Z. Hu, B. Shen, Q. Zhou, J. R. Lombardi and D. M. Lindsay, in preparation.

5. Z. Hu, B. Shen, Q. Zhou, S. Deosaran, J. R. Lombardi and D. M. Lindsay, *Optical Absorption and Raman Spectra of Mass-selected Niobium Dimers in Argon Matrices*, SPIE Proceedings Vol. **1599**, XXXX (1992), in press.

6. Z. Hu, B. Shen, Q. Zhou, S. Deosaran, J. R. Lombardi, D. M. Lindsay and W. Harbich, J. Chem. Phys, **95**, 2206 (1991).

7. R. Keller in *The Physics and Technology of Ion Sources* , edited by I. G. Brown (Wiley, New York, 1989), Chapter 7.

8. M. D. Morse, Chem. Rev. **86**, 1049 (1986).

9. E. M. Spain and M. D. Morse, Int. J. Mass Spec. and Ion Phys. **102**, 183 (1990).

10. We use units of nA-h (the product of the current times the deposition time in hours) where 1 nA-h = 2.25 x 10^{13} particles.

11. J. K. Bates and D. M. Gruen, High Temp. Sci. **10**, 27 (1978); C. Steinbrüchel and D. M. Gruen, J. Chem. Phys. **74**, 205 (1981).

12. W. Klotzbücher and G.A. Ozin, Inorg. Chem. **19**, 3767 (1980).

13. Wavelengths in this paper pertain to measurements in air, whereas energies are given in vacuum wavenumbers.

14. *Handbook of Physics and Chemistry*, ed. by R. C. Weast (Chemical Rubber, Cleveland, 1975).

15. C. Cossé, M. Fouassier, T. Mejean, M. Tranquille, D. P. DiLella and M. Moskovits, J. Chem. Phys. **73**, 6076 (1980).

16. K. Balasubramanian and Ch. Ravimohan, J. Chem. Phys. **92**, 3659 (1990).

18. C. W. Bauschlicher, private communication.

19. M. P. Andrews and G. A. Ozin, J. Phys. Chem. **90**, 2852 (1986).

PHASE TRANSITIONS AND COALESCENCE PHENOMENA IN Bi CLUSTERS

M. G. MITCH and J. S. LANNIN
Dept. of Physics
Penn State University
University Park, PA 16802

In-situ UHV Raman scattering and ex-situ transmission electron microscopy (TEM) measurements have been made on Bi clusters deposited on disordered C at 110K. Raman spectra show a structural phase transition from amorphous to nanocrystalline clusters at an equivalent film thickness of 8.5Å. Changes in TEM cluster size distributions at the transition thickness provide evidence for growth coalescence phenomena assisting the transition. A combined TEM and Raman analysis leads to an estimate of about 1300 atoms per cluster at the transition. Interaction of atomic H with amorphous clusters yields a similar transition as seen in the Raman spectra for thicknesses of 6.5-8Å. TEM distributions show evidence of H induced cluster motion, resulting in a mobility coalescence driven transition.

1. Introduction

It has recently been shown that Bi clusters undergo a phase transition from bulk-like rhombohedral, nanocrystalline order to an amorphous structure with decreasing film thickness.[1] Complementing this Raman scattering study, TEM has provided information regarding the size distribution of Bi clusters at and around the transition point. TEM studies have indicated that these clusters exhibit a distribution of sizes which may influence the nature of the observed 8.5Å transition. Such a distribution of sizes implies that particles with different structures may be present for a given film thickness. It is therefore useful to explore the relative importance of different size particles to the physical properties of the ultrathin film. In this work, more detailed information is provided that reveals the role of the size distribution and the particular importance of Bi cluster coalescence in enhancing the transition with increasing film thickness. A

P. Jena et al. (eds.), Physics and Chemistry of Finite Systems: From Clusters to Crystals, Vol. II, 977–982.
© 1992 *Kluwer Academic Publishers.*

combination of Raman scattering and TEM measurements are utilized to obtain information on coalescence effects as well as to estimate the approximate transition size for a single cluster to transform from an amorphous to nanocrystalline state.

Recent Raman scattering studies of low temperature Bi clusters have indicated an atomic H induced phase transition that is similar to that observed with increasing film thickness.[2] TEM and Raman scattering measurements are employed to study this transition, indicating the important role of H induced cluster motion at low temperatures.

2. Experiment

Ultrathin films of Bi were deposited in an ultrahigh vacuum chamber of base pressure 2×10^{-10} Torr onto a substrate held at 110K by liquid nitrogen cooling. DC magnetron sputtering at an Argon pressure of 1 mTorr provided a source of Bi atoms. Bi nucleation took place on substrates of disordered carbon (d-C) for both Raman and TEM measurements. These substrates were formed by high vacuum RF diode sputtering, and subsequently cleaned in UHV by 200 eV Ar bombardment from a Kaufman type ion source. Bi equivalent film thickness was determined by a calibrated quartz-crystal microbalance.

In-situ Raman scattering measurements were performed with a Spex Triplemate spectrograph at 6096Å excitation wavelength from an Ar ion laser pumped dye laser. Trilayer structures of Al, SiO_2, and d-C were designed for low (< 5 %) reflectance, yielding interference-enhanced Raman scattering (IERS).[3] Multichannel detection combined with IERS enabled the study of Bi films $\geq$ 1.5Å equivalent thickness.

Ex-situ TEM measurements were performed on a Phillips 420T. To protect the sample from oxidation and to reduce Bi cluster mobility, thin film overcoats of ~10Å thick a-Ge were deposited. XPS measurements confirmed the absence of any appreciable changes in cluster size upon heating from 110K to 300K.[1]

3. Results and Discussion

The presence of a phase transition in Bi clusters is clearly seen in the Raman spectra of Figure 1. For 8.5Å equivalent thickness, the spectral form is derivable from the dispersion curves of rhombohedral Bi with finite size effect, $k \neq 0$ wavevector admixing. The polarization dependence of this double peaked spectra is similar to that found in bulk Bi. In contrast, for 7Å Bi, the Raman spectra have a completely different form and polarization dependence, corresponding to an amorphous state in these clusters.[1]

Raman intensity analysis of weighted VH and HH polarization spectra allows an estimation of the size at which individual clusters transform from an amorphous to a nanocrystalline phase. In general, Raman scattering weights particles in proportion to their volume and cross section. It is observed that the depolarized VH component preserves integrated intensity at the transition. Therefore,clusters of the most probable volume will determine the form of the VH spectra. A more sensitive transformation indicator is given by the HH component, as its intensity increases by a factor of three across the transition from amorphous to nanocrystalline phase. Combining these Raman results with cluster volume distributions around the transition yields a transition size estimate of approximately 56Å, with ~1300 ±25% atoms per cluster.

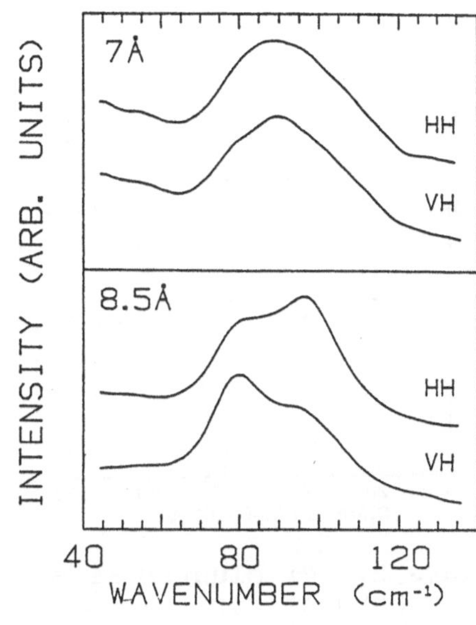

Figure 1. Raman spectra of Bi films of 7Å and 8.5Å thickness.

The key to understanding the mechanism of the phase transition seen in the Raman spectra is the cluster size distribution as a function of equivalent film thickness. Cluster size distributions were obtained by measurement of cluster average diameters on TEM micrographs. The latter clearly show films composed of isolated clusters for all thicknesses studied here ($\leq$8.5Å). In Figures 2a) and 3a), size distributions for clusters are shown for 7Å and 8.5Å Bi. Immediately apparent are the different forms of the distributions, showing a change from unimodal to bimodal with increasing thickness. It has been suggested theoretically that the distribution for clusters undergoing growth coalescence should have just the qualitative form shown in Figure 3a).[4] All depositions $\leq$7Å show a unimodal size distribution and a single peaked, amorphous phase Raman spectrum. The bimodal distribution at 8.5Å implies that certain sized clusters have grown and coalesced, forming larger clusters with a nanocrystalline structure which is shown by the Raman spectra.

The observed transition at a film thickness of 8.5Å is thus enhanced by the growth coalescence of a high density of Bi clusters formed at low temperature on the d-C substrate. Measurements on substrates with a lower nucleation density are

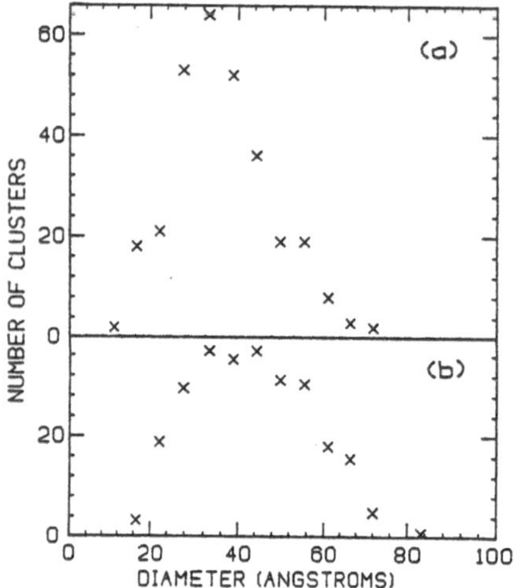

Figure 2. Bi cluster size
distributions for (a) 7Å
thickness and (b) 7Å
thickness with H exposure.

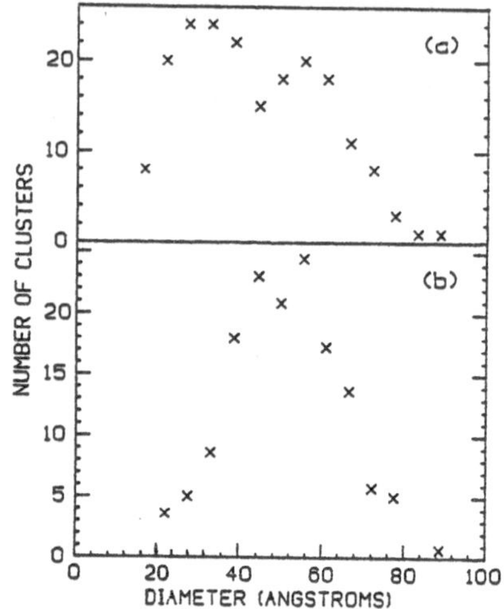

Figure 3. Bi cluster size
distributions for (a) 8.5Å
thickness and (b) 8.5Å
thickness with H exposure.

in progress to determine if the
estimated transition size is similar to that observed in the absence of appreciable coalescence.
In addition, the role of cluster
size on particle melting is under
study.

A phase transition is also
observed at 110K in the Raman
spectra when amorphous clusters
in ultrathin films of equivalent
thickness 6.5-8Å are exposed to
atomic H (made by passing H_2 over
a hot filament). As seen in Figure 4, the Raman spectrum changes
in form (and polarization) in an
analogous fashion to that from
the growth coalescence transformation, exhibiting amorphous to
nanocrystalline spectral changes.
To determine the precise nature
of the effect of H on Bi clusters, TEM and UPS studies were
performed on samples exposed to
atomic H after cluster deposition.

It has been shown that the

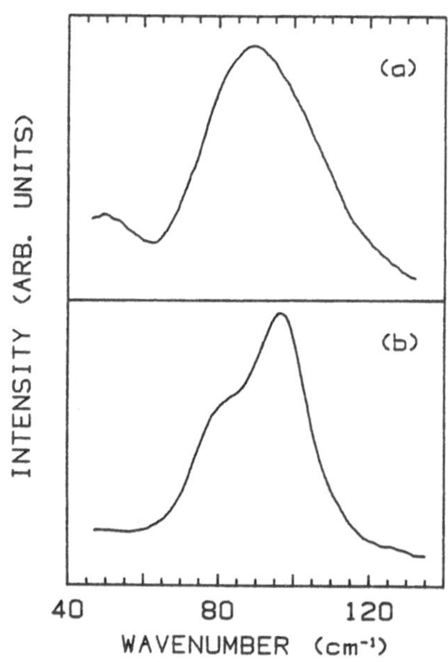

Figure 4. Bi HU Raman
spectra of (a) 7Å
thickness and (b) 7Å
thickness with H exposure.

effect of atomic H on a-Ge clusters is to satisfy surface dangling bonds, thus making the Raman spectrum more bulk-like in form. In addition, Raman and UPS measurements indicate Ge-H bond formation.[5] In contrast to a-Ge clusters, UPS and Raman measurements do not indicate evidence for Bi-H bond formation.

One key result of TEM studies is shown in Figure 3. In contrast to the Ge case, atomic H appears to induce cluster motion, even at 110K. The change in form of the cluster size distribution at 8.5Å with the addition of H is consistent with the presence of mobility coalescence phenomena. Also, a reduction in the number of clusters per unit area is seen upon H addition, further supporting the idea of mobility coalescence.

Given the observed increase in number of larger sized clusters at the expense of smaller ones, the atomic surface diffusion process of Ostwald ripening may seem plausible as an explanation for the observed H effects on cluster distributions. However, as seen in Figure 2, the shape change of the 7Å distribution with H is consistent with coalescence, rather than atomic surface diffusion processes.[6] The high diameter tail of the distribution shown in Figure 2b) contains large, nanocrystalline clusters. The application of transition size modeling to these distributions is fully consistent with the prediction of a nanocrystalline Raman spectrum for films exposed to H at 7Å thickness (see Figure 4b)).

The observed changes in the Bi size distribution due to atomic H are surprising, given the low temperature and the size of the clusters that are moving (hundreds of atoms). The density of Bi clusters and their average size at 7Å film thickness yields an average cluster separation of ~40Å. This implies that cluster motion over a scale of ~20Å is required for coalescence to occur, assuming both coalescing clusters move. It is possible that weak bonds of Bi clusters to d-C are removed by H bonding to substrate defect sites. Under such conditions, motion of Bi clusters over short distances at low temperature may be feasible. Such a chemisorbed gas layer has previously been suggested to explain the motion of Ag crystallites on graphite substrates.[7]

In summary, Bi clusters undergo phase transformations due to both growth coalescence and H-induced mobility coalescence processes. Using a combination of Raman scattering and TEM data, the size of clusters in an amorphous or nanocrystalline phase has been estimated.

4. Acknowledgements

This work was supported by U.S. DOE Grant No. DOE-FG02-84ER45095. The experimental assistance of S. Chase and Dr.

R. Yu in XPS and UPS and Professor P. Howell and Dr. W. Bacsa in TEM is gratefully acknowledged.

5. References

[1] Mitch, M.G., Chase, S.J., Fortner J., Yu, R.Q. and Lannin, J.S., (1991) 'Phase Transition in Ultrathin Bi Films', Phys. Rev. Lett., Vol. 67, No. 7, pp. 875-878.

[2] Chase, S.J., Mitch, M.G., Bacsa, W. and Lannin, J.S., (1991) 'Effect of Atomic H on Bi Nanostructures', Bull. American Phys. Soc., Vol.36, No.3, p.506.

[3] Connell, G.A.N., Nemanich, R.J. and Tsai, C.C., (1980) 'Interference Enhanced Raman Scattering from Very Thin Absorbing Films', Appl. Phys. Lett., Vol. 36, pp. 31-33.

[4] Venables, J.A., Spiller, G.D.T. and Hanbücken, M., (1984) 'Nucleation and Growth of Thin Films', Rep. Prog. Phys., Vol. 47, pp. 399-459.

[5] Yu, R.Q., Fortner, J., Chase, S.J. and Lannin, J.S., (1991) 'The Physical Properties of Ge Clusters and Ultrathin Films', in Averback, R.S., Bernholc, J. and Nelson, D.L. (eds.), Clusters and Cluster-Assembled Materials, Mat. Res. Soc. Symp. Proc., Vol. 206, pp. 403-408.

[6] Granqvist, C.G. and Buhrman, R.A., (1976) 'Size Distributions for Supported Metal Catalysts', J. Catal., Vol. 42, pp. 477-479.

[7] Sears, G.W. and Hudson, J.B., (1963) 'Mobility of Silver Crystallites on Surfaces of MoS_2 and Graphite', J. Appl. Phys., Vol. 44, pp. 2380-2381.

ROLE OF SELF-INTERACTION CORRECTIONS IN THE PHOTOABSORPTION CROSS SECTION OF SMALL METAL CLUSTERS

J. M. PACHECO and W. EKARDT
Fritz-Haber Institut der Max-Planck Gessellschaft
Faradayweg 4-6, 1000 Berlin 33, GERMANY

ABSTRACT.   The Time Dependent Local Density Approximation (TDLDA) is modified to incorporate the Self-Interaction Correction (SIC) of Perdew and Zunger. It is then applied to study the static polarizabilities and phoabsorption cross sections of small, spherical, neutral metal particles with 8, 20, and 40 Sodium atoms.

1. Introduction

The LDA of Density Functional Theory (DFT) and its extension[1] - TDLDA - to describe excited states of many-electron systems are self-interacting approximations. The success of the Self-Interaction Correction[2] (SIC) in describing the Ground State (GS) properties of diverse many-electron systems[3], prompts for the study of its implications in the determination of their excitation spectra. We shall tackle this question within the field of metal clusters, by calculating their photoresponse.

The SIC leads to a departure from mean-field theory, so that the general TDLDA formalism developed in ref.[1] is no longer applicable. Therefore, in this paper we reformulate TDLDA to make it suitable for inclusion of SIC (we shall henceforth denote this formalism as SIC-TDLDA). The paper is organized as follows. In section 2 we summarize the SIC corrections and discuss its implications in the ground-state properties of metal clusters; section 3 is devoted to the study of their photoabsorption spectrum and conclusions are collected in section 4.

2. Gound state properties

In the jellium model, the SIC-LDA single-particle potentials read[2]

$$V_{SIC}^{(i)}(\vec{r}) = V_I(\vec{r}) + e^2 \int \frac{[n(\vec{r}\,') - n_i(\vec{r}\,')]d\vec{r}\,'}{|\vec{r} - \vec{r}\,'|} + V_{xc}[n(\vec{r})] - V_{xc}[n_i(\vec{r})] \qquad (1)$$

where $n(\vec{r}) = \sum_{i=1}^{N} |\psi_i^{(i)}(\vec{r})|^2$, $n_i(\vec{r}) = |\psi_i^{(i)}(\vec{r})|^2$, N denotes the total number of valence electrons in the cluster, and $V_I(\vec{r})$ the jellium potential of N positive ions (Wigner-Seitz radius $r_s = 4\, a_0$). For V_{xc} we take the parametrization of ref.[4]. We

P. Jena et al. (eds.), Physics and Chemistry of Finite Systems: From Clusters to Crystals, Vol. II, 983–988.
© 1992 *Kluwer Academic Publishers.*

denote by $\psi_j^{(i)}(\vec{r})$ the eigenfunction and by $\epsilon_j^{(i)}$ the associated eigenvalue with quantum numbers j, solutions of the "Kohn-Sham"-like equation for potential $V_{SIC}^{(i)}(\vec{r})$,

$$\left[\frac{-\hbar^2}{2m}\Delta + V_{SIC}^{(i)}(\vec{r})\right]\psi_j^{(i)}(\vec{r}) = \epsilon_j^{(i)}\psi_j^{(i)}(\vec{r})\,. \tag{2}$$

The results for the self-consistent solution of equations (1),(2) are shown in fig.1 for Na_{20}, in comparison with standard LDA results.

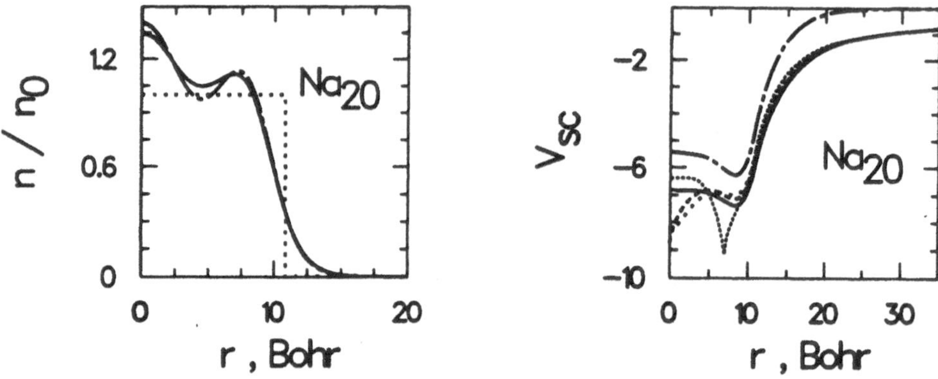

FIG 1. **LEFT:** The LDA radial density (in units of the jellium density n_0) is drawn with a bar dashed line, whereas the SIC-LDA density corresponds to the full drawn curve. **RIGHT:** The LDA mean field potential is drawn with a bar-dashed line; The other four curves correspond to the "1s", "1p", "1d" and "2s" orbital SIC-LDA potentials, drawn with full curve, thick dashed curve, dotted curve and thin dotted curve.

Although hardly discernible, the SIC-LDA density extends effectively more outside of the jellium edge, as can be inferred from the results of table 1 for the static polarizabilities.

N	$TDLDA$	$SIC - TDLDA$	EXP
8	1.41	1.52	1.72 $\pm$ 0.03
20	1.34	1.46	1.58 $\pm$ 0.04
40	1.30	1.41	1.56 $\pm$ 0.03

table 1. Static polarizabilities of Na_8 Na_{20} and Na_{40}, in units of $R_0^3 = Nr_s^3$ as calculated with TDLDA and SIC-TDLDA. EXP stands for the experimental values taken from ref.[5].

In the right panel, the four different potentials corresponding to the four orbital quantum numbers of occupied states in Na_{20} are significantly deeper than the single mean-field LDA potential and display a common, physically correct Coulomb tail, into which all merge somewhere outside of the jellium radius, in contrast with the tail of the LDA potential, determined essentially by V_{xc} in eq.(1).

3. Photoabsorption cross section

The standard expression for the LDA independent particle susceptibility reads[1]

$$\chi_0(\vec{r},\vec{r}\,';\hbar\omega) = \sum_i^{occ} \psi_i^*(\vec{r})\psi_i(\vec{r})[G(\vec{r},\vec{r}\,',\epsilon_i+\hbar\omega)+G^*(\vec{r},\vec{r}\,',\epsilon_i-\hbar\omega)] \ . \qquad (3)$$

The sum extends over all occupied single particle states. The Green's functions include all (infinite) electronic transitions from each of the occupied single particle states, and the 2 terms in square brackets ensure the exact cancellation of the transitions which explicitly violate the Pauli principle. In SIC-LDA, however, from each of the occupied single particle states, to which corresponds a specific single particle potential, there is an infinite number of possible electronic transitions, which should promote the electron into a state which is also SIC corrected. The natural choice is therefore to compute the orbitals into which the electron is to be promoted with the potential appropriate for the initial state of the electron[2, 6]. This, in turn, amounts to define orbital dependent Green's functions, in a straightforward correspondence with the originating orbital dependent potential. However, care must be exercised, so that the Pauli principle is not violated, which amounts to subtract explicitly the finite number of violating terms. This leads to,

$$\chi_0(\vec{r},\vec{r}\,';\hbar\omega) = \sum_i^{occ} \psi_i^{(i)*}(\vec{r})\psi_i^{(i)}(\vec{r}\,')\left[G^{(i)}(\vec{r},\vec{r}\,',\epsilon_i^{(i)}+\hbar\omega)-\sum_j^{occ}\frac{\psi_j^{(i)}(\vec{r})\psi_j^{(i)*}(\vec{r}\,')}{\epsilon_i^{(i)}+\hbar\omega-\epsilon_j^{(i)}+i\delta}\right]$$

$$+ \ \sum_i^{occ}\psi_i^{(i)}(\vec{r})\psi_i^{(i)*}(\vec{r}\,')\left[G^{(i)*}(\vec{r},\vec{r}\,',\epsilon_i^{(i)}-\hbar\omega)-\sum_j^{occ}\frac{\psi_j^{(i)*}(\vec{r})\psi_j^{(i)}(\vec{r}\,')}{\epsilon_i^{(i)}-\hbar\omega-\epsilon_j^{(i)}-i\delta}\right], \qquad (4)$$

the sums being restricted to the set of single-particle quantum numbers corresponding to occupied orbitals in the GS.

Once $\chi_0(\vec{r},\vec{r}\,';\hbar\omega)$ is determined the SIC-TDLDA susceptibility is now calculated as the solution of the following integral equation[7]:

$$\chi(\vec{r},\vec{r}\,';\hbar\omega) = \chi_0(\vec{r},\vec{r}\,';\hbar\omega) + \int d\vec{r}\,''d\vec{r}\,'''\chi_0(\vec{r},\vec{r}\,'';\hbar\omega)K(\vec{r}\,'',\vec{r}\,''')\chi(\vec{r}\,''',\vec{r}\,';\hbar\omega) \ , \qquad (5)$$

where the screening kernel reads (we do not perform SIC in $K(\vec{r},\vec{r}\,')$),

$$K(\vec{r},\vec{r}\,') = \frac{e^2}{|\vec{r}-\vec{r}\,'|} + \frac{dV_{xc}}{dn}\delta(\vec{r}-\vec{r}\,') \ . \qquad (6)$$

Due to the spherical symmetry of the clusters we consider here, the dipole polarizability can then be written(for details, cf. ref.[7]),

$$\alpha(\hbar\omega) = -e^2\frac{4\pi}{3}\int_0^{+\infty} dr\, r^3 \int_0^{+\infty} dr'\, r'^3 \chi_{L=1}(r,r',\hbar\omega)\,. \tag{7}$$

Finally, the photoabsorption cross section $\sigma(\hbar\omega)$ is simply related to the imaginary part of the polarizability by $\sigma(\hbar\omega) = (4\pi\omega/c)Im[\alpha(\hbar\omega)]$.

The calculated values for the static polarizabilities of Na_8, Na_{20} and Na_{40} are collected in table 1, together with the experimental values taken from ref.[5] and the TDLDA results. The SIC-TDLDA provides a systematic improvement with respect to TDLDA[8], towards the experimental values.

In fig.2 the results of the SIC-TDLDA dynamical response are depicted with solid lines; dashed lines show the previously calculated TDLDA results[7]. The SIC-TDLDA response is systematically red-shifted with respect to the TDLDA, in direct relation with the change in the static polarizabilities. Furthermore, the pattern of Landau fragmentation is not significantly changed from TDLDA to SIC-TDLDA, apart from the general trend of SIC-TDLDA to spread the collective response over a wider energy interval. In particular, one observes the systematic appearance of small humps in the ultraviolet(UV) part of the spectra, a feature which is directly related to the (now) physically correct Coulomb tail of the SIC-LDA potentials.

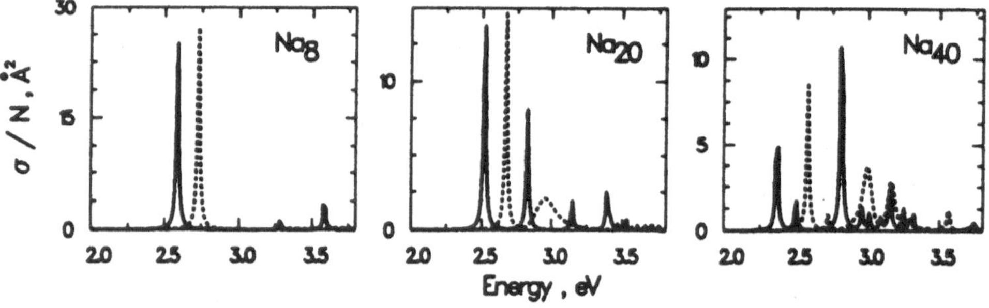

FIG 2. Photoabsorption cross sections per atom for Na_8, Na_{20}, and Na_{40}, in $\overset{\circ}{A}^2$, calculated making use of the SIC-TDLDA (full lines) and TDLDA(dashed lines). In all cases the value used for the infinitesimal delta (cf. eq.(4) and ref.[7]) was 10 meV.

In sharp contrast with the LDA case, the SIC-LDA potentials accomodate an infinite number of bound states. However, and because their overall depth (see fig.1) is significantly lowered, the wave-functions of the occupied states are very similar. Remaining well localized, the occupied states will not overlap significantly with the loosely bound Rydberg states, preserving the essential featues of the Landau fragmentation pattern. Note finally, that since the SIC-TDLDA conserves the Thomas-Reiche-Kuhn (TRK) sum rule the occurrence of the UV humps implies a reduction of total strength in the visible region.

Comparison with experiment is carried out in fig.3 for Na_8, for which there is the most detailed experimental information. To this end, the damping mechanisms discussed in ref.[9] and not present in SIC-TDLDA were simulated by folding the results of fig.2 with normalized Lorentzian functions, including the appropriate damping ratio ($\Gamma/\hbar\omega = 0.1$, in accord with the findings in ref.ref.[9]).

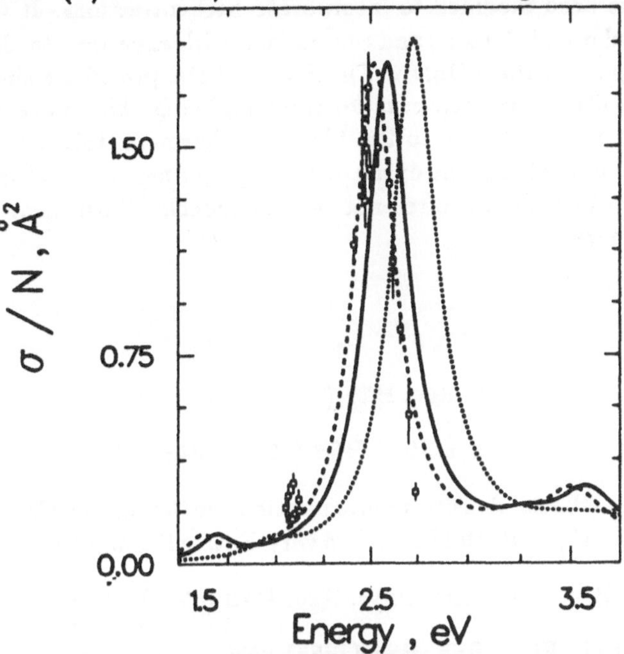

FIG 3. Experimental values for the photoabsorption cross section per atom for Na_8, in $\overset{\circ}{A}^2$, taken from ref.[10] are drawn with small squares. The results of the present SIC-TDLDA calculation correspond to the full drawn curve. The dashed curve corresponds to red-shifting the SIC-TDLDA results by 3% so as to bring the calculated plasmon peak to its most probable position. Finally, the thin dotted curve represents the TDLDA result for Na_8.

The SIC-TDLDA results (full line) display a better agreement with experiment than TDLDA (dotted line). They are, however, $\approx$ 3% blue-shifted with respect to the experimental peaks. For comparing the folded SIC-TDLDA lineshapes with the experimental results, we purposely shifted the dynamical response by 3%, the resulting curve being displayed with a dashed line. While the overall features of the data seem to emerge from the present formalism, we note that the SIC-TDLDA predicts a sizeable acummulation of strength in a region of the absorption spectrum not yet measured. It would be interesting to resolve this part of the spectrum experimentally, in order to confirm whether or not the sum rule strength which seems to be missing in present-day experimental results, is indeed to be found in the UV region of the spectrum, as predicted by the SIC-TDLDA.

4. Conclusion

The TDLDA has been modified to incorporate SIC corrections. It was applied specifically to neutral metal clusters, and comparison with experimental data favours this model with respect to the TDLDA. The SIC-TDLDA provides a simple framework which requires little extra computational effort, being able to overcome many of the unpleasant features intrinsic to the LDA. Furthermore, this new framework opens the possibility of studying the excitation spectra of negatively charged many-electron systems, to which LDA is often not even applicable. Work along these lines will be published elsewhere.

References

[1] A. Zangwill, P. Soven, Phys. Rev. **B21** (1980) 1561;

[2] J. P. Perdew, A. Zunger, Phys. Rev. **B23** (1981) 5048;

[3] J. P. Perdew, in "Local density approximations in Quantum Chemistry and Solid State Physics", J. P. Dahl and J. Avery Edts.,(Plenum 1982), pag. 173;

[4] O. Gunnarsson, B. I. Lundqvist, Phys. Rev. **B13** (1976) 4274;

[5] W. D. Knight et al., Phys. Rev. **B31** (1985) 2539;

[6] W. J. Hunt, W. A. Goddard, Chem. Phys. Lett. **3** (1969) 414;

[7] W. Ekardt, Phys. Rev. Lett. **52** (1984) 1925; Phys. Rev. **B31** (1985) 6360;

[8] Results for static polarizabilities, making use of the modified Sternheimer equation and using a different SIC model have been reported before in P.Stampfli and K. H. Benemann, Phys. Rev. **B39** (1989) 1007;

[9] J. M. Pacheco, R. A. Broglia, Phys. Rev. Lett. **62** (1989) 1400.

[10] K. Selby et al., Phys. Rev. **B43** (1991) 4565;

NONLINEAR OPTICAL TRANSMISSION OF NANOMETRIC GOLD COLLOIDS AND ITS DEPENDENCE ON PARTICLE SIZE

K.Puech, F.Z.Henari and W.Blau,
Department of Pure and Applied Physics,
Trinity College,
Dublin 2, Ireland.

D.J.Cardin,
Chemistry Department,
Trinity College,
Dublin 2, Ireland.

D.Duff * and P.P.Edwards#
University Chemical Laboratory,
Lensfield Road, Cambridge CB2 1EW
England.
*Now at, Technisch-Chemisches Labor.,ETH-Zentrum, CH-8092 Zurich, Switzerland.
#Now at Department of Chemistry, University of Birmingham, Birmingham, England.

ABSTRACT. Nanometric gold colloids in water, varying in size from 20-4000Å possess a large third order nonlinear optical response at wavelengths of 612nm and 651nm. Near the surface plasmon resonance frequency it is reported to be very strongly resonantly enhanced[1,2]. We have performed experiments to determine values for the imaginary part of the nonlinear susceptibility. They are found to be of the orders $10^{-17}m^2/V^2$ (10^{-9} esu). According to the free electron theory $\chi^{(3)}$ should vary as the inverse third power of the particle radius[2]. In our experiments, however, we find no relationship between nonlinearity and particle size. The experimental curves are fitted using a theoretical rate equation model. From this we can obtain the excited state properties of the particles quantitatively.

INTRODUCTION

The nonlinear optical properties of small metal and semiconductor particles, in a dielectric medium, are of fundamental interest. These particles are three dimensional quantum confined electron systems. Materials such as gold colloids in water have been studied on earlier occasions[3]; experiments were performed at the plasmon resonance frequency of the colloids and it is found that the nonlinear response is very strongly enhanced at this frequency[2].

P. Jena et al. (eds.), Physics and Chemistry of Finite Systems: From Clusters to Crystals, Vol. II, 989–993.
© 1992 Kluwer Academic Publishers.

In this letter we present results of nonlinear optical measurements on gold colloids at wavelengths of 612nm and 651nm, in a range of samples of greatly varying size.

It has been reported that the width of the absorption band of such colloids, which is roughly proportional to the imaginary part of the susceptibility, increases with a decrease in particle size[2,3]. This effect was accounted for classically in terms of a limited mean free electron path and quantum mechanically in terms of the quantum size effect.

In our experiments we found no dependence of width of absorption band on nonlinearity .

EXPERIMENTAL SET-UP

Most of the colloidal sols were prepared according to the method of Frens[4] - reduction of chloroauric acid solution by sodium citrate, which can produce the most monodisperse form of divided metal known in the colloidal size regime.

All the gold hydrosols were additionally concentrated by removal of solvent (water) before measurement of NLO properties. For the 15 Å sol this was performed under vacuum without addition of polymer, room temperature, to minimise sol coarsening. For all the other sols, PVP_{40K} was added, to a concentration of about 2g per millimole of metal atoms, before boiling down using a hotplate, either in a stream of Nitrogen or otherwise in air.

An example of a high resolution transmission electron microscope micrograph of a typical sample is depicted in figure 1

Figure 1: TEM Micrograph of a 167Å gold colloid

Nonlinear Optical measurements were carried out using the optical saturation method.The experimental apparatus is as shown in figure 2[5,6].

Figure 2: Experimental setup for NLO measurements

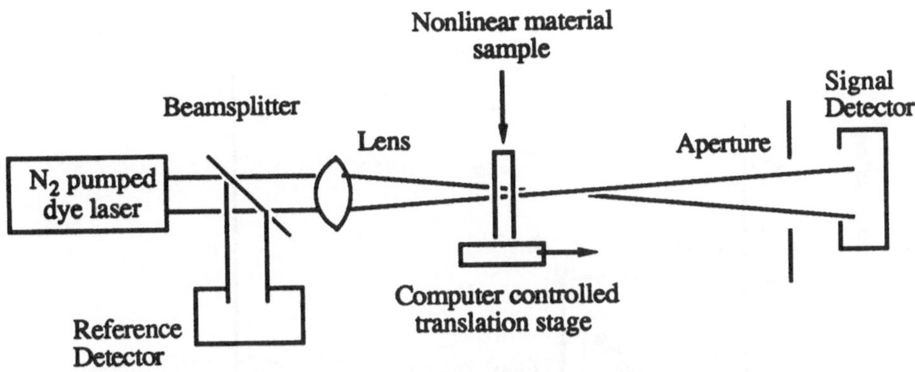

Using a Gaussian beam in a tight focus limiting geometry, we measured the transmission of the gold colloids as a function of their position, measured with respect to the focal plane. From such a trace one can determine the ground state and the excited state absorption cross sections and also Isat, the intensity required to saturate the ground state. For a three level system in the steady state, the imaginary part of the susceptibility is given by,

$$\chi^{(3)} = \frac{n_0 \, \varepsilon_0 \, C' \, \lambda \, N_A \, C \, (\sigma_{ex} - 2\sigma_0)}{16\pi \, I_{sat}}$$

ε_0 = Permittivity of free space C' = concentration of the solution

C = Velocity of light λ = Wavelength N_A = Avogadros no.

n_0 = Refractive index of sample σ_0 = Ground state absorption cross section

σ_{ex} = Excited state absorption cross section

RESULTS AND DISCUSSION

We measured a range of samples varying in size from 15Å to 4000Å. All the colloids exhibit bleaching with increasing intensity. A typical saturation curve is shown in figure 3.

Figure 3: Intensity dependent transmission of 315Å gold colloid at 651nm

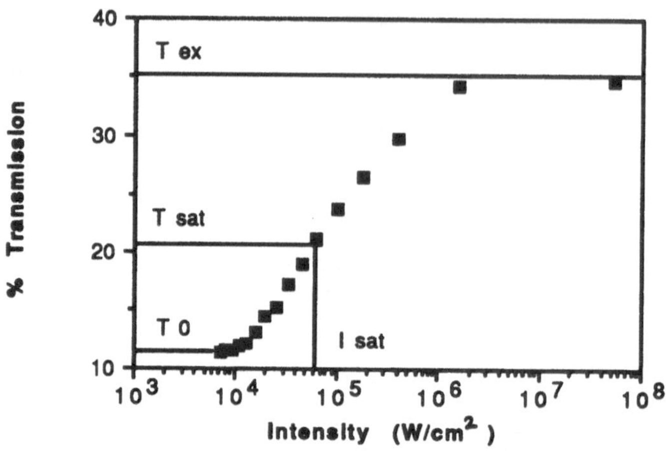

Using the values for σ_0 and σ_{ex} obtained from such a curve we calculated values for the imaginary part of the susceptibility for each of the colloids tested. It can be seen that there is no dependence between $\chi^{(3)}$ and particle size. This is illustrated in figure 4 where $\chi^{(3)}$ measurements are plotted against the diameter of the of the particle at a wavelength of 612nm.

Figure 4: Dependence of $\chi^{(3)}$ on particle size

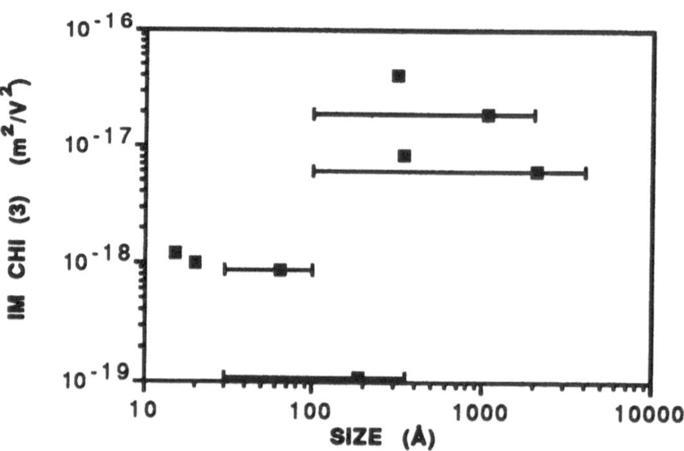

The results at a wavelength of 651 nm are summarized in Table 1.

Table 1 : Overview of experimental results

Size (Å)	σ_0 (cm^2)	I_{SAT} (W/cm^2)	$\chi^{(3)}$ (m^2/V^2)
100-4000	1.396×10^{-17}	8×10^6	-6.14×10^{-18}
100-2000	2.180×10^{-17}	6×10^6	-1.90×10^{-17}
315	2.383×10^{-17}	4×10^6	-4.09×10^{-17}
30-100	2.005×10^{-17}	1.5×10^7	-8.83×10^{-19}
20	2.250×10^{-17}	1.4×10^7	-1.01×10^{-18}
341	2.865×10^{-17}	2.5×10^7	-8.36×10^{-18}
30-350	5.929×10^{-18}	8.5×10^7	-1.11×10^{-19}
15	2.302×10^{-17}	3×10^7	-1.21×10^{-18}

Thus, we conclude from this that the free electron theory for electrons in a quantum box is invalid in its present state.

REFERENCES

1. Ricard, D., Roussignol, P. and Flytzanis, C. (1985) " Surface mediated enhancement of Optical phase conjugation in metal colloids ", Optics Letters 10, 511-513.
2. Hache,F., Ricard, D. and Flytzanis, C. (1986) " Optical nonlinearities of small metal particles: surface mediated response and Quantum size effects ", J. Opt. Soc. Am. B 3, 1647-1654.
3. Doremus, R.H. (1964), J. Chem. Phys. 40, 2389-
4. Frens, G. (1973), Nature Phys. Sci. 241, 20-22.
5.Sheik-bahae, M., Said, A. and Van Stryland, E.(1989) " High Sensitivity, Single Beam n_2 measurements ", Optics Letters 14, 955-957
6. Sheik-bahae, M., Said, A.,Wei, T., Hagan, D. and Van Stryland, E. (1990) " Sensitive measurement of Optical Nonlinearities Using a Single Beam", IEEE Journal of Quantum Electronics 26, 760-769.

DEPENDENCE OF THE PLASMON FREQUENCY OF CLUSTER EMBEDDED IN A MATRIX VERSUS THE CLUSTER SIZE

R.J. TARENTO[1], P. JOYES[2], J. VAN DE WALLE[2]

[1] CNRS Laboratoire de Physique des Matériaux, 1 Place Aristide Briand, 92195 Meudon Cédex, France.

[2] Laboratoire de Physique des Solides,Bât. 510, Université Paris–Sud, 91405 Orsay Cédex (France)

ABSTRACT

In the present contribution, by using the hydrodynamic model proposed by Bloch, we have derived the dependence versus the size of the plasmon frequencies for the case of the coupling between the bulk (surface) aggregate plasmon and the surface matrix plasmon.

1. INTRODUCTION

The study of cluster has attracted increasing works in the latter years. The phenomenon is the consequence of the great interest for aggregates ; either for industrial applications (i.e. in chemistry : catalyst...) or for fundamental point of view. But nevertheless some buffling problems have been pointed out for a long time and have not received a satisfactory answer. One of them deals with the cluster photoaborption spectra and the link with the plasmon frequency. Since the pioneer calculation done by Mie, some theoretical works have been devoted to the subject of plasmon in aggregate. For small clusters (radius < 40 Å) quantum effects have not to be neglected, however for large clusters, a classical treatment is quite accurate. Let us pay some attention to the case of free aggregate. For the onedimensional case (plane surface), assuming a zero electric field for the unperturbated system, Ritchie [1,2] has derived the dispersion equation. From this dispersion equation (established for an infinite surface) one may deduce the plasmon frequency of aggregates and the trend is that the larger the cluster the smaller the plasmon frequency. Bennett [2,3] has implemented the previous solution, i.e. the positive background is not locally compensated by the electronic charge within a small layer and therefore an

P. Jena et al. (eds.), Physics and Chemistry of Finite Systems: From Clusters to Crystals, Vol. II, 995–999.
© 1992 *Kluwer Academic Publishers.*

electric field is created : for larger cluster Bennett's results are in agreement with Ritchie's one, but however for smaller clusters his conclusion is the reverse. An alternative method has been applied by Ruffin [4] to the case of spherical aggregates by using the hydrodynamic model proposed by Bloch, and the results are in agreement with Ritchie's ones. However experimental results, both photoabsorption and electron loss spectra confirm Bennett calculations. Recently the case of aggregate in a matrix has attracted a lot of works, chiefly linked to the spectroscopy of electronic loss energy. The situation is less trivial. What is sure, is the importance of the neighbourhood on the behaviour. One problem which worths to be studied is the coupling between the matrix and the aggregate. Such coupling has been noticed in particular with gas bubble in Aluminium [5]. The present article examines two cases of coupling (i.e. the coupling between the bulk or surface aggregate plasmon and the surface matrix plasmon. The first part deals with the basic principle of the hydrodynamics model.

2. HYDRODYNAMICS MODEL OF PLASMON MODES.

The present work is based on the Bloch works who proposed a hydrodynamic model to study the state of a continuous electron fluid. In this picture the hydrodynamic pressure of the electrons is responsible for the spatial dispersion i.e. : for the variation of the mode frequencies with the wave vector. Let assume that the fluid has a charge density $e(n + \Delta n)$ (and mass density $m(n + \Delta n)$) ; the fluid is moving in a positive immobile background of charge density - en. Let $\xi(r)$ be the displacement from equilibrium. With respect to the fundamental equations of dynamics and the Poisson equation, the excess of charge Δn and the displacement $\xi(r)$ are linked :

$$\Delta n = - n \operatorname{div} \xi \tag{1}$$

If we consider the electronic fluid as a classical fluid, then the deviation ΔP of the hydrodynamic pressure from its equilibrium value is :

$$\Delta P = - n\, m\, \beta^2 \operatorname{div} \xi \tag{2}$$

where β would be the sound velocity if the medium were neutral. β gives the strength of the dispersion. For the bulk, β is given by :

$$\beta = \sqrt{\frac{2}{3}}\, V_F$$

Let E be the electric field inside the plasma zone, it derives from the electric potential χ, therefore the Poisson's equation is equal to :

$$\nabla^2 \Phi = 4\pi\, ne \operatorname{div} \xi \tag{4}$$

Within the present frame, and for normal modes of frequency Ω, the equation of motion is :

$$- \Omega^2 m\, \xi = - \operatorname{grad} (e\chi - m\, \beta^2 \operatorname{div} \xi) \tag{5}$$

In fact instead of working with the displacement ξ, it is classical to introduce a displacement potential χ :

$$\xi = - \, \mathrm{grad} \; \chi \tag{6}$$

Hence the relation between the electronic and displacement potential :

$$\Phi = - \left(\frac{m}{e}\right)\left[\Omega^2 + \beta^2 \, \nabla^2\right]\chi \tag{7}$$

Operating on equ.(7) with the operator ∇^2, we obtain the following tractable equation :

$$\nabla^2(\Omega^2 - \omega_p^2 + \beta^2 \, \nabla^2 \, \chi) = 0 \tag{8}$$

where ω_p is the bulk plasmon frequency ($\omega_p = \left(\frac{4\pi n_0 e^2}{m}\right)^{1/2}$). In our calculation two zones are considered :

– the spherical cluster zone described by β_1 and the bulk plasmon frequency ω_{p1},
– the matrix zone with a bulk plasmon frequency ω_{p2} and the parameter β_2.

The complete determination of the system need to specify the boundary equations. The first condition is the continuity both of the displacement and of the electrical potential and their derivatives at the cluster–matrix interface (BC condition). Mathematicaly to solve the problem need to know the solution of equ.(8). The displacement potential $\chi|r,\theta,\varphi|$ is for each zone a linear combination of solutions of equations :

$$\nabla^2\chi = \lambda\chi \;\; \text{with} \;\; \lambda = 0 \;\; \text{or} \;\; \lambda = \frac{\omega_p^2 - \Omega^2}{\beta^2} \tag{9}$$

Owing to the spherical symmetry, the eigenfunctions are researched as :

$$\chi(r,\theta,\varphi) = \sum_\ell f_\ell(r) \, Y_{\ell,m}(\theta,\varphi) \tag{10}$$

where $Y_{\ell,m}(\theta,\varphi)$ is the spherical harmonics and $f_\ell(r)$ depends on the orbital number ℓ and verifies the differential equation :

$$\left(\frac{d^2}{dr^2} - \frac{\ell(\ell+1)}{r^2} - \lambda\right)(r \, f_\ell(r)) = 0 \tag{11}$$

Let us set $\lambda = \pm \, k^2$ (+ for surfacewise plasmon, – for bulkwise plasmon). The classical solutions of (11) are :

998

$$\lambda = 0 \qquad f_\ell(r) = r^\ell \text{ or } r^{-\ell-1}$$

$$\lambda = -k^2 \quad f_\ell(r) = \sqrt{\frac{\pi}{kr}}\, J_{\ell+1/2}(kr) = j_\ell(kr) \text{ or } \sqrt{\frac{\pi}{kr}}\, J_{-\ell-1/2}(kr) = \eta_\ell(kr) \qquad (12)$$

$$\lambda = +k^2 \quad f_\ell(r) = \sqrt{\frac{\pi}{kr}}\, I_{\ell+1/2}(kr) = i_\ell(kr) \text{ or } \sqrt{\frac{\pi}{kr}}\, K_{\ell+1/2}(kr) = k_\ell(kr)$$

where J, K, I are the usual Bessel functions.

This displacement potential $\chi(r,\theta,\varphi)$ have to be expanded over the solutions of (12) for both zones. Two cases have been investigated, the plasmon coupling between the surface matrix plasmon and the bulk or surface aggregate plasmon. The BC conditions lead to the following relations to determine the plasmon frequency Ω of the system coupling between the bulk surface aggregate plasmon and the surface matrix plasmon :

$$\begin{vmatrix} \Omega^2 R^\ell & \omega_{P1}^2 f_\ell(k_1 R) & -\Omega^2 R^{-\ell-1} & -\omega_{P2}^2 k_\ell(k_2 R) \\[2mm] \Omega^2 \ell R^{\ell-1} & \omega_{P1}^2 k_1 \dfrac{\partial f_\ell(k_1 R)}{\partial r} & \Omega^2(\ell+1) R^{-\ell-2} & -\omega_{P2}^2 k_2 \dfrac{\partial k_\ell(k_2 R)}{\partial r} \\[2mm] \ell R^{\ell-1} & k_1 \dfrac{\partial f_\ell(k_1 R)}{\partial r} & (\ell+1) R^{-\ell-2} & -k_2 \dfrac{\partial k_\ell(k_2 R)}{\partial r} \\[2mm] \ell(\ell-1)R^{\ell-2} & k_1^2 \dfrac{\partial^2 f_\ell(k_1 R)}{\partial r^2} & -(\ell+1)(\ell+2)R^{-\ell-3} & -k_2^2 \dfrac{\partial^2 k_\ell(k_2 R)}{\partial r^2} \end{vmatrix} = 0 \qquad (13)$$

where R is the aggregate radius, and f_ℓ are the function $j_\ell(i_\ell)$ respectively in the bulk (surface) aggregate plasmon case.

3. Application to the case of a spherical potassium cluster in a sodium matrix.

The model developed in the part 2 has been investigated for the case of a spherical potassium cluster (i.e. $\omega_{P1} = 6.08 \ 10^{15}$ Hz, $\beta_1 = 0.70 \ 10^6$ m/s) embedded in a sodium matrix (i.e. $\omega_{P2} = 9.42 \ 10^{15}$ Hz, $\beta_2 = 1.05 \ 10^6$ m/s). The figures 1(2) display the plasmon frequency dependence versus the aggregate size respectively for the coupling between the matrix plasmon and the surface (volume) plasmon for $\ell = 2$. The results can be understood as the following. For large R the two media are equivalent and the matrix and the aggregate behave like two coupled oscillators whose the eigen frequencies are Ω_1 and Ω_2, so the frequencies are nearly $\dfrac{\Omega_1 + \Omega_2}{2}$ and $\dfrac{|\Omega_1 - \Omega_2|}{2}$ where Ω_1 is equal to $\omega_{P1}\left(\dfrac{\omega_{P1}}{\sqrt{3}}\right)$ for the volume (surface) aggregate case and Ω_2 is equal to $\dfrac{\omega_{P2}}{\sqrt{3}}$. For smaller cluster, the coupling is weaker and when the radius goes to zero, the plasmon frequencies reaches Ω_2 which can be understood that for R smaller than a characteristic length the aggregate plasmon cannot be excited. These two limit cases (large and small R) are observed in our solution. The article gives some insight for the determination of an accurate microanalysis by plasmon energy loss spectroscopy.

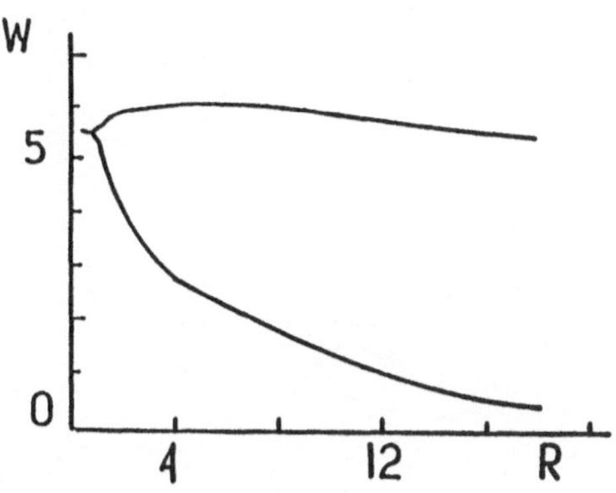

Figure 1. Plasmon frequency $\omega(\times 10^{15}$ Hz) dependence versus the size R (nm) given by the coupling between the bulk aggregate plasmon and the surface matrix plasmon.

Figure 2. Plasmon frequency $\omega(\times 10^{15}$ Hz) dependence versus the size given by the coupling between the surface aggregate plasmon R(nm) and the surface matrix plasmon.

References

[1] R.H. Ritchie, Progr. Theo. Phys. 29 (1963) 607.

[2] P. Joyes, Les agrégats inorganiques élémentaires : edited by "Les Editions de Physique" (1990).

[3] A.J. Bennett, Phys. Rev. B1 (1970) 203.

[4] R. Ruffin, J. Phys. Chem. Sol. 39 (1978) 233.

[5] P. Henoc and L. Henry, J. Phys. C1-31 (1970) 55.

Collective Excitations in Silver Cluster Ions

J. Tiggesbäumker, L. Köller, H.O. Lutz, K.H. Meiwes-Broer
Fakultät für Physik, Universität Bielefeld, 4800 Bielefeld 1, F.R.G.

Abstract: Silver cluster ions are produced by sputtering of solid silver by 25 keV Xe^+ bombardment. After mass selection, collinear ion beam depletion technique is used to measure absolute photofragmentation cross-sections in the photon energy range from $\hbar\omega = 2.3$ to 5.7 eV. Giant resonances are found which can be interpreted in terms of a collective electron oscillation. The optical spectra of spherical Ag_9^+ and Ag_{21}^+ show a blue shift of the resonance energy, in contrast to an expected behaviour within the jellium model. For Ag_{11}^+, a splitting of the giant resonance is found.

Introduction

The pioneering theoretical work of Ekardt[1] and experimental results of de Heer et al.[2] revealed that the optical activity of very small metal particles with only about ten atoms can — to a first approximation — be understood as a simple collective excitation of the valence electrons. Classically, such an excitation is equivalent to the first excitation mode in the Mie theory. It has its counterpart in the giant dipole resonance of nuclear physics[3]. Very recently, optical spectra of free neutral and charged alkali clusters showed that only a few atoms are needed to establish a giant resonance[4-8].

In this contribution we concentrate on the optical response of silver clusters. Plasmon excitations have been found in silver particles of the size 10–100 nm in the near UV-region[9-15]. Red- as well as blue-shifts produced some confusion in the past (for details see ref. 14). Up to now no such experiments have been done on mass selected silver clusters in the gas phase. Here we present optical spectra of the closed-shell clusters Ag_9^+ and Ag_{21}^+ and the open-shell cluster Ag_{11}^+.

P. Jena et al. (eds.), Physics and Chemistry of Finite Systems: From Clusters to Crystals, Vol. II, 1001–1006.
© 1992 *Kluwer Academic Publishers.*

Experimental setup

The experimental setup for production of mass selected metal cluster ions is described in detail elsewhere[16]. In short: Clusters are generated by bombarding a silver target with 25 keV Xe^+. After acceleration to 1.8 keV and mass selection by a Wien filter, a monodispersed cluster ion beam is subject to experimental investigation. Typical count rates without depletion are $5 \cdot 10^4$/sec for Ag_9^+, and $1.2 \cdot 10^4$/sec for Ag_{11}^+ and Ag_{21}^+, respectively. Two collimators (1 mm diameter, distance 10 cm) confine the ion beam and define the interaction region. Here the clusters are irradiated collinearly with light from a pulsed excimer-pumped dye laser (Lambda Physik LPX 110, FL2002). In order to extend the photon energy region, second harmonic generation is achieved via frequency-doubling in BBO1 and BBO2. Additionally, excimer laser lines (248 nm, 308 nm) are used. Behind the interaction region the cluster beam is steered out of the laser beam axis with a static quadrupol ion deflector which acts as an energy selector field. Photodissociated clusters therefore cannot reach the detector. Additionally, a grid can be used to repell cluster fragments. Care is taken to get optimal overlap of cluster and laser beam. A pyroelectrical detector serves to measure the laser intensity (typically $0.05 - 1.5$ mJ/cm^2 per pulse) before and after each run. Using this technique, depletion spectra are recorded for different photon energies $E = \hbar\omega = 2.3 \ldots 5.7$ eV. Beer's law is then used to extract absolute photofragment cross sections.

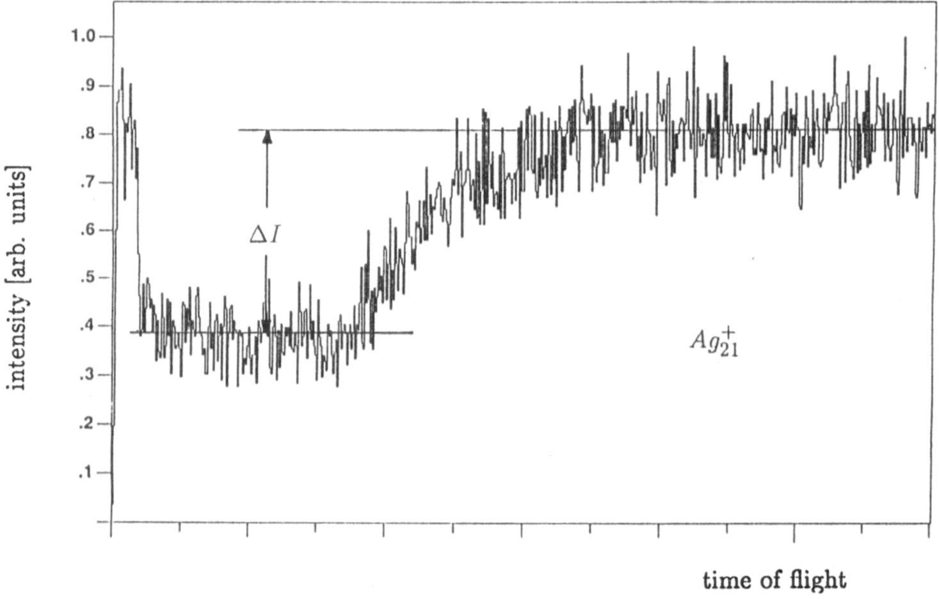

Fig. 1: Ion depletion spectrum of Ag_{21}^+ as illustration of a typical experimental signal. At a photon energy of 3.7 eV and 250 μJ/cm^2 per pulse, a fragmentation cross section of 15 $\mathring{A}^2$ is calculated for this ΔI.

Results and Discussion

Figure 2 compiles the photofragmentation cross sections of Ag_9^+, Ag_{11}^+ and Ag_{21}^+. In the case of 8 and 20 electron systems, single peaks dominate the spectra. As is indicated by the error bars, there might possibly be further structure which is not resolved. Additionally, a systematic trend is found for an increased "background" signal at higher photon energies. Nevertheless, there is no doubt about the existence and overall shapes of these giant resonances. In the case of Ag_9^+ the peak maximum is shifted towards higher energies (blue shift) with respect to Ag_{21}^+. The linewidth of Ag_{21}^+ appears to be smaller than that of Ag_9^+. The giant resonances are well fitted by single Lorentzians.

The classical Mie theory[17] assumes that all of the oscillator strength is exhausted by the surface plasma resonance; i.e., no single particle excitations are regarded, the system is treated as ideal metallic particle. In the limit of very small particles, only dipole exitations are supposed to contribute to the optical activity and induce simple absorption profiles in a relatively narrow energy range. In this situation, dipole sum rules can be used[3], which yield a linear dependence of the integrated photoabsorption cross section on the number of valence electrons in the cluster. The oscillator strength OS (normalized to N_e) derived from our measurements are 0.94 for Ag_9^+ and 0.64 for Ag_{21}^+, thus showing that most of OS is indeed exhausted by the giant resonance. At the same time, we observe a trend of an apparently decreasing oscillator strength with increasing cluster size. This might be due to the experimental method used; note that we measure photofragmentation instead of photoabsorption. Previous experiments showed that clusters can store a significant amount of energy for a long time[18]. The storage time increases with N and may lead to a decrease of the measured apparent photoabsorption cross section in the experimental time window. On the other side, the decrease of OS in silver might also be due to the excitation of a s-surface plasmon which should be located outside our spectral range. In this case its excitation probability would increase with N. Finally, a thorough consideration by, e.g., RPA, or CI calculations might lead to the conclusion that not one single state exhausts all the oscillator strength but that there are further significant contributions outside the main line.

In contrast to the 8- and 20-electron systems, the optical spectrum of Ag_{11}^+ (10 valence electrons) shows two well-separated peaks at $\hbar\omega_{01}=3.44$ eV and $\hbar\omega_{02}=4.22$ eV, see Fig. 2. Such splitting arise from collective excitations in deformed clusters where — in the simplest case of an ellipsoid — two peaks are expected according to different resonance energies for the two main axes. Within the ellipsoidal shell model[4,19] a 10-electron system undergoes prolate deformation which leads to an intensity distribution of 1/3 and 2/3 for the low and high energy peaks, respectively. For Ag_{11}^+ we find from the integrals of the two Lorentzians an intensity ratio of 0.24 : 0.76. The discrepancy from the calculated values might originate from shape isomerism due to the high internal energy[20]. We find a deformation parameter[21] of 0.33 for Ag_{11}^+.

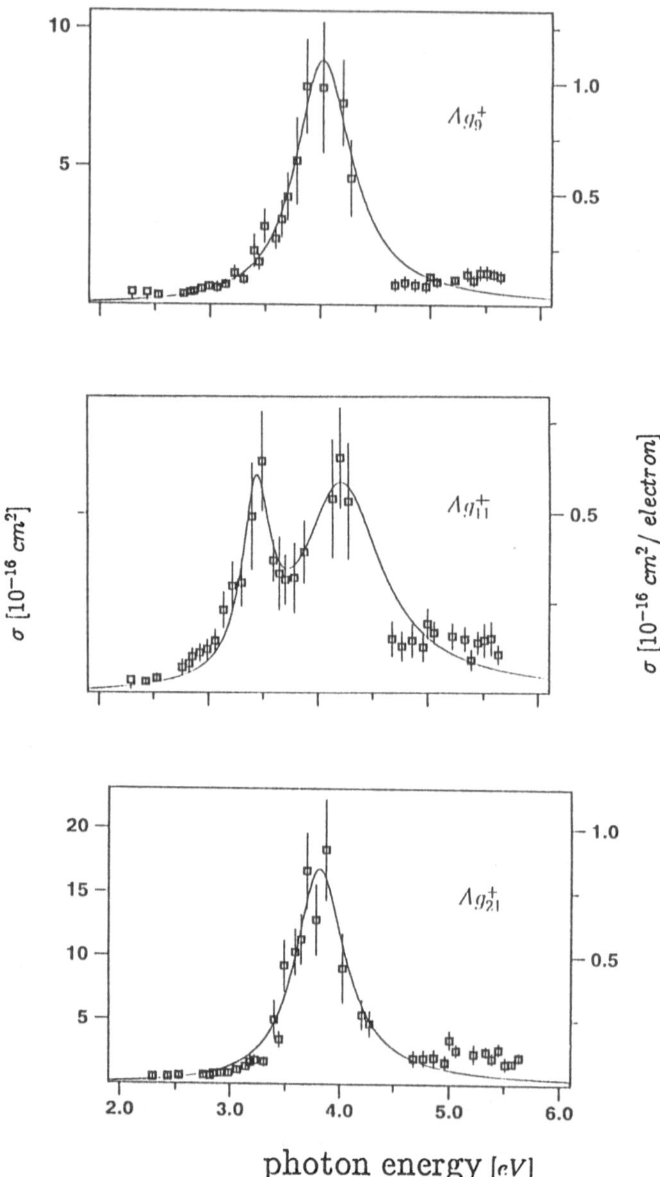

Fig. 2: Absolute photofragmentation cross section profiles of sputtered silver cluster ions. Lorentz curves are used to fit the experimental values. Fit parameters for:

Ag_9^+: $\hbar\omega_0 = 4.02$ eV, $\Gamma = 0.62$ eV, $\sigma = 8.84$ Å^2

Ag_{11}^+: $\hbar\omega_{01} = 3.44$ eV, $\Gamma_1 = 0.30$ eV, $\sigma_1 = 4.93$ Å^2, $\hbar\omega_{02} = 4.22$ eV, $\Gamma_2 = 0.87$ eV, $\sigma_2 = 5.65$ Å^2

Ag_{21}^+: $\hbar\omega_0 = 3.82$ eV, $\Gamma = 0.56$ eV, $\sigma = 16.79$ Å^2

For the energy of the giant resonance $\hbar\omega_0$, we first consider the optical activity of metallic particles as described by the Mie theory. The position of the dipolar plasma resonance $\hbar\omega_{Mie}$ of a metallic sphere in vacuum can be calculated in the quasistatic approximation[17] and is given by

$$\varepsilon(\omega_{Mie}) - 2 = 0 \tag{1}$$

with $\varepsilon(\omega)$ the dielectric function of the metal with bulk plasmon energy $\hbar\omega_p$. Note that for free electron metals this condition is equivalent to

$$\hbar\omega_{Mie} = \frac{\hbar\omega_p}{\sqrt{3}} \quad , \quad \omega_p^2 = \frac{4\pi n e^2}{m} \tag{2}$$

with n the electron density, m the electron mass. In terms of the static polarizablity α one finds:

$$\omega_{Mie}^2 = \frac{Ne^2}{m\alpha} \tag{3}$$

with $\alpha = Nr_s^3$, r_s the Wigner Seitz radius.

Eqs. 2 and 3 are commonly used to calculate ω_{Mie}, yielding $\hbar\omega_{Mie} = 5.2$ eV for silver. The value found with optical data[22] and eq. 1 ($\hbar\omega_0 = 3.5$ eV), however, differs significantly from that of the free electron model (eq. 2,3). The deviation is thought to originate from the d-electrons in silver.

For decreasing size, the influence of the electron spill-out increases α, thus leading to a red shift of the resonance energy, as is calculated in the jellium model[1,23]. This prediction appears to be in rough agreement with measurements on alkali clusters. For Ag_N^+ we find a clear blue-shift which cannot be explained in the jellium model in its present form.

The semiempirical approach from Apell et al.[24] considers a smooth variation of the dielectric function in the surface region and the non-local response of the electrons. A new quantity in units of a length is introduced which measures the center of gravity of the induced surface charge. The contributions from d-electrons are accounted for within this model through the optical data of silver[22]. We find values of 3.93 eV for Ag_9^+ and 3.83 eV for Ag_{21}^+ in excellent agreement with the observed data. A complete discussion of this approach and the overall blue shift found in all investigated clusters up to Ag_{70}^+ will be given in a further publication[25].

In conclusion, giant resonances govern the optical spectra of small silver clusters ions. The observed spectra directly reveal the mean cluster shape. For the resonance frequencies a semiempirical approach of Apell et al. and the use of the optical constants of silver turn out to yield values close to the measured ones. A satisfying and more detailed description based, e.g., on the jellium model, has still to be developed.

Acknowledgement

Financial support by the Deutsche Forschungsgemeinschaft is gratefully acknowledged.

References

[1] W. Ekardt, Phys. Rev. B 31 (1985) 6360
[2] W.A. de Heer, K. Selby, V. Kresin, J. Masui, M. Vollmer, A. Châtelain, W.D. Knight, Phys. Rev. Lett. 59 (1987) 1805
[3] J. Specht, A. v. d. Woude, Rep. Prog. Phys. 44 (1981) 719
[4] K. Selby, M. Vollmer, J. Masui, V. Kresin, W.A. de Heer, W.D. Knight, Phys. Rev. B 40 (1989) 5417
[5] C.R.C. Wang, S. Pollack, J. Hunter, G. Alameddin, T. Hoover, D. Cameron, S. Liu, M.M. Kappes, Z. Phys. D 19 (1991) 13
[6] H. Fallgren, T.P. Martin, Chem. Phys. Lett. 168 (1990) 233
[7] C. Bréchignac, Ph. Cahuzac, F. Carlier, J. Leygnier, Chem. Phys. Lett. 164 (1989) 433
[8] J. Blanc, M. Broyer, J. Chevaleyre, Ph. Dugourd, H. Kühling, P. Labastie, M. Ulbricht, J.P. Wolf, L. Wöste, Z. Phys. D 19 (1991) 7
[9] J.D. Eversole, H.P. Broida, Phys. Rev. B 15 (1977) 1644
[10] Y. Borensztein, P. de Andrès, R. Monreal, T. Lopez-Rios, F. Flores, Phys. Rev. B 33 (1986) 2828
[11] L. Genzel, T.P. Martin, U. Kreibig, Z. Phys. B 21 (1975) 339
[12] K.-P. Charlé, W. Schulze, B. Winter, Z. Phys. D 12 (1989) 471
[13] J.R. Heath, Phys. Rev. B 40 (1989) 9982
[14] U. Kreibig, L. Genzel, Surf. Sci. 156 (1985) 678
[15] A. Henglein, Chem. Phys. Lett. 154 (1989) 473
[16] W. Begemann, K.H. Meiwes-Broer, H.O. Lutz, Z. Phys. D 3 (1986) 183
[17] G. Mie, Ann. d. Phys. 25 (1908) 377
[18] W. Begemann, K.H. Meiwes-Broer, H.O. Lutz, Phys. Rev. Lett. 56 (1986) 2248
[19] K. Clemenger, Phys. Rev. B 32 (1985) 1359
[20] G. Lauritsch, P.G. Reinhard, J. Meyer, M. Brack, to be published
[21] E. Lipparini, S. Stringari, Z. Phys. D 18 (1991) 193
[22] P.B. Johnson, R.W. Christy, Phys. Rev. B 6 (1972) 4370
[23] D.E. Beck, Phys. Rev. B 35 (1987) 7325
[24] P. Apell, Å. Ljungbert, Sol. St. Com. 44 (1982) 1367
[25] J. Tiggesbäumker, L. Köller, K.H. Meiwes-Broer, to be published

^{195}Pt NMR OF Pt$_{309}$PHEN$^*_{36}$O$_{30}$ METALLIC CLUSTERS

D. van der Putten, H.B. Brom, L.J.de Jongh, G.Schmid*

Kamerlingh Onnes Laboratory, Leiden University, P.O. Box 9506,
2300 RA Leiden, The Netherlands

* Institut für Anorganische Chemie, Universität Essen,
Universitätsstrasse 5-7, D-4300 Essen 1, Germany

ABSTRACT. ^{195}Pt NMR lineshape and spin-lattice relaxation measurements at 77 K are presented for the polynuclear metallic cluster compound Pt$_{309}$Phen$^*_{36}$O$_{30}$. Two peaks in the NMR lineshape are found centered at H_0/ν_0=1.110 G/kHz and 1.096 G/kHz. The high field peak is attributed to Pt spins located in the interior of the cluster which are in a metallic environment. The low field peak has its origin in platinum atoms that constitute the surface shell of the cluster. These atoms are involved in chemical bonds with the ligands.

1 Introduction

The study of small metal particles is of fundamental interest in solid state physics and chemistry [1,2]. It is concerned with the mesoscopic region between atom and bulk. Physical properties of these systems strongly depend on the size distribution of the particles. It is therefore of crucial importance to reduce the spread in particle diameters as much as possible. Ideal candidates with respect to the size distribution problem are the polynuclear metal clusters [2]. They are composed of a well defined metal core surrounded by a ligand shell and thus offer the possibility of a systematic study of physical properties as a function of cluster size [3].

A number of ^{195}Pt NMR experiments on small platinum clusters have appeared in the literature [4-13]. Platinum exhibits one of the largest Knight shifts (K) of any metal. This offers the possibility to study the changes of electronic properties going from bulk to small particles. Small particle NMR lines are often extremely broad, which is attributed to the fact that Pt nuclei located at the surface do not exhibit any observable Knight shift, while the nuclei located more towards the center of the cluster have a more bulk like Knight shift. In order to explain such a spatial variation of K, Makowka et al. [11] have proposed a model in which the local density of states (LDOS) at the Fermi level of the 5d electrons is an exponentially decreasing function of distance when going from the center of the particle to the surface. The 6s-LDOS was shown to exhibit large oscillations due to the geometric boundary conditions of the spherical Bessel functions, which were used for the 6s electron eigenfunctions. Experimental evidence for the existence of these spatial Knight shift oscillations was given by Rhodes et al. [7] who obtained anomalously large

P. Jena et al. (eds.), Physics and Chemistry of Finite Systems: From Clusters to Crystals, Vol. II, 1007–1012.
© 1992 *Kluwer Academic Publishers.*

Knight shift gradients from the slow beats in the decay of the spin-echo intensities. In contrast, Bucher and van der Klink [12] found that the $6s$-LDOS is relatively constant in small platinum particles.

The surface nuclei of Pt catalysts do have rather short spin-lattice relaxation times (T_1) [7,12], typical for a metal, although K for these surface atoms is very small or zero. This is explained by a two-band model [11], in which the negative contribution to the Knight shift from the $5d$ band cancels the positive contribution from the $6s$ band. T_1 is inversely proportional to the sum of both $5d$ and $6s$ electron densities and can thus still be short.

In this paper we present NMR experiments on the polynuclear cluster $Pt_{309}Phen^*_{36}O_{30}$. The 309 Pt atoms constituting the metal core have a cuboctahedral structure (Fig.1) the volume of which is represented by a sphere with a radius of 10.4 Å.

Figure 1: Cuboctahedral structure of a cluster of 309 Pt atoms.

The metal core is embedded in a dielectric matrix formed by the ligands. The outer shell of the metal core consists of 162 atoms. The inner shells of the remaining 147 Pt atoms are built from 1, 12, 42 and 92 atoms.

2   Experimental results

A home built pulsed NMR spectrometer was used in a frequency range between 82-87 MHz. The applied magnetic field was 9.393 T. Typical 90° pulse lengths were 2 μs. The NMR lineshape was recorded by measuring the integrated intensity of the spin-echo as a function of frequency. The echo intensities were corrected for both T_1 and T_2 effects. For the determination of the spin-lattice relaxation time T_1 we used the saturation technique which consists of a train of short-spaced pulses followed by a spin-echo sequence. The intensity of the echo is measured as a function of delay time between the saturating pulse train and the echo sequence. Instead of the linear delay time increment usually employed we used a logarithmic time scale.

The lineshape at 77 K as a function of H_0/ν_0 is presented in Fig.2. It clearly shows the absence of a bulk signal which should have been located at $H_0/\nu_0=1.138$ G/kHz. Two separate peaks can be distinguished. The low field peak at 1.096 G/kHz is located in the region where the chemical shift values of ^{195}Pt atoms in insulating compounds are found. Similar values of low field peaks have been reported by Makowka et al. [11]. The full width at half maximum of the surface peak is about 9.1×10^3 ppm relative to H_2PtI_6.

The high field peak at 1.110 G/kHz is attributed to platinum atoms in a metallic environment. This is inferred from the spin-lattice relaxation time (T_1) (Fig.3), which is of the order of 1 ms. Similar values of T_1 as a function of H_0/ν_0 on platinum catalysts have been reported in Refs.[7,12].

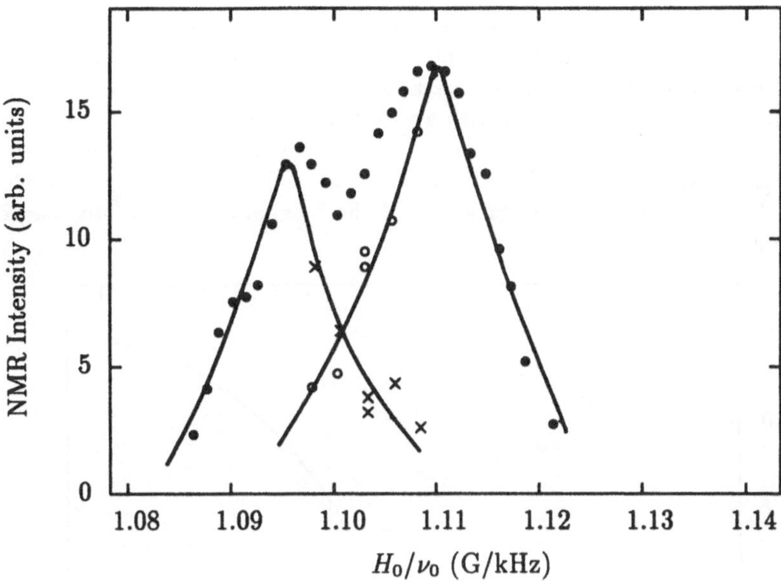

Figure 2: ^{195}Pt NMR lineshape at 77 K. •: spin- echo intensity corrected for T_1 and T_2. o: intensity of the short T_1 component. ×: intensity of the long T_1 component. Note the absence of a bulk peak at H_0/ν_0=1.138 G/kHz.

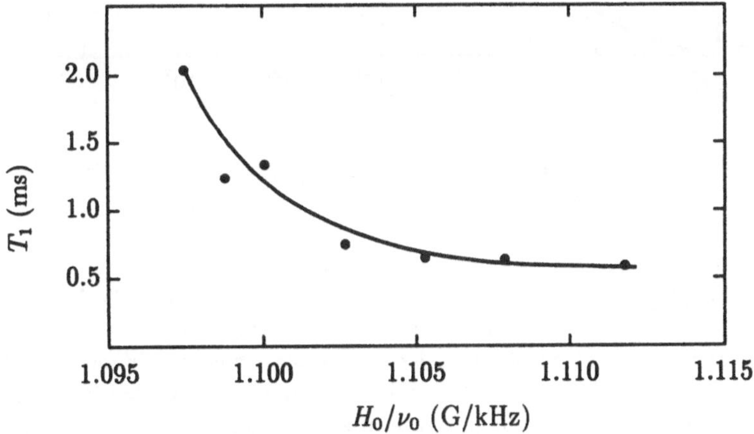

Figure 3: Spin-lattice relaxation time T_1 of the short component of the relaxation of the longitudinal magnetization as a function of H_0/ν_0. These values compare very well with the ones given in Refs.7 and 12.

The relaxation behavior of the platinum atoms is peculiar. Starting at the high field side of the metallic peak, the relaxation is in the form of a single exponential with a T_1 of about 0.6 ms (Fig.3). Moving further down field, T_1 slowly increases as expected but now a second component in the relaxation starts to appear. The relaxation of this second component is clearly non-exponential, with a very long mean T_1.

The intensity of the non-exponential component increases when one moves further down field, until finally at the low field side of the low field peak the relaxation is completely non-exponential. At this point equilibrium values of the magnetization are not reached even after 100 s (Fig.4).

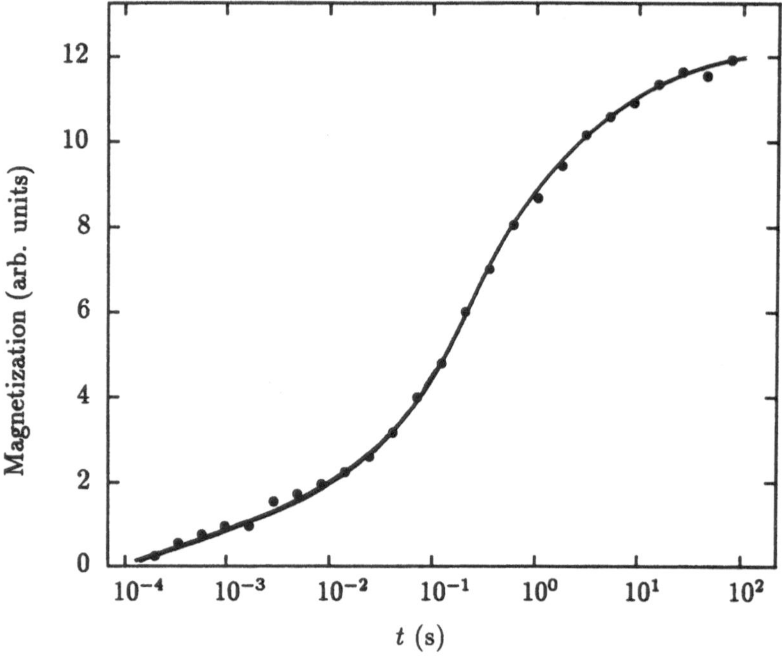

Figure 4: Non-exponential spin-lattice relaxation at $H_0/\nu_0=1.0897$ G/kHz. The solid line is the best fit to $M(t) = M_\infty\{1 - \exp-[(t/T_1)^\beta]\}$, with $\beta=0.43$ and $T_1=0.69$ s.

A fit of the experimental data points of Fig.4 to the stretched exponential $M(t) = M_\infty\{1 - \exp[-(t/T_1)^\beta]\}$, in which β is a measure of the non- exponential behavior, yields $T_1=0.69$ s and $\beta=0.43$. In view of this very long relaxation time we attribute the low field peak to platinum atoms on the metal core surface which are involved in bonds to the surrounding ligands and therefore have a low itinerant electron density. A relaxation mechanism might be provided by the presence of a paramagnetic spin, e.g. a free radical in one of the ligand molecules. Slow spin diffusion towards the paramagnetic center causes the relaxation to be non-exponential. The relative intensities of the two components in the relaxation are used to separate the metallic from the surface peak. The integrated intensities of the two peaks show that roughly 60 % of the platinum atoms behave as in a metallic environment.

We also observed the slow beats in the decay of the spin-echo with a frequency of 4.8 kHz independent of the position in the metallic peak. In platinum alloys a frequency of 4.2 kHz was found [14]

3 Discussion

In contrast to previous NMR results on small platinum particles [7,11,12] T_1 of the surface atoms is extremely long in our experiment. The short T_1 around $K=0$ was attributed to the cancellation of the positive Knight shift K_s due to a finite density of states of the $6s$ electrons with the negative shift K_d of the $5d$ electrons. $1/T_1$ is proportional to $K_s^2 + K_d^2$. We therefore conclude that a fraction of the surface atoms have a low itinerant electron density. Since the platinum atoms responsible for the low field peak do not experience any metallic behavior we can subtract 40 % of the Pt atoms from the metal core. This yields a metallic particle whose volume is represented by a sphere with a radius of about 8.7 Å. From the location of the metallic peak it is concluded that the density of states is reduced to a large extent with respect to bulk platinum. With a suitable choice of parameters the position and width of the metallic peak can be reproduced by the model proposed by Makowka et al. [11]. However, the simpler model of Bucher and van der Klink [12] fits equally well. The fact that we did not observe any dependence on line position of the slow-beat frequency in the spin-echo decay speaks in favor of the latter model, i.e. a constant LDOS of $6s$ electrons and a $5d$-LDOS which decreases exponentially towards the surface. These slow beats arise from the exchange coupling between the nuclear spins. The coupling constant depends on the $6s$-LDOS. Consequently, the constant beat frequency implies a constant $6s$-LDOS.

Experiments such as the present are of value in establishing the minimal size needed for a transition metal cluster to show "metallic behavior", since a Knight shift in the NMR is considered to be one of the hallmarks of metallicity. What is needed is in fact a density of energy levels around the Fermi energy that is sufficiently high to yield level spacings comparable to or smaller than the thermal energy $k_B T$. A rough estimate of the level spacing can be obtained from the ratio $2E_F/N$, where the Fermi energy would be a few eV for a large Pt cluster and N is the number of valence electrons in the cluster (10 for Pt). This yields a spacing of order 4 meV (taking $E_F \simeq 6$ eV and $N \simeq 3000$), to be compared with the thermal energy of about 7 meV at liquid N_2 temperatures. Evidently, several mechanisms of level broadening could be involved, including a finite transfer integral between neighboring cluster molecules. Another factor which plays a role is the high (fcc-packing) symmetry of the core. Some of the levels will be grouped together resulting in a reduced level splitting. It is therefore of interest to pursue these experiments to lower temperatures, which is what we are doing at present. Lastly, we mention that the conclusions of this paper are corroborated by parallel studies of the magnetic susceptibility [15] and low-temperature specific heat [16] of the giant metal cluster compound $Pd_{561}Phen_{36}O_{200}$.

Acknowledgement: J.T.Moonen is acknowledged for his contribution to the experiments. This work is part of the research program of the Leiden Materials Center and is supported by the "Stichting voor Fundamenteel Onderzoek der Materie" (FOM), which is sponsored by the "Nederlandse Organisatie voor Wetenschappelijk Onderzoek" (NWO). One of us (DvdP) is financially supported by an IOP grant (BP203).

References

[1] W.P.Halperin, Rev. Mod. Phys. $\underline{58}$, 533 (1986)

[2] G.Schmid, Structure and Bonding $\underline{62}$, 51 (1985); Endeavour, New Series $\underline{14}$ (Pergamon Press, Oxford, 1990); *Aspects of Homogeneous Catalysis*, $\underline{7}$, 1 (1990), Ed. R.Ugo (Kluwer Academic, Dordrecht)

[3] L.J.de Jongh, J.A.O.de Aguiar, H.B.Brom, G.Longoni, J.M.van Ruitenbeek, G.Schmid, H.H.A.Smit, M.P.J.van Staveren and R.C.Thiel, Z.Phys.D $\underline{12}$, 445 (1989) and references cited therein.

[4] I.Yu, A.A.V.Gibson, E.R.Hunt and W.P.Halperin, Phys. Rev. Lett. $\underline{44}$, 348 (1980)

[5] I.Yu and W.P.Halperin, J.Low Temp. Phys. $\underline{45}$, 189 (1981)

[6] H.E.Rhodes, P.-K.Wang, H.T.Stokes, C.P.Slichter and J.H.Sinfelt, Phys. Rev. B $\underline{26}$, 3559 (1982)

[7] H.E.Rhodes, P.-K.Wang, C.D.Makowka, S.L.Rudaz, H.T.Stokes, C.P.Slichter and J.H.Sinfelt, Phys. Rev. B $\underline{26}$, 3569 (1982)

[8] T.Stokes, H.E.Rhodes, P.-K.Wang, C.P.Slichter and J.H.Sinfelt, Phys. Rev. B $\underline{26}$, 3575 (1982)

[9] J.J.van der Klink, J.Buttet and M.Graetzel, Phys. Rev. B $\underline{29}$, 6352 (1984)

[10] B.J.Pronk, H.B.Brom, A.Ceriotti and G.Longoni, Solid State Comm. $\underline{64}$, 7 (1987)

[11] C.D.Makowka, C.P.Slichter and J.H.Sinfelt, Phys. Rev. B $\underline{31}$, 5663 (1985)

[12] J.P.Bucher and J.J.van der Klink, Phys. Rev. B $\underline{38}$, 11038 (1988)

[13] J.P.Bucher, J.Buttet and J.J.van der Klink, Surf. Sci. $\underline{214}$, 347 (1989)

[14] C.Froidevaux and M.Weger, Phys. Rev. Lett. $\underline{12}$, 123 (1964)

[15] J.M. van Ruitenbeek, M.J.G.M.Jurgens, G.Schmid, D.A.van Leeuwen, H.W.Zandbergen and L.J.de Jongh, Z. Phys. D $\underline{19}$, 267 (1991)

[16] J.Baak, H.B.Brom, L.J.de Jongh and G.Schmid, This conference.

OBSERVATION OF WAVE LOCALIZATION IN PENROSE LATTICES

Chumin Wang, O. Navarro[1], M. Cruz, *and* R. Fuentes,
Instituto de Investigaciones en Materiales, UNAM
Apartado Postal 70-360, 04510, México D.F., MEXICO

and

R.A. Barrio
Instituto de Física, UNAM
Apartado Postal 20-364, 01000, México D.F., MEXICO

ABSTRACT. Experimental measurements of wave interference made in a Penrose lattice of LC electric oscillators are reported. The measured spectrum of injection currents as a function of frequency is analyzed theoretically by solving Kirchhoff's laws with a dissipative term. This problem could be mapped to the case of finding the solution of the Schrödinger's equation for isotropic phonons, or s electrons in these quasicrystals. Measurements of the spatial distribution of amplitudes were performed in this network, and the results reveal key features concerning the controversial topic of localization of waves in quasicrystals. There are truly extended quasicrystalline *eigen*-states at low frequencies, and critically localized states when the wavelength becomes comparable to the lattice parameter.

The critical nature of the *eigen*-functions of excitations in one- and two-dimensional (2D) quasiperiodic lattices [1-3] has been of much current interest, due to the expected unusual transport properties [4] derived from such spectra of excitations, although measurements made in real quasicrystals have failed to resolve the structure. The spectrum of electrons in Penrose tiles shows an array of alternating bands and gaps [5], although the nature of the spatial decay of the wave functions is a matter of controversy. There are some very fine theoretical studies [6] about localization of wave functions in Penrose models and much effort made recently to test the predictions in artificially produced quasiperiodic systems. Amongst them, it is worthwhile mentioning the experiment made in a Josephson junction array tiled as a Penrose lattice [2]. Another experiment, more relevant to visualizing the spatial distribution of wave amplitudes, is the one made with tuning forks situated in the centers of the Penrose rhombuses, suitably connected and excited with sound waves [3]. The measurement of the amplitude of sound waves in this experiment needs extremely sophisticated equipment.

In this paper, we propose a clearer way of visualizing the interference of waves, making electric measurements on a network of LC oscillators linked together as a Penrose

---

[1] Permanent address: Escuela de Ciencias Físico Matemáticas de la Universidad Autónoma de Sinaloa, Apartado Postal 1872, Culiacán, Sin., MEXICO.

1013

P. Jena et al. (eds.), Physics and Chemistry of Finite Systems: From Clusters to Crystals, Vol. II, 1013–1018.
© 1992 *Kluwer Academic Publishers.*

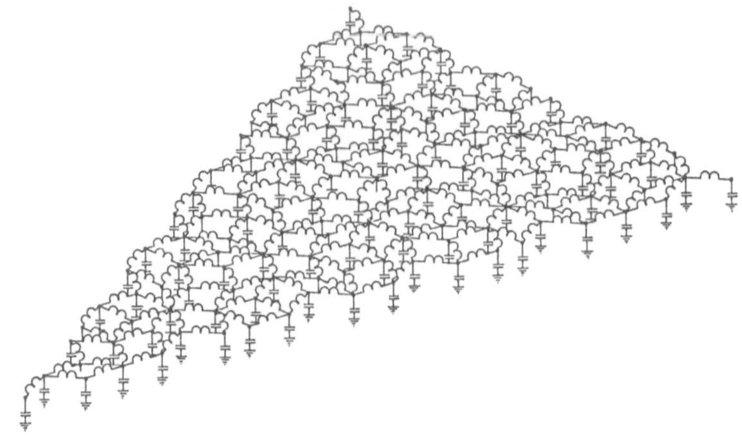

Figure 1: A schematic drawing of the Penrose network of oscillators. The bonds of the Penrose lattice are replaced by inductances and in each vertex there is a capacitor with one end connected to earth and the another to the inductances.

lattice. With this model one can easily calculate the measured quantities, thus allowing an exact analysis of the physical response of the system, since one is measuring the phase coherence of a monochromatic AC wave with a desired frequency. We show that this system simulates extremely well the behaviour of isotropic electrons and phonons in real Penrose quasicrystals, which is not surprising, since all these problems are equivalent in the tight-binding language.

Figure 1 is a sketch of the actual network. A Penrose lattice of generation $n =13$ was drawn on a fiber glass board with a grounded copper plate on one surface, according to the inflation procedure developed by Wang and Barrio [5]. This lattice contains 137 vertices and 240 bonds of length 5 cm. Each vertex consists of a capacitor of 1.0 $\mu F\pm10\%$ with one grounded end. The bonds of the Penrose tile are toroidal inductances of 7.8 $mH\pm5\%$, which are welded in the vertices to the free end of the capacitors. Therefore, the array can be visualized as a 2D quasiperiodic network of transmission lines.

In this array there are dissipative effects and undesired mutual inductances, which vary with frequency. In fact, the toroidal form was chosen to minimize the mutual inductance problem, and the toroids were made of ferrite cores 30 mm in external diameter, 18.5 mm internal diameter and 8 mm in width, with a relative magnetic permeability $\mu/\mu_0 \simeq 1500$ and coiled with 90 turns of copper wire of 1.1 mm in diameter in order to lower the resistance. The number of turns was adjusted in each component to yield the same inductance at $\nu = 1KHz$. All coiled components were separately characterized as a function of ν to reduce differences between them to less than 5%. The dissipative behaviour was tested experimentally, and the results could be adjusted to a smooth continuous function of the frequency and therefore could be taken into account by introducing a complex inductance L, with imaginary part equal to $-R(\omega)/\omega$. When fit to experimental data, a function $R(\omega) = 0.006\omega^{0.7}$ was obtained.

It is possible to excite this system in various ways, for instance, one might irradiate with RF and measure the absorption spectrum, or one could keep the rms voltage constant at one or several vertices, so as to change boundary conditions, or alternatively, keep the current flow constant at a given path. All these experiments can be readily made. However, for the sake of clarity, in this work an AC voltage source was connected to the extreme of the acute angle of the triangle in Fig. 1, and varying the frequency.

The theoretical treatment of such a situation can be made by analyzing Kirchhoff's laws, written in the following convenient form:

$$i\omega C V_n + \frac{1}{i\omega L}\sum_m (V_n - V_m) = 0, \tag{1}$$

where the capacitance C and the inductance L have been taken constant for all elements. Here V_n is the voltage measured on site n when a stationary wave of frequency ω is excited in the system. The summation spans all the m sites directly connected to n. The solution of equation (1) is easily found and it is the same as the solution of the secular equation

$$- M\omega^2 U_n + K \sum_m (U_n - U_m) = 0 \tag{2}$$

for the displacements U_n of masses M on each site connected together with "isotropic springs" with force constant K. Or, alternatively, equation (2) could be identified as the usual tight binding equations of motion for s electrons. The boundary conditions depend on how the excitation is introduced. In our case, if one injects on site $n = 1$, the first equation has an inhomogeneous current term I(ω).

The results of measuring the current I(ω) arriving at the site on the extreme left in Fig. 1, are shown in Fig. 2 for $V_1 = 0.1 volts$. The data are compared with the appropriate solution of equation (1) with complex L. Observe that the agreement of the frequency distribution of modes is reasonable, without any fitting parameters, besides R(ω). The frequencies of the normal modes are independent of the value of R(ω), although the relative intensities of the high frequency peaks are lowered by the dissipative effect of R(ω).

The amplitude of the waves can be measured for any state (ω) at all sites. In Figure 3 we show the spatial distribution of amplitudes for the frequencies of the first 6 peaks of Fig. 2. Notice that the peak at $80 Hz$ corresponds to an acoustic mode of wave length four times larger than the size of the system. This is to be expected, since for these low frequencies the wave does not "see" the discreteness of the lattice. The subsequent modes can be understood as analogous to the excited states of a continuous membrane, although the presence of the discrete quasicrystalline lattice is increasingly felt. Observe that the envelope stationary wave is not periodic, for instance, along the front border of Fig. 3c, at $\nu = 700 Hz$, the ratio of the distance between subsequent nodes is approximately the golden ratio 1.618.

As one increases the frequency of the mode, the spatial distribution starts being more complicated, but the amplitude starts being localized in certain regions. The effects of the border are more important in a quasiperiodic system than in an ordered lattice [7], and the quasiperiodicity is expected to affect the high frequency modes, when the wave length

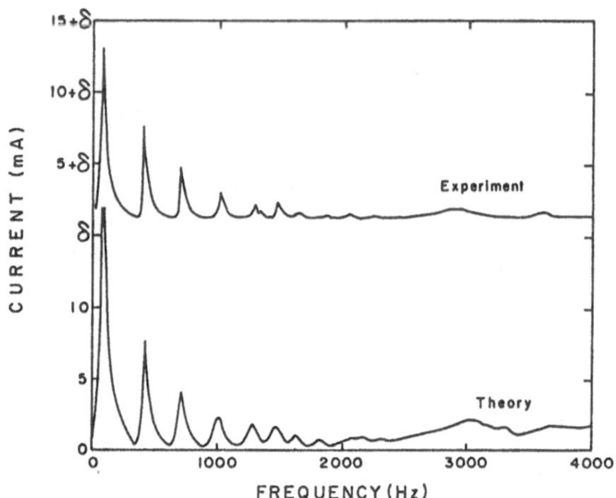

Figure 2: Comparison between the theoretical and experimental current spectrum. The AC voltage source was injected at the site on the acute angle of the triangle of Fig. 1, and the current was measured by placing an amperimeter in series. The vertical scale for the experimental data has been shifted by δ.

becomes comparable to the size of the bonds. This is evidenced by the large boundary amplitudes observed in Fig. 3f. The effect of localization of the amplitude is not always seen at the boundary, but also around some special sites in the interior of the lattice, therefore, it is expected that this effect should be present in an infinite lattice.

It is worthwhile pointing out that the localization seen is different from the usual concept of localization around an impurity, or that which arises due to random disorder, since the fluctuations in the amplitude of the wave function are present throughout the network, which is not disordered. In the theory of quasiperiodic functions one is able to predict the existence of isolated states, which in a way resemble the nature of certain non-dispersive states in periodic solids. One simple example is given by the pure p states in a four coordinated lattice with the Weaire and Thorpe Hamiltonian [8], which are localized along chains. There are theoretical analyses of wave functions for the tight-binding s-electron band in the Penrose vertex problem [9], in particular, the amplitude of the central peak is confined in regions separated by chains of sites.

There are in the literature numerous studies about the spatial distribution of the amplitude of the wave functions of electrons in quasiperiodic systems [6], and usually the pictures are very complicated. These can be compared with our high frequency maps, which show a lot of structure due to the incommensurability of the wave with the lattice. The important point is that it does not matter how complicated the plot should appear, the underlying theory is able to predict the exact form of the wave function.

In summary, we have presented a new method of examining wave propagation in 2D real quasicrystalline lattices. This device provides a precise way to measure directly the amplitude of the *eigen*-states at any point in the system. The model is versatile and

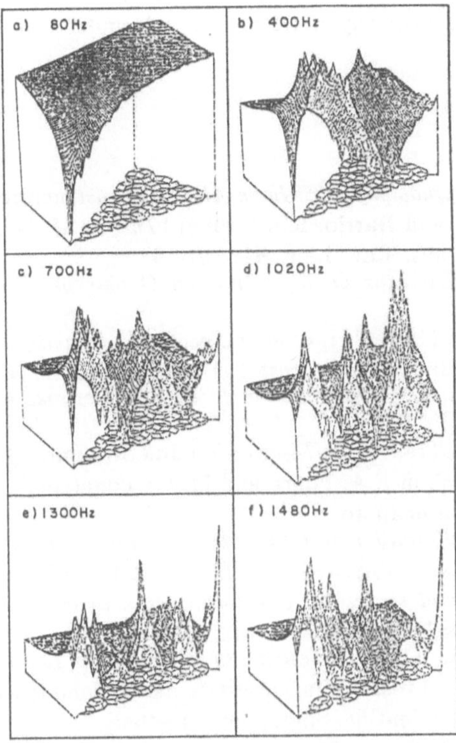

Figure 3: The amplitude distribution for the first 6 *eigen*-states. The frequencies of these states are indicated in the figures. The vertical axis in these figures measures the voltages at the vertices.

could be used in a number of different experiments, as selective conductance, anisotropy of current flow, radio frequency resonance phenomena, or the possibility of rising non-linear response. In particular, an experiment with RF excitation should be very interesting, since the analogy to thermal excitation of s electrons is quite close. These experiments are currently under way in our laboratory.

Although the network we examined is relatively small, the results show the effects of quasiperiodicity on the *eigen*-modes of the system, and reveal the mechanism of localization in quasiperiodic lattices. It is important to emphasize that the simplicity of the network allows to isolate the quasicrystalline characteristics of the system, and therefore a good correspondence between theory and experiment is attained, a situation that is not usually true in real materials. This gives us great confidence in the results.

Similar modeling methods could also be used to investigate the situation when there are defects or mistakes in the quasiperiodicity of the lattice, as phasons [10]. Obviously other systems, as octagonal or dodecagonal quasicrystals [11] and fractal networks [12], could be devised.

We thank R. Alvárez, E. Amano, J. Andrade, and E. Martínez, for valuable help. This work was partially supported by UNAM, DGAPA under contract No. IN-100-289-UNAM.

REFERENCES.

[1] Merlin, R., *et. al* (1985) '*Quasiperiodic GaAs-AlAs Heterostructures*', Phys. Rev. Lett. **55**, 1768-1770; Wang, C. and Barrio, R.A. (1988) '*Theory of the Raman Response in Fibonacci Superlattices*' Phys. Rev. Lett. **61**, 191-194.

[2] Behrooz, A., *et. al* (1986) '*Flux Quantization on Quasicrystalline Networks*', Phys. Rev. Lett. **57**, 368-371.

[3] He, S. and Maynard, J.D. (1989) '*Eigenvalue Spectrum, Density of States, and Eigenfunctión in a Two-Dimensional Quasicrystal*', Phys. Rev. Lett. **62**, 1888-1891.

[4] Tsunetsugu, H. and Ueda, K. (1988) '*Conductance of a Penrose tiling*', Phys. Rev. B **38**, 10109-10112.

[5] Wang, C. and Barrio, R.A. (1991) '*the Electronic Band Structure of Penrose Lattices: A Renormalization Approach*', in F.A. Ponce and M. Cardona(eds.), Lectures on Surface Science, World Scientific, Singapore.

[6] Sutherland, B. (1986) '*Self-similar ground-state wave function for electrons on a two-dimensional Penrose lattice*', Phys. Rev. B **34**, 3904-3909; Ma, P. and Liu, Y. (1989) '*Inflation rules, band structure, and localization of electronic states in a two-dimensional Penrose lattice*', Phys. Rev. B **39**, 9904-9911.

[7] Barrio, R.A. and Wang, C. (1990) '*Electron Localization in Large Fibonacci Chains*', in M. José Yacamán, *et. al* (eds.), Quasicrystals and Incommensurate Structures in Condensed Matter, World Scientific, Singapore, 448-464.

[8] Ziman, J.M. (1979) *Models of Disorder*, Cambridge University Press, Cambridge, p. 456.

[9] Wang, C. and Barrio, R.B. *to be published*.

[10] Horn, P.M., *et. al* (1986) '*Systematics of Disorder in Quasiperiodic Material*', Phys. Rev. Lett. **57**, 1444-1447; Strandburg, K.J., Tang, L.-H., and Jarić, M.V. (1989) '*Phason Elasticity in Entropic Quasicrystals*', Phys. Rev. Lett. **63**, 314-317.

[11] Socolar, J.E.S. (1989) '*Simple octagonal and dodecagonal quasicrystals*', Phys. Rev. B **39**, 10519-10551.

[12] Liu, S.H. (1986) '*Fractals and Their Applications in Condensed Matter Physics*', Solid State Physics **39**, 207-273.

PHOTOELECTRON SPECTROSCOPY OF WEAKLY-BOUND ELECTRONS IN SODIUM CHLORIDE CLUSTER ANIONS

P. XIA, A. J. COX, Y. A. YANG, AND L. A. BLOOMFIELD
Department of Physics
University of Virginia
Charlottesville, Virginia 22901
United States

ABSTRACT. Photoelectron spectra are reported for stoichiometric sodium chloride anions containing a single excess electron. Using a magnetic bottle time of flight spectrometer, photoelectron spectra and binding energies of electrons were obtained for the mass-selected negative clusters $(NaCl)_n^-$ ($n=1$-13, 22, except $n=6$, 12). From the spectra of energies observed, we obtained information about the localizations of the excess electrons in the anion clusters and the spatial rearrangement of their host clusters.

1. Introduction

Atomic and molecular clusters are remarkable systems in which to study the interactions of excess electrons with materials. Alkali-halide clusters are particularly interesting both experimentally and theoretically[1-9] because they remain ionically bound, even as very tiny particles. Thus, the anion clusters of stoichiometric species containing a single excess electron provide a microscopic laboratory in which to study the localizations of the excess electrons and their effects on spatial and electronic structures of their host particles.

The localizations of excess electrons attached to alkali-halide clusters have been predicted as follows[3,4,6,7]: An excess electron may (i) localize in an internal halogen anion vacancy, as in the case of an F center defect in the bulk alkali halide crystal; (ii) be bound on a cluster surface; (iii) localize about a specific alkali-metal ion; or (iv) be trapped in a dipole potential energy minima.[5] Furthermore, the electron localization in a particular cluster is strongly dependent on the cluster's structure, which changes with different temperatures and formation processes.[4,10]

In this paper, we report the photoelectron spectra of $(NaCl)_n^-$ anions ($n=1$-13, 22, except $n=6$, 12) using a magnetic bottle time of flight spectrometer. Using time of flight, we can measure the electron's kinetic energy and obtain its binding energy by subtracting that kinetic energy from the photon energy. From the spectra of energies observed, we obtain information about the localization of the excess electron in the negative cluster and the spatial rearrangement of its host cluster induced by the electron attachment.

P. Jena et al. (eds.), Physics and Chemistry of Finite Systems: From Clusters to Crystals, Vol. II, 1019–1024.
© *1992 Kluwer Academic Publishers.*

2. Experimental

Fig.1 is a schematic drawing of our magnetic bottle time-of-flight photoelectron spectrometer. A pulsed negative cluster ion beam is generated by the laser vaporization cluster source described previously.[11] In brief, a sample disk of NaCl crystal is vaporized by the 15-ns pulse of a focused ArF excimer laser. The incident energy density at 193nm is about $5J/cm^2$ over a 1mm × 2mm focal spot. The laser pulse coincides with a pulse of Helium gas (1-10 bars) which blows across the sample, cooling and entraining the material leaving the surface of the sample. The supersonic molecular beam containing neutral and charged particles travels through a molecular beam skimmer, and then the negative clusters present in the beam are accelerated so that when they exit the tiltable Wiley-McLaren acceleration plates, their kinetic energy is about 1350 eV. Whenever necessary, the cluster beam is decelerated and focused by an Einzel lens and a lens assembly.

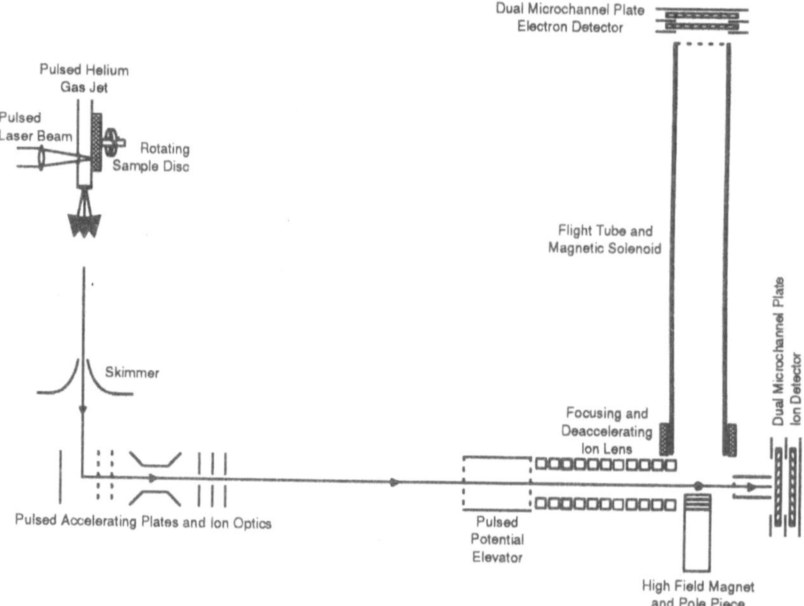

Figure 1. Schematic diagram of experimental apparatus.

As the clusters drift through the photodetachment region, an 8ns pulse of 532nm radiation (the second harmonic from a Nd:YAG laser) is used to photodetach electrons from the clusters. The photodetachment region of the spectrometer is positioned between the magnetic time-of flight tube and the magnetic pole piece, and centered along their symmetry axes. The high magnetic field is produced by a pair of permanent magnets(Nd:Fe:B) each 1 inch in diameter and 1/8 inch thickness, mounted on a 3 inch long soft iron pole piece. At the laser interaction region, the magnetic field is about 600-700 G. In the magnetic field, the photoelectrons are tightly constrained to follow the local field lines. In order to channel these electron trajectories up the time-of-flight drift tube, a weak (3-5 G) static field is produced by a long solenoid magnet mounted coaxially along the full length of the flight tube. At the end of the flight tube, the electrons pass through a grounded high transmission copper mesh and are accelerated towards a dual microchannel plate detector.

After photodetachment, the charged clusters and the resultant neutral clusters enter a set of coaxial 4 inches tubes: the inner one, a 0.5 inch diameter copper tube, is biased at $+2000$ V, while the outer one, a 0.75 inch diameter stainless steel tube coated with graphite, is grounded. A dual microchannel plate detector with its front plate biased at $+2000$ eV is mounted 0.2 inch behind of the end of the tubes so that the charged clusters are accelerated into the detector and the resultant neutral clusters are not. With this design, we can accurately determine the particular cluster being photodetached by subtracting alternative shots with the photodetachment laser on and off to yield a photodepletion peak and daughter ion peaks if photofragmentation happened. All the photodepletion peaks show there is no significant photofragmentation occurring with the detachment laser intensity of about 6-8mJ/pulse at 532 nm. The advantage of the photoelectron spectroscopy is that the detector is directly collecting the photoelectrons rather than photoionized neutral particles.

3. Results and Discussion

Fig. 2 presents the electron binding energy spectra for the negative stoichiometric sodium chloride clusters $(NaCl)_n{}^-$ with n=1-13,22, except n=6, 12. Each spectrum is calibrated by the known binding energy values of negative ion K^- and Na^-.

Fig. 3 shows the the abundance of the clusters of $(NaCl)_n^-$ (n=1-22). The clusters $(NaCl)_{13}{}^-$ and $(NaCl)_{22}{}^-$ have relatively high abundance since the clusters $(NaCl)_{13}Cl^-$ and $(NaCl)_{22}Cl^-$ have stable cubic structures.[10,12] The excess electrons in these two clusters apparently occupy chlorine ion vacancies to form stable cubic structures. The electron binding energies of both clusters are relatively high, 1.4 eV for the cluster $(NaCl)_{13}{}^-$ and 1.65 eV for the cluster $(NaCl)_{22}{}^-$. This is one type of localization of the excess electron, which falls into the class of an F-center defect in the bulk alkali halide crystal.

F-center localization of the excess electron has also been observed in neutral clusters such as Na_nF_{n-1} with the binding energies as high as 3.8 eV.[6,7] The difference in the electron binding energy between neutral and negative cluster both containing an excess electron is due to the different interaction between the electron and ions when the electron leaves its host cluster. In the neutral cluster, the detached electron leaves behind a positive cluster ion and must work against the long-ranged Coulomb interaction. In the negative ion, the electron leaves behind a neutral cluster and experiences only short ranged forces. It is noted that there is a small peak with binding energy around 0.25 eV in each photoelectron energy spectrum of clusters $(NaCl)_{13}^-$ and $(NaCl)_{22}^-$. This is probably due to an isomer induced by the excess electron. These isomers are small in comparison to the dominant peaks.

In the cluster $(NaCl)_4{}^-$, the electrons could be bound at the surface of a $2 \times 2 \times 2$ cube, localized in a chlorine ion vacancy of the 3×3 configuration, or both, because the neutral cluster $(NaCl)_4$ forms a stable cubic ionic structure $(2 \times 2 \times 2)$ and the negative cluster $(NaCl)_4Cl^-$ forms a 3×3 stable structure.[12] In the binding energy spectrum of the cluster of $(NaCl)_4{}^-$, only one peak appears with the binding energy about 1.4 eV. The calculated vertical binding energy of $(NaCl)_4{}^-$ for the 3×3 planar structure with the electron in one of the corners is about 1.8 eV.[13] Landman *et al.* explored mechanisms of the cluster isomerization induced by electron attachment to neutral $(NaCl)_4$.[5] They predicted that the electron in $(NaCl)_4^-$ is bound on the surface of a neutral cluster with the $2 \times 2 \times 2$ configuration or at one of the corners of a 3×3 plane, depending on the internal temperature of the clusters. The internal temperature of clusters produced in our source is unknown right now, but work is in progress on a source which can measure the clusters' internal temperature precisely. If the theoretical predictions are correct, we expect to see the temperature transition between the different isomers.

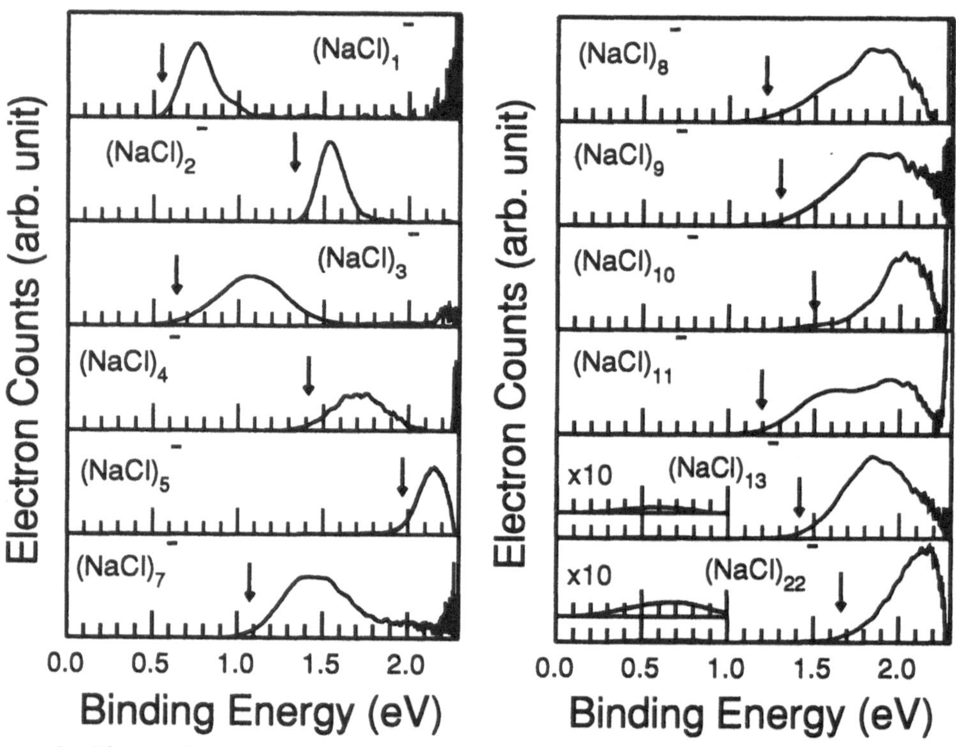

Figure 2. Electron binding energy spectra for stoichiometric clusters $(NaCl)_n^-$ for n=1-13, 22, except n=6, 12. The arrows approximately indicate the binding energy of the electron in its host cluster. For n=13 and n=22, the small peak at 0.25 eV is magnified by a factor of 10.

The electron binding energy for $(NaCl)_2^-$ is relatively high at about 1.35 eV. From theoretical calculations,[14] the most stable structure of neutral $(NaCl)_2$ is cyclic with zero dipole moments, but the calculation failed to yield a stable anion for a cyclic dimer. The linear dimer has a vertical binding energy of about 1.3 eV, based on the calculations.[14] The experimental result is in fairly close agreement with the previous experimental [9] and theoretical results. Therefore, with the addition of an electron, the $(NaCl)_2$ dimer becomes a bent chain and the electron is located before the first sodium. Thus the electron appears to be bound in the dipole field of a highly distorted $(NaCl)_2$ dimer molecule.

Although the electron in $(NaCl)_1^-$ is bound in the dipole field of a Na^+ ion and a Cl^- ion, the binding energy of this electron is low, 0.53 eV. The electron affinity of the NaCl molecule is 0.72 eV. The electron binding energy in $(NaCl)_3^-$ is low also, 0.6 eV. The neutral trimer is a ring of six ions with alternating charge.[12] The extra electron may be trapped inside the ring by a weak dipole field. Nevertheless, a detailed interpretation of these photoelectron spectra will require extensive comparison to electronic structure calculations of the clusters.

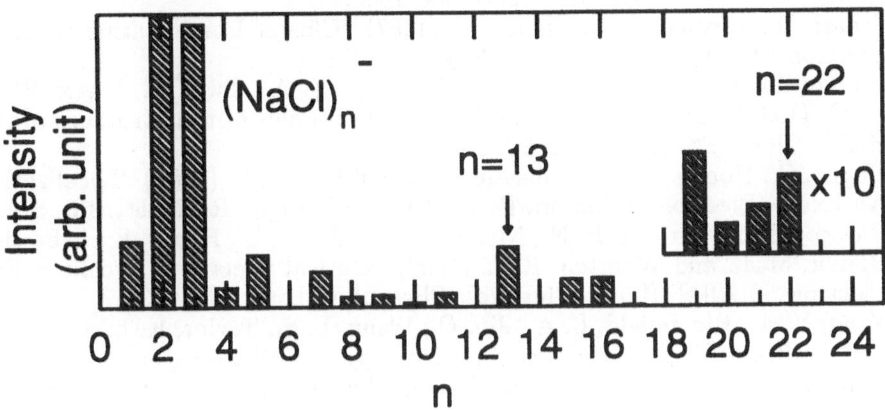

Figure 3. Relative cluster abundance for $NaCl_n^-$. The peaks for $n \geq 18$ are magnified by a factor of 10.

4. Conclusion

In the present study of sodium chloride clusters, the stoichiometric cluster anions $(NaCl)_n^-$ were produced in a laser vaporization supersonic nozzle source. We have studied the excess electrons in these clusters in the size range of n=1-13, 22 by photoelectron spectroscopy. Using a magnetic bottle time of flight spectrometer, the photoelectron spectra and the binding energies of electrons were obtained. The photoelectron spectra show that the binding energies of electrons is below 2.3 eV. Three types of possible electron localizations have been observed as following: the electrons in $(NaCl)_{13}^-$ and $(NaCl)_{22}^-$ occupy a chlorine ion vacancy to form stable cubic structures; the electron in $(NaCl)_3^-$ is trapped at the surface of the cluster with the binding energy below 0.7 eV; the electrons in the other clusters are trapped in a dipole potential energy minima. Also the configuration of cluster $(NaCl)_2^-$ is linear, while the neutral cluster of $(NaCl)_2$ is cyclic. The distortion of the neutral stable structure due to the attachment of the excess electron is exhibited. The localization of the electron in the cluster $(NaCl)_4^-$ needs more experimental measurements as a function of the cluster's internal temperature to observe the temperature transition between the isomers. The complete interpretation of all these photoelectron spectra require more electronic structure calculations.

The authors gratefully acknowledge technical help from D. Fatemi. Acknowledgment is made to the donors of The Petroleum Research Fund, administered by the ACS, for support of this research.

5. References

1. Rajagopal, G., Barnett, R. N., Landnam, U. (1991) "Metallization of Ionic Clusters", Phys. Rev. Lett. **67** , 727.
2. Pollack, S., Wang, C. R. and Kappes, M. M. (1990) "On the Optical Absorption Spectrum of Na_2Cl", Chem. Phys. Lett. **175**, 209.
3. Landman, U., Scharf, D. and Jortner, J. (1985) "Electron Localization in Alkali-Halide Clusters" , Phys. Rev. Lett. **54**, 1860.
4. Scharf, D., Jortner, J., Landman, U. (1987) "Cluster Isomerization Induced by Electron Attachment", J. Chem. Phys. **87**, 2716.
5. Bloomfield, L. A., Conover, C. W. S., Yang, Y. A., Twu, Y. J. and Phillips, N. G. (1991) "Experimental and Theoretical Studies of the Structure of Alkali Halide Clusters", Z. Phys. D **20**, 93.
6. Honea, C., Homer, M. L., Labastie, P. and Whetten, L. (1989) "Localization of An Excess Electron in Sodium Halide Clusters", Phys. Rev. Lett. **63**, 394.
7. Rajagopal, G., Barnett, R. N., Nitzan, A., Landman, U., Honea, E., Labastie, P., Homer, M. L. and Whetten, R. L. (1990) "Optical Spectra of Localized Excess Electrons in Alkali Halide Clusters", Phys. Rev. Lett. **64** , 2933.
8. Yang, Y. A., Bloomfield, L. A., Jin, C., Wang, L. S., Taylor, K. L. and Smalley, R.E. "Ultraviolet Photoelectron Spectroscopy and Photofragmentation Studies of Excess Electrons in Potassium Iodide Cluster Anions", J. Chem. Phys., in press.
9. Yang, Y. A., Conover, C. W. S. and Bloomfield, L. A. (1989) "Production and Photodetachment of Stoichiometric Sodium Chloride Cluster Anions", Chem. Phys. Lett. **158**, 279.
10. Diefenbach, J. and Martin, T. P. (1985) "Model Calculations for Alkali Halide Clusters", J. Chem. Phys. **83**, 4585.
11. Twu, Y. J., Conover, C. W. S., Yang, Y. A. and Bloomfield, L. A. (1990) "Alkali-Halide Cluster Ions Produced by Laser Vaporization of Solids", Phys. Rev. B **42**, 5306.
12. Phillips, N. G., Conover, C. W. S. and Bloomfield, L. A. (1991) "Calculations of the Binding Energies and Structures of Sodium Chloride Clusters and Cluster Ions", J. Chem. Phys.**94**,4980.
13. Conover, C. W. S. (1990) "Studies of the Structure, Binding properties and Defects in Alkali-Halide Microclusters", Ph. D. Thesis at the University of Virginia.
14. Sunil, K. K. and Jordan, K. D. (1987) "Negative Ion Formation in Alkali Halide Clusters", J. Phys. Chem. **91**, 1710.

THE UNIQUE NATURE OF METAL CLUSTER OXIDATION

JAMES L. GOLE
High Temperature Laboratory
Center for Optical Science and Engineering
and School of Physics
Georgia Institute of Technology
Atlanta, Georgia 30332

ABSTRACT. Oven based entrainment flow and supersonic expansion techniques are used to form small metal and metalloid molecules and to study their unusual oxidation behavior. These metal and metalloid molecules oxidize to form not only a distinct class of metal atom grouped cluster oxides and halides under kinetically as opposed to thermodynamically controlled conditions but also unusual (unexpected) excited electronic state product distributions. We exemplify this behavior with a focus on copper cluster oxidation, contrasting the emission spectra generated for the asymmetric and symmetric isomers of Cu_2O and comparing these with theory. A subset of the oxidation studies touches on the sodium trimer - halogen atom reactions, $(Na_3 - X(Cl,Br,I)$, which create a continuous electronic population inversion based on the chemical pumping of Na_2 with laser amplifiers throughout the visible and the extension of these studies to copper and magnesium trimer halogenation.

INTRODUCTION

The cluster oxidation process, the unique products of metal cluster oxidative transformation, and the optical signatures of the products of oxidation may have important implications for the extension of concepts in molecular reaction dynamics and the fingerprinting of local surface reactive environments. The study of cluster oxidation provides a unique opportunity to address an unusual and multicentered reaction chemistry and can furnish information on grouped metal atom-oxidant interactions which can serve as an important component in the modeling of chemically significant systems on a much larger macroscopic scale.

At present, the internal mode structure and dynamics associated not only with bare metal and metalloid clusters[1] but also with their kinetically controlled oxidation[2-5] is still largely neglected. The limited information which is available demonstrates a unique oxidation behavior and virtually unexpected reactive branching.[5] In approaching the study of the oxidation dynamics which a number of metal (and metalloid) clusters undergo as they form a distinct class of metal atom grouped cluster oxides[2] and halides[3] under <u>kinetically</u> as opposed to thermodynami-

1025

P. Jena et al. (eds.), Physics and Chemistry of Finite Systems: From Clusters to Crystals, Vol. II, 1025–1237.
© 1992 *Kluwer Academic Publishers.*

cally controlled conditions. We have been concerned with the characterization of the internal mode structure of the product metal clustered oxides and halides. In developing these studies, we have analyzed the first vibrationally resolved optical signatures[2] for several "asymmetric" metal cluster oxides, and demonstrated the first visible chemical laser amplifier from a metal cluster oxidation process.[4] These studies graphically illustrate the dramatic and unexpected oxidation behavior characteristic of small metal cluster reactions and point to the potential for developing new insights on the nature of chemical reactivity. We consider several examples.

"Oxidation of Metal and Metalloid Cluster Systems"

We have developed entrainment flow devices[6,7] which facilitate the generation of a substantial and usable continuous flux of small metal clusters. Using these devices, it has been possible to record the optical signatures of the reaction products associated with the oxidation of small boron (B_x+NO_2,N_2O),[2(a)-(c)] copper (Cu_x+O_3,Cl_2,Cl,F_2,F),[2(a),8,9] silver (Ag_x+O_3,Cl_2),[2(a),(d)] manganese (Mn_x+O_3),[2(b),(e)] chromium (Cr_x+O_3,F_2),[2,10] and magnesium (Mg_x+F,Cl) clusters. With the generation of a sufficient metal cluster flow, subsequently oxidized, we obtain quantal information on the energy levels of several asymmetric metal clustered oxides, M_nO_y ($n\geq2$) and preliminary data on the copper, chromium, and magnesium clustered chlorides and fluorides. Thusfar, this work has led to three general observations concerning cluster reactions.

(1) Cluster oxidations, through a multicentered reaction capability (excluding fluorine and chlorine atom rxns.), often yield product molecules in higher energy states than have been accessible to the corresponding atomic reactions.

(2) The products of metal cluster reactions encompass metal rich molecules. In many cases, metal-metal bonds are present and behave with an unusual fluxionality in these product molecules.

(3) Kinetics rather than thermodynamics most often controls the nature of the initially formed products of metal cluster oxidation.

Quantum level probes of the products of metal cluster oxidation are being developed with a current emphasis on two distinct source configurations. In one configuration a stream of metal clusters formed through the "supersonic expansion" of the metallic element of interest is made to intersect a selected oxidant (modified beam-gas configuration), the products of reaction being studied using a combination of chemiluminescent and laser fluorescent techniques. This configuration is being used to study the sodium trimer-halogen atom reactions.[4] The second, more versatile source configuration to which we have alluded to above lies intermediate to a low pressure molecular beam and high pressure flow device.[6,7] Clusters are formed from a high metal flux source and further agglomerated by an entraining argon or helium flow maintained at room to liquid nitrogen temperature. These clusters are subsequently oxidized by a concentric oxidant flow, the resulting products being passed to a small mass spectrometer system. Using this agglomeration flow source, we have successfully obtained the first quantal information on the energy levels and optical signatures of several metal cluster oxides and select halides (M_nO, M_nX). In these studies, it is possible to **combine the energetic constraints of the chemiluminescent process, distinct optical signatures, and mass spectrometry for product identification.**[6]

"Copper Cluster Oxidation"

The reaction exothermicity for the process

$$Cu + O_3 \rightarrow CuO + O_2 \tag{1}$$

is 1.76 ± 0.05 eV. This process is energetic enough to populate the v' = 2 level of the CuO $\delta^2\Sigma^+$ state. The chemiluminescent spectrum observed for reaction (1) under near single collision conditions with atomic copper dominant is in excellent agreement with the available energy for the atomic oxidation process.

If the energy liberated in a multicentered copper cluster reaction is largely collected in one of the reaction product molecules, a considerable enhancement of the available energy to populate product molecule excited electronic states can be obtained. The oxidation of copper clusters produced in an entrained copper flow (copper flux $\geq 10^{18}$ /cm^2-sec) with ozone under multiple collision conditions produces emission (Figure 1) which originates from the $D^2\Delta$, $C^2\Pi$, $A^2\Sigma$, $B^2\Delta$ and $A'^2\Sigma$

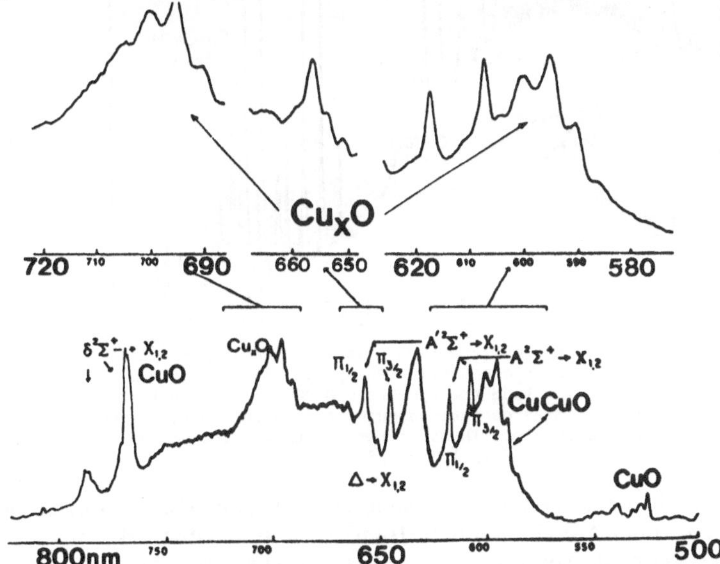

Figure 1: Chemiluminescent spectrum resulting from the multiple collision ($P_{Tot} \approx 600$ mTorr) oxidation of small copper molecules and clusters. Clusters are formed in a moderate copper agglomeration mode and oxidized with ozone. The emission spectrum, taken with an EMI 9808 phototube at a resolution of 0.8 nm is dominated by CuO, and Cu$_x$O (x $\geq$ 2) emission features where the polyatomic emitters correspond to the copper clustered oxides.

states of CuO. These emission features are not observed under conditions in which one simply increases the copper atom flux promoting a substantial increase in the CuO $\delta^2\Sigma^+$ emission. The population of the higher-lying states in CuO requires more energy than is available from the copper atom reaction. This increased exothermicity could be provided by the reaction of vibrationally hot copper dimer or, equally likely with moderate to high metal agglomeration, by the reaction of ground state copper trimer via the process

$$Cu_3 + O_3 \rightarrow CuO^* + Cu_2O_2 \ . \tag{2}$$

Under a variety of experimental conditions, the first emission spectra for both the asymmetric copper clustered oxides and the symmetric CuOCu molecule have now been successfully generated.[8] A selection of the data obtained for these copper oxides is indicated in Figures 1 and 2 where emission spectra for both symmetric and asymmetric Cu_2O as well as the higher order copper clustered, Cu_xO ($x \geq 2$) complexes[6] are displayed.

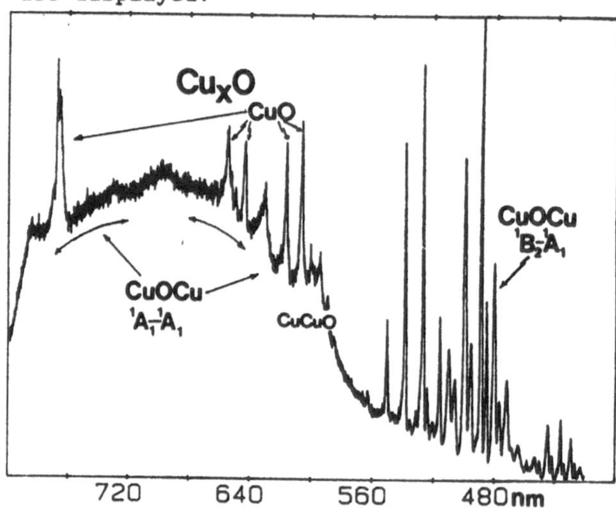

Figure 2: Chemiluminescent spectrum taken under multiple collision conditions in a non-effusive agglomeration mode dominated by emission from the symmetric CuOCu molecule but also showing emission corresponding to CuO and to the asymmetric CuCuO molecule. The spectrum was taken with an EMI 9808 phototube at a resolution of 0.5 nm.

Significant differences between the electronic spectra and energy levels for the CuCuO and CuOCu molecules are evidenced not only by the distinctly different location of emission features but also by the appearance, extent, and intensity of these features. The spectral features for the CuCuO molecule (Fig. 1) at $\sim$ 600 nm appear to correspond to a short progression in a 132 cm^{-1} bending mode whereas the observed structured emission spectrum for the CuOCu $^1B_2$ - $^1A_1$ transition appears to be dominated by moderate progressions in the ground and

excited state symmetric stretching modes of the symmetric isomer·
The spectral features recorded for CuOCu bear a strong correlation with
recent quantum chemical calculations by Bauschlicher, Langhoff, and
Siegbahn[11] (Table I). These authors have estimated bond lengths, bond
angles, transition moments, and electronic state locations for the
ground and several excited states of the CuOCu molecule. The
calculations are

Table I

Calculated Properties of Ground and Excited States of Symmetrical Cu_2O

STATE	r(Cu-O) (Angstroms)	BOND ANGLE (Degrees)	T_e(APPROX.) (cm^{-1})
$^1A_1$ (Ground)	1.793	105.7	0
$^1B_1$ (0.3)[a]	1.964	81.4	14221
$^1A_1$ (1.6)[a]	1.921	180.0	17181
$^1B_2$ (1.0)[a]	1.956	77.3	20336
$^1A_2$ (0.1)[a]	1.942	123.3	22133

[a]Transition Moment in Atomic Units.

currently undergoing further refinement, especially in correlation with
the data derived from the emission spectra depicted in Figure 2.
However, the relative intensities of the CuOCu $^1A_1$ and $^1B_2$ transitions
depicted in Figure 2 are in very good agreement with the calculated
relative transition moments. Further, the change in bond angle and bond
length predicted for the $^1A_1$ - $^1A_1$ transition (θ - a substantial 75°)
is commensurate with a long virtually unresolved progression dominated
by the CuOCu bending mode, much like that characterizing a similar
transition in the water molecule at the fringes of the vacuum ultra-
violet region.[12] For the $^1B_2$ - $^1A_1$ transition, the experimental data
indicates a ground state symmetric stretch frequency of   640 cm^{-1}
($v'' = 640 \pm 10$ cm^{-1}) and an excited state symmetric stretch normal mode
frequency of 409 ± 10 cm^{-1}.

The spectroscopy of the copper cluster oxides is now the focus of
considerable further study. This effort includes laser induced fluorescence at moderate and higher resolution to facilitate the evaluation
of further vibrational mode structure and the determination of both the
CuCuO and CuOCu ground state geometries. Potential functions now
generated for CuOCu and $CuCuO^{14}$ indicate that these molecules are quite
bent although they are also characterized by very flat bending mode
potentials. It appears that CuCuO can undergo a facile isomerization to
CuOCu with a high probability for interconversion of the two isomers.

A study of the isomer interconversion may bear some relevance to the assessment of the role which copper oxide plays in high T_c superconductors.[13] Here the movement of the copper and oxygen atoms as dictated by the Cu_xO potential function, especially the vibrational modes associated with out-of-plane bending, may play a role in the high T_c mechanism.

A Continuous Visible Chemical Laser Amplifier from a Metal Cluster Oxidation Reaction"

The alkali metals are particularly amenable to detection with transition probabilities among alkali metal atom or dimer electronic states being among the largest recorded.[14] Therefore, with a desire to initiate efforts focused on the study of metal cluster oxidation, it seemed appropriate to study what were felt to be the simplest metal cluster oxidation reactions, the alkali trimer-halogen atom reactions, producing reaction products with well defined and characterized electronic transitions. In retrospect, the study of alkali trimer - halogen atom reactive encounters has demonstrated several surprises.

The high cross section, highly exothermic $Na_3 - X$ (Cl,Br,I) reactions form Na_2^* in several of the sodium dimer excited electronic states[15] indicated schematically in Figure 3. The energetics of the

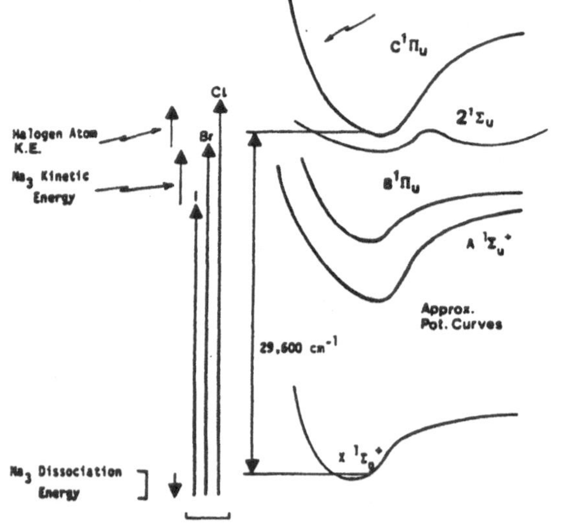

Figure 3: Energetics associated with the formation of Na_2 produced by the Na_3 - X (Cl,Br,I) chemiluminescent reaction. Potential curves are drawn approximately.

reactive processes of interest are indicated in the figure. The Na_3-Cl and Na_3-Br reactions are sufficiently exothermic to readily populate the A $^1\Sigma_u^+$, B $^1\Pi_u$, C' (2 $^1\Sigma_u^+$ - X $^1\Sigma_g^+$) and C $^1\Pi_u$ states of sodium dimer. The available energy results from the formation of a moderately strong sodium halide bond and the rupture of a weak sodium trimer bond. The Na_3-I reaction is much less exothermic. However, the contribution of the Na_3 kinetic energy and the reaction of halogen atoms, generated from a 1500 K source, in the high energy tail of their kinetic energy distribution allows the population of a few vibrational levels in the Na_2 C $^1\Pi_u$ state and several levels of the double minimum C' state. The optical signatures for the processes

$$Na_3 + Cl,Br,I \rightarrow Na_2^* + NaX \quad (X = Cl,Br,I) \tag{3}$$

encompass emission from a limited number of Na_2 band systems including the A, B, C, and C' states. Surprisingly, the observed emission is characterized by sharp, well defined, emission regions[15] (Figures 4,5)

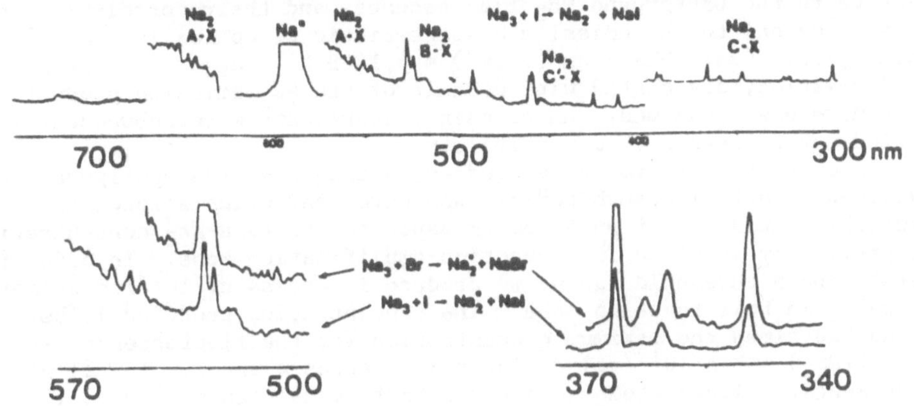

Figure 4: Chemiluminescent emission from the $Na_3 + Br$ and $Na_3 + I$ reactions forming excited states of Na_2, whose optical signature dominates the observed emission, and the sodium halide. Sharp emission features superimposed on a broad background are apparent including those at 527, 492, 460.5, 436, and 426 nm. Spectral resolution is 0.6 nm. See text for discussion.

Figure 5: Comparison of (a) observed and (b) calculated emission spectra for the Na_2 B-X emission system. The experimental spectrum corresponds to chemiluminescence from the Na_3-Br reaction. The calculated spectrum, which was obtained for a rotational temperature, $T_{Rot} \approx 1000K$, represents an estimate of _effective_ rotational temperatures for Na_2 product formation under near single collision conditions and therefore not at equilibrium. Relative vibrational populations input for Na_2 B-X, v' = 0-6 were in the ratio 1.00:1.17:1.33: 1.50:1.58:1.67:1.54. The location of contributions from vibrational levels v' = 6 (__), 5(---), and 4(_._) of the Na_2 B state in transition to vibrational levels v'' = 14-9, v'' = 13-8, and v'' = 12-7 of the Na_2 ground state are indicated.

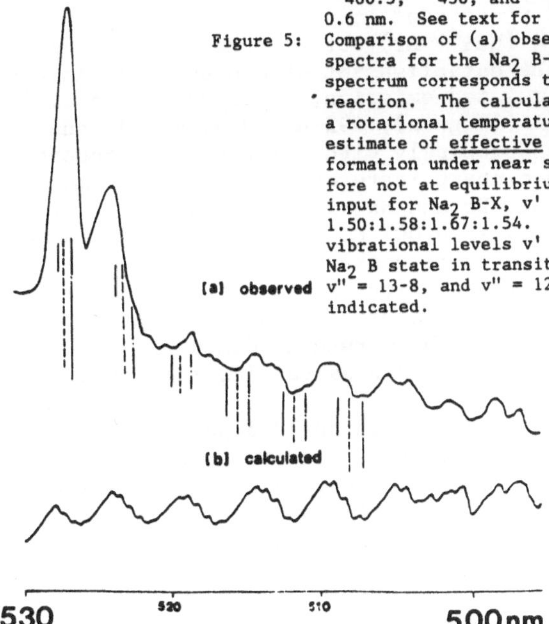

superimposed on a much weaker but perceptible Na_2 dominated background. As Fig. 5 demonstrates, through comparison with the best fit calculated sodium dimer spectrum, these sharp emission features are not readily explained by invoking a purely fluorescent process involving sodium dimer.

The sharp nature of several of the B-X, C-X, and C'-X Na_2 emission features (Fig. 4), their near exponential growth with Na_3 concentration relative to the background Na_2 fluorescence, and their correlation in certain regions to the emission characteristic of optically pumped Na_2 laser systems (ex: 528.2 nm (v',v'') = 6,14(B-X)) suggested that stimulated emission, associated with certain of the Na_2 emission products might have been observed. Laser gain measurements were subsequently carried out to assess this possibility.

In order to perform these studies, a unique source configuration, discussed in detail elsewhere[6,15], was developed which allowed the supersonic expansion of pure sodium vapor to create a Na_3 concentration not previously attained in a reaction-amplification zone. In order to operate above threshold (gain) we produce in excess of 10^{13}/cc trimer molecules in this reaction zone. The concentration produced is between 10 and 500 times the maximum concentration for the fluorescence experiments ($[Na_3] \sim 3 \times 10^{12}$/cc) depicted in Figures 4 and 5. In all of the experiments, halogen atoms were produced from halogen molecules in a high temperature carbon furnace[16] with $\sim 95\%$ efficiency. These halogen atoms exit the furnace into the reaction zone initiating the trimer-halogen atom chemical reaction. The halogen flow is adjusted to optimize gain.[15(b)]

Laser gain and hence amplification is found to be characteristic of several of the sharper and more pronounced emission features apparent in the spectra depicted in Figure 4 (Na_3 - Br) corresponding to a stimulated emission process and to the establishment of a population inversion. Optical gain through stimulated emission, measured at 0.5 cm^{-1} resolution in the regions 527 nm (1% gain), 492 nm (0.3% gain), and 460.5 nm (0.8% gain), correlates precisely with the reactive process and the relative intensities of those sharp features observed while monitoring the light emitted from the Na_3-Br and Na_3-I reactions (Figs. 4,5). High resolution ring dye laser scans (0.007 cm^{-1}) in the 527 nm region indicate that the gain for the system is close to 3.8% for an individual rovibronic transition with approximately four to five transitions near 527 nm showing gain. At 459.8 nm, a gain of 2.3% has been measured for an individual rovibronic transition. These results demonstrate the continuous amplifying medium for a visible chemical laser in at least three wavelength regions.[15]

The gain observed in these systems is strongly dependent on both the Na_3 and halogen atom concentration. As the probe laser power was changed, the amplification signal was found to vary linearly indicating that the measurements were taken in the small signal gain regime. In order to further verify that these gain observations were not artifacts which could be observed at any wavelength including those corresponding to an intense spectral feature, the gain experiment was repeated at several wavelengths including the sodium D-line wavelength at 588.9 nm. No gain was observed at any other wavelength in the 420-600 nm region.

At the Na D-line wavelength the signal was seen to decrease significantly, indicating considerable scattering or absorption of the probe
laser photons.

Because of the low Na_3 ionization potential and the high halogen
electron affinities,[17] the Na_3-halogen atom reactions are expected to
proceed via an electron jump mechanism with extremely high cross sections,[18] producing substantial Na_2 excited state populations. The
created population inversions monitored thusfar are thought to be
sustained by (1) the large number of free halogen atoms reacting with
Na_2 molecules in those ground state levels on which the transitions
emanating from the Na_2 excited states terminate and (2) collisional
relaxation of ground state sodium molecules.[15] The cross section for
reaction of vibrationally excited ground state Na_2 is expected to be
substantial relative to that corresponding to collision induced vibrational deactivation of the Na_2 manifold. Extremely efficient reactions
thus greatly aid the rapid depletion of the lower state levels in this
system allowing one to sustain a continuous population inversion. The
current results certainly indicate an unusual and unexpected oxidation
behavior which may be manifest in a number of metal cluster reactions.
This suggests that the study of these processes will aid the development
of new insights on chemical reactivity.

"Extensions of Metal Cluster Oxidation Studies"

The oxidation studies which we have exemplified for the copper and
sodium systems and briefly outlined for other reactive combinations are
now being extended to additional metal and metalloid based constit-
uencies. A source which we have developed to study the $Bi_2 + F$
reaction[19] is now being used to investigate the fluorine atom oxidation
of copper[11] and magnesium[11] clusters. The fluorine atom oxidations of
small copper molecules are of particular interest as we consider the
analogy between the single valence s electron copper and sodium
oxidation reactions. These copper studies also correlate strongly with
the elegant work of Parson,[20] Sadeghi,[21] and coworkers.

We are completing studies of the CuF chemiluminescence from the
copper dimer - fluorine atom reaction. The effort is being extended to
form the electronically excited copper cluster fluorides $Cu_{x-1}F$ from Cu_x
($x \geq 3$) + F reactive encounters. Also, in an attempt to generate the
Mg_2 excimer analogs of the Na_2 laser amplifiers discussed previously, we
are attempting to form magnesium clusters and observe the excited state
products of their oxidation with F and Cl atoms. Surprisingly, preliminary results on this system signal the formation of excited state Mg_xF
(Fig. 6) and Mg_xCl charge transfer complexes where x is most likely
two. The chemiluminescent spectra depicted in Fig. 6 are thought to
result in part from the magnesium trimer reaction which, under the
experimental conditions surveyed thusfar, appears to produce minimal
Mg_2^* fluorescence. These systems are now the subject of continued study
in our laboratory.

Our initial observations[6,7] suggest that larger cluster oxide
species with significant metal-metal bonding can be made and explored
using combined chemiluminescent and LIF techniques. It may be possible

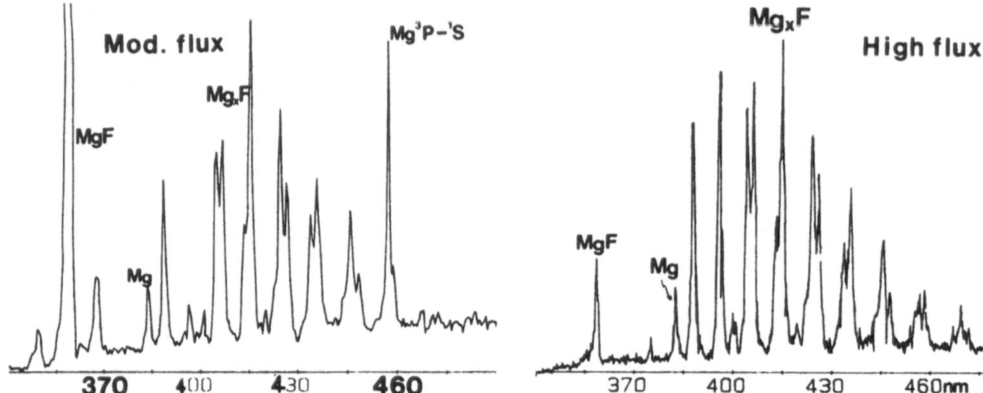

Figure 6: Chemiluminescent spectra resulting from the reaction of small magnesium molecules (Mg$_2$, Mg$_3$), formed in a high metal mole fraction agglomeration flow of dry ice cooled helium, with fluorine atoms. The observed spectra correspond to MgF (A$^2\Pi$ - X^2 +), Mg ^{3}P - ^{1}S, ^{3}S - ^{3}P, and ^{3}D - ^{3}P atomic, and Mg$_x$F emission features, which vary in intensity with increasing mole fraction of magnesium. The progression of bands associated with the Mg$_x$F emitter (charge transfer complex), where x is likely 2, appear to correspond to a progression in the MgF stretch on which is superimposed features associated with an Mg$_2$+ moiety. Preliminary studies indicate that MgF a$^2\Pi$ and Mg ^{3}P are formed from the Mg$_2$ + F reaction, Mg ^{3}D and Mg ^{3}S are formed from an energy pooling process involving Mg ^{3}P and MgF A$^2\Pi$ and the Mg$_x$F charge transfer emission results from the Mg$_3$-F reaction. The spectra are taken with an RCA 1P28 phototube at 1 nm resolution.

to generate species, the fingerprints of which are relevant to the detailed microscopic description of those properties which can contribute to the catalytic behavior of an oxidized metal surface or, in a future study of silicon cluster halogenation, to the nature and quality of a surface etch (Si$_n$ + X (X=Cl,Cl$_2$,F,F$_2$). As an ultimate goal, we wish to develop a detailed description of the intimate environment associated with the metal cluster-oxygen or metal cluster-halogen interaction, (1) determining how small clusters of metal atoms interact with the oxygen or halogen atom and (2) considering the dynamic behavior which these clustered atoms may exhibit as they move about the oxygen or halogen atom. Using a combination of chemiluminescent and laser fluorescent probes of the metal clustered oxides, it should be possible to establish structures and determine, through bond angle and vibrational frequency evaluation, the manner in which these small metal cluster groupings interact with an oxygen atom when they are formed in a unique kinetically controlled environment. It is precisely this information which can provide the productive tension between experiment and theory required for the development of systematically constructed and meaningful model systems describing the nature of ligand-metal surface interactions.

ACKNOWLEDGEMENT

We thank the National Science Foundation, the ACS-PRF, the Eastman Kodak Company, the Georgia Tech Foundation through a grant from Mrs. Betty Peterman Gole, and the Army Research Office and the Air Force Office of Scientific Research. This work would not have been possible without the capable assistance of J. R. Woodward, S. H. Cobb, K. X. He, M. McQuaid, K. K. Shen, C. B. Winstead, T. R. Burkholder, D. Grantier, R. Kahlscheuer and J. Bray, and collaboration with T. C. Devore, D. A. Dixon, S. Langhoff, and C. W. Bauschlicher.

REFERENCES

1. Na_3: (a) A. Herrmann, M. Hofmann, S. Leutwyler, E. Schumacher, and
   L. Woste, Chem. Phys. Lett. 62 (1979) 216; (b) J. L. Gole, G. J.
   Green, S. A. Pace, and D. R. Preuss, J. Chem. Phys. 76 (1982) 2247;
   (c) G. Delacretaz, E. R. Grant, R. L. Whetten, L. Woste, and J. W.
   Zwanziger, Phys. Rev. Lett. 56 (1986) 2598.
   Cu_3: (d) M. D. Morse, J. B. Hopkins, P. R. R. Langridge-Smith, and
   R. E. Smalley, J. Chem. Phys. 79 (1983) 5316; (e) W. H. Crumley,
   J. S. Hayden, and J. L. Gole, J. Chem. Phys. 84 (1986) 5250;
   (f) E. A. Rohlfing and J. J. Valentini, Chem. Phys. Lett. 126 (1986)
   113.
   Ag_3: (g) P. Y. Cheng and M. A. Duncan, Chem. Phys. Lett. 152 (1988)
   341.
   Al_3: (h) Z. Fu, G. W. Lemire, Y. M. Hamrick, S. Taylor, J.-C. Shui,
   and M. D. Morse, J. Chem. Phys. 88 (1988) 3524.
   Ni_3: (i) K. M. Ervin, J. Ho, and W. C. Lineberger, J. Chem. Phys.
   89 (1988) 4514. (j) R. W. Woodward, S. H. Cobb, and J. L. Gole,
   J. Phys. Chem. 92 (1988) 1404.
   Cu_4^+: (k) M. F. Jarrold and K. M. Creegan, Chem. Phys. Lett. 166
   (1990) 116.
   C_x: (1) P. F. Bernath, K. H. Kinkle, and J. J. Keady, Science 244
   (1989) 562; (m) J. Heath, A. L. Cooksy, M. H. W. Gruebele, C. A.
   Schumuttermaer, and R. J. Saykally, Science 244 (1989) 564;
   (n) N. Mozzen-Ahmadi, A. R. W. McKellar, and T. Amano, J. Chem.
   Phys. 91 (1989) 2140.
   Si_x: (o) T. N. Kitsopoulos, C. J. Chick, A. Weaver, and D. M.
   Neumark, "Vibrationally Resolved Photoelectron Spectra of Si_3^- and
   Si_4^-", J. Chem. Phys. 93 (1990) 6108. (p) C. B. Winstead, K. X. He,
   T. Hammond, and J. L. Gole, "Electric Field Enhanced Laser Induced
   Plasma Spectroscopy of Jet Cooled Silicon Trimer", Chem. Phys.
   Letts. 181 (1991) 222.
2. (a) R. W. Woodward, P. N. Le, M. Temmen, and J. L. Gole, J. Phys.
   Chem. 91 (1987) 2637; (b) J. L. Gole, "Quantum Level Probes of Small
   Metal Clusters and Their Oxidations", American Institute of Physics
   Conference Proceedings, No. 160, Advances in Laser Science II -
   Optical Science and Engineering Series 8, pg. 439; (c) T. C. Devore,
   R. W. Woodward and J. L. Gole, J. Phys. Chem. 92 (1988) 6919; (d)
   R. W. Woodward, P. N. Le, T. C. Devore, D. A. Dixon, and J. L. Gole,
   J. Phys. Chem. 94 (1990) 756; (e) T. C. Devore, J. R. Woodward, and
   J. L. Gole, J. Phys. Chem. 93 (1989) 4920; (f) M. J. McQuaid and
   J. L. Gole, "Stability and Oxidation of Metal Based CO and CO_2
   Complexes", Proceedings of the Fourth International Laser Science
   Conference, A.I.P. Conf. Proc. No. 191, Optical Science and
   Engineering Series 10, pg. 687.
3. T. C. Devore, M. McQuaid, and J. L. Gole, High Temp. Science 29
   (1990) 1.
4. S. H. Cobb, J. R. Woodward, and J. L. Gole, Chem. Phys. Lett. 143
   (1988) 205; 156 (1989) 197.
5. T. C. Devore and J. L. Gole, "Oxidation of Small Metal Clusters",
   Proceedings of the Sixth International Conference on High

Temperature Materials, High Temperature Science 27 (1989) 49.
6. "The Unique Dynamics of Metal Cluster Oxidation and Complexation", to appear in <u>Advances in Metal and Semiconductor Clusters</u>, Vol. I, Spectroscopy and Dynamics, ed. M. A. Duncan, JAI Press, in press.
7. Note that this present approach bears some resemblance to the liquid nitrogen entrainment - induced agglomeration of metal clusters used by Stein and coworkers and Solliard. These authors formed much larger aggregates which they studied using electron diffraction techniques. See for example, (a) B. G. de Boer and G. D. Stein, Surf. Sci. 106 (1981) 84. (b) C. Solliard, Ph.D. Thesis, Ecole Polytechnique Federal de Lausanne, Switzerland, 1983. (c) See also Surf. Sci. 106 (1981) 58. J. de Phys. C2 (1977) 167.
8. (a) T. C. Devore, C. W. Bauschlicher, Jr., S. R. Langhoff, Per E. M. Siegbahn, M. Sulkes and J. L. Gole, Formation, Electronic Spectra, and Electronic Structure of the Low-Lying Singlet States of Symmetrical Cu_2O, to be submitted for publication. (b) T. C. Devore, C. W. Bauschlicher, Jr., T. Burkeholder and J. L. Gole, A Comparative Study of the Oxidation of Atomic Copper and Higher Copper Clusters Under a Single and Multiple Collision Conditions; Electronic Structure of the Asymmetric Copper Clustered Oxides, Cu_xO, to be submitted for publication.
9. K. K. Shen, C. Winstead, L. Brock, K. Dulaney, T. Devore and J. L. Gole, work in progress.
10. T. C. Devore and J. L. Gole, Chemical Physics 133 (1989) 95.
11. To be published - see reference 8.
12. See G. Herzberg, <u>Spectra of Polyatomic Molecules</u>, Van Nostrand Reinhold, 1977.
13. D. A. Dixon, private communication.
14. W. C. Stwalley, private communication.
15. (a) W. H. Crumley, J. L. Gole and D. A. Dixon, J. Chem. Phys. 76 (1982) 6439. (b) S. H. Cobb, J. R. Woodward, and J. L. Gole, Chem. Phys. Lett. 143 (1988) 205. (c) S. H. Cobb, J. R. Woodward, and J. L. Gole, Chem. Phys. Lett. 157 (1989) 197. (d) S. H. Cobb, J. R. Woodward and J. L. Gole, "Continuous Chemical Laser Amplifiers in the Visible Region", Proceedings of the Fourth International Laser Science Conference, A.I.P. Conf. Proc. No. 191, Optical Science and Engineering Series 10, pg. 68.
16. (a) W. H. Crumley, Ph.D. Thesis, Georgia Institute of Technology, 1985. (b) S. H. Cobb. Ph.D. Thesis, Georgia Institute of Technology, 1988.
17. See for example, (a) R. S. Berry and C. W. Reimann, J. Chem. Phys. 38 (1963) 1540, (b) R. S. Berry, J. Chem. Phys. 27 (1957) 1288, (c) W. S. Struve, J. R. Krenos, D. L. McFadden and D. R. Herschbach, J. Chem. Phys. 62 (1975) 404. (d) R. C. Oldenborg, J. L. Gole and R. N. Zare, J. Chem. Phys. 60 (1974) 4032.
18. Given Na_2 and Na_3 ionization potentials of 4.87 and 3.97 eV (A. Hermann, E. Schumacher, and L. Woste, J. Chem. Phys. 68 (1978) 2327, and an electron affinity of 3.3363 eV for atomic bromine, we determine a very substantial electron jump cross section $\sigma = \pi$

$(14.38/3.97-3.36) = 1746$ A^2 $(1.75 \times 10^{-13}$ cm^2) for the Na$_3$ - Br reaction and $\sigma = \pi$ $(14.38/4.87-3.36)) = 285$ A^2 $(2.85 \times 10^{-14}$ cm^2) for the Na$_2$ - Br reaction.

19. "On the BiF Bond Dissociation Energy and Evaluation of the BiF Red Emission Band Systems", with T. C. Devore, L. Brock, and K. Dulaney, Chemical Physics, in press.
20. R. W. Schwenz and J. M. Parson, J. Chem. Phys. 73 (1980) 259.
21. P. Baltayan, F. Hartmann, J. C. Pebay-Peyroula, and N. Sadeghi, Chem. Phys. Lett. 120 (1988) 123. N. Sadeghi - private communication of unpublished results.

ENERGETICS OF ATOMIC AND CLUSTER PROCESSES ON SURFACES

TIEN T. TSONG[*]
Institute of Physics, Academia Sinica
Nankang, Taipei, Taiwan 11529 RoC

C. L. CHEN, J. LIU and C. W. WU
Physics Department, The Pennsylvania State University
University Park, Pennsylvania 16802 USA

ABSTRACT. The energies involved in various surface atomic and cluster processes, such as interaction between two surface atoms, dissociation of an atom from a cluster, a surface layer or a crystal, and cohesion of small surface clusters etc., have been measured with the FIM and the atom-probe FIM. Structure transformations of small surface clusters have been observed and the differences in the cohesive energies of different cluster structures have been measured also.

1. INTRODUCTION

One of the subjects of considerable current interest is how atoms interact to form a cluster and to grow in size to eventually become a crystal. In this growth process, atoms are condensed on a small surface by vapor deposition. These atoms will interact with one another as well as with the surface. Often, before they are locked into stable sites, they will diffuse on the surface, encounter with one another and form small surface atomic clusters.[1] Behavior of these small atomic clusters can be very interesting and is often puzzling also. One may consider a cluster on a surface a complex system which has a finite system (the cluster) coupled to a very large system (the surface). Using the field ion microscope (FIM) and the atom-probe FIM, it is possible to study the mechanisms and energetics of many of these surface atomic and cluster processes. We will report some of our recent studies related to the subject of this conference here. It is most encouraging to note that recent developments in theoretical techniques make possible to calculate of the energetics of surface atomic and cluster processes.[2]

2. SURFACE ATOMS

2.1. Diffusion

Using the field ion microscope (FIM),[3] single metal and semiconductor atoms on metal surfaces can be directly imaged. In particle theories of random walk diffusion, essential data are the atomic jump directions, the jump-length distributions, the displacement distributions and the mean square displacements, and their temperature dependences. All these data can be derived directly from FIM

P. Jena et al. (eds.), Physics and Chemistry of Finite Systems: From Clusters to Crystals, Vol. II, 1039–1045.
© 1992 *Kluwer Academic Publishers.*

images. Most FIM data of surface diffusion of single adsorbed atoms can be interpreted to occur by atomic hopping over the saddle points of the atom–surface interaction as illustrated in Fig. 1(a). The diffusion coefficient D_s is related to the pre-exponential factor D_0, which is related to the mean square displacement according to the Einstein relation, and the activation energy E_d according to $D_s = D_0 \exp(-E_d/kT)$. The activation energy is found to be very specific to the diffusion system. For metal and Si adatoms on various surfaces of Ni, Rh, W, Ir, and Pt etc., it ranges from about 0.1 to 1.0 eV. The pre-exponential factor, which is related to the frequency of small oscillation of the diffusing atom, on the other hand, is not sensitive to the diffusion system. It is always in the range of $\sim 10^{-3}$ cm^2/s.

Another mechanism of surface diffusion is the atomic replacement mechanism as illustrated in Fig. 1(b).[4] Atomic-replacement was thought to be a viable adatom diffusion mechanism for all systems. However, for a hetero-system, once a foreign adatom goes into a substitutional site in the surface, it is stable and cannot be replaced out to an adatom site again. Thus the atomic replacement is a viable mechanism for self-diffusion. For hetero-systems, it is the earliest stage of surface alloying, not a viable mechanism for adatom diffusion. The energy needed for the atomic replacement is quite low, only less than 1 eV for refractory metal adatoms on fcc {001} surface of refractory metals.

2.2. Binding with the Substrate

The binding energy of an atom with the substrate should depend on the atomic configuration of the site the surface atom sits. In low temperature field evaporation, atoms are field evaporated one by one from kink sites of the surface. By measuring the kinetic energy of the most energetic ions in low temperature field evaporation, the binding energy of kink site atoms can be determined. Using an ultra-high resolution pulsed-laser time-of-flight atom-probe FIM, we have measured the binding energies of kink site atoms of several metals which all agree with the cohesive energies of these metals to within 0.2 eV.[5] It can be easily argued that the binding energy of kink site atoms should indeed be equal to the cohesive energy.

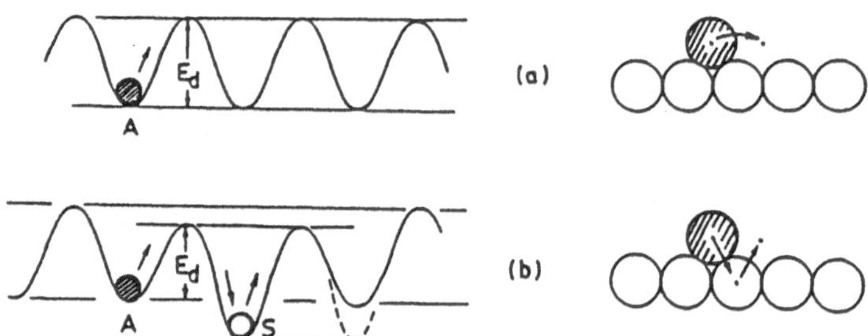

Fig. 1 (a) Random walk diffusion by atomic hopping. (b) Random walk self-diffusion by atomic replacement.

The binding energy of self-adsorbed atoms can be determined from a measurement of the dissociation energy of kink site atoms, at the edges of a surface layer, to the terrace sites. This measurement has been done for the Ir {001} surface.[6] The binding energy is found to be 6.38 eV which is smaller than the binding energy of kink site atoms, 6.94 eV, by 0.56 eV.

3. SURFACE CLUSTERS

3.1. Cluster Cohesion

At a sufficiently high temperature, condensed atoms will diffuse on the surface. In their diffusion motion, they may encounter with one another and may associate into clusters of different sizes. The stability of a cluster depends on the cohesion of atoms in the cluster as mediated by the substrate. The substrate usually exerts a very strong effect on these atoms. Adatoms behave completely differently from the atomic state. For an example, adatom-adatom interactions on metal surfaces are surprisingly weak, often have a strength of only about one tenth of a chemical bond, or only on the order of 0.1 eV for most adatom pairs on metal surfaces.[7] Adatom-adatom interactions are long-ranged and may exhibit small oscillatory tails. In a few exceptional cases such as Ir-Ir pairs on the Ir {110} surface, the cohesion is quite strong, or 1.10±0.11 eV at the closest bond separation of ~2.8 Å.[6] This is comparable to the strength of a typical chemical bond.

3.2. Cluster Diffusion

Similar to surface diffusion of single atoms, diffusion of small atomic clusters can occur by hopping of individual atoms or by atomic replacement in a concerted motion of all the atoms involved. W, Re and Ir adatoms in the adjacent atomic channels of the W {112} surface tend to form chain-like clusters with the atoms assuming only two bond lengths, presumably 4.47 Å and 5.32 Å. Diffusion of such chain-like structures can be directly seen to occur by random but coupled hopping of individual atoms.[8] A detailed study of diffusion of W-dimers on the W {110} surface also conclude that the two atoms hop nearly independently by stretching the bond during the hopping. The activation energy of dimer diffusion should then be approximately equal to that of single atoms plus the difference in the potential energy of the dimer at the equilibrium bond separation and at the stretched bond length, or should be different from single atom diffusion by only 0.1 to 0.3 eV. Fig. 2 shows field ion images of how a three atom Ir cluster is formed by gradual combination of three Ir adatoms from three different surface channels of an Ir (110) surface.

Atomic replacement surface cluster diffusion has been studied for Ir-dimers on the non-reconstructed Ir {110} surface. The activation energy is about 0.40 eV higher than that of single Ir adatoms, or about 1.15 eV vs 0.75 eV.

3.3. Cluster Structures and Structure Transformation

A cluster often can assume more than one structure.[8] For an example, small Ir and Pt clusters on the Ir and Pt {111} and {001} surfaces can assume either one-dimensional or two-dimensional structures.[9,10] Also, whereas one normally would

1042

think that a cluster of an fcc metal on its surfaces should form a structure with the closest possible bond configuration, thus the cluster should have a 2-D or a 3-D structure, for Ir and Pt atoms on their {001} surfaces, linear-chain structures are often more stable than 2-D island structures if the cluster size is very small.[11] For Pt-clusters on the Pt {001} surface, the stability of the 1-D vs. 2-D structures oscillates with the number of atoms in the cluster.[12] Fig. 3 shows how three Ir-dimers, formed from six single Ir adatoms, combine to form a six-atom cluster and how this cluster changes from a 2-D to a 1-D structure and vice versa.

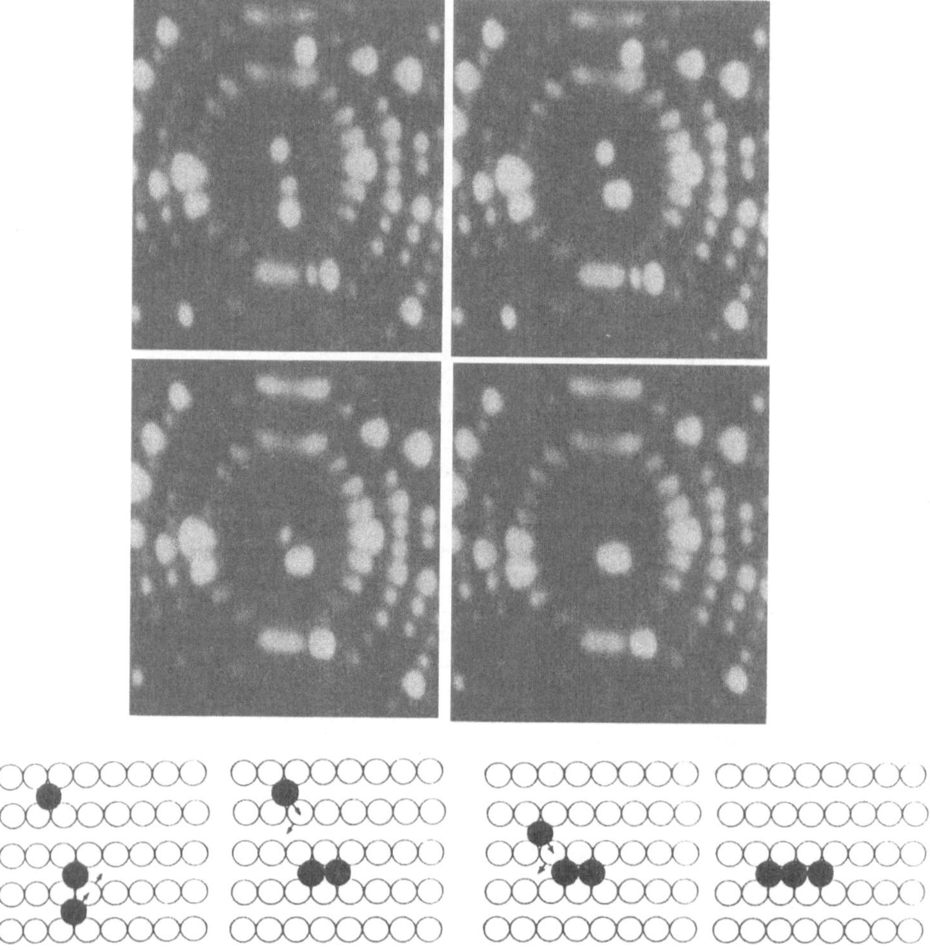

Fig. 2 Field ion images and diagrams showing how a 3-atom Ir cluster is formed by gradual association of three Ir adatoms from different surface channels of an Ir (110) surface.

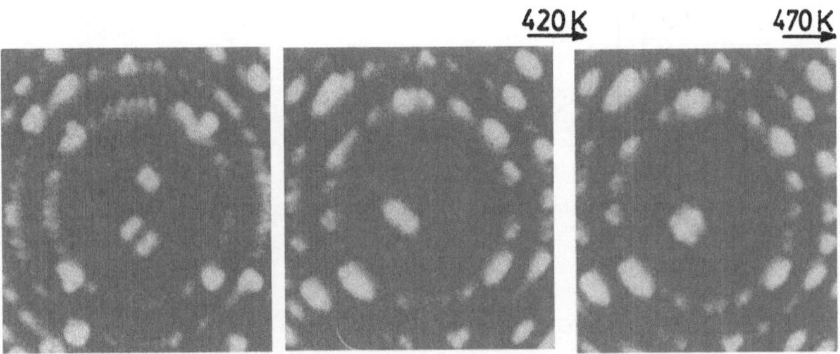

Fig. 3 Field ion images showing how three Ir-dimers, formed from vapor deposited Ir adatoms, on an Ir (001) surface combine into a six-atom cluster and how this atomic cluster transform between a 1-D and a 2-D structure.

A measurement of the difference in the internal binding energies of 3-atom Ir clusters on the Ir {111} and {001} surfaces has recently been made.[17] On the {111} surface, 3-atom Ir clusters with the closely-packed 2-D equilateral triangular structure is more stable than the 1-D linear-chain structure (see Fig. 4). The difference in the internal binding energies of the two structures derived from the data shown in Fig. 5(a) is about 0.10 eV. On the {001} surface, the situation is reversed, or the 1-D linear-chain structure is more stable with a lower binding energy of 0.335 eV as shown in Fig. 5(b). In addition, there is a significant

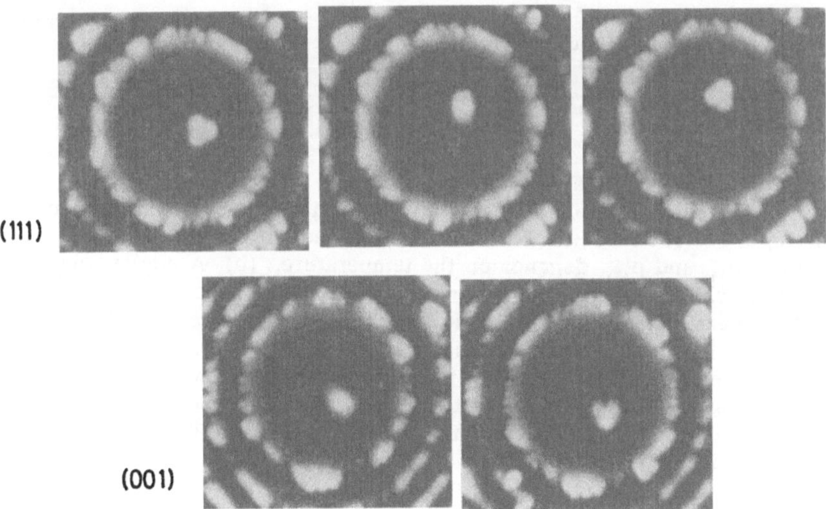

Fig. 4 Field ion images showing the 1-D and 2-D structures of 3-atom Ir clusters on the Ir {111} and {001} surfaces.

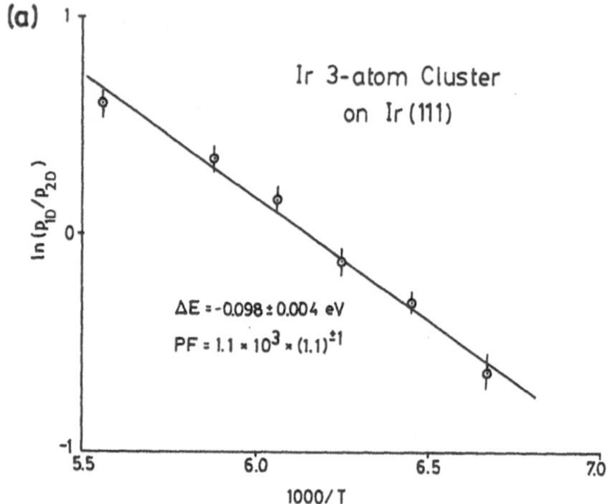

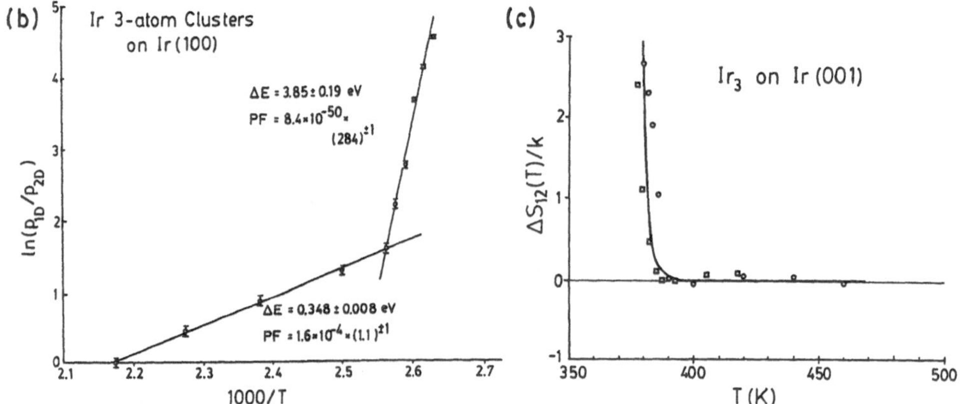

Fig. 5 (a) An Arrhenius-like plot showing how the ratio of the probabilities of observing a 3-atom Ir cluster on an Ir (111) surface in the 1-D and 2-D structures, p_{1D} and p_{2D}, depends on the temperature. (b) A similar plot for a 3-atom Ir cluster on an Ir (001) surface. Note the significant deviation from the linear behavior below 385 K which can be interpreted to have a temperature dependent entropy factor of the 3-atom cluster. (c) The entropy factor, or $k\ell n(W_{1D}/W_{2D})$, is found to be temperature dependent for 3-atom Ir cluster on the Ir (001) surface. W_{1D} and W_{2D} represent the statistical weights in phase space of the 1-D and 2-D structures. $\Delta S_{12}(0) = S_1 - S_2$ is $(7.0 \pm 2.5)k$ on the (111) and $(-8.3 \pm 1.5)k$ on the (001).

deviation from the linear behavior in the Arrhenius plot which can be interpreted to have a temperature dependent entropy term for this cluster as shown in Fig. 5(c). We believe the sudden deviation at a threshold temperature of about 385 K is an indication of a structure phase transition. Although phase transition can occur only for a system with a large number of interacting particles and should not be expected for a cluster with only three atoms, one has to recognize that these 3 atoms not only interact among themselves, but also interact with substrate atoms by inducing an elastic deformation as well as an electronic charge re-distribution of the surface around the 3-atom cluster. In other words, a 3-atom cluster on a surface is not an isolated system with only three atoms. A structure phase transformation having a size effect might be invoked to explain the sudden deviation of the data from a simple linear Arrhenius behavior around 385 K.

Work reported in this paper was done under the support of NSF.

References

* On leave from the Pennsylvania State University. Holder of a distinguished chair of the NSC of RoC.

1. a. G. Ehrlich and K. Stolt, Ann. Rev. Phys. Chem. 31, 603 (1980); b. T. T. Tsong, Rpt. Progress Phys. 51, 759 (1988); c. papers by D. W. Bassett, and T. T. Tsong in Proc. NATO Adv. Summer Inst. on Surface Mobilities of Solid Materials, Vol. B36, 1983, Plenum, V. T. Binh ed.

2. P. J. Feibelman, Ann. Rev. Phys. Chem. 40, 261 (1989); Y. Li, M. R. Press, S. N. Khanna and P. Jena, Phys. Rev. B41, 4930 (1990).

3. See for an example T. T. Tsong, Atom-Probe Field Ion Microscopy, Cambridge Univ. Press, Cambridge, New York (1990).

4. C. L. Chen and T. T. Tsong, Phys. Rev. Lett. 64, 3147 (1990); G. L. Kellogg and P. J. Feibelman, Phys. Rev. Lett. 64, 3143 (1990); P. J. Feibelman, Phys. Rev. Lett. 65, 729 (1990).

5. J. Liu, C. W. Wu and T. T. Tsong, Phys. Reb. B43, 11595 (1991).

6. C. L. Chen and T. T. Tsong, Phys. Rev. B41, 12403 (1990).

7. T. T. Tsong and R. Casanova, Phys. Rev. B24, 3063 (1981).

8. T. T. Tsong, Phys. Rev. B6, 407 (1972).

9. P. R. Schroebel and G. L. Kellogg, Phys. Rev. Lett. 61, 578 (1988).

10. C. L. Chen and T. T. Tsong, Appl. Phys. A51, 405 (1990).

11. C. L. Chen and T. T. Tsong, Phys. Rev. B42, 1464 (1990); Appl. Phys. A51, 405 (1990).

THE D_2+Ni_{13} REACTION: MODE-SPECIFIC AND STRUCTURE-SPECIFIC FEATURES

J. JELLINEK AND Z. B. GÜVENÇ

Chemistry Division, Argonne National Laboratory, Argonne, Illinois 60439 USA

ABSTRACT. Results of a quasiclassical simulation study pertaining to the qualitative and quantitative aspects of the mode-selectivity and structure-reactivity correlation in the dissociative adsorption of a D_2 molecule on a Ni_{13} cluster are presented. The role of resonances (molecularly adsorbed precursor states) in the reaction is discussed, and the resonances are characterized in terms of their formation probabilities and lifetimes.

1. Introduction

Interactions of atomic and molecular clusters with individual atoms and molecules is a fascinating area of the physics and chemistry of finite systems. Historically, theoretical studies of these interactions emerged as a way of modelling gas - surface interfaces. In the last two decades, however, it became widely accepted that clusters form a new form of matter, which is of paramount interest and importance in its own right. From the fundamental point of view, understanding of cluster properties and of mechanisms of their interactions is necessary to fill in the conceptual gap between atoms and molecules (gas phase) and solids/liquids (condensed phase). The practical interest in clusters and in their interactions stems from the role they play in a variety of natural phenomena (e.g., in the upper atmosphere and in the soil), as well as in different technologically important processes (e.g., in heterogeneous catalysis). From the point of view of the applications, the metal cluster - molecule systems are of particular interest.

Interactions of metal clusters with molecules have been studied experimentally quite intensively in the last few years [1]. General characteristics, such as kinetics and energetics of these interactions, as well as more cluster-specific features, such as structure - reactivity correlation [2], have been probed. The accumulated data provide a challenge to the theory. Most theoretical studies focused on metal surface - molecule interactions, using clusters as models of the surfaces [3]. The clusters were prepared as slabs of crystal lattices and were often assumed to be rigid. Only in a limited number of studies have the electronic properties and the nuclear dynamics of metal cluster - ligand systems been considered for the sake of investigation of cluster interactions per se. Electronic structure calculations [4] yield preferred binding sites and binding energies for adatoms and admolecules, and at least a partial mapping of the system potential energy surface (PES) along selected paths. Dynamical studies [5-8] use model and semiempirical potentials to describe the motion of the constituent nuclei (atoms) and to extract characteristic quantities, such as probabilities, cross sections, rate constants and activation energies of the different processes that can take place.

The first theoretical dynamical treatment of the reaction of a D_2 molecule with Ni_n, n = 4-13, clusters was carried out by Raghavan, Stave and DePristo [5] (RSD) under the assumption that the clusters are rigid. A major finding of this study is that the reactivity of a

P. Jena et al. (eds.), Physics and Chemistry of Finite Systems: From Clusters to Crystals, Vol. II, 1047–1056.
© 1992 *Kluwer Academic Publishers.*

given cluster may strongly depend on its structure. In subsequent treatments [7-8] the restriction of rigidity of the clusters has been lifted. We have performed a detailed quasi-classical simulation study of the $D_2 + Ni_{13}$ collision system. All the degrees of freedom of the system were allowed to evolve in time, and all the possible channels of the interaction (dissociative adsorption, molecular adsorption, and inelastic scattering) were included into consideration. The different processes were studied as functions of the initial rovibrational state of D_2, the collision energy, the structure and, in a limited way, the temperature of Ni_{13}. A partial account of the results has been given [7]. Here we present a further discussion of the reactive interaction of D_2 with Ni_{13} that results in dissociative adsorption of the molecule on the cluster. The theoretical background and the computational details are given in the next section. The results and their discussion with a focus on the structure-reactivity correlation, the roles of the initial state of the molecule and of the molecularly adsorbed precursor (resonance) states are presented in Section 3. Section 4 contains a summary.

2. Theoretical Background and Computational Procedure

The technique we used is quasiclassical trajectory simulations. The time evolution of the $D_2 + Ni_{13}$ collision systems was obtained by solving Hamilton's equations of motion for all the degrees of freedom. The D_2 molecule was prepared in a specified quantized rovibrational state (v_i, j_i) using a quasiclassical prescription [9]; v_i and j_i are the initial vibrational and rotational quantum numbers, respectively. The molecule was placed asymptotically far (8.5 Å) from the center of mass of a nonrotating and nontranslating Ni_{13} and sent with a specified impact parameter b and initial relative translational (collision) energy E_{tr}^i towards the cluster. The cluster was prepared initially in an icosahedral (ico), cuboctahedral (cubo) or hexagonal close-packed (hcp) geometry. Since the two latter structures are metastable, we chose them at $T = 0$ K, where the temperature T is defined as

$$T = \frac{2<E_k>}{(3n-6)k} \; , \tag{1}$$

E_k is the total kinetic energy of the cluster, k is the Boltzmann constant, and $<>$ stands for time-average. The icosahedral cluster was prepared initially at $T = 298$ K; test runs at $T = 0$ K indicated no change in the reactivity of this cluster in the 0-298 K temperature range [7a].

The phase space trajectories were generated using a fourth-order variable step-size predictor-corrector propagator. With the maximal step-size of $5 \cdot 10^{-16}$s the total energy and the total momenta (linear and angular) were conserved within 0.03% and 0.15%, respectively. The forces defining the time evolution of the system were calculated from a potential of the form

$$V = V_{EA} + V_{LEPS} \; , \tag{2}$$

where V_{EA} is an embedded-atom potential describing the interaction of the Ni atoms in the cluster, and V_{LEPS} is a LEPS (London-Eyring-Polanyi-Sato) potential representing the intramolecular (D-D) and the molecule - cluster interactions. We have used Voter and Chen's [10] parameterization of V_{EA}, which, in addition to the nickel bulk properties, builds in also the equilibrium bond length (2.2Å) and bond energy (1.95 eV) of the nickel dimer. The V_{LEPS} is the PESII of RSD [5], modified by a smoothing function of the form used by Truong et al. [11] (the details will be given elsewhere). The parameters of the

V_{LEPS} [5] are such that this potential reproduces the values of the binding energy and of the binding height of an H atom on a threefold site of an icosahedral Ni_{13}, as calculated using the corrected effective medium approach.

In general, each trajectory was run until one of the two events took place: (1) the molecule adsorbed dissociatively on the cluster - reactive event; (2) the molecule scattered from the cluster and departed into asymptotic region (8.5 Å from the center of mass of the cluster) - nonreactive event. The trajectory was qualified as reactive and terminated if and when the D-D distance reached the value of $3d_0$, where $d_0 = 0.741$Å is the equilibrium bond length of the D_2 molecule. Thus, possible recombinations and subsequent desorptions of the molecule were precluded. It was found that each of the channels, the reactive and the nonreactive, could be realized either through a direct process or via a molecularly adsorbed intermediate ("precursor") state. To quantify the processes, sweeps of N = 500 trajectories were run for each set of initial conditions specified by the values of v_i, j_i, E_{tr}^i, b, T and the structure of the cluster. Each trajectory in a sweep corresponded to different initial orientations of the molecule and of the cluster and to a different initial phase of the D_2 oscillator; these were chosen randomly. The b- and E_{tr}^i-dependent probabilities $P_{v_i,j_i,T,...}(b,E_{tr}^i)$ of the different processes for a given initial and, in the case of the nonreactive channel, also final [7b] state of the molecule were calculated as

$$P_{v_i,j_i,T,...}(b,E_{tr}^i) = \frac{\tilde{N}_{v_i,j_i,T,...}(b,E_{tr}^i)}{N} , \tag{3}$$

where $\tilde{N}...$ is the number of trajectories which resulted in the process of interest. The E_{tr}^i-dependent state-resolved cross sections $\sigma...$ were calculated using the expression

$$\sigma_{v_i,j_i,T,...}(E_{tr}^i) = 2\pi \int_0^{b_{max}} bP_{v_i,j_i,T,...}(b,E_{tr}^i)db, \tag{4}$$

where b_{max} is the maximal impact parameter of relevance for the process of interest. In the next section we present results for the dissociative adsorption cross sections.

One of the intricate aspects of the reaction is the role of the precursor (classical resonance) states and of their lifetimes. These lifetimes can be extracted from the simulations in the following way. Let $\tilde{N}^R...(b,E_{tr}^i)$ be the total number of reactive trajectories and $\tilde{N}^o...(b,E_{tr}^i)$ - the total number of reactive precursor states (i.e., resonances which eventually dissociate) in a given sweep; it is implied that $\tilde{N}^R...(b,E_{tr}^i)$ is obtained by running the trajectories long enough to meet the criteria for their termination mentioned above. Sweeps with shorter times of propagation will yield a smaller number of reactive trajectories $\tilde{N}^R...(t|b,E_{tr}^i)$ because the longer-lived precursor states will survive and will not contribute to $\tilde{N}^R...(t|b,E_{tr}^i)$. If the time t is measured from the instant of formation of the resonances, i.e., $t = t_r-t_{ar}$, where t_r is the fixed time of propagation of each trajectory in the sweep and t_{ar} is the "arrival time" needed for the molecule to reach the cluster (t_{ar} can be viewed as dependent only on b and E_{tr}^i and, thus, constant for a given sweep) then

$$\tilde{N}^R...(t|b,E_{tr}^i) = \tilde{N}^{DR}...(b,E_{tr}^i) + \tilde{N}^o...(b,E_{tr}^i)\left[1 - e^{-\kappa^d...(b,E_{tr}^i)t}\right]$$

$$= \tilde{N}^R...(b,E_{tr}^i) - \tilde{N}^o...(b,E_{tr}^i)e^{-\kappa^d...(b,E_{tr}^i)t} , \tag{5}$$

where $\kappa^d_{\ldots}(b,E^i_{tr})$ is the reactive resonance decay rate constant and $N^{DR}_{\ldots}(b,E^i_{tr})$ is the number of the direct reaction events in the sweep. Dividing Eq. (5) by N and introducing the corresponding probabilities P, one can rewrite it in the form

$$\ln\left[P^R_{\ldots}(b,E^i_{tr}) - P^R_{\ldots}(t|b,E^i_{tr})\right] = -\kappa^d_{\ldots}(b,E^i_{tr})t + \ln P^\circ_{\ldots}(b,E^i_{tr}) . \tag{6}$$

Repeating sweeps with different t_r's, one calculates the left-hand side (lhs) of Eq. (6) on a grid of t's. To correlate the t's with t_r's, one has to know the value of $t_{ar}(b,E^i_{tr})$. This latter can be calculated by introducing an arrival criterion. The procedure we used is based on the following consideration. As discussed above, both the direct dissociation and the reactive resonances, contribute to $P^R_{\ldots}(t|b,E^i_{tr})$. Sweeps with decreasing propagation times result in decreasing reaction probabilities since the longer-lived resonances don't dissociate on the time scale of the runs. The rate of change of the reaction probability with time is $\kappa^d_{\ldots}(b,E^i_{tr})P^\circ_{\ldots}(b,E^i_{tr})\exp[-\kappa^d_{\ldots}(b,E^i_{tr})t]$, i.e., it is changing smoothly at all $t > 0$. At $t = 0$, i.e., at $t_r = t_{ar}$, $P^R_{\ldots}(t_r=t_{ar}|b,E^i_{tr}) = P^{DR}_{\ldots}(b,E^i_{tr})$, where $P^{DR}_{\ldots}$ is the probability of the direct reaction. For propagation times $t_r < t_{ar}$, $P^R_{\ldots}(t_r<t_{ar}|b,E^i_{tr}) = 0$. Thus, t_{ar} is that value of t_r at which $P^R_{\ldots}(t_r|b,E^i_{tr})$ changes its value discontinuously from $P^{DR}_{\ldots}$ to zero. If $P^{DR}_{\ldots} = 0$, $P^R_{\ldots}(t_r|b,E^i_{tr})$ approaches zero at $t_r = t_{ar}$ in a smooth way. In order to determine t_{ar}, and consequently t, one has to find that largest value of t_r at which $P^R_{\ldots}(t_r|b,E^i_{tr}) = 0$. In numerical calculations the value of t_{ar} may be slightly affected by the choice of the dissociation criterion. A careful choice of this criterion, however, makes the uncertainty in t_{ar} negligible. Fitting the calculated values of the lhs of Eq. (6) to a linear function of t one obtains $\kappa^d_{\ldots}(b,E^i_{tr})$ (and consequently the reactive resonance lifetime $\tau_{\ldots} = 1/\kappa^d_{\ldots}$), as well as the probability $P^\circ_{\ldots}(b,E^i_{tr})$ of formation of reactive resonances. If $P^{DR}_{\ldots}(t|b,E^i_{tr}) = 0$, $P^\circ_{\ldots}(b,E^i_{tr}) = P^R_{\ldots}(b,E^i_{tr})$. Using Eq. (4) and assuming that $\kappa^d_{\ldots}$ does not depend on b (denote this b-independent rate constant $\bar{\kappa}^d_{\ldots}$), one can rewrite Eq. (6) in terms of the corresponding cross sections σ:

$$\ln\left[(\sigma^R_{\ldots}(E^i_{tr}) - \sigma^R_{\ldots}(t|E^i_{tr})\right] = -\bar{\kappa}^d_{\ldots}(E^i_{tr})t + \ln\sigma^\circ_{\ldots}(E^i_{tr}) . \tag{7}$$

In general, however, $\bar{\kappa}^d_{\ldots}$ does depend on b (see Section 3) and Eq. (7) should be regarded as the definition of an effective, averaged over b, rate constant $\bar{\kappa}^d_{\ldots}(E^i_{tr})$. To obtain the value of $\bar{\kappa}^d_{\ldots}$, one fits the lhs of Eq. (7) calculated on a grid of t's to a straight line.

3. Results and Discussion

In this section we present and discuss results pertaining to the reactive interaction of a D_2 molecule with a Ni_{13} cluster. The embedded-atom potential defines the icosahedral geometry as the most stable structure of Ni_{13}; its metastable cuboctahedral and hcp isomers are almost degenerate energetically (the details will be given elsewhere). In Fig. 1 the cross sections of the dissociative adsorption of the molecule on a room-temperature icosahedral Ni_{13} are shown as functions of the collision energy for different initial rovibrational states of the molecule. Apart from the ground ($v_i = 0$, $j_i = 0$) state, we considered also ($v_i = 0$, $j_i = 3$), the most populated at room temperature, and the ($v_i = 0$, $j_i = 10$) and ($v_i = 1, j_i = 0$) states, which are almost degenerate energetically. For all the (v_i, j_i) states the cross sections increase monotonically with the collision energy and display state-dependent thresholds. The higher the initial rovibrational energy of the molecule, the larger the corresponding cross sections and the smaller the threshold to the reaction. We have determined [7]

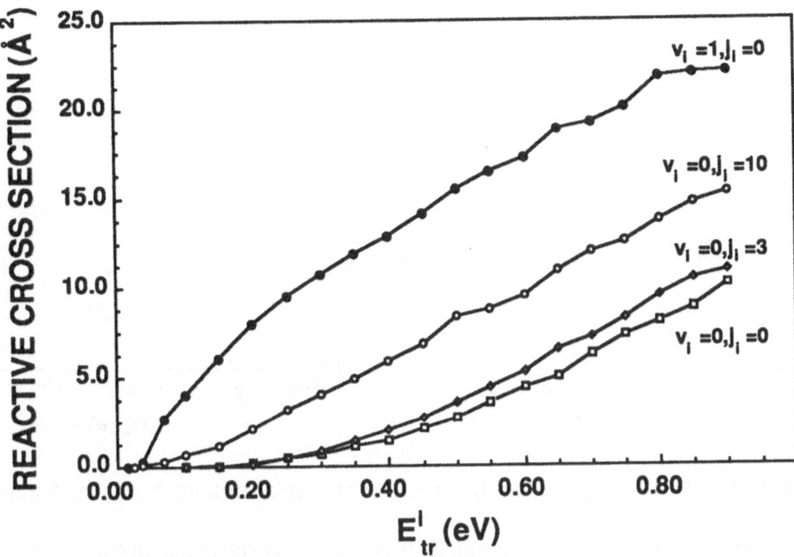

Figure 1. Dissociative chemisorption cross sections as functions of the collision energy E^i_{tr} for a room-temperature icosahedral Ni_{13} and different initial rovibrational (v_i, j_i) states of D_2.

that the threshold values of E^i_{tr} are measures of the (v_i, j_i)-dependent dynamical barriers for the dissociative adsorption of the molecule on the cluster. The large difference in the cross sections for the $(v_i = 0, j_i = 10)$ and $(v_i = 1, j_i = 0)$ degenerate states indicates a strong mode-selectivity of the reaction: an initial vibrational excitation of the molecule results in a much more efficient bond breaking than an energetically equal initial rotational excitation. Figure 2 displays the reactive cross sections for zero-temperature cuboctahedral and hcp Ni_{13}. Similarly to the case of the icosahedral cluster, higher initial rovibrational energies of the molecule result in larger cross sections and the mode-selectivity of the reaction is even more pronounced. But together with the similarities, the cuboctahedral and hcp isomers exhibit reactivity properties that are different from those of the icosahedral Ni_{13}. One of these is the absence of thresholds to the reaction. The most striking dissimilarity, however, is the nonmonotonic behavior of the cross sections as functions of the collision energy. At very low E^i_{tr}'s the cross sections display (v_i, j_i)-dependent maxima. These are especially high and sharp for $(v_i = 1, j_i = 0)$. One notices that the corresponding cross sections for the cuboctahedral and hcp structures are similar not only qualitatively, but also quantitatively, with a tendency of the hcp isomer to be somewhat less reactive; both structures are more reactive than the icosahedral isomer. This quantitative aspect of the structure - reactivity correlation in Ni_{13} agrees with what has been found by RSD in their study of the reactivity of rigid nickel clusters. Experimentally, the dependence of the reactivity on the isomer structure has been explored for niobium clusters [12]. Here we want to emphasize the qualitative differences in the cross sections, which indicate differences in the reaction mechanisms of the cuboctahedral and hcp structures, on the one hand, and the icosahedral geometry, on the other. The qualitative differences (presence or absence of thresholds and peaks) are particularly pronounced at low collision energies and

1052

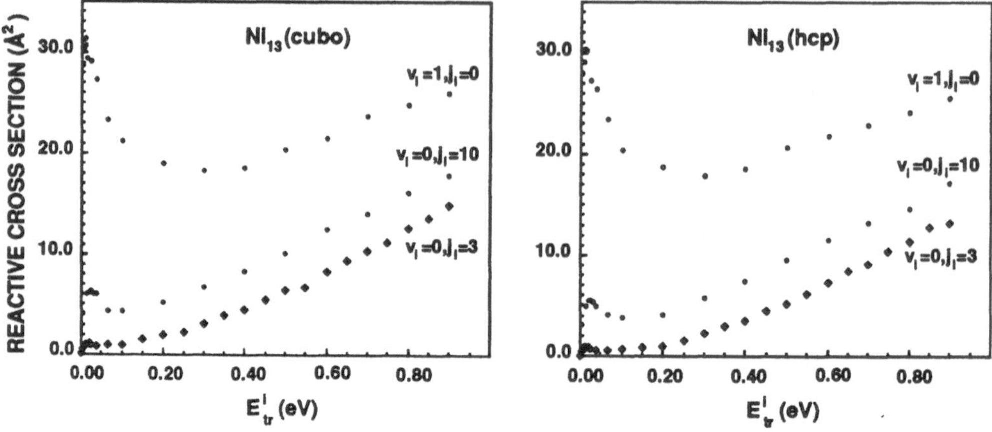

Figure 2. The same as Fig. 1 but for zero-temperature cuboctahedral and hcp Ni$_{13}$.

diminish at higher energies. This suggests that it is the differences in the surface topology of the different isomers of Ni$_{13}$ that cause the cross sections to behave so differently; at low E_{tr}^i's the molecule stays longer in the vicinity of the cluster and thus is more sensitive to the structural details of its surface. The surfaces of the cuboctahedral and hcp isomers are indeed quite different from that of the icosahedron. The former are comprised of triangles and squares, the latter - only of triangles. We have found that in the low collision energy range the reaction on the cuboctahedral and hcp clusters is dominated by the indirect process. The peaks in the cross sections are consequences of the high probability of formation of reactive resonances. As the collision energy increases, this probability decreases and so do the cross sections. Eventually, the direct process takes over and the cross sections become monotonically increasing.

We turn now to the discussion of the intriguing phenomenon of formation and decay of reactive resonances. In Fig. 3 the reaction probabilities $P^R...(t_r|E_{tr}^i)$ are shown as functions of the propagation time t_r for the case of the ($v_i = 1$, $j_i = 0$) state of the molecule and zero-temperature cuboctahedral Ni$_{13}$; three impact parameters and four values of E_{tr}^i corresponding to the peak range of the cross section were considered. It is seen that as t_r decreases, all the probabilities decay to zero smoothly. This means that, within the accuracy of our simulations, the direct dissociation channel is closed at the collision energies considered and, consequently, the reactive resonance formation probabilities $P^o...(b,E_{tr}^i) = P^R...(b,E_{tr}^i)$. The values of t_r at which the different $P^R...(t_r|b,E_{tr}^i)$ become zero are the corresponding arrival times.

Now that we have established the probabilities of formation of reactive resonances for the b's and E_{tr}^i's considered, the only remaining unknowns are the decay rates or, alternatively, the lifetimes. In order to find these we have used that $P^o... = P^R...$ and calculated

$$\zeta = \ln \left\{ \left[P^R...(b,E_{tr}^i) - P^R...(t_r|b,E_{tr}^i) \right] / P^R...(b,E_{tr}^i) \right\} \tag{8}$$

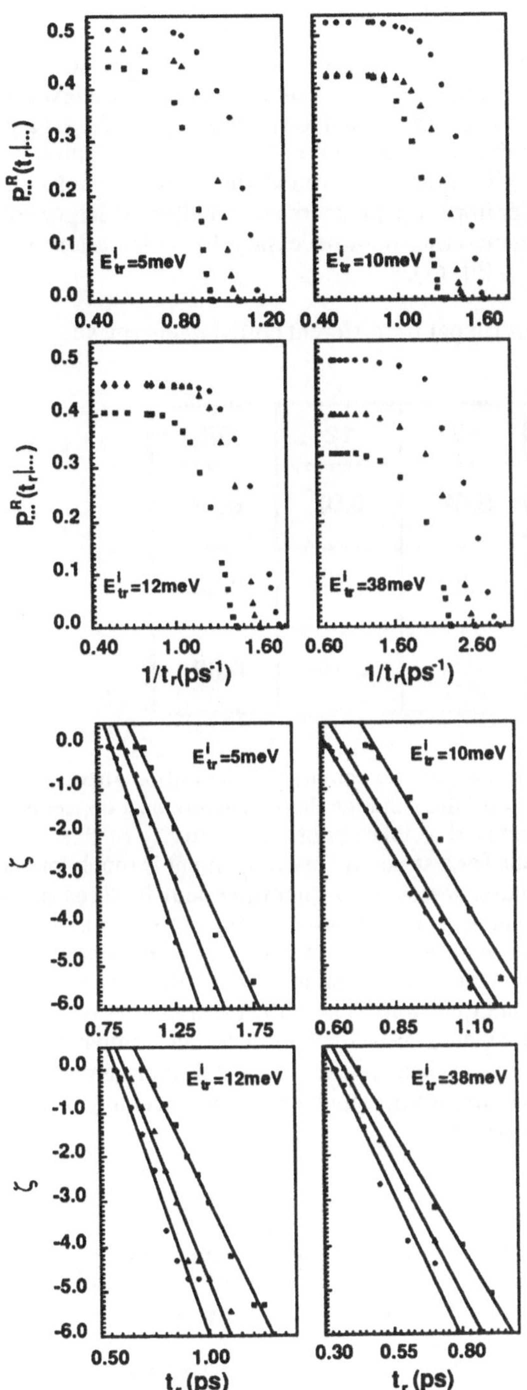

Figure 3. The time-dependent reaction probabilities as functions of the inverse propagation time for four different collision energies E_{tr}^i. The different symbols correspond to different values of b: •0.25 Å; ▲ 2.75 Å; ■ 3.75 Å.

Figure 4. ζ, Eq. (8), as a function of t_r for the same collision energies and impact parameters as in Fig. 3. The different symbols correspond to the same impact parameters as in Fig. 3.

on a grid of t_r's. The slopes of the fits of the ζ values to straight lines are the corresponding rate constants; of course, the slopes do not depend on whether one considers $\zeta(t_r)$ or $\zeta(t)$. The calculated points and their fits are shown in Fig. 4. The straight lines approximate the calculated data quite well, with the possible exception of a few points at the high-t_r end of the graphs. These points correspond to small values of the argument of E_{tr}^i -dependent lifetimes, calculated from the slopes of the fits, are listed in Table 1. They vary from 0.07 to 0.13 ps (~7 to 13 vibrations of a ground state free D_2 molecule) and show a tendency to be larger for larger impact parameters and smaller for larger collision energies. Work is in progress on generating data necessary for calculation of the b-averaged decay rate constants $\bar{\kappa}^d_{...}(E_{tr}^i)$, Eq. (7).

Table 1. Reactive resonance lifetimes (in ps) for different collision energies E_{tr}^i (in meV) and impact parameters b (in Å).

b \ E_{tr}^i	5	10	12	38
0.25	0.09	0.09	0.07	0.07
2.75	0.11	0.09	0.08	0.08
3.75	0.13	0.09	0.10	0.09

We conclude our discussion with a few general remarks. The cuboctahedral and hcp isomers of Ni_{13} are highly metastable, and they change their structure as a consequence of the interaction with D_2. It is still justified to talk about the reactivity of this cluster in different geometries since the time scale for a structural rearrangement is much longer than the reaction time. For measuring the state-resolved quantities and features discussed above, cluster beam experiments will have to be performed. The comparison of detailed theoretical and experimental data is necessary for improving the description of the interaction potentials. Finally, the resonance phenomenon we described is a classical and not quantal effect. One of the important questions is how the quantal effects would alter the classical results. This can be answered only through a direct comparison of the outcomes of the two approaches. Quantum mechanical dynamical treatments of clusters will have to involve quite substantial simplifying approximations because of the large number of degrees of freedom of these systems.

4. Summary

Results of a quasiclassical simulation study of the dissociative adsorption of a D_2 molecule on a Ni_{13} cluster were presented and discussed. Effects of the initial quantized rovibrational state of the molecule and of the cluster structure on the reaction were examined. A general trend, common to all three cluster isomers considered (icosahedron, cuboctahedron and hcp), is that higher initial rovibrational states of the molecule give larger reaction cross sections. The bond breaking process is, however, mode selective: an initial vibrational excitation of the molecule results in a substantially larger reaction cross section than an energetically equal initial rotational excitation. In addition to the quantitative aspect of the structure - reactivity correlation (the cuboctahedral and hcp isomers are more reactive than the icosahedral), the study also revealed intriguing qualitative correlations. The cross sections for the dissociative adsorption on the icosahedron are characterized by thresholds (indicating dynamical barriers to the reaction) and increase monotonically with the collision energy. No such thresholds have been obtained for the cuboctahedral and hcp isomers. Moreover, the reaction cross sections for these two structures change with the collision energy in a nonmonotonic fashion. The major difference between the reactivity of the icosahedron, on the one hand, and the cuboctahedral and hcp isomers, on the other, appears at low collision energies, when the molecule has more time to "sense" the surface topology of the cluster. At these energies the cuboctahedral and the hcp Ni_{13} produce maxima in the reaction cross sections, which depend on the initial rovibrational state of the molecule. As we have shown for the case of $(v_i = 1, j_i = 0)$, the maxima are generated by reactive resonances - molecularly adsorbed precursor states. An analysis of the resonances has been given in terms of the probabilities of their formation and the lifetimes.

Acknowledgments

Work performed under the auspices of the Office of Basic Energy Sciences, Division of Chemical Science, US-DOE under contract number W-31-109-ENG-38.

References

1. See, for example, J. L. Elkind, F. D. Weiss, J. M. Alford, R. T. Laaksonen, and R. E. Smalley, J. Chem. Phys. **88**, 5215 (1988); W. D. Reents, Jr. and M. L. Mandich, J. Phys. Chem. **92**, 2908 (1988); S. A. Ruatta and S. L. Anderson, J. Chem. Phys. **89**, 273 (1988); W. F. Hoffman, E. K. Parks, and S. J. Riley, J. Chem. Phys. **90**, 1526 (1989); Y. M. Hamrick, and M. D. Morse, J. Phys. Chem. **93**, 6494 (1989); S. K. Loh, L. Lian, and P. B. Armentrout, J. Chem. Phys. **91**, 6148 (1989); S. Nonose, Y. Sone, K. Onodera, S. Sudo, and K. Kaya, Chem. Phys. Lett. **164**, 427 (1989); P. Fayet, A. Kaldor, and D. M. Cox, J. Chem. Phys. **92**, 254 (1990); K. M. Creegan, and M. F. Jarrold, J. Am. Chem. Soc. **112**, 3768 (1990); R. E. Leuchtner, A. C. Harms, and A. W. Castleman, Jr., J. Chem. Phys. **92**, 6527 (1990); and references therein.
2. T. D. Klots, B. J. Winter, E. K. Parks and S. J. Riley, J. Chem. Phys. **92**, 2110 (1990).
3. See, for example, J. Tully, J. Chem. Phys. **73**, 633 (1980); P. E. M. Siegbahn, M. R. A. Blomberg, and C. W. Bauschlicher, J. Chem. Phys. **81**, 2103 (1984); C.-Y. Lee and A. E. DePristo, J. Chem. Phys. **85**, 4161 (1986); I. Panas, J. Schüle, P. Siegbahn, and U. Wahlgren, Chem. Phys. Lett. **149**, 265 (1988); G.-Q. Xu, S. L. Bernasek, and J. Tully, J. Chem. Phys. **88**, 3376 (1988); J. Harris, Surf. Sci. **221**, 335 (1989); U. Nielsen, D. Halstead, S. Holloway, and J. K. Nørskov, J. Chem. Phys. **93**, 2879 (1990); A. Chattopadhyay, H. Yang, and J. L. Whitten; J. Phys. Chem. **94**, 6379 (1990); E. A. Colbourn, in *Catalysis*, vol. 8, Royal Society

of Chemistry, Cambridge, 1989, p. 42; A. J. Cruz and B. Jackson, J. Chem. Phys. <u>94</u>, 5715 (1991); and references therein.

4. See, for example, T. H. Upton, Phys. Rev. Lett. <u>56</u>, 2168 (1986); T. H. Upton, D. M. Cox, and A. Kaldor, in *Physics and Chemistry of Small Clusters*, P. Jena, B. K. Rao and S. N. Khanna (Eds.), Plenum Press, NY, 1987, p. 755; K. Przybylski, J. Koutecký, V. Bonacić-Koutecký, P. von Ragué-Schleyer, and M. F. Guest, J. Chem. Phys. <u>94</u>, 5533 (1991); C. Mijoule, Y. Bouteiller, and D. R. Salahub, Surf. Sci. <u>253</u>, 375 (1991); and references therein.

5. K. Raghavan, M. S. Stave, and A. E. DePristo, Chem. Phys. Lett. <u>149</u>, 89 (1988); J. Chem. Phys. <u>91</u>, 1904 (1989).

6. J. E. Adams, J. Chem. Phys. <u>92</u>, 1849 (1990).

7. (a) J. Jellinek and Z. B. Güvenç, Z. Phys. D, <u>19</u>, 371 (1991); (b) J. Jellinek and Z. B. Güvenç, in *Proceedings of the 24th Jerusalem Symposium on Quantum Chemistry: Mode Selective Chemistry*, Kluwer Academic Publishers, Dordrecht, 1991 (in press).

8. R. Fournier, M. S. Stave, and A. D. DePristo, J. Chem. Phys. (to be published).

9. R. N. Porter, L. M. Raff, and W. H. Miller, J. Chem. Phys. <u>63</u>, 2214 (1975).

10. A. F. Voter and S. F. Chen, Mater. Res. Soc. Symp. <u>82</u>, 175 (1987).

11. T. N. Truong, D. G. Truhlar, and B. C. Garett, J. Phys. Chem. <u>93</u>, 8227 (1989).

12. M. B. Knickelbein and S. Yang, J. Chem. Phys. <u>93</u>, 1476 (1990); ibid <u>93</u>, 5760 (1990); and references therein.

Charge Transfer Involving Clusters

FRANK TRÄGER
Fachbereich Physik der Universität Kassel
Heinrich-Plett-Str. 40, D-3500 Kassel
Germany

Introduction

It has been discussed in numerous papers that the generation of neutral mass selected clusters is of great importance in a variety of aspects (see e.g. [1]). The main reasons for the interest in neutral monodisperse clusters are as follows: The cluster sources that are presently available generate mixtures of clusters with relatively broad mass distributions, the exact composition of which is unknown. In addition, fragmentation processes may easily occur if the clusters are ionized for mass spectrometry, i.e. size identification. Still, only few experimental techniques have been proposed and tested to produce size-selected neutral beams. An obvious approach to narrow the mass distribution of gas phase clusters is by choosing appropriate nucleation conditions of the source. Since the size of the clusters is strongly dependent on the temperature and pressure of the carrier gas, an influence on the mass distribution is possible (see e.g. [2]). Unfortunately, the generated mass distributions are still broad. Considerably more efficient mass separation of neutral clusters is possible by deflection of the cluster beam through collisions with a gas jet (see e.g. [3]). Experiments along these lines have been carried out on different cluster species, the main goal being the investigation of dissociation processes. Presently, it seems that neutralization of mass-selected cluster ions by charge exchange reactions has the largest potential for obtaining a high flux of neutral particles and for making the mass distribution as narrow as possible, even truly monodisperse.

P. Jena et al. (eds.), Physics and Chemistry of Finite Systems: From Clusters to Crystals, Vol. II, 1057–1064.

Charge Exchange

In the past, the technique of using charge transfer has been employed to produce fast atomic beams [4]. Application of the method to clusters has already been proposed as early as 1985 [5,6]. Recent experiments have illustrated [7-9] the large potential of the method and demonstrated that truly monodisperse cluster beams can indeed be produced. In principle, positively as well as negatively charged size-selected cluster ions can be neutralized.

Positively charged, mass-selected cluster ions X_n^+ can be neutralized in an atomic or molecular target gas Y by reactions of the type:

$$X_n^+ + Y \rightarrow X_n^{(*)} + Y^+ + \Delta E$$

The asterisk indicates that the neutralized clusters $X_n^{(*)}$ are not necessarily produced in the electronic ground state but can also be generated in an excited state. ΔE stands for the energy defect of the reaction and is given by the difference of the internal energy of the initial and final states of the reaction. If only electronic ground states are involved, ΔE is simply the difference of the ionization potentials of the collision partners.

If the energy defect ΔE is zero or very small the charge transfer is called resonant. The most simple reaction of this type is the symmetric charge transfer between identical species $A^+ + A \rightarrow A + A^+$. Resonant or near-resonant charge exchange can also be realized for reactions between different particles where ΔE is "accidentally" zero or very small. The cross section σ_n for resonant charge transfer decays as a function of the relative velocity of the collision partners [10]. Typical values of σ_n are on the order of 10^{-15} - 10^{-14} cm^2. In the case of nonresonant charge transfer, i.e. $\Delta E \neq 0$, the cross section σ_n increases rapidly for small energies, reaches a broad maximum and finally decreases as a function of kinetic energy. A theoretical description is complicated and has to rely on approximations [10].

Usually the cross section for electron capture is largest into that final state, for which the energy defect is minimal, i.e. the system tends to make the charge transfer as resonant as possible [11]. Resonant charge exchange takes place with practically no transfer of momentum. Therefore, a well collimated beam of neutralized clusters can be produced with a spatial divergence, which exceeds that of the original ion beam by very little, provided collisions between the particles are avoided. However, charge exchange results in a neutral, monodisperse cluster beam only, if the particles do not undergo extensive fragmentation. In general, the products of the neutralization reaction can include contributions from the following processes:

- Elastic collisions that modify the trajectories of the clusters.
- Dissociative charge transfer, i.e. electron capture in an unbound state or in an electronically excited state with subsequent deexcitation and fragmentation.
- Dissociative (inelastic) collisions of the type
  $C_n^+ + A \rightarrow C_{n-k}^+ + C_k + A$ between the cluster ion C_n^+ and the target atom or molecule.

A major difficulty in generating neutral mass separated cluster beams is to control the charge exchange in such a way that dissociation can be avoided or minimized. Essential in determining the outcome of the charge transfer are the conditions for energy resonance in the charge exchange reaction and the accessibility of bound product neutral levels in vertical transitions [12]. A reasonable probability exists to avoid fragmentation if the neutralized clusters carry as little internal energy as possible. Therefore, the target gas should be chosen such that energy resonance with the ground state of the neutral cluster exists and the electron is captured preferably in this level. Even then, however, fragmentation cannot be ruled out completely, since reconstruction processes with a release of energy are possible. In any case, the constituents of the neutralized beam have to be analyzed with great care in order to examine and avoid contamination with particles of different mass.

To detect possible fragmentation unambiguously, a technique known as translational spectroscopy (see e.g. [13]) can be applied profitably to analyze the neutralized cluster beam [7-9]. The principle of the method is as follows. If fragmentation occurs, internal energy is converted into kinetic energy of the fragments. Consequently, additional transverse momentum appears in the beam which causes a spatial deflection of the particles and a broadening of the neutral beam profile as compared to that of the ion beam. This broadening is measured with a position sensitive detector and interpreted in terms of the energy released in the fragmentation process. The measured beam profiles can also be compared with Monte Carlo simulations of the trajectories of the fragments [9]. Thus, possible fragmentation channels and energies can be extracted.

The experimental set-up for charge transfer and translational spectroscopy on clusters involves a gas aggregation source, a differential pumping stage, ionization of the clusters by electron bombardment and mass-selection in a quadrupole filter. The mass separated ions are directed into a charge exchange cell where a certain pressure of an atomic or molecular vapor is maintained. A transverse electric field behind the charge exchange region allows us to separate the neutrals from the ions. The clusters pass a drift region before the profiles of both the ions and the neutrals are measured.

Results

With the method of charge exchange mass-selected clusters could be produced for the first time. In the following several examples are presented that illustrate neutralization of mass separated sulfur and lead cluster ions under different conditions, i.e. non-resonant electron transfer (with respect to the neutral cluster ground state), where dissociation may occur, and near-resonant charge exchange, in which the electron is captured preferably in the neutral cluster ground state. As outlined above, fragmentation can be avoided in the latter case. Sodium and zinc were chosen as target gases for neutralization of S_n^+. Sodium has an ionization potential of 5.1 eV and zinc of 9.4 eV. The appearance potentials of sulfur clusters S_n with $2 \leq n \leq 8$ range from about 9 to 10 eV. Therefore, strongly non-resonant and near-resonant charge exchange can be realized with Na and Zn, respectively. Comparison of the ionic and neutral beam profiles of e.g. S_n^+ neutralized in Na shows, that a considerable broadening occurs. The ionization potentials of the two collision partners being different by 4.3 eV, this broadening is interpreted as electron transfer into an electronically excited cluster state, whose energy is released by subsequent fragmentation and converted into kinetic energy of the products. A near-resonant transfer of the electron into the neutral cluster ground state can be realized with Zn as target vapor. In this case only a small broadening of the neutral beam profile is observed. It is attributed to elastic scattering. This interpretation is supported by measurements of the cross sections for gaskinetic collisions and for charge transfer reactions as a function of cluster size. Therefore, fragmentation can be excluded for neutralization of S_6 and a number of other sulfur aggregates in Zn vapor.

More recently, metal clusters have also been investigated by charge exchange. Measurements of the rate of neutralized particles and transmitted ions have been performed with lead clusters Pb_6^+ and sodium target gas densities of 10^{12} up to several 10^{13} cm^{-2}, and Ar target gas densities ranging from 10^{13} to 10^{16} cm^{-2}. In the latter case, where a strongly nonresonant reaction occurs, only about 1 to 2% of the incident ions leave the charge transfer region as neutrals. Resonant neutralization in atomic sodium, on the other hand, converts as much as 30% of the cluster ions into neutrals. Typically, the rate of neutrals is on the order of 10^6 counts/s. The most essential result of the beam profile measurements is that no significant difference between the ionic and neutral profiles is observed for near-resonant charge exchange of Pb_n^+ (with n ranging from 2 to 12) in Na vapor. Non-resonant neutralization in Ar, however, causes a distinct broadening of the profiles. This observation was made independent of the cluster size. The measured charge exchange cross sections with sodium as target increase from 15 Å^2 for Pb_1^+ to 40 Å^2 for Pb_4^+ and 60 Å^2 for Pb_{12}^+. Such large values are obtained only for near-resonant neutralization, the ionization potentials of lead clusters [12] being very similar to the ionization energy of atomic Na. For non-resonant neutralization, values as low as several Å^2 are measured, e.g. for Ar or CS_2 as target gases. Also, the beams generated with these materials contain a large number of fragments. For Na, on the other hand, the quasi-resonant character of the charge exchange makes electron capture in the ground state of the neutralized cluster by far the dominating process. This minimizes transfer of energy so that fragmentation due to charge exchange is negligible.

It follows from the above discussion, that continuous mass-selected neutral sulphur and lead clusters Pb_n ($3 \leq n \leq 12$) have been generated.

Recently, the idea to generate neutral, size-selected cluster beams by charge transfer has been applied also by other groups, e.g. to alkali clusters like Na_n with $n \leq 21$ that were neutralized in Cs vapor [14]. A time-of-flight arrangement for the mass separation of the cluster ions was used and the neutralized clusters could be reionized for analysis of the neutral components of the beam. While the spectrum without reionization contains ion fragments of different mass, the reionized neutral products show very little fragmentation. Other work involving charge exchange was reported on Si_n clusters [15], and Rb_n clusters generated with a liquid metal ion source [16]. In both experiments, however, the constituents of the neutralized beam have not been identified. Multiply charged Au clusters were also studied [17]. It was found that the cross sections for fragmentation and charge transfer are of comparable magnitude. In a theoretical paper on charge exchange of alkali clusters the importance of the magic numbers on the cross sections was emphasized [18]. The theoretical treatment shows that the excess energy ΔE for Na_n with $2 < n \leq 8$ is smaller than the energy necessary to dissociate the cluster. This means that the neutral products have the same size as their parents. For larger clusters with $8 < n \leq 21$ the evaporation of monomers and dimers from the parent cluster results from the high stability of the magic numbers with complete shell filling.

In addition to the neutralization of positively charged clusters, photodetachment of an electron from a negatively charged cluster ion seems to offer interesting perspectives. Such a process should only weakly modify the cluster structure so that reconstruction with release of energy is minimized. However, photodetachment often has to compete with photofragmentation which is reflected in the mass distribution of the neutralized clusters. New experiments have demonstrated that photodetachment can indeed be used to generate neutral size-separated and collimated carbon cluster beams [19].

Conclusions

With the described experimental technique it could be demonstrated that neutral, mass-selected cluster beams can be generated. Such beams provide the basis for novel experiments on clusters under well defined conditions. Particularly interesting are, for example, electron- or photon-induced fragmentation or optical excitation with laser light. Also, the deposition of neutral clusters of a single size on well characterized surfaces offers exciting perspectives for future experiments.

This work was supported by the Deutsche Forschungsgemeinschaft.

References

[1] J. Kowalski, T. Stehlin, M. Vollmer, F. Träger, in "Physics of Clusters and Nanophase Materials", edited by M.S. Multani and V.K. Wadhawan, Phase Transitions **24-26**, 737 (1990)

[2] E. Choi, R.P. Andres, in Physics and Chemistry of Small Clusters, p. 61, eds. P. Jena, B.K. Rao, S.N. Khanna (Plenum, New York, 1987)

[3] U. Buck in Proc. International Summer School of Physics "Enrico Fermi", Course CVII, ed. G. Scoles, "The Chemical Physics of Atomic and Molecular Clusters", p. 543, North Holland (1990)

[4] H.S.W.Massey, H.B. Gilbody, Electronic and Ionic Impact Phenomena Vol. 4, (Oxford, Clarendon Press, 1974)

[5] M. Arnold, J. Kowalski, G. zu Putlitz, T. Stehlin, F. Träger Surf. Sci. **156**, 149 (1985)

[6] M. Arnold, J. Kowalski, G. zu Putlitz, T. Stehlin, F. Träger, Z. Phys. A **322**, 179 (1985)

[7] M. Abshagen, J. Kowalski, M. Meyberg, G. zu Putlitz, F. Träger, J. Well, Europhys. Lett. **5**, 13 (1988)

[8] M. Abshagen, T. Fischer, J. Kowalski, M. Meyberg, G. zu Putlitz, T. Stehlin, F. Träger, J. Well, J. Phys. **50-C2**, 169 (1989)

[9] M. Abshagen, J. Kowalski, M. Meyberg, G. zu Putlitz, J. Slaby, F. Träger,

Chem. Phys. Lett. **174**, 455 (1990)

[10] N.F. Mott, H.S.W. Massey, The Theory of Atomic Collisions,
(Oxford University Press, 1965)

[11] J. Perel, H.L. Daley, Phys. Rev. **A4**, 162 (1971)

[12] T.F. Moran, in "Charge-Transfer Reactions", in Electron-Molecule
Interactions and the Applications, (Academic Press, New York, 1984)

[13] D.P. de Bruijn, J. Los, Rev. Sci. Instrum. **53**, 1020 (1982)

[14] C. Bréchignac, Ph. Cahuzac, J. Leygnier, R. Pflaum, J. Weiner,
Phys. Rev. Lett. **61**, 314 (1988); C. Bréchignac, Ph. Cahuzac, F. Carlier, J.
Leygnier, I.V. Hertel, Z. Phys. D **17**, 61 (1990)

[15] W. Begemann W., R. Hector, Y.Y. Liu, J. Tiggesbäumker, K.H.
Meiwes-Broer, H.O. Lutz, Z. Phys. **D12**, 229 (1989)

[16] N.D. Bhaskar, R.P. Frueholz, C.M. Klimcak, R.A. Cook,
Chem. Phys. Lett. **154**, 175 (1989)

[17] W.A. Saunders, Phys. Rev. Lett. **62**, 1037 (1989)

[18] S.N. Khanna, P. Jena, B.K. Rao, Phys. Rev. **B 40**, 966 (1989)

[19] S. Suzuki, T. Wakabayashi, H. Shiromaru, C. Kittaka, Y. Achiba,
Chem. Phys. Lett. **182**, 12 (1991)

INTERACTION OF CLUSTERS WITH SUBSTRATES

B. F. CONSTANCE, B. K. RAO, and P. JENA
*Physics Department, Virginia Commonwealth University, Richmond,
VA 23284-2000, USA*

ABSTRACT. The equilibrium geometry, binding energy, and electronic structure
of a triatomic lithium cluster interacting with the (001) surface of lithium metal has
been studied using unrestricted Hartree-Fock theory. The surface was modeled by
clusters of 5 and 13 atoms of lithium whose positions were held fixed at their normal
crystallographic positions. Optimizing the geometry of the trimer while it interacted
with the surface showed that the trimer disintegrated and became a part of the
modeled bulk.

Introduction

The interaction of surfaces with adsorbates continues to be of great interest to
experimentalists and theoreticians alike because understanding these interactions is
of tremendous importance to the industries [1,2]. For example, the interaction of
oxygen with silicon surfaces guides the manufacture of microelectronic devices.
Catalysis depends completely on the chemisorptive dissociation of molecules on
surfaces. Due to these reasons many researchers have been working in this field to
obtain complete understanding of this phenomenon.

Recently the interest in these interactions has been even more increased due
to the introduction of the atomic clusters [3] into the arena. It is now well known
[4-8] that the atomic clusters have special properties which do not have much in
common with the properties of bulk materials. In the case of very small atomic
clusters the surface to bulk ratio is very high and almost all the atoms are suface
atoms. However, with the increase in the size of the atomic clusters the bulk-like
atoms grow in number. Thus one can fine-tune the surface to bulk ratio by
specifying the exact number of atoms in the clusters. One also knows [3,5] that the
properties of the clusters are dependent on the size and they do not change
monotonously to reach the bulk values. These lead scientists to believe that one can
produce atomic clusters with tailor-made properties and thus create cluster-materials
to be used for specific purposes.

The production of cluster materials for any use will require that the atomic
clusters be deposited on some surface because it is almost impossible to find
practical use for free floating clusters. Immediately the question arises as to
whether the interaction of the clusters with the surface changes the properties of the
clusters themselves. If this change is significant then the "cluster material" cannot

P. Jena et al. (eds.), Physics and Chemistry of Finite Systems: From Clusters to Crystals, Vol. II, 1065–1070.

be used because the property for which the cluster was initially designed is already lost. On the other hand, if the clusters do have absolutely no interaction with the surface then they will not stay on the surface and no cluster material will result. Thus, an ideal situation is when the clusters interact weakly with the substrate on which they are deposited and do not lose most of their original properties. Almost no theoretical work is available in the literature where the effect of substrates with clusters is studied in detail. With this in view we have studied the interaction of an Li_3 cluster with the (001) surface of Li metal. The lithium cluster was chosen because of its simple structure and the study of lithium on lithium eliminates some of the complications presented by a heterogenous sustem. In addition it is expected that the study of the energetics of this system may also contribute to the understanding of crystal growth.

Choice of Model

The study of the interaction at the surface should ideally include two related points. The first one is the influence of the adsorbates on the surface - how the wavefunction of the surface is affected by the presence of the adsorbate [2,9]. This can be answered by finding if the presence of the adsorbate causes a restructuring of the surface. The second point is related to the influence of the surface on the adsorbate. In the present study we shall start with the second point. The importance of this point has been justified in the last section.

Many techniques have been used in the past to represent the interaction of adsorbates at surfaces. Surfaces have been modeled as a semi-infinite jellium [10,11] and then with the application of density functional methods [12] good information have been obtained on the electron charge density, the dissappearance of clean surface states, and the chemisorbed energy. Another technique has been to represent the surface by a cluster of atoms restricted to sites representing a fragment [13-19] of the bulk. This approach allows the use of self-consistent field (SCF) techniques [20-22].

It is well known that the geometries of fragments of bulk crystals stabilize at bulk values for only a few atoms in the cluster [18,19]. This fact has enabled previous researchers to use limited models to obtain valuable information on the interaction between surfaces and molecules. For example, it has been shown by Kunz [13] that an approaching atom needs to be within 5 Å from the surface in order that any noticeable interaction may occur. This implies that such interactions are local and involve only a few atoms at the site of the reaction. Therefore, it is reasonable to represent the interaction between a surface and a small cluster by the interaction of two clusters, one representing the bulk surface and the other representing the cluster. In this case one can apply *ab-initio* methods such as the SCF-LCAO-MO technique [20,21] which uses the atomic wave functions of the system to create molecular orbitals and then self-consistently calculates the binding energies, optimum geometries, and other properties.

There is a disadvantage connected with the use of this method. It arises due to the limits put on the number of atoms which may be accomodated by this technique. The choice of approach is, however, redeemed by the facts previously stated concerning the localized interaction of the process under investigation. With all of these in consideration the lithium surface was initially modeled by five lithium atoms arranged at lattice sites for a bcc lithium crystal. Four of these atoms were arranged at the four corners of the top surface of the cube and the fifth atom was placed at the body-center position of the cube. The Li_3 trimer was allowed to interact with this model. As the next step in the investigation of the effect of the finite number of atoms representing the bulk surface, a 13 atom cluster was used to represent the bulk surface. These atoms were arranged at the lattice sites of the first and second layers of four unit cells as shown in Fig. 1. This model is a fair representation of the surface as seen by a small cluster (Li_3) because the boundaries of the Li_{13} cluster are too far from the center of the cluster to influence it [13].

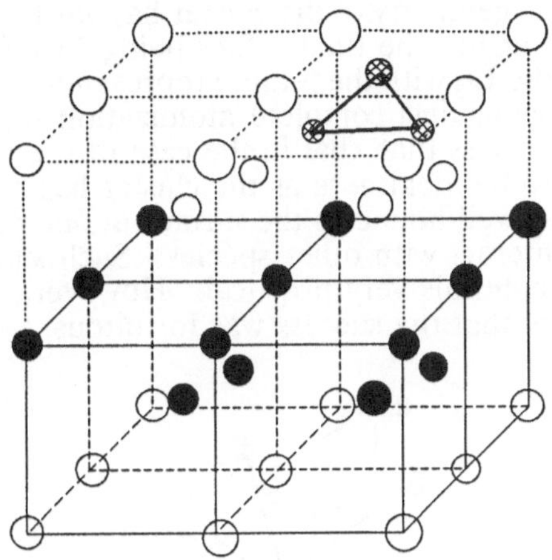

Fig.1   Model for the representation of the lithium metal surface by a 13 atom cluster. The other atoms in the lattice are represented by empty circles to aid observation. The Li_3 cluster approaching the (001) surface is shown by shaded circles.

Procedure

Using the SCF-LCAO-MO method within the Hartree-Fock approximation scheme to represent the effects of exchange the total energies of the chosen models were calculated using the GAUSSIAN-88 software [23]. For this purpose the STO-6G [20] Gaussian basis set was used. Although this is not of a double zela quality, Rao and Jena [4] have proved that it describes cluster geometries quite well. Rao et al. [18] have used this basis to study the geometries and properties of bulk lithium with success. Therefore, for the present work it was considered to be quite acceptable. In all the calculations the cluster representing the bulk was kept fixed at the initial geometry within the specified symmetry. The atom-atom distance was kept fixed for the corresponding bulk value. For this purpose the lattice constant was chosen to be 3.491 Å. The lithium trimer was allowed settle on its surface. While the Li_3 was approaching the surface its bond lengths and bond angles were varied in response to the energetics of the interaction starting with the configuration of an optimized trimer as obtained by Rao and Jena [4]. For each model, a complete geometry optimization for the trimer was done using the energy gradient method while searching for the optimum distance of the cluster atoms from the surface.

Results and Discussions

Before starting the discussions one needs to know the geometry of the free Li_3 cluster. It is an isosceles triangle [4] with an apex angle of 72.1° and the two equal bond lengths are 2.69 Å. When the geometry of such a cluster was optimized while interacting with the (001) surface of the five-atom model, the lowest energy configuration showed an interesting behavior. The configuration for the lowest energy is given in Fig. 2. It is clear that the Li_3 cluster has not lost its identity in the ensuing interaction. The geometry of the cluster has hardly changed from the initial bond lengths and bond angles the final values being 2.69 Å and 71.5° respectively. The binding energy of the Li_3 with the 5-atom representation of the bulk is only 1.76 eV. The binding energy against complete atomization in the case of the $Li_3 + Li_5$ system is 4.1 eV which is less than that in the case of the free Li_8 cluster (5.4 eV). The situation appears to be ideal here as the cluster has not lost its characteristics although it is reasonably well bound to the surface at one point leaving the two end atoms almost free to interact with other species. Such situations would make the production of cluster materials very practical. However, the following discussion will, unfortunately, show that the success was fortuitous.

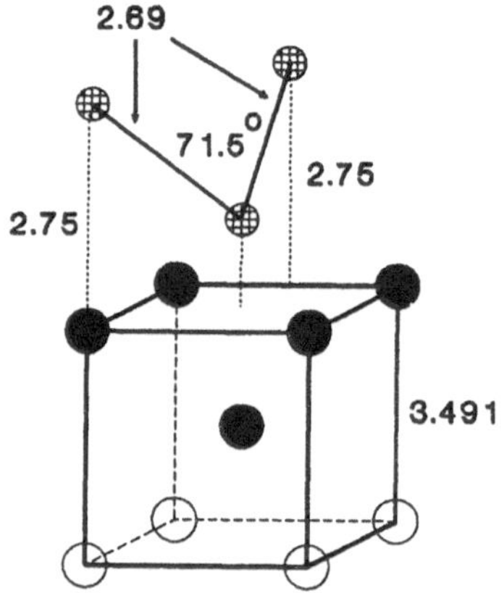

Fig. 2 The five-atom cluster representing a model bcc lithium metal surface. The filled circles show which atoms are included in the calculation. The empty circles are given to show the details of the model. The optimized Li_3 trimer is represented by the shaded circles. All bond lengths are given in Å. The positions of the trimer atoms are indicated by dotted lines.

The calculations performed with the 13-atom "bulk surface" showed a completely different behavior. When the cluster was at a distance of 5 Å above the surface it did not bind with the surface and its geometry stayed unchanged. However, in the process of energy minimization it migrated closer to the surface and then started responding to the effects of the surface by shifting and rotating. This was also accompanied by stretching of the bonds and changes in the bond angles. During the process of global minimization of the energy sometimes the bonds shrank and the trimer seemed to go have a tendency of going back to its original geometry. However, after a few more cycles this trend was reversed again. Obviously while traversing the Born-Oppenheimer surface, the trimer went across slight activation barriers. Towards the end there was a steady decrease in the energy of the system before finally stabilizing in the configuration showed in Fig. 3. Further investigation with different initial conditions did not reveal any lower energy in the Born-Oppenheimer surface. Therefore, it can be safe to assume that the configuration in Fig. 3 is of the lowest energy. The positions of the trimer atoms appear to be close to the body-center sites of an imaginary layer of unit cells above the surface. The distances between pairs of the two nearest atoms are 3.21 Å and the bond angle is approximately 94°. These atoms are between 1.3 and 1.6 Å above the simulated surface. If they would have occupied the exact body center positions corresponding to the present situation, the bond lengths would have been 3.491 Å and the bond angle would be 90°. The atoms would then have stayed at a uniform height of 1.7455 Å above the surface. However, with a Hartree-Fock calculation this trend appears to be closely followed. Thus the characteristics of the trimer are completely lost and the trimer atoms have been assimilated in the bulk.

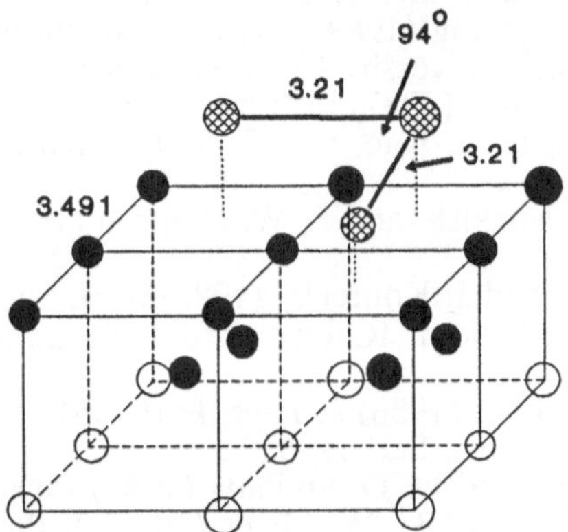

Fig. 3 Optimized $Li_{13} + Li_3$ cluster geometry showing where the trimer atoms have finally stabilized. The corresponding bond lengths and bond angles are given in Å and degrees respectively.

A similar study with a linear Li_3 chain showed a binding energy of 0.8 eV. It bound with the surface at a height of 2.63 Å. The end atoms of the chain did not go to any site corresponding to the bcc lattice. This observation is in contrast with the observations made of Ir clusters deposited on Ir surface [24]. More studies are in progress with other substrates.

Acknowledgements

This work was supported in part by a grant from the U. S. Department of Energy (DE-FG05-87ER45316). A part of the computations was conducted using the Cornell National Supercomputer Facility, a resource of the Cornell Theory Center, which is funded in part by the National Science Foundation, New York State, the IBM Corporation and members of the Center's Corporate Research Institute.

References

1. M. F. Jarrold, U. Ray, and K. M. Creegan (1990) J. Chem. Phys. 93, 1.
2. T. Stace (1990) Nature 346, 23. See also A. Zanguill (1988) "Physics at Surfaces" (Cambridge University Press, New York), pp. 292 and 208.
3. B. K. Rao, S. N. Khanna, and P. Jena (1990) Phase Transitions 24-26, 35.
4. B. K. Rao and P. Jena (1985) Phys. Rev. B 32, 2058.
5. B. K. Rao, S. N. Khanna, and P. Jena (1987), Phys. Rev. B 36, 953.
6. P. Jena, B. K. Rao, and S. N. Khanna, (1987) Eds., "Physics and Chemistry of Small Clusters" (Plenum, New York).
7. M. L. Cohen and W. D. Knight (1990) Physics Today 43, 42.
8. W. A. de Heer, W. D. Knight, M. Y. Chou, and M. L. Cohen (1987) Solid State Physics 40, 93.
9. S. K. Ramchurn, D. M. Bird, and D. W. Bullet (1990) J. Phys. C 2, 7435.
10. T. B. Grimley (1979) in "The Nature of the Surface Chemical Bond" edited by T. N. Rhodin and G. Ertl (North-Holland, New York)
11. R. G. Parr and W. Yang (1989) "Density Functional Theory of Atoms and Molecules" (Oxford University Press, London).
12. S. Saito and S. Ohnishi (1987) in "Physics and Chemistry of Small Clusters", edited by P. Jena, B. K. Rao, and S. N. Khanna (Plenum, New York), pp. 115.
13. A. B. Kunz, D. J. Mickish, and P. W. Deutsch (1973) Solid St. Commun. 13, 35.
14. H. -O. Beckmann and J. Koutecky (1982) Surf. Sci. 120, 127.
15. P. Fantucci, G. Pacchioni, R. Ugo, and J. Fernandez-Rico (1981) J. Mol. Catalysis 13, 339.
16. B. K. Rao and P. Jena (1986) J. Phys. F 16, 461.
17. J. B. Moffat (1979) Surf. Sci. 84, 65.
18. B. K. Rao, P. Jena, and D. D. Shillady (1984) Phys. Rev. B 30, 7293.
19. A. K. Ray, J. L. Fry, and C. W. Myles (1985) J. Phys. B 18, 381.
20. W. J. Hehre, L. Radom, P. v. R. Schleyer, and J. A. Pople (1986) "Ab initio Molecular Orbital Theory" (John Wiley, New York).
21. A. Szabo and N. S. Ostlund (1982) "Modern Quantum Chemistry: Introduction to Advanced Electronic Structure Theory" (McMillan, New York).
22. R. P. Messmer (1979) in "The Nature of the Surface Chemical Bond" edited by T. N. Rhodin and G. Ertl (North-Holland, New York).
23. M. J. Frisch, M. Head-Gordon, H. B. Schlegel, K Raghavachari, J. S. Binkley, C. Gonzalez, D. J. DeFrees, D. J. Fox, R. A. Whiteside, R. Seeger, C. F. Melius, J. Baker, R. L. Martin, L. R. Kahn, J. P. Stewart, E. M. Fluder, S. Topiol, and J. A. Pople (1988) 'GAUSSIAN 88' (Gaussian Inc., Pittsburgh).
24. C. Chen and T. T. Tsong (1990) Appl. Phys. A 51, 405.

DETECTION AND CHARACTERIZATION OF SILVER PARTICLES IN SODIUM SILICATE
GLASSES BY MEANS OF HIGH RESOLUTION ELECTRON MICROSCOPY

M. DUBIEL (a), H. HOFMEISTER (b) and ST. THIEL (a)
(a) Martin-Luther-University Halle-Wittenberg, Department of
     Physics, O-4010 Halle, Germany
(b) Institute of Solid State Physics and Electron Microscopy,
     O-4050 Halle, Germany

ABSTRACT. The present experiments were performed to study nanocrystalline silver precipitations in sodium silicate glasses. The structure of
these particles should exhibit a strong size dependence. The high resolution electron microscopy HREM was used to image the crystalline lattice. It was possible to exclude any influences both of the preparation
procedure and the electron beam on the particle behaviour during the
experiments. HREM images show the (111) and (200) lattice plane fringes
of the silver fcc stucture. It could be observed a strong contraction of
the lattice parameters in comparison to isolated silver particles. The
results indicate the existence of an additional external pressure induced by the thermal treatment of the glass.

1. INTRODUCTION

Small crystalline particles play an important role in the field between
molecular and solid state of matter. Drastic changes of the properties
may occur in dependence on the size of the particles. Comparing small
particles with the bulk material one finds diminished lattice parameters
and melting temperatures or changes in the electronic structure (see for
example Monot (1984)). Therefore the macroscopic properties of ceramics,
catalysts, other heterogeneous materials or glasses containing small
precipitated particles are influenced by their size and structure.

The aim of the present paper is to investigate the structure of
silver precipitations in sodium silicate glasses. Usually such precipitations are examined by optical spectroscopy giving informations averaged over a large number of particles (Kreibig (1986)). Moreover no
informations about the crystalline structure of the particles are available. In order to study size, size distribution and shape of single
particles includung their crystalline structure the application of electron microscopical methods is required by all means. For electron microscopical observation of small particles in bulk materials, however, a
special preparation technique is needed to obtain thin enough specimens.
Furthermore one has to exclude structural changes caused both by the
preparation procedure and by the electron beam during electron micro-

P. Jena et al. (eds.), Physics and Chemistry of Finite Systems: From Clusters to Crystals, Vol. II, 1071–1076.
© 1992 *Kluwer Academic Publishers.*

scopical observations. The results concerning form and crystallographic structure of the particles should be discussed in comparison with data measured on isolated particles, that means they are formed without any matrix interaction. In this way conclusions about the influence of the precipitation process and the surrounding glass matrix on the particle configuration could be derived. For these investigations a large range of different particle sizes would be advantageous.

2. EXPERIMENTAL

2. 1. Sample preparation

For the experiments sodium silicate glasses containing 0.13 weight% Fe_2O_3 were used. The silver precipitations were formed by a sodium-silver ion exchange and additional thermal treatments at 600°C. The temperature of the ion exchange was 330°C which is well below the glass transformation temperature. It is generally assumed that the precipitations are formed by the diffusion of silver ions into the glass matrix followed by the reduction to neutral atoms and the aggregation to metallic precipitations. In the present case the reducing agent consists of iron ions. The penetration depth of the silver particles as well as their size and size distribution depend on the thermal treatment of the glass. Optical spectroscopy was used to measure the mean size of the particles. With these results appropriate sample regions could be choosen for electron microscopical examination.

2. 2. Electron microscopical preparation

The transmission electron microscopical (TEM) observations were performed with a microscope JEM 100C operating at 100 kV and a JEM 4000 EX operating at 400 kV, respectively. A direct imaging of precipitations by TEM is possible provided that the samples are thin enough. This could be realized by mechanical grinding and subsequent ion beam thinning using Argon ions. In order to reduce the sample temperature during this procedure a minimized beam current (< 50 uA) and a relatively low incidence angle of the Argon ions (12...15°) were used. To avoid possible influences of the ion thinning procedure on the particle structure the results of TEM have been compared with experiments at particles prepared by chemical etching of glass with hydrofluoric acid. Both methods showed the same results.

2. 3. Electron beam effects

Inelastic interactions of the electrons with the sample during TEM observations act as a source of additional structural transformations. For example, it could be shown that an intense electron beam leads to the formation of oxygen bubbles as a consequence of the destruction of the bonds in the network of the investigated silicate glasses. Besides of small reorientations of the crystalline particles also coagulation and crystal growth processes of the silver particles could be induced during the investigations (see fig. 1). These effects, however, could be avoided using reduced electron beam currents and observation times.

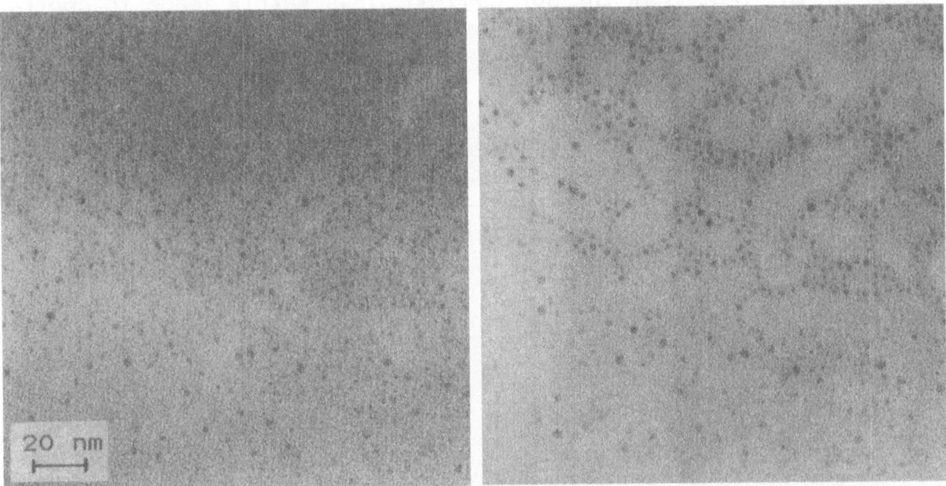

Figure 1. Effect of an intense electron beam irradiation using 100 keV electrons. Representation of the same glass region before (left) and after (right) irradiation.

3. CHARACTERIZATION OF ISOLATED SILVER PARTICLES

Because of the large portion of surface atoms in small particles the structural behaviour differs from that of bulk material. A size dependent lattice contraction induced by the surface stress f is one of the consesequences. The size dependence of the parameters given by the work which is necessary to increase the surface by dA as f dA. With that the volume V and the internal pressure p are changed. That leads to

$$f \; dA = dp \; dV. \tag{1}$$

Assuming spherical shape of the particles with a fcc structure the measured lattice contractions of small particles with the radius r can be characterized by the surface stress and the change of the internal pressure (Mays et al. (1968)) as follows

$$f = - \; 3r \; da \; / \; 2a \; K \tag{2}$$

and

$$dp = 2f \; / \; r, \tag{3}$$

where a is the lattice constant of the bulk material and da the reduction of this value in the particles; K represents the compressibility. The results of several authors using electron diffraction (Wassermann and Vermaak (1970)) and EXAFS (Montano et al. (1984)) experiments lead to values of the surface stress between 1 and 3 N / m.

4. EMBEDDED SILVER PARTICLES

In spite of the fact that the surrounding glass matrix leads to contrast
deteriorations the (111) and (200) lattice plane fringes could be observed using HREM. Two examples are shown in fig. 2 and 3. The resulting

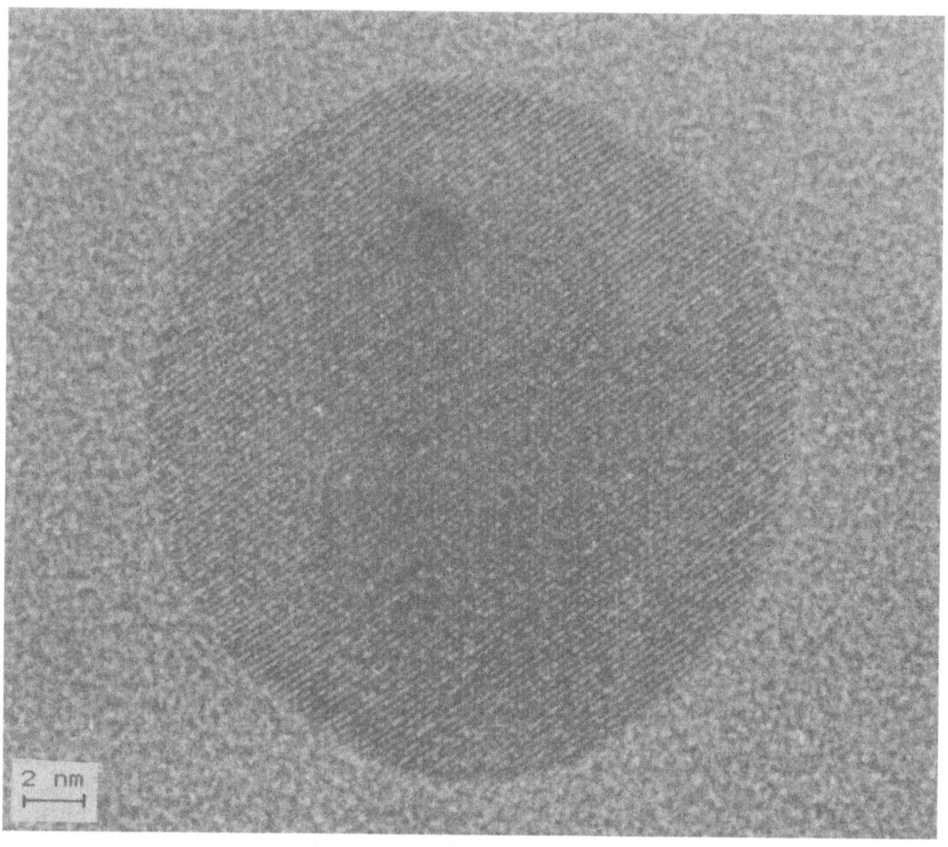

Figure 2. High resolution electron micrograph showing a silver precipitation with a diameter of 22 nm

change of the lattice constant in dependence on the particle size is
represented in fig. 4. The data indicate an increased lattice contraction in comparison to isolated particles. The formal description with
equ. (2) gives a f' value of 57.1 N / m. In this case the stress was
designated by f' because of the transition from the surface of isolated
particles to the interface between particle and glass matrix. Usually
this transition yields a diminished stress value. But the present experiments suggest the presence of an additional external pressure, which
acts in addition to the interface stress.

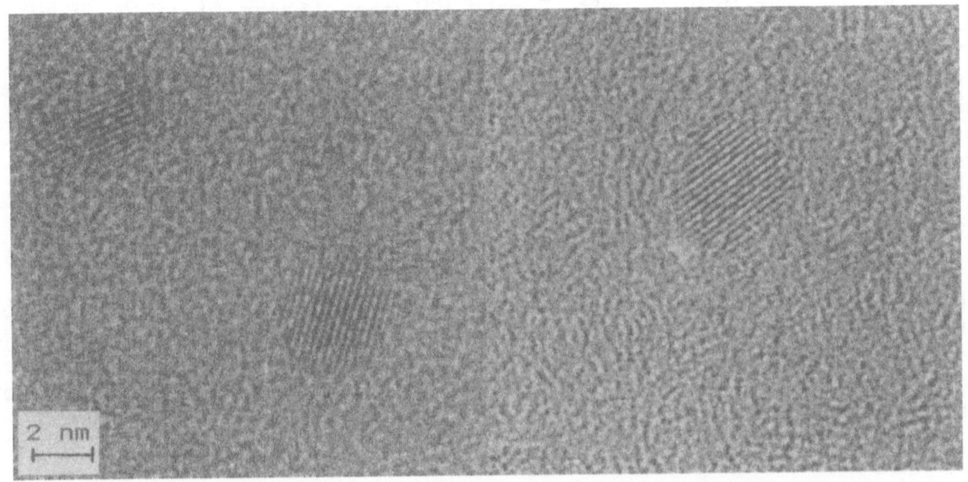

Figure 3. (111) lattice plane fringes of small silver particles.

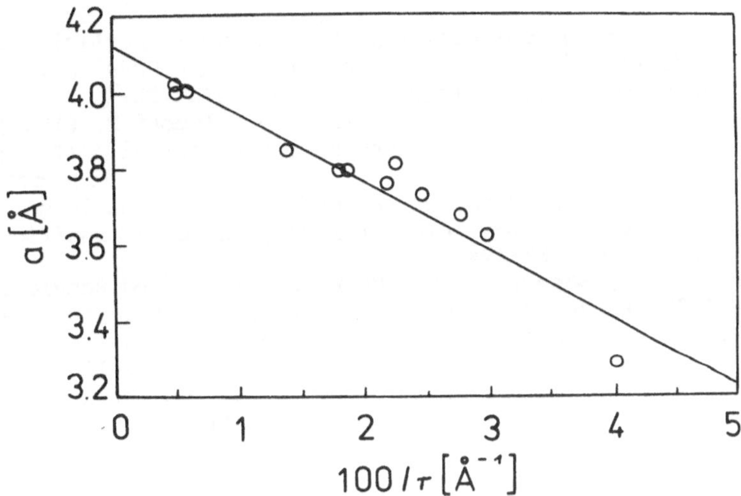

Figure 4. Lattice constant contraction of silver precipitations in a sodium silicate glass measured at room temperature.

In order to explain the experimental results it was assumed that the additional pressure is caused by the temperature difference between the formation of precipitations and the HREM examination. The particles are formed at 600°C and the HREM observations were carried out nearly at room temperature. Therefore it is suggested that the increased pressure could be induced both by the reduced particle radius at rooom temperature without matrix interactions and by different expansion coefficients of silver particles and glass matrix. The first contribution can be estimated approximately from the expansion coefficient of bulk silver material. The resulting value can be neglected. The increased external pressure dp' as a result of the different expansion behaviour one can calculate from the corresponding expansion coefficients β_i and compressibilities K_i of glass (g) and silver (Ag) by

$$dp' = -dT \left(\beta_g / K_g - \beta_{Ag} / K_{Ag} \right). \tag{4}$$

That leads to a constant value of dp' independent on the particle size. But the description of the experimental lattice parameters requires a size dependence of the silver particle expansion coefficient. However such experimental data are not available. That will be the subject of further investigations.

5. REFERENCES

Kreibig, U. (1986) "Optical properties of small particles in insulating matrices", in Contribution of Cluster Physics to Materials, Science and Technology, Dordrecht, 373-423

Mays, C. W., Vermaak, J. S. and Kuhlmann-Wilsdorf D. (1968) "On surface stress and surface tension", Surface Sci., 12, 134-140

Monot, R. (1984) "Noble metal clusters", in A. Baldereschi et al. (eds.), Proc. Int. Symp. on the Physics of latent Image Formation in Silver Halides (Triest 1983), World Scientific Publ. Co., Singapore, 157-194

Montano, R. A., Shenoy, G. K., Morrison, T. I. and Schulze, W. (1984) "EXAFS and Mössbauer study of small metal clusters isolated in rare-gas solids", in K. O. Hodgson et al. (eds.), EXAFS and Near Edge Structure III, Springer-Verlag, Berlin, 231-233

Wassermann, H. J. and Vermaak, J. S. (1970) "On the determination of a lattice contraction in very small silver particles", Surface Sci., 22, 164-172

Reactions and Thermochemistry of Mixed Cluster Ions

M. Samy El-Shall and George M. Daly
Department of Chemistry
Virginia Commonwealth University
Richmond, Virginia 23284-2006, USA

ABSTRACT.   Protonated clusters of the type $H^+(CH_3OH)_n(CH_3CN)_m$ show a predominant sequence with the composition $H^+(CH_3OH)_n(CH_3CN)_2$, where the solvent shells are completed by blocking groups on both ends [4]. Even though ionization involves large excess energies, the observed cluster distributions are sensitive to thermochemical differences as small as 1-3 kcal/mole. Nucleophilic substitution reactions have been observed following the ionization of p-difluorobenzene-methanol mixed clusters. The generation of p-fluoroanisole ion exhibits a strong dependence on the degree of solvation of the cluster. The results for different alcohols will be discussed.

1. Introduction

Cluster ions are subject to intensive current research [1-3]. Much of the current activity concerns the ionization of van-der-Waals complexes formed in supersonic beam expansions. Ionization can lead to partial evaporation and in some cases to intra-cluster reactions. In many cases, series of clusters are produced with a distribution where the abundance of some clusters suggests the occurrence of "magic numbers". Both the intra-cluster chemistry and product distributions are fundamentally related to cluster energetics. In this paper, we are concerned with the following questions: (1) how sensitive are cluster compositions to energetics? and (2) how does chemical reactivity change as a function of cluster size? In order to shed more light on these questions, we chose the mixed clusters of methanol, acetonitrile and water as model systems for investigating the sensitivity of cluster distribution to energetics and the mixed clusters of p-difluorobenzene-alcohol as examples of intra-cluster nucleophilic substitution reactions.

2. Experimental

Mixed clusters, generated by pulsed adiabatic expansion in a supersonic cluster beam apparatus, were ionized by EI and subsequently mass analyzed in a quadrupole mass spectrometer [4]. Equilibrium measurements were carried out using the NIST pulsed high-pressure mass spectrometer [5].

P. Jena et al. (eds.), Physics and Chemistry of Finite Systems: From Clusters to Crystals, Vol. II, 1077–1082.
© 1992 Kluwer Academic Publishers.

3. Results and Discussion

Figure 1 displays a typical 70 eV EI mass spectrum observed from the ionization of a cluster beam generated from the expansion of a 1:1 vapor mixture of CH_3CN and CH_3OH in He carrier gas. The prominent peaks in the mass spectrum are due to the sequence $H^+M_n A_2$ (denoted as a_n) where A and M are acetonitrile and methanol, respectively. This manifold is predominant over a wide range of vapor composition (30%-80% A with respect to M) and stagnation pressures (1-6 Bar) and is strongly independent of the electron impact energy (20-100 eV), indicating that the cluster ions $H^+M_n A_2$ have particularly stable structures independent of the distribution of the neutral clusters produced. The structure of the simplest ion in this series can be pictured as a central protonated methanol with two terminal acetonitrile molecules as shown below.

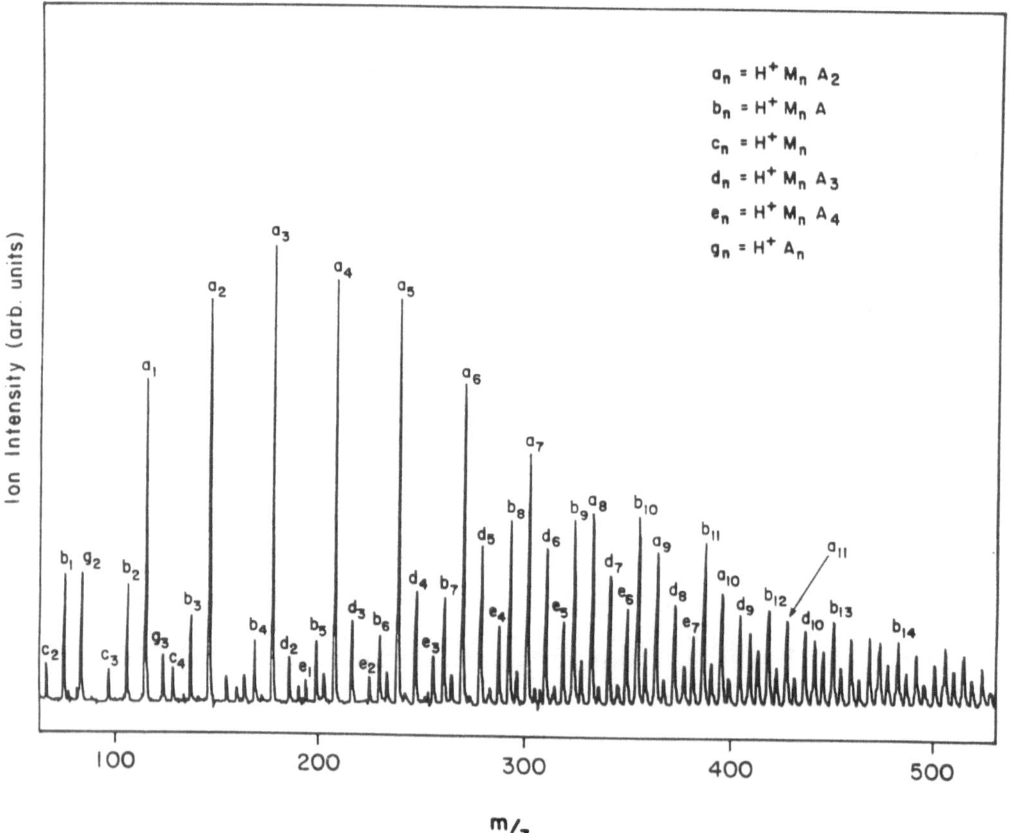

Figure 1. Cluster spectrum of acetonitrile-methanol mixture, 1:1 vapor mixture, in He carrier gas.

The results suggest that cluster ions with n < 9 prefer to lose methanol to generate the sequence $H^+A_2M_n$, while for larger clusters with n > 9 the loss of acetonitrile is the dominate process, thus generating the H+AM$_n$ series.

In order to define the thermochemistry associated with the cluster sequences, we measured the enthalpies (ΔH^o) of the various clustering and exchange equilibria using pulsed high-pressure mass spectrometry. The results for clusters containing up to four molecules are shown in Figure 2. The results show that the switching of CH3CN for CH3OH is favorable until the clusters incorporate two CH3CN molecules and become blocked at both ends of the CH3OH chain. Past this, further exchange becomes endothermic.

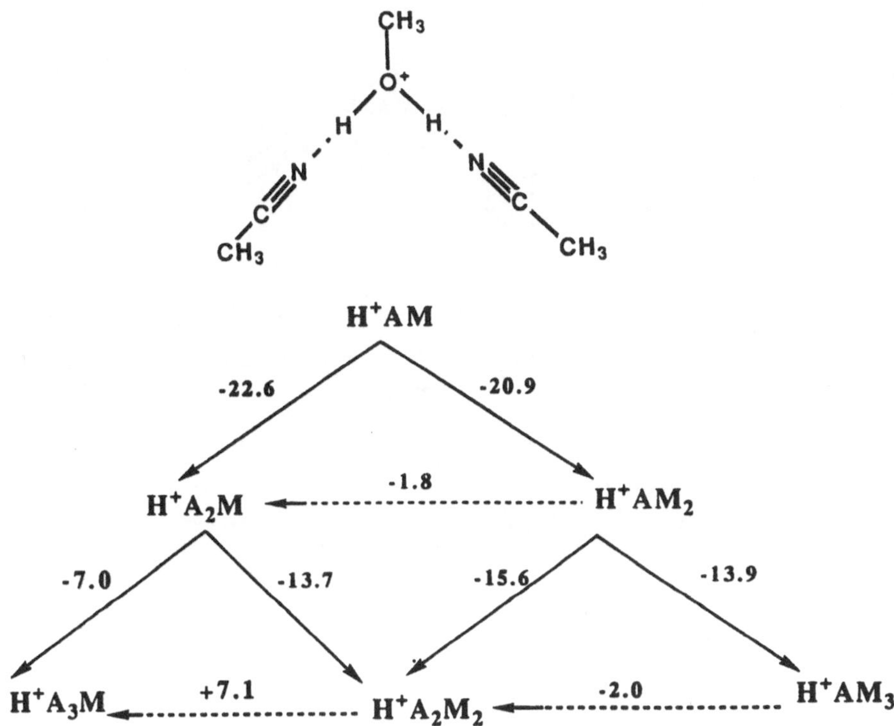

Figure 2. ΔH^o (kcal/mol) for clustering and exchange equilbria.

In the three component clusters $H^+(CH_3CN)_n$ $(CH_3OH)_m$ $(H_2O)_k$,the predominant sequence observed (as shown in Figure 3) has the composition H_3O^+ $(CH_3OH)_m$ $(CH_3CN)_3$. This is consistent with shell filling structures with three acetonitrile molecules around a protonated core ion H_3O^+. These structures allow the formation of higher clusters through the incorporation of hydrogen-bonded methanol chains to the protic sites of the H_3O^+ core ion as shown in Figure 4.

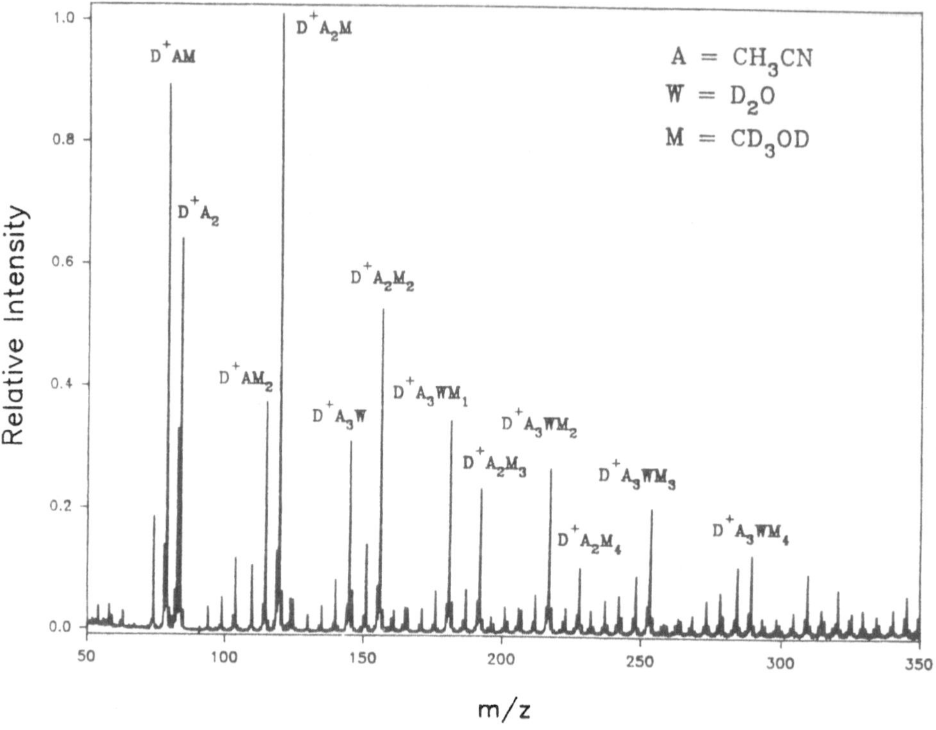

Figure 3. Cluster spectrum for acetonitrile-methanol-water mixture (1:1:1).

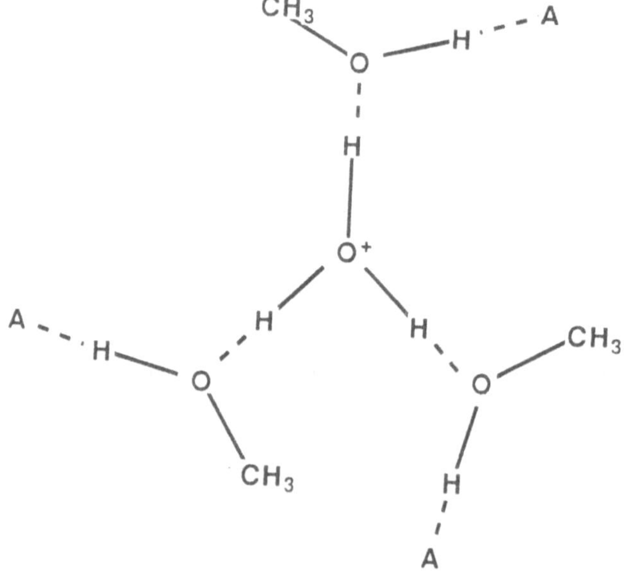

Figure 4. Structure of $H_3O^+(CH_3OH)_3(CH_3CN)_3$ cluster.

Figure 5 shows the mass spectrum observed from the EI ionization of p-difluorobenzene-methanol mixed clusters. In addition to methanol, p-difluorobenzene and their mixed clusters,the spectrum shows two peaks corresponding to p-fluoroanisole ion and p-fluoroanisole-methanol dimer ion which result from the intra-cluster nucleophilic substitution reaction between p-difluorobenzene radical cation and methanol molecules according to

$$[C_6H_4F_2^+ + (CH_3OH)_2](CH_3OH)_{n-2} \longrightarrow$$
$$[C_6H_4FOCH_3]^+(CH_3OH)_{n-2} + CH_3OH + HF$$

Similar results were obtained for other alcohols such as ethanol and isopropanol. Isotopic and pressure studies indicate that in all cases, the 1:1 p-difluorobenzene-alcohol dimer does not participate in the reaction in agreement with Brutschy et al.[6,7] who used resonant two-photon ionization to study these intra-cluster reactions. This result is consistent with the failure to observe the analogous ion-molecule

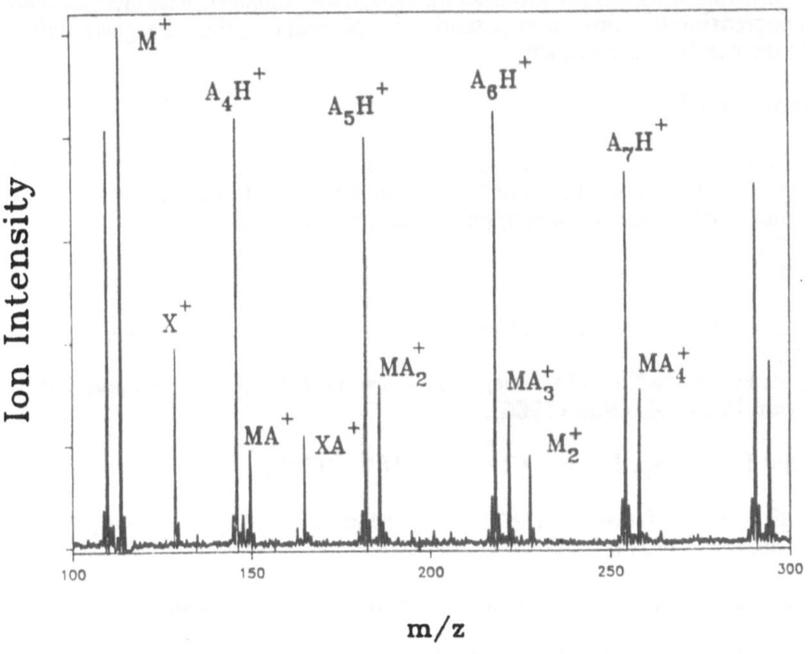

Figure 5, Cluster spectrum of p-difluorobenzene (M) and methanol-d4 (A). The mixed clusters are designated MA_n and the nucleophilic substitution product, p-fluoroanisole, is designated X. The monomer intensity of p-difluorobenzene is divided by 10.

reaction in the gas phase despite the large exothermicity of about 100 kJ/mol [6]. This exothermicity can result in extensive evaporation of alcohol molecules from the cluster, essentially stripping the cluster. Our results indicate that the reactivity of different alcohols towards nucleophilic aromatic substitution varies according to isopropanol > ethanol > methanol. This trend correlates with the ionization potential of the nucleophile.

4. Conclusions

The main conclusions that can be drawn from this study are as follows: 1. The cluster ion distributions from beam expansion are sensitive to thermochemical differences as small as 1-3 kcal/mol. The sensitivity to thermochemistry is observed even when ionization and intra-cluster proton transfer can deposit large amounts of energy. This suggests that the final evaporation steps occur from clusters with little excess energy. Thermochemical data make it possible to correlate the cluster distributions with energetics, and will be useful also in calculating the rate constants for the evaporation processes.
2. Intra-cluster nucleophilic substitution reactions have been observed following EI ionization of mixed p-difluorobenzene-alcohol clusters. The alcohol reactivity increases with decreasing the ionization potential which suggests that charge-transfer plays a crucial role in the reaction mechanism.

5. Acknowledgement

Acknowledgement is made to the donors of the Petroleum Research Fund, administered by the American Chemical Society and to the Thomas F. and Kate Miller Jeffress Memorial Trust for the partial support of this research.

6. References

1. S. Wei; W. B. Tzeng; A. W. Castleman, Jr., J. Chem. Phys. 92, 332 (1990).

2. C.A. Deakyne, M. Meot-Ner (Mautner), C.L. Campbell, M.G. Hughes and S.P. Murphy, J. Chem. Phys. 84, 4958 (1986).

3. C. Lifshitz and F. Louage, J. Phys. Chem. 93, 5633 (1989).

4. M. S. El-Shall, S. R. Olafsdottir, M. Meot-Ner (Mautner) and L. W. Sieck, Chem. Phys. Letters, 185, 193 (1991).

5. M. S. El-Shall and M. Meot-Ner (Mautner), J. Phys. Chem. 91, 1088 (1987).

6. B. Brutschy, J. Phys. Chem.94, 8637 (1990).

7. B. Brutschy, J. Eggert, C. Janes and H. Baumgartel, J. Phys. Chem.95, 5041 (1991).

Ionic Polymerization Within Van der Waals Clusters

M. Samy El-Shall
Department of Chemistry
Virginia Commonwealth University
Richmond, Virginia 23284-2006, USA

ABSTRACT. Consecutive bimolecular ion-molecule addition and elimination reactions have been observed within Van der Waals clusters of isoprene and vinyl chloride following EI ionization at pressures of 10^{-8} torr. Clusters provide a feasible and valuable approach for understanding the mechanism of ionic polymerization, and how the size of polymer chains is controlled in such a process.

1. Introduction

Cationic gas phase polymerization has been known for several years and is a subject of growing interest. Findings in this field will contribute to fundamental understanding of a wide range of problems, including the mechanism of polymerization, the production of large molecules in interstellar media, nucleation on polymers in the vapor phase, soot formation in hydrocarbon flames and photoemmision from polymeric ions generated from highly exothermic processes [1-3].

Gas phase polymeric ions can be formed through consecutive condensation reactions between undersaturated parent ions with their precursor compounds. One interesting possibility that has not been explored yet is the study of ionic polymerization within vdw clusters of the monomer molecules. The reaction can be initiated following the ionization of a neutral cluster beam formed in a supersonic expansion. Addition and elimination reactions can take place within the ionized clusters resulting in a product ion distribution which reflects both the stability of the polymeric ions and the kinetics of the reaction. The survival of a condensation ion formed in a highly exothermic process within the cluster proceeds via evaporative loss where the cluster boils off its vdw bonded molecules leaving a bare polymeric ion. Another mechanism for energy dissipation involves photoemission from the excited ions. The processes involved in the cluster ionic polymerization are illustrated in Figure 1.

Recently, we have studied cationic polymerization in isoprene and vinyl chloride clusters following electron impact ionization [4,5]. In this paper we summarize our results which show interesting chemistry, with unique size effects and an unexpected degree of sensitivity to thermochemistry.

P. Jena et al. (eds.), Physics and Chemistry of Finite Systems: From Clusters to Crystals, Vol. II, 1083–1088.
© 1992 *Kluwer Academic Publishers.*

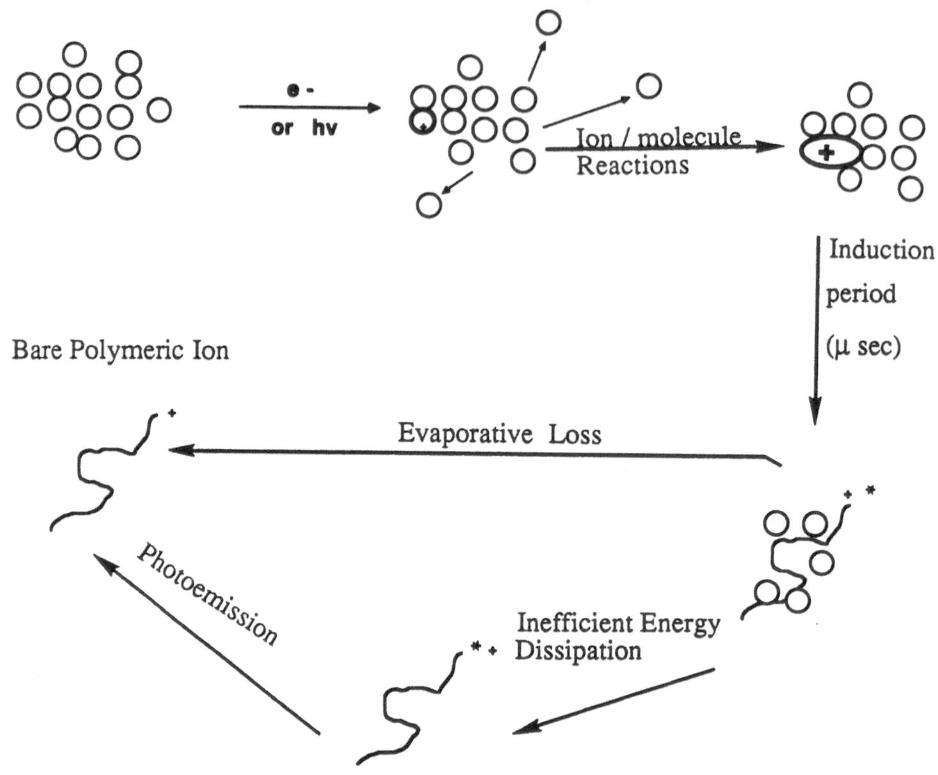

Figure 1. Processes involved in the cluster ionic polymerization.

2.Experimental

Neutral clusters, generated by pulsed adiabatic expansion in a supersonic cluster beam apparatus, were ionized by EI and subsequently mass analyzed in a quadrupole mass spectrometer [4,5].

3. Results and Discussion

Figure 2 displays segments of the mass spectrum observed for a cluster beam of isoprene. The ions observed are consistent with their formation by eliminative ion-molecule reactions between isoprene radical cation and neutral isoprene within the clusters.These ions are similar to those observed by Kaschers and Cooks from their ion-molecule reactions after 300 ms reaction time in an ion trap [6]. These ions are the characteristic fragment ions observed in the daughter spectrum of limonene, which is formed by a Diels - Alder reaction of isoprene radical cation with its neutral molecule according to :

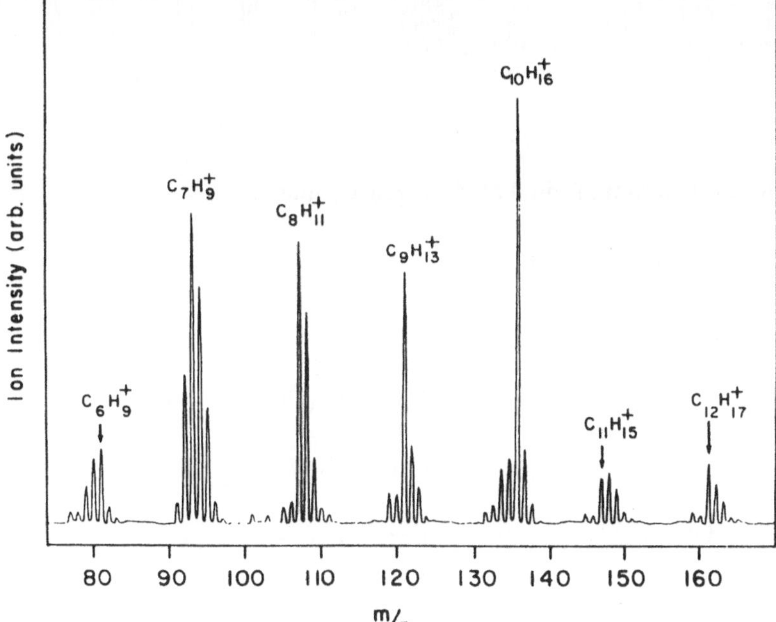

A magic number has been observed for $C_{14}H_{21}^+$ ion and attributed to a stable cyclic structure which can be derived from a series of ion-molecule consecutive addition-elimination reactions. We believe that the stability of this ion is associated with its closed shell electronic structure (even number of electrons) and we propose the cyclic structure (b) shown below. It must be noted that other cyclic isomers such as (a) and (c) are also possible.

Our results suggest that cationic polymerization of isoprene proceeds via the formation of the stable dimer $C_{10}H_{16}^+$ which, most likely, has a structure similar to limonene and can undergo further addition reactions to generate larger ions.

Figure 2. Mass spectrum of isoprene cluster ions.

Figure 3 displays a representative segment of a typical 70 eV EI mass spectrum of a cluster beam of vinyl chloride. The precursor ions $C_2H_3Cl^+$ and $C_2H_3^+$ undergo condensation ion-molecule reactions with neighboring C_2H_3Cl molecules in the cluster with subsequent elimination of neutral species such as HCl, Cl, CH_2Cl and C_2H_2.

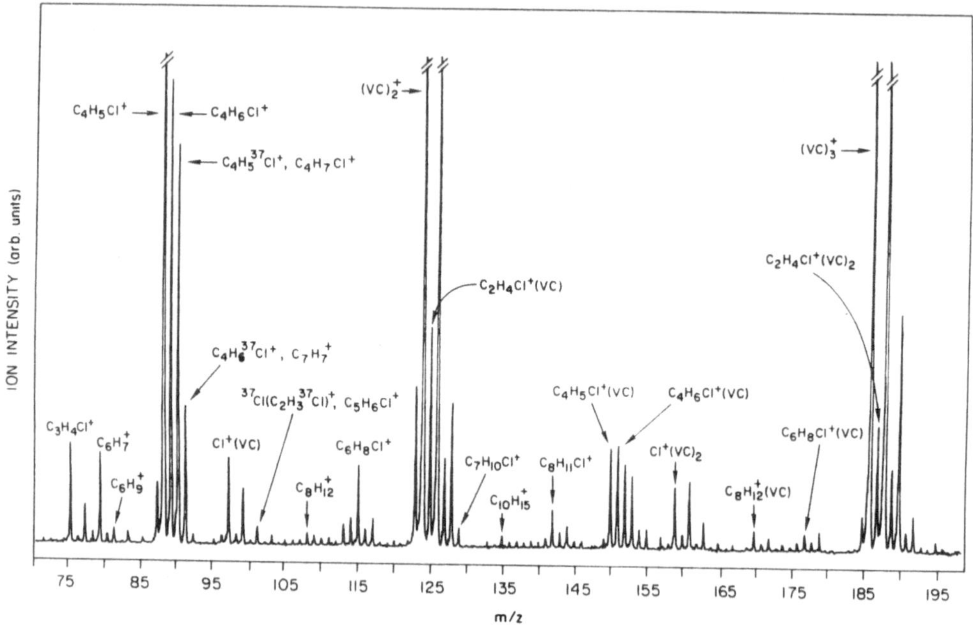

Figure 3. Mass spectrum of vinyl chloride cluster ions.

The observed reactions are as follows:

$$(C_2H_3Cl^+ + C_2H_3Cl)\,(VC)_{n-2} \longrightarrow C_4H_5Cl^+\,(VC)_{n-2} + HCl \tag{1}$$

$$C_4H_6Cl^+\,(VC)_{n-2} + Cl \tag{2}$$

$$C_3H_4Cl^+\,(VC)_{n-2} + CH_2Cl \tag{3}$$

$$(C_2H_3^+ + C_2H_3Cl)\,(VC)_{n-2} \longrightarrow C_2H_4Cl^+\,(VC)_{n-2} + C_2H_2 \tag{4}$$

$$C_4H_5^+\,(VC)_{n-2} + HCl \tag{5}$$

$$C_4H_6^+\,(VC)_{n-2} + Cl \tag{6}$$

Reactions (1-6) constitute a general scheme for the first stages of the cationic chemistry observed in vinyl chloride. Interestingly, this scheme explains most of the processes which we observe to occur within the ionized VC clusters. The ion $C_3H_4Cl^+$ is not observed in any clustering sequence with VC. It must be noted that the ion/molecule reaction analogies of reactions (1-3) are exothermic by 41, 21-14 and 5 ±7 kcal/mol, respectively. Therefore the elimination of CH_2Cl could be barely allowed energetically for the reaction of the unsolvated $C_2H_3Cl^+$ ion. The fact that the sequence $C_3H_4Cl^+(VC)_{n-2}$ is not observed for any other cluster with n larger than 2 may indicate that ion solvation renders the reaction less exothermic. Perhaps at larger clusters following ionization there is a hinderance to the elimination of CH_2Cl and thus this reaction channel is closed. We also observe that the elimination reactions initiated by $C_2H_3Cl^+$ terminate after three successive steps; each involving elimination of HCl or Cl. The product ions are of the general formula $C_8H_kCl^+$, k=9-13. These ions can rearrange to some stable structures and this may account for the lack of further elimination reactions. Figure 4 is a representative reaction scheme for the $C_2H_3Cl^+$ ion which takes into account the three processes thought to be important namely; elimination, addition and decomposition reactions within the clusters. Similar schemes can be used to account for all possible combinations of elimination reactions and also for the reactions of $C_2H_3^+$ within the clusters. This overall reaction mechanism is capable of accounting for all the observed mass peaks in VC clusters.

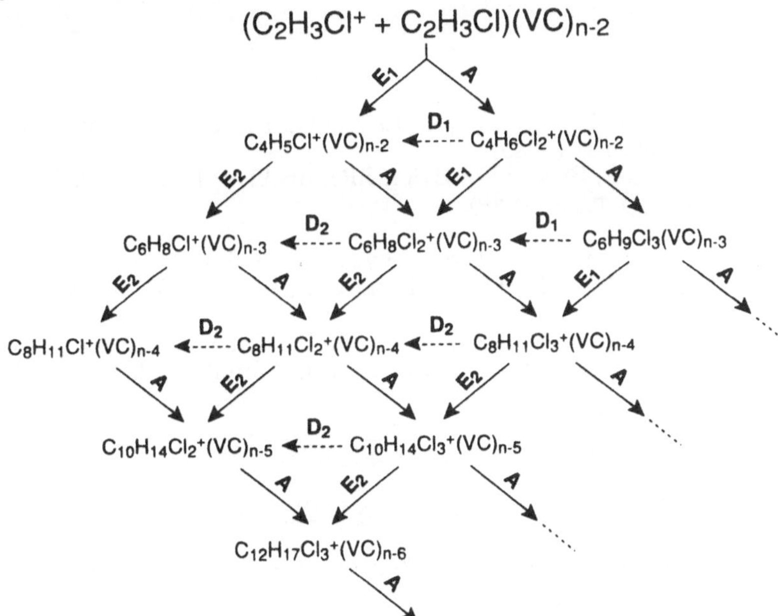

Figure 4. A representative scheme for the reactions of $C_2H_3Cl^+$ within vinyl chloride clusters : A = addition of C_2H_3Cl , E1 = addition / elimination of HCl , E2 = addition / elimination of Cl , D1 = loss of HCl , D2 = loss of Cl.

4. Conclusions

The main conclusions that can be drawn from this study are as follows :

1. Sequential polymerization reactions with several condensation steps can occur in ionized Van der Waals clusters. This occurs despite the fact that condensation reactions are exothermic, and the energy released could dissociate the cluster.The competition between the condensation reactions and monomer evaporation can control the ultimate size that the polymer can reach in the cluster. It is therefore important to understand the kinetic factors that control this competition.

2. The formation of stable ionic species (cyclic structure) interrputs the general pattern of successive addition reactions. A clear example in isoprene study is the observation of a magic number corresponding to $C_{14}H_{21}^{+}$ ion. It is interesting that similar cyclic structures have been found in polyisoprene formed by bulk cationic polymerization of isoprene.

Acknowledgement

Acknowledgement is made to the donors of the Petroleum Research Fund, administered by the American Chemical Society and to the Thomas F. and Kate Miller Jeffress Memorial Trust for the partial support of this research.

References

1.See for example:
   (a) Buckley, T. J.; Sieck, L. W.; Metz, R.; Lias, S.G.; Liebman, J.F. Int. J. Mass Spectrom. Ion Proc.,65, 181 (1985).
   (b) Oztwik, F.; Moini, M.; Brill, F.W.; Eyler, J.R.; Buckley, T.J.; Lias S.G.; Ausloos, P.J. J. Phys. Chem., 93,4038 (1989).

2. Dalgarno, A.; Black, J.H. Rep. Prog. Phys., 39, 573 (1976) Huntress, W.T. Chem.Soc. Rev., 6, 295 (1977); Herbst, E.; Klemperer, W. Phys.Today, 29, 32 (1976).

3. Calcote, H.F. Flame, 42, 215 (1981); Olson, D.B.; Calcote, H.F. in "Particulate Carbon: Formation During Combustion" Siegle, D.C.; Smith, G.W. (Eds.), Plenum Press, New York,1981, p. 177.

4. El-Shall,M.S. and Marks,C., J. Phys. Chem.,95, 4932 (1991).

5. El-Shall,M.S. and Schriver,K.E., J. Chem. Phys., 95, 3001 (1991).

6. Kascheres,C. and Cooks,R.G., Analytica Chimica Acta, 215, 22 (1988).

METASTABLE FRAGMENTATION OF RARE GAS CLUSTER IONS INITIATED BY EXCIMER DECAY

M. FOLTIN, G. WALDER, T.D. MÄRK
Institut für Ionenphysik, Universität Innsbruck
Technikerstrasse 25, A 6020 Innsbruck, Austria

ABSTRACT. A new unusual metastable fragmentation channel of argon and neon cluster ions produced by the electron impact ionization of a neutral cluster beam has been studied with a double focusing sector field mass spectrometer. In the case of argon cluster ions it leads to evaporation of up to 10 monomers in the μs time regime. It is shown that this fragmentation is initiated by the radiative decay of a metastable excimer embeded in the cluster ion. Formation of the excimer is induced by multiple electron scattering within the cluster.

1. Introduction

Metastable fragmentation studies of ionized van der Waals clusters are of considerable importance for the elucidation of energy acquisition, storage and disposal processes in finite atomic and molecular systems, representing an intermediate stage between the gaseous and solid state of matter. In the last decade the occurrence of several distinct types of metastable fragmentation of cluster ions has been reported (see e.g. [1]). These types differ in the way how excess energy is deposited by electron or photon impact ionization and how this energy is released subsequently via metastable decay reactions.

In most cases the excess energy gained during the electron impact ionization is randomized over intermolecular van der Waals modes and then it is subsequently released giving rise to sequential evaporation of up to two monomers in the metastable time regime [2]. This process (often referred to as the metastable fragmentation due to the vibrational predissociacion) is well described by the evaporative ensemble model of Klots [3]. A different metastable fragmentation pattern is shown in the case of nitrogen cluster ions $(N_2)_n^+$, where the evaporation of more than two monomers can be observed and the dissociation probability plotted versus the number of desorbed N_2 monomers exhibits a characteristic oscillatory pattern [4]. This was explained by delayed radiationless relaxation of vibrationally excited N_2 molecules in the cluster followed by an abrupt energy release via evaporation [4].

Recently, we have discovered a rather unusual metastable fragmentation channel occurring in argon cluster ions [5]. In contrast to the well known case of single monomer evaporation due to vibrational predissociation, in the new decay reaction the number of ejected Ar monomers rises from 2 for Ar_4^+ up to 10 for Ar_{30}^+. Similar metastable fragmentation channels were observed

P. Jena et al. (eds.), Physics and Chemistry of Finite Systems: From Clusters to Crystals, Vol. II, 1089–1094.
© *1992 Kluwer Academic Publishers.*

also in neon cluster ions [6]. In this paper we summarize our studies of the dependence of the metastable fractions on (i) the electron energy, (ii) parent cluster size and the number of ejected monomers, and (iii) the time since ion formation to dissociation. This allows us to conclude that a metastable excimer $Ar_2^*(^3\Sigma_u^+)$ (or $Ne_2^*(^3\Sigma_u^+)$) localized inside the cluster ion is responsible for the observed unusual decay pattern. The *radiative* decay of this excimer leads to the repulsion of Ar (Ne) atoms in the ground state $Ar_2(^1\Sigma_g^+)$ (or $Ne_2(^1\Sigma_g^+)$) and to subsequent disintegration of the cluster. Formation of the excimer is induced by an electron which undergoes multiple scattering (i.e. ionization and excitation) within the same cluster.

This work provides a clear evidence for possible coexistence of an ionized "chromophore" and a localized long-living electronically excited state inside the cluster of the size of several atoms to several tens of atoms. The coexistence of an ionized and a long-living excited species within the cluster was already reported by Gspann and Vollmar [7] for ionized helium clusters of 10^6 to 10^8 atoms. In this work the same phenomenon is shown for the first time to occur also in small cluster sizes. Moreover, we are able to detect and follow the decay of electronically excited state in the metastable time regime.

2. Experimental

The supersonic beam/electron-impact ionization source/mass spectrometer system used has been described in detail earlier [8]. Neutral Ar clusters are formed by expanding 2.5 bar of Ar at -160 °C through a 20 µm nozzle. (Neutral Ne clusters are produced by expanding 9 bar of Ne at -192 °C). The ensuing cluster beam proceeds through a skimmer into a differentially pumped inner chamber and is crossed 10 cm downstream from the nozzle at right angles by an electron beam of variable energy. Cluster ions thereby produced are (i) extracted at right angles from the ionization region, (ii) accelerated by an acceleration voltage (between 1.5 and 3 kV), (iii) analyzed in a double-focusing sector field mass spectrometer, and (iv) detected after postacceleration with a secondary electron multiplier. A possible metastable decay of an ion produced in the ion source can be detected either in the first field free region between the accelerating region and the magnetic sector, or in the second field free region between the magnetic sector and the electric sector by proper tuning of both sector fields. The dependence of the metastable fraction on the time since ion formation can be obtained by varying the acceleration voltage.

3. Results and Discussion

Fig.1 shows the dependence of the metastable fraction (which is the ratio between the fragment and the parent ion current) on the number of evaporated monomers for Ar_{10}^+, Ar_{16}^+ and Ar_{25}^+ parent cluster ions, decaying in the first field-free region of our mass spectrometer. As expected, the strongest metastable fragmentation channel is the usual single-monomer evaporation. Suprisingly, it is shown for the first time that there exists also an additional metastable fragmentation channel, leading to evaporation of more than two monomers. In the case of Ar_{10}^+ the number of Ar monomers evaporated in this new unusual fragmentation channel is 3 to 6 with the strongest evaporation of 4 monomers (Fig. 1). In the case of Ar_{16}^+ this new channel is bimodal with 2 to 6 monomers evaporated in the first mode and 6 to 10 monomers evaporated in the second mode (see Fig. 1). For Ar_{25}^+ the first mode has disappeared and the number of monomers evaporated in the second mode is 8 to 11 (Fig. 1).

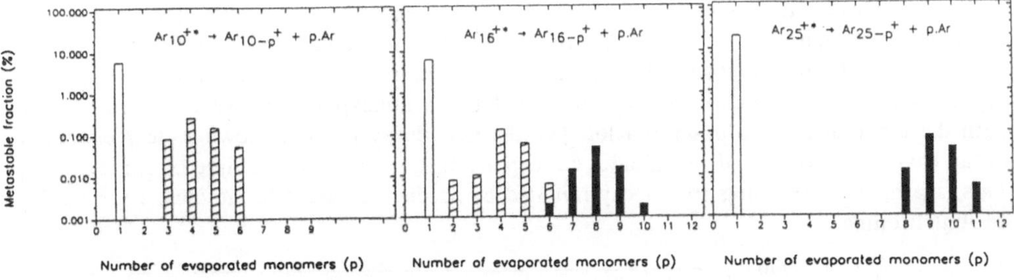

Fig. 1. The dependence of the metastable fraction on the number of evaporated monomers p for Ar$_{10}^{+}$, Ar$_{16}^{+}$ and Ar$_{25}^{+}$ parent cluster ions.

The average number of evaporated monomers in these two new modes increases with the parent cluster size according to Fig. 2. Fragmentation in the first mode is observable only for small cluster ions (with size n<19), while the second mode is appearing for biger cluster sizes (for n>10). Local minima in the dependence of the average number of evaporated monomers on the parent cluster size correspond to fragmentation of parent ions with a rather stable geometry (e.g. Ar$_{13}^{+}$ and Ar$_{19}^{+}$). The asymptotic value of the number p of evaporated monomers for cluster ions of the size n>>20 is approx. p=10. Comparing this value with the number of monomers evaporated during the photofragmentation of argon cluster ions (where the energy deposited to system is known) it is possible to estimate the energy needed for this evaporation process to be approx. 0.9 eV for p=10. Delayed release of such an amount of energy in the metastable time regime can be explained only by the existence of a localized long-livinig excited state inside the cluster.

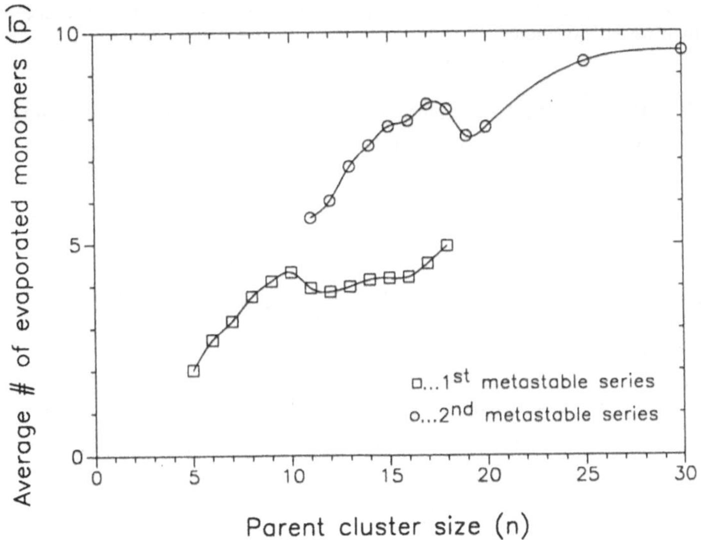

Fig. 2. Dependence of the average number of evaporated monomers $\bar{p}$ on the parent cluster ion size n for the new metastable fragmentation channel Ar$_{n}^{+}$ → Ar$_{n-p}^{+}$ + p.Ar

This conclusion is further substantiated by another experimental observation. In Fig. 3 we show the dependence of the metastable fraction on the time since ionization to fragmentation, compared for the newly observed metastable decay channel and for the usual single monomer evaporation. Whereas monomer evaporation exhibits a nonexponential behavior in accordance with the theoretical description of Klots [3], the new decay process shows a clear exponential decay (over three orders of magnitude) corresponding to an average lifetime of approx. 1.2 μs. This, again, strongly points to a decay initiated by species embeded in the cluster ion having a definite lifetime.

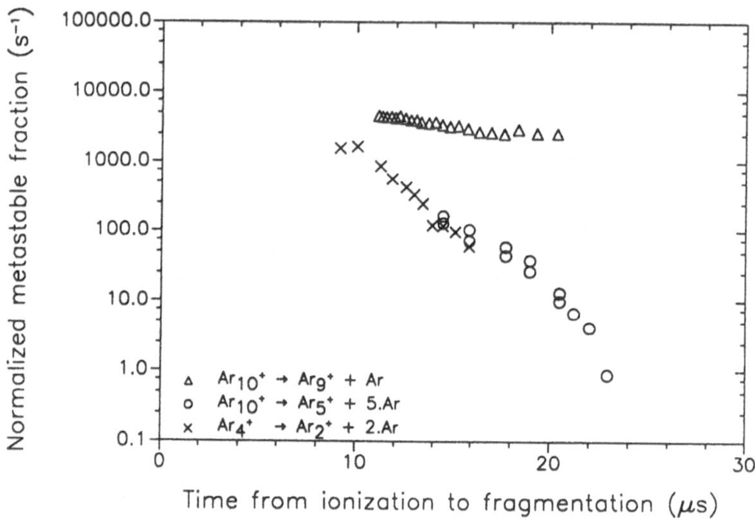

Fig. 3. Normalized metastable fraction as a function of the time since ionization to entering the first field free region of the mass spectrometer.

In order to find the type of a localized metastable state initiating the fragmentation, we studied the dependence of the fragment ion current on the electron energy. In Fig. 4, this dependence is shown for the metastable decay

$$Ar_4^+ \rightarrow Ar_2^+ + 2.Ar \qquad (1)$$

For comparison, the electron energy dependence of the Ar_4^+ parent ion current is also shown. Linear extrapolation in the onset region shows, that the appearance energy (AE) of the metastable decay (1) is 26.9 ± 0.4 eV, which is approximately 11.9 eV more than the AE of the parent ion Ar_4^+ (in this measurement AE (Ar_4^+) = 15.0 ± 0.3 eV). The lowest metastable electronically excited states of the Ar^+ ion are lying more than 15 eV above the ground ionic state. The lowest excited states of the neutral Ar are the $(3p^5 4s)$ $^3P_2$, $^3P_0$ metastable states and the $^3P_1$, $^1P_1$ resonant states, lying 11.55 to 11.83 eV above the ground state. Due to the solvation the electron energies needed for the population of first excited states in clusters are generally slightly higher than those in isolated atoms [9]. It was found that also all the other metastable decays belonging to the new metastable decay channel show electron energy dependences similar to that observed in the metastable decay (1).

Taking into consideration the measured AE of the metastable fragmentation (1), we have to conclude therefore that electronically excited states of *neutral* Ar are involved in the fragmentation process (1), populated by an electron which undergo multiple scattering (i.e. the ionization of one atom and the excitation of an another atom) inside the cluster.

In order to further support our conclusions, the electron energy dependences of the parent and the fragment ion currents were measured also for a new metastable fragmentation of the *neon* tetramer ion:

$$Ne_4^+ \rightarrow Ne_2^+ + 2.Ne, \qquad (2)$$

which was recently observed in our laboratory [6] (Fig. 4). The appearance energy of the metastable decay (2) is 37.2 ± 0.8 eV, which is approximately 16.4 eV more than the AE of the parent ion Ne_4^+ (in this measurement AE $(Ne_4^+) = 20.8 \pm 0.3$ eV). The lowest excited states of the neutral Ne are the $(2p^5 3s)$ $^3P_{0,2}$ metastable states and the $^{1,3}P_1$ resonant states, lying 16.62 to 16.85 eV above the ground state. As in the *argon* case, the difference between the appearance energy of the metastable decay reaction (2) and the appearance energy of the parent ion Ne_4^+ corresponds quantitatively very well to the value of the excitation energy for the lowest lying excited states of *neutral* Ne.

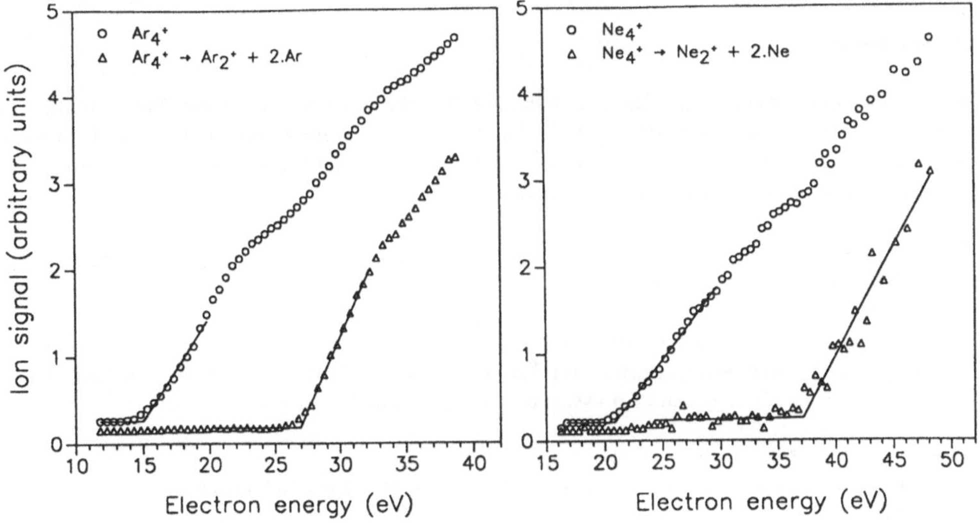

Fig. 4. Left part: Ion current as a function of the electron energy for the parent ion Ar_4^+ (o) and for the fragment ion Ar_2^+ (Δ) produced by the metastable decay reaction (1). Right part: Ion current as a function of the electron energy for the parent ion Ne_4^+ (o) and for the fragment ion Ne_2^+ (Δ) produced by the metastable decay reaction (2).

Fluorescence studies in liquid and solid argon and neon indicate [10-12] that after electron or photon impact excitation and subsequent cascades of fast nonradiative transitions, the excimer dimer Ar_2^* (Ne_2^*) in the $0_u^-, 1_u$ $(^3\Sigma_u^+)$ or 0_u^+ $(^1\Sigma_u^+)$ bonding state is formed. The second excited 0_u^+ state (correlated with the $^3P_1$ atomic level) is radiatively coupled with the 0_g^+ $(^1\Sigma_g^+)$ ground electronic state and therefore it is short living (lifetime of several ns). The first excited, nearly degenerate $0_u^-, 1_u$ $(^3\Sigma_u^+)$ state is metastable, but it's radiative lifetime is considerably reduced with respect to the corresponding atomic state $^3P_2$ and lies in the microsecond region.

Appearance energy measurements (see above) allow us to propose, that the metastable $Ar_2^*(\Sigma_u^+)$ (or $Ne_2^*(^3\Sigma_u^+)$) excimer is also formed in the presently investigated cluster ions. It's *radiative* decay followed by repulsion of Ar (or Ne) atoms in the ground $Ar_2(^1\Sigma_g^+)$ $(Ne_2(^1\Sigma_g^+))$ state would lead to fragmentation of the cluster ion in the microsecond region, observable in our mass spectrometer. In the case of *argon* cluster ions the corresponding fragmentation lifetime of approx. 1.2 μs deduced from the dependence of the metastable fraction on the time since ionization to fragmentation (see above) is perfectly agreeing with the radiative lifetime of the Ar_2^* $(^3\Sigma_u^+)$ excimer in solid and liquid argon, respectively [10]. In the case of *neon* cluster ions we can see the metastable decay (2) not only in the first field free region but also in the second field free region, i.e. approx. 17 μs after the ionization which is again consistent with the fluorescence measurements, providing radiative lifetimes between 5 μs ([11], excimer in solid) and 11 μs ([12], isolated excimer). No similar metastable decay was observed in krypton and xenon cluster ions, because the correponding Σ_u^+ states have only lifetimes in the ns range and therefore will not be observable in the μs time regime of our experimental time window.

4. Conclusions

The present work provides unambiguous evidence that the unusual metastable fragmentation of rare gas cluster ions can be initiated by radiative decay of an excimer embeded in the cluster ion. Appearance energy measurements represent quantitative support for this model, while time dependence measurements support it qualitatively.

Acknowledgements

Work partially supported by the Österreichischer Fonds zur Förderung der Wissenschaftlichen Forschung and the Bundesministerium für Wissenschaft und Forschung, Wien, Austria. It is a pleasure to thank Dr. Paul Scheier for assistance in the production of Ne clusters.

References
1. T.D. Märk and O. Echt, in Clusters of Atoms and Molecules (H. Haberland, Ed.) Springer, Heidelberg (1990)
2. P. Scheier and T.D. Märk, Phys. Rev. Lett. **59**, 1813 (1987)
3. C.E. Klots, J. Chem. Phys. **83**, 5454 (1985); Z. Phys. D **5**, 83 (1987), J. Phys. Chem. **92**, 5864 (1988)
4. T. Leisner, O. Echt, O. Kandler, X.J. Yan, and E. Recknagel, Chem. Phys. Lett. **148**, 386 (1988); P. Scheier and T.D. Märk, Chem. Phys. Lett. **148**, 393 (1988); T.F. Magnera, D.E. David and J. Michl, J. Chem. Soc. Faraday
5. M. Foltin, G. Walder, A.W. Castleman, Jr., and T.D. Märk, J. Chem. Phys. **94**, 810 (1991)
6. M. Foltin and T.D. Märk, Chem. Phys. Lett. **180**, 317 (1991)
7. J. Gspann and H. Vollmar, J. Chem. Phys. **73**, 1657 (1980)
8. T.D. Märk, P. Scheier, K. Leiter, W. Ritter, K. Stephan and A. Stamatovic, Int. J. Mass Spectrom. Ion Proc. **74**, 281 (1986)
9. A. Burose, C. Becker, and A. Ding, Symposium on Atomic and Surface Physics 90 (T. D. Märk, F. Howorka, Ed.), Obertraun 1990, p. 302
10. M. Joppien, F. Grotelüschen, T. Kloiber, M. Lengen, T. Möller, J. Wörmer, G. Zimmerer, J. Keto, M. Kykta and M.C. Castex, J. Luminescence, in print (1991)
11. R. Gaethke, P. Gürtler, R. Kink, E. Roick, and G. Zimmerer, phys. stat. sol. (b) **124**, 335 (1984)
12. B. Schneider and J.S. Cohen, J. Chem. Phys. **61**, 3230 (1974)

ELECTRONIC STATES and H-ADSORPTION of Co and Co-V MICROCLUSTERS

N. FUJIMA and T. YAMAGUCHI
Faculty of Engineering, Shizuoka University,
Hamamatsu 432, Japan

ABSTRACT   By using the DV-Xα-LCAO method, electronic states are calculated for icosahedral Co_{13}, $Co_{12}V$ clusters and for $Co_{13}H_2$, $Co_{12}VH_2$ clusters of C_{2v} symmetry. The gross feature of the electronic structure near the 12 exterior Co atoms of the Co_{13} cluster is similar to that of the $Co_{12}V$ cluster. However, the density-of-states around the highest occupied level is larger in the Co_{13} cluster than in the $Co_{12}V$ cluster. For the $Co_{13}H_2$ and $Co_{12}VH_2$ clusters, electronic structures of the levels which contain much $1s$ atomic orbital of H are different from each other: the $1s$ orbital of the H atom interacts strongly with the $3d$ orbital of the exterior Co atoms and make a bonding orbital in $Co_{13}H_2$ cluster, but it interacts weakly and makes a non-bonding orbital in the $Co_{12}VH_2$ cluster. This explains the experimental result, that is , H_2 molecules are easily adsorbed to Co_{13} clusters and hardly adsorbed to $Co_{12}V$ clusters.

1   Introduction

Experimental and theoretical studies in microclusters have revealed remarkable properties of microclusters as a new material. Reaction studies have been done recently. These studies revealed that reactivities of metal clusters to molecules have anomalous size dependences [1]-[7]. Nonose *et al.* have observed, in the cluster beam experiments using the laser vaporization method, that a $Co_{12}V$ cluster is stable to H_2-adsorption while Co_{13}, $Co_{11}V_2$ and Co_NV ($N < 12$) clusters are unstable [7]. They also observed similar stablities of Co_NV ($N = 13 \sim 18$) clusters, although it was not so clear as the $Co_{12}V$ cluster [8]. These results suggest that the clusters which have a V atom and consist of more than 13 atoms are stable to the adsorption. In addition, they suggest that the V atom is located at an interior site of the cluster since clusters which consist of more than 13 atoms are likely to have interior sites. Then, the interior atom seems to have an important role in reaction with the H_2 molecule.

El-Batanouny *et al.*   have found in photoemission experiments that a H atom adsorbs well both on the clean Nb(110) surface and on the surface with a multilayer of Pd, but does not adsorb on Nb(110) surface with a monolayer of Pd [9]. The fact that the H atom does not adsorb on the surface with a monolayer of Pd is due to the decrease of the density-of-states (DOS) of d electrons near the Fermi level such as the DOS of a noble metal. A similar behavior has been observed in the experiment of CO-adsorption [10]. We expect that these anomalous reductions of molecular adsorption on the surface resemble the stability of the

P. Jena et al. (eds.), Physics and Chemistry of Finite Systems: From Clusters to Crystals, Vol. II, 1095–1100.
© 1992 *Kluwer Academic Publishers.*

Co$_{12}$V cluster to H-adsorption in the sense of molecular adsorption on a late transition-metal monolayer with an early transition-metal substrate, that is, the interior V atom of Co$_{12}$V cluster corresponds to the Nb(110) surface and the exterior 12 Co atoms correspond to the monolayer of Pd.

Kumar and Bennemann have calculated, by using a self-consistent tight-binding scheme, the electronic structure of the Pd-overlayer on Nb(100) surface [11]. Estimating the interface binding energy between the Pd-layer and the Nb substrate, they expect that the H-adsorption on surfaces depends on the interface interaction.

2 Calculations

We calculate the electronic states of icosahedral (Ih) Co$_{13}$ and Co$_{12}$V clusters in Section 3 and Co$_{13}$H$_2$ and Co$_{12}$VH$_2$ clusters of C$_{2v}$ symmetry in Section 4. All calculations are performed by using the self-consistent-charge DV-Xα-LCAO method with the $1s{\sim}4s$, $2p{\sim}4p$ and $3d$ orbitals of transition-metal atoms and the $1s$ orbital of the H atom.

Figure 1 shows the atomic structure of the cluster whose electronic states are calculated. The bare cluster of 13 atoms has Ih-symmetry. Since the atomic distance between the exterior atoms of the icosahedron is 5 % longer than that between the exterior and center atoms, the atomic distances of the clusters are chosen such that the root-mean-square error to the distance of the bulk crystal is minimum: 4.57 a.u. for the Co-Co distance and 4.62 a.u. for the V-Co distance between the center and exterior atoms. We assume that two H atoms are located directly on the exterior Co atoms with distance of the sum of atomic radii, 3.00 a.u. The symmetry of the H-added cluster is C$_{2v}$. The levels of only the A$_1$ and B$_1$ representations of C$_{2v}$ symmetry have the $1s$ component of the H atom.

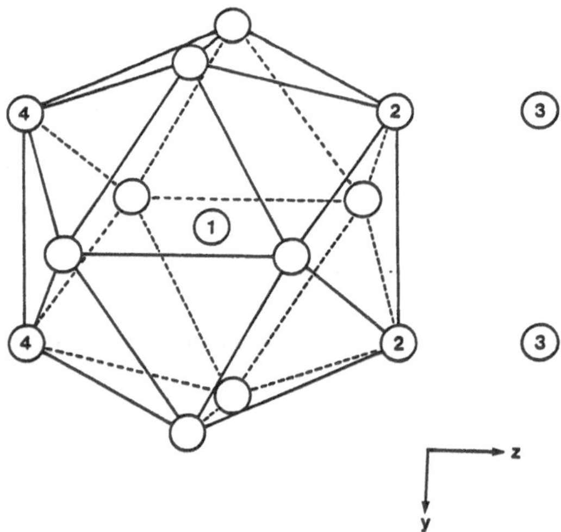

Fig. 1. Atomic structure of the Co$_{12}$MH$_2$ clusters. The atom sites designated by numbers are located on the y-z plane. The atom site (3) represents an added hydrogen atom. The atom sites (2) and (4) represent exterior Co atoms on the plane. The atom site (1) is the center site where a Co or V atom is placed.

3   Electronic states of Co_{13} and $Co_{12}V$ clusters

Figure 2 shows the energy level diagrams of the Co_{13} and $Co_{12}V$ clusters. In the figure, the solid and broken curves indicate the partial density-of-states (PDOS) of $3d$ orbitals of the whole cluster (multiplied by a factor 0.1) and that of the $3d$ orbital of the center atom (1) in Fig.1, respectively. Here, the PDOS is obtained by plotting the magnitude of the Mulliken charge of atomic orbitals for each levels and by replacing the discrete level by a Gaussian with FWHM of 0.02 a.u. The dotted line points to the highest occupied level.

The $3d$-PDOS of the whole cluster has two large peaks at -0.26 and -0.19 a.u. They are mainly composed of the $3d$ orbitals of the exterior Co atoms. The PDOS near the highest occupied level of the Co_{13} cluster is 15 % larger than that of the $Co_{12}V$ cluster. The higher peak of the Co_{13} cluster has higher energy than that of the $Co_{12}V$ cluster. These facts correspond to the DOS of the monolayer of Pd on Nb(110) surface in the photoemission spectra [9,10].

The $3d$-PDOS of the center Co atom in the Co_{13} cluster has a large peak at -0.35 a.u., which belongs to a 5-fold H_g representation in Ih-symmetry. On the other hand, the $3d$-PDOS of the V atom in the $Co_{12}V$ cluster has two large peaks between which the highest occupied level exists. This suggests that the V atom interacts with the exterior Co atoms more strongly than the center Co atom of the Co_{13} cluster.

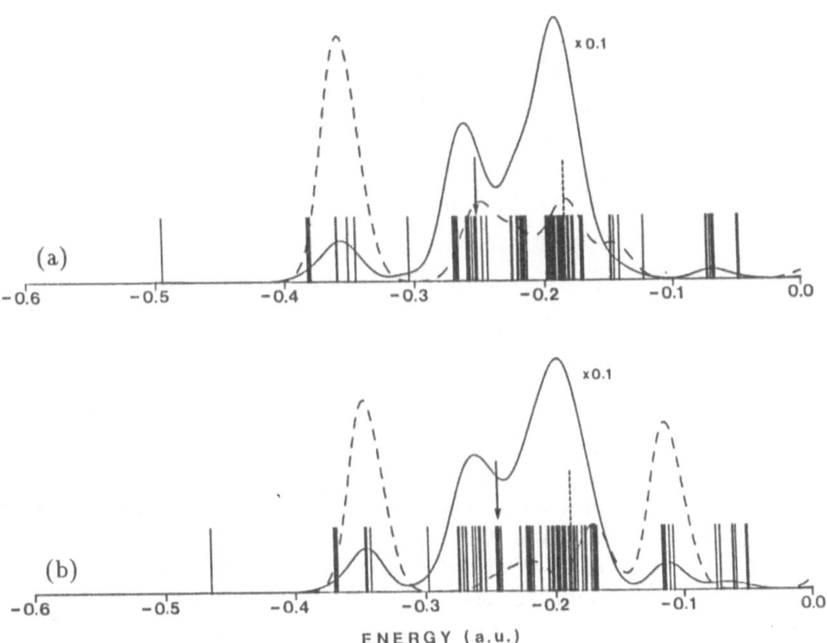

Fig. 2. Energy level diagrams in a.u. (a) Co_{13} and (b) $Co_{12}V$ clusters.

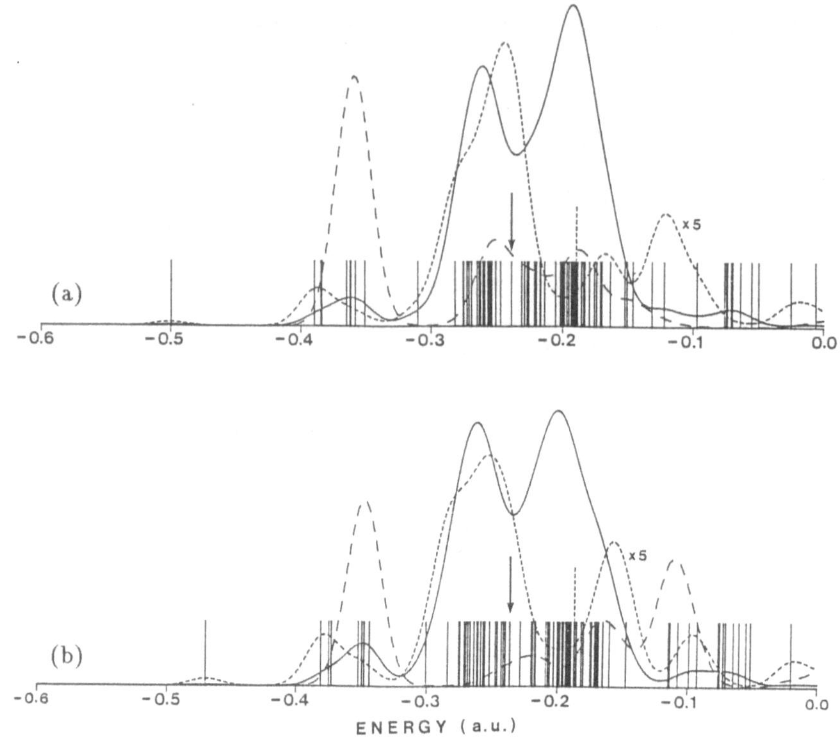

Fig. 3. Energy level diagrams. (a) $Co_{13}H_2$ and (b) $Co_{12}VH_2$ clusters.

4 Electronic states of $Co_{13}H_2$ and $Co_{12}VH_2$ clusters

Figure 3 shows the energy level diagrams of $Co_{13}H_2$ and $Co_{12}VH_2$ clusters. The solid, broken and dotted curves indicate the $3d$-PDOS of Co (2) atoms adjoining H atoms, the $3d$-PDOS of the center Co (1) or V(1) atom and the $1s$-PDOS of H (3) atoms, respectively. The $1s$-PDOS is multiplied by a factor 5. The solid curves of both clusters have similar shapes except the region near the highest occupied level. The broken curve has a large peak at -0.36 a.u. in Fig.3(a), and two peaks at -0.35 and -0.11 a.u. in Fig.3 (b). Similarly to the discussion in the previous section, the $3d$ electron of the V atom interacts strongly with the exterior Co atoms. The dotted curves of both clusters have two large peaks above and below the highest occupied level. The energy separation of two peaks of the $Co_{13}H_2$ cluster is larger than that of the $Co_{12}VH_2$ cluster. This means that the exterior Co-H interaction is larger in the $Co_{13}H_2$ cluster.

Figure 4 shows the contour map of electron density on the y-z plane including a center atom site (1), four exterior sites(2, 4) and H-sites(3) in Fig.1. Here, Fig.4(a) shows the electron density of the B_1 level at -0.239 a.u. of the $Co_{13}H_2$ cluster and Fig.4(b) shows that of the B_1 level at -0.236 a.u. of the $Co_{12}VH_2$ cluster. These levels are designated by arrows in Fig.3. In the central region of both clusters, the charges are mainly composed of the $3d_{yz}$

orbital of the center atom (1). In the region between the exterior Co (2) and H (3) atoms, the charge in Fig.4(a) is much larger than that of Fig.4(b). It seems that the orbital is bonding in Fig.4(a) and non-bonding in Fig.4(b). This agrees with the adsorption experiments [7] that the Co_{13} cluster is unstable and the $Co_{12}V$ cluster is stable to H-adsorption.

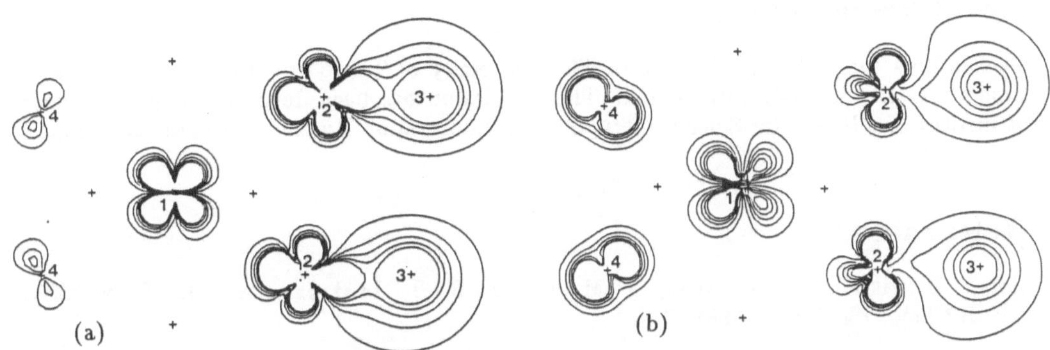

Fig. 4. Contour map of electron density on the y-z plane (arbitrary units). (a) B_1 level (-0.239 a.u.) of the $Co_{13}H_2$ cluster. (b) B_1 level (-0.236 a.u.) of the $Co_{12}VH_2$ cluster.

5 Discussion

In the previous section, we showed that the exterior Co-H interaction depends on the sort of the center atom, *i.e.* it is large and small when the center atom is Co and V, respectively. In addition, the center-exterior interaction is small and large when the center atom is Co and V. Calculation shows that the $1s$-PDOS of the H atom in the $Co_{12}H_2$ cluster whose center site is vacant is similar to that of the $Co_{13}H_2$ cluster. Then, the existence of the center Co atom does not affect the exterior Co-H interaction. The large (small) interaction between the center and exterior atoms causes the small (large) interaction between the exterior and added atoms. This explains the anomalous stablity of the $Co_{12}V$ cluster to H-adsorption.

Next, we discuss the magnitude of the interaction between the center-exterior atoms. The magnitude of the $3d$-$3d$ interaction is expected to be derived from the radius of the atomic orbital relative to the inter-atomic distance: we estimate the ζ value of the $3d$ orbital in the single ζ-STO $r^2 e^{-\zeta r}$ from the numerical orbital of the DV-Xα calculation: 3.93 in a.u. for Co and 2.80 for V atom. The radius of the V $3d$ orbital is larger than that of the Co $3d$ orbital. Note that the observed atomic radius in the crystal is 2.53 a.u. for V and 2.42 a.u. for Co. Then, we expect the center-exterior interaction which is small for the Co_{13} cluster and large for the $Co_{12}V$ cluster.

Calculation shows that the electronic structure of the A_1 level is similar for each cluster. The A_1 level contains the $3d_{x^2-y^2}$ and $3d_{3z^2-r^2}$ orbitals of the center atom, but does not contain the $3d_{yz}$ orbital which interacts strongly with the exterior atoms (2, 4).

In conclusion, the calculated electronic states of the $Co_{13}H_2$ and $Co_{12}VH_2$ clusters are qualitatively consistent with the experimental results of the stability of the $Co_{12}V$ cluster to

H-adsorption and the reactivity of the Co_{13} cluster. In future work, we will calculate the electronic states of the $Co_{13}H_2$ and $Co_{12}VH_2$ clusters where H atoms are adsorbed on other sites such as the hollow site with C_3 rotational symmetry. Furthermore, we will calculate electronic states of $Co_{12}MH_2$ (M=Fe, Mn and Cr) clusters and discuss the stabilty of these clusters to H-adsorption.

Acknowledgment

The authors thank Dr. S. Nonose of The University of Tokyo and Professor K. Kaya of Keio University for useful discussion. The calculation has been done at Computer Center of Institute for Molecular Science, Okagaki, and Information Processing Center of Shizuoka University.

References

[1] J. L. Elkind, F. D. Weiss, J. M. Alford, R. T. Laaksonen and R. E. Smalley: J. Chem. Phys. **88** (1988) 5215.

[2] E. K. Parks, G. C. Nieman, L. G. Pobo, and S. J. Riley: J. Chem. Phys. **88** (1988) 6260.

[3] M. R. Zakin, D. M. Cox, R. O. Brickman and A. Kaldor: J. Phys. Chem. **93** (1989) 6823.

[4] K. Fuke, S. Nonose, N. Kikuchi and K. Kaya: Chem. Phys. Lett. **147** (1988) 479.

[5] S. Nonose, Y. Sone, N. Kikuchi, K. Fuke and K. Kaya: Chem. Phys. Lett. **158** (1989) 152.

[6] M. D. Geusic, S. C. Morse and R. E. Smalley: J. Chem. Phys. **82** (1985) 590.

[7] S. Nonose, Y. Sone, K. Onodera, S. Sudo and K. Kaya: J. Phys. Chem. **94** (1990) 2744.

[8] K. Kaya: private communications.

[9] M. El-Batanouny, M. Strongen, G. P. Williams and J. Colbert: Phys. Rev. Lett. **46** (1981) 269.

[10] M. W. Ruckman and M. Strongin: Phys. Rev. **B29** (1984) 7105.

[11] V. Kumar and K. H. Bennemann: Phys. Rev. **B28** (1983) 3138.

CHEMISTRY WITHIN VAN DER WAALS CLUSTERS OF UNSATURATED MOLECULES: OBSERVATION OF CATIONIC POLYMERIZATION

J. F. GARVEY, M. T. COOLBAUGH, S. G. WHITNEY, W. R. PEIFER,
and G. VAIDYANATHAN
Dept. of Chemistry, State University of New York at Buffalo, Buffalo, NY 14214

ABSTRACT.   We present in this paper the observation of intracluster ion-molecule polymerization reactions within clusters containing molecules which possess double or triple C-C bonds. This effect is borne out by the observation of unusual magic numbers in the cluster ion distribution only under expansion conditions which favor the production of large neutral clusters. These magic numbers are rationalized in terms of the productio of covalently bonded cyclic molecular ions which terminate the cationic polymerization process.

1. Introduction

A subject of great interest in recent years has been the study of the physics of weakly bound van der Waals clusters. These species have been probed in a variety of ways to gain an understanding of their formation and to determine their various physical properties. However, the study of chemical reactions within clusters is especially intriguing since clusters can conceptionally bridge the disparate fields of bimolecular gas-phase reaction dynamics and solution chemistry. The ultimate goal of these studies is to obtain an understanding of the factors that govern reactions in solution but which are absent in gas phase processes. By concentrating on the chemistry within these cluster systems one can directly observe how the behavior of the system changes as a function of stepwise solvation.

Most of the recent work in this area consists of utilizing the neutral cluster as one of the reagents for a bimolecular reaction[1-8] with the product cluster ion being directly detected via conventional mass spectrometric techniques. Apart from the observation of protonated clusters {i.e., $(H_2O)_nH^+$,[9] $(NH_3)_nH^+$,[10,11] $(CH_3COCH_3)_nH^+$,[12] $(CH_3OH)_nH^+$,[13] $(CH_3OCH_3)_nH^+$,[13] and a few herocluster ions such as $[(ROH)_n(H_2O)_m]H^{+}$[14,15] and $[(CH_3OH)_n(H_2O)_m]H^+$,[16] } there are few reported cases of chemical reactions taking place within the cluster ion itself.[12,16-20] Yet it has been known that electron impact ionization of clusters leads to ions that closely resemble many of the intermediates found in

1101

P. Jena et al. (eds.), Physics and Chemistry of Finite Systems: From Clusters to Crystals, Vol. II, 1101–1107.
© 1992 *Kluwer Academic Publishers.*

bimolecular ion-molecule reactions.[21,22]

In addition to this typical unimolecular and bimolecular gas-phase chemistry already studied within clusters, our group has recently observed the generation of new cluster product ions which cannot be explained by either of these two known processes. That is, we observe product ion formation that has absolutely no counterpart with gas-phase bimolecular reactions and which only occurs within a van der Waals cluster[23]. These new processes, which we have begun to document in the past four years at SUNY, include the generation of $(C_2H_4F_2)_{n\geq4}H^+$ ions from 1,1-difluoroethane clusters[24], the generation of $(CH_3OCH_3)_nH_3O^+$ & $(CH_3OCH_3)_nCH_3OH_2^+$ ions from dimethyl ether clusters[25,26], the generation of $(NH_3)_nN_2H_5^+$ ions from ammonia clusters[27] and the photo-generation of MoO^+ and MoO_2^+ ions from van der Waals clusters of molybdenum hexacarbonyls[28]. The observation of these new chemical processes which occur <u>only</u> within a cluster, is particularly exciting for chemists in that we may now utilize clusters as a novel "crock-pot" in which to produce new molecules, which could not be produced be any other means.

The majority of our experiments consist of generating a beam of neutral van der Waals clusters and then, by electron impact, performing mass spectroscopy on the ion cluster species generated within the molecular beam. Though the cation within the cluster is rapidly generated ($\sim10^{-14}$ s), it takes microseconds before the generated cluster ion is mass selected by the quadrupole filter. On such a lengthy time scale the cation within the cluster may rid itself of its excess energy either by fragmentation, evaporation of neutral monomers or the solvated cation may chemically react with one (or more!) of the solvating neutrals. In any case, a new product cluster ion has been generated, which is then detected via mass spectroscopy.

One may therefore visualize the electron impact ionizer of our mass spectrometer as a *'reaction cell'* in which the precursor cluster ion is generated and allowed to *'incubate'* for microseconds. After this time period the newly generated product cluster ions are subsequently analyzed via mass spectroscopy. By observing the distribution of product cluster ions in the mass spectra, we can then deduce the ion-molecule chemistry which is occurring within the bulk cluster, and monitor how this chemistry changes as a function of cluster size.

2. Results and Discussion

2.1 ETHYLENE CLUSTERS[29]

The ethylene cluster mass spectrum is quite simple in that it is composed of a single sequence of cluster ions with the formula $(C_2H_4)_n^+$, which one would naturally attribute to the unreacted parent cluster ion. We observe that at decreased nozzle temperature (or increased stagnation pressure, 1.5-3.5 atm)[30] a pronounced ion intensity (i.e., magic number) appears for the empirical formula $(C_2H_4)_4^+$, such that it dominates the entire spectrum.

This observation of n=4 being a magic number only under certain expansion conditions is difficult to explain solely in terms of stabilities of either the neutral or ionic parent ethene clusters. A closed solvent shell, hydrogen bonded to a central cation, is typically the driving force for the appearance of a magic number in a molecular cluster. In a system, such as ethene, where hydrogen bonding does not play a significant role, the observation of pronounced magic numbers is therefore not expected since the distribution of neutral clusters is produced by processes which are essentially statistical in nature (i.e., why should the $(C_2H_4)_4^+$ cluster ion be so stable?). There would seem to be no *a priori* reason to predict that any particular ethene cluster size should be extraordinarily stable. High pressure mass spectrometry[30] has shown that the $C_2H_4^+$ cation undergoes an exothermic condensation reaction with a neutral C_2H_4 molecule to form a branched $C_4H_8^+$ cation. This new $C_4H_8^+$ cation can then undergo successive reactions with additional C_2H_4 molecules to form a larger and larger branched cation with the general formula $C_{2m}H_{4m}^+$. The rate of each successive condensation reaction decreases rapidly as the extent of the branching of the product cation increases, This drop in the reaction rate has been attributed to steric effects (i.e., highly branched $C_{2m}H_{4m}^+$ product cations being less reactive)[30].

We now speculate that the ethene cluster ions we observe represent the products of a similar series of <u>intracluster</u> condensation reactions. That is, the observation of the magic number at n = 4 (under expansion conditions which create extensive clustering) is due to the formation of the $C_8H_{16}^+$ molecular ion generated via a series of successive ion-molecule reactions within the cluster. Since these condensation reactions are exothermic[31] we would then also expect extensive evaporation of unreacted monomers (as seen before with the ammonia system).

In summary, neutral $(C_2H_4)_n$ clusters with n > 4, following ionization via electron impact, can react within the cluster to give primarily the $C_8H_{16}^+$ cation as shown in reaction (1). The intensity of these ions would then be

$$(C_2H_4)_{n-1}C_2H_4^+ \;\rightarrow\; (C_2H_4)_{n-2}[C_4H_8^+]^* \;\rightarrow\; (C_2H_4)_{n-3}[C_6H_{12}^+]^*$$
$$\rightarrow\; C_8H_{16}^+ + n\text{-}4\,(C_2H_4) \qquad (1)$$

expected to increase as the distribution of the neutral ethene clusters grows larger (i.e., as either $T_o \downarrow$ or $P_o \uparrow$: n $\uparrow$). The prominent peak at n = 4 represents a balance between two competing effects: the larger the cluster the greater the probability of reaction 1, however, as the cluster size increases the rate constants of each individual step decrease. This is a new interpretation in that this particular magic number is attributed not to a central cation solvated by neutral monomers but rather due to the generation of a new molecular ion. We have recently additional studies of clusters containing ethene, 1,1-difluoroethene and propene and have again observed similar phenomena: observation of magic numbers which are consistent with the generation of cyclic molecular ions[32]. These results indicate that cluster cationic polymerization is indeed a common reaction for the family of olefinic van der Waals clusters.

2.2 ACETYLENE/ACETONE CLUSTERS[33]

Under the expansion conditions utilized, three classes of 'cluster' ions may be observed in the acetylene/acetone cluster mass spectra: 1) neat acetylene ions, 2) neat acetone ions, and 3) herocluster ions (i.e., $A_nT_m^+$ where $A = C_2H_2$ and $T = CH_3COCH_3$). The behavior of the first two classes of ions is consistent with previous reports[12,15] and will only be briefly considered here. At the highest stagnation pressure utilized in the present experiments (2.0 atm.), it was not possible to observe neat acetylene cluster ions higher than the trimer which may be attributed to the production of benzene cations[34,35] within the acetylene clusters followed by evaporative losses of the remaining monomers as shown in reaction (2).

$$A_n^+ \rightarrow C_6H_6^+ + (n\text{-}3)\, A \qquad (2)$$

Two main series of herocluster ion were observed, A_nT^+ and $A_nC_2H_3O^+$. All of the $A_nT_m^+$ clusters displayed very similar behavior with the $A_2T_m^+$ ions always becoming very prominent as the expansion pressure was increased for $m = 1\text{-}4$. We also note that the $A_2C_2H_3O^+$ ions are the most intense ions under all of the expansion conditions investigated with very little intensity observed beyond $n = 3$.

We feel, in light of our previous work, that the A_2T^+ and $A_2C_2H_3O^+$ ions represent the covalently bonded molecular ions $C_7H_{10}O^+$ and $C_6H_7O^+$, respectively via intracluster ion-molecule association reactions. The pressure behavior exhibited by the data is very reminiscent of that obtained for neat olefin clusters as indicated in section 2.1. We propose that ionization of acetylene/acetone clusters leads to the production of $C_7H_{10}O^+$ and $C_6H_7O^+$ ions The fact that the magic number occurs for $n = 2$ is also suggestive of the formation of a molecular ion in this system since this corresponds to replacement of one acetylene by an acetone (or acetyl) during the ionic reaction (i.e., 2 C_2H_2 + $CH_3COCH_3^+$ or $2C_2H_2 + CH_3CO^+$).

Unfortunately the present mass spectral data do not allow us to unequivocally assign the structures of the $C_7H_{10}O^+$ and $C_6H_7O^+$ ions but previous investigations of the ion-molecule chemistry of acetone[36] as well as the polymer chemistry of carbonyl compounds[37], suggest that a reaction across the C=O bond should give rise to an ether linkage (i.e., a C-O-C bond). If the reaction proceeds by initial production of an acetone ion within the cluster, the product ion is formed via reaction (3).

$$A_nT_m + e^- \rightarrow A_nT_{m\text{-}1}CH_3COCH_3^+ + 2\,e^- \rightarrow C_7H_{10}O^+ + (n\text{-}2)\, A + (m\text{-}2)\, T \qquad (3)$$

If this reaction is halted by production of a six-membered ring, as appears to be the case for neat acetylene clusters, the 2,2-dimethyl-2H-pyran radical ion would appear to be the expected product.

If the reaction proceeds by initial production of an acetyl ion (CH_3CO^+) within the cluster, the reaction will proceed as shown in reaction (4)

$$A_nT_m + e^- \rightarrow A_nT_{m-1}CH_3CO^+ + CH_3 + 2\,e^- \rightarrow C_6H_7O^+ + (n\text{-}2)\,A + (m\text{-}2)\,T \quad (4)$$

If the reaction is halted by production of a six-membered ring, then the expected ion product will be the 2-methylpyrylium ion ($C_6H_7O^+$). The fact that the $A_nC_2H_3O^+$ ion intensity distribution shows anomalous features under all expansion conditions, and very little intensity beyond n = 3, may indicate that reaction sequence (4) is not responsible for the exclusive production of the $C_6H_7O^+$ ions and may also arise from fragmentation of excited ions.

The present results suggest that it is possible to synthesize relatively complex molecules starting from simple molecular 'building blocks' within van der Waals clusters through ion-molecule reactions. Because of the differing stabilities and reactivities of ionic and neutral molecules, this method may be useful in production of ions of unusual structure, as well as providing insight into the basic mechanisms of ionic polymerizations and is a source of continuing study within our group.

3. Future Directions

The examples just shown represent the beginning of a new distinctive chemistry which can occur within the environs of a gas phase cluster. New experimental techniques which our group hopes to employ within the near future include the use of mass selected cluster beams to directly observe reaction dynamics, and the application of spectroscopic probes, such as laser induced fluorescence, to probe the internal states of the radical product generated via these cluster reaction.

Acknowledgements

This research was supported by the Office of Naval Research which is hereby gratefully acknowledged. JFG also acknowledges the Alfred P. Sloan Foundation for a Research Fellowship (1991-1993).

References

(1) Whitehead, J. C.; Grice, R. *Faraday Discuss. Chem. Soc.* **1973**, *55*, 320.

(2) King, D. L.; Dixon, D. A.; Herschbach, D. R. *J. Am. Chem. Soc.* **1974**, *96*, 3328.

(3) Gonzalez Urena, A.; Bernstein, R. B.; Phillips, G. R. *J. Chem. Phys.* **1975**, *62*, 1818.

(4) Behrens, R. B., Jr.; Freedman, A.; Herm, R. R.; Parr, T. P. *J. Chem. Phys.* **1975**, *63*, 4622.

(5) Wren, D. J.; Menzinger, M. *Chem. Phys.* **1982**, *66*, 85.

(6) Nieman, J.; Na'aman, R. *Chem. Phys.*, **1984**, *90*, 407.

(7) Morse, M. D.; Smalley, R. E. *Ber. Bunsenges. Phys. Chem.* **1984**, *88*, 208.

(8) Whetten, R. L.; Cox, D. M.; Trevor, D. J.; Kaldor, A. *Surf. Sci.* **1985**, *156*, 8.

(9) Hermann, V.; Kay, B. D.; Castleman, A. W., Jr.; *Chem. Phys.* **1982**, *72*, 185.

(10) Stephan, K.; Futrell, J. H.; Peterson, K. I.; Castleman, A. W., Jr.; Wagner, H. E.; Djuric, N.; Märk, T. D. *Int. J. Mass Spectrom. Ion Phys.* **1962**, *44*, 167.

(11) Echt, O.; Morgan, S.; Dao, P. D.; Stanley, R. J.; Castleman, A. W., Jr. *Ber. Bunsenges. Phys. Chem.* **1984**, *88*, 217.

(12) Stace, A. J.; Shukla, A. K. *J. Phys. Chem.* **1982**, *86*, 865.

(13) Grimsrud, E. P.; Kebarle, P. *J. Am. Chem. Soc.* **1973**, *95*, 7939. Morgan, S.; Castleman, A. W, Jr. *J. Phys. Chem.* **1989**, *93*, 4544. Ibed., *J. Am. Chem. Soc.*, **1987**, *109*, 2868.

(14) Stace, A. J.; Shukla, A. K. *J. Am. Chem. Soc.* **1982**, *82*, 5314.

(15) Stace, A. J.; Moore, C. *J. Am. Chem. Soc.* **1983**, *83*, 1814.

(16) Kenny, J. E.; Brumbaugh, D. V.; Levy, D. H. *J. Chem. Phys.* **1979**, *71*, 4757.

(17) Klots, C. E.; Compton, R. N. *J. Chem. Phys.* **1978**, *69*, 1644, Klots, C. E. *Radiat. Phys. Chem.* **1982**, *20*, 51. Ibed. *Kinetics of Ion-Molecule Reactions*; Ausloos, P., Ed.; Plenum: New York, 1979; p 69.

(18) Ono, Y.; Ng, C. Y. *J. Am. Chem. Soc.* **1982**, *104*, 4752.

(19) Nishi, N.; Yamamoto, K.; Shinohara, H.; Nagashima, U.; Okuyama, T. *Chem. Phys. Lett.* **1985**, *122*, 599.

(20) Stace, A. J. *J. Am. Chem. Soc.* **1985**, *107*, 755.

(21) Milne, T. A.; Beachey, J. E.; Greene, F. T. *J. Chem. Phys.* **1972**, *56*, 3007.

(22) Ceyer, S. T.; Tiedemann, P. W.; Ng. C. Y.; Mahan, B. H.; Lee, Y. T. *J. Chem. Phys.* **1979**, *70*, 2138.

(23) Garvey, J. F.; Peifer, W. R.; Coolbaugh, M. T.; *Accts. of Chem. Res.*, **1991**, *24*, 48.

(24) Coolbaugh, M. T.; Peifer, W. R.; Garvey, J. F. *J. Phys. Chem.* **1990**, *94*, 1619.

(25) Garvey, J. F.; Bernstein, R. B. *J. Am. Chem. Soc.* **1987**, *109*, 1921.

(26) Coolbaugh, M. T.; Peifer, W. R.; Garvey, J. F. *J. Am. Chem. Soc.* **1990**, *112*, 3692.

(27) Peifer, W. R.; Coolbaugh, M. T.; Garvey, J. F. *J. Chem. Phys.* **1989**, *91*, 6684.

(28) Peifer, W. R.; Garvey, J. F. *J. Phys. Chem.* **1989**, *93* , 5906. Ibed. *Int. J. of Mass Spectrom. Ion Proc.* **1990**, *102*, 1.

(29) Coolbaugh, M. T.; Peifer, W. R.; Garvey, J. F. *Chem. Phys. Lett.* **1990** *168*, 337,

(30) Kebarle, P; Haynes, R. M., *J. Chem. Phys.* **1967** *47*, 1676.

(31) for the reaction $C_2H_4^+ + C_2H_4 \rightarrow C_4H_8^+$, $\Delta H^o = -1.43$ to -2.60 eV depending on the structure of the product

(32) Coolbaugh, M. T.; Vaidyanathan, G.,Peifer, W. R.; Garvey, J. F. *J. Phys. Chem.* **1991** *95*, 8337.

(33) Whitney, S. G.; Coolbaugh, M. T.; Vaidyanathan, G.,Peifer, W. R.; Garvey, J. F. *J. Phys. Chem.* **1991** *95*, 9625.

(34) Ono,Y.; Ng, C. Y. *J. Am. Chem. Soc.* **1982**, *104*, 4752.

(35) Shinohara, H.; Sato, H.; Washida, N. *J. Phys Chem.* **1990**, *94*, 6718.

(36) Hiraoka, K.; Takimoto, H.; Morise, K.; Shoda, T.; Nakamura, S. *Bull. Chem. Soc. Jpn.* **1986**, *59*, 2247.

(37) Allcock, H. R.; Lampe, F. W. *Contemporary Polymer Chemistry* Prentice Hall: Engelwood Cliffs, NJ, **1981**.

PULSED LASER ABLATION AND CLUSTER ION FORMATION FROM GROUP VI ELEMENTS S, Se, Te, AND MIXTURES WITH Ni AND Cu

A. GIARDINI GUIDONI, A. MELE, S. PICCIRILLO, and G. PIZZELLA
Dipartimento di Chimica, Università degli Studi di Roma "La Sapienza",
P.le A. Moro 5
00185 ROMA
ITALY

R. TEGHIL
Dipartimento di Chimica, Università degli Studi della Basilicata,
Via N. Sauro, 85 -
85100 POTENZA
ITALY

ABSTRACT. Laser mass spectrometry of positive and negative cluster ions produced from S, Se and Te, pure as well as in mixture with Ni and Cu are reported. The ion intensity decreases with increasing cluster size, and this is combined with discontinuities likely related to the geometrical structures of the cluster.

1. Introduction

Deposition of thin films by pulsed laser ablation of a solid target has been demonstrated to be a useful technique which may compete with other procedures. High Tc superconducting films were obtained by laser evaporation of suitable target [1]. Diamond like carbon films were grown also by laser ablation of graphite and some organic compounds [2]. Thus far we have used laser mass spectrometry to study aggregation phenomena on pure compounds and mixed systems. The investigation on oxides and on carbon and organic material is a typical example [3,4]. The experimental results have shown some of the chemical pathways which follow laser evaporation and lead to nucleation and to the growth processes for thin film formation. Metal sulphides, selenides and tellurides have been indicated as products which may find application in the field of semiconductivity [5]. In the present study we have produced clusters by chemical combination of Ni and Cu with VI group elements by laser evaporation of suitable mixtures. The products of the reaction have been analyzed using a time of flight mass spectrometer. The intensity distribution of the observed peaks allows the identification of clusters of high and low stability which can be correlated with geometrical structure.

P. Jena et al. (eds.), Physics and Chemistry of Finite Systems: From Clusters to Crystals, Vol. II, 1109–1114.
© 1992 Kluwer Academic Publishers.

2. Experimental

The laser mass spectrometer consists of a Nd-YAG laser system quadrupled in frequency (λ=266 nm τ_{FWHM}=6 ns) combined with a reflectron time of flight mass spectometer (TOF). When the laser is focussed on the sample to a spot of ~ 10 μm diameter the power is ranging from 10^8 to 10^{10} W/cm^2. The procedures of cluster production and measurements have been previously described [3].

Reagent grade sulphur, selenium and tellurium were used as powder sample. Mixtures of these elements with Ni and Cu were obtained by direct combination of the pure elements finely grinded onto a Ni or Cu microscope grid. The grid was inserted into the sample stage and covered by a quartz slide. The spectra of positive and negative ions were recorded and averaged over several runs.

3. Results and discussion

TOF mass spectra of laser ablated samples of pure sulphur, selenium and tellurium powders show positive and negative cluster ion formation of formula S_n^+, Se_n^+ and Te_n^+ or S_n^-, Se_n^-, Te_n^- with n up to 32. The ion intensity of the clusters decreases with increasing size and this is accompanied by discontinuities of the mass distribution of positive and negative cluster ions. The positive and negative cluster size is larger for sulphur than for selenium and tellurium. In our experimental conditions, that is under direct laser induced desorption, the temperature at the solid gas interface is high enough for all three elements to produce mainly atoms and ionized atoms. The process of aggregation and growth of the observed clusters appears to depend likely only on the binding energy that is in the order S>Se>Te [6].

The analysis of mass spectra show that the distribution presents some alternation. It may be noted that in the spectra of the three pure elements, higher intensities at n=2,5,7,8 for sulphur, at n=2,5,7 for selenium and at n=2,3, for tellurium are observed.

It may be tempting to correlate these numbers with the experimental findings of the U.V. photoelectron studies of Se_n clusters that show characteristic differences between the spectra of Se_2 Se_5 and Se_7 with respect to Se_6 and Se_8. These differences have been attributed to different geometrical structure [7]. Photodissociation of tellurium clusters suggests also the existence of stable two, five and seven atom ion clusters which are produced preferentially [8]. Sulphur clusters stability may probably be correlated to what found for selenium and tellurium. S_8 structure is particularly stable also in a simple laser evaporation experiment.

The geometrical structures and stability of sulphur and some selenium and tellurium clusters have been estensively investigated. The most interesting aspect of these studies is that many of these clusters exhibit open or ring forms, whose energy has been calculated [10]. On the contrary the stability among clusters containing different number of atoms has not been determined.

Laser ablation of mixtures of Ni-S, Ni-Se and Ni-Te produces positive and negative cluster ions. Sequences of the formula NiS_n^+, $NiSe_n^+$, $NiTe_n^+$ are predominant; n is ranging up to 22 for sulphur and up to 11 for selenium and tellurium. Other pure VI group element ion clusters are also observed together with Ni^+ and Ni_2^+. In fig. 1 the low resolution spectrum of a mixture of nickel and sulphur is reported, in the mass range covering up to m/e 500.

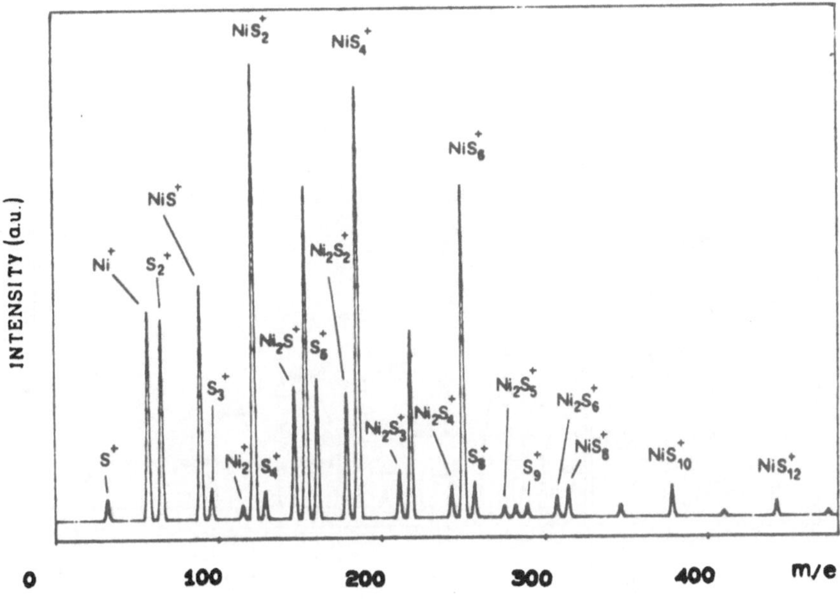

Fig. 1 - Laser time-of-flight mass spectrum of a mixture of nickel and sulphur.

S_n^+ cluster ions can be seen together some less intense structures to the left of the principal sequence NiS_n^+. These peaks can be attributed to a $Ni_2S_n^+$ ions. For what concern the predominant series it may be noted that higher intensities are observed at n=2,4,6,8,10,12,14. This higher stability for even values of n is not found in the spectra of the bare sulphur clusters. The difference between the spectra of the bare elements and the mixtures cannot be interpreted in terms of a simple molecular orbital theory as in the case of carbon clusters [3].

Similar sequences $NiSe_n$ and $NiTe_n$ of positive and negative cluster ions are found in the mass spectra of nickel selenium and nickel tellurium mixtures. The peak intensity continuously decrease by increasing n as shown in fig. 2 for the Ni-Se mixture. These mixed cluster ions do not show any alternation. Other less intense sequences containing two nickel atoms $Ni_2Se_n^+$ are observed in the spectra up to n=8. Very weak peaks at m/e corresponding at $Ni_3Se_n^+$ with n ranging up to 4 were also observed. This trend is typical of a building up of an ionic crystal structure. Similar structures for this type of compounds have been previously discussed [9, 11, 12].

Cu-Se and Cu-Te mixtures also produces positive and negative cluster ions sequences $Cu Se_n$, Cu_2Se_n and $Cu Te_n$ as shown in fig. 3, where the TOF mass spectrum of a mixture of copper and tellurium is reported.

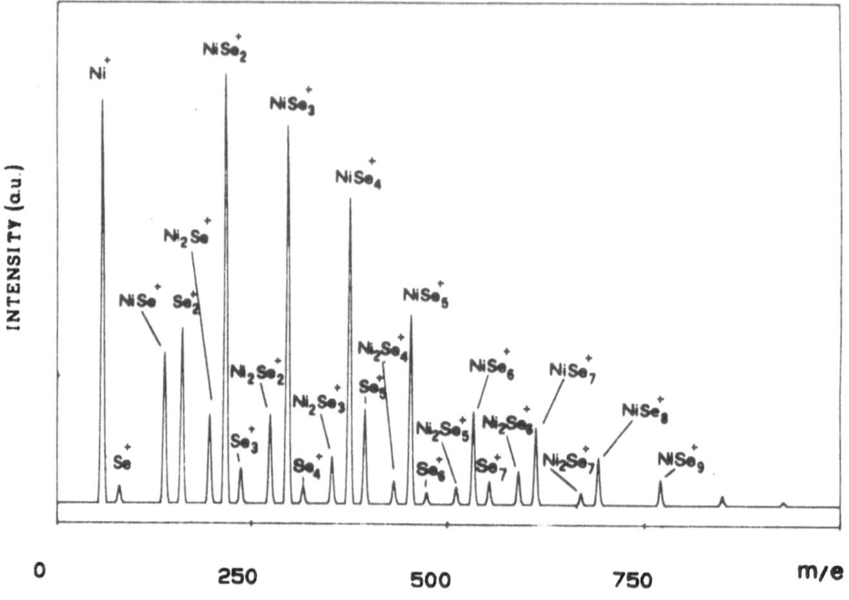

Fig. 2 - Laser time-of-flight mass spectrum of a mixture of nickel and selenium.

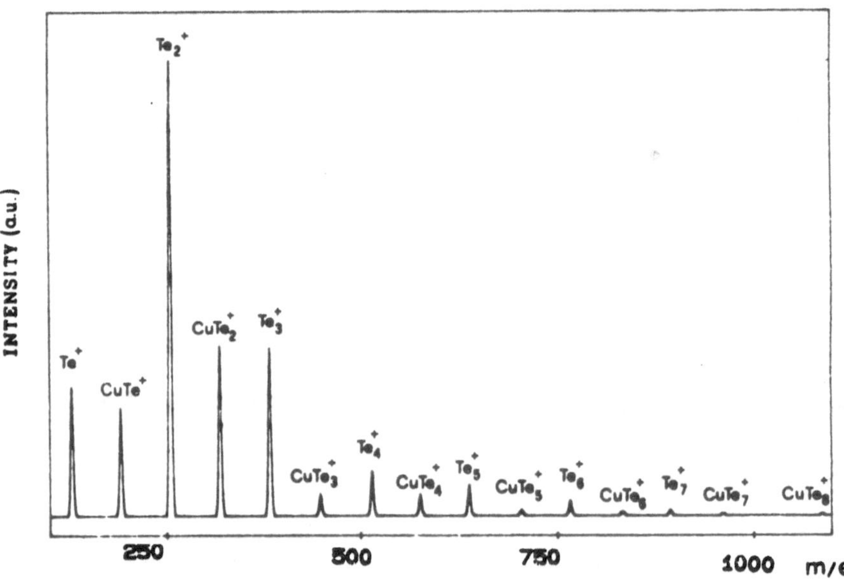

Fig. 3 - Laser time of flight mass spectrum of a mixture of copper and tellurium.

The intensity of the peaks Te_n^+ and $Cu\,Te_n^+$ is strongly decreasing and no intensity anomalies are observed. The mixed copper tellurium ions in which the copper dimer is present are coincident in one mass unit with the tellurium peak. Cu-S mixtures produce also $Cu\,S_n$ positive and negative cluster ions. The overlapping of m/e at one mass unit for Cu and S_2 makes the analysis of these spectra very difficult.

4. Summary

From these results it appears that direct laser evaporation is a very simple technique to produce from binary mixtures polysulphides, polyselenides and polytellurides clusters and is an applicable source for the production of thin films.

Clusters formed at the solid gas interface provide the beam of material that will be deposited. The composition of the target, the laser peak power density, the pulse rate are all factors affecting the nature of ablated products and hence determine the nature of sulphide selenide and telluride film being formed.

5. Acknowledgements

This work has been partially supported by CNR P.F. Tecnologie Elettroottiche and P.F. Chimica Fine.

6. References

1)   a) Venkatesan T., Wu H.D., Inam A., and Wachtman J.B. (1988), Appl. Phys. Lett. 52, 1193; b) Balsoch M., Olander D.R., and Russo R.E. (1989), Appl. Phys. Lett. 52 (2), 197; c) Giardini-Guidoni A., Teghil R., Morone A., Ambrico M., (1991) Less J. Comm. Met..

2)   a) Athwal I.S., Mele A., and Ogryrlo E.A. (1991), Proceeding of Diamond Films '91 Nice Franc. Diamond and Related Materials; b) Mehta B.R. and Ogryalo E.A. (1990), Surf. and Coating Tech. 43, 80.

3)   a) Mele A., Consalvo D., Stranges D., Giardini Guidoni A. and Teghil R. (1990), Int. J. Mass Spectrom Ion Processes 95, 359; b) Giardini Guidoni A., Mele A., Pizzella G. e Teghil R. (1991), 2 Phys. D-atoms, Molecules and Clusters 20, 89.

4)   Lineman D.N., Somayajula K.V., Sharkey A.G., and Hercules D.M. (1989), J. Phys. Chem. 93, 5026.

5)   Keller S.P. (1980), Handbook on semiconductors, North Holland Publishing. Amsterdam.

6)   Mills K.C. (1974), Thermodynamic Data for Inorganic Sulphides, Selenides and Tellurides, Butterworths, London.

7) Becker J., Rademan K., and Hensel F. (1991), Z. Phys. D - Atoms, Molecules and Clusters $\underline{19}$, 229.

8) Willey K.F., Cheng P.Y., Taylor T.G., Bishop M.B., and Duncan M.A. (1990), J. Phys. Chem. $\underline{94}$, 1544.

9) Martin T.P. (1984), J. Chem. Phys. $\underline{81}$, 4426.

10) Raghavachari K., Rohlfing C.M., and Binkley J.S. (1990), J. Chem. Phys., $\underline{93}$ (8), 5862.

11) Janz G.J., Roduner E., Coutts J.W. and Downey Jr. J.R. (1976), Inorg. Chem. $\underline{15}$, 1751.

12) Giardini-Guidoni A. and Mele A. (1991), Laser Chem..

CLUSTERDEPOSITION AND CLUSTEREROSION

J. GSPANN
Institut für Mikrostrukturtechnik der Universität
und des Kernforschungszentrums Karlsruhe
Postfach 3640, W-7500 Karlsruhe, Federal Republic of Germany

ABSTRACT. Modification of microtechnical surfaces through cluster impact can be done either constructively by cluster deposition or abrasively by cluster erosion. The high mass flux achievable with ionized and electrically accelerated beams of not too small clusters is a major reason for attraction in both cases. In addition, the collective motion of the cluster material provides an important distinction from isolated atom impact. Rather slowly impinging, molten clusters will facilitate epitaxial growth without substrate damage while impact melting and compaction may well serve for metallization purposes. At higher cluster energies impact crater formation differs from conventional sputtering. Erosion rate reduction and surface smoothing observed in high flux high energy experiments apparently result from plasma shielding and superficial melting.

Introduction

Cluster beams resulting from condensation in nozzle expansions are distinguished from molecular beams by their high mass flux and high directivity. These properties have made them conceptually attractive for mass transfer applications of various kinds, first for fueling nuclear fusion devices /1/, then for depositing thin films at high deposition rates /2/, and more recently, for abrasive surface micromachining /3/. Ionization and electrical acceleration of the cluster ions can be used for achieving the optimum kinetic cluster energy for the respective purpose, i. e. about 10 keV/proton for fusion fueling, 1 to 10 eV/atom for film deposition, and more than about 100 eV/atom for micromachining. As these energies may be reached by accelerating clusters with only one elementary charge, space charge induced beam blowup is drastically reduced in comparison with atomic ion beams of the same mass flux. For these beam related advantages to become fully effective, the clusters should have got sufficient size, e.g. a thousand atoms per elementary charge. Furthermore, clusters are thought to provide additional advantages due to their collective degrees of freedom when interacting with a surface. These aspects will be discussed for cluster deposition as well as for abrasive interaction, called "clustererosion" in the following.

P. Jena et al. (eds.), Physics and Chemistry of Finite Systems: From Clusters to Crystals, Vol. II, 1115–1120.
© 1992 *Kluwer Academic Publishers.*

Clusterdeposition

According to the ionized cluster beam (ICB) concept /2/, the impact of a cluster onto a substrate leads to cluster disintegration and subsequent lateral diffusion of the cluster atoms, as illustrated in Figure 1a. This picture bases on the idea of "loosely bound" clusters, since only the clusters, and not the substrate, are thought to disintegrate. However, the internal binding of a cluster of about a thousand atoms does not differ so much from that of the bulk, and cluster and film material are the same in film formation, perhaps except for the first atomic layer. Hence, a more loose coupling in the clusters can only result from a much higher internal temperature of the clusters, compared to that of the substrate. Eventually, the clusters may even differ in phase, being in the liquid state /4/. Figure 1b illustrates this as a possibly more realistic case of surface interaction: A hot nanometer droplet impinges on a colder surface, spreads laterally without disintegration, and an intimate contact with the substrate leads to local substrate heating.

Up to now, only molecular dynamics calculations can give information on the fate of the impinging clusters. K.H. Müller has presented two-dimensional calculations of close-packed hexagonal clusters of 91 atoms impinging onto a single crystal substrate of the same material /5/. Using a Lennard-Jones 12-6 interatomic potential, he found epitaxial film growth with very little defects for a kinetic energy of the cluster atoms corresponding to 1.5 times the interatomic potential well depth ε. Clusters did not disintegrate but melted together to form the epitaxial layer. At lower energies, the clusters retained their shape and settled down without preferred orientation and with many open spacings between them. At higher energies, corresponding to 4ε per cluster atom, the clusters started to disintegrate but also first substrate defects were created.

Comparable 3-dimensional calculations of film formation seem not to exist yet. Single silver cluster impacts on carbon were calculated by Y. Yamamura to lead to a practically monatomic spreading at 6 eV/atom kinetic energy without much disintegration or lateral diffusion. Some silver atoms were found to have penetrated one atomic layer into the carbon. At higher impact energies, 10 eV/atom, more cluster disintegration and even reflection of silver atoms was observed, but again also more substrate damage /6/.

Cluster internal temperatures and surface melting.

The results of Müller were obtained with clusters and substrate being at $T=0$ initially. Raising these temperatures to 0.3 ε/k, corresponding to about half the melting temperature, was said to lead to more flattening of the clusters and smoother films. (Herein, k denotes the Boltzmann constant.) Cluster temperatures other, in particular appreciably higher, than the substrate temperature were not tested, apparently.

If, however, the clusters are formed in pure vapor expansion, they are expected to be rather hot. The reason is that the clusters are able to cool down by evaporation cooling only after having reached high vacuum since inside the vapor flow colder clusters will always be heated up by the latent heat released by further growth. Since the evaporation rate drastically falls with decreasing cluster temperature, the clusters reach in practical distances

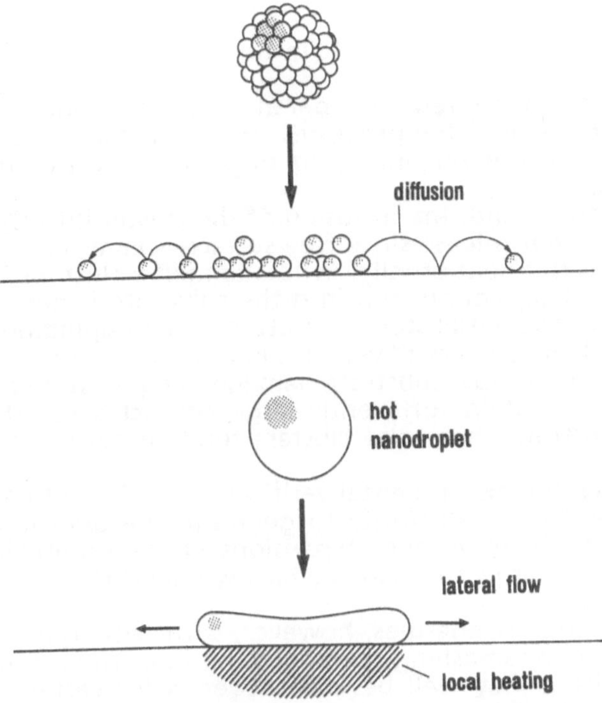

Figure 1: a) Model of cluster impact followed by atomic disintegration and subsequent lateral diffusion.
b) Alternative model of nanodroplet impact leading to lateral flow without disintegration.

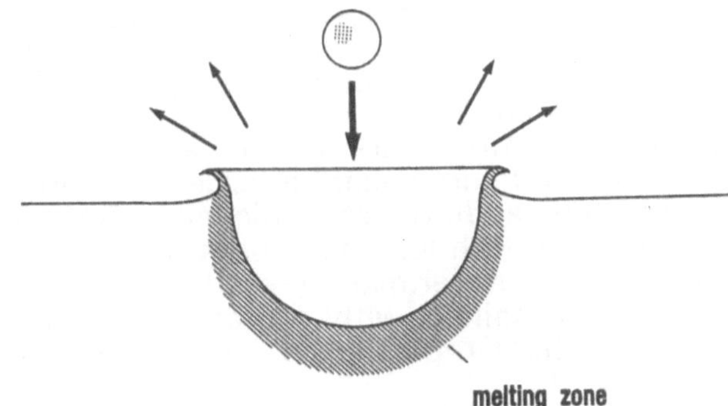

Figure 2: High velocity cluster impact leading to shockwave induced cratering and melt-compaction.

from the nozzle a final temperature T_f given by /7/

$$T_f = (10 \ln 10)\, u_0/k \qquad (1)$$

where u_0 denotes the latent heat per atom. For the higher boiling substances like metals these final temperatures are much higher than the respective melting temperatures /4/, justifying thus the notion of the impinging hot nanodroplets.

On the other hand, the inclusion of the cluster into the forming film always provides heat release since free surface is vanishing. One can calculate that for metals this heat usually suffices to melt a cluster of more than about 5000 atoms, if heat conduction into the substrate is neglected. Hence, one may argue that the initial state of matter of the impinging clusters does not matter so much. In the case of very low kinetic energy impact, however, which may be desirable for low substrate damage, a liquid state of the clusters may be favorable to obtain sufficiently large contact area. The calculations of Müller demonstrate that solid clusters tend to settle down in very loose packings.

Up to now, the experimental verification of the hot nanodroplet impact was hampered by the difficulty to generate the appropriate cluster sizes, except for Cs /8/, by pure vapor expansions. Quite recently however, we were able to generate large Zn clusters which will lend themselves for deposition studies.

At higher impact energies, however, sufficient contact is enforced also for solid clusters by smashing them into the film. The resulting densification and local melting may well be advantageous for certain applications, like metallization or via filling. Corresponding results have been reported recently by Haberland and coworkers /9/.

Clustererosion

At even higher impact energies, more substrate or film material will be removed than is deposited. Again, cluster impact differs from atom or atomic ion impact by the collective effects of the cluster material, providing a tendency for lateral motion instead of forward penetration or backward reflection. In order to underline this distinction, the term "cluster erosion" is used instead of sputtering.

In simulation calculations starting from atomic sputtering, the collective effects are usually called "nonlinear" in order to express that cluster impact means more than the addition of single atom impacts. Up to now, only calculations for clusters with less than about 50 atoms have been presented /10/. On the other hand, experimental results on abrasive micromachining of surfaces have been obtained with clusters of a thousand carbondioxide molecules /11/. For these rather large clusters one can try to use a more macroscopic treatment.

High velocity impact cratering always follows the same pattern, be the projectiles meteorites of 10^{12} g mass and 60 m diameter, like the one having generated the Barringer crater in Arizona, or iron microparticles of 10^{-12} g and 10^{-6} m diameter /12/. If the impact velocity is a multiple of the sound velocity in the involved materials, shock waves originate from the locus of im-

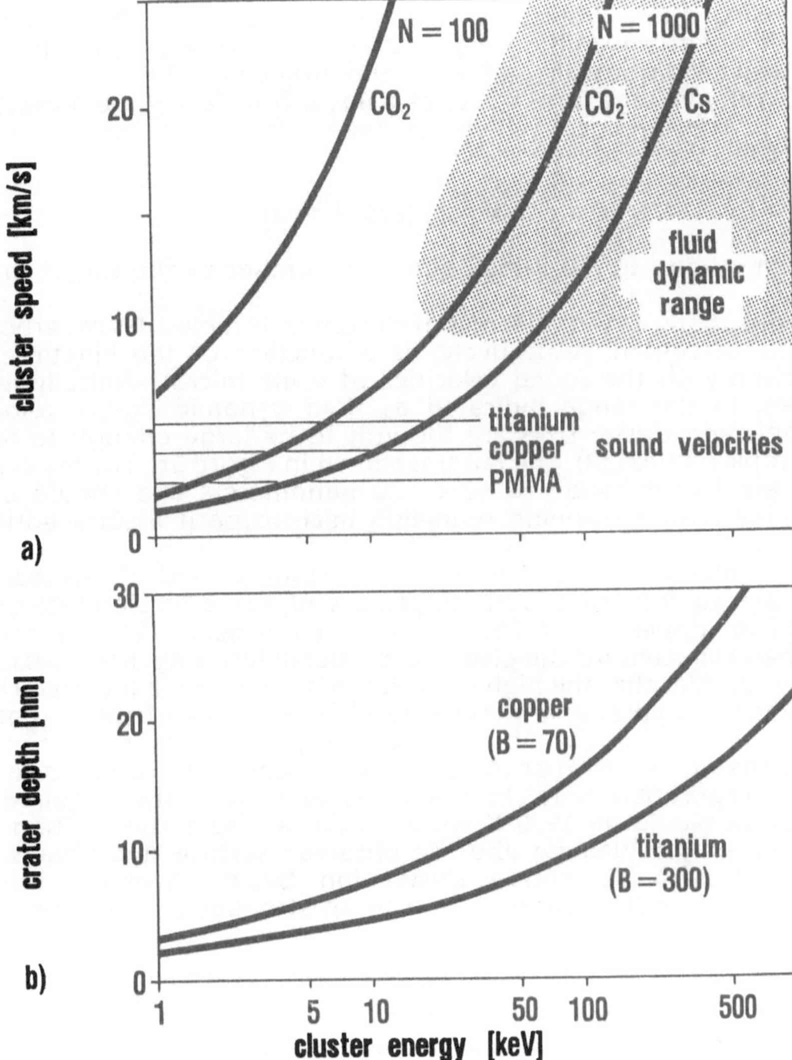

Figure 3:   a) Velocities of clusters of N atoms or molecules as a function of the cluster energy in comparison with the sound velocities in some microtechnical materials.
b) Corresponding crater depths for materials of Brinell hardness B

pact, running into the target as well as into the projectile, melting and compressing these bodies to enormous pressures, which then lead to explosive material expulsion. Figure 2 shows the typical hemispherical crater, giving also a realistic scaling of crater and projectile /13 /.

The crater volume turns out to be a linear function of the projectile kinetic energy E. The empirical relation between the crater depth T and impact energy E is

$$T = 1.45 \, (E/B)^{1/3} \, nm \qquad (2)$$

with E in eV and B, the Brinell hardness number of the target material, in kg/mm^2.

Figure 3a shows the velocities of clusters of N molecules of carbondioxide, or atoms of cesium, respectively, as a function of the kinetic energy, in comparison with the sound velocities of some microtechnically interesting materials. In the range indicated as fluid dynamic cluster velocities are supersonic and cluster sizes are thought to be large enough to follow the macroscopic relation (2). The latter is shown in Figure 3b. The expected crater depths are in the range of 10 to 20 nanometers and should be readily measurable with a scanning tunneling microscope if a sufficiently smooth target is used.

The cumulative cluster erosion after some period of bombardment is much easier to determine. Surprisingly, the reported erosion rate of 70 nm/s obtained in copper with a focussed cluster ion beam /11/ is a factor 1000 lower than expected for the given cluster beam intensity. For an explanation, one may consider that the high mass flux achieved with a focussed cluster ion beam leads to appreciable plasma densities in front of the target. Such a plasma is indeed visually observed. Clusters may not survive the passage through the plasma so that the low rate of abrasion would correspond to thermal evaporation only. In the considered case the required surface temperature would be 1500 K which is 150 K above the melting point of copper, possibly explaining also the observed surface smoothness. Current experiments with defocussed cluster ion beams, avoiding the plasma shielding, seem indeed to indicate a much stronger erosion due to cluster impact /14 /.

References

/1/   E.W. Becker, Laser and Particle Beams 7 (1989) 743

/2/   T. Takagi, Ionized-cluster beam deposition and epitaxy, Noyes Publications, New Jersey, 1988

/3/   R. Beuhler, L. Friedman, Chem. Rev. 86 (1986) 521

/4/   J. Gspann, Z. Phys. D3 (1986) 143

/5/   K.H. Müller, J.Appl. Phys. 61 (1987) 2516

/6/   Y. Yamamura, Nucl. Instr. Methods B 45 (1990) 707;

/7/   J. Gspann, in Physics of Electronic and Atomic Collisions, S. Datz ed., North-Holland 1982, 79

/8/   J. Gspann, Z. Phys. D 20 (1991) 421

/9/   H. Haberland, M. Karrais, M. Mall, Z. Phys. D 20 (1991) 413

/10/   P. Sigmund, J. Physique 50 (1989), C2-175

/11/   P.R.W. Henkes, R. Klingelhöfer, J. Physique 50 (1989) C2-159

/12/   V. Rudolph, Z. Naturforsch. 24a (1969) 326

/13/   J.J.W. Gehring, in High-Velocity Impact Phenomena, R. Kinslow ed., Acad. Press, New York 1970, 463

/14/   P.R.W. Henkes, B. Krevet, private communication

MAGIC NUMBERS OF SODIUM/ANTIMONY-CLUSTERS PRODUCED BY GAS AGGREGATION AND SUBSEQUENT PHOTOIONIZATION

A. HARTMANN AND A. W. CASTLEMAN, JR.
Department of Chemistry
The Pennsylvania State University
University Park, PA 16802

ABSTRACT. The gas aggregation technique is used to produce Na_xSb_y clusters containing up to 20 antimony atoms. Magic numbers in the small size range can be explained by assuming highly ionic bonds in the binary Zintl-type cluster ions. Larger clusters display a distinct odd/even alternation which is affected by the ionization conditions.

1. Introduction

Intermetallic compounds, containing alkali metals and electronegative maingroup elements, are known to form isolated ionic units both in the liquid and solid state [1,2]. Furthermore, various gaseous so-called Zintl ions are known in equilibrium and nonequilibrium vapors [3,4,5,6]. In the present investigation, attention is focused on sodium/antimony clusters. As will be seen in this study, several stoichiometries which correspond to Zintl ions in the condensed phase show up as magic numbers in the mass spectra. By varying the wavelength of the ionizing laser, we reveal a change in the observed cluster ion distribution.

2. Experimental

Details of the apparatus are described elsewhere [7] and only those features pertinent to the present experiment are briefly outlined here. Zintl cluster species were produced by a gas aggregation source under well-defined free evaporation conditions. Alloys of the composition NaSb are prepared in situ by annealing equiatomic amounts of sodium and antimony in a boronnitride crucible at 600 K. The temperature was then increased until a sufficient cluster ion intensity could be observed. A chromel-alumel thermocouple was used to monitor the temperature of the crucible during the evaporation process. A feedback circuit allows us to maintain a constant evaporation temperature during an experiment within the limits of ±1K. High purity helium carrier gas (typical flow: 1.55 standard liters per minute) is used to achieve cooling and cluster growth. The clusters were subjected to multiphoton ionization (unfocused high fluence KrF- or ArF-excimer laser radiation) and time-of-flight mass analysis.

P. Jena et al. (eds.), Physics and Chemistry of Finite Systems: From Clusters to Crystals, Vol. II, 1121–1124.
© 1992 *Kluwer Academic Publishers.*

3. Results and Discussion

Figures 1a and 1b show the distribution of various Na_xSb_y cluster ions at a crucible temperature of 890 K.

Neither a change in the evaporation temperature (860-1020 K) nor a variation in the inert gas flow (0.8-3.5 slm) affects the observed signal pattern.

Magic numbers corresponding to species which display especially high abundances are observed in the small binary sodium/antimony clusters: these can be explained by assuming ion formation due to direct ionization- or photofragmentation processes.

In the size range where x+y is equal to or less than 12, especially prominent peaks appear which are similar to the known polyanions Sb_2^{2-}, Sb_3^{3-}, Sb_4^{2-}, and Sb_7^{3-}. For example, both $Na_3Sb_3^+$ and $Na_4Sb_3^+$ represent the stable Zintl-ion Sb_3^{3-} formed directly from a neutral precursor or as a fragmentation product after the ionization.

In those cases where x+y is equal or greater than 20, the prominent mass species display a distinct odd/even alternation. Clusters containing an <u>odd</u> number of antimony atoms, bond only an <u>even</u> number of sodium atoms and vice versa. The valence electron count of these clusters yield, by considering the unit positive charge on each cation, always an <u>even</u> number of total electrons.

Gaseous species in this size range have no isolated counterparts in condensed material [2]. The sequence of abundance in which the large clusters appear (no strong magic numbers) is different from what we observe in the case of smaller Na_xSb_y-cluster ions. While sodium atoms act as electron donors in Zintl-type clusters, it is possible that in larger aggregates the more metal-like properties of sodium dominate.

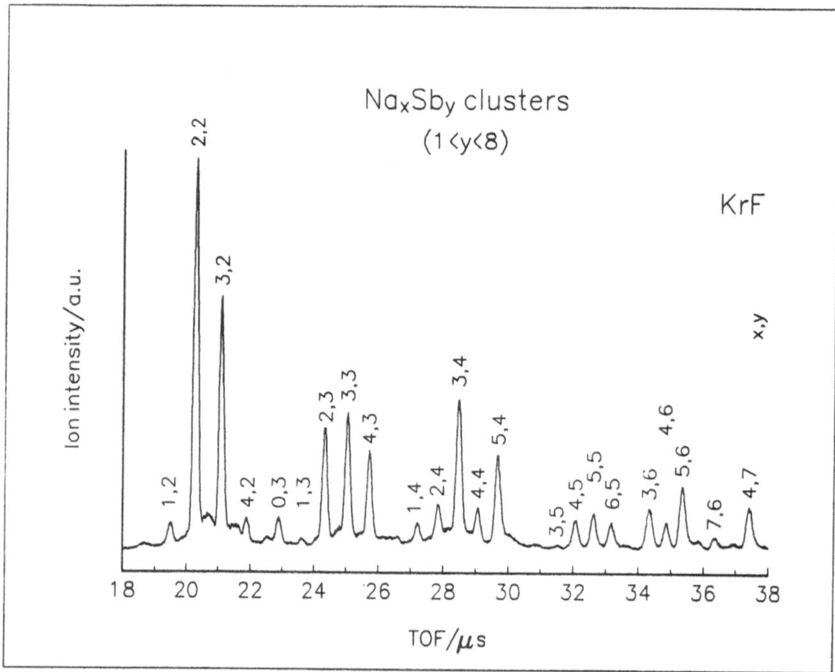

Figure 1a. Mass spectrum of Na_xSb_y-clusters containing up to seven antimony atoms (T=890 K). KrF-excimer laser photons (4.98 eV) were used to achieve ionization.

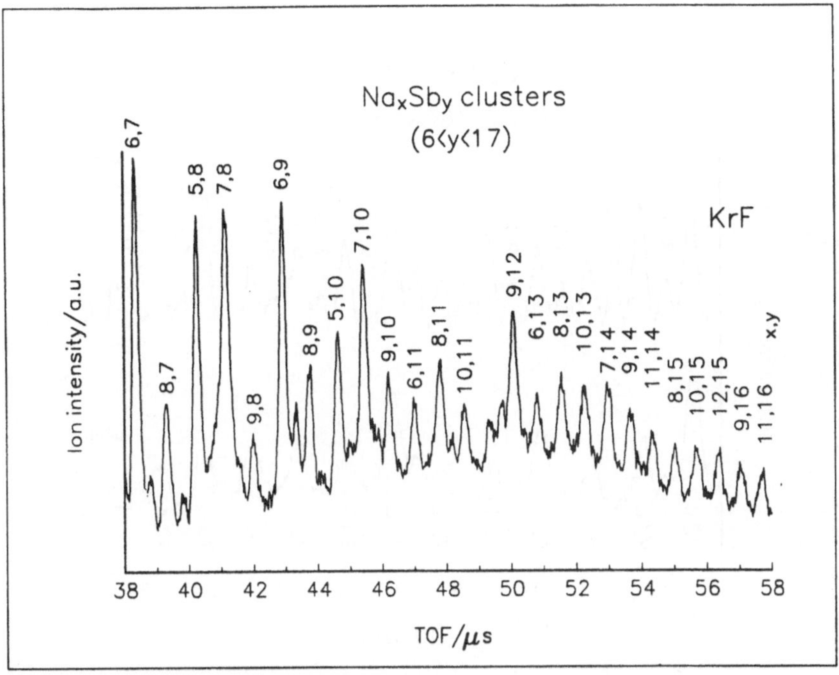

Figure 1b.   Mass spectrum of Na$_x$Sb$_y$-clusters containing seven to sixteen antimony atoms (T=890 K).   Same ionization conditions as mentioned in 1a.

A detailed comparison of a part of the mass spectrum taken with KrF and ArF laser radiation is shown in Figure 2.   It is observed, especially in the small size range, that the overall signal pattern is not dramatically affected by using the lower wavelength radiation of the ArF-excimer laser.   However, in the displayed mass range it can be clearly seen that additional peaks arise upon using the 193 nm radiation.   By performing an electron count of these ions in the same way as mentioned above, one always ends up with an <u>odd</u> number of total electrons.   This behavior points to the lower stability of these clusters which could be related to a composition dependent ionization mechanism.   A nonresonant multiphoton process at 248 nm (KrF, 4.98 eV) could lead to extensive fragmentation of unstable clusters into stable ions while a threshold ionization at 193 nm (ArF, 6.4 eV) might produce ions which are related to neutral precursors.

This consideration follows upon comparing the appearance energy for the known binary ion NaSb+ (6.1 ± 0.3 eV) [8] with the used photon energy of the respective laser radiation.

Comparing results of this work with mass spectra obtained by the Knudsen effusion technique shows that most of the known equilibrium species (NaSb, Na$_2$Sb$_2$, Na$_2$Sb$_4$, and Na$_6$Sb$_4$) [8] could be also detected in our experiments.

It is remarkable that the cluster spectra shown are not influenced by contributions from pure antimony clusters which are known in a size range up to 240 antimony atoms [9].   Furthermore, there is no evidence that stable (Sb$_4$)$_n$-clusters directly affect the observed binary cluster ion distribution.

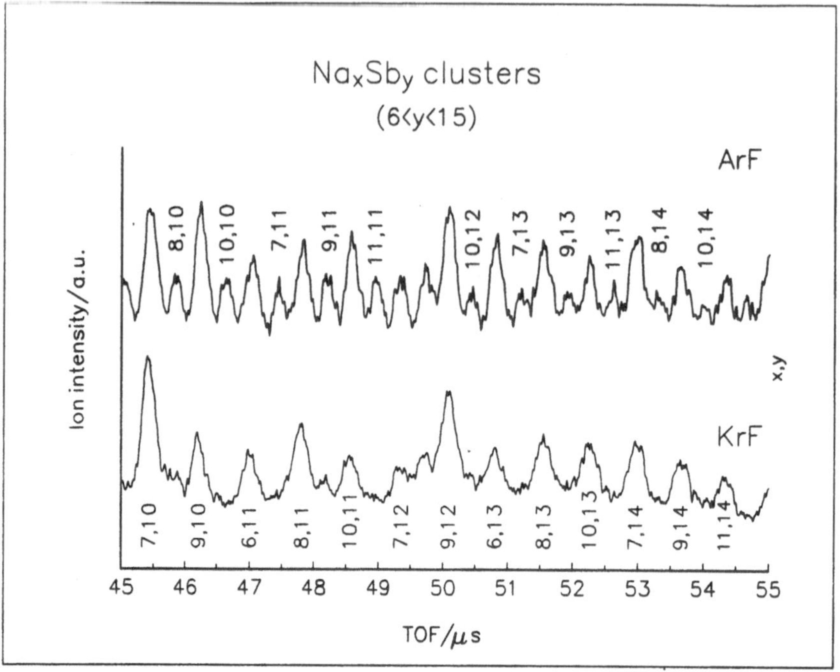

Figure 2. Mass spectra of Na$_x$Sb$_y$-clusters in the size range 6<y<17. The spectra were obtained by using KrF (4.98 eV)- and ArF (6.4 eV)-excimer laser photons. Both spectra were taken at the same evaporation temperature (890K).

Acknowledgements

Financial support by the U.S. Department of Energy, Grant No. DE-FG02-88-ER60668, is gratefully acknowledged. One of the authors (A. Hartmann, Feodor Lynen-Fellow) acknowledges support by the Alexander von Humboldt Foundation, Germany.

References

1. E. Zintl, J. Goubeau and W. Dullenkopf, *Z. Phys. Chem.*, Abt. A., **154**, 1 (1931).
2. J. D. Corbett, *Chem. Rev.*, **85**, 383-397 (1985).
3. T. P. Martin, *Angew. Chem. Int. Ed. Engl.*, **25**, 197-211 (1986).
4. A. Hartmann and K. G. Weil, *Angew. Chem. Int. Ed. Engl.*, **27**, 1091 (1988).
5. R. W. Farley and A. W. Castleman, Jr., *J. Am. Chem. Soc.*, **111**, 2734 (1989).
6. D. Schild, R. Pflaum, G. Riefer, and E. Recknagel, *Z. Phys. D*, **12**, 235-236 (1989).
7. R. W. Farley, P. Ziemann, and A. W. Castleman, Jr., *Z. Phys. D*, **14**, 353 (1989).
8. T. Scheuring and K. G. Weil, *Surf. Sci.*, **156**, 475 (1985).
9. K. Sattler, J. Muehlbach, P. Pfau, and E. Recknagel, *Phys. Lett.*, **87A**, 418 (1982).

COLLISION–INDUCED REACTIONS OF MOLECULAR CLUSTER IONS

S. NONOSE, J. HIROKAWA, M. ICHIHASHI, M. SAKAMOTO,
T. TAHARA and T. KONDOW
Department of Chemistry, Faculty of Science,
The University of Tokyo, Bunkyo–ku, Tokyo 113,
Japan

ABSTRACT. The collision–induced reactions of $(CO_2)_n^+$ $(n = 1 - 15)$ and Ar_n^+ $(n = 1 - 15)$ with NH_3 and Kr were studied by use of a tandem mass–spectrometer equipped with octapole ion guides (OPIG). The results showed that evaporation, charge transfer and fusion were dominant pathways in the reactions. The cross sections and the branching ratios were obtained as a function of the size of the parent cluster ion and the collision energy. The fusion pathway was found to be sizable in the collision of a large cluster ion with a target having a collision energies below 0.2 eV.

1. Introduction

Collision–induced reactions of cluster ions have attracted much attention, because of the characteristic properties of the cluster ions, which critically depend on their cluster size. In particular, a specific nature of cluster collision could be demonstrated in collision processes involving ions of van der Waals clusters, because these cluster ions are often made of a tightly bound ion core and a weakly bound solvation shell. This characteristic structure provides us a unique opportunity to explore reaction processes which are scarcely encountered in ordinary collision processes. In this regard, we investigated the collision reactions of the $(CO_2)_n^+$ + NH_3 and the Ar_n^+ + Kr systems as typical examples.

2. Experimental

An experimental setup is shown in Fig. 1. The cluster ions produced by electron impact on neutral clusters of CO_2 and Ar were mass-selected by a quadrupole mass spectrometer (Extrel, 162–8), and were allowed to collide with the target molecules in a collision region of OPIG. The product ions were mass-analyzed by a sector–magnet mass spectrometer (JEOL, JMS–D300). The ion source, the quadrupole mass spectrometer, and the collision region were floated up to 1 kV against the

1125

P. Jena et al. (eds.), Physics and Chemistry of Finite Systems: From Clusters to Crystals, Vol. II, 1125–1130.
© 1992 *Kluwer Academic Publishers.*

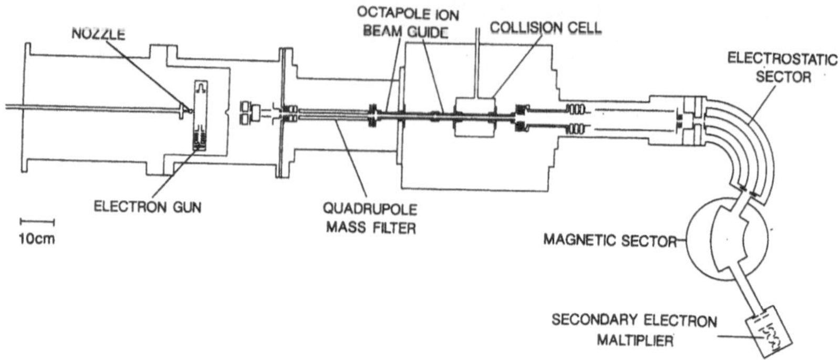

Fig. 1: An experimental setup.

ground. The total cross section were measured by varying the pressure of the target gas in the range of $10^{-5} - 10^{-6}$ Torr; the background pressure was attained to be less than 5×10^{-7} Torr. The branching ratio of a given product ion was estimated from the ion intensities.

3. Results and Discussion

3-1. $(CO_2)_n^+$ + NH_3 system

Reaction Pathways

The measurements of the product ions show that the reaction of $(CO_2)_n^+$ with NH_3 proceeds as follows:

$$(CO_2)_n^+ + NH_3 \rightarrow NH_3^+ + nCO_2 \qquad \text{(charge transfer)} \quad (1)$$
$$(CO_2)_n^+ + NH_3 \rightarrow (CO_2)_{n'}^+ + (n-n')CO_2 + NH_3 \qquad \text{(evaporation)} \quad (2)$$
$$(CO_2)_n^+ + NH_3 \rightarrow (CO_2)_{n'}NH_3^+ + (n-n')CO_2 \qquad \text{(fusion)}. \quad (3)$$

Dependences on Collision Energy

Figure 2 shows the collision–energy dependence of the total cross section, the cross sections for pathways (1)+(3) and pathway (2), for the $(CO_2)_{12}^+$ + NH_3 reaction system. As shown in Fig. 2, the total cross section decreases gradually while that of pathway (2) increases slightly, as the collision energy increases. The cross section for pathways (1)+(3) decreases rapidly with the collision energy. In particular, large daughter ions, such as $(CO_2)_4NH_3^+$, almost vanish at the collision ener–

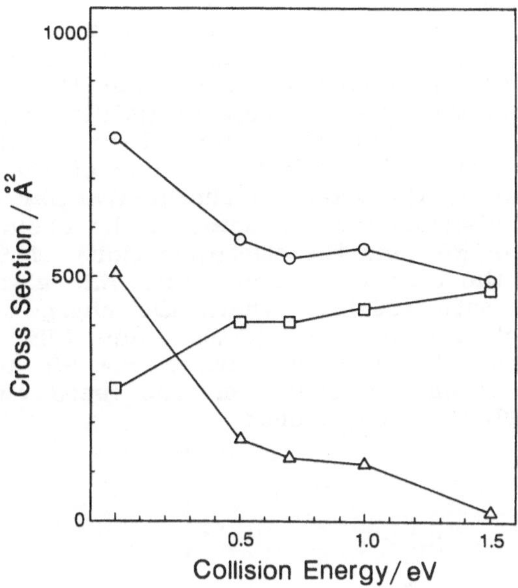

Fig. 2: The total cross section, the cross sections of pathways (1)+(3) and pathway (2) are plotted as a function of the collision energy in the reaction of $(CO_2)_{12}^+$ with NH_3; open circles, open triangles, and open squares represent the total cross section, the cross sections of pathways (1)+(3) and pathway (2), respectively.

gies of more than 0.5 eV, whereas the smallest product ion, NH_3^+, remains even at 1.0 eV. Evidently, the fusion is influenced more greatly by the increase of the collision energy. There seems to be no appreciable activation energy for the fusion process since it occurs efficiently in small collision energies.

Dependences on Cluster Size

The total cross section was roughly proportional to the cluster size, n, and reached 1500 $Å^2$ at n = 15. In the collision energies of 0 ~ 0.2 eV, at most 4 CO_2 molecules were found to evaporate, irrespective of n.

A particular abundance of NH_3^+ in the range of n = 1 − 5 indicates that in this range the charge transfer supersedes the fusion (see Fig. 3). Accordingly, the sharp decrease of the cross section for pathways (1)+(3) with the cluster size implies that pathway (1) diminishes above n = 6. It is likely that the solvation core of $(CO_2)_n^+$ prevents the target NH_3 molecule from reaching into the region where the charge transfer is able to occur efficiently. The effective radius of the shell of $(CO_2)_6^+$ can be approximated by the effective charge transfer distance between CO_2^+ and NH_3; the distance was estimated to be about 4.5 Å from the ionization potential of NH_3 and the electron affinity of CO_2^+. Seemingly, in a large cluster, the charge exchange with NH_3 always proceeds via the fusion of NH_3 with $(CO_2)_n^+$, where the charge exchange occurs inside the fused cluster made of $(CO_2)_n^+$ and NH_3. As the excess energy associated with the electron transfer from NH_3 to CO_2^+ is 3.0 eV, several CO_2 molecules should leave from the fused cluster ion, and a daughter ion, $(CO_2)_{n'}NH_3^+$, is produced.

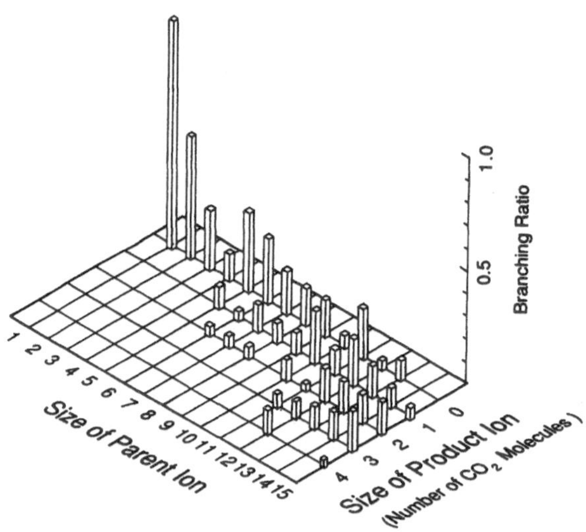

Fig. 3: The branching ratio of pathways (1)+(3) in the reaction between $(CO_2)_n^+$ and NH_3. The symbols, n and n', represent the sizes of the parent cluster ion and the daughter ion, $(CO_2)_{n'}NH_3^+$, respectively. The collision energy is 0 ~ 0.2 eV.

3-2. Ar_n^+ + Kr system

Similar phenomena were observed in the collision-induced reaction of Ar_n^+ with Kr; the product ions were Kr^+, $Ar_{n'}^+$ (n' < n), and $Ar_{n'}Kr^+$ (n'=1, 2 and 3), which are considered to be produced from the charge transfer (pathway(1)), the evaporation (pathway (2)), and the fusion (pathway (3)), respectively, as observed in the $(CO_2)_n^+$ + NH_3 reaction system.

Dependences on Collision Energy

With the increase of the collision energy, from 0.1 to 10 eV, the daughter ions produced from the pathways (1) and (3) disappeared, while the number of the evaporated Ar atoms and the size distribution of the daughter ions produced by the evaporation was not significantly influenced by the collision energy.

Dependences on Cluster Size

The branching ratio of pathways (1)+(3) is shown as a function of the cluster size, n. (see Fig. 4). The branching ratio decreases smoothly, and levels off at n ~ 6 with the increase of n. The result can be

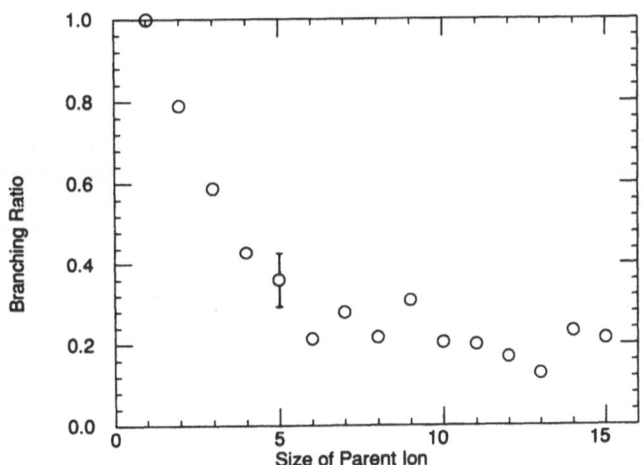

Fig. 4: The branching ratio of pathways (1)+(3) is shown as a function of the size of the cluster size, n, in the reaction of Ar_n^+ with Kr. The collision energy is 1.0 eV.

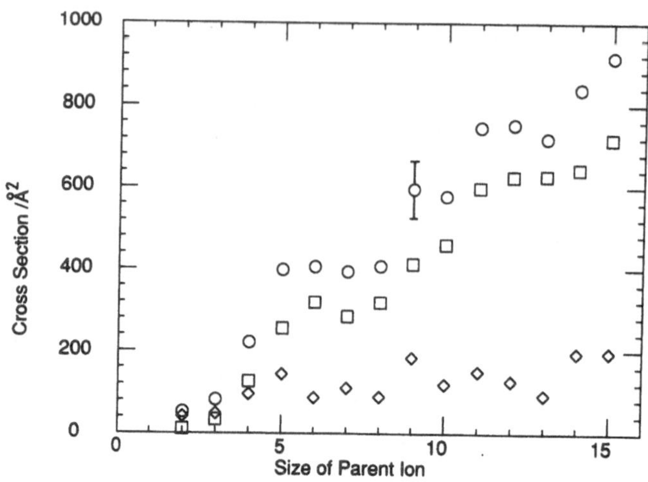

Fig. 5: The total cross section, the cross sections of pathways (1)+(3) and pathway (2) are shown as a function of the cluster size, n, in the reaction of Ar_n^+ with Kr; open circles, open diamonds, and open squares represent the total cross section, the cross sections of pathways (1)+(3) and pathway (2), respectively. The collision energy is 1.0 eV.

explained by charge transfer mechanism similar to the $(CO_2)_n^+$ + NH_3 system. The total cross section, the cross section for pathways (1)+(3) and pathway(2) are plotted against the cluster size, n, as shown in Fig. 5. The total cross section increases in proportion to n, and reaches 900 $Å^2$ at n = 15. The cross section for pathway (2) increases monotonously with the increase of n, while the cross section for pathways (1)+(3) does not change much with n. It was also found that the average number of Ar atoms evaporated from Ar_n^+ by 1.0 eV collision was 0.3n.

FORMATION AND REACTIVITY OF SILICON CARBIDE CLUSTER CATIONS

DENISE C. PARENT
Chemistry Division/Code 6113
Naval Research Laboratory
Washington, D. C. 20375-5000
U. S. A.

ABSTRACT. Direct laser vaporization of a mixture of silicon and graphite powders produced silicon carbide cluster cations containing 1-3 silicon atoms and 1-12 carbon atoms. Gas-phase ion-molecule reactions with acetylene and other gases were studied in the FTMS. With C_2H_2 the cations with 1 silicon atom behave very similarly to the carbon cluster cations, indicating a linear structure for these ions. The cations with 2 silicon atoms exhibit unusual reactivity. The initial reaction to add CH_2 is very slow. With increasing time, the reaction becomes faster and forms the adduct. ^{13}C labelling studies revealed a process of carbon exchange, which may be linked to the "activation" process just described. A structural or electronic state isomerization is believed to be the cause. Cleavage of carbon-carbon multiple bonds, dehydrogenation, deoxygenation and dehalogenation reactions of $Si_2C_2^+$ are also reported. Charge transfer reactions of the initial $Si_2C_2^+$ ion population were used to bracket the ionization potential of the corresponding neutral.

1. Introduction

Silicon carbide is an important technological material for the electronics industry and in material science. Small silicon carbide molecules have also been observed in interstellar space. The pure carbon and silicon clusters have been extensively studied, with reports on their formation and stability, physical properties, chemistry, and structure. There are surprisingly few experimental investigations of the mixed silicon/carbon clusters, however. Some spectroscopic studies and theoretical calculations of the structures of the smaller neutral clusters have been done. In this article I will describe the formation of gas-phase silicon carbide cluster cations and focus on their chemistry. A more complete treatment will be given in [1].

2. Experimental

All experiments were performed with a Fourier transform ion cyclotron resonance (ICR) mass spectrometer (FTMS) equipped with a 3 Tesla superconducting magnet, fully described in [2]. The silicon carbide cluster cations were produced by direct laser vaporization of pressed pellets of mixed silicon and graphite powders, using the focussed output of a frequency-doubled Nd:YAG laser. The ions were trapped in the ICR cell of the FTMS for further study.

P. Jena et al. (eds.), Physics and Chemistry of Finite Systems: From Clusters to Crystals, Vol. II, 1131–1136.
© 1992 *Kluwer Academic Publishers.*

Reactions were studied by first mass-selecting the reactant ion and then allowing it to react with the neutral reagent gas, which was held at a static pressure of 5-30 x 10^{-8} torr. Rate constants and product branching ratios were measured by varying the reaction time.

3. Results and Discussion

3.1. CLUSTER CATION FORMATION

Silicon carbide cations are only formed when using low laser power (oscillator only), which suggests they are not as stable as the carbon cluster cations. Three dominant ion series are observed: C_y^+ (y=3-28); SiC_y^+ (y=2-12); and $Si_2C_y^+$ (y=0-11). In some spectra one also sees $Si_3C_y^+$ (y=0-2) and Si_4^+, but the intensity of these ions is generally very weak.

The relative abundances of the SiC_y^+ ions show even/odd alternations (even y peaks more intense) up to y=9, with a monotonic decrease in abundance for larger y. The $Si_2C_y^+$ ions exhibit the same even/odd alternation in intensity up to y=5, followed by a monotonic decrease. The most abundant ions are SiC_2^+, Si_2^+ and $Si_2C_2^+$, as well as some carbon cluster ions.

3.2. REACTIONS WITH ACETYLENE

The SiC_y^+ and $Si_2C_y^+$ ions were reacted with acetylene, for comparison with the carbon cluster ions [3]. The variation in rate constant with cluster ion size is plotted in Figure 1.

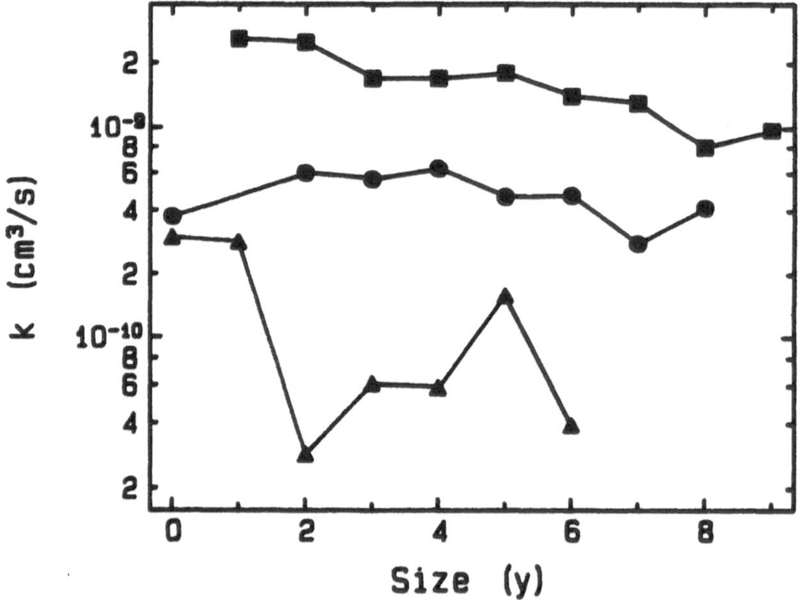

Figure 1. The rate constants for the reactions of (■) C_y^+, (●) SiC_y^+ and (▲) $Si_2C_y^+$ with acetylene, as a function of the number of carbon atoms. Results for C_y^+ (y>2) are from [3]. (SiC^+ is not formed.) For $Si_2C_y^+$ the rate constant of the long time process (see text) is plotted.

The major products are as follows. For C_y^+, C_2H addition predominates; some adduct formation with or without loss of C_3 becomes important for the larger ions. C_2H addition is also the major product for the SiC_y^+ ions; adduct formation is only important for $y=7$. The $Si_2C_y^+$ ions with 0 or 1 carbons also show mainly C_2H addition. The larger ions in this series react in a most unusual manner. Initially the ion reacts slowly to add CH_2 (loss of C); with increasing reaction time the reaction becomes more rapid and the ion forms the adduct. In Figure 1 it is the rate constant for this latter process which is plotted.

The cluster ions with one Si atom react similarly to the carbon clusters C_y^+: both predominantly add C_2H, and the k *vs.* y curve for SiC_y^+ shows the same size variation as the C_y^+ curve when it is shifted to $y+1$ (to account for the Si atom). The rate constants are uniformly ~ 3 times smaller for SiC_y^+. These results strongly support a linear structure for the SiC_y^+ formed in the FTMS (analogous to the smaller C_y^+ ions), but with only one reactive carbene end. The cluster ions containing two Si atoms react very differently, suggesting a completely different structure.

These results bring up several questions. What is the source of the C lost in the short time process? (See next section on Labelling Studies.) What is responsible for the unusual kinetics? Is it excess energy, either kinetic or internal, or are we observing a structural isomerization? (See the section on Activation Studies below.)

3.3. ^{13}C LABELLING STUDIES

Using amorphous ^{13}C graphite in the pellet, labelled silicon carbide ions were formed. In the reaction with acetylene, the following carbon exchange reactions were observed to occur:

$$Si_2{}^{13}C_y{}^+ + {}^{12}C_2H_2 \rightarrow Si_2{}^{13}C_{y-1}{}^{12}C^+ + {}^{12}C^{13}CH_2, \tag{1}$$

$$Si_2{}^{13}C_y{}^+ + {}^{12}C_2H_2 \rightarrow Si_2{}^{13}C_{y-2}{}^{12}C_2{}^+ + {}^{13}C_2H_2. \tag{2}$$

(The neutral product is assumed to be acetylene since the second-lowest energy product channel lies almost 6 eV higher.) The same phenomena is observed when using ^{13}C labelled acetylene.

This observation of carbon exchange was totally unexpected, and is rather surprising. The ions as initially formed in our instrument are able to catalyze cleavage of the carbon-carbon triple bond! Carbon exchange was observed for the $Si_2C_y^+$ ($y=1-5$) ions but not for any of the SiC_y^+ ions. The double exchange product is formed directly from the reactant ion in reaction 2, as well as by a two-step mechanism from the product of reaction 1. Carbon exchange is initially faster than any other reaction.

The presence of the carbon exchange products obscures the picture when attempting to determine the source of the carbon lost from the adduct product. However, it is clear that scrambling of the carbons occurs on the same time scale as reaction.

3.4. "ACTIVATION" STUDIES

Figure 2 shows one example of the "activation" experiments performed. The kinetics of the reaction of $Si_2C_2^+$ at a constant low pressure of acetylene are plotted. The effect of added argon gas is also shown. Similar experiments were done with SF_6, a good quencher of rotational and vibrational energy, as the collision gas. For comparison, different pressures of acetylene were also used. The efficiency of the collision gas in "activating" $Si_2C_2^+$ was calculated.

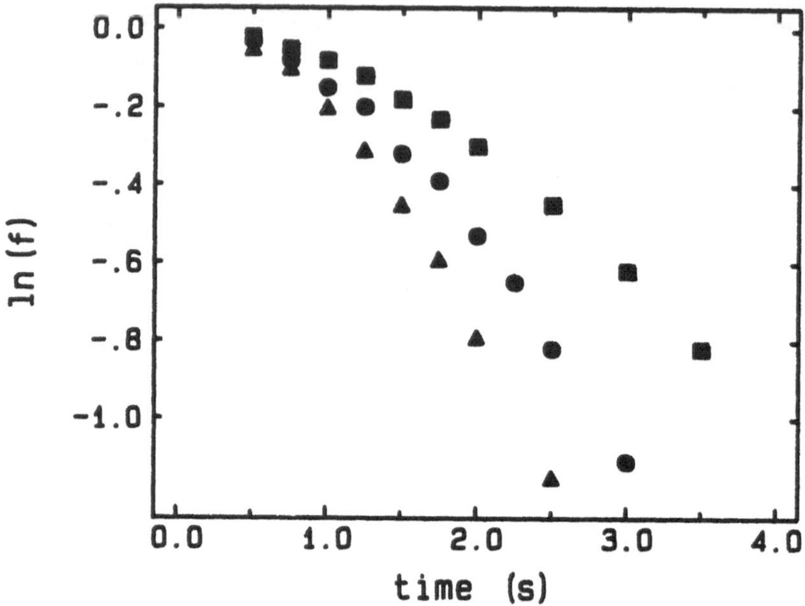

Figure 2. First-order kinetic plot (extent of reaction *vs.* reaction time) for the reaction of $Si_2C_2^+$ with acetylene ($\blacksquare$ $P=6\times10^{-8}$ torr). The effect of added argon buffer gas ($\bullet$ $P=10\times10^{-8}$ torr and $\blacktriangle$ $P=35\times10^{-8}$ torr) is also shown.

Ar and SF_6 show the same "activating" efficiency, and are ≈ 3 times less efficient than C_2H_2. This result rules out excess rotational or vibrational energy in the ion as the cause of the observed kinetic behavior. The expected trend in kinetic energy partitioning ($SF_6 > Ar > C_2H_2$) is not seen, which suggests excess kinetic energy is not the cause either.

There are many calculations on the structures of small <u>neutral</u> silicon carbide species. Of interest here are those of Fitzgerald and Bartlett [4] and Lammertsma and Güner [5] on various structural isomers of Si_2C_2. They found 4 minima on the singlet potential energy hypersurface. The singlet ground state geometry is a rhombus. The distorted trapezoid, linear Si-C-C-Si chain, and distorted rhombus geometries are found at progressively higher energies. Lammertsma and Güner also considered triplet forms of Si_2C_2. Again 4 minima were found. The lowest triplet isomer, the linear Si-C-C-Si configuration, is isoenergetic with the lowest singlet isomer, the rhombus. Conversion from one to the other requires bond breaking and rearrangement of the atoms in the molecule.

Perhaps what we are seeing is such an isomerization. The higher "activation" efficiency of acetylene is presumably related to the carbon exchange reaction. This suggests a transition between two structures or electronic states. The unusual kinetics of $Si_2C_y^+$ may arise from initial formation of a geometric or electronic structure of low reactivity, which isomerizes to a more reactive form of the ion.

3.5. REACTIONS OF $Si_2C_2^+$ WITH UNSATURATED HYDROCARBONS

The unusual reactivity of $Si_2C_2^+$ with acetylene prompted the investigation of reactions with other unsaturated hydrocarbons. The results are briefly summarized in Table 1.

TABLE 1. A summary of the reaction products observed in the reactions of $Si_2C_2^+$ with the listed neutrals.

Neutral	Ion Products
$HC \equiv CH$	A, A-C, C exchange
$H_2C = CH_2$	$Si_2C_2H_2^+$, C exchange
$H_2C = C = CH_2$	+C, +CH$_2$, +H, A-H, CT, CT-H
$HC \equiv C\text{-}CH_3$	+CH$_2$, $SiC_3H_3^+$, A-H, +C$_2$H
$HC \equiv C\text{-}CF_3$	Si_2F^+, SiF^+, +F, A-SiF$_2$

Notes: A = adduct ion. CT = charge transfer.

The reaction with acetylene was discussed above in detail. Similarly, carbon exchange was also observed in the reaction with ethylene. This result, along with the formation of the +C and +CH$_2$ products in the allene reaction, indicates that carbon-carbon double bonds are also cleaved. Cleavage of carbon-carbon triple bonds is much less favorable. In the reaction with propyne, labelling experiments revealed that attack occurs about equally at the acetylenic C-H and C-C bonds. There was only weak evidence for attack at the $C \equiv C$ bond.

The dehydrogenation reaction with ethylene is another example of the interesting chemistry of these silicon carbide ions. Unsurprisingly perhaps, reaction with HC_3F_3 shows that formation of silicon fluoride is favored over $Si_xC_yH_z$ species.

3.6. REACTIONS OF $Si_2C_2^+$ WITH AROMATIC NEUTRALS

Bohme and coworkers [6], studying the reactions of Si^+ with large aromatics, observed adduct formation and unusual reactivity of the adduct ion. Reactions of various cluster cations with aromatic compounds have been used by Bach and Eyler [7] to determine the ionization potentials of the corresponding cluster neutrals. The reactions of $Si_2C_2^+$ with aromatic neutrals were studied for comparison with previous work, and the results are briefly summarized in Table 2.

Charge transfer reactions of the initial $Si_2C_2^+$ population bracket the ionization potential of the corresponding neutral at 9.55 ± 0.3 eV. This is significantly higher than the value of 8.2 eV found by Drowart *et al.* [8]. The more reactive form of the ion gives a lower ionization potential for the neutral, but the exact value is still uncertain.

The deoxygenation reaction with nitrobenzene is rather remarkable, and is analogous to the deoxygenation reactions of Si^+ with NO_2 and N_2O [6]. Dehalogenation of chlorinated and fluorinated compounds is also observed.

TABLE 2. A summary of the reactivity of $Si_2C_2^+$ with various aromatic neutrals, with an emphasis on charge transfer.

Neutral	IP	CT	Other Ion Products
aniline	7.72	Y	A-H
p-xylene	8.44	Y	$A-CH_3$, N^+-H
toluene	8.82	Y	$A-CH_3$, $A-C_2H_5$, $A-H_2$
1,2,4-Cl-benzene	9.04	Y	$Si_xC_yCl_z^+$
benzene	9.25	Y	$+C_4H_4$, A, $+C_6H_5$, $+C_2H$
nitrobenzene	9.86	N	N^+-O, N^+-O_2
hexa-F-benzene	9.91	?	SiF^+ (very slow)

Notes: IP in eV. CT=charge transfer, Y=yes and N=no. A=adduct ion. N=neutral.

4. References

[1.] Parent, D. C. (1991) manuscripts in preparation.

[2.] Parent, D. C. and McElvany, S. W. (1989) 'Investigations of small carbon cluster ion structures by reactions with HCN', J. Am. Chem. Soc. 111, 2393-2401.

[3.] McElvany, S. W. (1988) 'Reactions of carbon cluster ions with small hydrocarbons', J. Chem. Phys. 89, 2063-2075.

[4.] Fitzgerald, G. B. and Bartlett, R. J. (1990) 'Optimum structures and vibrational frequencies of $(SiC)_2$ clusters', Int. J. Quant. Chem. 38, 121-128.

[5.] Lammertsma, K. and Güner, O. F. (1988) 'Structures and energies of disilicon dicarbide, C_2Si_2', J. Am. Chem. Soc. 110, 5239-5245.

[6.] Bohme, D. K. (1990) 'Chemistry initiated by atomic silicon ions in the gas phase: formation of silicon-bearing ions and molecules', Int. J. Mass Spectrom. Ion Proc. 100, 719-736.

[7.] Bach, S. B. H. and Eyler, J. R. (1990) 'Determination of carbon cluster ionization potentials via charge transfer reactions', J. Chem. Phys. 92, 358-363.

[8.] Drowart, J., De Maria, G. and Inghram, M. G. (1958) 'Thermodynamic study of SiC utilizing a mass spectrometer', J. Chem. Phys. 29, 1015-1021.

ENCAGEMENT OF ALKALI METAL IONS IN GAS-PHASE WATER CLUSTERS

A. SELINGER AND A. W. CASTLEMAN, JR.
Department of Chemistry
The Pennsylvania State University
University Park, PA 16802

ABSTRACT. A fast-flow reactor was used to produce and investigate ionic water clusters with the general composition $M^+(H_2O)_n$; where M is one of the alkali metals Li, Na, K or Cs and n is up to 45. A series of magic numbers could be observed, especially pronounced for $K^+(H_2O)_{20}$ and $Cs^+(H_2O)_{20}$ and to some degree for Lithium, and for some other cluster sizes. Evidence is presented that these stable cluster ions are gas phase hydrates in analogy to the well-known clathrate hydrates in the condensed phase.

1. Introduction

It has been known for a long time that water crystallizes in structures different from ice under the presence of substances as different as xenon or tetrahydrofuran [1]. Despite the variety of the enclosed species, the vast majority of those mixed crystals, called clathrates, are based on only two different water lattice structures. Only recently considerable new interest in those clathrates has arisen, as they are known or believed to play an important role in natural phenomena like methane storage in ocean sediments, clogging of gas pipelines, and formation of atmospheric aerosols [2,3].

There are three known examples where gas phase cluster ions are believed to form similar clathrate structures. In all three cases 20 water molecules form a cage, in which the ions H_3O^+ [4], OH^- [5] and NH_4^+ [6] are enclosed. In this study we present evidence for the formation of gas phase clathrates with encaged alkali ions.

2. Experimental

The apparatus has been described before [7] and was only slightly modified for the present study. The cluster ions are formed in a cooled thermionic filament source, thermalized in a fast flow tube, and detected in a quadrupole mass spectrometer.

The source is comprised of a heated filament exposed to the flow of a helium-water mixture at about 32 torr (4200 Pa) where the partial pressure of water is less then 0.5 % of helium. The filament is covered with a glassy-like bead of the approximated stoichiometry $M_2O \cdot Al_2O_3 \cdot 2SiO_2$, formed with the usual procedure. The ions are carried through a 3 mm orifice from the source into a fast-flow tube which is operated at 0.3 Torr (40 Pa) and serves in these experiments mainly to ensure the thermalization of the formed cluster ions. Therefore, it was operated for the present study at temperatures between -100 and -120°C. After thermalization, the ions are sampled by a quadrupole mass filter located at the exit of the flow tube. The presented spectra are obtained by mass scanning for three minutes each.

P. Jena et al. (eds.), Physics and Chemistry of Finite Systems: From Clusters to Crystals, Vol. II, 1137–1140.
© 1992 *Kluwer Academic Publishers.*

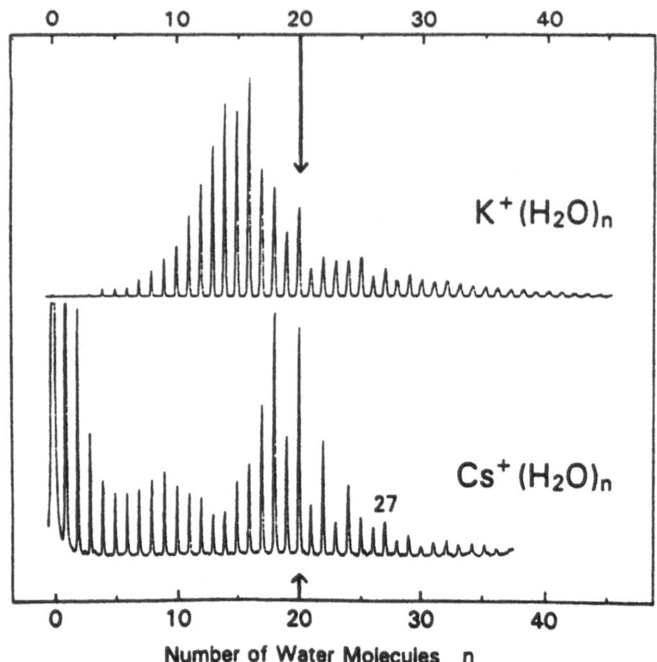

Figure 1.   Mass spectra of thermalized alkali metal cation-water clusters
(a) $K^+(H_2O)_n$ at -110 °C ;   (b) $Cs^+(H_2O)_n$ at -117 °C

3. Results and Discussion

Figures 1(a) and (b) show mass spectra of cluster distributions $M^+(H_2O)_n$ with M=K and Cs, respectively.

Potassium ions with up to 45 attached water molecules could be detected at -110°C. The strongest signal in this spectrum is due to the cluster $K^+(H_2O)_{16}$. This maximum can be shifted to higher or lower masses by minor changes in the source and flow tube temperature or by adjustment of the water vapor flow. However, "magic numbers", that is signals with enhanced intensities in respect to the neighboring intensities of higher mass, proved to be independent of the overall maximum of the mass spectrum. The signal preceding the largest intensity drop is caused by $K^+(H_2O)_{20}$. Weaker, but still completely reproducible features occur at n = 16, 18, 25, 27 and 29.

The cesium/water spectrum Fig. 1b, obtained at -117°C, shows signals of clusters up to $Cs^+(H_2O)_{37}$, and has obvious structural similarities to Fig. 1a. Due to the low binding energy of water molecules to the cesium ion, there is a strong signal for Cs⁺ with zero, one or two attached water molecules. Again, 20 attached water molecules form an especially stable Ion: $Cs^+(H_2O)_{20}$. As in the case of potassium, this magic number is surrounded by a set of weaker ones, namely for n = 18, 22, 24, 27 and 29.

Surprisingly, preliminary results for lithium and sodium indicate a magic number for $Li^+(H_2O)_{20}$, but none for sodium/water clusters. In agreement with our results, no magic numbers have been found for sodium/water clusters by Hertel et al in a different experimental approach [8].

It is rather evident from these findings that 20 water molecules form an especially stable cluster when stabilized by certain alkali ions. The most reasonable explanation assumes the formation of a pentagonal dodecahedron with oxygen atoms in the corners and hydrogen bonds forming the edges as displayed in figure 2. This structure has recently been established in a study of the cluster ion $H_3O^+(H_2O)_{20}$ [9]. The same dodecahedron is also part of the crystal structures of solid phase clathrates type I and II. The formation of those clathrates is more a matter of the size of the enclosed particle than of any other physical or chemical property, although ions are not known to form clathrates in the condensed phase.

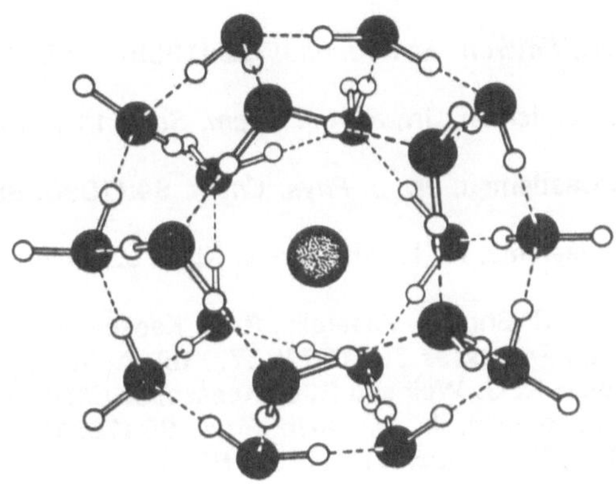

Figure 2. Proposed dodecahedral structure for $M^+(H_2O)_{20}$. White spheres, hydrogen; black sheres, oxygen; gray sphere, alkali ion; dashed lines, hydrogen bonds.

The other observed magic numbers can be explained in terms of different, in some cases distorted, water shell structures, as discussed somewhere else [10, 11].

4. Conclusions

In view of the present results for atomic ions in water clusters, we believe we have shown further evidence for the existence of gas phase clathrates of a so far unknown variety. However, as not all atomic ions form a stable cluster with 20 water molecules, further investigations are necessary to reveal which properties of ions are essential criteria for the formation of those clathrates.

Acknowledgements

Financial support by the U.S. Department of Energy, Grant No. DE-FGO2-88ER60648 and the National Science Foundation, Grant No. ATM-90-15855, is gratefully acknowledged. One of the authors (A. Selinger, Feodor Lynen-Fellow) acknowledges support by the Alexander von Humboldt-Foundation, Germany.

References

[1] D. W. Davidson in *Water, A Comprehensive Treatise;* Franks, F., Ed.; Plenum Press: New York, 1973; Vol. 2, Chapter 3.

[2] T. Appenzeller, *Science* **252** (1991) 1790.

[3] A. W. Castleman, Jr., *Environ. Sci. Technol.* **22** (1988) 1265.

[4] X. Yang and A. W. Castleman, Jr., *J. Am. Chem. Soc.* **111** (1989) 6845.

[5] X. Yang and A. W. Castleman, Jr., *J. Phys. Chem.* **94** (1990) 8500.

[6] H. Shinohara, U. Nagashima, H. Tanaka and N. Nishi, *J. Chem. Phys.* **83** (1985) 4183.

[7] (a) B. L. Upschulte, R. J. Shul, R. Pasarella, R. G. Keesee and A. W. Castleman, Jr., *Int. J. Mass Spectrom. Ion Processes* **75** (1987) 27.   (b) A. W. Castleman, Jr., S. Sigsworth, R. E. Leuchtner, K. G. Weil and R. G. Keesee, *J. Chem. Phys.* **86** (1987) 3829. (c) X. Yang and A. W. Castleman, Jr., *J. Chem. Phys.* **95** (1991) 130. (d) X. Yang and A. W. Castleman, Jr., *J. Am. Chem. Soc.* **113** (1991) 6776.

[8] I. V. Hertel, C. Hüglin, C. Nitsch and C. P. Schulz, *Phys. Rev. Lett.* **67** (1991) 1767.

[9] S. Wei, Z. Shi and A. W. Castleman, Jr., *J. Chem. Phys.* **94** (1991) 3268.

[10] A. Selinger and A. W. Castleman, Jr., *J. Phys. Chem.* **95** (1991) 8442.

[11] A. Selinger and A. W. Castleman, Jr., to be published.

DEPOSITION OF LEAD CLUSTERS ON COLD SILVER AND PHOTOEMISSION BY SYNCHROTRON RADIATION

H.R. SIEKMANN, E. HOLUB-KRAPPE*, BU. WRENGER,
CH. PETTENKOFER*, K.H. MEIWES-BROER

Fakultät für Physik, Universität Bielefeld,
4800 Bielefeld 1, F.R.G
* Bereich S, Hahn-Meitner-Institut,
1000 Berlin 39, F.R.G.

ABSTRACT: Lead clusters grown in a pulsed arc cluster ion source (PACIS) are deposited under UHV conditions on LN_2-cooled polycristalline silver targets. Photoemission experiments with light from the synchrotron BESSY serve to investigate the electronic level structure of the clusters. The Pb 5d core levels clearly shift with the cluster size, yielding — for small clusters with below about 100 atoms each — a shift of - 0.33 eV relative to the bulk value. The same maximum shift is obtained in the case of atom deposition. Born-Haber cycles and the equivalent core approximation are used to explain the observed shifts.

There is fundamental interest in exploring the size dependent electronic structure of supported metal clusters. During the past decades, most photoemission work has been done on clusters grown on the surface of materials with a defined amount of adsorption sites, mainly amorphous carbon [1–3]. Also, several attempts have been made to deposit mass-selected clusters [4–6], but so far only very small ones have been deposited from a beam. Core level photoemission from deposits on C or SiO_2 is dominated by the effect of final state charging. In addition, size distributions possibly cause line broadening and mask size effects. We have performed photoemission studies of lead clusters grown in a Pulsed Arc Cluster Ion Source (PACIS) [7–8] and deposited onto cold Ag. In this contribution, the influence of cluster size on the Pb 5d core levels will be presented and discussed.

The source produces beams with intensities high enough to deposit clusters with a small size distribution. E.g. for lead, even 2 m downstream the source, 0.05 ML/s is an average deposition rate. The "heart" of the PACIS is a pulsed high current arc fired between two electrically isolated electrodes in the presence of a carrier gas [7–8]. The resulting metal cluster pulse (length $> 100\mu s$) is cooled in the expansion into high vacuum. After skimming and differential pumping, the charged part of the beam can be analyzed by time-of-flight mass spectroscopy. A typical mass spectrum for Pb_n^- is shown in Fig. 1. It should be pointed out that the real maximum of the distribution is located at considerably higher n, as the detection probability decreases with increasing cluster size, i.e. mass.

P. Jena et al. (eds.), Physics and Chemistry of Finite Systems: From Clusters to Crystals, Vol. II, 1141–1146.
© 1992 Kluwer Academic Publishers.

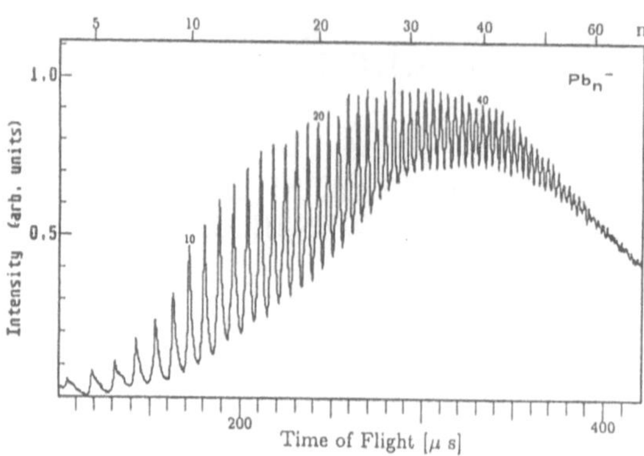

Fig. 1: Mass spectrum of lead cluster anions as emerging directly from the PACIS, measured by time-of-flight mass spectroscopy. Due to technical reasons only small clusters are recorded here. From deposition studies we know that the size distribution extends to n > 1000.

For deposition, the cluster ions are accelerated and mass-separated by a Wien filter. A LN_2 cooled silver target mounted within UHV can be biased in order to achieve soft landing. Deposition times range from a few minutes (neutrals) to about 1 hour (mass-selected ions). The sample will be exposed to light from a toroidal grating monochromator (TGM7, 20-120 eV) at the Berlin electron storage ring BESSY. A 150° hemispherical analyzer serves to record the photoelectron spectra. The overall energy resolution including monochromator and analyzer reaches 190 meV at $h\nu = 50$ eV. All photoemission spectra are recorded in an angle-resolved mode at normal emission.

First, we deposit about 0.1 ML Pb atoms on polycristalline Ag at 160 K. This yields a small core level intensity centered at 17.67 eV. The corresponding core level shift of - 0.33 eV with respect to the bulk value can be fully accounted for by a Born-Haber cycle, as we demonstrate below. Then the PACIS is tuned to yield mainly small clusters ($n_{ave} < 100$), which are now deposited at a target temperature of 230 K. The resulting peak positions for the Pb $5d_{5/2}$ level as well as the respective linewidths are compiled in Fig. 2, see square symbols. Rutherford backscattering (RBS) analysis [9] under comparable experimental conditions serves to estimate the coverage (and may be wrong by a factor of 2). For the lowest coverage, a total shift of 0.33 eV towards lower binding energy (BE) is measured (relative to the bulk value of 18.0 eV below ϵ_F [10–12]). Further deposition increases the BE until the bulk value is reached at a coverage of about 10^{16} atoms/cm^2. Deposition on Cu under identical conditions leads to a similar maximum shift of - 0.31 eV.

Thus we find that, first, under the conditions of small cluster ($n_{ave} < 100$) deposition at 230 K, the minimum BE is the same than that of lead atoms on Ag. So, these small clusters either show no distinct shift with respect to the atom, or, in the case of deposition at 160 K, the atoms aggregate to small clusters. We present arguments for the former possibility below. Second, we observe a steady shift of the core levels towards the bulk value. As the cooling is fairly mild (230 K), the mobility of the clusters seems to be large enough to induce aggregation to larger clusters with increasing coverage, leading to a coverage-dependent shift. Further evidence for the temperature effect on mobility comes, e.g., from observations of cluster size by SEM analysis in the Pb/Si-system with clusters from the PACIS [9].

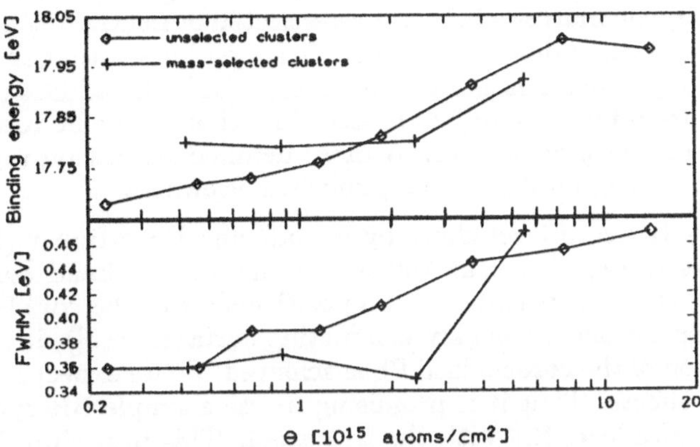

Fig. 2: Comparison of Pb 5d core electron binding energy and linewidth for small unselected Pb_n on Ag (230 K) (squares), and mass-selected clusters on cold Ag (160 K) (crosses), versus the coverage Θ. Mass-selected Pb-clusters exhibit a certain core level binding energy which — up to a critical coverage — turns out to be independent on the coverage. Further deposition leads to a shift towards the bulk value. At the same time, the line broadens significantly.

By using the Wien filter to mass-select a distinct mass range of the ionic part of the beam, clusters with defined size are now deposited on an Ag target at 160 K. For n ranging approximately between 100 and 1000, the Pb $5d_{5/2}$ BE shifts from 17.74 eV to 17.92 eV. For a given cluster size, as Pb_n coverage increases, the positions of the core levels remain centered around a constant value (within $\pm$ 0.02 eV). Only beyond a critical coverage Θ_{cr}, the BE shifts towards the bulk value. This is shown in the top of Fig. 2 and compared to the unselected and high-temperature case. The bottom part of the figure illustrates the variation of the linewidth with coverage for the two cases. While small cluster deposition at 230 K leads to a steady increase in width with increasing coverage, the mass selected clusters only show line broadening beyond Θ_{cr}. The maximum width of 0.47 eV is somewhat larger than data from other authors, 0.40 [10] and 0.41 eV [12] in the bulk case. The minimum linewidth as we determined for $\Theta < \Theta_{cr}$ (here 0.35 eV) is below that of other cluster depositions [1–5]. In the case of metal clusters on C or SiO_2 substrates, finite linewidths at the lowest coverages have been attributed to phonon broadening [5] or size distribution [2]. Even on metals broader lines have been observed [1], although, as is the case for alloys, one would not expect line broadening. Mason [2] has pointed out the similarity between clusters and alloys. Thus we can conclude that the size distribution of the deposited clusters remains rather narrow, presumably similar to the preformed size. Furthermore, our measurements clearly show that as long as the surface is only partially covered, the BE shift depends mainly on cluster size.

In spite of elaborate valence band studies, it is not possible to reveal distinct structures being derived from the atomic p-levels of lead. From photoelectron spectroscopy of free Pb_n we know that pronounced and well-resolved narrow lines are characteristic for these clusters [13]. Instead, the Ag sp-valence band gains unstructured intensity, no matter whether mass-selected or neutral deposition. This is similar to the case of atom deposition, where it is well known that the sp valence band on top of another metal is hardly to be seen [12]. In addition, resonance tunneling is known to induce severe line broadening, thus might be responsible for a smearing out of the sharp structures. A more detailed discussion together with results from studies of Au_n on Ag will be published elsewhere.

We now discuss the core level shifts by introducing Born-Haber (BH) cycles, starting with the atom deposition at 160 K. As the valence band studies reveal strong interaction with the substrate and no positive shift of the core levels occurs (as is the case of depositions on weakly conducting surfaces [1–5]), it is clear that efficient neutralization of the core ionized Pb is achieved. So we expect no significant final state charging effects. Thus it is promising to use a simple BH cycle [1,12,14] which connects the core level BE with the free atom. This procedure is discussed in detail by Johansson et al. [14].

In short, if an atom with atomic number Z (here $Z = 82$ for Pb) is desorbed from the surface by applying the desorption energy $E_{AD}(Z)$ and is then core-ionized $(Z^* + e)$ by applying $E_{B,Atom}$ (the BE of the free atom), the equivalent core approximation can be used. This yields an equivalence between the two configurations: $(Z^* + e) \simeq (Z + 1)^+ + e$. Then the valence ionized $(Z + 1)^+$-atom is neutralized gaining $I(Z + 1)$. Finally the $(Z + 1)$-atom (Bi) is adsorbed, thereby releasing $E_{AD}(Z + 1)$. The result is equivalent to a core ionized, fully screened atom Z, thus the difference in the cycle is assumed to be the core level BE $E_B(Z)$ referred to the Fermi edge:

$$E_B(Z) = E_{AD}(Z) + E_{B,Atom} - I(Z + 1) - E_{AD}(Z + 1) \qquad (1)$$

We use the data for E_{AD} from the semi-empirical model introduced by Miedema and Dorleijn [15] and $E_{B,Atom}$ from gas phase measurements of atomic lead [16]. The result together with the data used is shown in Table 1.

System	Atom	$E_{B,Atom}$	I_1	E_{AD}	$E_{B,calc}$	$E_{B,exp}$
Errors				±0.025	±0.05	±0.02
Pb/Ag	Pb	25.27	—	2.236	17.72	17.67
	Bi	—	7.287	2.496		
Pb/Cu	Pb	25.27	—	2.860	17.67	17.69
	Bi	—	7.287	3.172		

Table 1: Atomic (gas phase) binding energy (BE) $E_{B,Atom}$ [16], first ionization potential I_1, adsorption energies E_{AD} [15], calculated ($E_{B,calc}$) and measured ($E_{B,exp}$) $Pb\ 5d_{5/2}$ core level BE referred to the Fermi edge. All values in (eV).

The agreement between calculation and experiment is extremely good, as is the case for the Pb/Ni-system [12].

For an explanation of the Pb 5d core level shifts of the deposited clusters, we use BH cycles as have been applied by Wertheim [1]. First, we consider small, two-dimensional clusters, denoted by an index "n". Here, every atom in the cluster is in contact with the surface. Introducing the cohesive interaction in the Pb_n cluster, $E_{AGR}(Pb_n)$ [17], and applying this energy to a Pb_n on Ag, we are left with a Pb atom on Ag and may follow the same cycle as in (1). Next, we need to solute a Bi atom in the Pb_n by gaining $E_{SOL}^{Bi}(Pb_n) = E_{AGR}(Bi_n) + E_{IMP}$, E_{IMP} being the implantation energy to transfer a Bi from Bi_n to Pb_n. If we neglect this usually small term [1] we get instead of (1):

$$E_B(Pb_n) = E_{AGR}(Pb_n) + E_{AD}^{Pb}(Ag) + E_{B,Atom} - IP^{Bi} - E_{AD}^{Bi}(Ag) - E_{AGR}(Bi_n) \quad (2)$$

Second, we consider large, three-dimensional clusters, denoted by an index "N". Then photoelectrons mainly stem from lead atoms only in contact with the Pb_N-cluster and we should replace adsorption energies by cohesive energies of Pb_N, $E_{COH}(Pb_N)$. Again neglecting E_{IMP}, the BH cycle becomes

$$E_B(Pb_N) = E_{COH}(Pb_N) + E_{B,Atom} - IP^{Bi} - E_{COH}(Bi_N) \quad (3)$$

If we now compare the BE of adsorbed clusters and adsorbed atoms, we get for the relative shift

$$\Delta E_B(Pb_n) = E_B(Pb_n) - E_B(Pb_1) = E_{AGR}(Pb_n) - E_{AGR}(Bi_n) = \Delta E_{AGR}^n \quad (4)$$

and

$$\begin{aligned}
\Delta E_B(Pb_N) &= E_B(Pb_N) - E_B(Pb_1) \\
&= E_{COH}(Pb_N) - E_{COH}(Bi_N) - E_{AD}^{Pb}(Ag) + E_{AD}^{Bi}(Ag) \quad (5) \\
&= \Delta E_{COH}^N - \Delta E_{AD}(Ag)
\end{aligned}$$

For $N \to \infty$ we arrive at the bulk and/or surface values. Now we remember that the difference in cohesive energies between Pb and Bi is very small [15]. Relative to the bulk, the cohesive energy in a cluster containing 100-1000 atoms reduces to about 80–90%, respectively [18]. If we suppose that this reduction is similar in Pb and Bi, both ΔE_{AGR}^n and ΔE_{COH}^N can be considered to be small. Thus $\Delta E_B(Pb_N)$ (5) will be governed largely by ΔE_{AD}, while $\Delta E_B(Pb_n)$ (4) is supposed to be small. This means that small n-clusters possibly experience little shift with respect to the monomer, as is supposed earlier in the discussion. Indeed, if we look carefully at the data of Gürtler and Jacobi on the Pb/Ni-System [12], where a coverage-dependent shift reflects the size variation of the growing two-dimensional clusters, we recognize a pronounced shift only beyond a coverage of $\Theta \simeq 0.5$ ML. Another example is the deposition of Sn on Metglas, where no shift is observed between 3×10^{13} and 2×10^{14} atoms/cm^2 [1].

For the large N-clusters, we quickly should arrive at the bulk/surface-value for the BE, as the dominating adsorption term in (5) pushes the core line to the bulk value. Using the experimental data from Table 1, we get $\Delta E_{AD}(Ag) \simeq -0.31$ eV and arrive at $E_B \simeq 17.98$ eV, i.e. the bulk value.

Clusters of sizes in between "n" and "N" will exhibit both the influence of adsorption and cohesive energies, leading to a size dependence of the BE.

In summary, we have shown that the PACIS is suitable for mass-selected cluster deposition for medium-sized clusters with narrow size distributions. Pb 5d-core levels experience shifts depending mainly on cluster size as long as the surface is only partially covered. Shifts for atoms, small and large clusters can be explained by Born-Haber cycles involving cohesive and adsorption energies.

Acknowledgements

We thank J. Lehmann and the staff of BESSY for technical assistance. Helpful discussions with M. Bronold and A. Klein are gratefully acknowledged.

References

1. G.K. Wertheim, Z. Phys. D 12, 319 (1989); Z. Phys. B 66, 53 (1987); G.K. Wertheim, S.B. Di Cenzo, Phys. Rev. B 37, 844 (1988)
2. M.G. Mason, Phys. Rev. B 27, 748 (1983)
3. S.L. Qiu, X. Pan, M. Strongin and P.H. Citrin, Phys. Rev. B 36, 1292 (1987)
4. S.B. Di Cenzo, S.D. Berry and E.H. Hartford, Jr., Phys. Rev. B 38, 8465 (1988)
5. W. Eberhard, P. Fayet, D. Cox, Z. Fu, A. Kaldor, R. Sherwood and D. Sondericker, Phys. Scri. 41, 892 (1990); Phys. Rev. Lett. 64, 780 (1990)
6. P. Fayet, Th. Detzel, H.V. Roy, F. Patthey and W.D. Schneider, to be published
7. G. Ganteför, H.R. Siekmann, H.O. Lutz and K.H. Meiwes-Broer, Chem. Phys. Lett. 165, 293 (1990)
8. H.R. Siekmann, Ch. Lüder, J. Faehrmann, H.O. Lutz and K.H. Meiwes-Broer, Z. Phys. D 20, 417 (1991)
9. P. Jonk, H.R. Siekmann, T. Brammer, B. Wrenger, K.H. Meiwes-Broer, this volume, and to be published
10. A.W. Potts, P.J. Bridgen, D. S. Law and E.P.F. Lee, J. El. Spec. Rel. Phen. 24, 267 (1981)
11. D. Chadwick and M.A. Karolewski, Appl. Surf. Sci. 9, 98 (1981)
12. K. Gürtler and K. Jacobi, Surf. Sci. 134, 309 (1983)
13. G. Ganteför, M. Gausa, K.H. Meiwes-Broer and H.O. Lutz, Z. Phys. D 12, 405 (1989)
14. B. Johansson and N. Mårtensson, Phys. Rev. B 21, 4427 (1980)
15. A.R. Miedema and J.W.F. Dorleijn, Surf. Sci. 95, 447 (1980)
16. N. Sandner, V. Schmidt, W. Mehlhorn, F. Wuilleumier, M.Y. Adam and J.P. Desclaux, J. Phys. B 13, 2937 (1980)
17. The index "AGR" is chosen, because in ref. [1] E_{AGR} describes the aggregation energy on the surface.
18. L. Skala and H. Müller in: PDMS and Clusters, edited by E.R. Hilf, F. Kammer and K. Wien, Springer-Verlag Berlin, Heidelberg, 144 (1987)

Intracluster Anionic Polymerization of $(CH_2=CXCN)_n$ (X = H, D, CH_3, and Cl) in collision with high-Rydberg rare gas atoms and electrons

T. TSUKUDA and T. KONDOW
Department of Chemistry, Faculty of Science,
The University of Tokyo, Bunkyo-ku, Tokyo 113, Japan

ABSTRACT. Intracluster polymerization in the gas-phase clusters, $(CH_2=CXCN)_n$ [X=H (AN), D (AN-d_1), CH_3 (MAN), and Cl (CAN)], following collisional electron transfer from Kr^{**} was investigated by mass spectrometry. Cluster anions, such as $[(AN-d_1)_n-D_2]^-$, $[(MAN)_n-n'\cdot HCN]^-$, and $[(CAN)_n-n'\cdot HCl]^-$, were observed, where $[(AN-d_1)_n-D_2]^-$ is produced by abstracting a D_2 molecule from $(AN-d_1)_n^-$. These product anions indicate that polymerization proceeds in the clusters by the electron attachment. Extraordinary large abundances of the AN cluster anions with the size of n=3k are explained in such a manner that the intracluster polymerization is constrained by a ring geometry of three AN molecules in the AN clusters. The effect of the X moiety on the product anions suggests that the intracluster polymerization is explained by the anionic mechanism.

1. Introduction

Recently, ionic polymerization in the gas-phase clusters has been explored by use of mass spectrometry [1-4]. The propagation of polymerization is much influenced by the intermolecular distances between the propagating end and the adjacent monomer. It is conceivable, therefore, that spatial arrangements of the constituent molecules in the clusters influence on the behaviors of the intracluster polymerization. In this connection, a hydrogen-bonded cluster appears to be suitable for the investigation of such a geometrical effect because the anisotropic hydrogen bond gives specific geometrical arrangements. In reality, structure-constrained polymerization has been discovered in the electron attachment to acrylonitrile (AN) clusters [4]. In the intracluster polymerization, the excess energy generated by the exothermic reaction is sufficiently large that several molecules could be liberated from the cluster system by the rupture of the chemical bonds. The loss of the molecules enables us to investigate such intracluster polymerization by observing the product cluster anions.

In this report, we described a systematic study of anionic polymerizations in the clusters of $CH_2=CXCN$ (X = H, D, CH_3, and Cl) induced

P. Jena et al. (eds.), Physics and Chemistry of Finite Systems: From Clusters to Crystals, Vol. II, 1147–1152.
© 1992 *Kluwer Academic Publishers.*

by electron attachment. In order to minimize the disturbance due to the ionization and to enhance the ionization efficiency, the slow outermost electron of Rydberg Kr atoms (Kr**) were collisionally transferred to the clusters [5].

2. Experimental

The experimental apparatus has been described in detail elsewhere [6]. The apparatus consists of a supersonic nozzle beam source, a triple-grid ion source, and a quadrupole mass spectrometer. A seeded sample gas was prepared by flowing helium gas at a pressure of 1–5 atm over a liquid sample placed in a reservoir. The temperature of the reservoir was kept at 70 °C during the measurements. The seeded gas was expanded through a nozzle of 0.05 mm diameter. A beam of a neutral cluster, after passing through a skimmer and a collimator, was allowed to enter the ion source consisting of filaments and three concentric cylindrical grids. The cluster was then ionized either by high-Rydberg atom impact (RAI) or electron impact (EI) in the central region of the ion source.

In the RAI ionization, Kr gas was admitted into the ion source at a pushing pressure of 0.2 Torr and was excited to Kr** by impact of 50-eV electrons. The principal quantum number of Kr** was estimated to be in the range of 25–40 by use of field ionization [6]. In the EI ionization, the energy of electrons was varied in the range of 1–10 eV. The energy spread (fwhw) was determined to be about 1 eV with reference to the cross section for electron attachment to SF_6.

Cluster anions thus produced were mass-analyzed and detected by a Ceratron with an ion conversion dynode. The signal was counted by a multichannel scaler based on a microcomputer. The mass-to-charge ratios of the detected anions were calibrated against those of $(CO_2)_n^-$ observed simultaneously.

3. Results and Discussion

3–1. Acrylonitrile clusters
In the collisional electron transfer from Kr** to acrylonitrile (AN) clusters, the following cluster anions were observed;

(a)	$(AN)_n^-$	with $n \geq 1$,
(b)	$[(AN)_n–H]^-$	with $n \geq 2$,
(c)	$[(AN)_n–H_2]^-$	with $n \geq 2$,
(d)	$[(AN)_n–H_2–HCN]^-$	with $n \geq 3$, etc.,

where, for example, $[(AN)_n–H]^-$ denotes the cluster anion produced by abstracting one hydrogen atom from $(AN)_n^-$. The population of each ion was obtained by normalizing the ion intensity with the total intensity of the cluster anions observed. The populations of ion (a)–(d) thus obtained are plotted against the cluster size, n, in Fig. 1. The populations

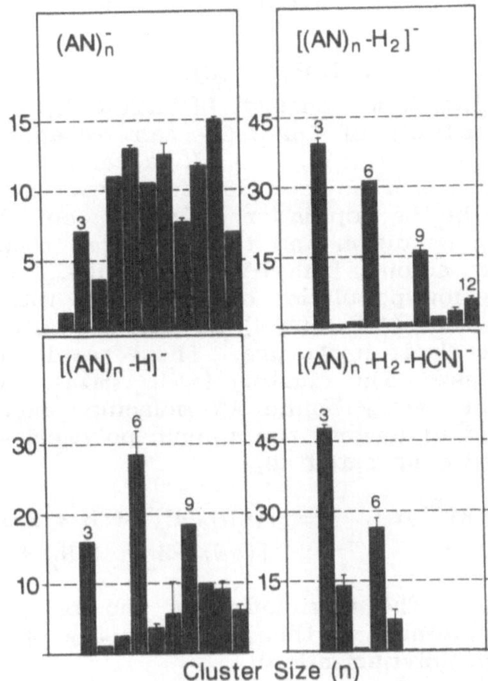

Fig. 1 The size distributions of the cluster anions, $(AN)_n^-$ and $[(AN)_n-Y]^-$ (Y=H, H₂, H₂+HCN), produced by collisional electron transfer from Kr** to acrylonitrile (AN) clusters.

of ion (b)–(d) with n=3k are prominent and otherwise very small. Ion (b)–(d) with n<3 is scarcely populated. Because the trimer anions are the smallest ions observed, at least three AN molecules in (AN)ₙ must participate in the formation of ion (b)–(d). Further, in the electron attachment to a deuterated cluster, $(AN-d_1)_n$, cluster anions, such as $(AN-d_1)_3^-$, $[(AN-d_1)_3-D]^-$, $[(AN-d_1)_3-D_2]^-$, and $[(AN-d_1)_3-D_2-HCN]^-$, were observed in the region where trimer ions are observed. This result indicates that chemical bonds are formed between two or more constituent molecules in the formation of ion (c) and (d), because the deuterium atoms are abstracted from different component molecules. In the EI ionization, the populations of ion (b)–(d) with only n=3 increase as the electron energy increases, whereas ions with n≠3k remain scarcely populated. The populations of ion (b)–(d) with n<3 do not change with the electron energy up to 10 eV. Conceivably, strong bonds, such as covalent bonds, are formed within the trimer anions because they are not fragmented to smaller cluster anions by introduction of much energy. The consideration of the energetics also supports of the intracluster polymerization. If a single molecule captures an electron in (AN)₃, the heat of reaction for the process,

$$(AN)_3 + Kr^{**} \longrightarrow [AN-H]^-(AN)_2 + H + Kr^+,$$

is given by

$$\Delta H \approx D[CH_2CCN-H] - EA[CH_2=C\cdot CN],$$

where the bond dissociation energy, $D[CH_2CCN-H]$, is reported to be 4 eV and the electron affinity of $CH_2=C\cdot CN$ can be approximated by that of cyanomethyl radical (1.507 eV). As ΔH is calculated to be +2.5 eV, the process is endothermic if no intracluster polymerization proceeds.

An increase in the population of larger neutral clusters by changing the stagnation condition and the seed gas resulted in the enrichment of the trimer anions, $[(AN)_3-Y]^-$ (Y=H, H$_2$). This finding together with the extremely low population of ion (b)-(d) with n=4 and 5 leads us to the conclusion that $(AN)_m$ (m≥4) as well as $(AN)_3$ give rise to these trimer ions by the electron capture. These results can be explained by the following process: The cluster, $(AN)_m$ (m>4), consisting of a trimer unit and additional weakly bound AN molecules captures an electron in the trimer unit and undergoes the anionic polymerization. The additional AN molecules are evaporated as,

$$(AN)_3(AN)_k + Kr^{**}/e \longrightarrow [(AN)_3-H]^- + H + k\cdot AN + Kr^+,$$
$$[(AN)_3-H_2]^- + H_2 + k\cdot AN + Kr^+,$$

where k = 1, 2,.... The termination of the polymerization within the trimer unit is attributable to the ring geometry of the unit (structure-constrained anionic polymerization).

3-2. 2-Chloroacrylonitrile and methacrylonitrile clusters

In the electron attachment to 2-chloroacrylonitrile (CAN) and methacrylonitrile (MAN) clusters, the following cluster anions were produced:

(e) $(CAN)_n^-$		with n≥1,
(f) $(MAN)_n^-$		with n≥2,
(g) $[(CAN)_n-n'\cdot HCl]^-$		with n≥2,
(h) $[(MAN)_n-n'\cdot HCN]^-$		with n=3-4.

The size distributions of the observed anions are shown in Figs. 2(A) and (B). Two or more constituent molecules should take part in the formation of ion (g) and (h) because more than two HCl and HCN molecules are abstracted from the cluster system in the ionization. This finding leads us to conclude that ion (g) and (h) are produced via intracluster anionic polymerization following the electron attachment. If no intracluster reaction, such as polymerization, occurred, the process for the formation of $[(CAN)_2-HCl]^-$ from $(CAN)_2$ would be endothermic by +1.5 eV. Thus, the intracluster polymerization should occur even when one HCl and HCN molecule are released from the cluster after the electron attachment. The abstraction of HCl and HCN from CAN^- or MAN^- is considered to be either endothermic; this conjecture is supported by no observation of $[CAN-HCl]^-$ and $[MAN-HCN]^-$ in the mass spectra. It is

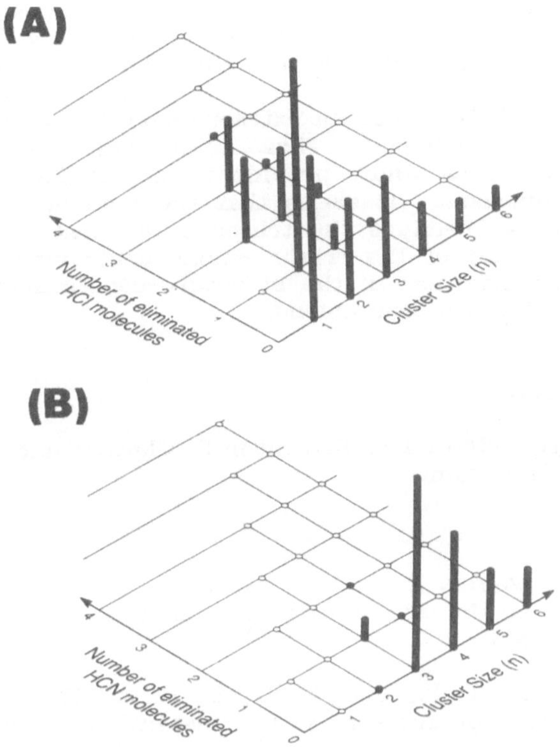

Fig. 2   The size distributions   of the cluster anions, $[(CAN)_n-n'\cdot HCl]^-$ and $[(MAN)_n-n'\cdot HCN]^-$ produced in the collision of Kr^{**} with the 2-chloroacrylonitrile (CAN) clusters (panel (A)) and the methacrylonitrile (MAN) clusters (panel (B)).

likely that an excess energy generated in the polymerization is dissipated by liberating HCl and HCN molecules from the nascent polymer anions and that the product ion (g) and (h) have polyene structures. Similar thermal degradation of the CAN and MAN polymers occurs by releasing HCl and HCN molecules, respectively [7, 8].

The dependence of the $(MAN)_n^-$ (n=2–4) intensities on the stagnation pressure was almost the same as those of $(MAN)_n^+$ (n=2–4), respectively. The agreement of these stagnation pressure dependences implies that the extensive evaporation is unlikely in the formation of intact cluster anions, (e) and (f), so that the electron is simply captured in a constituent molecule or molecules in them. Thus, the intensity ratio of ion (g)and (h) to the corresponding intact anions, (e) and (f), respectively, can be used as a measure for the branching fraction of the polymerization in a given cluster. In the ionization of $(CAN)_n$, ion (g) is produced in a comparable amount with the intact ion (ion (e)), whereas in the ionization of $(MAN)_n$, ion (h) is not produced efficiently (see Figs. 2 (A) and (B)). These findings are interpreted in such a manner that the intracluster polymerization is enhanced by introduction of electron–

withdrawing group, Cl, while the reaction is suppressed by an electron-releasing group, CH3. This substituent effect on the polymerization seems to favor the anionic mechanism for the intracluster polymerization of $(CH_2=CXCN)_m$.

Ion (g), $[(MAN)_3-HCN]^-$, has an extraordinary large abundance, as observed in the AN trimer anion. These trimer anions are probably 6-membered cyclic structures which provide more stability than polymer anions with the other sizes. It seems that the intracluster polymerization terminates so as to produce the trimer or at most tetramer anions. The particular abundance of the trimer anions could originate from the dynamics of the polymerization in the structure-constrained environment as well as stability of the trimer anions.

Acknowledgements

The authors are indebted to Professor R. Okazaki and Dr. T. Nagata for their valuable discussion.

References

[1] Coolbaugh, M. T., Peifer, W. R., and Garvey, J. F. (1990) 'Observation of a magic number in the ion distribution of ethene clusters', Chem. Phys. Lett., 168, 337-344.

[2] Morita, H., Freitas, J. E., and El-Sayed, M. A. (1991) 'Laser ionic multiphoton dissociation of some acrylate clusters', J. Phys. Chem., 95, 1664-1667.

[3] El-Shall, M. S. and Marks, C. (1991) 'Cationic polymerization within gas-phase clusters of isoprene', J. Phys. Chem., 95, 4932-4935.

[4] Tsukuda, T. and Kondow, T. (1991) 'Structure-constrained anionic polymerization in hydrogen bonded acrylonitrile clusters', J. Chem. Phys., in press.

[5] Mitsuke, K., Kondow, T., and Kuchitsu, K. (1986) 'Formation of negative cluster ions in collision of SF_6 clusters with krypton Rydberg atoms', J. Phys. Chem., 90, 1552-1556.

[6] Kondow, T. (1987) 'Ionization of clusters in collision with high-Rydberg rare gas atoms', J. Phys. Chem., 91, 1307-1316.

[7] Grassie, N. and Grant, E. M., (1967) 'Thermal degradation of poly(α-chloroacrylonitrile)', J. Polym. Sci., C, 591-599.

[8] Bell, J. W. and Mulchandani, R. K. (1965) 'Thermal degradation of courtelle acrylic fibers. I. chemical changes induced by heat', J. Soc. Dyers Colourists, 81, 16-21.

FORMATION OF NEGATIVE CLUSTER IONS BY ELECTRON CAPTURE.

A.A.VOSTRIKOV, D.Yu.DUBOV, and I.V.SAMOILOV
Institute of Thermophysics
Siberian Branch of the USSR Acad.Sci.
Novosibirsk 630090
U.S.S.R.

ABSTRACT. The formation of long-lived negative ions has been studied by the method of crossing beams of $(N_2)_N$, $(CO_2)_N$, $(N_2O)_N$, and $(H_2O)_N$ clusters and electrons. Clustered beams are formed from a central stream-line of a free jet. The absolute cross sections for the negative ion formation σ^- in dependence on electron energy E_e and mean cluser size $\bar{N}$ in the range $0 < E_e < 100$ eV and $10 < \bar{N} < 2 \cdot 10^3$ were received. The non-dissociative attachment of electron at E_e 0 eV was observed for all clusters; and also for $(CO_2)_N$, $(N_2O)_N$, $(H_2O)_N$ clusters the presence of resonance peaks of dissociative attachment was detected.

It was found for $(N_2)_N$ clusters, that long-lived negative ions appear not only at $E_e \approx 0$ eV, but also at those values of energy E_e, at which in the experiments on electron scattering the formation of short-lived N_2^- ions was observed. Especially strong attachment was found at $E_e = 2$ eV. The appearance of resonance peaks on curves $\sigma^-(E_e)$ for N_2 may be explained by the fact, that the form and the position of potential curves of N_2^- ion in cluster change in such a way, that the electron affinity of molecule becomes positive. Simultaneously it is supposed, that the delay of non-stable states of negative ions N_2, CO_2, N_2O, and H_2O, related with energy losses by an electron, may be accompanied with electron capture by a group of molecules in the cluster.

1.INTRODUCTON

Cluster anions have an important role in atmospheric, plasma- and gasdynamic, and biophysical processes. Systematic studies of cluster anion formation in crossing cluster and electron beams began about ten year ago [1,2]. Cluster beams are formed from a central stream-line of a supersonic jet. At definite gas pressure P_0 and temperature T_0 in nozzle source, due to adiabatic gas cooling, in the jet the condensation initiates, i.e. clusters are forming [3].

This paper presents the results of molecular beam studies of long-lived negative ion formation at the collision of low-energy electrons $(0 < E_e$, eV $< 100)$ with CO_2, N_2O, H_2O, and N_2 clusters, containing from two to several thousand molecules per cluster. The study was carried out

P. Jena et al. (eds.), Physics and Chemistry of Finite Systems: From Clusters to Crystals, Vol. II, 1153–1158.
© 1992 *Kluwer Academic Publishers.*

by the crossing beam method. The absolute cross sections for the negative ion formation were obtained in dependence on electron energy E_e and mean cluster size $\bar{N}$.

2. EXPERIMENTAL

The description of the setup and diagnostical methods are given in [4]. Gases and superheated vapour were coming through a sonic nozzle with diameter $d_* = 1$ mm or 0.3 mm in the case of N_2 expansion. The special attention was paid for nitrogen cleaning from admixtures of water and CO_2.

The cross section measuring technique for negative cluster formation is described in [5]. Electron beam was formed from an oxide cathod. The control of electron energy in a beam was carried out on the peak of resonance electron attachment to SF_6 molecule and by the retarding potential technique. Working with CO_2, N_2O, and H_2O we had the FWHM of electron energy in a beam, ΔE_e, $\lesssim 0.3$ eV; working with N_2, it was $\Delta E_e \lesssim 1$ eV. Ions were registrated by a Faraday cup collector in the regime of cluster beam modulation by a mechanic interrupter. Time of ion moving in a beam to collector was of 10^{-4} s order. Mean cluster size and full cluster size distribution function were measured by the retarding potential technique.

Current I^- and intensity J are related with specific cross section of negative ion formation (per one molecule in a cluster) σ_o^- as: $I^- = i \cdot L \cdot \sigma_o^-(\bar{N}, E_e) \cdot \sum_{N=2}^{\infty} n_N$, where i – electron current, L – the length of interaction of electrons with cluster beam, n_N – density of clusters with size N.

The beam intensity J, is equal to $A \cdot v \cdot \bar{N} \cdot \sum_{N=2}^{\infty} n_N$, where A – constant, v – particle velocity in a beam. Normalizing ion current I^- on J, we receive for specific cross section σ_o^-

$$\sigma_o^-(\bar{N}, E_e) = A \cdot v/(L \cdot i) \cdot (I^-/J) \tag{1}$$

Since the cross section of electron attachment to isolated CO_2, N_2O, and H_2O molecules is known [6], that dividing I^- on J in the region of P_o, where the condensation in a jet is absent (to the left from the arrow in [5]) and making this value equal to the known cross section of attachment to isolated molecule σ_1^-, we receive [5]:

$$\sigma_o^-(\bar{N}, E_e) = \sigma_1^- \cdot I^-(\bar{N}, E_e)/J(\bar{N}) \cdot J(1)/I^-(1, E_e) \quad . \tag{2}$$

3. RESULTS AND DISCUSSION

Figure 1 shows the cross sections for negative ion formation in dependence on electron energy E_e for clusters of investigated substances (mean cluster sizes $\bar{N}$ are marked near the curves) and for molecules ($N = 1$). To reveal the peculiarities of behaviour of these dependences , the curves $\sigma^-(E_e)$ are converged in their maxima at $E_e \approx 0$ eV for CO_2 and H_2O clusters, and for N_2O clusters – on the most abrupt region at $E_e = 3$ eV. For N_2 clusters the cross sections are given in absolute units. This figure presents also the dependences of the cross sections for dissociative attachment of electron to isolated molecules (in relative units) [6]. These curves are marked by letter "a". Smaller peaks $\sigma^-(E_e)$ of dissocia-

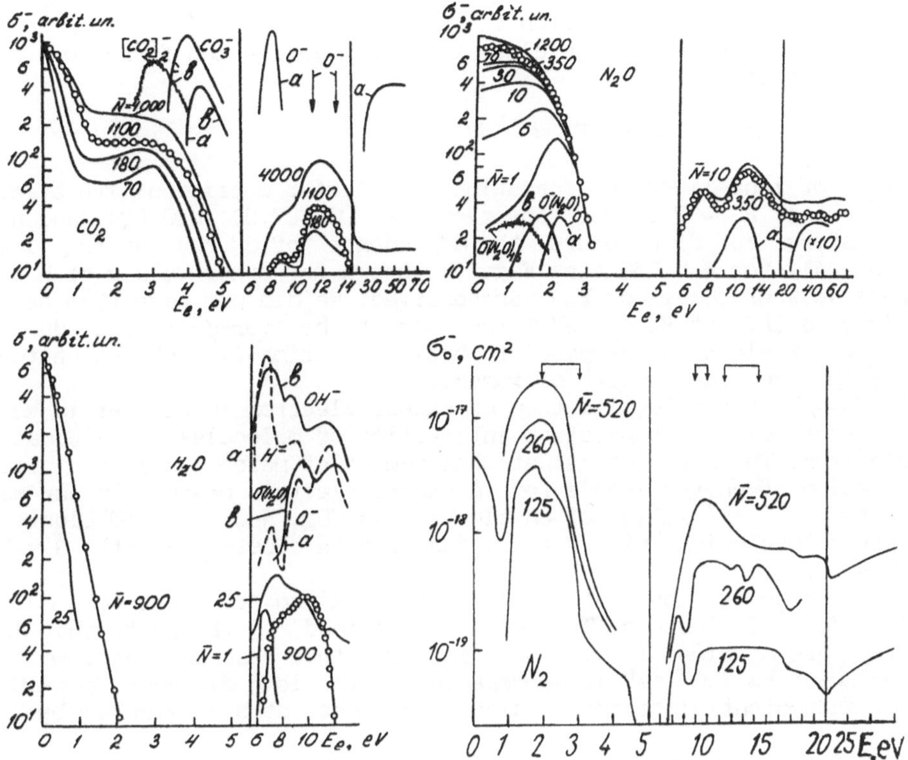

Figure 1. Cross sections for the formation of negative cluster ions versus electron energy E_e.

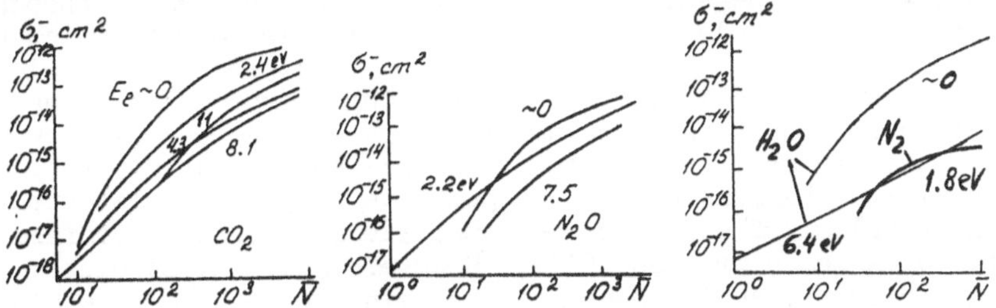

Figure 2. Cross sections for the formation of negative cluster ions versus mean cluster size $\bar{N}$.

tive attachment to isolated molecule CO_2 are shown in the figure by arrows. The figure also shows (in arbitrary units) the result of mass-spectrometric measurements of cluster ion composition forming in a cluster beam ($N < 30$ molecules) by electron capture [1,2]. These curves are marked by letter "b".

Figure 2 presents the dependences of total cross sections for negative ion formation $\sigma^- = \bar{N} \cdot \sigma_o^-$ for investigated clusters on mean cluster size $\bar{N}$ at some values of electron energy E_e.

Let us discuss general peculiarities of negative ion formation, revealed in this study.

3.1. Electron attachment at $E_e \approx 0$ eV.

The effect of long-lived ion formation at cluster collision with electrons at energy $E_e \approx 0$ eV was observed for CO_2 [7,5], N_2O [8], and H_2O [8,9] clusters. For N_2 clusters the dependence of $\sigma^-(E_e)$ at $E_e < 1$ eV is given only for $\bar{N} = 260$ (see Fig.1), though for other values of $\bar{N}$ the same behaviour of curves $\sigma^-(E_e)$ was observed. We did not give them because values $\sigma^-(E_e)$ at $E_e \approx 1$ eV turned out to be over-decreased due to large width of electron energy distribution function (~ 1 eV) which took place at the experiments with nitrogen.

At $E_e \approx 0$ eV the localization of excess electron in cluster is seen to be the result of cooperative interaction of molecules in a cluster with electron. The question is: at what number of molecules in a cluster the formation of energy levels for an excess electron begins. In our measurements the localization of an electron in CO_2, N_2O, and H_2O clusters is observed, beginning with $\bar{N} = 10$, and for N_2 clusters - with $\bar{N} = 25$ molecules.

Figure 2 shows, that at $E_e \approx 0$ eV the maximum cross section σ^- is revealed for H_2O clusters. This may be explained by the fact, that H_2O molecules are dipoles. It was shown in [10], that resonance capture of electron with $E_e < 0.1$ eV is accompanied with long-distance potental capture, and potential capture begins to dominate at $N > 36$ molecules.

3.2. Negative ion formation at $E_e > U_i$ - ionization energy.

Figure 1 shows that for CO_2, N_2O, and N_2 clusters the negative ion output depending weakly on energy E_e, are observed. For CO_2 and N_2O molecules the existence of negative ion current at $E_e > U_i$ has a threshold character and may be explained by reaction of polar dissociation [6]: $AB + e^- \longrightarrow AB^{**} \longrightarrow A^- + B^+ + e^-$. To realize this process in a cluster an ion of one sign should leave the cluster. Our measurements of negative ion distribution function, forming at $E_e > U_i$ and $E_e \approx 0$ eV show that the cluster size are practically the same. This means that separation of charges in cluster happens mainly by positive fragment outlet from the cluster. The results of our mass-spectrometric measurements confirm the possibility of such a process [11]. However, polar dissociation is not the only process, and perhaps not the main one for negative cluster ion formation at $E_e > U_i$. At least for N_2 clusters the possibility of this process is not evident.

3.3. Negative ion formation at the energy E_e, corresponding to dissociative attachment of an electron to molecules.

It is known that in two-body collision of CO_2 molecule with electron the O^- ions are mainly formed. The most intensive peaks of $\sigma^-(E_e)$ are observed at the energy $E_e = 4.3$ and 8.2 eV, less intensive at $E_e = 11.3$ and 13 eV [6]. At collision of electrons with clusters ($N < 30$) in the peaks of dissociative attachment the ions $O^-(CO_2)_{N>0}$, $O_2^-(CO_2)_{N>0}$ and $(CO_2)_N^-{}_{<25}$ are detected (see Fig.1) [1,2]. At $N > 6$ the number of $(CO_2)_N^-$ ions is observed to become the main ions. In [10] we suppose that the formation of $(CO_2)_N^-$ is related with electron localization in a cluster by a group of molecules after autodetachment with large energy losses by an electron. Figure 2 shows that the cross section of electron capture at the energy $E_e < 0.3$ eV by clusters with the size $\bar{N} > 10$ is more than a cross section of dissociative attachment. The shift of the first peak $\sigma^-(E_e)$ is determined by the decrease of dissociative attachment energy, approximately on the value of O^- ion solvatation in cluster.

Let us analyze the peculiarities of negative ion formation in N_2O clusters. The molecule of N_2O possesses positive electron affinity ($\lesssim$ 0.3 eV). However, neutral N_2O and ion N_2O^- have different geometry. This prevents the formation of ion N_2O^- in two-body collision, but leads to the strong dependence of attachment cross section on vibration excitation of molecule [6]. In cluster molecules are not vibrationally excited, however the geometry of N_2O may become more similar to the geometry of N_2O^- ion. A superposition of resonance peaks for dissociative attachment and direct "cooperative" capture at $E_e \approx 0$ eV gave the pattern of $\sigma^-(E_e)$ presented at $E_e < 3$ eV in Fig.1.

We observed the existence of autodetached state of N_2O^- with peak in $\sigma^-(E_e)$ at $E_e = 7.5$ eV. In the case of isolated molecule the decay of this state may lead to the formation of NO^- ion. However, because of smallness of electron affinity (~ 0.03 eV [6]) this ion is not observed in the gas phase. In the case of clusters, detachment of an electron from NO^- ion leads with high probability to $(N_2O)_N^-$ formation.

For isolated molecules of H_2O, it was found that in the dissociative attachment with maximum cross section σ^- at $E_e = 6.4$ and 8.8 eV ion H^- forms, and $E_e = 11.4$ eV $-$ O^- ion. In cluster in these peaks $\sigma^-(E_e)$ ions OH^- are mainly formed [2]. Their formation is explained in [2] through ion-molecular reactions of ions H^- and O^- with molecules H_2O in cluster. It is seen in Fig.1, that at transition from isolated molecule to cluster with increase of $\bar{N}$ the more rapid growth of the cross section σ^- at $E_e = 11.4$ eV occurs. Probably it is related with the fact, that formation of H^- ion in a cluster is accompanied by essential transformation of cluster structure, because the hydrogen bonds of molecules H_2O in clusters are characterized by redistribution of negative charge on oxigen atom.

As far as N_2 molecules are concerned, the following facts are known [6]. Long-lived ions N_2^- and N^- do not appear at the collision of electron with N_2 molecule. However, in the experiments on scattering of electrons, a whole serie of short-lived negative ions $N_2^-{}^*$ was registrated; parent molecules for these ions are electronically excited states of N_2 molecule and N_2^+ ion. For example, the formation of $N_2^-{}^*$ according to the form resonance was observed at excitation of N_2 into states $B^3\Pi_g$, $A^3\Sigma_u^+$, $C^3\Pi_u$, and also into ground and excited states of N_2^+ ion. But energy level of a low state of $N_2^-{}^*$ ion is 1.4-2 eV higher than the initial state. Electron capture according to electronically excited Feshbach resonance, which appear at excitation of N_2 into the state $E^3\Sigma_g$ at energy $E_e = 11.5$ eV, corresponds to the positive electron affinity about 0.4 eV. We should note the existence of resonance states of $N_2^-{}^*$ ion in the energy area $E_e = 20\text{-}24$ eV [6]. The ground state of N_2^- ion is appro-

1158

ximately 2 eV higher than for the ground state of N_2 molecule. The lifetime of this state is order 10^{-15} s.

In Fig.1 for N_2 all values of energy E_e, where resonance states N_2^{-*}
are observed, are marked by vertical arrows. It is seen, that in clusters at these values of energy E_e we observe long-lived N_2^- ions. Probably at transition of N_2 molecules into the condensed state in clusters
the form and the position of potential curves are changing in such a way,
that the resonance of the form transfers into the Feshbach resonances.
By this, the unfavourable overlap of the potential curves in the Franck-
Condon region leads, finally, to the long-lived N_2^- ion.

For the ground state of N_2^- ion the shift of potential energy curve
causes the formation of vibrationally excited Feshbach resonance with
several stable vibration levels, situated below the ground state of N_2
molecule (outside the Franck-Condon region), similar to the behaviour of
O_2^- ion [6]. Dissipation of vibrational energy of N_2^- ion among many inner degrees of freedom in a cluster prevents the autodetachment of electron.

4. REFERENCES

1. Klots, C.E. and Compton, R.N. (1978) 'Electron attachment to van
   der Waals polymers of carbon dioxide and nitrous oxide', J.Chem.
   Phys. 69, 1636-1643.
2. Klots, C.E. and Compton, R.N. (1978) 'Electron attachment to van
   der Waals polymers of water', J.Chem.Phys. 69, 1644-1646.
3. Vostrikov, A.A. and Dubov, D.Yu. (1991) 'Cluster generation in a
   free jet for molecular beam studies', in A.E.Beylich (ed.), Rare-
   fied Gas Dynamics, VCH, Weinheim, pp.1156-1163.
4. Vostrikov, A.A., Mironov, S.G., and Semyachkin, B.E. (1984) 'Investigation of nonequilibrium phenomena by molecular beam method',
   Fluid Mech.- Sov.Res. 11, 98-121.
5. Vostrikov, A.A. and Predtechenskiy, M.R. (1985) 'Interaction of
   electrons with CO_2 van der Waals clusters', Sov.Phys.-Tech.Phys.
   30, 529-534.
6. Massey, H. (1976) 'Negative ions', Cambridge University Press, Cambridge - London - New York - Melbourne.
7. Stamatovic, A., Leiter, K., Ritter, W., Stephan, K., and Maerk, T.D.
   (1985) 'Electron attachment to carbon dioxide clusters at very
   low electron energies', J.Chem.Phys. 83, 2942-2946.
8. Vostrikov, A.A., Dubov, D.Yu., and Predtechenskij, M.R. (1986)
   'Characteristics of the formation of negative H_2O, N_2O, and CO_2
   cluster ions', Sov.Phys.-Tech.Phys.(USA) 31, 825-826.
9. Haberland, H., Ludewight, C., Schindler, H.-G., and Worsnop, D.R.
   (1984) 'Experimental observation of the negatively charged water
   dimer and other small $(H_2O)_n$ clusters', J.Chem.Phys. 81, 3742-
   - 3744.
10. Vostrikov, A.A., Dubov, D.Yu., and Predtechenskij, M.R. (1987)
    'Water clusters: electron attachment, ionization, electrification
    upon dissociation', Sov.Phys.-Tech.Phys.(USA) 32, 459-466.
11. Vostrikov, A.A., Gilyova, V.P., and Dubov, D.Yu. (1991) 'Cluster
    size effect on electron-induced luminescence', Z.Phys.D - Atoms,
    Molecules and Clusters 20, 205-208.

SCANNING TUNNELING MICROSCOPY STUDY OF THE INTERACTION BETWEEN ADSORBED CLUSTERS AND GRAPHITE SUBSTRATES

J. Xhie, K. Sattler, M. Ge and N. Venkateswaran
Department of Physics and Astronomy
University of Hawaii at Manoa
Honolulu, Hawaii 96822, USA

In previous studies we have observed ten different superstructures on graphite near platinum clusters using scanning tunneling microscopy. Additional types of such structures were recently found for clusters of cobalt as well. The occurrence of the structures was independent of the size and shape of the clusters. They have the same periodicity as for platinum clusters and can be described by periodic charge-density modulations (PCDM) on the graphite lattice. The PCDM may be explained by surface stress produced by the clusters.

Introduction

Graphite is a popular substrate for use with the scanning tunneling microscope (STM). It can be easily cleaved to give atomically flat surfaces, exhibiting a perfect lattice over large areas. In the presence of defects such as steps, holes and adsorbed particles, superstructures of graphite have been observed frequently in the surrounding regions[1-7]. It is believed that these superstructures are a result of a surface charge-density redistribution due to perturbations by defects.

In a previous paper, we reported ten types of superstructures observed on graphite near adsorbed platinum particles and proposed a model of periodic charge-density modulation (PCDM)[1]. We ascribed the structures to a superposition of the graphite lattice and a localized periodic charge-density modulation induced by the adsorbed platinum particles. The PCDM components have a period of 1.5a (a = 0.245 nm, the graphite lattice constant) and a preferred orientation with 30° rotation relative to the underlying graphite lattice. The superstructures can be produced by superimposing up to three sets of the PCDM components in three symmetry directions on the graphite lattice.

In this paper, we report four types of new superstructures found recently on graphite near cobalt clusters. We show that they have the same periodicity and orientation as in the case of platinum and can also be described by the PCDM model.

P. Jena et al. (eds.), Physics and Chemistry of Finite Systems: From Clusters to Crystals, Vol. II, 1159–1163.
© 1992 *Kluwer Academic Publishers.*

Experiment

The samples were prepared in a high vacuum (10^{-7} Torr) chamber by vapor deposition of cobalt on freshly cleaved highly-oriented pyrolytic graphite substrates. After evaporation, the samples were kept in vacuum overnight with vacuum pumps turned off to allow slow oxidization before exposition to air. This process has been shown to be critical in forming stable and conducting cobalt particles on the graphite surface[8]. Analysis by x-ray photoelectron spectroscopy (XPS) showed a typical 5% coverage of a cobalt monolayer on the graphite surface.

The sample was then transferred to a scanning tunneling microscope operated at atmospheric conditions[9]. At room temperature, the STM images were taken in constant current mode where the tip-sample distances were kept constant and the variations of the z motion of the tip were recorded. The tunneling current was 1.5 nA at a bias voltage of 80 mV. Oxidized cobalt clusters sizing from 1 nm to 6 nm were observed to be randomly distributed on the surface of graphite.

Results and Discussion

Superstructures of graphite were found near small (1 nm - 3 nm) oxidized clusters of cobalt. Although most of them are of the same types as those observed previously on graphite near adsorbed platinum particles, additional structures were found for cobalt samples.

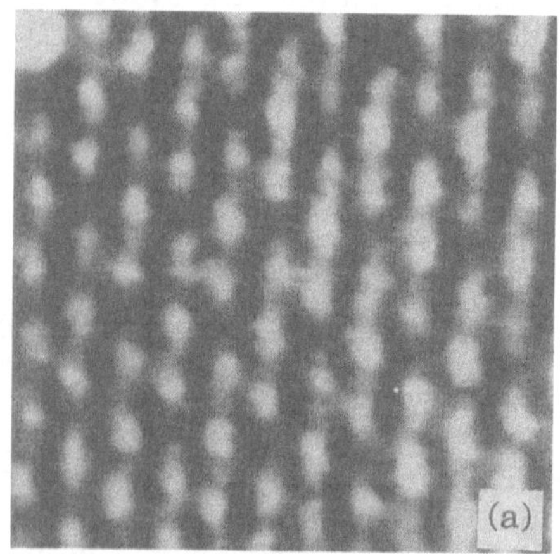

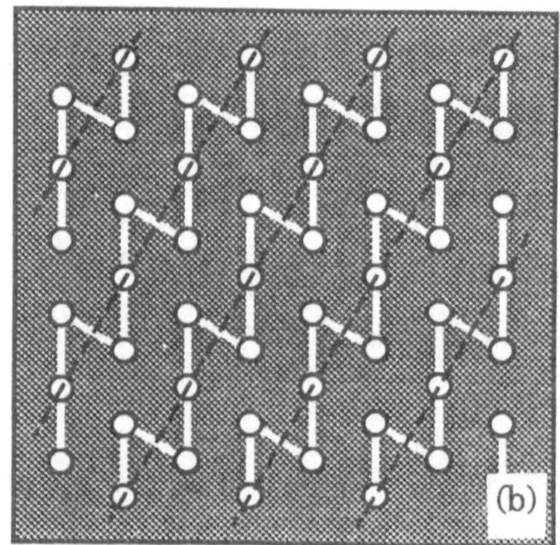

Figure 1. (a) STM image of a superstructure with a PCDM superimposing on the triangular graphite lattice. (b) Schematic model corresponding to the superstructure in (a). The circles represent the graphite lattice and the dashed lines the maxima of the PCDM components.

Figure 1(a) shows a STM image of a superstructure on graphite near an oxidized cobalt cluster. Figure 1(b) is the corresponding schematic model of the structure, where the circles represent the graphite triangular β-sites[10] and the dashed lines the maxima of the PCDM component. The superstructure is formed by superimposing a PCDM component on the graphite triangular lattice. The PCDM has a period of 1.5a and orient in a direction rotated 30° with respect to the graphite lattice.

Figure 2(a) displays another superstructure where the underlying graphite lattice is honeycomb, showing both the α-sites and the β-sites. This honeycomb lattice is modulated by a PCDM component shown as dashed lines in Figure 2(b).

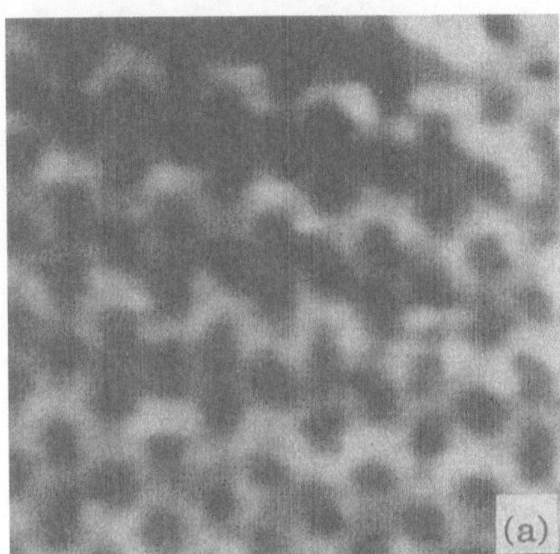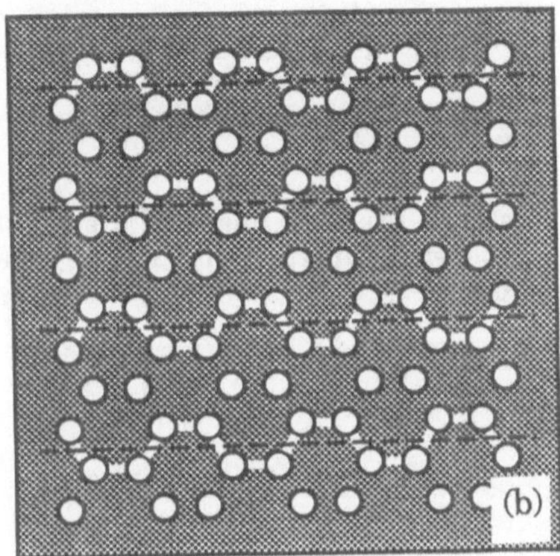

Figure 2. (a) STM image of a superstructure with the PCDM superimposing on the honeycomb graphite lattice. (b) Schematic model.

Figure 3 illustrates a superstructure where the underlying honeycomb lattice is modulated by three sets of PCDM components, each having a period of 1.5a rotated 30° relative to the graphite and 60° relative to each other. As a result, the graphite atoms located at the maxima of the PCDMs are highlighted as bright spots forming a large honeycomb structure.

A more complicated structure combining two superstructures is displayed in Figure 4(a). A large triangular lattice is formed from the bright spots at the bonds between a α-site and a neighboring β-site. In between are less bright chains of connected hexagons. With the PCDM model, this structure can be produced by superimposing three sets of PCDM components on the honeycomb graphite lattice. As illustrated in Figure 4(b), two sets of the PCDMs have their vertices on the highlighted bonds forming the large triangular lattice. The third set of the PCDM component shown in horizontal dashed lines has a smaller intensity and therefore highlights the less bright hexagon chains. This suggests that the three PCDM components can have different intensities and phases and act rather independently.

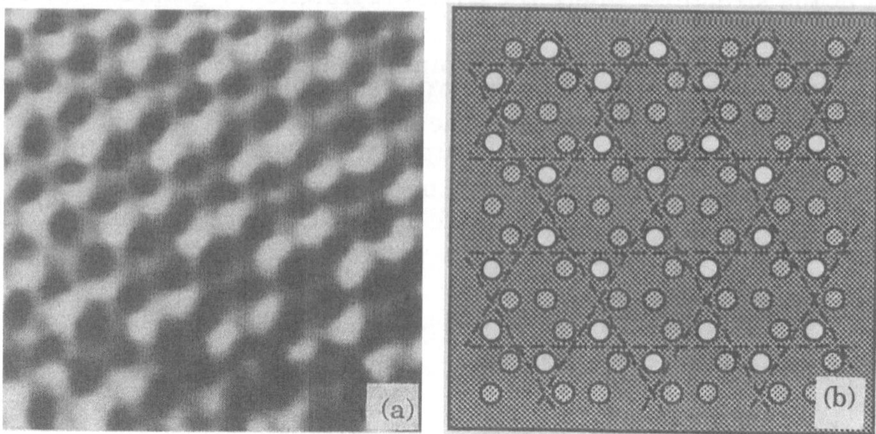

Figure 3. (a) STM image of a superstructure with three sets of the PCDM components superimposing on the honeycomb graphite lattice. (b) Schematic model.

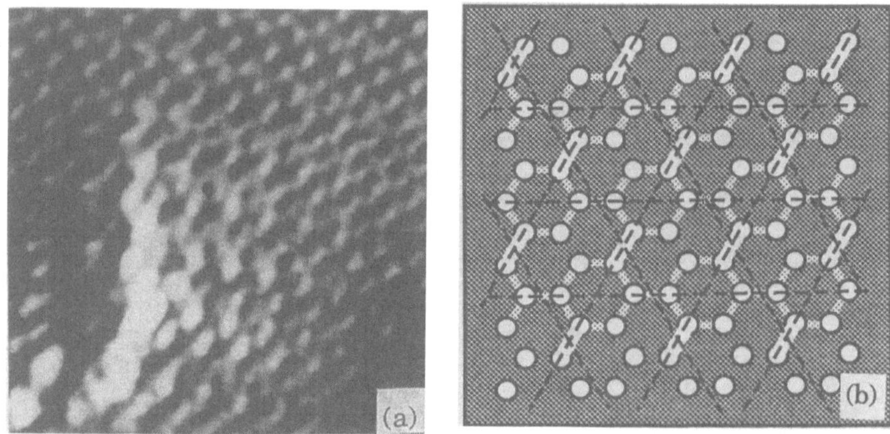

Figure 4. (a). STM image of a superstructure with three sets of the PCDM components superimposing on the honeycomb graphite lattice. (b) Schematic model.

Although the effect of the PCDMs on larger adsorbed clusters is insignificant, the atoms of small clusters have a tendency to be bound to the maxima of the PCDMs. The dimer in the lower left corner of the Figure 4(a) shows an example of such an effect.

The physical origin of the PCDMs is not well understood. Surface stress may be a possible reason. The stress exerted by the adsorbed clusters may induce a distortion of the Fermi surface and, as a consequence, a charge-density redistribution. This redistribution, especially its preferred periodicity and orientation needs to be explained theoretically.

Conclusion

New types of superstructures have been found near oxidized cobalt clusters on graphite. They have the same periodicity and orientation as those observed previously and can be explained by the PCDM model. The superstructures are produced by superimposing up to three sets of PCDM components on the graphite lattice. They have a period of 1.5a and are oriented in three symmetry directions, each rotated 30° relative to graphite.

Acknowledgements

We are grateful to A. Rizzetti, D. Yamamoto and W. Pong, who prepared some of the samples. Financial support from the Office of Technology Transfer and Economic Development (OTTED) and from the Research Council of the University of Hawaii is gratefully acknowledged.

References

1 J. Xhie, K. Sattler, U. Müller, N. Venkateswaran, and G. Raina, Phys. Rev. B **43**, 8917 (1991).
2 Physics Today **41**, 129 (1988).
3 T. R. Albrecht, H. A. Mizes, J. Nogami, S.-I. Park and C. F. Quate, Appl. Phys. Lett. **52**, 362 (1988).
4 J. P. Rabe, M. Sano, D. Batchelder, and A. A. Kalatchev, J. of Microsc. (Oxford) **152**, 573 (1988).
5 H. A. Mizes and J. S. Foster, Science **244**. 559 (1989).
6 E. D. Ganz, "Scanning Tunneling Microscopy of Metals on Graphite", Ph.D. thesis, University of California at Berkeley (1988).
7 H. A. Mizes and W. A. Harrison, J. Vac. Sci. Tech. A.**6**, 300 (1988).
8 A. Rizzetti, J. Xhie, K. Sattler, D. Yamamoto, and W. Pong, "XPS and STM Studies of Oxidized Cobalt Particles on Graphite", J. of Electron Spectrosc. Relat. Phenom. (in print).
9 Digital Instruments, Inc., Santa Barbara, CA, USA.
10 D. Tomanek, S. Louie, H. J. Mamin, D. W. Abraham, R. E. Thomson, E. Ganz, and J. Clarke, Phys. Rev. B **35**, 7790 (1987).

MAGIC NUMBERS OF METAL AND METAL ALLOY CLUSTERS AND THEIR CHEMICAL REACTIVITY

Y. YAMADA[1] AND A. W. CASTLEMAN, JR.
Department of Chemistry,
Pennsylvania State University,
University Park, PA 16802

ABSTRACT. Pure metal and metal alloy clusters such as Cu_n, Ag_n, Cu_nAg_m, Cu_nAl_m, Cu_nIn_m, Ag_nAl_m, and Ag_nIn_m are produced by a gas aggregation source. Ionized clusters show magic numbers at jellium shell closings. From the reaction product analysis of the clusters, it is shown that the neutral clusters corresponding to jellium shell closing are particularly unreactive.

1. Introduction

Currently there is considerable interest in the electronic nature of metal clusters and their influence on mass spectral abundances in cluster distributions commonly referred to as magic numbers. In the case of simple free electron systems, these are often observed to obey the jellium shell model [1] in terms of their influence on the mass spectral abundances of particular species. In recent work, we have focused our attention on an investigation of metals beyond the simple alkali metal type, but still on ones that are expected to display appreciable free electron behavior. The purpose of the present investigation is to ascertain the nature of the magic numbers in these systems, with attention to alloys among species whose individual components correspond to the electronic shell closings predicted by the jellium model.

2. Experimental

A gas aggregation source is used in the present work to form species of selected composition [2]. The pure metal sample or metal mixture is heated in a crucible and evaporated into a flow of He or Ar gas. In the flow reactor tube, neutral metal clusters are reacted with reactant gas to investigate the influence of cluster size and electronic character on chemical reactions of the evolving species. The clusters are ionized by a KrF excimer laser (4.98 eV) and detected using a time-of-flight mass spectrometer system for product identification and analysis.

---

[1] Japan Atomic Energy Research Institute, Tokai-mura, Naka-gun, Ibaraki-ken, JAPAN

P. Jena et al. (eds.), *Physics and Chemistry of Finite Systems: From Clusters to Crystals, Vol. II*, 1165–1169.
© 1992 *Kluwer Academic Publishers.*

3. Results and discussion

3.1. MAGIC NUMBERS IN THE METAL AND METAL ALLOY SYSTEM

Recently, the major focus of attention has been on copper, silver, and on these species in combination with each other and with other metal alloying partners. Initial experiments were done to investigate magic numbers in the individual pure metal systems such as Cu_n and Ag_n to elucidate their comparison with the predictions of the jellium shell model. Magic numbers reflecting electronic structure in coinage metal clusters were first observed in mass spectra of sputtered charged clusters [3]. In the spectra of Cu_n^+ (Fig. 1.), smaller clusters have higher intensities and show exponential like distributions up to Cu_n; n=65. Besides the general distributions, we can clearly see even-odd alternations in the smaller mass range, and magic numbers are observed at certain size. These magic numbers reveal that ionization and fragmentation results in the ionized species which for the single electron systems display magic numbers, for example at Cu_n^+, n=9 and 21. The mass spectra of Ag_n^+ show the same pattern as Cu_n^+ clusters, displaying the same magic numbers. The larger magic numbers predicted from the jellium shell closing model are not clearly observed in the larger mass range. Magic numbers become evident when cluster stability plays a role in either the mechanism of formation or in the dynamics of dissociation following ionization and our observations suggest that neither of these process are important for larger clusters in the present study.

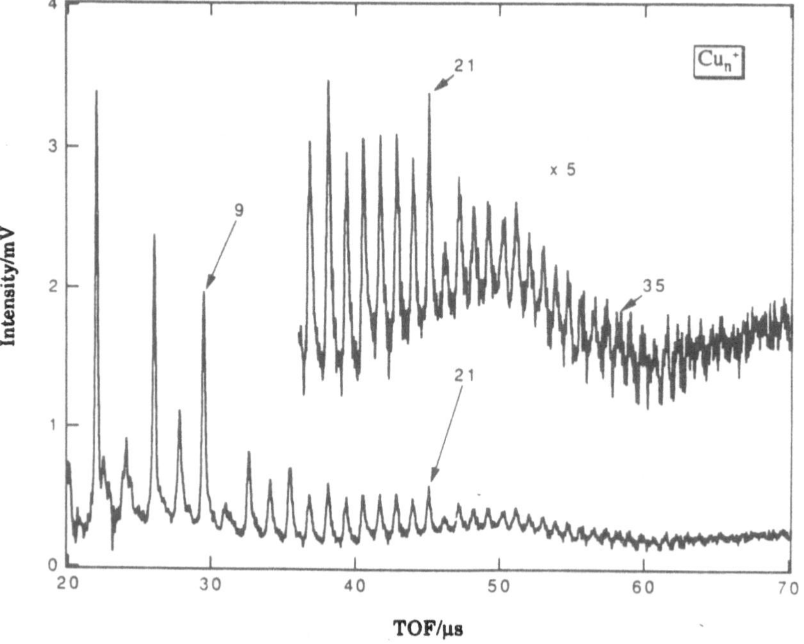

Fig. 1. Time-of-flight spectrum of Cu_n^+ clusters.

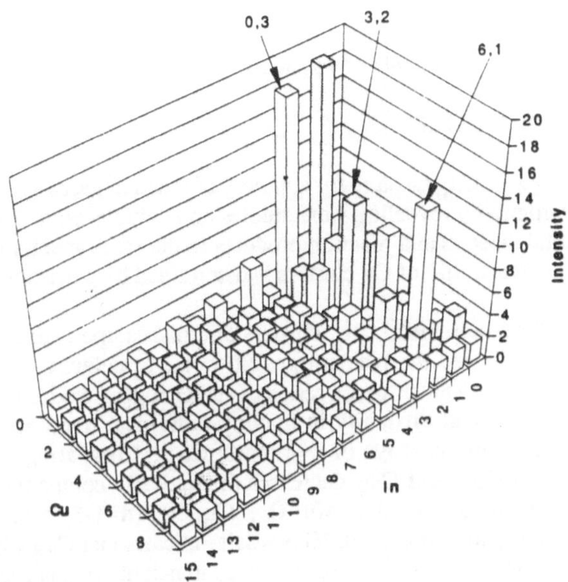

Fig. 2. Signal intensity of $Cu_nIn_m^+$ clusters.

The magic numbers are observed at (n,m)=(0,3),(3,2),(6,1).

We measured the alloys of these systems $Cu_nAg_m^+$ to investigate the influence the substitution of the other metal. Cu and Ag mix easily to form alloy clusters with various mixture ratios observed in a mass spectrum. Similar to the general tendency of the pure Cu_n^+ or Ag_n^+ clusters, smaller clusters have high intensity in this alloy system. Counting the total number of valence electrons in a cluster n+m; Cu_nAg_m, even-odd alternation is also observed. The clusters with even number of valence electrons n+m-1 have higher intensity than neighboring clusters with odd number of electrons. The magic numbers are clearly seen where n+m has the same value as those of the pure components with shell closings; n+m-1=8,20.

It is interesting to investigate the alloy clusters composed with metals having different valence electron to determine the importance of the electronic contribution to stability [4]. We employed Al and In which is normally a trivalent metal, to form alloy clusters with Cu or Ag. We vaporized the component metals in a single crucible and observed the spectra with various mixture ratios employed in different experiments. In the case of the Cu_nAl_m and Cu_nIn_m system, clusters with various ratios were observed in the same spectrum. Counting the total number of valence electron in a cluster n+3m-1; Cu_nAl_m, we can see even-odd alternation and magic numbers at n+3m-1=8,20. Similar to the Cu_nAl_m system, the Cu_nIn_m system displays the same pattern (Fig. 2.). In the spectra of the Ag_nAl_m system, these metals do not mix very well and the only signal we can observe is Ag_nAl where only one Al atom attaches to Ag_n clusters. In this series, a magic number is clearly observed at Ag_6Al which has 8 valence electrons in a cluster. As the difference between the mass of Ag and In are small, peaks in observed mass spectra overlap each other and it is difficult to measure the intensity of the individual clusters, but still we can see the even-odd alternation.

It is demonstrated that the jellium shell model is a good guide to the mass spectral abundances of the cations of these metal and metal alloy systems.

3.2. REACTIVITY OF CLUSTERS

We observed reactions of Cu_n clusters for a variety of reactant gases such as O_2, H_2S, CO, CO_2, HCl, and HBr. Particularly revealing were studies on reactions with O_2 and H_2S which show both unreacted metal clusters and reaction products in the same spectra after the reaction. It was reported that chemical stability of Cu clusters is determined by electronic and geometric structure [5].

The intensities of the pure metal Cu_n clusters diminishes upon adding the O_2 gas to the flow tube (Fig. 3.). We adjusted the reactant gas flow rate to detect the small amount of reaction products. When we add too much gas, a series of stable copper oxides are formed which leads to extensive mass overlaps and difficulty in making individual mass assignments. The reaction products were observed in the range of n=4 to 12; Cu_n. Comparing the intensity of pure Cu_n peaks, the intensity of Cu_7 and Cu_9 decreases dramatically compared to the intensity of Cu_8. Reaction products were observed at Cu_7O_n, Cu_9O_n (n=1-3) while no reaction products corresponding to Cu_8O_n were observed. This result indicates that Cu_8 which is stable as predicted from shell closing model is very unreactive compared to neighboring clusters; Cu_7 and Cu_9.

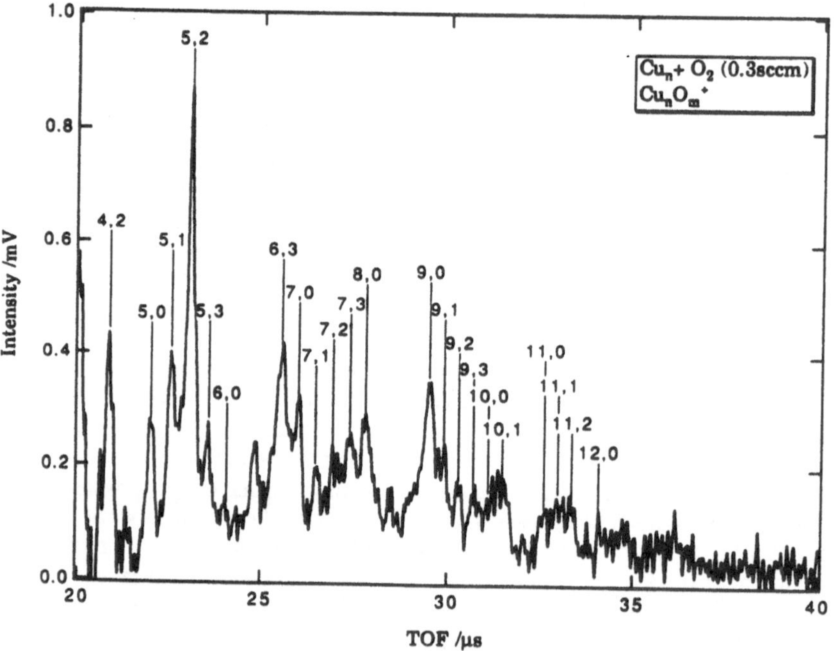

Fig. 3. Time-of-flight spectrum of Cu_n^+ clusters and their reaction products; 0.3 sccm of O_2. Indicated numbers correspond to Cu_nO_m; n, m.

Besides the O_2 gas, H_2S was also reactive with Cu_n clusters and reaction products such as Cu_nS were observed. The peak intensity of the Cu_8 decreases less compared to neighboring clusters Cu_7, Cu_9 in the spectra after reaction.

In these experiments, neutral clusters are ionized with nonresonant ionization process and the spectra may contain substantial fragmentation [6]. Nevertheless, the difference of intensities after reaction gives information on reactivities. Neutral precursors of clusters corresponding to jellium shell closings were found to be particularly unreactive compared to clusters of neighboring size. This is an especially exciting and interesting finding in view of the expectation that chemical reactions are dramatically influenced by electronic effects, particularly shell closing.

Acknowledgement

Financial support by the U.S. Department of Energy, Grant No. DE-FGO2-88-ER60668, is gratefully acknowledged.

References

1. W. D. Knight, K. Clemenger,W. de Heer, W. Saunders, M. Chou, M. Cohen, *Phys. Rev. Lett.*, **5 2**, 2141 (1984).
2. R. W. Farley, P. Ziemann, and A. W. Castleman, Jr., *Z. Phys. D.-Atoms, Molecules and Clusters* , **1 4**, 353 (1989).
3. I. Katakuse, T. Ichihara, Y. Fujita, T. Matsuo, T. Sakurai, and H. Matsuda, *Int. J. Mass Spectrom. Ion Proc.*, **6 7**, 229 (1985).
4. M. M. Kappes, *Chem. Rev.*, 88, 382 (1988).
5. B. J. Winter, E. K. Parks, and S. J. Riley, *J. Chem. Phys.*, **9 4**, 15 (1991).
6. D. E. Powers, S. G. Hansen, M. E. Geusic, D. L. Michalopoulos, and R. E. Smalley, *J. Chem. Phys.*, **7 8**, 2866 (1983).

SCANNING TUNNELING MICROSCOPY OF TRIS-(1,10-PHENANTHROLINE) RUTHENIUM (II) CHLORIDE ON GRAPHITE, COPPER AND GALLIUM ARSENIDE

K.C. YUNG, T.M. VESS, AND M.L. MYRICK

Environmental Sciences Division
Lawrence Livermore National Laboratories
P.O. Box 808, L-524
Livermore, CA 94550

ABSTRACT.   An inorganic complex, $[Ru(1,10\text{-phenanthroline})_3]Cl_2$ (RP3), was applied to highly ordered pyrolytic graphite (HOPG), copper and n-doped GaAs.   Scanning tunneling microscopy (STM) investigations of the complex on three substrates yielded direct evidence of the molecular layer on Cu and GaAs, but not on HOPG.   The results suggest that substrates can be chosen to "tune" observations of non-conductive ultrasmall moieties with STM.

1. Introduction

Several probe microscopies have been employed in recent years to image sorbed molecules. These probe microscopies have in common a sharp tip that is traced across a surface with some form of feedback information (e.g., tunneling current) used to conform to surface topography. Of these microscopies, scanning tunneling microscopy (STM) appears best suited to the study of electronic phenomena because it monitors the conduction of electrons by materials deposited on a surface. In general, molecular species are non-conductive. However, numerous STM observations of molecules support the existence of conduction mechanisms through molecules on the time scale of the tunneling event (assumed to be near 10^{-16} sec). A resonance tunneling mechanism has been suggested for this molecular conduction.[1-2] However, little work has been performed aimed at taking advantage of this phenomenon to image molecules. Beyond simple imaging, however, resonance tunneling is the basis of research into some new types of semiconductor devices (e.g., resonance tunneling diodes and lasers), and could form the basis of electronic devices operating with single-molecule-scale switching. This article presents indirect experimental evidence in support of resonance tunneling processes through molecules in the STM tunneling gap.

2. Experimental

2.1   SUBSTRATE PREPARATION

An HOPG monochromator was obtained from Union Carbide. It was attached to an aluminum platten with silver paste, and cleaved for measurements by peeling off the top layers of graphite with adhesive tape. $[Ru(phen)_3]Cl_2$ (RP3, phen = 1,10-phenanthroline) was added to the surface by drying a methanolic solution.

P. Jena et al. (eds.), Physics and Chemistry of Finite Systems: From Clusters to Crystals, Vol. II, 1171–1176.
© 1992 *Kluwer Academic Publishers.*

Gold colloidal particles were obtained from E•Y Industries. The colloids used were 0.1% w/w Au, with 20 Å Au particles. They were deposited on HOPG by drying drops of the colloidal suspension in methanol on the graphite block. After exposure to RP3 in solution, the colloids were centrifuged from suspension and resuspended in fresh methanol three times to remove excess RP3. Drops of the Au colloid with surface-adsorbed RP3 were then added to the graphite surface as with the uncoated colloid.

Copper plates were prepared by polishing to an optical finish using a final polish of 0.3 μm grade Imperial Lapping Film. RP3 was added to the surface by drying, as for HOPG.

GaAs samples were also mechanically polished. These samples were heavily n-doped for higher conductivity. RP3 was added to the surfaces of the GaAs samples by drying, as above.

RP3 was obtained from Alfa Chemical Company and used without further purification. Solutions were 10-4 M, corresponding to an average surface coverage of approximately 50 molecular layers. The drops left no visible residue except at the edges; given the concentration of residue at the edges, the coverage in the central sampling region of the drops was likely significantly less than 50 molecular layers.

2.2   MICROSCOPE

The microscope used for these measurements was constructed at LLNL, and consisted of a single-tube scanning head with a platinum-rhodium tip and an electronics package operated by an LSI-11 microcomputer. This system has been described elsewhere.[3]

2.3   RESONANCE RAMAN

Resonance Raman scattering measurements were performed on the samples as prepared for STM. These measurements were performed with the 457.9-nm line of an Ar^+ laser, and detected with a $N_2(l)$-cooled CCD array (Princeton Instruments). The front half of a SPEX Model 1401 double monochromator was used for detection, with a Physical Optics, Inc. hoographic filter to reject the Rayleigh component of scattering.

3.   Results

Resonance Raman measurements performed on each of the substrates (except the Au-RP3 complex) showed vibrational frequencies consistent with those found in single-crystal RP3. These data support a micro-crystalline or near-crystalline formation of RP3 on each of the surfaces, in which most RP3 molecules are surrounded by similar molecules, rather than bound to the surface or strongly perturbed by the presence of the surface.

Images obtained on HOPG showed no evidence for the presence of RP3, regardless of tunneling bias voltage, current, of estimated depth of coverage.

Au particles were used in an effort to bind RP3 molecules to the surface more tightly. It is known from surface-enhanced Raman measurements that RP3 binds tightly to gold. Images of the HOPG covered 20 Å Au colloidal particles obtained *before* complexation with RP3 clearly showed evidence of the Au particles, indicating that these particles are massive enough to resist being pushed aside during imaging. These particles typically aggregated into clumps, with each particle well defined.

Following complexation with RP3, no Au particles were found in repeated scanning. Evidence for intermittent contact was present in the form of numerous anomalous current spikes during the scan, frequently associated with a specific region of the surface area being studied. This is likely an indication that the particles were present, but that the RP3 coating of the particles prevented good electrical conduction through them.

The copper plates provided the first clear evidence of RP3 in these studies. The surface morphology of the coated copper was distinct from that of uncoated Cu prepared the same way. Uncoated Cu showed residual scratches several hundred angstroms wide from the polishing process in long parallel rows. RP3 on Cu appeared as ill-defined cloudy deposits. The best resolution obtained was 100 Å; molecular resolution was not approached. However, the new coating did have distinct physical properties from the uncoated Cu. Uncoated Cu was impervious to a mild voltage pulse (4 V, 2 μsec). However, the coating was pierced by this voltage pulse, to an apparent depth near 200 Å, consistent with the depth of the overlayer.

On GaAs, the layer was also immediately obvious from a change in surface morphology. In this case, the surface of the polished GaAs was also covered with residual scratches, but the chemical debris removal performed following the polishing process had edtched all the sharper edges, so that no sharp feature was observable. The overlayer had a clear microcrystalline form, evidencing sharp edges and striations indicative of a "growth direction". Again, the uncoated GaAs was not perturbed by the 4V, 2 μsec pulse used above. However, the overlayer was pierced immediately by this pulse, again to a depth of near 200 Å, consistent with the likely thickness of the overlayer.

Examination of the overlayer on GaAs indicated that the best resolution obtainable for this layer was considerably better than 100 Å. It was frequently possible to observe features on the 10 Å scale or finer. As an example, Figure 1 shows a scan area of 80 X 80 Å, with what appear to be individual 10-Å spheres corresponding to individual molecules.

4.   Discussion

Molecules are difficult to image with STM. Most imaging is done on flat substrates such as highly ordered pyrolytic graphite (HOPG) and single-crystal supports. This is because STM is incapable at present of clearly delineating what topographic features are due to molecules and which are not. Very flat substrates have been viewed as necessary for the identification of molecules on the assumption that surface extrema will be molecular in origin. This limits the number of substrates that are applicable to imaging in air with the STM, because most metals and semiconductors form rough oxide layers. This has meant that the substrate of choice for most molecular imaging in air has been HOPG, with some reports of gold-coated supports.

The major difficulty with obtaining images of molecules on these substrates is that molecules are frequently swept from the field of view by the STM tip. The Fermi energy of HOPG and gold are both near 5 eV from vacuum; most molecules do not have orbitals at or near this energy. This is readily shown by considering that the reversible redox potentials of most molecules are not within one volt of the standard hydrogen potential, corresponding to approximately -4.6 eV vs. vacuum.[4] Hence, one would expect that, if resonance tunneling is the dominant mechanism by which molecules conduct in STM, most molecules would not assist the passage of electrons very well on these substrates.

In this study, three substrates are used with distinctly different Fermi energies. Figure 2 shows how these energies compare to the energies of the orbitals in RP3. For GaAs, the energy

Figure 1

Energy vs. Vacuum
[Ru(phen)3]2+ orbitals
LUMO
-3.4 eV
-4.1 eV (GaAs)0
-4.6 eV (Cu)0
-5.0 eV (HOPG)0
-5.1 eV (Au)0
HOMO
-6.2 eV

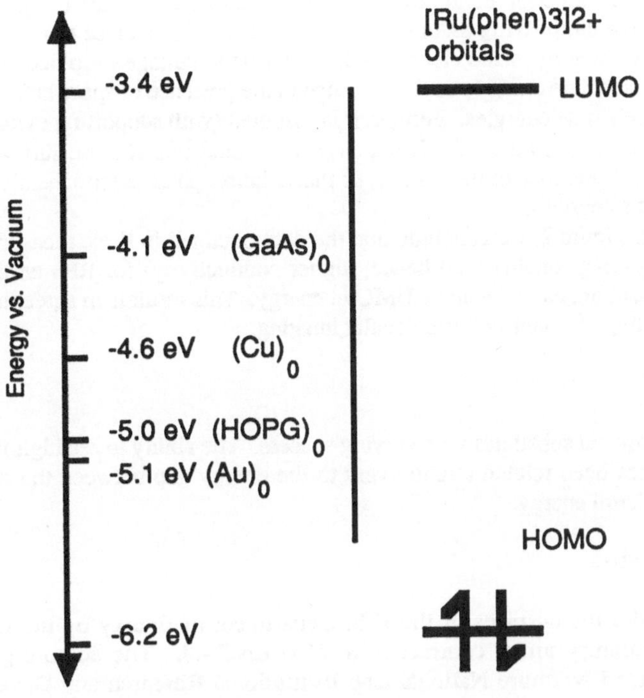

Figure 2

given is the energy of the conduction band edge at the surface in the absence of space charge effects.[5] In each case, the simplifying assumption is made that RP3 is close to the surface of these conductors that they share the same electric field effects (the voltage drop occurs over the vacuum gap), so that their absolute energy positions change in the presence of space charge, but not the relative positions of their state energies. Further, it is assumed (with supporting evidence from surface-enhanced Raman scattering) that RP3 is not chemically modified at these surfaces, and is not strongly perturbed by the presence of the surface, so that solution-phase electrochemistry can be used to estimate bound state energies.

Comparing results to Figure 2, we conclude that the data obtained in these measurements is consistent with improving image quality (and hence, higher conductivity) for RP3 on substrates with Fermi energies approaching its molecular LUMO in energy. This result is in agreement with a dominant resonance tunneling mechanism for molecular imaging.

5.   Conclusions

RP3 has been imaged on several substrates with varying success. The ability to see high-resolution images of this complex has been related qualitatively to the energy gap between the molecular LUMO and the substrate Fermi energy.

6.   Acknowledgements

Work was performed under the auspices of the U.S. Department of Energy by the Lawrence Livermore National Laboratory under contract # W-7405-ENG-48. The authors gratefully acknowledge the Lawrence Livermore National Lab Institutional Research and Development (IR&D) fund and Lynn Anspaugh of the Environmental Sciences Division for supporting this research. The authors would like to express their appreciation to Dr. Raymond Mariella of the Electrical Engineering Department for supplying GaAs wafers, and to Drs. Wigbert Siekhaus of the Chemistry and Material Sciences Division and Rodney Balhorn of the Biomedical Division at LLNL for the use of their STMs.

7.   References

1.   Lindsay, S.; Sankey, O.; Li, Y.; Herbst, C.; Rupprecht, A. (1990) "Pressure and Resonance Effects in Scanning Tunneling Microscopy of Molecular Adsorbates" J. Phys. Chem. 94, 4655-60.
2.   Yuan, J.; Shao, Z. (1990) "Simple Model of Image Formation by Scanning Tunneling Microscopy of Non-Conducting Materials" Ultramicroscopy 34, 223-6.
3.   Beebe, T., Wilson, T., Ogletree, D., Katz, J., Balhorn, R., Salmeron, M., Siekhaus, W. (1989) "Direct Observaton of Native DNA Structures with the Scanning Tunneling Microscope" Sicnece 243, 370-2.
4.   Lohmann, F. (1967) "Fermi-Niveau und Flachbandpotential von Molekulkristallen Aromatischer Kohlenwasserstoffe" Z. Naturforsch., Teil A, 22, 843-4.
5.   Gobelli, G.; Allen, F. (1965) "Photoelectric Properties of Cleaved GaAs, GaSb, InAs, and InSb Surfaces; Comparison with Si and Ge" Phys.Rev. 137, A245-54.

CLUSTER QUANTUM-CHEMICAL STUDY OF DIHYDROGEN, METHANE AND WATER MOLECULES: INTERACTIONS WITH PURE AND LITHIUM DOPED MAGNESIUM OXIDE

N.U. ZHANPEISOV and G.M. ZHIDOMIROV
Institute of Catalysis
Novosibirsk 630090
USSR

ABSTRACT. In the framework of supermolecular approach using MINDO/3 method the various channels of dissociative chemisorption of dihydrogen, methane and water molecules on pure and lithium doped magnesium oxide surfaces are considered. On the basis of obtained calculation results the possible ways of these molecules activation on MgO and Li/MgO , the mechanism of destructive interaction of water with magnesia, the nature of surface hydroxyl groups and some other problems are discussed.

1. Introduction

Chemical activity of magnesium oxide in catalytic reactions is often connected with low - coordinated magnesium and oxygen ions (Mg_{LC}^{2+} and O_{LC}^{2-}) of various surface irregularities – faces, edges, corners, etc. [1,2]. In spite of the large quantity of the experimental and theoretical investigations in this field up to now there are no satisfactory coordinated ideas about the molecular structure, relative concentration and the mechanism of catalytic action of these centres. Undoubtly, the quantum – chemical investigation will play an important role in the development of these ideas. The present paper when studying the reactions of dissociative chemisorption of dihydrogen, methane and water molecules on MgO and Li/MgO discusses the semiempirical quantum-chemical MINDO/3 method which is especially parameterized for studying these systems. On the basis of obtained calculation results the possible ways of these molecules activation on MgO and Li/MgO , the mechanism of destructive interaction of water with magnesia, the nature of surface hydroxyl groups and some other problems are discussed.

P. Jena et al. (eds.), Physics and Chemistry of Finite Systems: From Clusters to Crystals, Vol. II, 1177–1182.
© 1992 *Kluwer Academic Publishers.*

2.   Calculation method and surface model

The quantum-chemical cluster calculations were performed in the framework of MINDO/3 method whose parameterization was extended for studying Li and Mg containing compounds [3,4]. Magnesium oxide surface was modelled by the clusters of Mg_4O_4 , Mg_9O_9 , $Mg_{12}O_{12}$, $Mg_{18}O_{18}$ and $Mg_{32}O_{32}$. As it was shown [4] the growth of cluster size does not lead to the large change of geometry and charge state of the similar surface parts – the corners, the edges, the faces. Note that in the framework of such supermolecular approach the character of the charge changes on the Mg_{LC}^{2+} and O_{LC}^{2-} ions with their coordination number variations, the relaxation of various Mg_{LC}^{2+} and O_{LC}^{2-} ions and their influence on energetics of the considered chemisorptional processes are correctly described [4]. In the calculations full optimization of geometry of the chemisorption complexes was carried out. The calculations of molecular clusters corresponding to Li/MgO promoted by lithium were carried out by isomorphic substitution of one of Mg_{LC}^{2+} ion on Li_{LC}^{+} cation with further compensation of the excessive charge by H^+ connected with O_{3C}^{2-}. In calculations Mg_8O_8LiOH and $Mg_{11}O_{11}LiOH$ clusters were used. Here also full optimization of local structure of this substitution was performed. Calculations of the O-D stretching vibrational frequencies ν_{OD} of various surface OD-groups on MgO are carried out in the framework of a harmonic oscillator approach in the locality of the minimum of total energy by a parabola at the optimized distance of $R_{OH} \pm 0.005 \ A^O$.

3. Results and discussions

3.1. DISSOCIATIVE CHEMISORPTION OF DIHYDROGEN , METHANE AND WATER MOLECULES ON PURE MAGNESIUM OXIDE

It is a generally accepted idea that at the interaction with the dehydroxylated MgO surface many molecules chemisorb dissociatively [1,2,4]. Taking this into account various forms of dissociative chemisorption of these molecules on MgO are studied. Calculation results of dissociative chemisorption energy of dihydrogen (methane) molecules on various pairs of AC of Mg_9O_9, $Mg_{12}O_{12}$ and $Mg_{32}O_{32}$ are given in Table 1 and for water one – in Table 2

TABLE 1. Calculated MINDO/3 dissociative chemisorption energy (ΔE, kcal/mol) of dihydrogen (methane) molecules on various pairs of AC of pure magnesium oxide.

AC pair	$\Delta E,^*$ kcal/mol		
	Mg_{3C}^{2+}	Mg_{4C}^{2+}	Mg_{5C}^{2+}
O_{3C}^{2-}	20 (15)	4 (− 3)	−11 (− 12)
O_{4C}^{2-}	− 7 (− 10)	− 23 (− 30)	−38 (− 35)
O_{5C}^{2-}	− 22 (− 27)	− 39 (− 47)	−53 (− 56)

*Sign (+) corresponds to stabilization under chemisorption.

We note that the transition from Mg_9O_9 cluster to more extended ones practically does not influence on the structural and energetical characteristics of chemisorption complexes realized on the similar surface parts. As it was expected with increasing LC both magnesium cations and oxygen anions the energetic effect of dissociative chemisorption decreases. Maximum gain in chemisorption energy takes place when two O_{3C}^{2-} and Mg_{3C}^{2+} ions act as the centres of localization of dissociated molecule fragments. For methane this is the only state of heterolytic adsorption with the gain of energy (Table 1).

TABLE 2. Calculated MINDO/3 dissociative chemisorption energy (ΔE, kcal/mol) of water molecule on MgO.

AC pair	Mg_{3C}^{2+}	Mg_{4C}^{2+}	Mg_{5C}^{2+}
O_{3C}^{2-}	50	31	−
O_{4C}^{2-}	25	6	− 6
O_{5C}^{2-}	−	− 11	− 23

Evidently, the quantity of such pair centres on the surface must be very small. Adsorption on the statistically preferable ion pairs $O_{3C}^{2-} - Mg_{4C}^{2+}$ or $O_{4C}^{2-} - Mg_{3C}^{2+}$ only in the case of H_2 leads to stabilization. Perhaps this is connected with the circumstances when the dissociative adsorption of CH_4 on the contrast to H_2 proceeds only at high temperatures. In contrast to heterolytic dissociation of dihydrogen and methane molecules the dissociative chemisorption of water molecule can proceed not only with participation of three-coordinated acid-base centres but also with the participation of four-coordinated AC. These results are in good agreement with the available experimental data [2,5].

3.2. DISSOCIATIVE CHEMISORPTION OF DIHYDROGEN AND METHANE MOLECULES ON LITHIUM DOPED MAGNESIUM OXIDE

It is worth noting that in our case we use the molecular model which by its nature is neutral but not radical. This approach is in contrast to Lunsford et al. [6] who proposed the existence of $[Li^+ - O^-]$ centres in lithium promoted Li/MgO . As it was shown below by the using such unradical molecular model for Li/MgO one can rather fully describe the observed sharp increasing of activity of Li/MgO in comparison with pure MgO. Calculation results of dissociative chemisorption energy of dihydrogen (methane) molecule on statistically more active pairs of acid-base centres are given in Table 3. In contrast to the analogous results for pure MgO surface (Table 1) dissociative chemisorption of both H_2 and CH_4 on Li/MgO can proceed

TABLE 3. Calculated MINDO/3 dissocative chemsorption energy (ΔE, kcal/mol) of dihydrogen (methane) molecule on Li/MgO.

AC pair	LC = 5	4	3
$O^{2-}_{3C} - Mg^{2+}_{4C}$	13 (8)	11 (12)	4 (5)
$O^{2-}_{4C} - Mg^{2+}_{3C}$	19 (17)	19 (18)	12 (9)
$O^{2-}_{4C} - Mg^{2+}_{4C}$	3 (− 9)	5 (−18)	1 (−15)

not only with participation of three − coordinated acid − base centres, but also with participation of a pair of four-coordinated AC (placed on the edges of oxide lattice). This result corresponds to numerous experimental data proving that promotion of magnesium oxide by lithium leads to sharp increase of activity of the first in comparison with non-promoted MgO. The qualitative picture of the activation of H_2 and CH_4 on Li/MgO is the same as for pure magnesium oxide.

3.3. MECHANISM OF DESTRUCTIVE INTERACTION OF WATER WITH MAGNESIA.

Now let us consider the mechanism of hydroxylation of magnesium oxide. As it was shown [2] that treatment of MgO in water vapours leads to a considerable growth of quantity of three – coordinated sites and keeping of a considerable amount of MgO powder in a fixed quantity of H_2O vapours does not lead to any changes in the specific surface area by BET method despite considerable changes in surface structure. These experimental results may be explained within the limits of the data obtained by us in the following way. Let us consider a pairs of Mg_{3C}^{2+} and O_{4C}^{2-}. For the first water molecule the dissociative chemisorption with the formation of the fragment Mg–OH and OH-group is possible. Hereafter, if the second water molecule is adsorbed on the same LAS but with participation of another neighbour basic O_{4C}^{2-} site, the embryo of a new phase – $Mg(OH)_2$ – is formed. If now this embryo is excluded with the formation of a vacancy in Mg_{3C}^{2+} this process turns to be profitable (the calculated value of energy of this process is 21 kcal/mol). Removal of the cations Mg_{4C}^{2+} and Mg_{5C}^{2+} from the edge and (001) face turns to be unprofitable by −24 and −74 kcal/mol respectively. This probably explains the growth of three-coordinated sites, but unlike [2], not the ionic pairs are removed primarily but basically the Mg_{3C}^{2+} ions. The growth of the specific surface area by BET method may be explained by the next decomposition of the formed new $Mg(OH)_2$ phase with isolation of MgO of smaller sizes.

3.4. NATURE OF THE SURFACE HYDROXYL GROUPS.

The surface hydroxyl groups play an important role in catalytic activity of magnesium oxide and can behave as AC for many catalytic reactions. In this connection the O–D stretching vibrational frequency ν_{OD} of various surface OD – group on MgO have been performed. These OD – groups differs from each other by the coordination degree of hydroxyl by metal atoms in first and by oxygen one in the second coordination spheres . Calculated MINDO/3 results of ν_{OD} corrected with using a scaling factor $f = 0.924$

1182

and a well - known correlation of $\nu_{OH} / \nu_{OD} = 1.35$ are the following: $Mg_{5C} - O_1D$ (2763), $Mg_{4C} - O_1D$ (2765), $Mg_{3C} - O_1D$ (2779), $O_{3C}D$ (2699), $O_{4C}D$ (2635), $O_{5C}D$ (2584). All ν_{OD} values in cm^{-1}. On the basis of these data we can drawn the following conclusions. First, in accordance with general ideas [7] the formation of coordination bond decreases the O-D stretching vibrational frequency, i.e. the OD- group in the more higher corrdination degree by metal atoms has the lowest band frequency. This is due to the fact that when increasing the coordination number of OD - group by Mg atoms its strength decreases and it leads to the lengthening of OH bond . The latter results in decreasing the electron density on OH bond. Second, the changing of the coordination number of Mg atom relatively weakly affects the ν_{O_1D} band frequency. Here in contrast to [8]

but in accordance to [9], the reverse dependence of ν_{O_1D} on coordnation number of Mg atom is observed, i.e. upon increasing the number of O atoms in the second coordination sphere the ν_{O_1D} band frequency decreases.

Accordingly, the OH bond order decreases and the corresponding OH bond length increases, the latter leading to increasing the electron density on OH bond.

4. References

1   Ito, T., Murakami, T. and Tokuda, T. J. Chem. Soc.,
    Faraday Trans. 1, **79** (1983) 913.
2   Stone, F.S., Garrone, E. and Zecchina, A. Mater. Chem.
    Phys., **13** (1985) 331.
3   Zhidomirov, G.M., Pelmenschikov, A.G., Zhanpeisov, N.U.
    and Grebenyuk, A.G. Kinet. Catal., **28** (1987) 86 (Russ.).
4   Zhanpeisov, N.U., Pelmenschikov A.G. and Zhidomirov, G.
    Kinet. Catal., **31** (1990) 563 (Russ.).
5   Jones, C.F., Reeve, R.A., Rigg, R., Segal, R.L., Smart,
    R.St.C. and Turner, P.S. J. Chem. Soc., Faraday Trans.1
    80 (1984) 2609.
6   Ito, T. and Lunsford, J.H. Nature , **314** (1985) 721.
7   Bellamy, L.J. Advances in IInfrared spectra of Complex
    Molecules, Moskva, Mir, 1971 (Russ.).
8   Shido, T., Asakura, K. and Iwasawa, Y. J. Chem. Soc.,
    Faraday Trans. 1, **85** (1989) 441.
9   Pelmenshchikov, A.G. and Zhidomirov, G.M. React. Kinet.
    Catal. Lett., **31** (1986) 85.

NANOSTRUCTURE FABRICATION USING BIOMOLECULAR TEMPLATES

K. Douglas, G. Devaud, M.K. Lyon, and N.A. Clark
University of Colorado
Department of Physics
Boulder, CO 80309-0390
USA

ABSTRACT. Metal structures with features on the 10nm length scale (nanostructures) have been fabricated in a parallel process employing two-dimensional monolayer protein crystals as patterning templates. These nanostructures have been assayed on thin film substrates using transmission electron microscopy (TEM) and on thick (bulk) substrates using atomic force microscopy (AFM). Compound structures (nanoheterostructures) have also been fabricated in which adsorbates selectively self-assemble onto a nanostructure. Such nanoheterostructures have been created using both nanometer scale clusters of metal atoms and biological macromolecules as selective adsorbates.

1. Introduction

The development of nanotechnology, i.e., materials and devices structured on the nanometer length scale is an important scientific and technical goal. Biomolecular crystals provide a means of fabricating nanostructures via parallel (as opposed to serial) processes. In our laboratory we have developed a process to fabricate nanometer-scale periodic arrays of holes in a metallic overlay using two-dimensional biomolecular crystals (S-layers) as templates (Figure 1) [1].

The technique of employing a crystalline surface as a patterning element for nanometer fabrication is motivated by its intrinsic parallelism. Mesoscopic crystalline monolayers are particularly attractive for nanometer scale fabrication because their periodicity provides a length scale amenable to structural study by a variety of established methods which represent a technology base for the development of new nanometer fabrication techniques. The present work is an extension of the metal decoration and shadowing methods employed in biomembrane structural investigations [2,3]. Additionally, the structural redundancy of periodic molecular arrays offers a significant advantage in that a single preparation yields many examples of the same process, enabling fluctuation effects, which will become of increasing importance as device size is reduced, to be effectively probed.

We have further demonstrated the fabrication of composite biomolecular/solid state and solid state/solid state heterostructures of nanometer dimension [4] (Figure 2). The site-specific incorporation of adsorbates onto nanostructured surfaces offers novel possibilities for applications requiring specific chemical, optical or electronic functions.

P. Jena et al. (eds.), Physics and Chemistry of Finite Systems: From Clusters to Crystals, Vol. II, 1183–1191.
© 1992 *Kluwer Academic Publishers.*

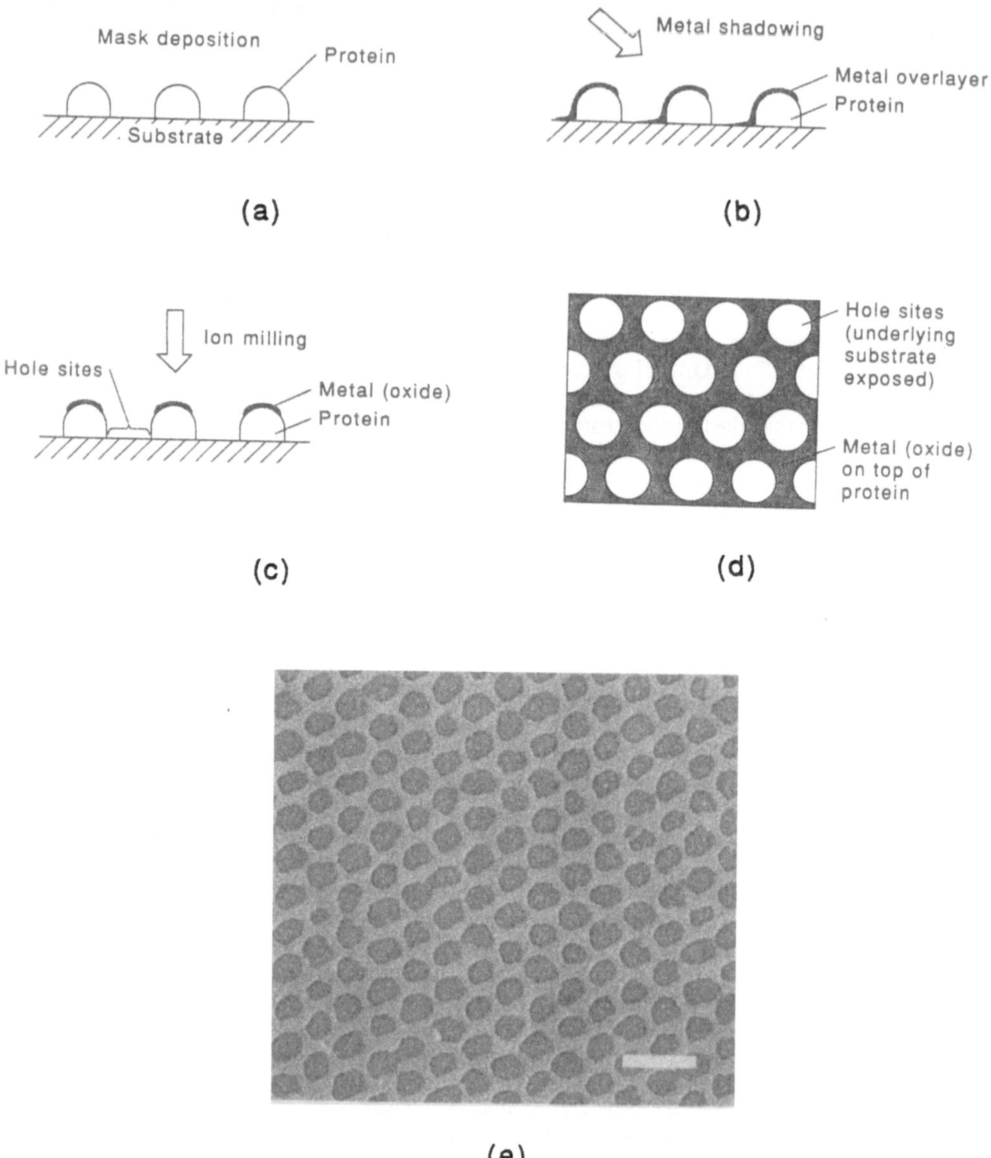

Figure 1: Processing steps to make nanostructures by templating operation. (a) Deposition of protein crystals onto substrate. (b) Shadow metallization of protein by electron beam vaporization of titanium (12 Å). (c) FAB milling to remove metal from substrate, leaving titanium coated protein. (d) Plan view of idealized nanostructure. (e) Image of metallized, milled nanostructure by transmission electron microscopy (TEM). (Template: *Sulfolobus solfataricus*.) (Scale bar = 40 nm).

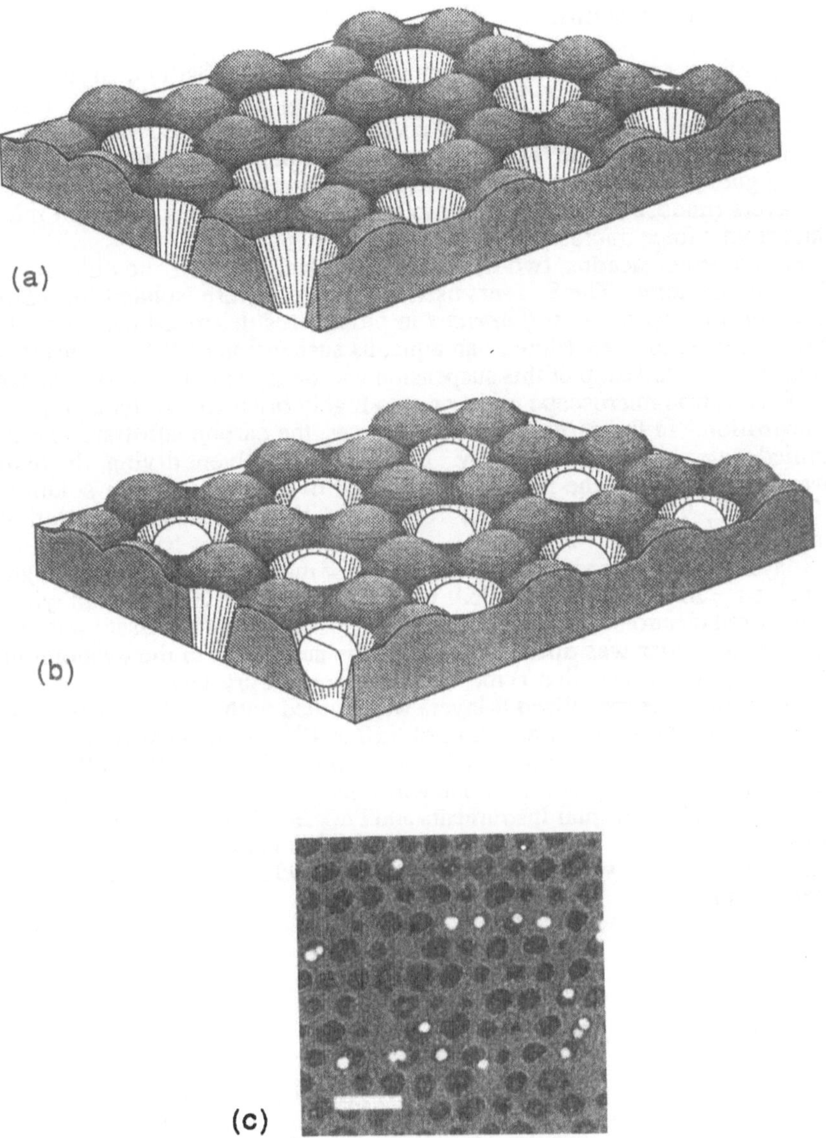

Figure 2: Nanoheterostructure formation from nanostructure. (a) Oblique view of idealized nanostructure showing six-fold symmetry of holes. Dark protrusions represent metallized protein. (b) Nanoheterostructure is formed by the selective placement of adsorbates into nanostructure hole sites. (c) TEM micrograph of colloidal gold selectively adsorbed to hole sites of nanostructure. Note high degree of preferential adsorption. (Scale bar = 50 nm)

2. Experimental Procedure

The process of nanostructure fabrication involves the deposition, drying and metal shadowing of the S-layer of *Sulfolobus* bacteria with 10±2 Å of metal, followed by normal incidence fast atom bombardment (FAB) milling. During FAB milling, metal is removed from the pores of the S-layer through a combination of sputtering and surface diffusion (Figure 1). Imaging of both metallized, unmilled S-layers and metallized, milled S-layers (nanostructures) is accomplished by transmission electron microscopy (TEM) and atomic force microscopy (AFM).

S-layer is a proteinaceous, two-dimensional crystal which is the outermost surface layer of certain bacteria. The S-layers used in this work were isolated from *Sulfolobus acidocaldarius* or *Sulfolobus solfataricus* in procedures described elsewhere [5]. The product of this isolation procedure is an aqueous suspension of S-layer sheets (average diameter ≤ 1µ). A 10ml drop of this suspension was deposited onto carbon coated 200 or 400 mesh Cu electron microscope grids, or onto highly oriented pyrolytic graphite for a 1 minute incubation. In the case of air-dried S-layer, the carbon substrates were washed with distilled water, and allowed to dry. In the case of solvent drying, the hydrated S-layers were dehydrated in the following sequence of alcohol in water solutions: 15%, 30%, 50%, 70%, 95%, 100%, 100%, 100%, and then dried by evaporation of the remaining solvent.

The S-layers were then metallized with 10±2 Å of metal at an angle of 50 degrees from normal incidence in an Edwards E306A vacuum deposition system employing a magnetically focussed electron beam gun. A steady deposition rate (1-2Å/sec) was established before a manual shutter was opened to expose the substrates to the evaporating metal. thicknesses were measured with a Temescal IC6000 quartz crystal monitor.

After deposition, the metallized S-layers were milled with 2 keV Ar atoms in a Gatan 600 Duomill for 45-60 seconds at a flux of 1±0.5x10^{14}/sec-cm^2 at room temperature. Transmission electron microscopy (TEM) was done with a JEOL 100C at 100 keV. Atomic force microscopy (AFM) was done with a Digital Instruments NanoScope II using cantilever tips from both Digital Instruments and Park Scientific (Microlevers, 0.6 µ) with a variety of force constants. We operated the AFM at zero net force (bias voltage set to the voltage at engagement). Typical scan speed was 8 or 19 Hz. All imaging was done in air at room temperature.

3. Results

3.1. SELECTION OF TEMPLATES

One source of fluctuation generation in the process of parallel nanostructuring results from fluctuations in the crystalline patterning template itself. As the bacteria grows, additional S-layer protein is produced and added to the growing surface [6]. Defect structures in *S. acidocaldarius* have been inferred from image analysis using high resolution TEM [7]. We have ourselves **directly imaged** twin boundaries in metallized S-layers from *Sulfolobus acidocaldarius* using atomic force microscopy (AFM) (Figure 3) [8]. We are pursuing several means of obtaining larger fluctuation-free patterning templates as will be described in the **Discussion** section.

3.2. DEPOSITION OF TEMPLATES

Another challenge in parallel fabricated nanostructures is structural distortion of the biological patterning templates [9]. We have improved the structure preservation of our protein patterning templates by changing from air drying samples prior to metallization and milling to solvent drying of samples. We have found that it is possible to improve structure preservation by exchanging the water in the samples with volatile solvents, e.g., ethanol and tertiary butanol, and then air drying from these alcohols. The alcohol is quite miscible with the water in the samples and once the samples have been dehydrated, air drying from these solvents produces much less deformation because they have a much lower surface tension than water.

Figure 3: Atomic force microscope (AFM) image of cytoplasmic surface of S-layer of *Sulfolobus acidocaldarius* which has been air-dried and metallized showing rotation twin boundary. The orientation of the triads of dimers of S-layer protein changes at a twin boundary as shown. (Scale bar = 45 nm.)

3.3. BEHAVIOR OF OVERLAYER MATERIAL

We have found a marked disparity in the pattern creation in overlayer films of different materials but otherwise identical experimental conditions. For example, Figures 4(a) and 4(b) show transmission electron micrographs of parallel fabricated nanostructures in which the deposited overlayer material was nickel and titanium and, respectively. We have performed electron beam deposition of numerous metal and metal alloy ultrathin films, both single and multilayer. For example, we have deposited ultrathin films of tantalum, platinum, chromium, vanadium, palladium, rhenium, silver, niobium, molybdenum, germanium, and platinum/iridium, as well as combinations of these materials [10]. We have found that titanium thin films are best suited for the fabrication and evaluation of nanostructures and nanoheterostructures. When we overlay S-layers with a titanium film using a conventional electron beam source and expose the film to the atmosphere prior to and after FAB milling, an oxide forms on the titanium. The oxidized titanium

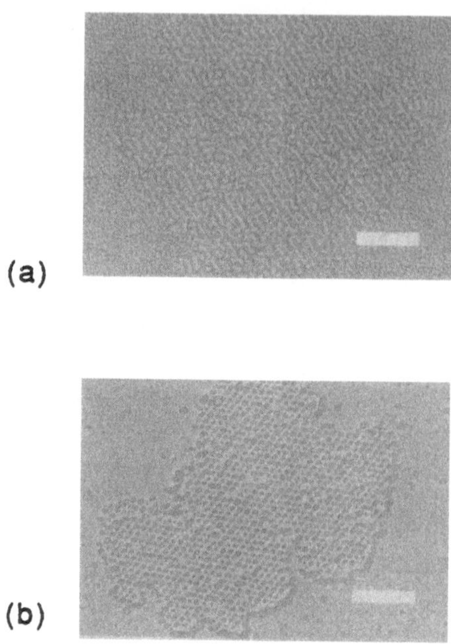

(a)

(b)

Figure 4: TEM micrographs of nanostructures formed from (a) nickel (oxidized) (Scale bar = 200 nm) and (b) titanium (oxidized). In both cases, the S-layer of *S. acidocaldarius* was shadowed with 12 Å of metal and FAB milled for 20 seconds. (Scale bar = 200 nm).

nanostructure is readily imaged by the AFM (Figure 5) [11]. However, these nanostructures do not have sufficiently high conductivity for acceptable scanning tunneling microscope (STM) imaging. Because the STM provides the opportunity to *manipulate* nanostructures by using high values of bias voltage, for example, it is desirable to fabricate nanostructures which can be imaged by the STM as well as by the AFM. In the **Discussion** section we will address this subject.

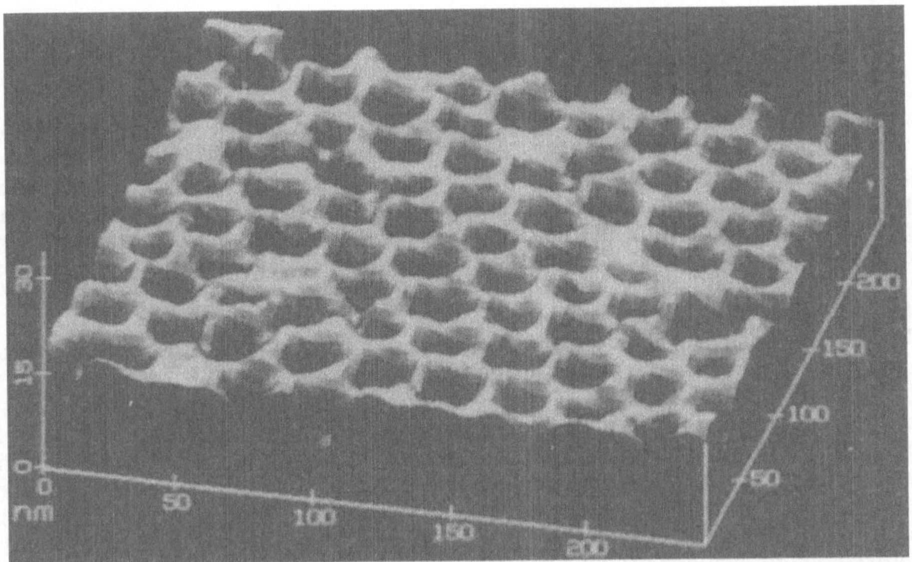

Figure 5: Atomic force microscope (AFM) image of nanostructure formed from metallized, milled S-layer of *S. acidocaldarius*. This image shows *unfiltered* data displayed in a surface plot mode.

3.4. NANOSTRUCTURE FORMATION

We have studied the time evolution of the milling process as well as the dependence of pattern formation on incident fast atom energies using transmission electron microscopy [12]. The initial step in the pattern formation during milling occurs within 5 seconds: the thinnest areas of metal are removed, creating crescent-shaped holes. These incipient holes are further enlarged over the next 30 seconds. The final step, occurring in the last 10 seconds of milling, is a smoothing of the jagged edges of the holes, resulting in a rounded appearance. This patterning process is not observed for incident energies below 1.5 keV: no incipient hole formation is observed. This sharp patterning threshold suggests that the initial effect of milling, the opening of the crescent shaped holes, is the result of a sputtering process. Subsequently the hole boundaries give the appearance of "flowing" to their final rounded shape in a continuous way. This suggests that the second two steps of the patterning process are predominantly the effect of surface diffusion.

3.5. NANOHETEROSTRUCTURES

Since we have moved from using Ta/W metal overlayers [4] to Ti overlayers, we have found that preferential adsorption is present in the milled (and oxidized) Ti films with no additional processing steps: such anionic adsorbates that do bind to the nanostructures bind exclusively to the exposed carbon sites and not to the metal film. Consequently, the focus of our recent nanoheterostructure experiments has been on obtaining saturation of the carbon sites is discussed in the following section.

4. Discussion

4.1. SELECTION OF TEMPLATES

One approach to obtaining larger fluctuation-free patterning templates is to induce recrystallization of S-layer subunits. With the appropriate combination of conditions (e.g., temperature, concentration of bivalent cations, etc.) recrystallized S-layer sheets with a greater than ten-fold increase in diameter (compared to native S-layer sheets) can be formed [13]. It is possible, moreover, to induce large sheet recrystallization which appears to be free of defects [14]. The availability of such larger patterning arrays could also be useful for applications experiments employing the nanostructures formed from these larger patterning arrays.

We are also exploring an alternative means of acquiring larger nanostructures and one that addresses the topic of the placement and orientation of patterning arrays. Clearly the applicability of our techniques will be greatly enhanced if the patterning arrays can be controlled to cover chosen areas and have chosen orientations. A possible scenario for ordering the arrays on the surface would be to coat the (smooth) surface to be patterned with a low molecular weight layer which could be chemically modified using an electron beam, such that it binds the patterning protein only where exposed, for example on a square or triangular grid of nanometer wide lines. The patterning protein arrays would be solubilized, yielding single proteins in a solution that would then be brought into contact with the surface to selectively adsorb. A gradual evaporation of the suspending medium would drive reconstitution on the surface, growing single crystals as nucleated by the grid pattern. The feasibility of the basic step in this process has already been demonstrated by Zingsheim's selective binding of ferritin to STEM written lines [15].

4.2. BEHAVIOR OF OVERLAYER MATERIAL/STM IMAGING

In order to fabricate nanostructures which have sufficient lateral conductivity and support a sufficient tunneling current to be suitable for STM imaging, one approach is to reduce the oxidized titanium to improve its conductivity. It may also be possible to plate out a very thin conducting layer of platinum or gold onto the titanium. Alternatively, coating the S-layer with titanium nitride rather than titanium may enable STM imaging because of the relatively high conductivity of TiN.

4. 3. NANOSTRUCTURE FORMATION

As discussed earlier, the formation of nanostructures appears to be a combined effect of sputtering and surface diffusion both produced by FAB milling. We will refine our understanding of the patterning process and the reproducibility and homogeneity of the resulting nanostructures by employing a newly acquired *Precision* Ion Milling System (PIMS). For example, with the PIMS both a stationary and a rastered beam are obtainable. The rastered beam has an adjustable scan rate as well as a continuously adjustable size and aspect ratio. The shape of milled features can depend on the beam raster (for the same total flux) because of effects attributable to a combination of redeposition and self-focusing [16]. Another milling parameter available on the PIMS is sample temperature which is continuously adjustable from -196°C to +90°C. This feature could be of particular importance in understanding and controlling that portion of the nanostructure fabrication process which is dominated by surface diffusion.

4.4. NANOHETEROSTRUCTURES

An important requirement for nanoheterostructure fabrication is complete coverage (site saturation) with the desired adsorbate of the exposed substrate binding sites in a nanostructure. Viewing the nanostructure as an ultramicroelectrode array (with the exposed carbon sites as the electrodes), one approach to fabricating nanoheterostructures might be electrochemical deposition [17]. Such a voltammetric experiment could be used to produce a solid-state/solid-state nanoheterostructure. The differing chemistries of the components of such heterometallic arrays might in turn be used to selectively anchor biomolecules to form biomolecular/solid-state nanoheterostructures.

5. Acknowledgements

We gratefully acknowledge the support of the Division of Advanced Energy Projects/Office of Basic Energy Sciences for Department of Energy Grant DE-FG02-89ER14077 to K. D.

6. References

1.   K. Douglas, N.A. Clark, and K.J. Rothschild, App. Phys. Let. *48*, 676 (1986).
2.   M.V. Nermut, Trends in Biochem. Sci. *8*, 303 (1983).
3.   J.E. Rash and C.S. Hudson (editors), <u>Freeze-Fracture: Methods, Artifacts and Interpretations</u>, Raven Press, New York (1979).
4.   K. Douglas, N.A. Clark, and K.J. Rothschild, Appl. Phys. Lett. *56*, 692 (1990).
5.   H. Michel, et al. in *Electron Microscopy at Molecular Dimensions,* W. Baumeister and W. Vogell (eds.), Springer-Verlag, Heidelberg, 1980, 34.
6.   J. Smit, "Protein Surface Layers of Bacteria," in *Bacterial Outer Membranes as Model Systems*, M. Inouye, ed. Wiley-Interscience, NY, 343 (1987).
7.   G. Lembcke, R. Durr, R. Hegerl, J. Microscopy *161*(2), 263 (1991).
8.   G. Devaud, P. Furcinitti, M.K. Lyon, and K. Douglas, submitted to Biophysical Journal.
9.   G. Devaud, K.Douglas, M.K. Lyon, and N.A. Clark, submitted to J. Vacuum Science A.
10. G. Devaud, and K. Douglas, submitted to Journal of Microscopy.
11. K. Douglas, G. Devaud, M.K. Lyon, and N.A. Clark, submitted to Science.
12. G. Devaud, K. Douglas, J. Fleming, M.K. Lyon, and N.A. Clark, Materials Research Society, Symposium Boston, MA, December 1991.
13. U.B. Sleytr and M. Sara, Appl. Microbiol. Biotechnol. *25*, 83 (1986).
14. U.B. Sleytr, private communication.
15. H.P. Zingsheim in <u>Scanning Electron Microscopy</u>, Vol. I, I.I.T. Research Institute, 357 (1977).
16. J. Melngailis, J. Vac. Sci. Technol. B *5*, 469 (1987).
17. R.M. Penner and C.R. Martin, Anal. Chem. *59*, 2625 (1987).

FILM DEPOSITION WITH CLUSTER BEAMS: AN ALTERNATE PATH TO EPITAXIAL, CRYSTALLINE FILMS

Isao Yamada
Ion Beam Engineering Experimental Laboratory
Kyoto University,
Sakyo, Kyoto 606-01 Japan

ABSTRACT Films deposited by beams containing a small number of large clusters differ markedly from those grown by conventional means with atomic beams. In many film-substrate combinations, ionized cluster beam deposition produces an extraordinarily uniform atomic arrangement in films deposited on large lattice misfit substrates. No significant chemical mixing between the film and substrate atoms is observed. Atomic scale observations of the film growth by tramsmission electron microscopy (TEM) and scanning tunneling microscopy (STM) have been made to understand the effects of neutral and ionized cluster beam bombardment on film formation processes. Examples of epitaxial growth of metal and ceramic films on large lattice misfit substrates are discussed.

1. Introduction

Studies of the properties of metal films grown by Ionized Cluster Beam (ICB) methods have found many examples of the formation of large-area epitaxial films on large lattice misfit substrates [1]. Such epitaxial films have been formed with Al on Si, Ge and sapphire substrates at room temperature. Metal films formed by ICB methods also have been found to have extraordinary atomic smoothness and to exhibit defect-free electrical characteristics [2]. Applications of these novel deposition techniques under development include formation of high-performance contacts and metallization for advanced IC devices [3] and high-reflectivity mirrors of excimer laser and soft x-ray optical systems [4].

ICB deposition has been investigated in this laboratory since the inception of the technique. While there is no question that high quality films can be made using ICB techniques; it is not clear that the presence of large clusters is the primary agent responsible. In order to

P. Jena et al. (eds.), Physics and Chemistry of Finite Systems: From Clusters to Crystals, Vol. II, 1193–1202.
© 1992 *Kluwer Academic Publishers.*

investigate the fundamentals of ICB deposition mechanisms, a "definitive" experiment for detection of clusters from an ICB source has been made [5]. The results show that the number of clusters of size 25 < n < 1600 was of the order one cluster per 10^7 atoms. Taking into account the cluster size, one atom in 5000 is contained in a cluster with between 25 and 1600 atoms.

The key question is how the relatively small number of atom clusters detected in an ICB beam might play so significant a role in nucleating the growth of films on a substrate and what is the role in film growth of the single-atom vapor ejected together with clusters through the nozzle. Atomic-scale observations of the early stages of film growth by transmission electron micicroscopy (TEM) and scanning tunneling microscopy (STM), combined with recent studies of the size distribution of clusters of atoms from ICB sources, have led to new understandings of origins of some of the extraordinary properties of thin film deposited by ICB methods. In studies of the growth of gold films on graphite substrates, the presence of a small fraction of atoms in the form of large clusters initiates stable nucleation sites by which immediate formation of uniform islands for film growth can be established. The growth of films from cluster-bombardment leads to fundamentally altered film properties. This paper will discuss bombarding effects of metal clusters and ionized metal clusters at the initial stages of film growth and resultant film characteristics.

2. TEM and STM observations of the film growth at the initial stages of the deposition

Atomic scale imaging by TEM and STM of the initial stages of film growth of Ag and Au on graphite substrates have been made.
Figure 1 shows TEM images of Ag deposition on graphite substrate from a cylindrical shape nozzle source (1 mm in diameter and 1 mm long) which produces an atomic beam containing some large atomic clusters, and from an open source (10 mm in diameter) for molecular beam. The mean layer thickness was varied from 0.4 to 4 Å at a deposition rate of 45 Å/min., where the nozzle source was at 1700 K and the open sources at 1460 K. Under these conditions, the Ag beam from the cylindrical nozzle source contains a large mass fraction of clusters. The open source emits evaporated beams consisting single atoms. The graphite substrate was kept at liquid nitrogen temperature and the deposited layers were encapsulated with a carbon overcoating to reduce the atomic migration and preserve the characteristic of the as-deposited Ag layer. Figure 2 shows the measured mean diameter of the Ag islands as a function of the mean layer thick-

0.5 s (0.38Å) 1 s (0.75Å) 1.5 s (1.13Å) 3 s (2.25Å) 5 s (3.75Å)

(a) NOZZLE(Tc=1700K, Ts=77K)

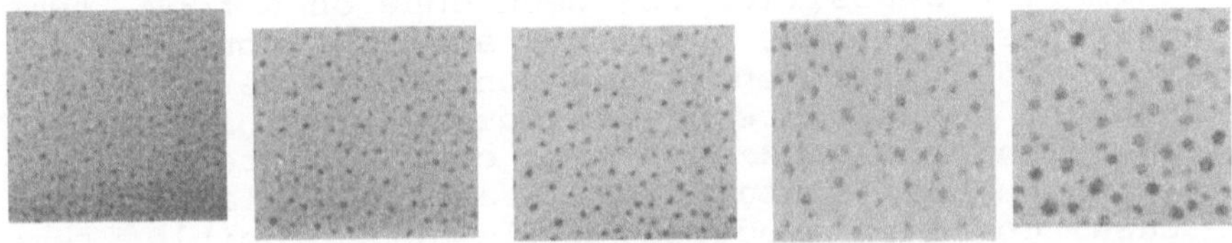

0.5 s (0.40Å) 1 s (0.78Å) 1.5 s (1.20Å) 3 s (2.44Å) 5 s (4.04Å)

(b) OPEN(Tc=1460k, Ts=77K)

Fig.1. TEM of the carbon-encapsulated silver deposited by the nozzle source (a) and by the open source (b).

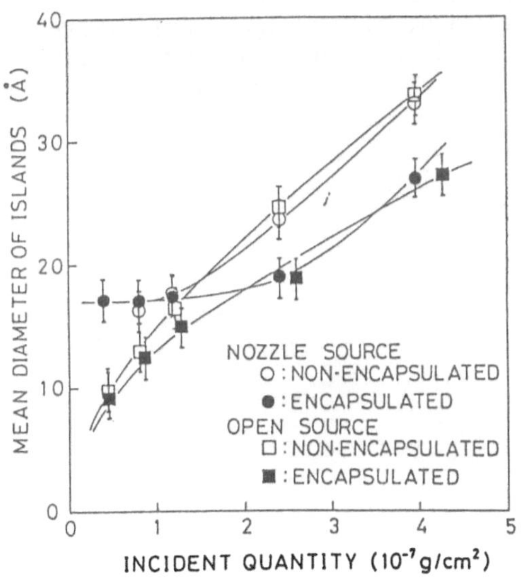

Fig.2. The mean diameter of silver islands deposited by the nozzle and the open source.

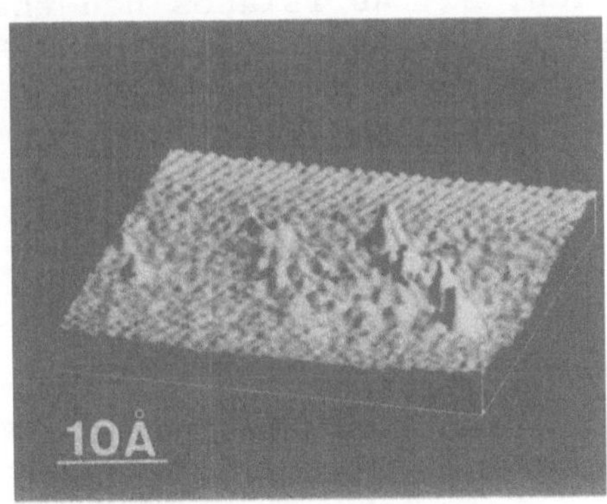

Fig.3. STM image of Au deposit on HOPG substrate by "MBE" at room temperature. The molecular beams were not ionized.

ness. In the open source case, the mean diameter of islands increases and number of islands decreases with increasing layer thickness. This result suggests that the film formation process involves formation of nucleation sites and growth of the nuclei by capturing migrating adatoms. On the other hand, for the cylindrical nozzle source case, the number of islands increases linearly with the mean layer thickness, while the mean island diameter is constant until the onset of coalescence of the islands. This results suggest that clusters contained in the ICB beam directly create stable nucleation sites.

In order to observe STM images at the initial stage of film formation deposition has been done on a room temperature graphite substrate under MBE and ICB conditions, respectively [6]. The deposition conditions were varied by adjusting the temperature of the source crucible, which had a 1 mm cylindrical nozzle. At a crucible temperature of 1920 K, the Au beam containing a fraction of large clusters (ICB case) could be produced, while at a crucible temperature of 1570 K, a single-atom (MBE case) beam of Au was obtained.

Figure 3 shows the typical STM images of Au island deposited under MBE conditions. The size and shape of Au islands were not uniform. Au adatoms deposited by the MBE method on the substrate often stuck to the tip of the STM during the observation, which indicates that adhesion of adatoms to the substrate was weak.

Figure 4 shows the STM images of Au deposited under ICB conditions for various deposition times. In the ICB deposition, all Au islands had the same size and shape throughout the early stage of the deposition that typical Au islands

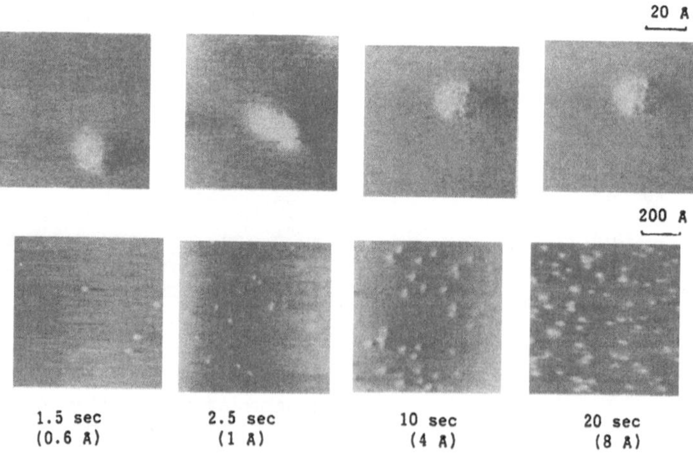

Fig.4. STM images of deposited at the initial stages of cluster beam deposition. (neutral beam, crucible temperature 1920 K)

were round in shape. The mean size of Au islands was about
20 Å in diameter and the number of atoms in a cluster was
estimated to be about 100 - 500 atoms. These results on the
size and shape of Au islands corresponds to the TEM results
for Ag beams. The number of Au islands increases linearly
with the deposition time at a rate of, 5.2×10^{10} cm^{-2} sec^{-1},
while the mean size of the islands remains constant during
the early stages of the deposition process. This suggests
that the islands formed at the initial stage of deposition
result from the direct impact of individual large clusters
from the nozzle source. Smaller islands, consisting of
several Au atoms, were not observed for the ICB deposition.
This suggests that the single atoms arriving to the substrate had re-evaporated and stable nuclei were not formed.

In order to study the effect of the kinetic energy of
cluster ions, we also observed by STM the initial stage of
the deposition process for accelerated beams.   Figure   5
shows typical Au islands deposited by ICB for (a) neutral
beams and (b) an acceleration voltage (Va) of 3 kV.   In the
case of neutral cluster beams, Au atoms in islands show no
orientation with respect to the HOPG substrate atoms.   On
the other hand, in the case of Va = 3 kV, about 10 percent
of the islands were flat and oriented to the substrate
lattice.   These oriented islands are considered to be formed
by the ionized and accelerated clusters.

STM results for the deposition of Au on carbon substrate
show that a film can be nucleated by arrival of the cluster
to the surface in the form of an "instant-island". However,
depending on the cohesiveness and mobility of the atoms in
the cluster, the film growth characteristics can be markedly
different.   In the case of Al deposition on SiO_2 and clean
silicon substrates, a large surface migration of adatoms is

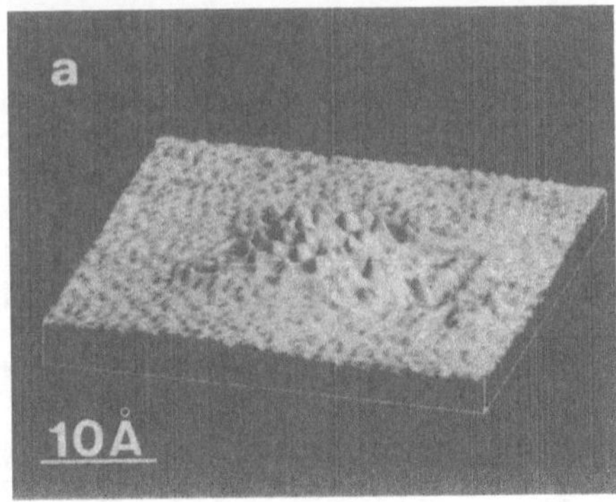

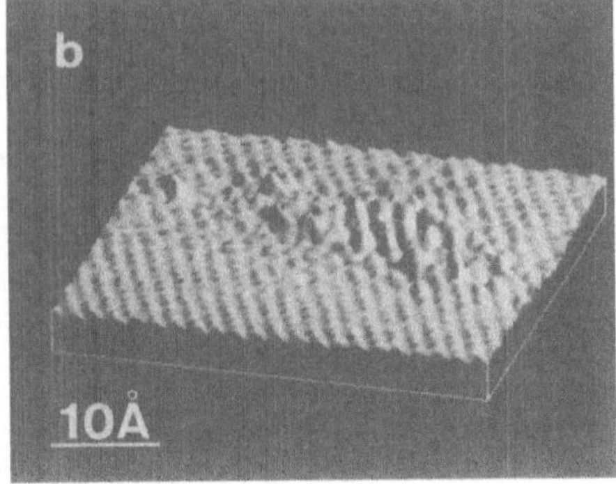

Fig.5. STM image of Au deposit on HOPG substrate by ICB at
neutral beams (a) and an acceleration voltage (Va) of 3 kV
(b).

observed. Figure 6 shows the result of the diffusion distance measurement of Al on SiO_2. The diffusion distance of Al increases as the acceleration voltage increases for low substrate temperatures. This tendency is also observed at 200 °C; but at 400 °C the order is reversed. Measurements on Si substrates indicated a similar tendency [7]. It is likely that this reversal is caused by an increased re-evaporation rate at high acceleration voltage and temperature.

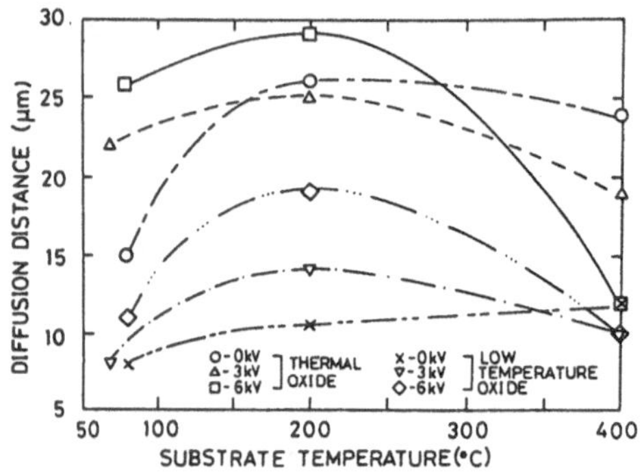

Fig.6. Surface diffusion distance of Al on two kinds of SiO_2 as a function of ion acceleration voltages and surface temperatures.

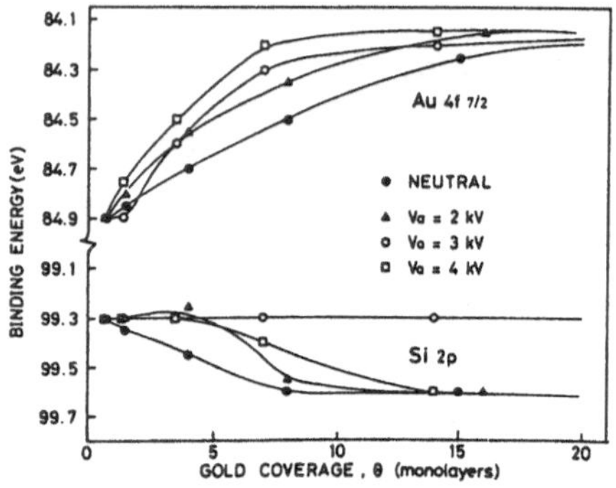

Fig.7. Si 2p and Au 4f core level shifts plotted as a function of Au coverage for different acceleration voltages.

3. Interface characteristics

The initial deposition stage plays an important role in the formation of interface. An in-situ XPS analysis of the interface characteristics was made during the ICB deposition of thin Au films on Si substrates [8]. Although the formation of a stable Au/Si interface is known to be very difficult [9], ICB bombardment is expected to have a remarkable influence on the interface characteristics.

Figure 7 compares the core level shift of Si 2p and Au 4f XPS peaks as a function of Au coverage for different ion acceleration voltages. Without the ion acceleration, Au 4f peak shifts towards lower binding energies and the Si 2p peak shifts towards higher binding energies with increasing Au coverage. This trend is in agreement with the result reported for conventional evaporation experiment [10], and is explained as the occurrence of Au-Si interaction at the initial stage of the film formation. When the acceleration voltage was applied, on the other hand, the Si 2p signal showed negligible peak shift and disappeared at relatively small Au coverage. Au 4f peak approaches more quickly to the bulk value. Figure 8 shows the energy separation between the two Au 5d peaks as a function of Au coverage. The energy separation of the films deposited without ion acceleration indicates that the Au atoms are in an isolated state up to few monolayers. With applying the acceleration voltage, however, the energy separation approaches the metallic value at one monolayer. These results suggest that the film-substrate interaction is suppressed and a uniform metallic Au film is formed at the initial stage by use of ICB.

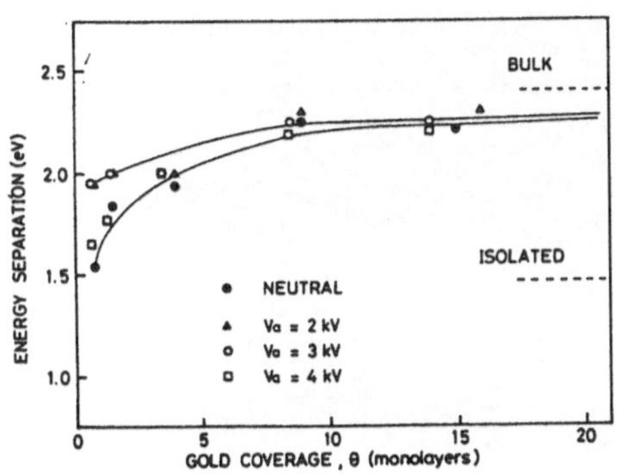

Fig.8. Energy separation between two main peaks of Au 5d state.

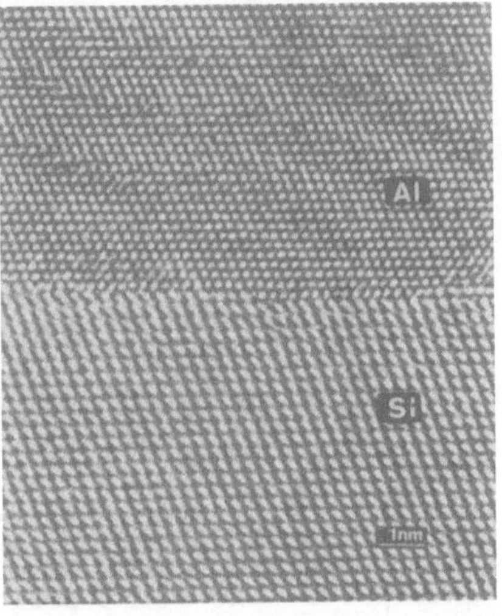

Fig.9. Cross sectional TEM image of the interface of Al film deposited on Si(111) at room temperature.

4. Film growth on large-misfit substrates

4.1. Metal films

Growth of epitaxial metal films has been studied by high-resolution transmission electron microscopy (TEM). This results show that the films have highly ordered atomic arrangements at the interface and grain boundaries [11]. Figure 9 shows a TEM micrograph at the interface of Si(111) substrate and an Al grown film. The micrograph shows the realization of the epitaxial growth of an Al film on a Si

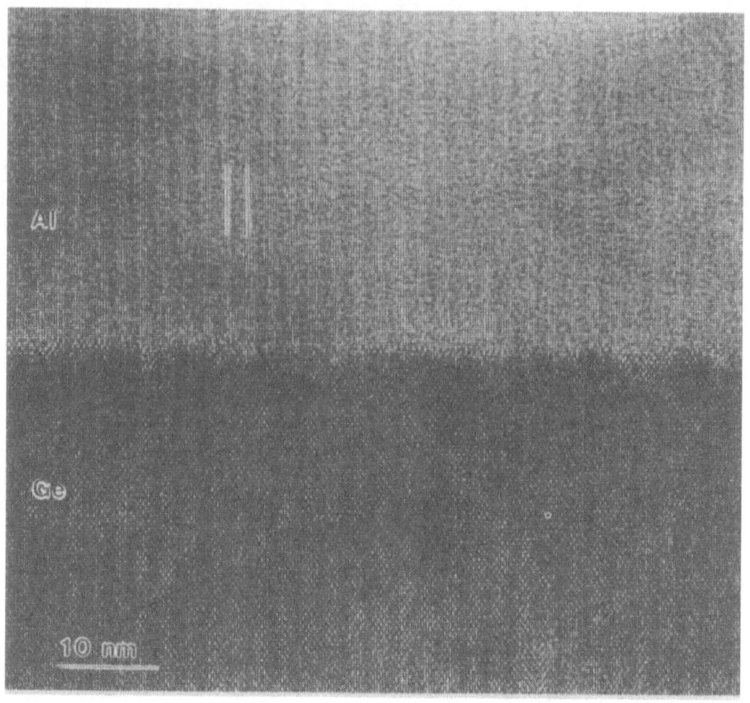

Fig.10. Moiré pattern produced by superimposing a high resolution micrograph of the Al/Ge interface with a standard net indicating the absence of atomic relaxations at the interface. The moiré period in the Al is marked.

substrate. The deposited film also has extraordinary atomic smoothness and exhibits defect-free electrical characteristics.

As seen by high resolution TEM pictures, Al/Ge interfaces are flat and there is no evidence of atomic interdiffusion between the film and the substrate. Figure 10 is a moiré pattern used to emphasize the pattern of elastic strain distribution [12]. The moiré analysis shows that there is very little distortion of the pattern in either the Al or Ge lattice, indicating no relaxation of the structure. Figure 11 shows schematic diagrams of the rigid, unrelaxed lattices with mismatch (a) , strained (b) and relaxed into periodic misfit dislocation array (c). In the epitaxial film growth on different lattice constant materials, Fig.11 (b) or (c) mostly occurs to release the strain energy at the interface. However, in the case of Al/Ge film by ICB deposition, the films formed a rigid, unrelaxed structure, as sketched in Fig.11 (a).

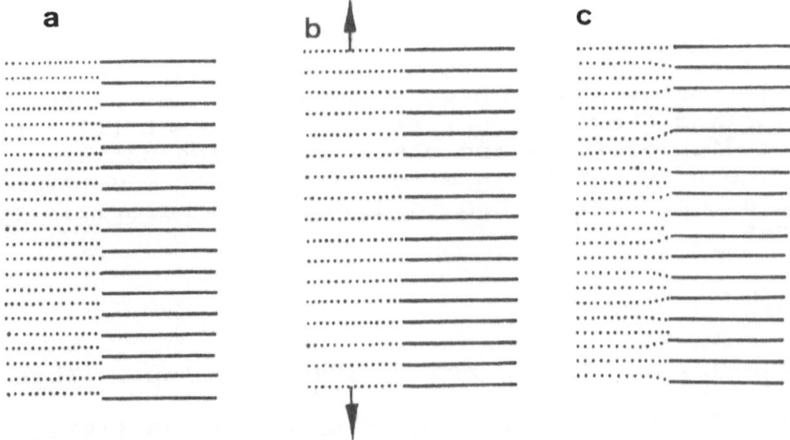

Fig.11. Schematic illustration of three possible interface structures; rigid (a), strained (b) and relaxed into periodic misfit dislocation array (c)

4.2 Ceramic films

Hetero-epitaxial growth of Al_2O_3 films on single crystal Al(111) films grown on Si(111) substrate has been made [13]. Al was used as a source material and the deposition was carried out in a O_2 atmospheres in the range of 10^{-4} to 10^{-5} Torr. The films were grown on Al/Si substrates with a 5 kV acceleration voltage and 200 °C substrate temperature. The deposited film was twin structure of alpha-Al_2O_3. The refractive index of the grown film was 1.763. This is almost equal to that of a sapphire crystal (n=1.765). In this case, the lattice mismatch is 16.8%. Since we have already shown that Al films could be epitaxially grown on sapphire substrates, these new results suggest that monolithic epitaxial metal/insulator/metal multi-layer structure could be formed over Si substrates.

Epitaxial TiO_2 films have been grown on sapphire(0001) and Al(111) substrates [14]. Ti depositions was made under O_2 gas atmospheres of 10^{-5} to 10^{-4} Torr. TiO_2(200) films grew parallel to the Al(111) surface. Lattice misfit is 13.3 % for this material combination.

5. Conclusions

Bombarding effects of ionized metal cluster beams have been studied. Bombarding effects for epitaxial growth were observed by TEM and STM observations during the initial stage of the deposition. The results also show the differences between ICB deposition and MBE. The TEM and STM results show that clusters produced during the expansion from small nozzle source have a remarkable influence on the nucleation and growth processes and resultant film properties. Studies of these and other material-substrate combinations are on going in an effort to further understand the fundamental mechanisms and to develop new materials by ICB techniques.

References

1. I.Yamada, H.Inokawa and T.Takagi, J. Appl. Phys. $\underline{56}$, 2746 (1984)
2. I.Yamada, Applied Surface Science, $\underline{43}$, 23 (1989)
3. I.Yamada, C.J.Palmstrøm,. E.Kennedy, J.W.Mayer, H.Inokawa and T.Takagi, Mat. Res. Soc. Symp. Proc., $\underline{Vol.37}$, 1227 (1984)
4. I.Yamada, G.H.Takaoka, H.Usui, F.Satou, Y.Itoh, K.Yamashita, S.Kitamoto, Y.Namba, Y.Hashimoto, Y.Maeyama and K.Machida, Nuc. Instr. Methods, $\underline{B55}$, 876 (1991)
5. W.L.Brown, R.L.McEachern, M.Sosnowski, G.Takaoka, H.Usui and I.Yamada, Nuc. Instr. Methods B, (1991) in press.
6. I.Yamada, G.H.Takaoka, H.Usui and S.K.Koh, Mat. Res. Soc. Symp. Proc., $\underline{Vol.206}$, 383 (1991)
7. L.L.Levenson, A.B.Swartzlander, H.Usui, I.Yamada and T.Takagi, Mat. Res. Soc. Symp. Proc., $\underline{Vol.128}$, 131 (1989)
8. I.Yamada, H.Usui, H.Harumoto, and T.Takagi, Mat. Res. Soc. Symp. Proc., $\underline{Vol.101}$, 172 (1988)
9. A.Hiraki, Surface Science, $\underline{168}$, 74 (1986)
10. L.Braicovich, C.M.Garner, P.R.Skeath, C.Y.Su, P.W.Chye, I.Lindau, and W.E.Spicer, Physical Review B, $\underline{20}$, 5131 (1979)
11. I.Yamada, H.Usui, S.Tanaka, U.Damen and K.H.Westmacott, J. Vac. Sci. Technol., $\underline{A8(13)}$, 1443 (1990)
12. U.Dahmen and K.H.Westmacott, Proc. Special Seminar on Ionized Cluster Beam Technology and its Application, in part of the Thirteen Symposium on Ion Sources and Ion-Assisted Technology, (1990) page 53.
13. H.Hirayama, G.H.Takaoka H.Usui and I.Yamada, Nuc. Instr. Methods B, (1991) in print
14. K.Fukushima and I.Yamada, submitted to the Seventh International Conference on Surface modification of Metals by Ion Beams. Washington, DC July 1991

APPLICATION OF ULTRA FINE PARTICLES

M.Oda, E.Fuchita, M.Tsuneizumi, S.Kashu, and C.Hayashi
UFP Division, Vacuum Metallurgical Co., Ltd.
516 Yokota, Sanbu-cho, Sanbu-gun, Chiba-pref., Japan 289-12

ABSTRACT. Ultra fine particles of organic and inorganic materials can be formed by the gas evaporation method(gas condensation method). An example of these UFP's application will be introduced. One of the applications is the gas deposition method. In the gas deposition method, particles formed in an evaporation chamber are carried to another chamber(a deposition chamber) through a pipe. Particles are accelerated in the pipe with a gas flow and come out of a nozzle located in the deposition chamber which is being evacuated down to 10^{-1} torr and deposited on a substrate to form UFP films. The final speed of the particles depends on the pressure difference between the evaporation chamber and the deposition chamber. The particles speed exceeds 1000m/s and adhesion strengths of the films reaches 500 Kgf/cm^2 as well as vacuum deposition films in the condition that the pressure of the evaporation chamber is at 4 atms. Patterns of spots and lines with 50 μm size and also wider films can be formed on a substrate without masking system. The process is used in producing wirings, electrodes and condensers on an industrial scale.

1.INTRODUCTIONS

Ultra fine particles are formed by the gas evaporation method(1). In the gas evaporation method, metal atoms are evaporated in an inert gas such as helium or argon, collide with gas atoms and cooled down condensing to particles. The particles sizes are controlled by changing the gas pressure or the evaporation temperature. Larger sized particles can be formed under conditions of higher pressures or higher temperatures.

Among the applications of ultra fine particles, the gas deposition method will be introduced. In the gas deposition method, the formed particles by the gas evaporation method are carried to another chamber through a pipe while in an aerosol state. The particles are accelerated in the pipe and sprayed on a substrate through a narrow nozzle and deposited on the substrate in the form of a UFP film. This method is called the gas deposition method(2). As for materials and forms of substrate, there are no restrictions. Glass, ceramics etc. can be used. In the near future, this method is expected to be used for forming electrodes or wirings because it has several advantages over

P. Jena et al. (eds.), Physics and Chemistry of Finite Systems: From Clusters to Crystals, Vol. II, 1203–1212.
© 1992 *Kluwer Academic Publishers.*

1204

conventional methods such as higher speed deposition precessing, direct write processing, low temperature processing and the ability to form uniformly mixed UFP films using more than two elements. Mixed UFP film formation process has succeeded to form high Tc-super conductors(3).

In this article, Ag films are formed for applications to wirings. Au films are formed for applications to microelectrodes. Adhesion strengths and electric conductivities of Ag films and patterning abilities using a multinozzle and a shutter system and adhesion strengths of Au films will be reported.

2.EXPERIMENTAL

The schematic diagram of the gas deposition apparatus is shown in fig.1. The gas deposition apparatus is composed of a UFP evaporation chamber, a deposition chamber(a spray chamber) and a transfer pipe. The chambers are evacuated down to 10^{-6} torr with a turbo molecular pump and a rotary pump and then helium gas is introduced to the evaporation chamber and the gas is carried to the deposition chamber through the transfer pipe. The gas pressure of the evaporation chamber is controlled from 0.13 atms(100torr) to 5 atms by changing the helium gas supplying speed.

Ag or Au metal atoms are evaporated from a resistance heating tungsten basket. Evaporated atoms are collided with gas atoms and cooled condensing into particles. The mean diameter of the particles was controlled to be 600Å by changing the electric power supplied to the basket, shown in fig.2, compensating for the effect of the evaporation chamber pressure on the particles diameters and the effect of the heat absorption of the tungsten basket with the higher pressure gas. The formed particles are carried with the gas flow and ejected out of the nozzle in the deposition chamber and deposited on a substrate. The deposition chamber is pumped down to be less than 0.3 torr by a mechanical booster pump and a rotary pump while the particles and the helium gas are being carried into.

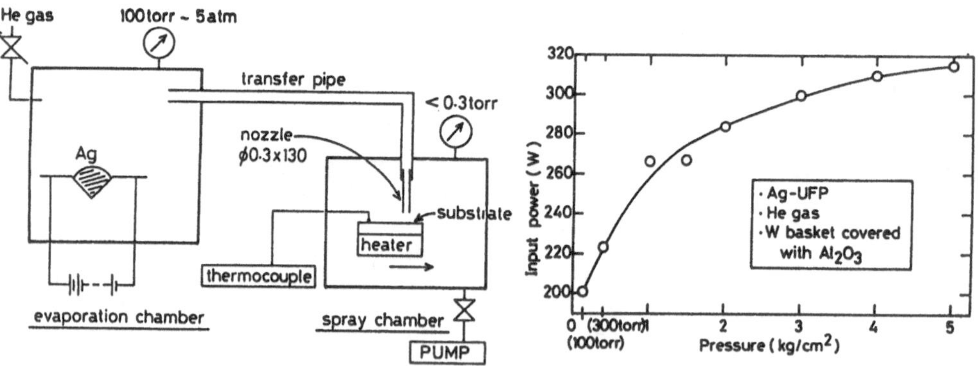

Fig.1 Schematic diagram of gas deposition apparatus

Fig.2 Evaporation chamber pressure of electric power supply to evaporation source

Two single nozzles and one multinozzle are used. The inner diameter of the first single nozzle is 0.3mm and the length of the nozzle is 130mm. The inner diameter of the second one is 0.1mm and the length of the nozzle is 30mm. The multinozzle has 10 nozzles with the inner diameter of 0.1mm aligned with each other with the period of 0.4mm and the length is 1.2mm, shown in fig.3. The distance between the outlet of the nozzles and the substrate are controlled to be from 0.7mm to 1.4mm. The first single nozzle is used for forming Ag films. The second one and the multinozzle are used for forming Au films. As the substrate, an Al_2O_3 plate is used for Ag film formations and a Si wafer covered with Ti(3000Å thick) film and Ni(3000Å thick) film is used for Au film formations.

A substrate holder is designed to be scanned along X direction and Y direction by a digital programable controller. The holder is constantly scanned at the speed of 125μm/sec for Ag film formations. For Au film formations, the holder is controlled to be stopped at the positions of the deposition and scanned to the next position at the speed of 125μm/sec. The shutter system has been developed to form spot patterns. This system is illustrated in fig.4. The shutter system functions as follows; when the deposition is finished at a position, V-1 valve is closed and V-2 valve and V-3 valve are opened simultaneously, the particles from the evaporation chamber are carried through V-2 to a vacuum pump, helium gas is introduced through V-3 to carry the particles located in between V-3 and the nozzle to the deposition chamber. The reaction time of the shutter system is 0.1 sec.

The formed films are observed with a scanning electron microscope(SEM). The adhesion strengths of the Ag films to the substrate are measured using a measuring equipment. A rod is glued onto the films and the adhesion strength is measured by pulling against each other. Electric conductivity is measured using the four wire method. The adhesion strengths of the Au films to the substrate are measured using a shearing force measuring equipment. The measuring equipment is illustrated in fig.5.

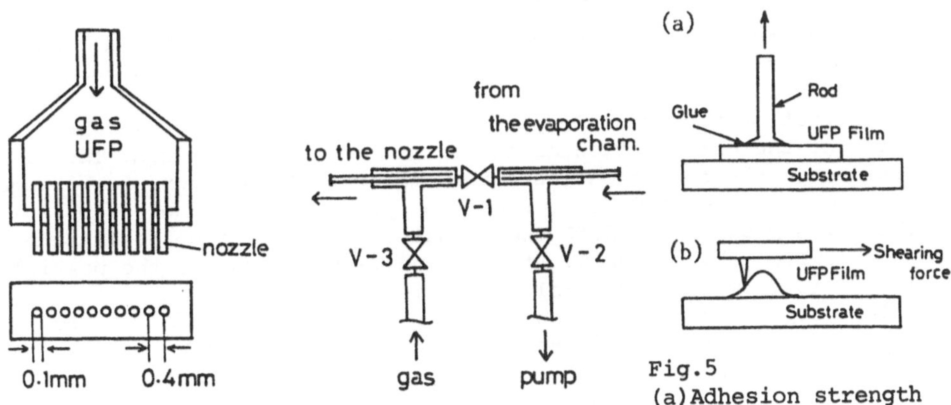

Fig.3 Schematic diagram of multinozzle

Fig.4 Schematic diagram of shutter system

Fig.5
(a) Adhesion strength
(b) Shearing force measuring equipment

3.RESULTS

3.1.Ag Film Formations for Wirings

3.1.1. Deposition Rate. The deposition rate of Ag films is 0.4 µm/sec using the nozzle with an inner diameter of 0.3mm at the scanning speed of 125µm/sec and at the evaporation chamber pressure of 5atms.

3.1.2. Effect of Evaporation Chamber Pressure to Adhesion Strengths. The SEM pictures of the surface and the crosssection of Ag films are shown in fig.6. The films are formed in the condition where the pressure in the evaporation chamber is changed from 0.13 atms and the substrate temperature is 30°C. The picture of 0.13 atms shows that most of the particles are sintered but with a lot of pores. But the other pictures show that the number of pores is reduced and the surface of the films become smooth with the pressure being increased.

The dependence of the adhesion strength of Ag films to the evaporation chamber pressure is illustrated in fig.7. The adhesion strength of 140kgf/cm^2 is obtained in the film of 1.0 atms. It increases to 550kgf/cm^2 as much as those of the vacuum deposition films.

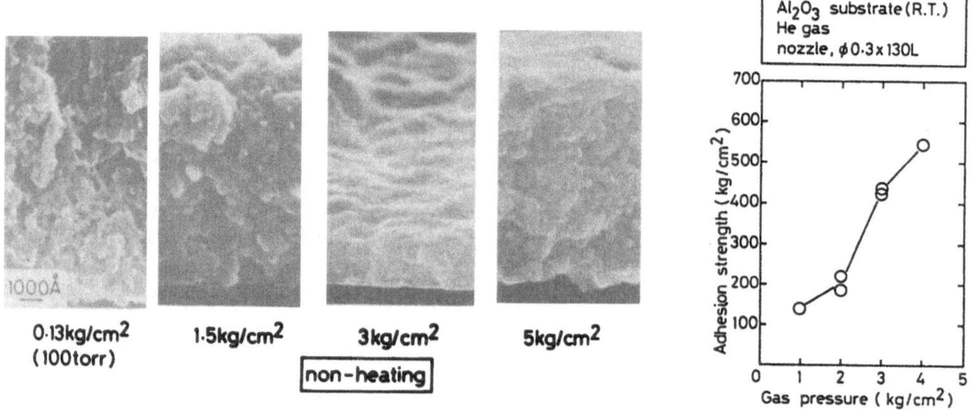

Fig.6 Scanning electron microscope picture
    of films obtained at from 0.13 atms
    and 30°C

Fig.7 Evaporation chamber
    pressure dependence of
    adhesion strength at 30°C

3.1.3. Effect of Substrate Temperature to Adhesion Strengths. The SEM pictures of the crosssection of the films are shown in fig.8. The films are formed in the condition where the substrate temperature is changed from 30°C to 400°C at 0.13 atms of the pressure in the evaporation chamber. The picture of 200°C shows that the number of pores is reduced compared with that of 30°C. At the condition of 400°C , the particles coalesced to grow up to several µm in diameter.

The dependence of the adhesion strengths of the films to the substrate temperature is illustrated in fig.9. The black circles and white circles indicate the results of the films obtained at the condition of 0.13 atms and 2 atms respectively. The adhesion strength of 80kg/cm^2 is obtained at the condition of 200°C and 0.13 atms. It increases to 310kg/cm^2 at 400°C. The comparison between the results of at 0.13 and 2 atms indicate that at the condition of 0.13 atms, a substrate temperature of 300°C is required to obtain the adhesion strength of the film obtained at 2 atms and 30°C.

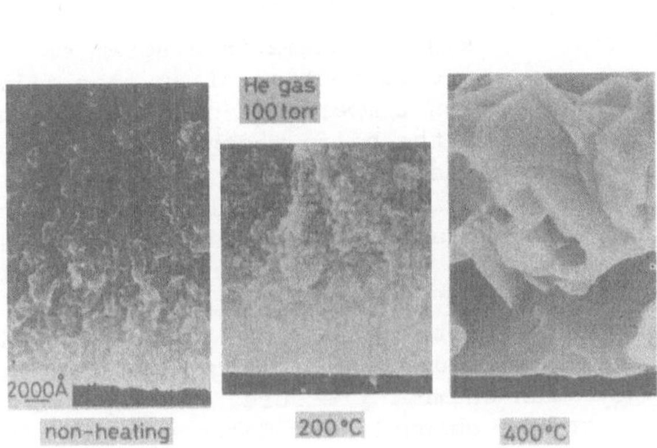

Fig.8 Scanning electron microscope picture of films obtained at from 30°C to 400°C at 0.13 atms

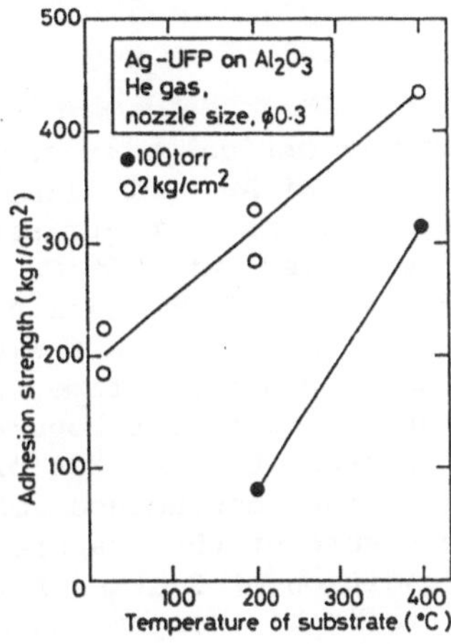

Fig.9 Substrate temperature dependence of adhesion strength of Ag films

3.1.4. Results of Specific Resistance. The dependence of the specific resistance of Ag films to the evaporaion chamber pressure is illustrated in fig.10. The white circles, the triangles and the black circles indicate the results of 30°C, 200°C and 400°C respectively. This indicates that the specific resistance of the films are reduced with the pressure in the evaporation chamber and the substrate temperature being increased. The specific resistance reaches less than 10 times that of Ag single crystal even at 30°C of the substrate temperature. It is low specific resistance enough for electrodes or wirings.

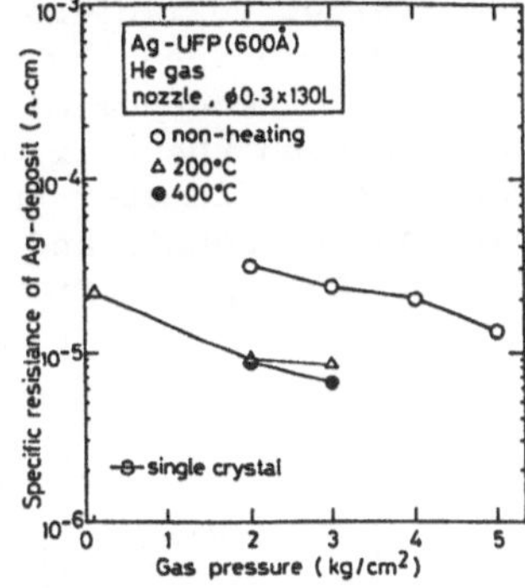

Fig.10 Evaporation chamber pressure dependence of specific resistance of Ag films

3.2. Au Film Formations for Microelectrodes

3.2.1. Shape of Au Spot Films and Deposition Rate. The optical microscope pictures of Au spot films are shown in fig.11. A crosssection of typical Au spot film is illustrated in fig.12. The shape is defined by the height, the half width, the bottom diameter and the tail diameter. The dependence of the height of Au spot films to the deposition time is illustrated in fig.13-a. At the condition of 2 atms of the evaporation chamber pressure, the deposition rate of Au

1208

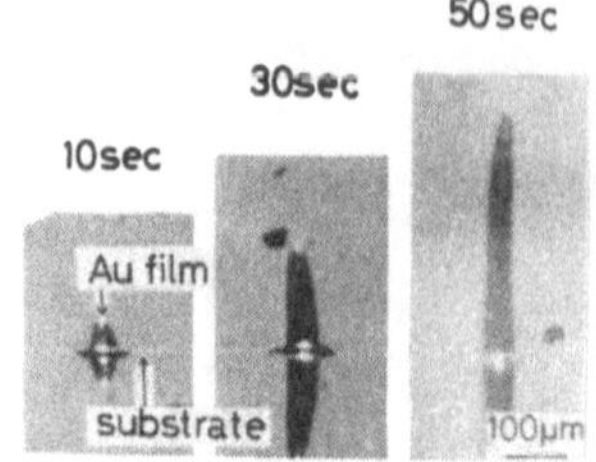

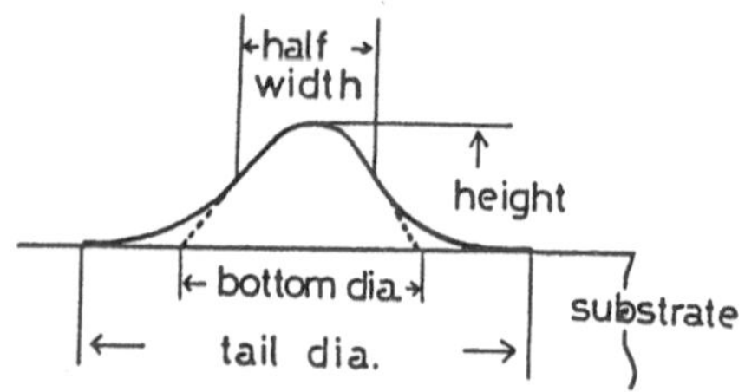

Fig.11 Optical microscope picture
       of Au spot films

Fig.12 Schematic diagram of
       crosssection of Au spot film

spot films is 0.5μm/sec at the evaporation temperature of 1380° C and
6.6 μm/sec at 1460°C. The dependence of the half width, the bottom
diameter and the tail diameter to the deposition time is illustrated in
fig.13-b. The half widths and the tail diameters are increased with
the deposition time while the bottom diameters are not much. The
bottom diameters bocome 50% of the nozzle inner diameter and the tail
diameters become 130% of it at the 50sec of the deposition time.

The dependence of the deposition rate of Au spot films to the
pressure of the evaporation chamber is illustrated in fig.13-c. At the
condition of 3 atms of the evaporation chamber pressure, the deposition
rate(both height change and volume change) is decreased at the
temperature of 1460°C because the evaporation rate is reduced much even
though the gas flow rate in the nozzle is increased. The dependence of
the bottom diameter and tail diameter to the pressure of the
evaporation chmber is illustrated in fig.13-d. At the condition of 3
atms, the average bottom diameter is increased by 50% of 2 atms while
the average tail diameter is not much, resulting that the height change
becomes more than the volume change.

The dependence of the bottom diameter and the tail diameter to the
gap between the nozzle and the substrate is illustrated in fig.13-e.
The bottom diameter and the tail diameter are increased when the gap
between the nozzle and the substrate is increased.
3.2.2. Au Spot Films with Multinozzles The optical microscope
pictures of Au spot films using a multinozzle is shown fig.14. All the
spot films are observed to be separated with each other but heights of
each spots are not uniform maybe because of the ununiformity in the
inner diameter of each nozzles.

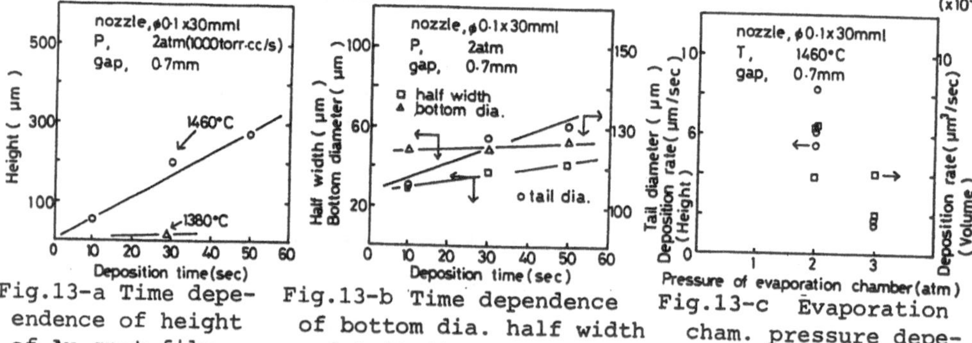

Fig.13-a Time depe-
endence of height
of Au spot film

Fig.13-b Time dependence
of bottom dia. half width
and tail dia. of Au film

Fig.13-c Evaporation
cham. pressure depe-
ndence of depo. rate

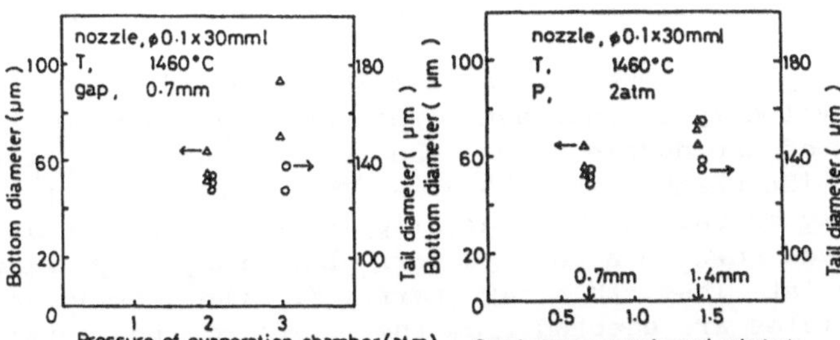

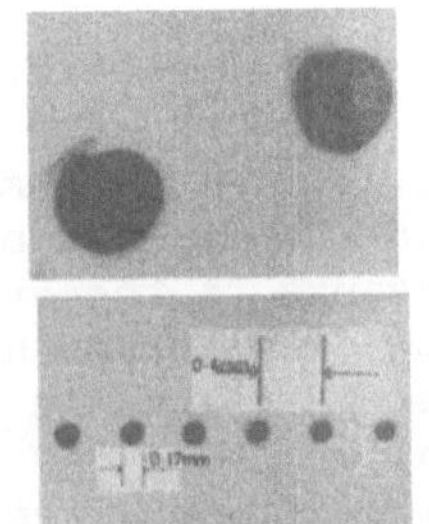

Fig.13-d Evaporation
Cham. pressure depe-
ndence of bottom dia.
and tail dia. of Au
spot film

Fig.13-e Gap between
nozzle and subtrate
dependence of bottom
dia. and tail dia. of
Au spot film

Fig.14 Optical microscope
picture of Au spot film
using a multinozzle

3.2.3. Results of Shutter System. Spot patterns of Au films using a shutter system is shown in fig.15. Deposition time is 5.0sec and the time between the depositions is 1.0sec. No tails can't be recognized between the spots.

3.2.4. Adhesion Strengths of Au Spot Films. The shearing strength of Au spot films are summerized in Tab.1. Au spot films for microelectrodes are usually formed by electroplatings. The average shearing strengths of Au films formed by electroplatings is about 10Kgf/mm^2. At the condition of 2atms of the evaporation chamber pressure and more than 200 °C of the substrate temperature, some of the gas deposition films have enough adhesion strengths for these applications.

Table 1. Shearing forces of
Au spot films

Sample No.	Substrate temp. (℃)	Pressure (atm)	Shearing Strength (kgf/mm^2)
1	150	3.0	0.1
2	150	3.0	0.3
3	150	2.0	0.3
4	200	2.0	27.0
5	200	2.0	8.4
6	200	2.0	7.2
7	250	2.0	15.0

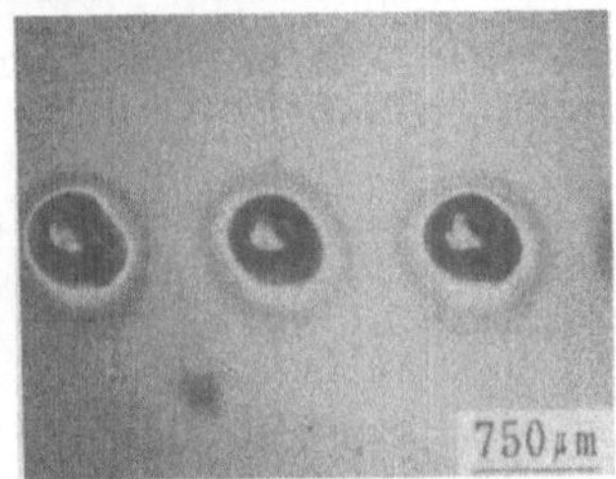

Fig.15 Optical microscope
picture of Au spot films
using shutter system

3. DISCUSSION

3.1.Ag Film Formation for Wirings

The experimental results indicate that the films which have higher densities and higher adhesion strengths can be obtained under the condition of higher pressure in the evaporation chamber and higher temperature of the substrate. Micrograins are left in the films obtained under the conditions of higher pressure and lower substrate temperature, even though the density of the films becomes higher. It is believed that the natural mechanisms of achieving higher densities and higher adhesion strengths in the films obtained under the condition of higher pressure in the evaporation chamber and lower substrate temperature are different from that of in the films obtained under the condition of lower pressure in the evaporation chamber and higher substrate temperature.

The average velocity of the particles is calculated and shown in fig.16. The helium gas velocity is increased exponentially reaching to Mach 1 at the outlet of the nozzle under the condition where the pressure of the evaporation chamber is 0.13 atms. The particles with a mean diameter of 600Å may be carried without slipping until the helium gas pressure is reduced down to around 40 torr but they may slip against the gas flow at less than 40 torr. As the result of calculations, the particles are ejected from the outlet of the nozzle at the velocity of about 300m/sec.

On the other hand, the particles can be carried without slipping to the outlet of the nozzle when the pressure of the evaporation chamber is 3.0 atms, because the pressure of the gas at the outlet of the nozzle is high enough to prevent slipping. The particles are estimated to be ejected from the outlet of the nozzle at the velocity of 970m/sec, that is Mach 1 of helium gas. It can not exceed Mach 1 of helium gas, because the gas flow velocity can not exceed the Mach 1 because of the mach wave resistance in a pipe.

It is estimated that the average velocity of the particles of 3 atms at the outlet of the nozzle is 3.23 times faster than that of 0.13 atms, this means, under these conditions. the kinetic energy is 10 times larger. This energy is estimated to be large enough for sintering the particles because the surface of the pictures are not oxidized. Under this condition, the particles do not bocome larger because excess energies are absorbed by the substrate kept at 30°C. On the other hand, under the condition of higher substrate temperature, the particles do become larger because the excess energies can be supplied from the substrate kept at over 200°C.

3.2.Au Film Formations for Microelectrodes

For an application to forming microelectrodes on the substrate such as 8 inhces Si wafer, specifications are as follows;
(1) Height of electrodes is about 20μm.
(2) The size of the electrodes is required to be 50 to 100μm.
(3) The pitch of the spot films is about 100 to 200μm.

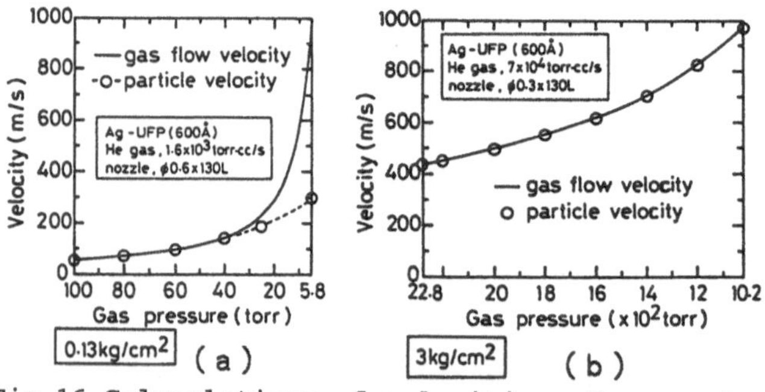

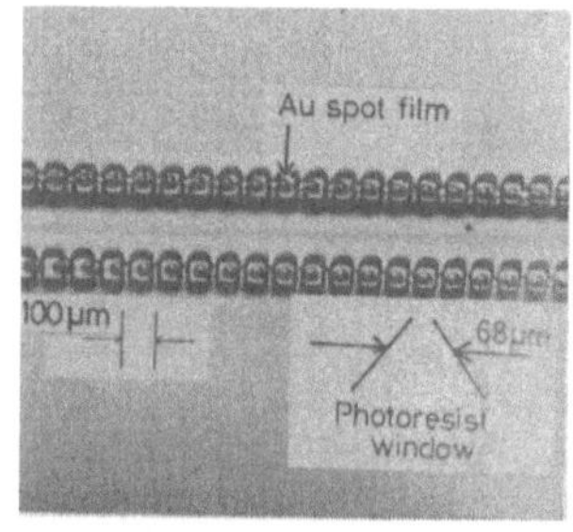

Fig.16 Caluculations of velocities of gas and particles in transfer pipe (a)Evaporation cham. pressure is 0.13 atms and (b)3 atms

Fig. 17 Optical microscope picture of film formed at windows of photoresist film

(4) About 30,000 spots must be formed within 30min(1800sec).
(5) The positioning error must be within 5µm.
(6) Strengths against shearing forces is more than 10kgf/mm^2.
In the present study, the deposition rate of 6.6µm/sec is obtained in the condition where the evaporation temperature is 1460 °C, the evaporation chamber pressur is 2 atm. At this deposition rate, about 3 sec is needed to form one spot film. About 1 to 2 sec is necessary for transfering to and positioning for the next spot. About 5 sec is necessary for both processes. At least, about 80 spots must be formed at one shot to form 30,000 spot films within 1800sec. A multinozzle with more than 80 nozzles must be developed for this application.

At the deposition rate of 6.6µm/sec, the tail diameter becomes more than 100µm to obtain a spot film with height of 20µm. The gap between the nozzle outlet and the substrate is required to be kept less than 0.5mm to reduce the tail diameters. In the case of the single nozzle, this condition may be possible to be obtained. But in the case of a multinozzle with 80 nozzles, it is difficult to be obtained because of surface roughnesses of the substrate. At the present stage, a process of patterning windows with photoresists in necessary to lift off the tail part of the films. As the example, the optical microscope picture of the films formed at the windows of a photoresist film is shown fig.17. Tail parts are thin enough to be lifted off.

The shearing strengths of some of the films reaches 10kgf/mm^2 at the condition that at more than 2 atms of the evaporation chamber pressure and at more than 200°C of the substrate temperature. This results roughly match with those of Ag films. The gas flow conductance of the pipe is proportional to d^4 at this pressure condition and the crosssection of the nozzle is proportional to d^2. As a result, the average gas velocity is proportional to d^2 until Mach 1. The gas velocities are distributed in a narrow pipe, that is, large at the center but small at close to the wall of the nozzle. In the case of Ag film formations, a nozzle with the inner diameter of 0.3mm is used, while a nozzle with 0.1mm is used in the case of Au spot films. The average gas velocity in the nozzle may be about 10 times reduced, compared with Ag film formations. It is estimated that only the particles carried in the center of the nozzle are contributed to the film formations. This matches with the results that the bottom diameter of the film is about 50% of the nozzle inner diameter at the condition of 2 atms, and the bottom diameter is increased at 3 atms.

4.CONCLUSION

In this article, we have shown that:
(1)The density and the adhesion strength of the Ag films increase with the increase of the pressure in the evaporation chamber.
(2)The adhesion strength and the specific resistance of the Ag films reach to 550kgf/cm^2 and 1.3 x 10^{5}Ωcm respectively even at 30°C of the substrate temperature. This film leaves the structure of micrograins.
(3)Au spot films with the bottom diameter of 50µm can be formed using a single nozzle with the inner diameter of 100µm.

(4)The multinozzle with 10 nozzles and the shutter system are effective for patternings.
(5)The shearing strength of some of the Au films is large enough to apply to microelectrodes.

References
1. M. Oda and N. Saegusa: Jpn.J.Appl.Phys.,Vol.24,No.9,L702(1985).
2. C. Hayashi: Jpn.J.Appl.Phys.,Vol.23,No.12,L910(1984).
3. K. Hatanaka, et al.: Proc. of 2nd International Symposium of Superconductivity,Nov.14(1989).

FULLERENE AND FULLERIDE FILMS

A. F. Hebard
A T & T Bell Laboratories
Murray Hill
New Jersey 07974

ABSTRACT. A brief description of the present status of research on fullerene and alkali-metal doped fulleride films is given. Deposition, metal doping, and characterization techniques are discussed. Measurement results including structure, stoichiometry, infrared and visible absorption, electrical transport, dielectric properties of the fullerenes, and superconducting properties of the fullerides are reviewed. Our present understanding of these novel materials has led to unexpected scientific insights together with considerations of how these materials might be used in technological applications.

1. Introduction

Carbon-sixty (C_{60}) with sixty carbon atoms symmetrically arrayed in a soccer ball configuration resembling the geodesic dome structures of inventor/architect Buckminster Fuller has the promise of being the ultimate "designer" cluster molecule to be employed in a wide variety of scientific studies and possible applications. The discovery of this unique molecule in gas phase[1] followed five years later by the announcement of a simple technique to produce macroscopic amounts[2] has stimulated strongly focused research efforts and some unexpected discoveries such as conductivity[3], superconductivity[4], and ferromagnetism[5]. The almost-spherical icosahedral symmetry of the C_{60} molecule, together with it's demonstrated ability to undergo reversible one-electron reductions[6] suggests a new chemistry with synthesis of compounds having unexpected and unusual properties.

It is appealing to consider C_{60} as a molecular building block not only because of it's promising chemistry but also because of it's uniformity of size and characteristic dimension. The 10 Å separation between close-packed C_{60} molecules represents an intermediate length scale, greater than typical atomic separations (~ 3 Å) and less than minimum feature dimensions (~ 100 Å) of fine-line lithography. Thus, nanometer-scale engineered structures with novel chemical and physical properties can be envisaged. Such structures might be fabricated on metal or semiconducting surfaces. It is likely that the technological foundations for such developments will derive from studies of fullerene and fulleride thin films rather than studies of bulk powders and crystals. The following brief review, therefore, focuses on thin films and although intended to be current, cannot be complete because of fast-breaking current research developments.

1213

P. Jena et al. (eds.), Physics and Chemistry of Finite Systems: From Clusters to Crystals, Vol. II, 1213–1220.
© 1992 *Kluwer Academic Publishers.*

2. Deposition

Published descriptions of fullerene film growth by sublimation[2] and laser vaporization of carbon in an inert atmosphere[7] contain the essential information embodied in present film-growth techniques. For reproducibility, sublimation in high vacuum is preferred[8,9]. One must only heat a crucible containing a fullerene powder at a controlled temperature to obtain a steady sublimation rate. Low temperature ($\sim 200°C$) outgassing of the solid fullerene source material before opening a shutter to expose the substrate will prevent high vapor pressure residues from reaching the substrate. The purity of the deposited film depends on the purity of the starting material. Purities better than 99.9% for separate C_{60} and C_{70} fractions have been shown to give correspondingly pure films. Mixed composition C_{60}/C_{70} films can be obtained by using fullerene powder which has not been separated into different mass fractions.

Solutions containing dissolved fullerenes can also be applied to a substrate leaving a solid film when the solvent evaporates. Although this is the simplest way to form a fullerene film, it suffers from two disadvantages: solvent residues are often left behind and the solid film has uncontrollable thickness variations.

Alkali-metal doped fulleride films have been obtained by exposing previously evaporated fullerene films to alkali metal vapor[3]. *In-situ* monitoring of the resistance can be used to determine stoichiometry. Alternatively, in ultra high vacuum systems, analysis of photoelectron emission spectra can also be used to infer stoichiometry[10,11]. It is also possible, though technically more difficult, to co-deposit both fullerene and alkali metal at the same time. This technique requires the precise calibration of independently controlled sources to obtain the desired stoichiometry. Properly executed, this technique should give films with uniform composition.

3. Physical Properties

C_{60} films have been successfully deposited on a variety of substrates including glass, Si, quartz, KBr, sapphire, and Si(111)[8,9]. On transparent substrates the films have a yellow cast, and by surviving the application and removal of Scotch tape[8] show remarkable adherence and robustness for a Van der Waals solid. Infrared transmission spectra can be used as a fingerprint test for phase purity and also to check for the presence of hydrocarbon contamination, revealed by C-H bend and stretch modes. Ultraviolet-visible absorption spectra, which result from the interband electronic transitions among the π-orbitals, can also be used to confirm phase purity. Optical absorption coefficients in the 200-400 nm range have values somewhat greater than Si but less than Ge. Photovoltage[12], photoconductivity[13,14], and measurements of a non-linear susceptibility χ_3 for C_{60} films[15] have been reported. Additional detailed optical characterization, such as the frequency dependence of the complex index of refraction[16] will place into perspective the applicability of these materials to optical and electro-optical applications.

Sublimated C_{60} films have a close-packed structure which has been indexed to face-centered cubic (fcc), in agreement with x-ray measurements on bulk powders[17]. Extent of crystallinity for films deposited on room-temperature substrates is only about 60 Å[8]. Larger crystalline grains might be obtained by sublimation onto lattice-matched substrates at elevated temperatures. Registry of C_{60} molecules with surface features, such as steps, on freshly-cleaved GaAs has been observed in Scanning Tunneling Microscopy (STM)[18]. These same measurements show layer-by-layer growth of close-packed crystalline plateaus that have a lateral extent similar to that seen by x-rays.

Band structure calculations of fcc C_{60} predict semiconducting behavior with a direct band gap near 1.5 eV[19]. The gap between the highest occupied molecular orbital and the lowest unoccupied molecular orbital is determined by photoemission studies to be 1.9 eV[20]. Transport measurements of $C60/C70$ films reveal a room-temperature resistivity of approximately 10^{14} Ωcm[21].

Capacitance of trilayer Al-C_{60}-Al structures has been studied as a function of C_{60} thickness and a relative permittivity $\kappa = 4.4$ obtained[8]. This is in agreement with refractive index measurements[16] and extrapolation of electron energy-loss spectroscopy data to zero frequency[22]. This polarizability is similar to that of SiO_x films, and represents a value at the low end of thin-film oxides. Electric-field breakdown strength is also low, near 10^5 V/cm, and may be due to the ease with which contaminant ions can diffuse through the interstices of the close-packed C_{60} structure.

4. Conducting Fulleride Films

Insulating fullerene films can be made conducting by exposure to alkali-metal vapor[3]. The alkali metal atoms are incorporated into the octahedral and tetrahedral sites of the face centered cubic (fcc) lattice where they donate electrons into the triply-degenerate lowest unoccupied molecular orbital (LUMO) of the C_{60} molecule. At half-band filling, corresponding to three donated electrons per C_{60} molecule, the conductivity is maximum and the stoichiometry is A_3C_{60} (A = K, Rb). Measurements of the Raman shift of the A_g pentagonal breathing mode confirm this picture of electron donation by showing that the mode shift to lower frequency for K_6C_{60} is twice that for K_3C_{60}[3,23]. Physically, the shift results because the donated electrons fill antibonding orbitals causing the molecule to expand slightly and vibrate at lower frequency.

Analogy with graphite intercalated compounds (GIC's) is useful but not complete. One important difference is that for GIC's the conductivity is anisotropic with maximum conductivity parallel to the planes whereas in fullerene solids the conductivity is isotropic. This difference arises because of the location of the intercalants: in GIC's the intercalants are layered between the graphite sheets whereas in fcc fullerene crystals the intercalants are arrayed symmetrically in the tetrahedral and octahedral sites surrounding each C_{60} molecule. Three dimensionality is favorable for metallic behavior since instabilities such as charge density waves which compete with metallic behavior tend to be favored in lower dimensions.

Doping films, powders, or crystals with alkali metals must be done in an inert environment. This environment can be either high vacuum or nonreactive gas such as helium. As a function of alkali metal stoichiometry x the resistivity of $A_x C_{60}$ (A=Li,Na,K,Rb,Cs) decreases rapidly from a value characteristic of an insulator, passes through a minimum, and then increases again. The resistivity minimum for $K_x C_{60}$ has been measured by Rutherford backscattering to occur at a stoichiometry $x = 3.00 \pm 0.05$ corresponding to half band filling[24]. The resistivity at the minimum of 2-3 mΩ cm for $K_3 C_{60}$ is smaller than the values obtained with other alkali intercalants[3]. Intercalant ions such as Cs are too large to fit into the tetrahedral sites of the fcc lattice and the smaller intercalant ions such as Li and Na are most likely asymmetrically located within their respective interstitial sites.

As doping proceeds towards $x = 3$ the color of the films changes from the yellow hue of the undoped films to a reddish and then grayish metallic color. In poor vacuum (10^{-5} Torr) doping to x=3 takes significantly longer than doping in ultra high vacuum. This is most likely due to diffusion barriers set up by the formation of contaminating oxides and hydroxides on the film surface.

Potassium appears to be a fast diffuser in C_{60} films, having a diffusion constant estimated[24] to be $10^{-12}\,cm^2\,sec^{-1}$ with an activation energy of 0.5eV[25]. This rapid diffusion is most likely due to the numerous grain boundaries associated with the small grains of the undoped film. Bulk powders with their larger micron size grains require significantly longer times (days) than films (seconds) to obtain uniform doping concentrations[26].

The electronic properties of *in situ* doped films have been extensively studied by photoemission and inverse photoemission[10,11,25,27]. Detailed information about molecular band structure in the undoped fullerenes and the occupation of a LUMO-derived conduction band arising from donated alkali atom electrons has led to a preliminary understanding of electronic structure. Intensity ratios give information about stoichiometry and lineshape analysis is consistent with the presence of discrete phases. The energy width of the half-filled conduction band of $K_3 C_{60}$ is approximately 0.5 eV.

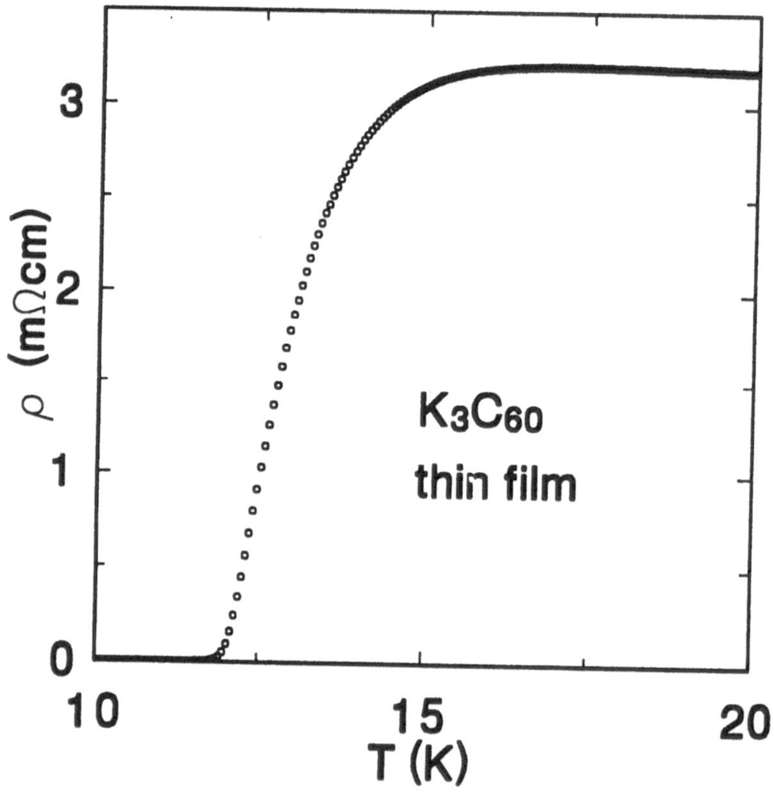

Figure 1: Resistance transition of a 1600 Å-thick $K_3 C_{60}$ thin film.

5. Superconductivity

K_3C_{60} and Rb_3C_{60} are superconducting with transition temperatures of 18 K[4] and 28 K[26,28] respectively. A resistance transition of a K_3C_{60} film[29] is shown in Fig. 1. The onset of superconductivity, defined as the temperature where the resistance noticeably begins to decrease, is at 15K with zero resistance occurring near 12 K. Using a 50% normal-state criterion, an upper critical field slope of 5.5T/K is found. The resulting values for the zero-temperature critical field of 50 T and coherence length of 26Å are in good agreement with magnetic measurements on bulk samples[30,31].

The reduction in transition temperature and width of the transition region have been attributed to a granular microstructure[29]. This granularity arises because of the coexistence of immiscible discrete phases. This picture is substantiated by photoemission[25,27], X-ray structure determinations of fcc C_{60}[17], fcc K_3C_{60}[32], bct K_4C_{60}[33], and bcc K_3C_{60}[34] phases, determination by C^{13} NMR for $x \leq 3$ of only two motionally narrowed lines corresponding to an admixture of C_{60} and K_3C_{60} phases, and observation of a maximum Meissner fraction for $x = 3$[26]. The temperature and stoichiometry dependence of the resistivity of K_xC_{60} is consistent with this evidence for immiscible phases[24]. Thus, as x increases from zero, conducting K_3C_{60} grains nucleate in an insulating C_{60} matrix and the temperature dependence of the resistance is activated with an energy determined by the electrostatic energy cost of creating charge carriers in an array of small (60 Å) grains. At $x = 3$ the grains coalesce, the charging energies become negligible, and a barely metallic resistivity of $3\,m\Omega\,cm$ obtains. At present it is not known if this value is intrinsic (intra-grain resistance) or extrinsic (inter-grain resistance).

6. Future Prospects

The demonstration that fullerene films can be made in purified form and then doped with alkali metals to achieve conducting and even superconducting behavior provides unique opportunities for scientific inquiry and possible applications. Understanding the mechanism for superconducting pairing, whether it be electron-phonon or electron-electron or a combination of both, could well lead to even higher transition temperatures in related compounds. Applications, however, might be seriously impeded by the extreme air sensitivity of the alkali-metal-doped compounds. The possibility of making chemical derivatives by incorporating metal ions within each fullerene cage or by replacing carbon atoms on the cage[36] greatly magnifies the opportunities. It is also worthwhile to think of fullerenes and their derivatives as building blocks for nanometer-scale structures. How fullerene molecules behave at interfaces and whether they can be patterned or assembled into mesoscopic structures are interesting questions. Use of fullerene and fulleride thin films in optical, opto-electronic, semiconducting, and superconducting applications may well take place in ways presently unimagined.

Acknowledgements

The author acknowledges useful discussions and fruitful collaboration with S. J. Duclos, A. T. Fiory, R. M. Fleming, S. H. Glarum, R. C. Haddon, G. P. Kochanski, A. R. Kortan, B. Miller, D. W. Murphy, T. T. M. Palstra, A. P. Ramirez, J. E. Rowe, M. J. Rosseinsky, R. Tycko, G. K. Wertheim, and W. L. Wilson.

References

[1] H. W. Kroto, J. R. Heath, S. C. O'Brien, R. F. Curl, and R. E. Smalley, 'C$_{60}$: Buckminsterfullerene', Nature **318**, 162(1985).

[2] W. Krätschmer, L. D. Lamb, K. Fostiropoulos, and D. R. Huffman, 'Solid C$_{60}$: A new form of carbon', Nature **347**, 354(1990).

[3] R. C. Haddon, A. F. Hebard, M. J. Rosseinsky, D. W. Murphy, S. J. Duclos, K. B. Lyons, B. Miller, J. M. Rosamilla, R. M. Fleming, A. R. Kortan, S. H. Glarum, A. V. Makhija, A. J. Muller, R. H. Eick, S. M. Zahurak, R. Tycko, G. Dabbagh, and F. A. Thiel, 'Conducting films of C$_{60}$ and C$_{70}$ by alkali-metal doping', Nature **350**, 320(1991).

[4] A. F. Hebard, M. J. Rosseinsky, R. C. Haddon, D. W. Murphy, S. H. Glarum, T. T. M. Palstra, A. P. Ramirez, and A. R. Kortan, 'Superconductivity at 18 K in potassium-doped C$_{60}$', Nature **350**, 600(1991).

[5] P. -M. Allemand, K. C. Khemani, A. Koch, F. Wudl, K. Holczer, S. Donovan, G. Gruner, and J. D. Thompson, 'Organic molecular soft ferromagnetism in a fullerene C$_{60}$', Science **253**, 301(1991).

[6] D. Dubois, K. M. Kadish, S. Flanigan, R. E. Haufler, L. P. F. Chibante, and L. J. Wilson, 'A spectroelectrochemical study of the C$_{60}$ and C$_{70}$ fullerenes and their mono-, di-, tri-, and tetraanions', J. Amer. Chem Soc. **113**, 4364(1991).

[7] G. Meijer and D. S. Bethune, 'Laser deposition of carbon clusters: A new approach to the study of fullerenes', J. Chem. Phys. **93**, 7800(1990).

[8] A. F. Hebard, R. C. Haddon, R. M. Fleming, and A. R. Kortan, 'Deposition and characterization of fullerene films', Appl. Phys. Lett. **59**, 2109(1991).

[9] W. M. Tong, D. A. A. Ohlberg, H. K. You, R. S. Williams, S. J. Anz, M. M. Alvarez, R. L. Whetten, Y. Rubin, and F. N. Diederich, 'X-ray diffraction and electron spectroscopy of epitaxial molecular C$_{60}$ films', J. Phys. Chem. **95**, 4709(1991).

[10] P. J. Benning, J. L. Martins, J. H. Weaver, L. P. F. Chibante, and R. E. Smalley, 'Electronic states of K$_x$C$_{60}$: Insulating, metallic, and superconducting character', Science **252**, 1417(1991).

[11] G. K. Wertheim, J. E. Rowe, D. N. E. Buchanan, E. E. Chaban, A. F. Hebard, A. R. Kortan, A. V. Makhija, and R. C. Haddon, 'Photoemission spectra and electronic properties of K$_x$C$_{60}$', Science **252**, 1419(1991).

[12] B. Miller, J. M. Rosamilia, G. Dabbagh, R. Tycko, R. C. Haddon, A. J. Muller, W. Wilson, D. W. Murphy, and A. F. Hebard, 'Photoelectrochemical behavior of C$_{60}$ films', J. Am. Chem. Soc. **113**, 6291(1991).

[13] J. Mort, K. Okumura, M. Machonkin, R. Ziolo, R. Huffman, and M. I. Ferguson, 'Photoconductivity in solid films of C$_{60}$/C$_{70}$', Chem. Phys. Lett. (in press), (1991).

[14] K. Pichler, S. Graham. O. M. Gelsen, R. H. Friend, W. J. Romanow, J. P. McCauley Jr., N. Coustel, J. E. Fischer, and A. B. Smith, 'Photophysical properties of solid films of fullerene, C$_{60}$', preprint, (1991).

[15] H. Hoshi, N. Nakamura, Y. Maruyama, T. Nakagawa, S. Suzuki, H. Shiromaru, and Y. Achiba, 'Optical second and third-harmonic generations in C_{60} film', Jpn. J. Appl. Phys. **30**, L1397(1991).

[16] S. L. Ren, Y. Wang, A. M. Rao, E. McRae, J. M. Holden, T. Hager, K. -A. Wang, W. -T. Lee, H. F. Ni, J. Selegue, and P. C. Eklund, 'Ellipsometric determination of the optical constants of C_{60} (Buckminsterfullerene) films', submitted to Appl. Phys. Lett.

[17] R. M. Fleming, T. Siegrist, P. M. Marsh, B. Hessen, A. R. Kortan, D. W. Murphy, R. C. Haddon, R. Tycko, G. Dabbagh, A. M. Mujsce, M. L. Kaplan, and S. M. Zahurek, 'Diffraction symmetry in crystalline close-packed C_{60}', Mat. Res. Soc. Symp. **206**, 691(1991).

[18] Y. Z. Li, M. Chander, J. C. Patrin, J. H. Weaver, L. P. F. Chibante, and R. E. Smalley, 'Order and Disorder in C_{60}, C_{70}, and $K_x C_{60}$ multilayers: Direct imaging with scanning tunneling microscopy', Science **253**, 429(1991).

[19] S. Saito and A. Oshiyama, 'Cohesive mechanism and energy bands of solid C_{60}', Phys. Rev. Lett. **66**, 2637(1991).

[20] J. Weaver, J. L. Martins, T. Komeda, Y. Chen, T. R. Ohno, G. H. Kroll, N. Troullier, R. E. Haufler, and R. E. Smalley, 'Electronic structure of solid C_{60}', Phys. Rev. Lett. **66**, 1741(1991).

[21] J. Mort, R. Ziolo, M. Machonkin, D. R. Huffman, and M. I. Ferguson, 'Electrical conductivity studies of undoped solid films of $C_{60/70}$', submitted to Chem. Phys. Lett.

[22] E. Sohmen, J. Fink, R. H. Baughman, and W. Krätschmer, 'Electron energy-loss spectroscopy studies on C_{60} and C_{70} fullerite', submitted to Z. Physik B.

[23] S. J. Duclos, R. C. Haddon, S. Glarum, A. F. Hebard, and K. B. Lyons, 'Raman Studies of alkali-metal doped $A_x C_{60}$ films (A = Na, K, Rb, Cs; x = 0, 3, 6)', Science, (in press).

[24] G. P. Kochanski, A. F. Hebard, R. C. Haddon, and A. T. Fiory, 'Electrical resistivity and stoichiometry of $K_x C_{60}$ films', Science (in press).

[25] P. J. Benning, D. M. Poirier, T. R. Ohno, Y. Chen, M. B. Jost, F. Stepniak, G. H. Kroll, and J. H. Weaver, 'C_{60} and C_{70} fullerenes and potassium fullerides', submitted to Phys. Rev. B.

[26] K. Holczer, O. Klein, S. -M. Huang, R. B. Kaner, K. -J. Fu, R. L. Whetten, and F. Diederich, 'Alkali-fulleride superconductors: synthesis, composition, and diamagnetic shielding', Science **252**, 1154(1991).

[27] C. T. Chen, L. H. Tjeng, P. Rudolf, G. Meigs, J. E. Rowe, J. Chen, J. P. McCauley, Jr., A. B. Smith III, A. R. McGhie, W. J. Romanow, and E. W. Plummer, 'Electronic states and phases of $K_x C_{60}$', Nature **352**, 603(1991).

[28] M. J. Rosseinsky, A. P. Ramirez, S. H. Glarum, D. W. Murphy, R. C. Haddon, A. F. Hebard, T. T. M. Palstra, A. R. Kortan, S. M. Zahurak, and A. V. Makhija, 'Superconductivity at 28 K in $Rb_x C_{60}$', Phys. Rev. Lett. **66**, 2830(1991).

[29] T. T. M. Palstra, R. C. Haddon, A. F. Hebard, and J. Zaanen, 'Electroic transport properties of $K_3 C_{60}$ films', submitted to Phys. Rev. Lett.

[30] K. Holczer, O. Klein, G. Grüner, J. D. Thompson, F. Diederich, and R. L. Whetten, Critical magnetic fields in the superconducting state of $K_3 C_{60}$´, Phys. Rev. Lett. **67**, 271(1991).

[31] A. P. Ramirez, M. J. Rosseinsky, D. W. Murphy, and R. C. Haddon, ´Density of electronic states in $K_{3-x} Rb_x C_{60}$´, submitted to Science.

[32] P. W. Stephens, L. Mihaly, P. L. Lee, R. L. Whetten, S. -M. Huang, R. Kaner, F. Diederich, and K. Holczer, ´Structure of single-phase superconducting $K_3 C_{60}$´, Nature **351**, 632(1991).

[33] R. M. Fleming, M. J. Rosseinsky, A. P. Ramirez, D. W. Murphy, J. C. Tully, R. C. Haddon, T. Siegrist, R. Tycko, S. H. Glarum, P. Marsh, G. Dabbagh, S. M. Zahurak, A. V. Makhija, and C. Hampton, ´Preparation and structure of the alkali-metal fulleride $A_4 C_{60}$´, Nature **352**, 701(1991).

[34] O. Zhou, J. E. Fischer, N. Coustel, S. Kycia, Q. Zhu, A. M. McGhie, W. J. Romanow, J. P. McCauley Jr., and A. B. Smith III, ´Structure and bonding in alkali metal-doped C_{60}´, Nature **351**, 462(1991).

[35] R. Tycko, G. Dabbagh, M. J. Rosseinsky, D. W. Murphy, R. M. Fleming, A. P. Ramirez, and J. C. Tully, ´^{13}C NMR Spectroscopy of $K_x C_{60}$: Phase separation, molecular dynamics, and metallic properties´, Science **353**, 884(1991).

[36] R. E. Smalley, ´Doping the fullerenes´, ACS symposium series, Large Carbon Clusters, edited by G. Hammond and V. Kuck, (1991).

NANOCRYSTALLINE AND ICOSAHEDRAL AMORPHOUS RF SPUTTERED CrNi(65:35) THIN FILMS

M.I.BIRJEGA, C.SARBU and M.ALEXE
Institute of Atomic Physics
Bucharest-Măgurele
P.O.Box MG-6, R-76900, ROMANIA

ABSTRACT. The effect of deposition parameters such as the substrate temperature T_d (300ºC, 250ºC, 200ºC) and argon gas pressure P, (600 mtorr, 300 mtorr, 150 mtorr, 20 mtorr) on the structure of 40 nm thick, RF sputtered CrNi(65:35) films onto freshly air-cleaved doped and electrically coloured alkali halide substrates, has been studied with TEM techniques. Nanocrystalline CrNi(65:35) thin films which contain the equilibrium phase separated state [α-Cr(bcc) and γ-Ni(fcc)] are obtained at T_d, 300ºC at high argon pres - sure [600 mtorr, 300 mtorr and 150 mtorr] and T_d, 250ºC and 200ºC only at the lowest argon pressure. Nanocrystalline CrNi(65:35) thin films which contain the metastable single phase extended α-Cr(bcc) solid solution are obtained at T_d 300ºC at the lowest argon pressure and at T_d, 250ºC at 150 mtorr argon pressure. Contrastless crack network microstructures of the CrNi(65 :35) thin films which contain the metastable icosahedral amorphous singlephase are obtained at T_d 200ºC and high argon pressure [300 mtorr, 150 mtorr]. CrNi(65:35) thin films which contain nanocrystalline of α-Cr(bcc) extended solid solution embedded in the icosahedral amorphous matrix are obtained at T_d 250ºC at high argon pressure [600 mtorr, 300 mtorr] and at T_d 200ºC at the highest argon pressure.

INTRODUCTION

Results of TEM and TED studies on vacuum deposited CrNi and Cr thin films and their relation to literature data have already been reported in several of our previous papers [1-6]. In the present work, we have concentrated on the influence of argon gas pressure in connection with substrate temperature on the phases and microstructure of RF sputtered CrNi (65:35) thin films.

P. Jena et al. (eds.), Physics and Chemistry of Finite Systems: From Clusters to Crystals, Vol. II, 1221–1226.
© 1992 *Kluwer Academic Publishers.*

EXPERIMENTAL PROCEDURE

The CrNi(65:35) thin films were deposited simultaneously onto freshly air-cleaved doped and electrically coloured alkali halide [KI(Ag°), KCl(Pb°), KCl(Ag°), KBr(Ag°)] substrates using a standard RF sputtering apparatus. Before deposition the substrates were heat cleaned at 300°C for 30 minutes and then cooled down to the deposition temperature T_d of 200°C, 250°C or 300°C. The residual gas pressure before deposition was 5×10^{-6} torr. Argon of 99.998 purity was used as sputtering gas. The sputtering of the CrNi (65:35) thin films was performed at pressures of 20 mtorr, 150 mtorr, 300 mtorr or 600 mtorr argon. The RF sputtering power was maintained constantly at 200 W. Also, the distance between the target-source and the substrate, as well as the deposition time were maintained constantly at 37 mm and 120 s, respectively. For structural investigations the thin CrNi(65:35) films were removed from the alkali halide substrates by floating them off in distilled water and were examined using a JEM-200 CX electron microscope working at 200 kV.

RESULTS AND DISCUSSIONS

As we can see from Table 1, fig.1 and fig.2 the TED and TEM data revealed a complex dependence of the resulting CrNi (65:35) thin films phases and microstructure on the variable deposition parameters chosen.

Table 1. Experimental data on the observed phases

T_d °C	P, argon mtorr	Stable α-Cr and γ-Ni solid solutions	Metastable	
			Extended α-Cr solid solution	Single icosahedral amorphous a_1
300	600; 300 150	Only	–	–
	20	–	Only	–
250	600; 300	–	Traces	Mostly
	150	–	Only	–
	20	Only	–	–
200	600	–	Traces	Mostly
	300; 150	–	–	Only
	20	Only	–	–

The equilibrium crystalline phase separated state[α-Cr(bcc) and γ-Ni(fcc)] is obtained at T_d 300°C, at high argon pres-

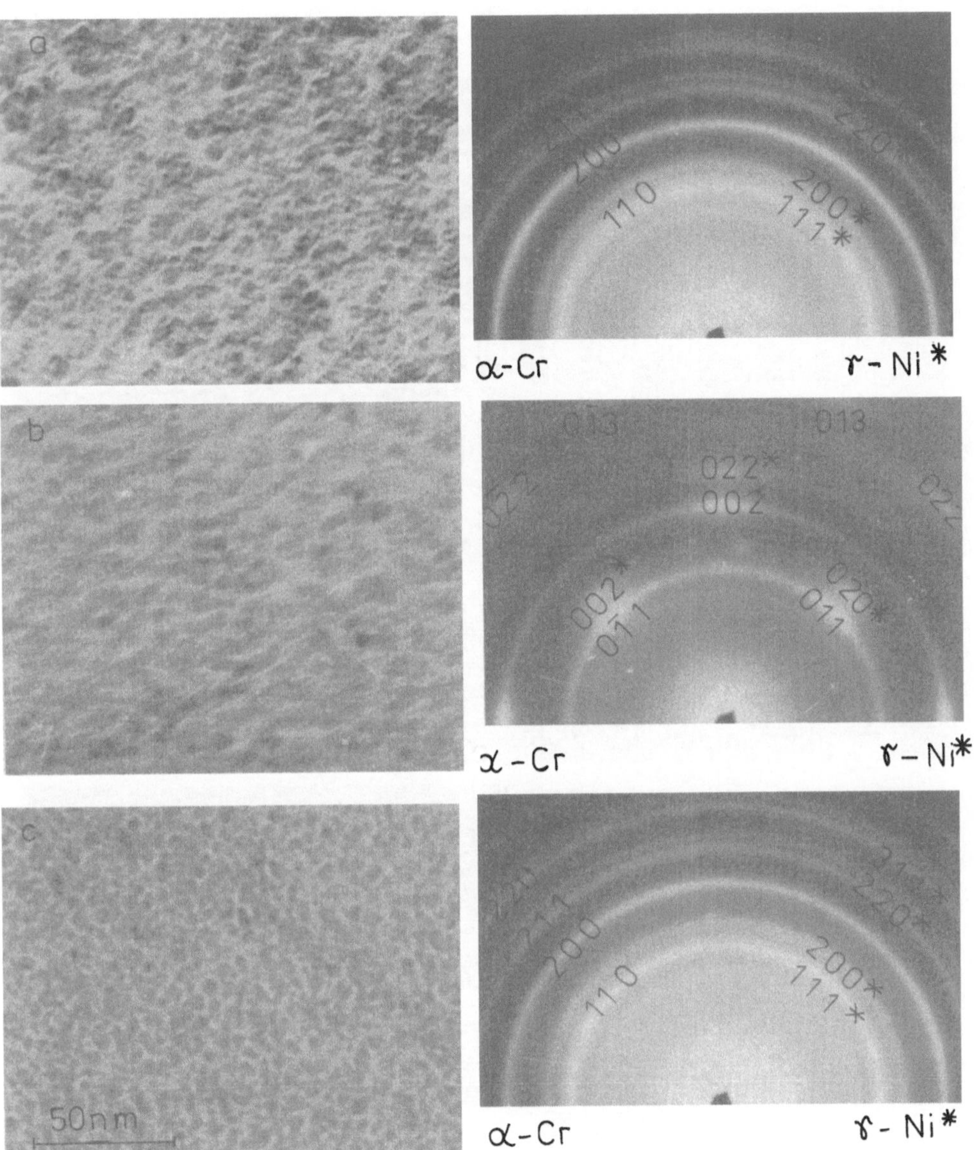

Fig.1. Electron micrographs and corresponding electron diffraction patterns of thin RF sputtered CuNi (65:35)
films on KI(Ag°) substrates at: a)(T_d = 300°C,
P=300 mtorr Ar);   b)(T_d = 300°C;   P = 150 mtorr Ar);
c) (T_d = 200°C;   P = 20 mtorr Ar).

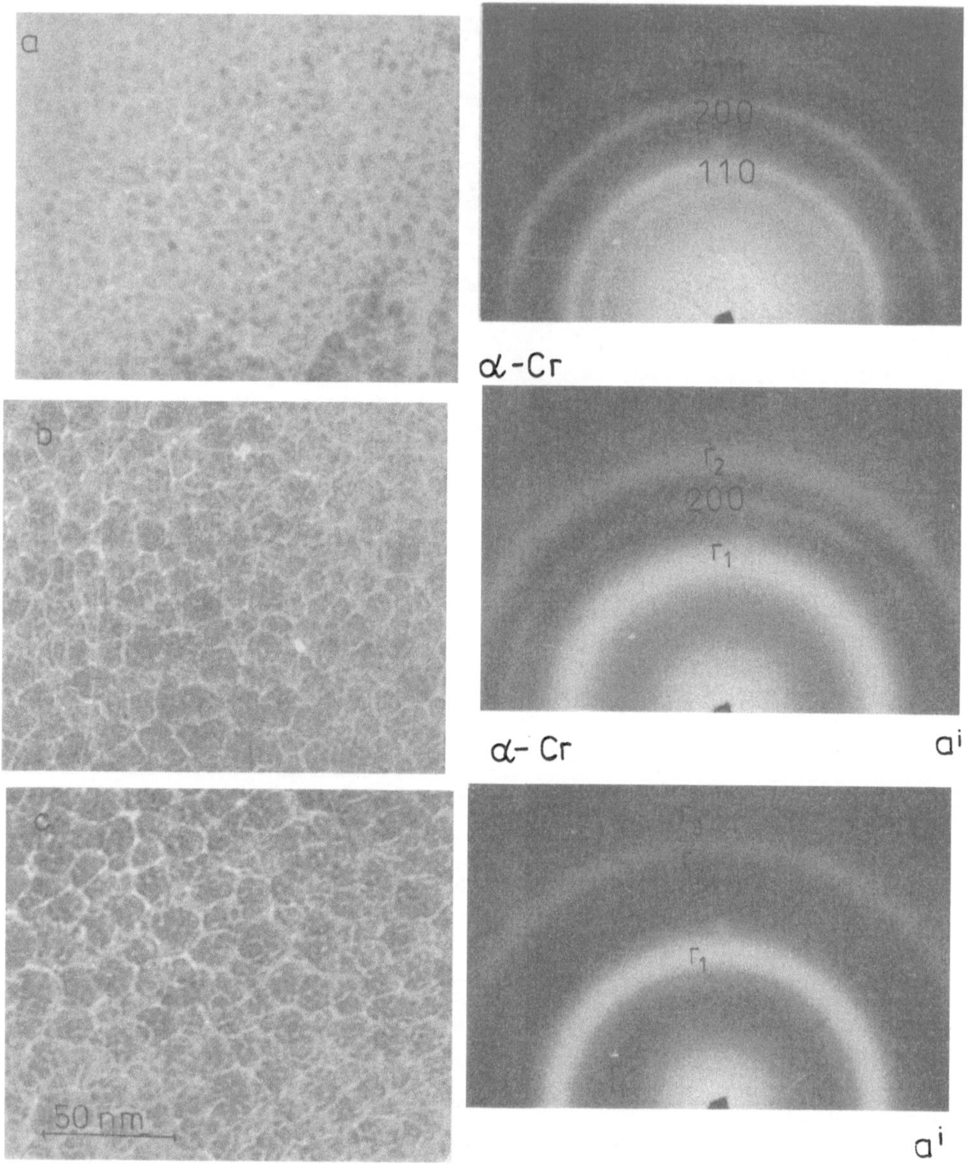

Fig.2. Electron micrographs and corresponding electron diffraction patterns of thin RF sputtered CrNi(65:35) films on KCl(Pb⁰) substrates at : a)(T_d = 300°C ; P = 20 mtorr Ar); b) (T_d = 250°C; P = 300 mtorr Ar); c) (T_d = 200°C; P = 300 mtorr Ar).

sure [600 mtorr, 300 mtorr and 150 mtorr] and at T_d 250°C and 200°C only at the lowest argon pressure (Table 1,fig. 1). As we can see from the electron microscope images (fig.1) though all of the thin CrNi(65:35) films contain nanocrystals belonging to the equilibrium phase separated state [1, 2, 4], their microstructures are different depending on the deposition parameters (T_d, P argon). One additional diffraction ring (0.24 nm) could be observed in the electron diffraction images (fig. 1, fig. 2a) and attributed to the cubic CrO (type NaCl) oxide.

Nanocrystalline CrNi(65:35) thin films which contain the metastable single phase extended α-Cr(bcc) solid solution and the CrO oxide are obtained at T_d 300°C at the lowest argon pressure and at T_d 250°C at 150 mtorr argon pressure (Table 1, fig. 2a).

The metastable icosahedral amorphous CrNi(65:35) single phase (icosahedral glass) is obtained at T_d 200°C and high argon pressure (300 mtorr, 150 mtorr) (Table 1, fig. 2c).The values of the peak positions in the structure factor of the amorphous CrNi(65:35) thin films are in good agreement with an assumed icosahedral short range order (SRO) [5, 6, 7].The contrastless electron microscope image (fig. 2c) of the as deposited CrNi(65:35) thin film reveals a crack network structure due to the existence of regions of slowly varying density or void formation during amorphous thin film growth [8].

The CrNi(65:35) thin films which contain nanocrystals of α-Cr(bcc) extended solid solution embedded in the icosahedral amorphous matrix, are obtained at T_d 250°C at high argon pressure (600 mtorr, 300 mtorr) and at T_d 200°C at the highest argon pressure (Table 1, fig. 2b). As we can see from the electron microscope images (fig. 2b, fig. 2c) the microstructure does not change considerably by the appearance of the nanocrystals.

Whereas further experimental data are necessary some primary considerations can be made.

It should thus be noted that the Frank-Kasper (δ, σ) crys - talline and quasicrystalline phases, which had been reported in both evaporated CrNi(65:35) thin films [1] and small CrNi(65:35) particles evaporated in an inert gas atmosphere [9, 10], do not occur. The formation of a single metastable icosahedral amorphous CrNi(65:35) phase at T_d 200°C, under protection of an argon atmosphere may hint to possible solid state reactions between the nuclei emerging and growing during (120 s) film deposition, which would be similar to those occurring for instance during the amorphization of NiZr powder or multilayers [11 - 13].

REFERENCES

[1] C.Sârbu, S.A.Rău, N.Popescu-Pogrion and M.I.Bîrjega, Thin Solid Films 23 (1975) 311.

[2] M.I.Bîrjega, N.Popescu-Pogrion, C.Sârbu and S.Rău, Thin Solid Films 34 (1976) 153.

[3] M.I.Bîrjega, N.Popescu-Pogrion, C.Sârbu and V.Topa, Thin Solid Films 58 (1979) 217.

[4] M.I.Bîrjega, C.A.Constantin, I.Th.Florescu and C.Sârbu, Thin Solid Films 92 (1982) 315.

[5] M.I.Bîrjega and N.Popescu-Pogrion, Thin Solid Films 171 (1989) 33.

[6] M.I.Bîrjega, N.Popescu-Pogrion and C.A.Constantin, Z.Phys.D20 (1991) 369.

[7] S.Sachdev and D.R.Nelson, Phys.Rev. B 32 (1985) 1480;

[8] R.Messier, J. Vac. Sci. Technol. A4 (1986) 490.

[9] N.Yukawa, Y.Fukano, M.Kawamura and T.Imura, Trans. Jpn. Inst. Met. 9 (1968) 272.

[10] T.Ishimasa, H.U.Nissen and Y.Fukano, Phys. Rev.Lett. 55(1985) 511.

[11] R.B.Schwarz and W.L.Johnson, Phys. Rev. Lett. 51 (1983) 415.

[12] K.Samwer, Phys. Rep. 161 (1988) 3.

[13] R.De Reus and F.W.Saris, Mat. Lett. 9 (1990) 487.

COMPOSITE CLUSTERS BY LASER VAPORIZATION

F.W. FROBEN
Free University Berlin
Institute for Experimental Physics
Arnimallee 14
1000 Berlin 33
Germany

ABSTRACT. Composite material-ceramics (boron nitride, aluminium oxyde, YBCO) is vaporized by pulse laser vaporization. Three different types of measurements are performed. Firstly the emission of the plasma above the solid surface is analysed by time resolved optical spectroscopy. Secondly, the clusters are condensed on a carbon film and measured by electron microscopy and finally the films formed on a clean surface are examined by mechanical and optical methods.

1. INTRODUCTION

Laser vaporization of solids - pure metals, mixed metal alloys, ceramics or other composite material - has been used since the early 80th for cluster formation [1-2] and to study there reactivity [3-4]. Mass spectroscopy, and optical spectroscopy in the gas phase as well as matrix isolation spectroscopy and thin film analysis after condensation helped to identify the species and to measure some of there physical properties [5-8].

Assumed that mostly neutral and ionized atoms are produced in the initial step and only very small molecules/clusters evaporate directly, the dynamics of evaporization and formation of larger clusters is not well understood, but this is important especially for applications like thin film formation [9-10].

Some of the questions to be asked are

1. Can optical spectroscopy with time and spatial resolution be used to optimize the cluster formation and understand some of the dynamics?
2. Is there an easy relationship between increasing cluster size and shifts in the optical spectra?
3. Is the thin film formation different if clusters instead of atoms or ions are condensed?
4. Is it possible to develop new high temperature material by pulse laser vaporization (PLV) and deposition?

Research on these questions especially on the application part is only in the beginning. Here only some preliminary results will be presented and it is hoped that further investigations along this line may clarify the problems.

P. Jena et al. (eds.), Physics and Chemistry of Finite Systems: From Clusters to Crystals, Vol. II, 1227–1231.
© 1992 Kluwer Academic Publishers.

2. EXPERIMENTAL

The clusters are produced by the IR or green Nd-YAG laser line with a pulse duration of 8 nsec. The original Smalley source [1] has been modified for easier interchange of the materials [11]. Between 10^{14} to 10^{16} atoms per laser pulse are evaporated and He, Ar or N_2 gas is used for beam formation. The optical emission in the plume above the vaporizing material is analysed perpendicular to the cluster beam [12] by time resolved spectroscopy, to get some idea about the dynamics of cluster formation.

The evaporated material is deposited on a 2-3 nm thick carbon film covering a copper grid for measurement of their size distribution by electron microscopy. In addition thin films are condensed under different experimental conditions on glass plates and examined for there mechanical stability and optical spectra taken in the visible and UV.

The material used is either obtained in the form of a rod or tube or can be sintered as tablets and then pressed together into a rod.

3. RESULTS AND DISCUSSION

The size distribution of clusters has been measured by electron microscopy (EMS). Fig. 1 shows YBCO, on the left site a 150 Å particle with structure and on the right site a 80 Å particle including power spectrum and computer assisted image reconstruction. The lattice distance observed corresponds to appr. 12 Å, very close to the c acis of YBCO. This is already observed for the deposition of clusters without further treatment. A 5-20% oxygen constant in the carrier gas does not change the results. For electron microscopy a number of precausions has to be observed to minimize coagulation after condensation.

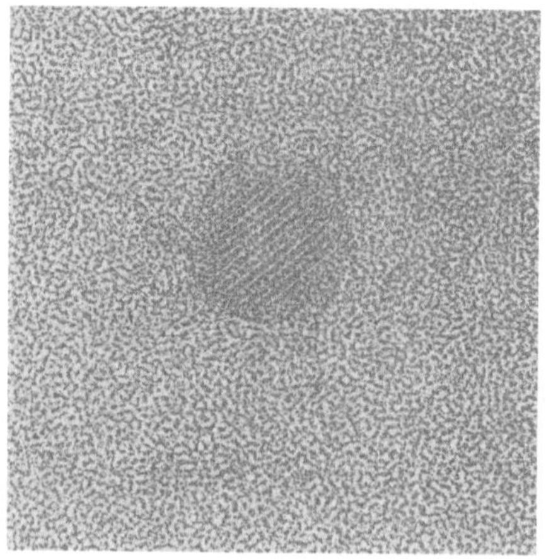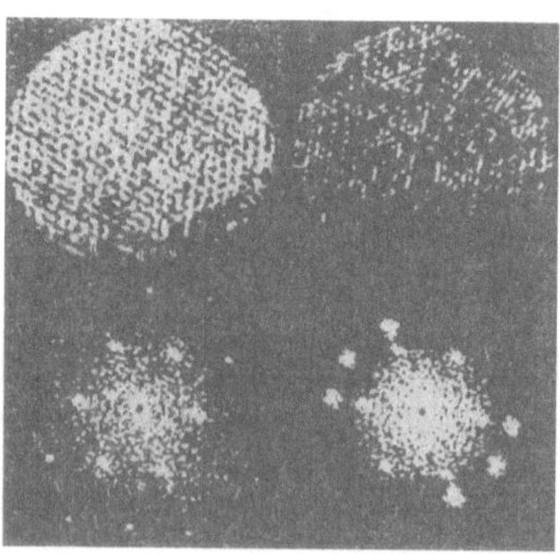

Figure 1.   EMS of YBCO on carbon film cluster of 150 Å diameter (left) and 80 Å diameter using computer assisted image reconstruction (right).

Time and spatial resolved optical spectroscopy in the plume above the vaporized material provides information on the energy in the plasma, the ratio of excited atomic neutrals and ions, the velocity of the beam and the formation of small clusters. The experiments have been performed using different gases, different pressure and gas pulse length for beam formation and different laser pulse energy to change the composition of the beam. The ratio of neutrals to ions can be changed from 10% to 80% ions and the velocity of the beam between 1 and 6 mach. The emission of the plume has been measures so far only very close to the surface [7] but it can be observed up to 50 or even 100 mm away from the vaporized material. Fig. 2 shows an example for $YBa_2Cu_3O_{7-\delta}$.

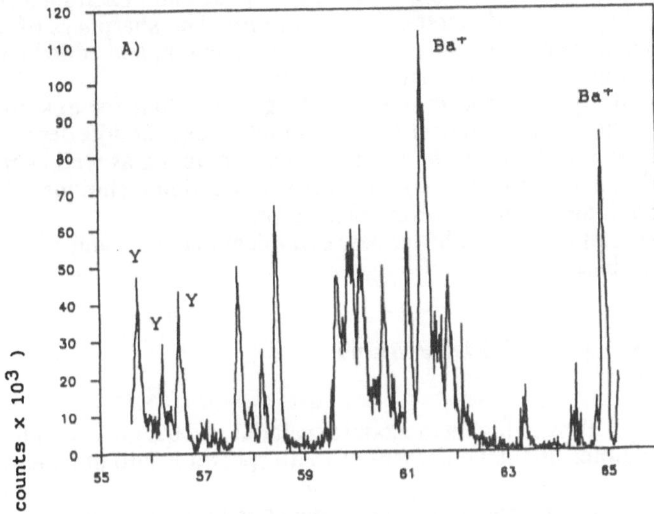

Figure 2.
Optical emission-spectra of laser-vaporized $YBa_2Cu_3O_{7-\delta}$ with oxygen carrier-gas (P = 0.07 Torr).

Signal 1-5µs (A) and

15-50µs (B) after lasershot. Viewpoint 30 mm from the target.

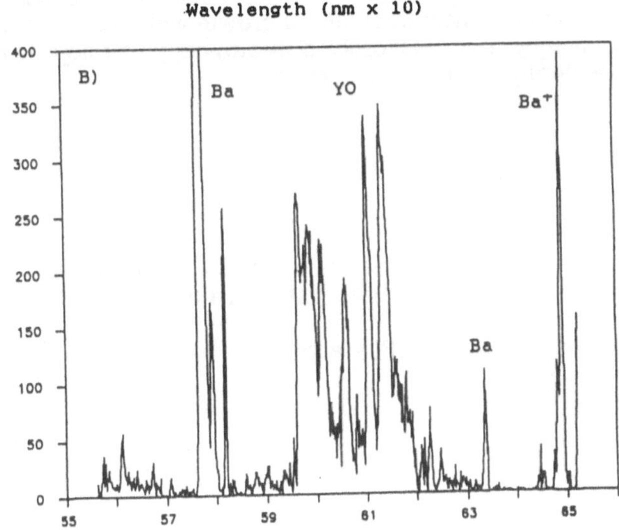

Mostly neutral and ionized atoms are observed and in addition the emission of diatomic YO, BaO, CuO plus a number of features of unknown origin. The energy of the plasma is high enough to allow excitations up to 24 eV. At equilibrium this would correspond to a temperature of around 10^5K. The emission intensity changes from non thermal within the first few microseconds (Fig. 2A) to thermalization after about 10 msec.(Fig. 2B). Comparison with pure metals shows that metal dimers are not excited in the spectra of YBCO. But the observation of the plasma emission shows only a part of the vaporized material. It is hoped to continue soon with laser induced fluorescence and to standardize some of the optical emissions for optimum cluster size distribution for condensation of thin films.

In addition, the vaporized material is deposited on glas plates and examined for mechanical stability, optical spectra and electrical conductivity. The sharpness of an image and the hardness depends very much on cluster size, ion content, and added gas. More than 50% ion content produces a more volatile coveradge.

This method of thin film production is much more general than for example chemical vapor depositions where at the moment fabricating integrated circuits depends on. Especially the carbide, and nitride films are expected to be improved as well as the semi-conductors and high T_c superconductors by the laser vaporization technique with subsequent cluster formation in the gas phase before deposition.

The technical part and implications for better new ceramical materials and surface coating will be reported elsewhere.

4. FUTURE ASPECTS OF DEVELOPMENT

The experiments on exotic thin films of high temperature material and the understanding of the cluster formation dynamics and development of there spectra is still at an early stage. Much of the practical purposes of thin film production, surface protection and modification can be certainly obtained by lucky coincidence, but a tailered construction will need a better inside view on the dynamics of cluster growth which can be obtained by different experimental techniques like time resolved spectroscopy. We hope to report some progress along this line during the next cluster meeting.

5. ACKNOWLEDGEMENT

The experimental results reported are obtained with the help of K. Möller, J. Kolenda, H. Yu and M. Ritz from the Free University and W. Schulze, C. Jackschath, I. Rabin, J. Tesche, C. Altenhein and J. Urban from the Fritz-Haber-Institute of the Max-Planck Gesellschaft.

The financial support by the German Research Foundation - DFG - through Sfb 337 is gratefully acknowledged.

6. References

[1]   Hopkins, J.B. Langridge-Smith, P.R.R., Morse, M.D. and Smalley, R.E., (1983) 'Supersonic metal cluster beams of refractory metals: Spectral investigation of ultra cold Mo_2' J.Chem. Phys. 78, 1627-1637.

[2]   Rohlfing, E.A., Cox, D.M., Kaldor, A. and Johnson, K.H. (1984), 'Photoionization spectra and electronic structure of small iron clusters', J. Chem. Phys. 81, 3846-3851.

[3]   Riley, S.J. (1989) 'Reactivity of free clusters' Z. Phys. D 12, 537-541.

[4]   Kaldor, A., Cox, D.M. and Zakin, M.R. (1988)'Reactions of gas-phase metal clusters' in Prigogine, I. and Rice, S.A. (eds.), Adv. Chem. Phys. Vol. LXX pp. 211-261.

[5]   Moskovits, M. 'Matrix-isolated metal clusters' (1989) in Andrews, L. and Moskovits, M. (eds.), Chemistry and Physics of Matrix isolated species Chapt. 3 p. 47-73, North Holland, Amsterdam.

[6]   Cai, M.F., Tsay, S.J., Dzugan T.P., Pak, K. and Bondybey, V.E. (1990) 'Laser vaporization of alloys: Fluorescence and electronic structure of AlCu', J. Phys. Chem. 94, 1313-1316.

[7]   Wu, X.D., Dutta, B., Hegde, M.S., Inam, A., Venkatesan, T., Chase, E.W., Chang, C.C. and Howard, R. (1989) 'Optical spectroscopy: An in situ diagnostic for pulsed laser deposition of high Tc superconducting thin films' Appl. Phys. Lett. 54, 179-181.

[8]   Urban III, F.K. and Bernstein, A. (1991)'Study of the ionized cluster beam technique' J. Vac. Sci. Technol A9, 537-541.

[9]   Allen, T.H., Beauchamp, W.T. and Hichwa, B.P. (1991) 'Enhanced optical thin-film materials', Photonics Spectra (3), 103-109.

[10]   Simon, R. (1989) 'High-temperature superconductors for microelectronics' Solid State Technology (9), 141-146.

[11]   Froben, F.W., Kolenda, J. and Möller, K. (1989), 'Fluorescence spectroscopy of laser vaporized species', Z. Phys. D 12, 485-487.

[12]   Froben, F.W. and Kolenda, J. (1990) 'Cluster production by laser material interaction with optical spectroscopic characterization', High Temperature Science 28, 15-20.

HIGH INTENSITY SILVER CLUSTER BEAMS

O. F. Hagena and G. Knop
Institut für Mikrostrukturtechnik
G. Linker
Institut für Nukleare Festkörperphysik
Kernforschungszentrum, PF 3640
D-7500 Karlsruhe, Germany

ABSTRACT. High-intensity silver cluster beams have been obtained from supersonic nozzle expansions of an Ar/Ag-mixture. This paper reports on cluster size and intensity as function of expansion characteristics, and discusses the formation of silver films by cluster beam deposition. Source parameters were: temperature T_o ≤ 2300 K, total pressure p_o ≤ 4000 hPa, silver partial pressure p_{Ag} ≤ 250 hPa, nozzle diameter d = 0.35 mm, nozzle cone angle 2α = 10°, and cone lengths l = 17 and 27 mm.
Silver beam intensity corresponded to deposition rates up to 75 nm/s at 0.3 m distance from the nozzle. This was a factor 20 higher compared to the maximum intensity from a sonic nozzle operated with the same silver mass flow. Increase of intensity was paralleled by an increase of mean cluster size up to 500 atoms/cluster, measured with the retarding field method. The low-mass region ≤ 1600 a.m.u., studied with a quadrupole mass spectrometer, exhibited multiply charged silver clusters, "magic numbers", and odd-even intensity variations known to be characteristic for metal-cluster ions. Thin silver films deposited onto room-temperature substrates were examined by REM and x-ray diffraction. In general, the diffraction patterns were those of polycrystalline fcc material. On glass substrates there was a preferred orientation of (111) and (200) planes parallel to the substrates surface. Minimum grain size was 60 nm.

1. Introduction

Cluster formation by condensation of the respective vapor in supersonic expansions is a well established method to obtain cluster beams from gaseous and low boiling-point materials [1,2]. For materials with high boiling points the formation of such "gasdynamic" clusters has been controversial [3,4,5]. We have shown, however, that clustering, e.g., of silver does occur if the expansion conditions satisfy the predictions of the scaling laws describing the formation of gasdynamic clusters [4]. Another confirmation of these scaling laws comes from recent experiments looking for clusters of Ga and Ag under flow conditions typical for the ionized cluster beam — "ICB" — concept [5]. As expected for the low values of the respective condensation parameter Γ^* < 200 [3], clustering was not

P. Jena et al. (eds.), Physics and Chemistry of Finite Systems: From Clusters to Crystals, Vol. II, 1233–1238.
© 1992 *Kluwer Academic Publishers.*

observed, and it was finally concluded, that the high quality of ICB films seems to be due to ion assisted deposition [5]. In this situation the effect of clusters for thin film formation is again an interesting question. Thus having established the conditions for silver cluster beams, we are now presenting data on the dependence of cluster size and intensity on the expansion characteristics and first results on the formation of thin silver films by cluster beam deposition.

2. Experimental Apparatus

A schematic of the experimental arrangement is shown in Fig. 1. It is based on a standard cluster beam apparatus with three differentially pumped vacuum chambers and is designed for expansions of a mixture of silver vapor and argon carrier gas. The nozzle source assembly has been operated at temperatures of up to 2300 K and total pressures of 4000 hPa. The gas mixture expands through a conical nozzle with 0.35 mm throat diameter and 10° full cone angle. The cone lenght was either 17 mm or 27 mm. The core of the supersonic flow enters the second stage through a heated skimmer, and the final cluster beam passes through a collimator into the third chamber, which contains the following devices for beam analysis: a) Rotating chopper disc for beam modulation. b) Quartz crystal rate meter for measuring silver beam intensity. c) Substrate holders for thin film deposition, mounted on the support of the rate meter. d) Ionization detector with retarding field [6] for measurement of beam density, beam velocity distribution (TOF), and cluster ion size. e) Quadrupole mass spectrometer for mass spectra up to 1600 amu.

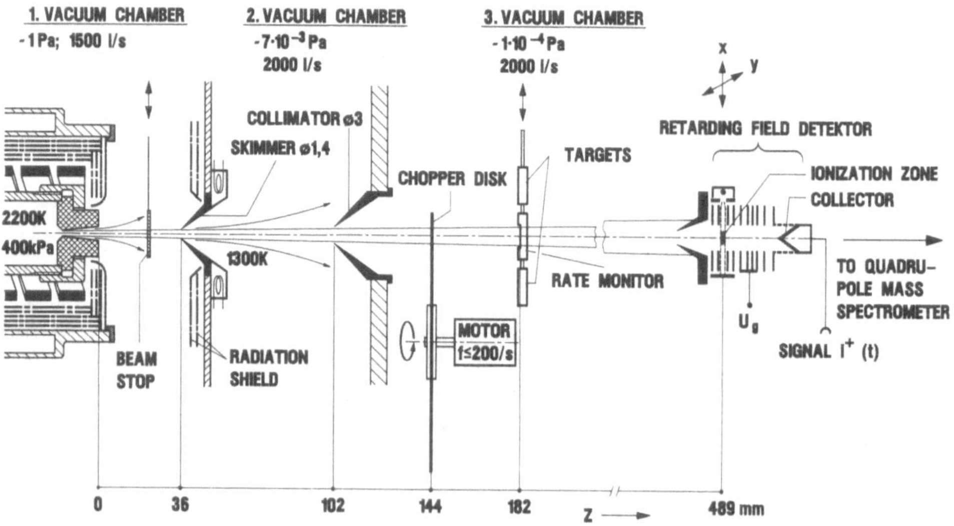

Figure 1.   Schematic of the cluster beam apparatus.

The total source pressure p_o of the expanding gas mixture is measured outside the nozzle source in the carrier gas supply line. The source temperature is determined with a pyrometer looking at the nozzle. The silver partial pressure p_{Ag} is calculated from measured carrier gas mass flow and the values of p_o and T_o. More details on the performance and construction of the apparatus are published in previous papers [4,7].

3. Results and Discussion

3.1 Cluster Size and Intensity

3.1.1 *Measurements with the quadrupole.* Figure 2 is a typical mass spectrum of silver cluster ions, obtained for T_o = 2100 K and p_o = 4000 hPa. It gives the relative intensities versus specific cluster size N/Z (N = silver atoms/cluster, Z = no. of elementary charges). The electron impact ionizer of the quadrupole was operated at 100 eV electron energy. Note that the ion distribution does not directly represent the respective distribution of the neutral clusters. This is due to fragmentation and multiple ionisation of the clusters upon electron impact [6] and to the mass-dependent transmission of the spectrometer [8]. The qualitative features of interest in the mass spectrum of Fig. 2 are
- sequence of mass peaks at integer values of N/Z with the odd-even variation of intensity characteristic for metal cluster ions.
- appearance of intermediate peaks at N/Z = 15/2, 17/2, 19/2.... and at N/Z = 37/3, 38/3, 40/3, 41/3 ... corresponding to doubly and triply ionized clusters
- "magic" numbers, i.e. peaks of relatively high or low intensity at N/Z = 9 (maximum) and 23/2 (minimum).

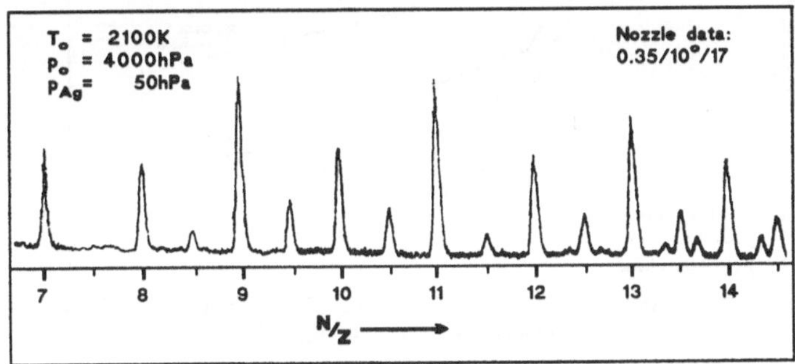

Figure 2. Cluster ion mass spectrum of a silver cluster beam. Source parameter: p_o = 4000hPa, T_o = 2100K, p_{Ag} = 50hPa. Nozzle parameter: d = 0.35mm, l = 17mm and 2α = $10°$.

The same features have been already observed for silver clusters produced by inert gas aggregation/electron impact ionization [9] and by laser evaporation/ multiphoton ionization [10], and they have been fully explained by the shell model for metal clusters [9,10]. Thus one can conclude that the different techniques of producing silver clusters do not affect these characteristics of the cluster ion spectrum.

Similar spectra were obtained when operating the nozzle source at higher temperatures: While the peak intensities increased, the intensity relations were almost the same. This suggests that the quadrupole mass range — for the conditions studied — contains primarily fragment ions which depend little on the original cluster size.

3.1.2 *Measurements with the retarding field detector.* Figure 3 shows representative measurements with the retarding field detector to determine cluster ion size. The diagram gives the detector ion current i^+, normalized by the ionizing electron current i^-, as function of the retarding field voltage U_R. Shown are four curves for different source temperatures T_o and thus silver partial pressures p_{Ag}. The decrease of ion signal for -200 V < U_R < 0 V (no retarding field) is due to the loss of atomic ions. Any intensity measured for 0 V < U_R is due to cluster ions for which the kinetic energy, $N \cdot m \cdot w^2/2$, where w is the flow velocity of the cluster beam, exceeds the energy loss in the retarding field, $Z \cdot e \cdot U_R$ [6]. For the present experimental conditions w was measured by TOF to be 1.5 km/s. This results in a monomer kinetic energy $m \cdot w^2/2 = 1$ eV. According to Fig. 3 there is no cluster signal for T_o= 2000 K, while at the higher temperatures an increasing cluster ion signal is observed. The respective half-intensity point — marked by an arrow — serves to define the characteristic cluster ion size <N/Z> which is used to characterize the respective cluster ion distribution [6]. Note that for 2100K ≤ T_o ≤ 2300 K both cluster size <N/Z> and silver pressure increase by about a factor of three.

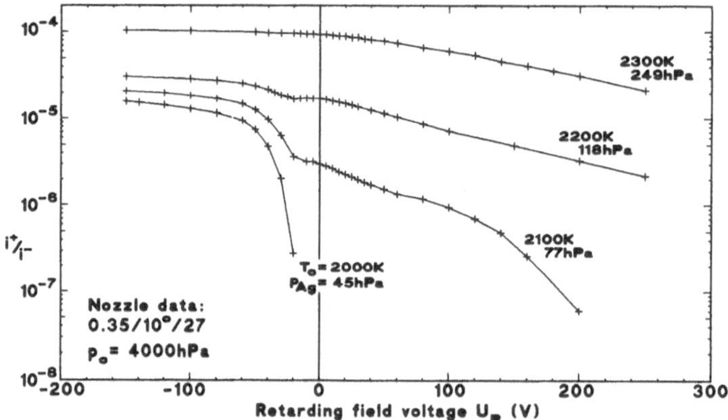

Figure 3.   Retarding field measurements: Normalized ion current i^+/i^- versus retarding field voltage U_R. The half-intensity points are marked by arrows. Nozzle parameter: d = 0.35mm, l = 27mm, 2α = 10°.

3.1.3 *Measurements with the rate meter.* Some results of deposition rate measurements are shown in Figure 4. It gives the silver cluster intensity, expressed as deposition rate dz/dt as function of silver partial pressure p_{Ag} for nozzle flows with constant source pressure p_o = 4000 hPa. Data are for two different nozzles. Silver pressure was increased by increasing T_o, and the respective temperatures are listed near the data points. For comparision the scale at the top of the figure gives the saturation temperatures T_s corresponding to the silver partial pressure p_{Ag}. For both nozzles silver intensity increases faster than silver pressure, and the rate of increase is higher for the longer nozzle. The maximum of 205nm/s corresponds to a rate of 75nm/s referred to a typical deposition distance of 0.3m. The respective silver consumption was m_{Ag} = 23 g/h. The measured rate of 75 nm/s exceeds that for an ideal *sonic* nozzle expansion, compared at the same silver consumption, by a factor 19. For ICB-conditions the highest rate reported for conditions with 21 g/h was only 1.1 nm/s [5]. Comparing the source temperatures T_o with the respective T_s shows that the silver vapour was always superheated, and that the superheating was more pronounced for the shorter nozzle. This may be explained by slightly different temperature gradients in the nozzle chamber.

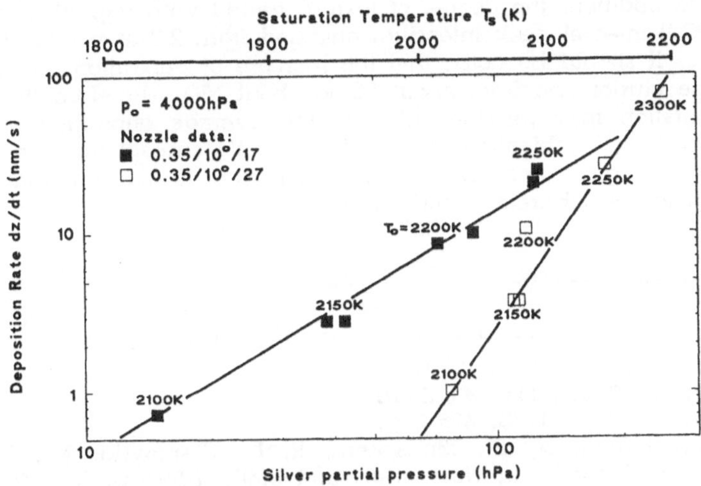

Figure 4.   Cluster beam intensity dz/dt as function of the silver partial pressure p_{Ag}. Source temperature T_o and saturation temperatur T_s are given. Nozzles parameter: d = 0.35mm, 2α = 10°, l = 17 respectively 27 mm.

3.2 Silver Film Properties

Silver cluster beams incident on room-temperature surfaces of e.g. the collimator and the chopper disc or of special substrates produced metallic films corresponding to the respective surface structure: Glass or Si wafer substrates produced smooth, highly specular films, whereas metal surfaces as well as the rough quartz crystal of the rate meter produced less bright films. These optical impressions where confirmed by REM investigations of the silver films: For all types of substrates and films of about 1000 nm thickness the microstructure of the silver surface was a close replica of the substrate. Especially films deposited on glass and Si-wafers were as smooth as the substrate. Adhesion of the films was limited: The films could be peeled off the structural elements of the apparatus, or lifted off the substrates by adhesive tape. Stronger adhesion is expected if the adsorbate layers covering the substrates are removed by proper cleaning procedures prior to silver cluster deposition.

Preliminary information on the crystallographic growth properties of these films was obtained by X-ray diffraction measurements. In general, the diffraction patterns were those of polycrystalline material revealing all lines of a fcc structure. For room-temperature deposition the peaks are remarkably narrow. The breadth of the (111) line, e.g., is $0.15°$ (FWHM in 2Θ) which indicates a minimum grain size of 60nm. The line breadths and thus also the grain sizes were found to be independent of deposition rate and substrate material (glass and (111) Si). An influence of the deposition parameters including substrates, however, was observed on the texture of the films examined by rocking curves. While the films on (111) Si revealed almost no texture, on glass a preferred orientation of (111) and (200) planes parallel to the substrate surface was observed with a broad distribution of the (111) grains and a narrower (few degrees FWHM) distribution for the (200) grains. In addition, the degree of texture varied with deposition rate; the ratio of (111) to (200) integral peak intensities changed from 2.3 at a rate of 5 nm/s to 0.7 at 47 nm/s. It should be noted that the increase of deposition rate involves an increase of the cluster size from about 50 to about 150. Therefore the change of crystallite orientation may be due either to the average deposition rate or to the cluster size, or to a combination of both factors. We suspect that the different growth behaviour on glass and (111) Si is rather due to the different surface quality finish than to the substrate material itself.

---

[1] Becker, E. W., Bier, K., and Henkes, W., Z. Phys. 146, 333 (1956)

[2] Hagena, O. F., Surf. Sci. 106, 101 (1981)

[3] Hagena, O. F., Z. Phys., D4, 291 (1987)

[4] Hagena, O. F., Z. Phys. D20, 425 (1991)

[5] Brown, W. L., Jarrold, M. F., McEachern, R. L., Sosnowski, M., Takaoka, G., Usui, H., and Yamada, I., Nucl. Instr. and Meth. B59/60, 182 (1991)

[6] Falter, H., Hagena, O. F., Henkes, W., and Wedel, H. V., Int. J. Mass Spectrom. Ion Phys., 4, 145 (1970)

[7] Hagena, O. F. and Knop, G., KfK Nachr. 23, 136 (1991)

[8] Dawson, P. H., Int. J. Mass Spectrom. Ion Phys., 14, 317 (1974)

[9] O. Kandler, K. Athanassenas, O. Echt, D. Kreisle, T. Leisner, and E. Recknagel Z. Phys. D 19, 151 (1991)

[10] I. Rabin, C. Jackschath, and W. Schulze, Z. Phys. D 19, 153 (1991)

OPTICAL CHARACTERISTICS OF Cu-NANOCLUSTER LAYERS ASSEMBLED BY ION IMPLANTATION

R. F. HAGLUND Jr.[1], R. H. MAGRUDER III[2], L. YANG[1],
J. E. WITTIG[2] and R. A. ZUHR[3]
[1]*Department of Physics and Astronomy and*
[2]*Department of Materials Science and Engineering*
Vanderbilt University, Nashville, TN 37235 U.S.A.
[3]*Solid-State Physics Division, Oak Ridge National Laboratory*
Oak Ridge, TN 37831 U.S.A.

ABSTRACT. We have generated Cu-nanocluster layers in solid insulating substrates by implanting Cu ions into fused silica, creating thin layers (~ 150 nm) of nanoclusters over a diameter of order 2 cm. Transmission electron microscopy shows that the size and size distribution can be controlled by the parameters of the ion implantation. We report measurements of the optical properties and nonlinear index of refraction on these unusual solid-phase cluster materials as a function of total implanted-ion dose.

1. Introduction and Motivation

Metallic clusters of nanometer dimensions have many desirable characteristics as optical materials: ultrafast switching times, high resistance to bulk and surface laser damage, and large third-order optical nonlinearities.[1] Recently we have made nanometer-size copper clusters in dense, solid-phase layers by implanting Cu ions in glass substrates.[2] Apart from the known nonlinear optical behavior of metal nanoclusters, we are motivated by the fact that ion implantation is already a standard technology for modifying the properties of semiconductors and forming optical waveguides in dielectric materials.

In this paper, we describe the creation of dense layers of nanometer-size clusters by ion implantation, and show that total ion dose significantly affects the size and size distribution of nanoclusters created by Cu-ion implantation in fused silica. We describe the optical properties of these cluster-assembled layers, including absorption, index of refraction, and nonlinear index, and consider briefly the physical origin of these optical properties.

2. Cluster Preparation and Characterization

Ion implantation is a well-established technique for creating materials which cannot be generated by ordinary glass-forming processes because they are constrained both by chemistry and by the thermodynamics of the glass transition.

As substrate materials we used optical quality silica (Spectrosil®) discs, 1mm thick x 20 mm diameter. The nominal cation impurity content of the Spectrosil® is less than 5 ppm by weight; the principal anion impurities are Cl (80 ppm) and OH (200 ppm). The

P. Jena et al. (eds.), Physics and Chemistry of Finite Systems: From Clusters to Crystals, Vol. II, 1239–1244.
© 1992 *Kluwer Academic Publishers.*

disks were implanted in a vacuum of 10^{-7} Torr with 160 keV singly-charged positive ions of Cu to doses of $6 \cdot 10^{16}$ (on both faces of the disc) and $12 \cdot 10^{16}$ ions·cm^{-2} (on a single side). The ion beam was electrostatically rastered to provide uniformity of implantation, and the current density was set at 10 μA·cm^{-2}, holding the "global" temperature of the substrate below 50 °C. However, this does not preclude the possibility of strong localized heating in the implanted layer. Since these doses are above saturation, compaction effects due to radiation damage were approximately the same for both samples.

The depth distribution of the implanted Cu was measured by He$^+$ ion backscattering The total dose calculated from the backscattering measurements was typically 90% of the integrated current measured during implantation, indicating minimal sample charging. The volume concentration of the Cu-layer at peak density is of order 0.1. The backscattering profiles of the implanted layers are bimodal: the larger of the two peaks is twice the size of the smaller, and appears 85 nm below the surface, while the smaller peak is at a distance of 170 nm from the surface; the implanted layer was some 150 nm thick FWHM.

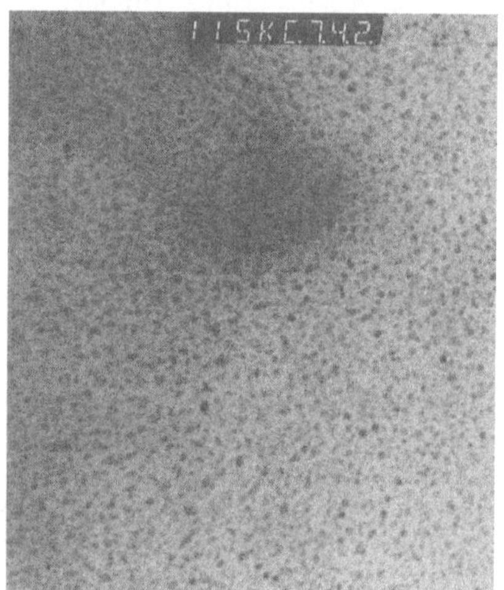

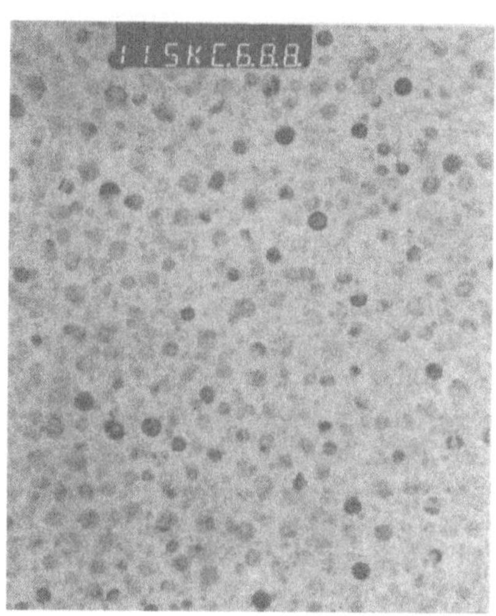

Figure 1. TEM micrograph of the layer implanted to a dose of $6 \cdot 10^{16}$ ions·cm.

Figure 2. TEM micrograph of the layer implanted to a dose of $12 \cdot 10^{16}$ ions·cm.

The size and size distribution of the Cu nanoclusters were characterized by transmission electron microscopy (TEM). Samples for TEM were prepared by cutting a 3-mm-diameter disc from the implanted wafers with a slurry drill, mechanically grinding the disc from the back side to a thicknesss of about 150 μm, and dimple grinding to approximately 50 μm from the implanted surface. Electron transparency was achieved by backthinning with an 5 keV Ar ion gun while cryogenically cooling the samples to minimize ion damage. The samples were then examined in a Philips CM20/T scanning transmission electron microscope operating at 200 kV. The high dose samples were stable in the microscope; the lower-dose samples, however, required carbon coating to eliminate charging effects.

Standard bright field (BF) imaging techniques were used to characterize the implanted layer. The BF images in Figures 1 and 2 show plane views of the ion-implanted layer at a magnification of $1.15 \cdot 10^5$. TEM scans made with a tilted sample revealed that the round particles embedded in the silica matrix are spherical with random crystallographic orientation. An electron diffraction pattern taken from a selected area within the displayed region exhibited the polycrystalline ring pattern characteristic of an fcc metallic copper structure, superimposed on the diffuse diffracted intensity from the amorphous silica matrix.

An area of each micrograph containing approximately 200 particles was selected as the basis for measuring particle size distribution. Individual particle diameters were measured using a digital vernier caliper and tabulated; particles touching the boundary of the selected area were weighted by a factor 0.5. By measuring the boundary surface and making use of both the photographic and TEM magnifications, a surface area was calculated. It was assumed that the depth of the analyzed region was equal to the FWHM depth of the implanted layer as measured by the ion backscattering profile, about 150 nm. This permits extraction of a volumetric density of the clusters. The density extracted by this means is in good agreement with the backscattering density for the $12 \cdot 10^{16}$ ions·cm^{-2} sample. For the $6 \cdot 10^{16}$ ions·cm^{-2} sample, the correlation is less strong, suggesting that a significant fraction of the nanoclusters may have diameters below the resolution limit of the TEM.

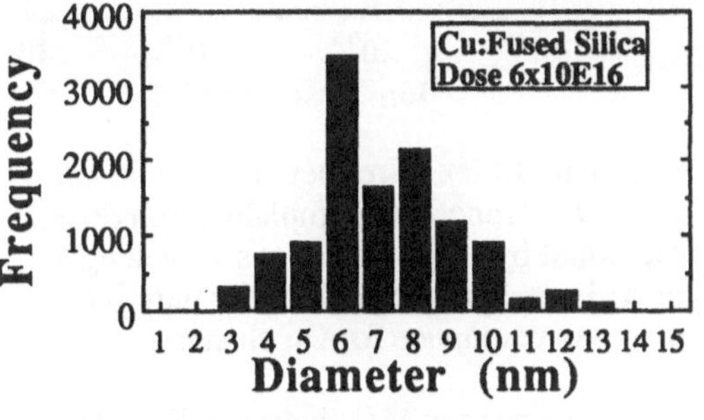

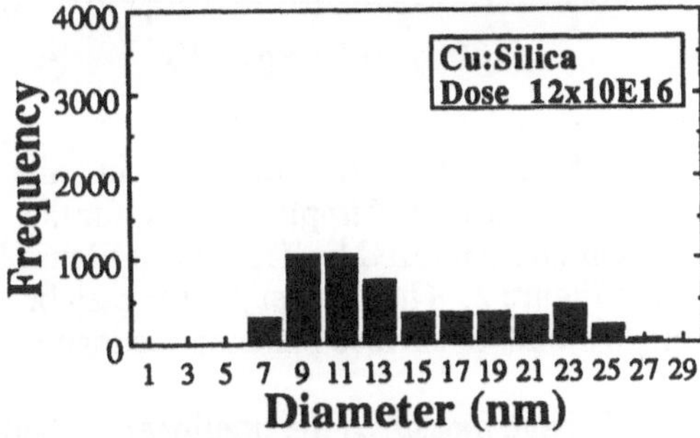

Figure 3. Distribution of cluster sizes in Cu:silica implanted to $6 \cdot 10^{16}$ ions·cm^{-2} using the protocol described above.

Figure 4. Distribution of cluster sizes in Cu:silica implanted to $12 \cdot 10^{16}$ ions·cm^{-2}. Note difference in scales from Fig. 3.

3. Linear and Nonlinear Optical Properties of the Nanocluster Layer

We have measured various optical properties of cluster-assembled layers, including linear absorption, linear index of refraction, and nonlinear index of refraction.

Optical absorption spectra were measured in a Cary dual-beam spectrophotometer, using an unimplanted sample as a reference. The optical absorption spectra in Figure 5 show a peak at 2.2 eV whose amplitude is dose dependent. The absorption in the 2.4 to 4.0 eV region increases with dose. In the 4 eV region the absorption per implanted ion is the same for both samples, while above 4.5 eV, the absorption per ion is again higher for the higher-dose sample. We attribute the larger absorption in the 2.4 to 4 eV region to the broader distribution of particle sizes in the higher-dose sample, as evident in the histogram of Figure 4. The peak at 2.2 eV may be due to the bulk surface plasmon resonance peak.

Ellipsometric measurements were carried out on Cu-implanted silica at a wavelength of 633 nm using a manual ellipsometer. Samples implanted with Cu over a wide range of doses were examined to determine their complex refractive index $n = n_0 + i\,k$. The data, shown in Figure 6, exhibited a sharp rise in index of refraction with increasing dose.[3] In this region of the spectrum, the complex refractive index increases due to optical excitation of the surface plasmon resonance of the Cu clusters.[4] Under these conditions, the imaginary part **k** of the refractive index increases rapidly and largely determines the index. Incidentally, our recent infrared reflectance measurements[5] show a red-shift in the location of the Si-O stretching mode peak with increasing ion dose. This behavior is consistent with breaking and reforming of the Si-O bonds as the nanoclusters grow in size.

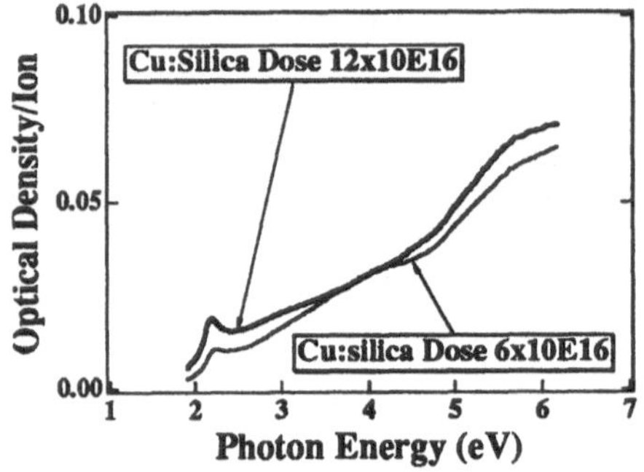

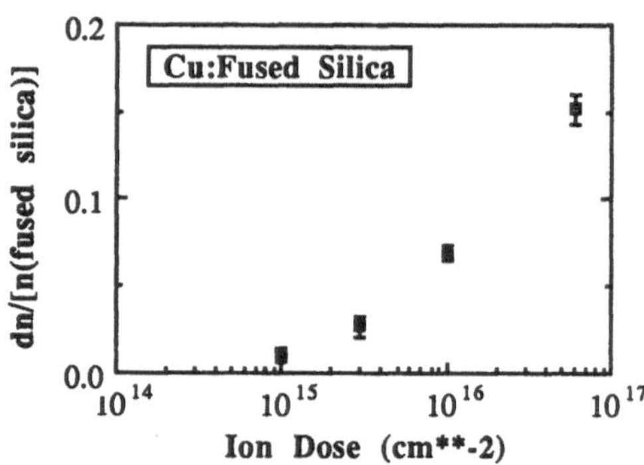

Figure 5. Optical absorption curve, in units of optical density per implanted ion, for the two samples analyzed by TEM from Figure 1 and Figure 2. The peak at 2.2 eV may be due to a bulk or surface plasmon resonance.

Figure 6. Index of refraction of Cu:silica layers as a function of implanted ion dose, measured by ellipsometry at a wavelength of 633 nm, near the peak of the putative plasmon resonance for Cu clusters.

We have measured the nonlinear refractive index of our samples to assess the potential of optically bistable materials of this type in nonlinear optical devices. The nonlinear index of refraction γ is defined in terms of the ordinary linear index n_0 and the (complex) third-order nonlinear dielectric susceptibility $\chi^{(3)}$ as follows (SI units):

$$n = n_o + \gamma \cdot I, \qquad \gamma = \frac{4\pi}{3n_o} 10^{-8} \cdot \mathbf{Re}\left[\chi^{(3)}\right] \tag{1}$$

where n is the total index of refraction and I is the laser intensity. The nonlinear index may arise from several mechanisms: electronic excitation, the thermo-optic effect, electrostriction, and ionic excitations. The mechanism which operates in any particular experiment will depend on the choice of laser pulse length, pulse repetition frequency, and laser wavelength in relation to the electronic structure and electron-phonon coupling in the material.

We measured the nonlinear index of refraction γ of Cu-implanted silica by observing variations in the far-field intensity of laser light transmitted through the sample as a function of laser intensity.[6] The laser was a continuous-wave, mode-locked, frequency-doubled Nd:YAG laser producing 100-ps pulses at a wavelength of 532 nm and a pulse-repetition frequency of 76 MHz. Peak irradiance in the focal spot was of order 10^{11} W·m^{-2}, below the threshold for thermal damage. The intensity was varied by translating the sample through the focal plane of a 150-mm focal length lens on a micrometer-driven translation stage. The transmitted intensity was monitored by a power meter located behind a beam-limiting aperture 90 cm from the focal plane, and normalized to the signal from a beamsplitter located before the focusing lens. Slow thermal drifts were compensated by averaging forward and backward scans. For a material with a positive nonlinear index of refraction, moving the sample toward, and then away, from the focal spot causes an initial decrease in the normalized far-field intensity, followed by a crossing through unity at the focal spot and then an increase in intensity.

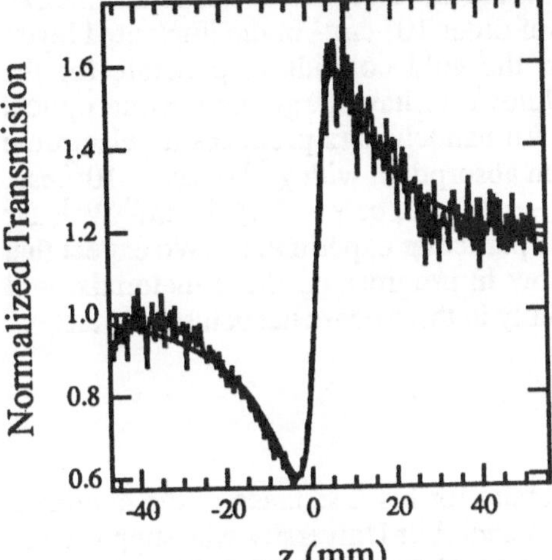

Figure 7. Far-field intensity as a function of sample distance from the focal plane for Cu:silica ($6 \cdot 10^{16}$ ions·cm^{-2}, two sides). The smooth curve drawn through the data is calculated from Eq. (2).

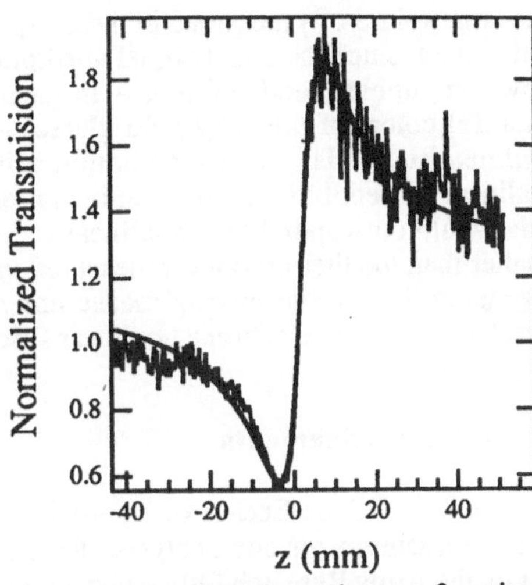

Figure 8. Far-field intensity as a function of sample distance from the focal plane for Cu:silica ($12 \cdot 10^{16}$ ions·cm^{-2}, one side). The smooth curve drawn through the data is calculated from Eq. (2).

Experimental results for the far-field intensity of laser light transmitted through the two Cu:silica samples are shown in Figures 7 and 8. For the thin, highly absorbing layers produced by ion implantation, the intensity-dependent change in refractive index $\gamma = \Delta n/I$ is extracted from the Z-scan measurement, and for moderate intensities turns out to be

$$\gamma = \frac{I_{max} - I_{min}}{I_{max} + I_{min}} \cdot \frac{n_o r_o^2}{z_o I_o L} \quad , \quad z_o \equiv \frac{\pi r_o^2}{\lambda} \tag{2}$$

where I_{max} and I_{min} are the maximum and minimum intensities recorded in the Z-scan; I_o is the laser peak intensity at the focal spot, r_o is the radius of the Gaussian beam profile at the focal plane, L is the thickness of the implanted layer, and z_o is the diffraction length. Note that the difference $\Delta I = I_{max} - I_{min}$ for the low-dose sample implanted on both sides (to a dose of $12 \cdot 10^{16}$ ions·cm^{-2}) is almost identical to that for the high-dose sample implanted on *one* side (to a dose of $12 \cdot 10^{16}$ ions·cm^{-2}), if one accounts for the Fresnel reflection at the two additional interfaces in the two-side-implanted sample.

We calibrated our results by measuring and reproducing the thermo-optic coefficient of CS_2, obtaining the accepted value. By fitting the Z-scan data to Eq. (2), we then found that $\gamma = 1.8 \cdot 10^{-7}$ cm^2/W. Because the extracted value for γ thus seems to be correlated with the total amount of implanted Cu, we conclude that the optical nonlinearity measured for these high-repetition-rate, 100-ps pulses is also probably due to the thermo-optic effect.

4. Discussion

The large values of γ measured in these experiments are almost certainly due to the thermo-optic effect, since the small-signal absorption is of order 10^4 cm^{-1} in the implanted layer. However, noble-metal colloids - in particular, the gold colloids responsible for the beautiful colors of the ruby-gold glasses - are known to have large picosecond optical nonlinearities.[7] The Kerr-type nonlinearity in Au nanoclusters produces an electronic nonlinear susceptibility dominated by hot-electron absorption, with $\chi^{(3)}$ of order 10^{-7}.esu. This would correspond to a nonlinear index $n_2 \sim 10^{-7}$ esu, or $\gamma \sim 2 \cdot 10^{-10}$ cm$^2 \cdot$W^{-1}, far smaller than the thermo-optic γ measured in these particular experiments. We expect that degenerate four-wave mixing measurements, now in progress on these materials, will reveal a picosecond electronic nonlinear susceptibility in these dense nanocluster layers.

5. Acknowledgements.

We thank Dr. Klaus Becker of Vanderbilt University for his assistance in the nonlinear optical measurements and analysis. Research at Vanderbilt University was supported in part by the Army Research Office under contract DAAL03-91G-0028. Oak Ridge National Laboratory is partially supported by the Division of Materials Science, U. S. Department of Energy, under contract DE-AC05-84OR21400 with Martin-Marietta Energy Systems, Inc.

6. References

1.   G. I. Stegemann and R. H. Stolen (1989) "Waveguides and fibers for nonlinear optics," J. Opt. Soc. B **6**, 652-662.
2.   K. Becker, Y. Li, R. F. Haglund, Jr., R. H. Magruder III, R. A. Weeks and R. A. Zuhr (1991) "Nonlinear and Fluorescence Properties of Ion-Implanted Fused Silica," Nucl. Instrum. Meth. in Phys. Res. B **59/60**, 1304-1307.
3.   R. F. Haglund, Jr., H. C. Mogul, R. A. Weeks and R. A. Zuhr (1991) "Changes in the Refractive index of fused silica due to implantation of transition-metal ions," J. Non-Cryst. Sol. **130**, 326-331.
4.   G. C. Papavassiliou (1976) "Optical Absorption Spectra of Silver, Gold, and Copper Thin Films Chemically Deposited on Quartz Plates," Z. Phys. Chemie (Leipzig) **257**, 241-248.
5.   R. F. Haglund, Jr., R. H. Magruder III, D. O. Henderson, S. H. Morgan, L. Yang and R. A. Zuhr (1992) "Nonlinear Index of Refraction in Cu- and Pb-Implanted Fused Silica," Nucl. Instrum. Meth. in Phys. Res. B, in press.
6.   M. Sheik-Bahae, A. A. Said, T. Wei, D. J. Hagan and E. W. VanStryland (1990) "Sensitive Measurement of Optical Nonlinearities Using a Single Beam," IEEE J. Quantum Elect. **26**, 760-769.
7.   F. Hache, D. Ricard, C. Flytzanis and U. Kreibig (1988) "The Optical Kerr Effect in Small Metal Particles and Metal Colloids: The Case of Gold," App. Phys. A **47**, 347-357.

THE STRUCTURE AND PROPERTIES OF NANO-SIZE CRYSTALLINE SILICON FILMS

Y. HE,[a] C. YIN,[a] W. TANG,[b] and T. GONG[b]

(a) Physics Dept., Nanjing University, Nanjing 210008, P. R. CHINA
(b) Lab for Infr. Phys., Shanghai Inst. Tech. Phys., Shanghai 200083, P. R. CHINA

ABSTRACT. We have fabricated the nano-size crystalline silicon films by using the high hydrogen diluted silane as the reactive gases and activated at r.f+d.c double power sources, in a conventional PECVD deposition system. The structure of growing films were detected by means of HREM, Raman scattering spectra, X-ray diffraction pattern, IR transmission spectra and ultra-violet ray analysis. The results show that there are a lot of novel structure performances and strange physical properties, e.g. the optical obsorption coefficient is higher than that of a-Si:H and uc-Si:H films, the room temperature conductivity σ_{rt} may reach 10^{-3}-$10^{-2}\Omega^{-1}cm^{-1}$, the hydrogen content C_H in nc-Si:H films is high than 30atm% etc. The nc-Si:H films have their unique features and different from both a-Si:H and uc-Si:H films.

1. INTRODUCTION

We have fabricated the nc-Si:H films by using the high hydrogen diluted silane as a reactant gases in a conventional PECVD deposition system. The structure of growing films were detected by means of HREM, Raman scattering spectra, X-ray diffraction pattern, IR transmission spectra and ultra-violet ray analysis. The results show that under the deposition conditions of our system, the growing films have the mean grain size about 3-4nm, the ratio of volume of fraction of crystalline region is around (50±5)%. These values coincide with the definite values of the nano crystalline materials by H.Gleiter.[1] The hydogen content in nc-Si:H films are as high as (20-40)atm%, these values are higher than that of a-Si:H and uc-Si:H films which deposited in the identical deposition system. The E_g^{opt} values of nc-Si:H films measured by means of ultra-violet spectra and estimated from Tauc curve[2] are 1.65-1.85 ev, which are lower than those of a-Si:H films, 1.9-2.1 ev. The HREM picture of nc-Si:H films shows that the size of smallest grain is about 2nm. Also, we can estimate that the thickness of interface between grains is 2-4 atomic layers from the volume fraction of the two components and the mean grain size. The orientation of grains in the network of nc-Si:H films is random. Since there is a novel morphology in the structure in nc-Si:H films, we measurd a lot of strange physical properties of these films.

2. THE TECHNOLOGICAL PARAMETERS OF NC-SI:H FILMS DEPOSITED BY PECVD METHOD

The nc-Si:H films are deposited at a conventional capacitance diode glow discharge system. The technological parameters are selected at T =300C, r.f power density

1245

P. Jena et al. (eds.), Physics and Chemistry of Finite Systems: From Clusters to Crystals, Vol. II, 1245–1250.
© 1992 *Kluwer Academic Publishers.*

0.44w/cm (13.56 MHz), the total pressure of reactant gases in deposition process 1.0-1.5 torr. Two prerequisite technological means are used to form the nano-phase morphology of growing films. First, the high hydrogen diluted silane are employed as the reactant gases, which can enhance the etching process at the meantime of deposition and promote the growing films nucleation and crystallization. Fig.1 gives a curve of optical energy gap Egopt of growing films versus the concentration ratio of reactant gases SiH_4 /SiH_4+H_2 and shows the value of E_g^{opt} appears a sharply decreasing at the ratio value< 2%, that there is an abrupt change in the structure of growing films.

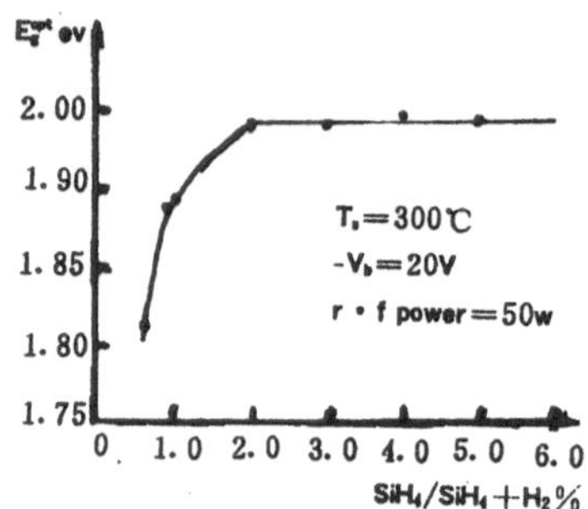

Fig 1. A curve of E_g^{opt} versus the ratio of reactant gases concentration of silane of growing films

Second, the important parameter of deposition nc-Si:H films is the negative substrate bias -V_b applied on the parallel electrodes of deposition system. During the growth of nc-Si:H film the negative bias decisively affects the energy and the nature of the species $[SiHn]^+$ and $[H]^+$ go towards the substrate surface. We consider that an appropriate substrate bias can affect the reaction rate of the gas-solid (substrate) interface, which may promote the nucleation rate and then enhance the densification of fine micro-crystallites in the growing films. Thus, We can control the ratio of volume fraction of crystallites, C=Ic/Ic+Ia, in nc-Si:H films by changing the substrate bias. Fig 2 shows the relationship of the C values versus with substrate bias -V_b . It indicates that the nano-phase silicon films appears in the growing film, i.e. the ratio of volume fraction of crystallites C reaches (50±5)%. Usually, we select the negative substrate bias in the range of 200v < -V_b < 300v in our deposition system. The C curve appears a maximum at -Vb = 250v, and then gradually decrease with the increasing of -V_b. We think that the -V_b is more than 250v, the radical species of $[SiHn]^+$ and $[H]^+$ have an excessive kinetic energy , the high energy species bombardment on the growing film limits the growth and nucleation of grains.

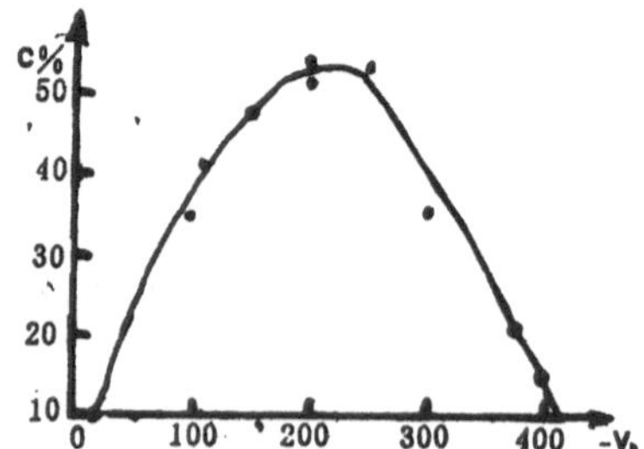

Fig 2 The C values versus with -V_b of deposited silicon films

3. THE STRUCTURE PERFORMANCES OF NC-SI:H FILMS

Fig 3 is a HREM picture of nc-Si:H film which we have deposited by means of PECVD method. From this picture we can see that there are a lot of fine and dense microcrystallites in the network. The orientation of these grains are completely random. We can estimate from the picture that the smallest grain is 2nm. The mean grains size is 3-4nm. We can also estimate the value of ratio of volume fraction of crystallites to be 47%. This value is corresponding to the value from Raman spectrum which is 45% (see Fig 4). According to the mean grain size and the ratio of volume fraction of interfaces, we may obtain the thichness of interface between grains is 2-4 atomic layers. Fig 4 gives a

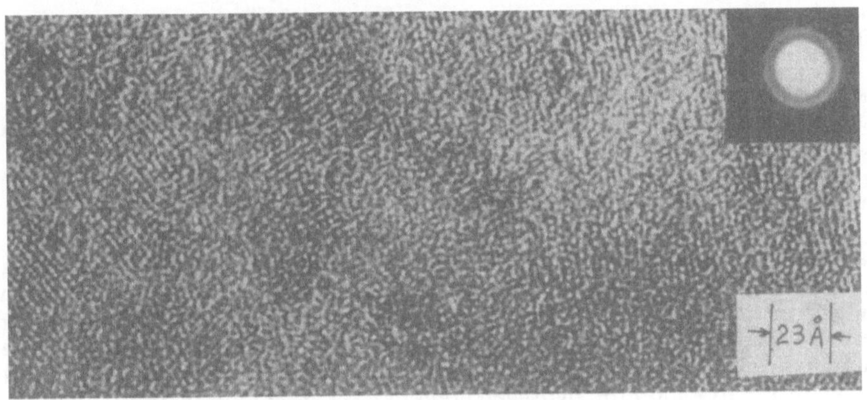

Fig 3 The HREM picture of nc-Si:H film (4,320,000)

typical Raman spectra TO modes of two nc-Si:H films. The curve 4A is of sample 1028, from the formula[3],

$$d=2\pi(B/\Delta\omega)^{1/2}$$

we may calculate that the mean grain size is 3.6nm, the ratio of volume fraction of crystalline $C=Ic/Ic+Ia$ is 46%. The curve 4B is of sample 201, of which $d\approx4nm$ and $C = 49\%$. Fig 5 is X-ray diffraction patterns of two nc-Si:H samples. The curve 5A is of sample 226, the mean grains size from FWHM of (111) peak calculated from Scherrer formula is 3.4nm. We can see there are many small and sharp peaks superimposing at the (111) peak and at lower diffration angles, which demonstrates the random nucleation and growth of grains. We can also see from Fig 5A, there appears a broad peak at $2\theta = 50°$, which is a merged peak of (220) and (311). Fig 5B is of sample 124, from this X-ray diffration pattern we can obtain the mean grain size is 5nm. The superimposed small peaks around (111) disappeared and the (220) and (311) peaks have been seperated individually. It is demonstrated that the micro-grains in 124 sample have been regularized[4].

4. SOME PHYSICAL PROPERTIES OF NC-SI:H FILMS

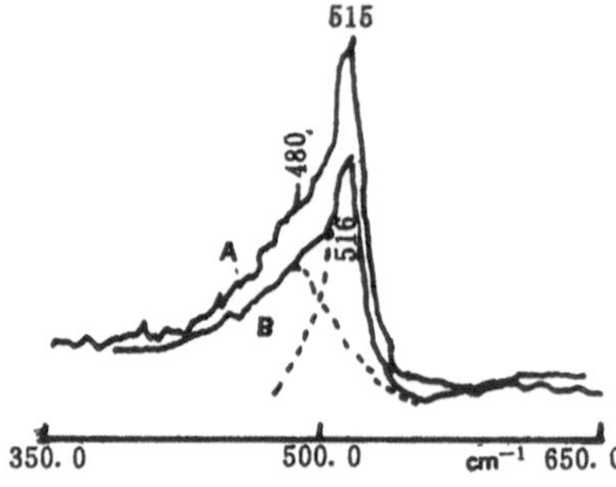

Fig 4 The Raman scattering
spectra of two nc-Si:H
films

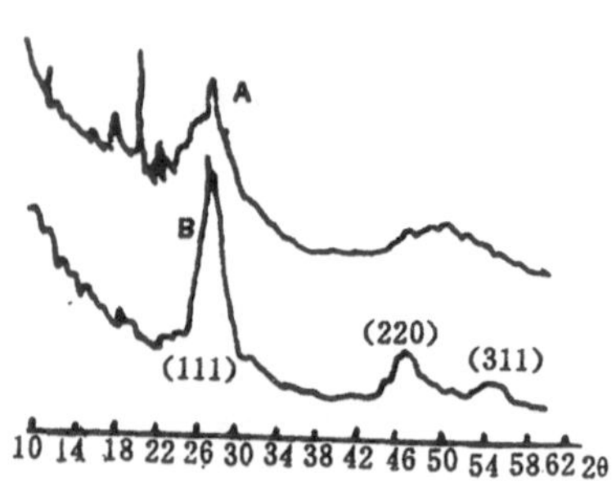

Fig 5 The X-ray diffrac-
tion pattern of two
nc-Si:H films

Because, there is a novel structure of nc-Si:H films, we have got a series of new properties which differed from both a-Si:H and uc-Si:H films. We may describe it as follows.

1.1 The hydrogen content and the bonding of silicon hydrogen

The total hydrogen content in the nc-Si:H films are obtained by integrating the area under the rocking mode at 640 cm⁻¹ of I.R spetra. Some of these correspond with the values which measured by resonant nucleation reaction analysis. The relationship of hydrogen content C_H with the ratio of volume fraction of crystallites C are shown in Fig 6. It is apparent, when the C value increases into the region of nano-phase structure (the shadow region in Fig 6) the hydrogen content C_H varied sharply from 20 atm% ~ 40 atm%. Also, we estimate the ratio of Si-H bonds and Si-H_2 bonds from the fine structure of 2089 cm⁻¹ and 840 cm⁻¹ peaks of I.R spectra, it is almost about 1:4 of these samples. So, we think that the major bonds are Si-H_2 in the nc-Si:H films.

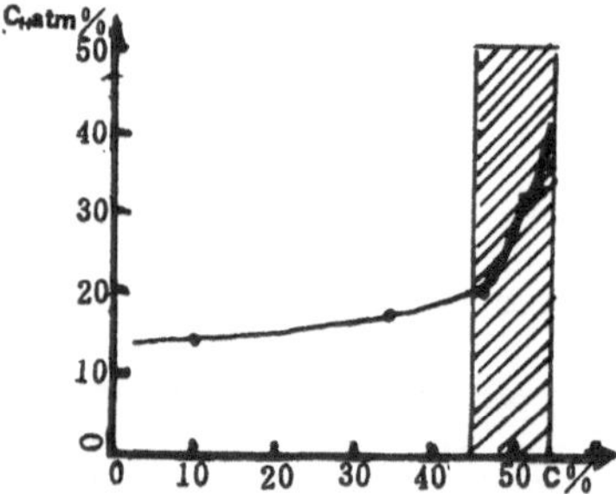

Fig 6 The hydrogen content CH versus with C values of silicon films

As mentioned above, an atmosphere of hydrogen atoms is concentrated at the interface region. The fine micro-crystallites are covered by dense hydrogen atoms, which limits the growth of grains. A large number of interface in nc-Si:H films, which have a ratio of volume fraction 50% in the films, may have a serious infuence on the structure and properties of films. This is a question that waits for further discussion. We find that the C_H content increased sharply with C values raising, however, the relative values of optical energy gap E_g^{opt} are decreased monotonously (Fig 7). This is also quite different from that a-Si:H films. As we know that the E_g^{opt} values are enhanced with increasing of hydrogen content in a-Si:H films.

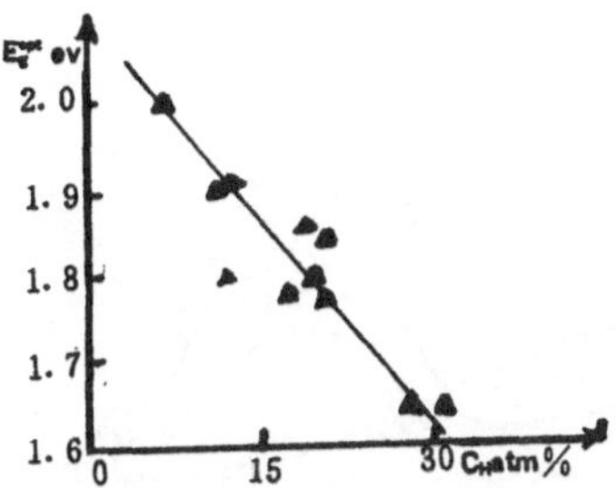

Fig 7 The E_g^{opt} varies with hydrogen content C_H in nc-S:H films

1.2 The optical absorption coefficient of nc-Si:H films

We measured the optical absorption coefficient α by means of photo-acoustic spectra (PAS) for some nc-Si:H, uc-Si:H and a-Si:H samples. It has shown that the α value of nc-Si:H films are higher than that of uc-Si:H and a-Si:H films, in the range of visible light (0.9-2.1 ev), which are deposited in a identical equipment, as shown in Fig 8.

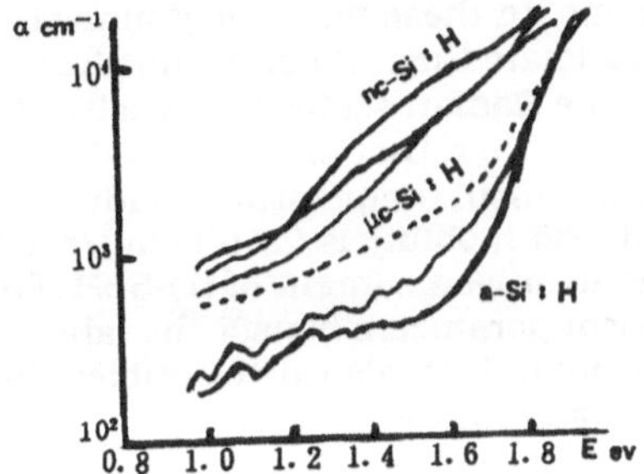

Fig 8 The α values of silicon films by means of PAS method

1.3 The photoluminescence spectra of nc-Si:H films

We measured the PL spectra of these films. The cw PL spectra were excited with the 5145 Å line of an argon ion laser and detected with a cooled photomultiplier. The spectra have been corrected for system response. The samples were placed in a temperature controlled croystat.

The measuring PL spectra of deposition films are shown in Fig 9, in which two a-SI:H samples with amorphous network and C≈0 , revealed a single and broad peak at 900-1000 nm (corresponding to 1.38-1.30 ev), this values are coincided with the literatures(7)(8). Moreover, there are two peaks at 900 nm (1.38 ev) and 1250 nm (0.99 ev) for each of the two uc-Si:H samples. The former one is due to amorphous network (a-peak) and the later one is corresponding to the location of the PL peak of c-Si (c-peak)(9). We consider that the c-peak results from the constituent crystallites of uc-Si:H network. But the width of c-peak are broading. It is interesting to note that there is only one PL peak at 1250 nm (c-peak) for both nc-Si:H samples, while we found no trace of a-peak. It may mean that there is no a-peak PL spectrum from random morphology in nc-Si:H network although the volume fraction of interfaces is about 50% at low temperature (T=15K).

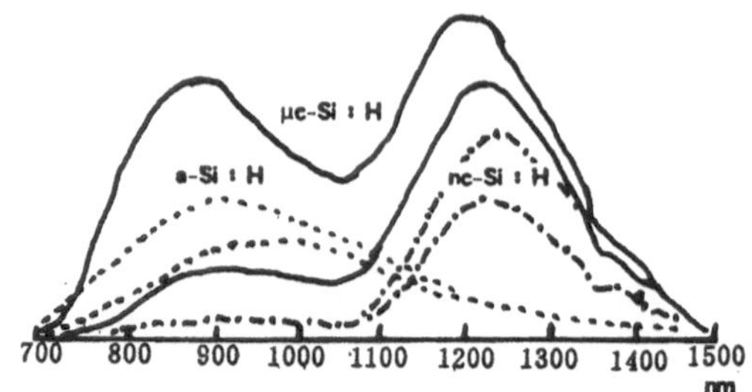

Fig 9 The PL spectra of deposited silicon films measured at T=15K
...... a-Si:H _______ uc-Si:H ------ nc-Si:H

5. CONCLUSION

In a conventional PECVD films deposition system, we employed the high hydrogen diluted silane as the reactant gases and at the conditions of precisely controlled the technological parameters have fabricated the silicon films, in which emerged nano-phase morphology, so-called nc-Si:H films. In these films the grains size is 3-5 nm, the smallest grain size is 2 nm and we can estimate the ratio of volume fraction of crystallites is the range of (46.6-55.3)% by using the Raman scattering spectra. A lot of physical properties of nc-Si:H films are better than that of a-Si:H and uc-Si:H films, such as the optical absorption coefficient α, the room temperature conductivity σ_{rt} may be reached 10^{-3}-$10^{-2}\Omega^{-1}cm^{-1}$, the Hall and drift mobility is (3-5) cm^2/sec·V, the photo-and thermal stability is very well, the hydrogen content of nc-Si:H films are high than 30atm%, and so on. These physical parametes benefit the fabrication of electronic devices, therefore, we think that not only it is a worthy subject for further research, but also it has potentiality for devices applications.

ACKNOWLEDGMENTS

We would like to thank Luchun Wang and Qi Li for HREM pecture,Kuanshi Cheng for Raman scattering measurement,Shuyi Zhang and Yiping Zang for PAS measurement,Qinghai Feng for samples deposition and we acknowledge Prof Hsiangna Liu and Hengnan Zhou for useful discussion. This reseach work was supported by The National Science Foundation of china.

REFERENC

[1] Glieter, H. [1989] Prog Materials Science 33, 223.
[2] Tauc, J. [1974] "Amorphous and Liquid Semiconductors" chap 4.
[3] Cheng Guangxu, Hua Xia et, al [1990] Phys State Solid (a) 118, K51-54.
[4] He Yuliang, Zhou Hengnan et, al(1990) Acta Physica Sinia 39, 1798.
[5] Cody, G. D. and Abeles, B. (1980) Solar Cell 2, 227.
[6] Fujii, Y. Moritani, A. and Nakai, J. (1981) Jpn. J. Appl Phys 20,361.
[7] Collins, R. W. Paesler,M.A.and Paul,W(1980) Solide State Commun 34,833.
[8] Campbell, I. H. Fauchet, P. M. and Lyon, S. A. (1990) Phys Rev B41,9871.

FIVEFOLD MULTIPLY-TWINNED CRYSTALLITES IN VAPOUR-DEPOSITED
AMORPHOUS THIN FILMS OF GERMANIUM STUDIED BY HREM

H. HOFMEISTER[1], P. WERNER[1] and T. JUNGHANNS[2]
[1]Inst. of Solid State Physics & Electron Microscopy
P.O. Box 250
O-4010 Halle (Saale)
Germany
[2]Inst. of Physics of the Ukrainian Academy of Sciences
Prospekt Nauki 46
252 650 Kiev
U.S.S.R.

ABSTRACT. Crystalline particles formed in amorphous thin films of Ge
during vapour deposition have been studied by high resolution electron microscopy. Multiply twinning is frequently observed in the particles. They show cyclic arranged microtwins the twin boundaries of
which meet in fivefold junctions parallel to the film normal. Nucleation
and growth of twinned structures as well as the formation of lattice
defects are discussed by means of various experimental findings.

1. INTRODUCTION

The investigation of structural peculiarities of amorphous semiconductors is of great interest not only with respect to the growth and
perfection of the corresponding crystalline phase, but also in the
frame of efforts to enlighten the basic processes of crystallization
from the amorphous phase.

Twinning and, in particular, multiply twinning is frequently observed in the crystal growth of diamond, silicon, and germanium thin
films [1-3]. Therewith, structures are formed which are composed of
cyclic arranged subunits, twin related to each other. Multiply twinned
structures cannot be attributed readily to one uniform mechanism of
formation. They are most intensively studied for small metal particles
having f.c.c. crystal lattice [4-6]. As a characteristic, the twin boundaries enclosing an angle of about 2 /5 meet at axes of fivefold
symmetry. Because of a lack in space filling, when composed of subunits with a perfect crystal structure, elastic strains are resulting
which may lead to the formation of lattice defects [6].

In order to elucidate the role of nuleation, growth, and twinning
in the formation of fivefold multiply-twinned crystallites, insight in
the structural details is required at an atomic scale of resolution. For
that high resolution electron microscopy (HREM) is a powerful tool. In

1251

P. Jena et al. (eds.), Physics and Chemistry of Finite Systems: From Clusters to Crystals, Vol. II, 1251–1256.

the present paper, dealing with a HREM study on amorphous thin
films of germanium which contain small crystallized regions (particles),
various stages of twinning and growth as well as of the formation of
lattice defects are reported and discussed.

2. EXPERIMENTAL

Amorphous thin films of Ge containing small crystalline particles,
about 5 to 30 nm in size, have been obtained by vapour deposition on
air cleaved NaCl crystals under a vacuum of nearly 5×10^{-5} Pa. The
deposition was carried out at substrate temperatures between 200°
and 300° C up to layer thicknesses between 10 and 20 nm at a rate
of about 0.9 nm min^{-1}. A resistance-heated Ta coil was used to evaporate the highly pure Ge (less than 2×10^{14} cm^{-3} impurities).
　　　　Electron microscopy examination of the Ge films detached from
the substrates and mounted on copper grids was done at a JEM 200
CX microscope operating at 200 kV and a JEM 4000 EX microscope
operating at 400 kV, respectively. HREM images of the particles, frequently situated in a <110> zone axis orientation were taken with the
objective lens appropriately defocused so to allow imaging of the
channels characteristic of this projection as bright dots.

3. RESULTS AND DISCUSSION

A considerable portion of the crystalline particles formed within the
amorphous Ge films during vapour deposition at appropriate substrate
temperatures (nearly 200° C or more) exhibit an internal structure
which is characterized by the presence of several microtwins. Typically the twins are arranged cyclic with the fivefold axis parallel to
the overall growth direction of the film. The large frequency of twinned crystallites found in the amorphous Ge films points to a preferred nucleation and growth of twinned strutures.

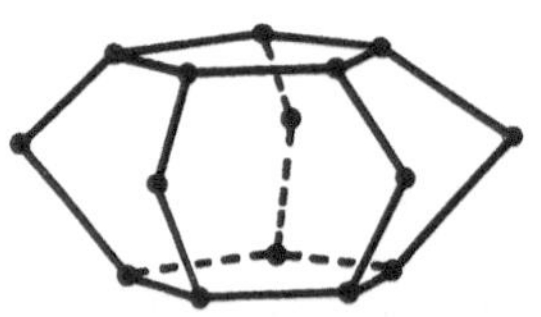

Fig.1: Twinned 15 atoms nucleus with fivefold symmetry

The smaller requirements of space
and the larger number of Ge-Ge bonds
per atom may be decisive in the nucleation of preformed twin structures.
Especially, this is valid for a configuration with fivefold symmetry
consisting of 15 atoms [7] as shown schematically in Fig.1.
　　　　At particles with only one fivefold twin junction lattice defects are to
be expected which allow stress relief. This can be recognized from
the distinct lattice distortions in the region of a twin boundary
(dashed) and the neighboured twin sector (encircled) at the particle
shown in Fig.2. A model of the corresponding defect structure is
given in Fig.3. The lattice distortion resulting from a stacking fault
bounded by partial dislocations A and B which produce a shear deformation in the lattice and a step at the twin boundary, respectively,

is supposed to involve atomic displacements by relaxation.

Fig.2: Particle with lattice distortions at a twin boundary step

Fig.3: Model of the defect structure of the particle of Fig.2

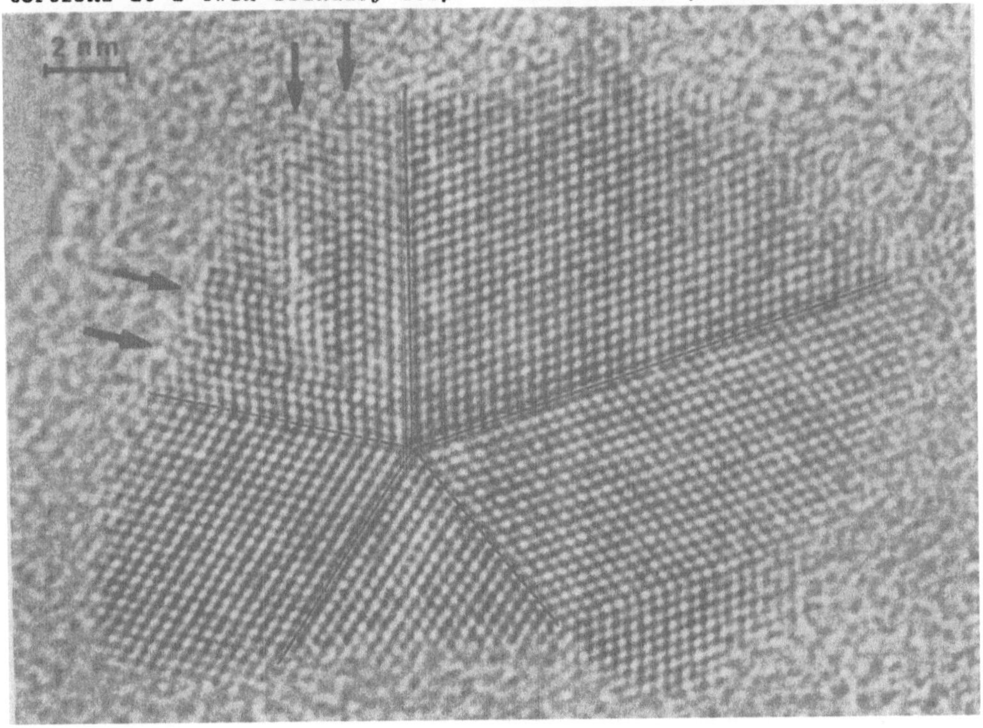

Fig.4: Particle with couples of stacking faults (arrowed) arranged
tetrahedrally in the area of one of the twin sectors

     As a typical defect in a little larger particles a tetrahedral
arrangement of stacking faults in one of the twins was found. The
particle shown in Fig.4 exhibits two couples of such stacking faults
the arrangement of which is schematically drawn in Fig.5. As can be
recognized from Fig.6 the tangential lattice dilatation achieved in this
way leads to an enlargement of the angle between the twin boundaries of the relevant twin.

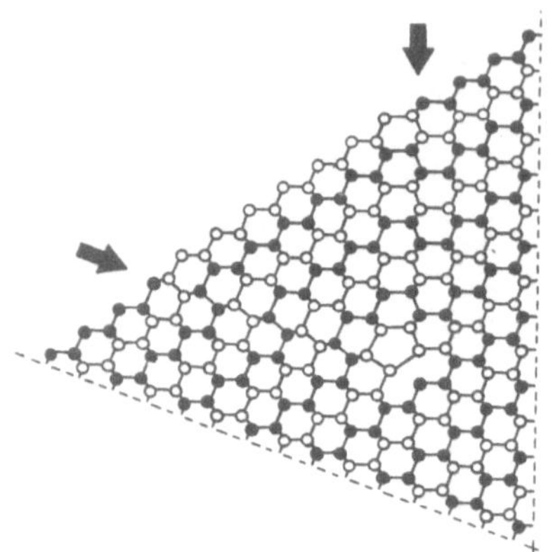

Fig.5: Model of a pair of tetrahedrally arranged stacking faults

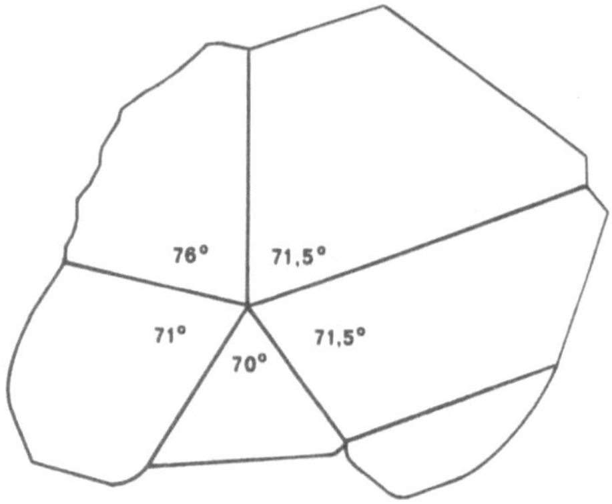

Fig.6: Schematic representation of
the particle of Fig.4

     During growth of the crystalline particles additional twin bands
may be formed by errors in the stacking of atoms. Successive growth
twinning on alternate twin planes may lead to configurations at which
the formation of fivefold twin junctions seems to be favoured energetically compared with the formation of grain boundaries. This may be

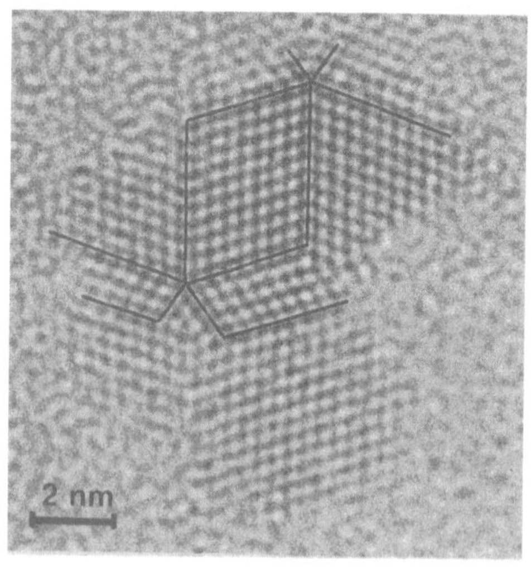

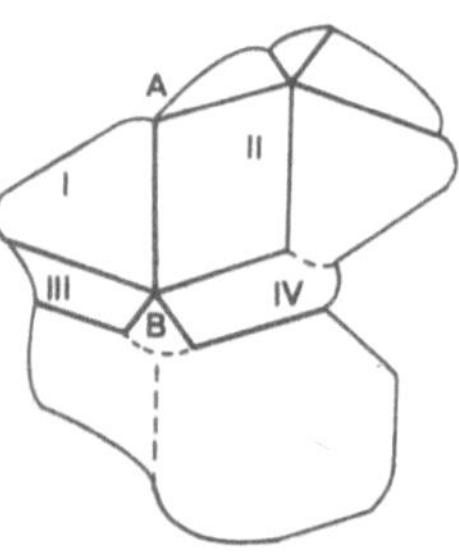

Fig.7: Structures formed by successive growth twinning

Fig.8: Schematic representation
of the particle of Fig.7

illustrated by the particle shown in Fig.7. The schematic representation given in Fig.8 as well as the model shown in Fig.9 clearly out line the configuration at which from a threefold junction of twin boundaries (B) of four twins (I – IV) by further growth of the twin bands III and IV preferably a fivefold twin junction may result instead of a grain boundary between the twins III and IV.

Additional twin bands may be formed during growth also by dislocation processes owing to elastic strains. This may lead to crossing of twins. Fig.10 gives an example of particles which contain several fivefold twin junctions within a network of twin bands which may result from cross twinning. As shown in Fig.11 by transformation of the distorted

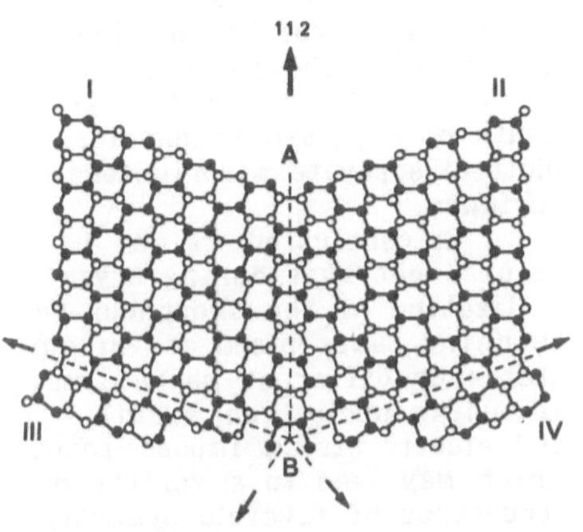

Fig.9: Model of a threefold twin junction during growth

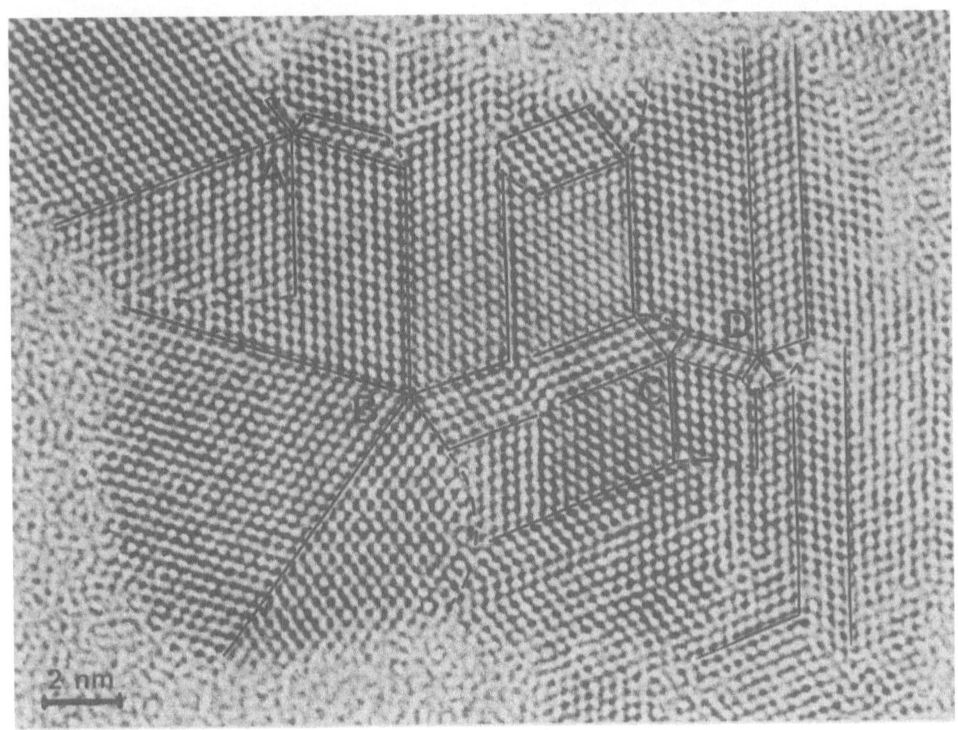

Fig.10: Particle with several fivefold twin junctions within a network of twin bands accompanied by subgrain boundaries (curved lines)

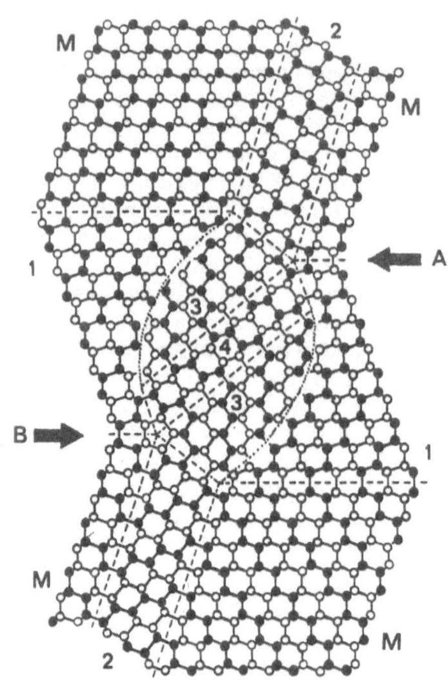

lattice in the crossing region [8] of the twin bands (1 & 2) additional twins (3 & 4) occur which form fivefold junctions (A & B) with the twins 1 and 2 as well as the matrix M. Segments of subgrain boundaries (dotted) separate misoriented regions.

In conclusion, it may be pointed out that during crystallization of the amorphous Ge twinning takes place at various stages by different mechanisms including nucleation, growth, and elastic strain impact, resp., which may lead to a variety of structures of fivefold symmetry stabilized by lattice defects.

Fig.11: Model of the formation of fivefold twin junctions by cross twinning

REFERENCES

[1] Williams, B.E., Glass, J.T., Davis, R.F. and Kobashi, K. (1990) "The analysis of defect structures and substrate/film interfaces of diamond thin films", J. Cryst. Growth 99,1168-1176.

[2] Sinclair, R. (1990) "In situ high-resolution electron microscopy", in L.D. Peachy and D.B. Williams (eds.),Proc. XIIth Int. Cong. Electr. Microsc., San Francisco Press, San Francisco, pp.512-513.

[3] Okabe, T., Kagawa, Y. and Takai, S. (1991) "High resolution electron microscopic observation of a pentagonal nucleus formed in amorphous germanium films", Phil. Mag. Lett. 63, 233-239.

[4] Ino, S. and Ogawa, S. (1967) "Multiply twinned particles at early stages of gold film formation on alkali halide crystals", J. Phys. Soc. Japan 22, 1365-1374.

[5] Gillet, M. (1977) "Structure of small metallic particles" Surf. Sci. 67, 139-157.

[6] Howie, A. and Marks, L.D. (1984) "Elastic strains and energy balance for multiply twinned particles", Phil. Mag. 49, 95-109.

[7] Douin, J., Dahmen, U. and Westmacott, K.H. (1991) "On the formation of twinned precipitates in Al-Ge alloys", Phil. Mag. B63, 867-890

[8] Dahmen, U., Westmacott,K.H., Pirouz, P. and Chaim, R. (1990) "The martensitic transformation in silicon — II. Crystallographic analysis" Acta metall. mater. 38, 323-328.

DISPERSION OF ULTRAFINE NICKEL PARTICLES ON ALUMINA FILMS AND THEIR
BEHAVIOUR IN OXYGEN AND HYDROGEN ATMOSPHERES

JITENDRA KUMAR and V. SUBRAMANIAN
Indian Institute of Technology
Materials Science Programme
Kanpur - 208016, India

ABSTRACT. Ultrafine nickel particles (typical diameter <10nm) have been
dispersed over alumina support films by thermal deposition technique in
vacuum ($\sim 10^{-5}$ torr) at substrate temperature of 400°C and studied with
regard to their size distribution, crystal structure and changes
occuring in oxygen and hydrogen atmospheres by transmission electron
microscopy. It is shown that deposits corresponding to mean thickness of
0.5 nm yield reasonable dispersion of particles with average size of
5.3nm and correspond to f.c.c. phase of nickel but undergo both physical
and chemical changes in oxygen and hydrogen atmospheres at 500°C. In
oxygen, they form NiO exhibiting a NaCl-type structure with a ~ 0.417nm.
Moreover, oxide particles get spread over the substrate and cause
increase in their average projected diameter due to overall expansion
and favourable changes in interfacial energies. In hydrogen, particles
show coarsening with simultaneous decrease in their number density and
faceting. Hydrogen insertion in general causes structural disorder and
turns nickel into hydride(s).

1. Introduction

Ultrafine metal particles dimensionally lie between atoms and the bulk.
They represent a highly dispersed state of metal and exhibit unique
structural, physical and chemical properties. As a consequence, they
have been receiving considerable attention in recent years both from
fundamental and application view points, particularly,to understand the
origin of their curious characteristics [1-5]. Of special interest are
particles of metals of group VIIIB and IB of periodical table which
when dispersed on suitable supports are used as catalyst for a number
of chemical reactions [5,6]. Numerous studies have been made on the
behaviour of some metals, but only a few are devoted to nickel. The
present investigation was therefore undertaken to disperse nickel
particles on alumina films and study their behaviour at the microscopic
level. Ideally, specimens should be amenable to experiment and yet
correspond to the real situations as closely as possible. Small metal
particles formed by vacuum evaporation technique have been considered
quite suitable for this purpose [1,5]. The method essentially involves
vacuum deposition of metal on a substrate and its subsequent heating in

P. Jena et al. (eds.), Physics and Chemistry of Finite Systems: From Clusters to Crystals, Vol. II, 1257–1262.
© 1992 Kluwer Academic Publishers.

hydrogen at elevated temperatures. Although the formation of particles does take place, yet, their purity is questionable. Therefore, an alternative method [3,4] has been adopted here, in which, particles are formed during the deposition step itself by maintaining the substrate at elevated temperature. This eliminates altogether the heating step and yields samples free of contamination for meaningful studies relating to their behaviour in oxygen and hydrogen atmospheres.

2. Experimental Details

Alumina support films of thickness ~13nm were prepared by anodization using 3 wt% tartaric acid solution of pH 5.5 as electrolyte, and transferred onto gold specimen grids [3]. Suitable amount of nickel for mean thickness of 0.5 nm was then deposited on alumina under vacuum (10^{-5} torr) at the substrate temperature of 400°C to obtain well separated particles. A Philips EM301 TEM was used at 80 kV to examine the morphology and structural crystallography of the resulting nickel particle. Samples were heat treated in each oxygen and hydrogen atmospheres at 500°C for various lengths of time. The gas flow rate was maintained at 45-50 ml/minute. The samples were examined in TEM again for associated changes and possible reactions.

3. Results and Discussion

3.1. FORMATION OF PARTICLES

The representative electron micrograph and diffraction pattern of nickel films of mass thickness 0.5 nm deposited on alumina support at 400°C are given in Fig.1. The micrograph clearly shows well separated particles with reasonable dispersion and shapes. The quantitative information in respect of size, percentage, average surface area, etc. as deduced from the micrograph are summarized in Table 1. Notice that the average size of particles is 5.3 nm. The indexing of pattern and interplanar (d) spacings of various diffraction rings match very well with the f.c.c. phase of nickel having a lattice parameter of 0.35238 nm [7]. In contrast, Arai, Ishikawa and Nishiyama [8] obtained round and dumbell shape particles of nickel in two step process. They first deposited a

Table 1. Dispersion parameters of particles in nickel-alumina system before (a) and after heat treatment in oxygen (b) and hydrogen (c) atmosphere at 500°C for 4 hours.

	(a)	(b)	(c)
Number of particles counted	1159	1104	672
Distribution ⎤ – particles %	28.6	18.6	16.1
peak ⎦ – size range (nm)	3.8-4.8	5.8-6.8	6.8-7.8
Median (nm)	4.0	5.1	6.7
Average particle size (nm)	5.3	7.7	8.9
Average surface area (m^2/g)	126.5	87.6	75.4

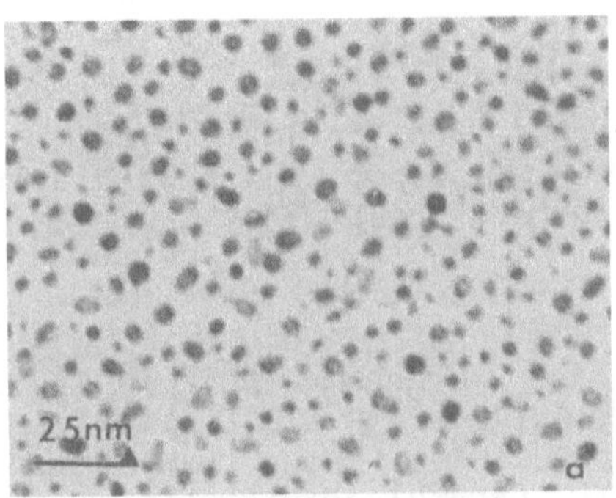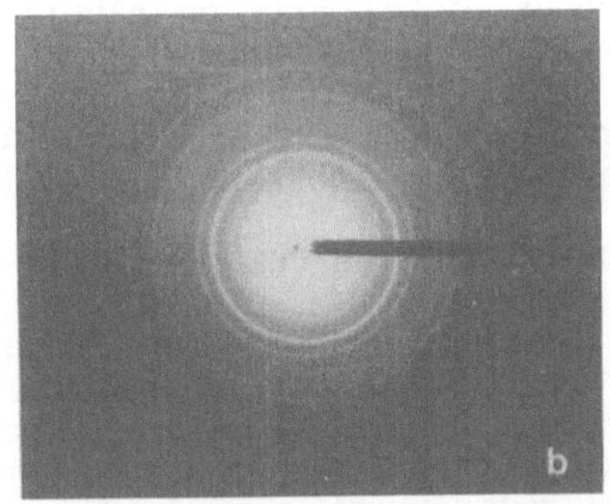

Figure 1. Electron micrograph a) and diffraction pattern b) of nickel particles dispersed over alumina at 400°C (mean thickenss 0.4 nm).

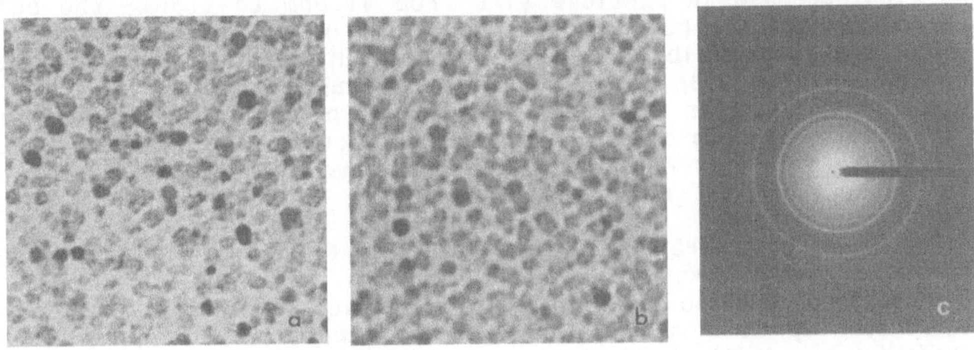

Figure 2. Nickel particles after heat treatment in oxygen at 500°C for a) 2h, b) 4h, and diffraction pattern c) corresponding to region a)

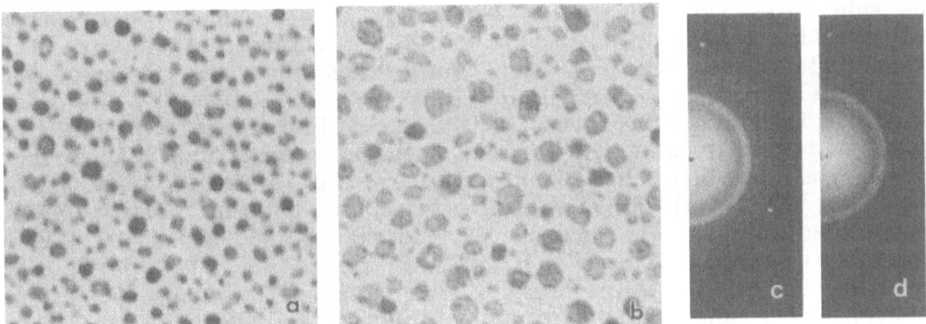

Figure 3. Electron micrographs and diffraction patterns of nickel particles after heat treatment in hydrogen at 500°C for 2h(a,c),4h(b,d).

continuous nickel film on alumina at room temperatre and then heated inside a TEM in the temperature range of 450° - 700°C at a pressure of 10^{-4} torr. However, the method adopted here is unique in the sense that nickel particles are formed during the deposition step itself by maintaining the substrate at elevated temperatures.

3.2. OXIDATION OF NICKEL PARTICLES

The microstructure and the corresponding diffaction pattern resulted after heat treatment of samples in oxygen at 500°C show marked differences in the (i) nature of particles and (ii) distribution and the intensity of diffraction rings (Fig.2). Careful interpretation of the pattern reveals presence of an f.c.c. phase. The crystal data match well with NaCl-type NiO having a lattice parameter of 0.4176 nm [7]. It means that the nickel particles have undergone a chemical change and got oxidized while being treated in oxygen atmosphere at 500°C. It is known that the lattice parameter of NiO increases [9] from 0.4178 nm to 0.4193 nm with decrease in particle size from 41.5nm to 9.5nm; the bulk parameter for NiO being 0.4176 nm. Some of the observations on morphology and distribution of particles after heat treatment in oxygen can be summarized as : a) increased coverage of substrate, b) no perceptible change in the number of particles per unit area, c) increase in average particle diameter, d) shift of normal size distribution and cumulative frequency curves towards higher particle diameters, and e) tailing of normal size distribution curves in the regime of large particle diameters.

The increase in particle size may result if (i) there is indeed a real growth of them by known processes (migration, collision and subsequent coalesence or Ostwald ripening), (ii) particles simply tend to wet the substrate due to favourable changes in interfacial energy following oxidation, or (iii) the resulting oxide assumes crystal structure that has larger volume per nickel atom in comparison to pure nickel (i.e., Pilling-Bedworth ratio Ø is larger than unity). In case of oxidation of nickel to NiO having NaCl-type structure, Ø takes a valve of 1.7. That means oxide occupies larger volume than that of metal utilized. It contributes to a 66% increase in volume and nearly 40% extra coverage of the substrate. Further, nickel oxide has lower surface energy than nickel and the energy of nickel-alumina interface decreases with oxidation [6]. These lead to lower contact angle for particles of NiO to be in equilibrium with the alumina substrate. Consequently, the oxide tries to wet the substrate. Both the above considerations contribute to apparent increase in particle size without decrease in their number density. The real growth of particles by coalescence or Ostwald ripening, on the otherhand, involves overall decrease in their number density. Also, heating in vacuum ~ 10^{-5} torr at 500°C, nickel particles show marginal growth without any chemical reaction. The particles smaller than the resolution of TEM (1 nm) can not be detected and so might eventually contribute to overall growth following heating at 500°C. However, since the substrate was maintained at 400°C during particle formation and their overall volume is found to match very well with the amount of nickel deposited, contribution of smaller size particles to growth is believed to be negligible. Therefore, the increase in particle size in oxygen atmosphere at 500°C is mainly due to enlargement of the unit cell following oxidation and spreading of the oxide on the substrate. Behaviour of nickel particles on Al_2O_3 has also been studied by Nakayama, Arai and Nishiyama [10] in oxygen atmospheres at 60 Torr. However, their initial nickel was in the form of a continuous layer that was heated in hydrogen for obtaining the particles. Nevertheless, they did observe the formation of annular, torous like or horse shoe shaped NiO particles on treating their samples in oxygen atmosphere. However, such studies can not possibly represent the true behaviour of nickel particles as they are likely to be contaminated by hydrogen in the preparation stage itself. A clear evidence for this observation is given in Section 3.3. In contrast, the results obtained in the present work do provide true picture as the particles have been generated during the deposition step without involving any treatment in hydrogen.

3.3. EVIDENCE FOR HYDRIDE FORMATION

Nickel particles prepared at substrate temperature of 400°C and treated in hydrogen atmosphere at 500°C resulted in microstructure and the diffraction pattern shown in Fig. 3. It can be noticed that diffraction rings get suppressed progressively with treatment in hydrogen. Moreover, rings first increase and then decrease in number and become somewhat diffuse with hydrogen uptake. It suggests that significant disorder is created by hydrogen insertion. In fact, survey of interplanar (d) spacings suggest the possibility of a chemical change involving hydride(s) formation. Initially, nickel rings also show up in the diffraction pattern at least upto 4h of treatment. Only three rings however, remain in 6h treated sample with none matching with nickel.

There are several reports [11-13] which deal with the existence of hydrides of nickel. However, only two hydrides reliably established [13] are the ß-phase with f.c.c. structure having a = 0.373 nm and the γ-phase of composition Ni_2H exhibiting hexagonal close-packed structure (CdI_2-type) with a =b = 0.266 nm, c=0.433 nm. While the ß-phase is unstable at room temperature and atmospheric pressure, the γ-phase is stable upto 350°C. Other hydrides reported include f.c.c. ß''-phase having a= 0.365 nm, $ß^{IV}$-phase with a=0.42 nm, hexagonal γ'-phase with a=b=0.463 nm, c=0.430 nm, hexagonal γ'''-phase having a=b=0.324 nm, c=0.520 nm, and hexagonal NiH with a=b=0.2645 nm, c=0.4312nm. As the diffraction patterns contain diffuse rings and in general are complex, it is difficult to index the pattern with certainty. Yet, they do show the presence of hydrides. The reaction is perhaps occuring as the particles of small size are usually very active [4,5]. The quantitative information as deduced in respect of size distribution parameters are given in Table 1. Some of the microscopic observations can be summarized as : a) significant decrease in number of particles per unit area, b) decreased coverage of substrate by particles and coalescence, c) faceting of particles, d) increase in average particle size, and e) shift of normal size distribution and cumalative frequency curves towards larger particles size. The observed faceting phenomenon suggests

that these particles have certain crystallographic planes of low
interfacial energies which allow their stability, even though they have
large surface areas. The particles are also seen touching each other
and forming neck(s) at places. It is known [11] that rate of increase
in size of nickel particles in hydrogen atmosphere at 600°C is
comparable to that under vacuum at the same temperature and sintering
proceeds via particle migration mechanism. The present observations can
not substantiate or disprove these findings. Nevertheless, it can be
pointed out that coarserning of nickel particles in hydrogen does
involve some sort of gas-metal interaction.

4. References

1   Hamilton, J.F. and Baetzold, R.C. (1979) "Catalysis by small metal
    clusters", Science 205, 1213-1220.
2   Granqvist,C.G. and Buhrman, R.A. (1976) "Ultrafine metal particles",
    J. Appl. Phys. 47, 2200 -2218
3   Kumar,J. and Palanisamy, R. (1987) "Formation of small particles of
    gold on alumina support films and their behaviour in oxygen and
    hydrogen atmospheres", Appl. Surface Sci. 29, 256-270.
4   Kumar,J. and Saxena, R. (1987) "Behaviour of platinum and palladium
    microclusters on alumina support films", in P. Jena, B.K. Rao and
    S.N. Khanna (eds.), Physics and Chemistry of Small Clusters, Plenum
    Press, New York, pp. 825-830.
5   Poppa,H. (1984) "Model studies in catalysis with UHV-deposited metal
    particles and clusters", Vacuum 34, 1081-1095.
6   Stevenson,S.A.,Dumesic,J.A.,Baker,R.T.K. and Ruckenstein, E. (eds.)
    (1987) Metal-Support Inteactions in Catalysis, Sintering and
    Redispersion, Van Nostrand Reinhold Company, New York.
7   Powder Diffraction File Nos. 4-0850, 4-0835.
8   Arai, M., Ishikawa, T. and Nishiyama, Y. (1982) "Surface migration
    of Ni on C, SiO_2, and Al_2O_3",J. Phys. Chem. 86, 577-581.
9   Fievet, F., Germi, P., De Bergevin, F. and Fig Larz, N. (1979)
    "Lattice parameters, microstructures and non-stoichiometry in NiO
    comparison between mosaic microcrystals and quasi-perfect single
    microcrystals", J. Appl. Cryst. 12, 387-394.
10 Nakayama, T., Arai, M and Nishiyama, Y. (1984) "Dispersion of nickel
    supported on alumina and silica in oxygen and hydrogen", J. Catal.
    87, 108-115
11 Kohdyrev, Yu. P., Baranova, R.V.,Imamov, R.M. and Semiletov, S.A.
    (1978) "Electron diffraction investigation of hexagonal nickel
    hydride", Sov, Phys. Crystallogr. 23, 405-408.
12 Kepka, J.B. and Czaputowicz, E.W. (1979) " Analysis of some results
    on Pd-H and Ni-H systems studied in high pressure-hydrogen
    conditions", Phys. Rev. B19,2414-16.
13 Baranova, R.V., Khodyrev, Yu. P. and Semiletov, S.A. (1982) nickel
    hydrides, nitrides and carbides",Sov. Phys. Crystallogr. 27, 554-558.

METALLIZATION OF Si(111) BY LEAD CLUSTER DEPOSITION

P. JONK, H.R. SIEKMANN, T. BRAMMER,
B. WRENGER, K.H. MEIWES-BROER
Fakultät für Physik
Universität Bielefeld
D-4800 Bielefeld 1
F.R.G.

ABSTRACT. Metallization of $Si(111)$ is achieved by deposition of neutral and charged lead clusters from a beam. While for small clusters deposits grow crystalline without additional acceleration of the cluster ions, acceleration is necessary for large clusters.

Semiconductor metallization at low substrate temperature is a growing field of interest. Ion assisted deposition processes in many cases are suited to induce epitaxy at room temperature. As a matter of fact, however, incorporation of rare gas atoms, high adatom mobilities and particle radiation damage set limits to these methods. Deposition of clusters from a beam [1] may overcome these problems: Due to the source conditions (acceleration in a seeding gas or ion acceleration in an electric field) the kinetic energy of the clusters ranges from about one hundred eV to several keV. During the impact this energy is converted to heat, thus locally influences the initial stage of film growth. At the same time the energy per atom is in the eV range as many atoms (100 – 1000) share the overall kinetic energy. Therefore, self-sputtering and surface damage should be avoided. So far there is still a discussion whether the experiments on "cluster deposition" really had been performed with reliable beams containing significant amounts of aggregates larger than a dimer. In order to stop this confusion we first work on cluster analysis in a beam (mass analysis, photoelectron spectroscopy). Then the investigated beams are used for the metallization of $Si(111)$.

First very recent results of this "real cluster" deposition are very promising [2-4]; they show that the optical character or the crystallinity of room temperature deposits can be adjusted by cluster mass and energy.

In this contribution we report on cluster beam deposition from a pulsed arc cluster ion source PACIS [5]. A pulsed discharge is fired between two metal rods.

1263

P. Jena et al. (eds.), Physics and Chemistry of Finite Systems: From Clusters to Crystals, Vol. II, 1263–1268.
© *1992 Kluwer Academic Publishers.*

The erroded (neutral and charged) material undergoes supersonic expansion in a He seeding gas. After differential pumping and eventually mass separation by a Wien filter, neutral as well as charged metal clusters are deposited with and without additional acceleration (up to $7keV$) of the cluster ions. In the "unaccelerated" case, the clusters still posses the energy gained from the supersonic expansion, typically about $1eV$ per atom. The deposits are performed on cleaned $Si(111)$ substrates at temperatures below $110°C$ in an UHV environment. For details of the cluster production and beam formation see ref. [6], this volume.

Figure 1 depicts the experimental arrangement. The deposition will be performed with clusters either from the beam axis or from an off axis position. In the first case, large clusters are selected (mean number of atoms $\bar{N} \sim 1500$). The side region contains smaller clusters, presumably $\bar{N} \sim 100 - 500$. By this, a rough mass selection is possible. In both cases, about 10 % of the beam content is charged.

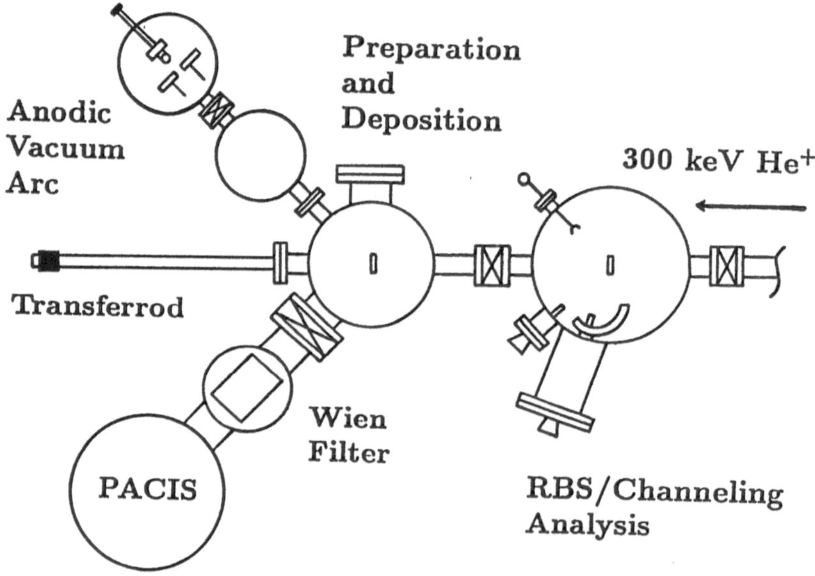

Fig. 1: Experimental arrangement for the investigation of semiconductor metallization. Neutral or charged metal clusters, produced in the PACIS, are deposited on $Si(111)$. RBS/Channeling serves to investigate the film crystallinity.

The crystalline quality of these deposits (thickness ranging between 10 and 120 Monolayer (ML)) is examined by RBS/Channeling with 300 $keV He^+$ (scattering angle $\theta = 123°$). While exact measurements on the layer thickness are limited by the depth resolution of the analyzing system to roughly ≥ 100 Å (for Pb and Sn), information on layer ordering, however, is still available down to about ten ML (from minimum yield studies, substrate shadowing, etc.). [7]

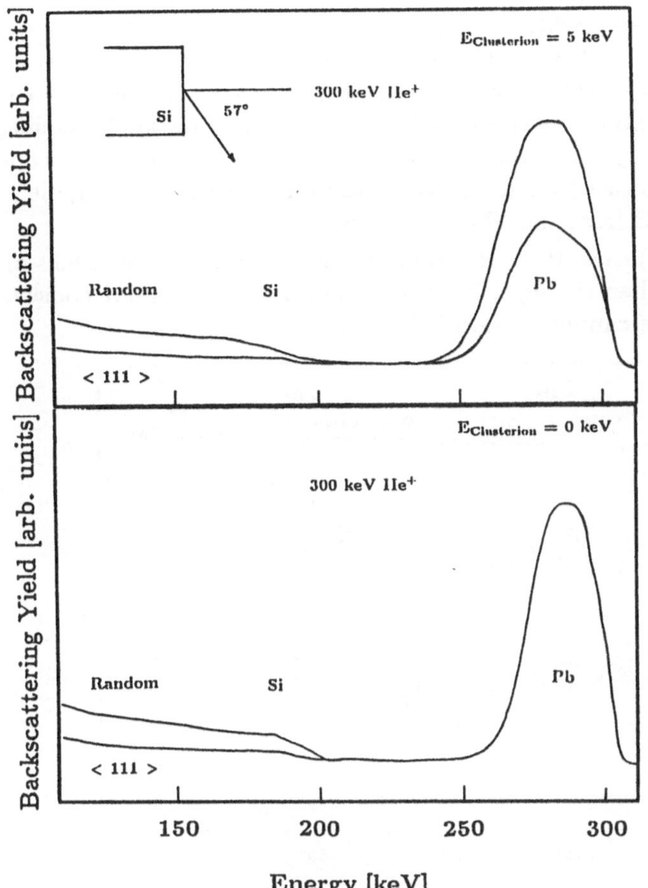

Fig. 2a:
Rutherford backscattering (RBS) spectra in random-direction (top curve) and < 111 > direction (bottom curve) of a *Pb* layer on *Si*(111). Here, the charged part of the *Pb* cluster beam has been accelerated to 5 *keV*.

Fig. 2b:
Same as figure 2a, but now without acceleration.

Figure 2a shows RBS/Channeling spectra for $\approx$ 120 ML large *Pb* clusters deposited (from the beam axis) on *Si*(111). Here the cluster ions are accelerated to 5*keV*. For the upper spectrum the incidence of the analyzing *He*$^+$ beam is chosen to be in random direction. On aligning the *He* beam to *Si* < 111 > the intensity of the backscattering yield from the *Pb* deposit decreases (bottom curve) which indicates an orientation of the *Pb* layer with respect to the *Si* < 111 > direction. This is not the case if the cluster ions are unaccelerated as it is shown in figure 2b: Now the backscattering yield from the metal layer remains unaffected by the alignment of the analyzing beam.

The decrease of the *Pb* signal upon alignment of the substrate is a first measure of the layer cristallinity. In the case of perfect epitaxial growth we expect a decrease of the signal to a few percent of its original value. Due to technical reasons this value is not obtained. Therefore, additional angular scans are performed where the $Si(111)$ plane is aligned parallel to the scattering plane. Upon rotating the substrate by an angle ϕ with respect to $<111>$, different crystal axes are explored. At tilt angles referring to certain crystal axes the RBS yield is reduced due to channeling in these directions. Fig. 3 compares three RBS angular scans: The first two scans originate from lead layers deposited from unaccelerated and accelerated large lead clusters, respectively, the third from the silicon substrate.

In the first (unaccelerated) case, the RBS signal turns out to be independent on ϕ. Such a behaviour occurs when the layer is either amorphous, or when it consists of unoriented cristallites. This cannot be decided at here.

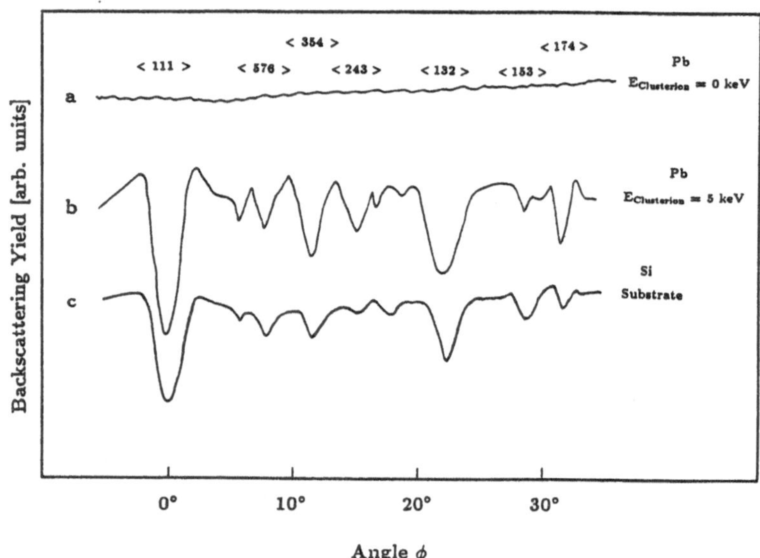

Fig. 3: Angular dependence of the backscattering yield from a) the Pb layer deposited from the unaccelerated cluster beam, b) Pb layer deposited from the accelerated beam, and c) from the Si substrate. The effect of ordering of the Pb layer upon cluster acceleration is obvious.

In the case of the accelerated beam, however, pronounced minima in the backscattering yield appear exactly at those angles where the silicon substrate axes are localized. Even without an exact analysis of the peak widths it is obvious that the accelerated large lead clusters form a layer having practically all cristalline axes of the silicon substrate. The assignment of the common axes is given in the figure. From this, the crystalline relation is found to be $Pb(111) \parallel Si(111)$, which is the same as for MBE grown Pb on $Si(111)$ [8,9].

In the case of off-axis deposition (small clusters), a similar degree of orientation is achieved already without additional acceleration. Here, the angular scans show the same crystalline axes as in figure 3 but with more pronounced minima indicating a better crystalline quality. Therefore, we conclude that the large Pb_N need an even higher acceleration voltage for a perfect growth.

There are several possible reasons for the different crystallization behaviour of large and small Pb_N clusters: From melting point studies of, e.g., Au, Pb and Sn particles it is known that generally melting temperatures appear to decrease with decreasing particle size [10-12]. For small clusters the energy gained from the supersonic expansion appears to be sufficient to raise the cluster temperature above the melting point during the impact. The larger clusters, on the other hand remain solid due to their higher melting point. During cooling the molten (smaller) clusters orientate with respect to the substrate, while the solid (bigger) clusters cannot accommodate due to their rigidity. For the latter ones an acceleration of the cluster ions is necessary to raise their temperature in order to induce orientation.

A further simple aspect of the influence of particle size could be the following: The accommodation to the substrate lattice structure should increase with the number of atoms coming into direct contact to the surface. Small clusters possess a large surface to volume ratio. Therefore, they should more likely grow in an ordered structure even without acceleration.

Our results on Sn cluster deposition seem to favour the melting point argument. Here, large as well as small cluster deposits possess an ordered structure even without additional acceleration, whereas MBE yields amorphous growth. In the above mentioned picture this could be due to the lower bulk melting point of Sn which means that the energy gain from the nozzle expansion is sufficient to melt the big clusters during the impact.

In conclusion, the experiments on Pb (and Sn) cluster beam deposition have shown that for small clusters deposits grow crystalline at a deposition energy of about 1 eV per atom. For large lead cluster, additional acceleration of the ionic part of the beam to several keV is necessary to induce cristalline growth. Without acceleration, the layer grows amorphous or polycristalline.

Acknowledgement

We thank Barbara Kessler and Walter Begemann for their help in the initial stage of this experiment. We are indebted to H.O. Lutz for his continous interest. Financial support by the Bundesministerium für Forschung und Technologie is greatfully acknowledged.

References:

[1] T. Takagi, I. Yamada, A. Sasaki, J. Vac. Sci. Technol. 12 (1975) 1128

[2] P. Jonk, H.R. Siekmann, T. Brammer, B. Wrenger, K.H. Meiwes-Broer, Proceedings of the 3rd Int. Symp. on Trends and New Applications in Thin Films, Strasbourg 1991

[3] G. Fuchs, P. Melinon, F. Santos Aires, M. Treilleux, B. Câbaud, A. Hoareau, Phys. Rev. B 44 (1991) 3926

[4] H. Haberland, M. Karrais, M. Mall, J. Vac. Sci. Technol., to be submitted

[5] G. Ganteför, H.R. Siekmann, H.O. Lutz, K.H. Meiwes-Broer, Chem. Phys. Lett. 165 (1990) 293;
H.R. Siekmann, Ch. Lüder, J. Faehrmann, H.O. Lutz, K.H. Meiwes-Broer, Z. Phys. D 20 (1991) 417

[6] H.R. Siekmann, E. Holub-Krappe, B. Wrenger, Ch. Pettenkofer, K.H. Meiwes-Broer, this volume

[7] L.C. Feldman, J.W. Mayer, S.T. Picraux, Material Analysis by Ion Channeling, Academic Press, New York 1982

[8] D.R. Heslinga, H.H. Weitering, D.P. van der Werf, T.M. Klapwijk, T. Hibma, Phys. Rev. Lett. 64 (1990) 1589

[9] G. Le Lay, J. Peretti, M. Hanbücken, W.S. Yang, Surf. Sci. 204 (1988) 57

[10] Ph. Buffat, J.P. Borel, Phys. Rev. A 13 (1976) 2287

[11] S.J. Peppiatt, J.R. Sambles, Proc. Roy. Soc. Lond. A 345 (1975) 387

[12] C.R.M. Wronski, Brit. J. Appl. Phys. 18 (1967) 1731

CRYSTALLINE NANOMETER-SCALE III-V CLUSTERS

W. A. SAUNDERS, P. C. SERCEL, H. A. ATWATER, & K. J. VAHALA
Deptarment of Applied Physics, M. S. 128-95
California Institute of Technology, Pasadena, CA 91125

R. C. FLAGAN
Deptartment of Chemical Engineering, M. S. 210-41
California Institute of Technology, Pasadena, CA 91125

ABSTRACT. We report the synthesis of crystalline nanometer-scale III-V clusters in the 5 to 50 nm size regime. The clusters are formed by homogeneous nucleation of a non-equilibrium vapor created by the explosive vaporization of a bulk sample in an inert atmosphere via a high voltage electrical discharge. Transmission electron micrographs and diffraction patterns for GaAs, InAs, and InP are presented.

In the hierarchy of reduced dimensionality semiconductor structures, quantum dots represent the extreme, "zero-dimensional," case.[1] The enhanced density of states of quantum dots, more akin to that of an atom than that of bulk, is predicted to lead to large enhancements in the performance of minority carrier devices, such as semiconductor lasers. Here, we discuss a vapor phase route to the production of III-V clusters in the quantum size regime as a first step toward the fabrication of quantum dots.

Several routes to the synthesis of nanometer scale GaAs clusters have been employed previously. Alivisatos *et al.*[2] and Uchida *et al.*[3] have synthesized crystalline GaAs clusters in quinoline solution. However, elemental analysis reveals that clusters produced in this manner incorporate significant impurities and, furthermore, have unequal Ga and As abundances.[2] Additionally, molecular species mask the optical properties of the GaAs clusters produced by this method.[5] In an alternate approach, Sandroff *et al.*[4], using molecular beam epitaxy (MBE), have grown nanometer scale GaAs clusters on silica substrates. This method produces stoichiometric and crystalline particles. However, particle shapes and sizes are highly nonuniform, and they are in intimate contact with the substrate.

Our apparatus[5] for producing the clusters is shown schematically in Fig. 1. The approach, adapted from techniques used to study the exploding wire phenomenon is useful for a wide variety of materials. Typically, GaAs

1269

P. Jena et al. (eds.), Physics and Chemistry of Finite Systems: From Clusters to Crystals, Vol. II, 1269–1273.
© 1992 *Kluwer Academic Publishers.*

wires, roughly 0.2 mm in diameter, are either etched from a photolithographically patterned wafer or cleaved from a wafer lapped to .1 to .2 mm thickness. A wire is carefully soldered into a holder in a vacuum chamber which is then pumped and back-filled with high purity gas (typically 99.999% Ar or He) to a pressure of 1 atm. After charging the capacitor to 6 to 12 kV, the spark gap is triggered, completing the circuit and vaporizing the wire.

The details of aerosol formation by exploding wires are not completely understood.[6] However, our results indicate that the background gas quenches the vapor and induces cluster nucleation before vapor-solid equilibrium is established. The role of the background gas is confirmed by experiments with different gases. For instance, we find clusters produced in H_2 do not have a zincblende crystal structure.

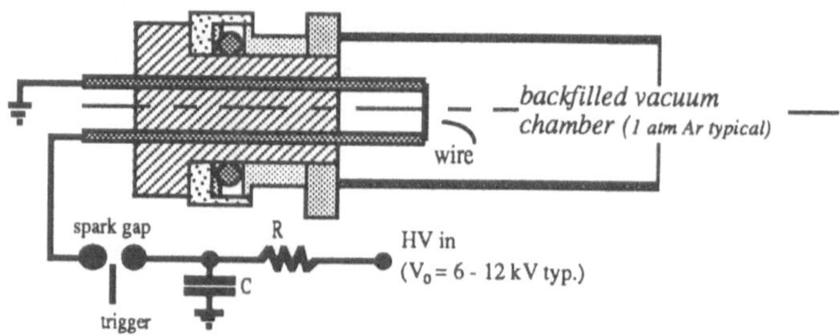

Figure 1. A schematic diagram of the exploding wire apparatus. By discharging the capacitor through a GaAs wire, a stoichiometric vapor of Ga and As is produced which subsequently nucleates to form crystalline GaAs clusters. Typically V is 6 - 12 kV and C = 1 μF. The stored energy corresponds to approximately 100 eV/atom.

A high resolution transmission electron micrograph of an agglomerate of the GaAs clusters, collected on a holey carbon substrate, is shown in Fig. 2. In general, the clusters produced by this method, when observed in the TEM tend to be agglomerated. This is largely due to the small experimental volume and long collection times employed in preparing a TEM sample. Previous work on aerosols produced in this manner suggests that agglomeration be effectively stopped by increasing the experimental volume.

Within several of the GaAs clusters, atomic rows are easily recognized. In addition, hexagonal features, indicative of faceting, are apparent. The clusters appear to have random relative orientations, which suggests that cluster nucleation occurs on a much shorter time scale than does agglomeration. An absence of lattice fringes which extend to the particle boundaries is due to a native oxide layer, approximately 1.5 nm thick, which surrounds each particle. This is attributed to the oxidation of collected particles during sample preparation for electron microscopy. Energy-dispersive x-ray analysis and electron energy loss spectroscopy both give results which are consistent with the clusters having a stoichiometric elemental composition. Electron diffraction confirms that the clusters have a zincblende crystal structure.

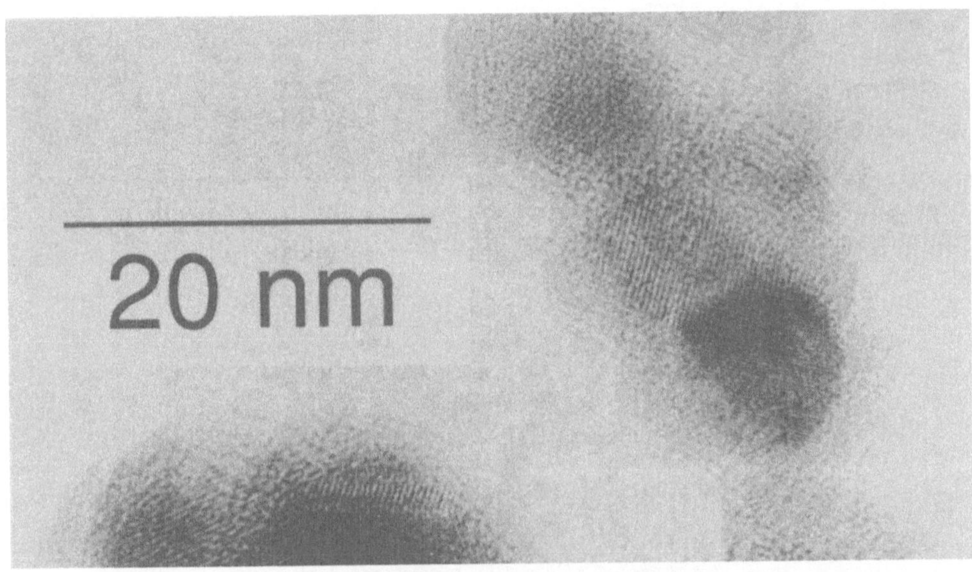

Figure 2. A high resolution transmission electron micrograph of GaAs clusters. Due to a high cluster density and small volume of the experimental apparatus, the clusters agglomerate into micron length chains. This has the advantage that the agglomerates may span holes in the carbon film, permitting measurements without interference from the carbon background. Lattice fringes and hexagonal features, suggesting faceting, are apparent.

Micrographs of InP clusters, generated by the exploding wire technique, are shown in Figure 3 . In Figure 3(a), the capacitor was charged to 7 kV, while in Figure 3(b), the capacitor was charged to 10kV, thus increasing the energy of the explosion by a factor of two. It is clear that the mean diameter of the clusters in the agglomerates decrease with increasing energy. Presumably, greater energy in the explosion leads to a faster propagation of the shock wave, rarifying the vapor more rapidly and leading to smaller cluster sizes. In both cases, electron diffraction revealed that the clusters are zincblende.

In Figure 5, a micrograph of an InAs cluster generated by the same technique is shown. The particle is approximately 20nm x 30 nm. Lattice

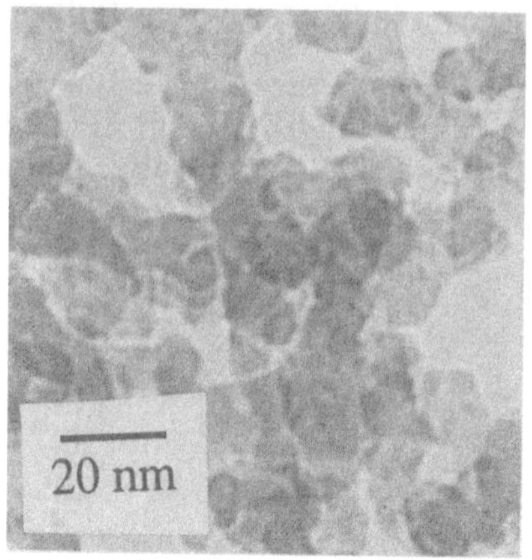
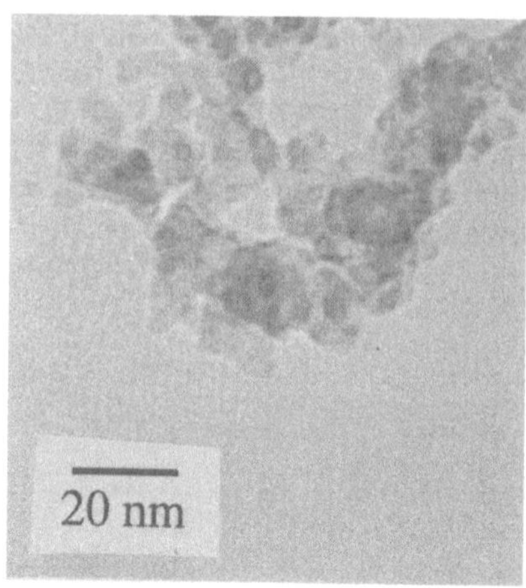

Figure 3. InP clusters preoduced at two different explosion energies. Increasing the energy of the explosion reduces the mean particles size. The discharge voltage is 7kV and 10kv in (a) and (b), respectively.

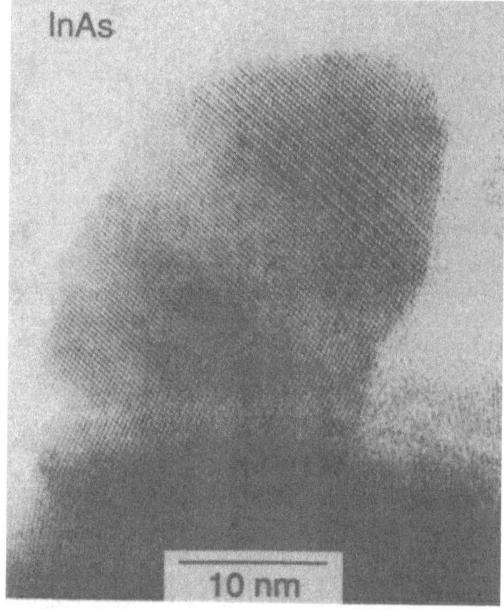

Figure 4. An InAs cluster produced by the exploding wire technique. A grain boundary is plainly visible.

fringes are plainly visible, and reveal the existence of a grain boundary in the cluster. In general, we rarely observe stacking faults or grain boundaries, and never observe them in clusters smaller than about 10 nm. This observation concurs with that of Sandroff *et al.* (Ref. 4) who observe stacking faults in GaAs particles in the size regime near 30 nm.

In summary, we have demonstrated the vapor phase synthesis of crystalline nanometer-scale III-V clusters. The method relies on the rapid condensation of a non-equilibrium vapor produced directly from the discharge of high electrical current through the wire. Electron microscopy and diffraction show that the cluster sizes fall well within the quantum size regime and that they have a bulk-like zincblende crystal structure.

A significant advantage of this method is its economy and the ease with which it can be adapted to different materials. We have performed studies on GaAs, InAs, InP, and GaSb and produced clusters in 5 to 40 nm size regimes. Studies on GaP and some II-VI materials are underway. We are presently characterizing source parameters with regard to cluster size and crystallinity. We are also actively investigating techniques to reduce cluster agglomeration.

We acknowledge support by NSF grant CTS-8912328 and the Powell Foundation. PCS acknowledges support by AT&T.

REFERENCES
1. L. E. Brus, J. Chem Phys. <u>80</u>, 4403-4409 (1984).
2. M. A. Olshavsky, A. N. Goldstein, and A. P. Alivisatos, J. Am. Chem. Soc <u>112</u>, 9438 -9440 (1990).
3. Hiroyuki Uchida, Calvin J. Curtis, and Arthur J. Nozik, J. Phys. Chem. <u>95</u>, 5382-5384 (1991).
4. C.J. Sandroff, J. P. Harbison, R. Ramesh, M. J. Andrejco, M. S. Hedge, D. M. Hwang, C. C. Chang, E. M. Vogel, Science <u>245</u>, 391-393,(1989).
5. W. A. Saunders, P. C. Sercel, H. A. Atwater, K. J. Vahala, and R. C. Flagan, (to be published)
6. T. Vijayan and V. K. Rohatgi, J. Appl. Phys. <u>63</u>, 2576-2582 (1988).

Incandescent Radiation from 3500 K Hot Clusters

R.SCHOLL, B.WEBER
Philips Research Laboratories
P.O.Box 1980
5100 Aachen
Germany

ABSTRACT. Hot refractory metal clusters have been produced by a novel experimental technique. A regenerative chemical cycle established in a high pressure microwave tungstenoxyhalide or rheniumoxide discharge leads to condensation of nanometer size tungsten or rhenium clusters with a mean temperature of about 3500 K or 4500 K, respectively. The clusters emit incandescent radiation with broad continuous spectra. The spectral shapes of the continua and the positions of their maxima (W 700 nm, Re 550 nm) are characteristic for the refractory metal. A first theoretical description of the clusters' radiation can be based on the Mie theory. The existence of clusters with spherical shape and a mean size of about 2 nanometers has been proven by a laser scattering experiment. The properties of the cluster radiation especially the continuous spectrum and the high luminous efficiency are attractive for lighting applications.

Introduction

The aim of the work presented in this paper is to produce hot (3500-4500 K) clusters and to investigate the applicability of their incandescent radiation for illumination. To understand the advantages of light production by clusters we have to examine other conventional artificial light sources : the incandescent lamp and the discharge lamp.

The incandescent lamp is a Planckian radiator with a continuous spectrum and therefore favourable colour rendering properties. The luminous efficiency of the incandescent lamp is limited by the maximal temperature of the filament of about 3400 K (some hundred Kelvins below the melting point of tungsten). Due to the Wien's displacement law a incandescent radiator of 3400 K emits mainly in the IR region. So low luminous efficiency but good light quality are characteristic for the incandescent lamp.

On the other hand we have the discharge lamp : electrons emitted by the electrodes stimulate atoms which emit light by spontaneous emission. The result is a atomic line or molecular band spectrum. Characteristic for discharge lamps are high luminous efficiencies but poor light quality.

It was our aim to combine the two advantages : high luminous efficiency **and** good light quality The idea is to create and to heat small particles (clusters) in a high pressure discharge and to use the thermal emission spectrum of these clusters for light generation. The thermal emission spectrum guarantees a continuous spectrum with good colour rendering properties, on the other hand it is possible to heat the clusters above their melting points and to shift the maximum of the emission into the visible range, the reason why the luminous efficiency increases. (There is also another effect due to the minuteness of the clusters which shifts the emission maximum into the visible range called Rayleigh effect).

P. Jena et al. (eds.), Physics and Chemistry of Finite Systems: From Clusters to Crystals, Vol. II, 1275–1280.
© 1992 *Kluwer Academic Publishers.*

1276

A cluster lamp has been realized at the Philips research laboratories Aachen (Germany) in form of a microwave excited high pressure discharge lamp ((1),(2)).

The cluster lamp consists of a cylindrical TM_{010} resonant microwave cavity in which via a couple antenna microwave power is coupled in (frequency 2.3-2.5 GHz). This microwave power is used to support a high pressure discharge in an electrodeless quartz vessel with a typical volume of 0.25 cm^3 (see fig.1). The experimental set-up has been described in more detail in references (3) and (4).

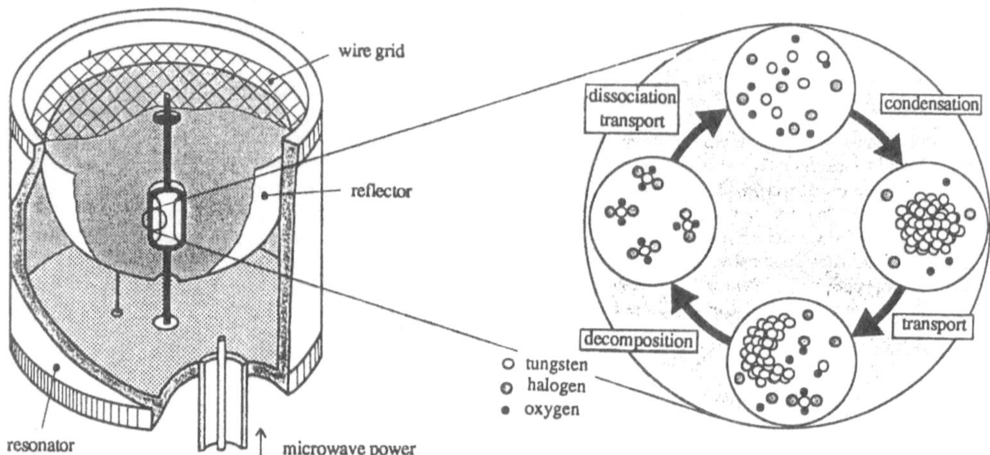

Fig.1:   Cluster lamp together with a schematic sketch showing the formation of tungsten clusters

The cluster lamp is based on the following principle. A compound like WO_2Br_2 is filled in an electrodeless quartz vessel . These WO_2Br_2 molecules evaporate completely at typical wall temperatures of 1000 K. In the center of the discharge at temperatures of about 4000 K these molecules dissociate and atomic tungsten is created in such high amounts that the partial pressure of tungsten exceeds its vapour pressure. So the tungsten vapour condenses in the hot region of the discharge and tungsten clusters are formed. The clusters are heated by the surrounding discharge and emit incandescent radiation until they are transported out of the condensation region. They are finally evaporated in hotter zones of the discharge or chemically dissolved in cooler parts (see fig.1). This cycle of cluster formation and decomposition proceeds stationary without any consumption of the substances filled into the discharge tube.

We use an electrodeless quartz vessel (and the excitation mechanism by microwaves) because WO_2Br_2 is a very aggressive compound which would destroy a tungsten electrode in a few minutes.

The Regenerative Chemical Cycle

Let us have a more detailed look to the condensation of tungsten. In the Philips research laboratories in Aachen a computer tool was developed which is able to calculate temperature-dependent chemical equilibria on the basis of thermodynamics ((5), (6)). The computer program minimize the Gibb's free energy ΔG for a given temperature and a total gas pressure as a function of the partial pressures of the dissociation products of the filled compounds. Fig.2 shows an example of such a calculation. The calculation includes 21 solid, 4 liquid and 23 gaseous species, but only the most important species are plotted. The filled compound WO_2Br_2 is stable up to about 2500 K. Going to higher temperatures we see that a first bromine atom is separated and the partial pressure of WO_2Br_1 increases. At about 3000 K the second bromine atom is separated and so on. At very high temperatures atomic tungsten dominates in the discharge. If we compare this partial pressure p_W of tungsten with its vapour pressure p^W_{vapour} (see fig.2: W1 vap.p.) we can see that in a broad temperature region $p_W > p^W_{vapour}$, that means that tungsten vapour condenses and tungsten clusters are created in the hot region of the discharge.

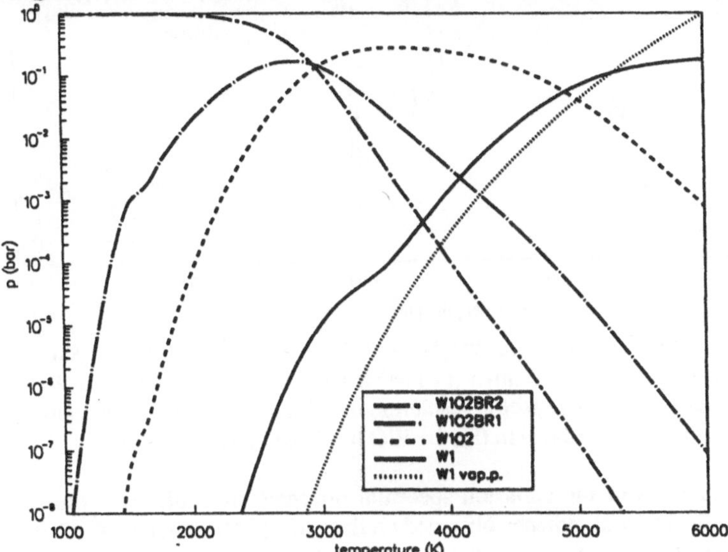

Fig.2 : Calculated chemical equilibrium composition in a WO_2Br_2 discharge

Experimental Results

The so created tungsten clusters have the following properties. The temperature of the clusters is about 3500 K, which could be detected by measuring the temperature of the surrounding plasma (a temperature of about 3500 K is also consistent with the position of the maximum of the clusters' emission).

We have proven the existence of clusters in the discharge by laser scattering. The light of an argon ion laser is focused into the discharge and the scattered laser light is detected under 90°. Light coming from spontaneous emission of the discharge or from laser light scattered by the vessel walls has been reduced by lock-in-techniques. The scattered laser light is linearly polarized perpendicular to the scattering plane, a fact from which we deduce that the clusters are spheres (7). From the scattering coefficient we estimate a mean radius of the clusters to about 2 nanometers.

In special operating modes, e.g. if the discharge power is reduced periodically for some milliseconds the clusters grow and are visible in a projection of the discharge.

The emission spectrum of the discharge consists of 3 contributions (see fig.3a) : identified atomic lines, molecular band radiation in the range 350-600 nm which originate from WO molecules (see fig.3b) and a broad dominating continuum with a maximum at about 700 nm which is the thermal emission spectrum of the hot tungsten clusters.

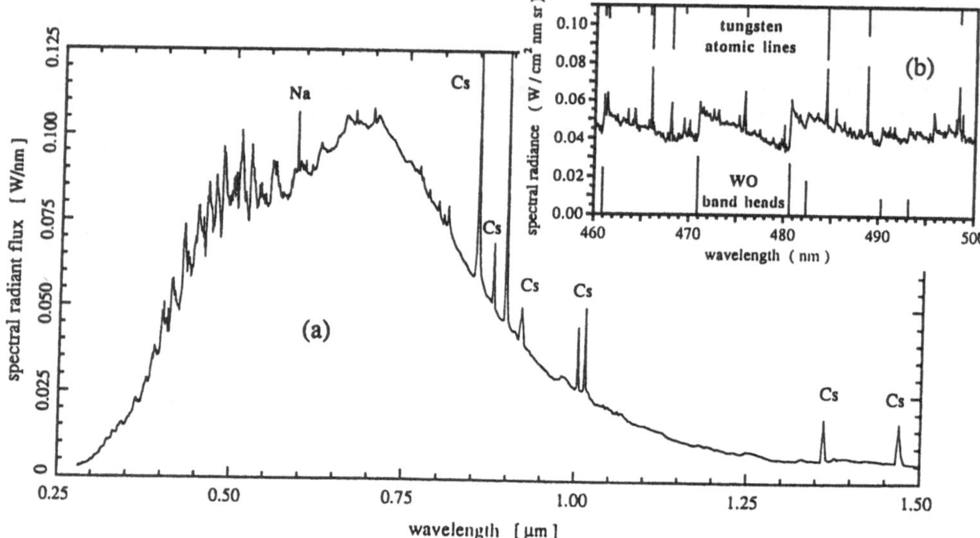

Fig.3 : (a) Measured emission spectrum of a WO$_2$Br$_2$ / CsBr / Ar discharge (5.3 µmol/cm^3 WO$_2$Br$_2$; 0.9 µmol/cm^3 CsBr; 13 mbar (r.t.) Ar) at a microwave power of 125 Watt. (b) Part of the spectrum with a spectral resolution of 0.1 nm. The spectral positions of tungsten atomic lines are tabulated in (8), the WO band head positions in (9).

The different contributions of the emission spectrum originate from different areas of the discharge. Fig.4 shows the spectral radiance observed on lines of sight through the discharge axis and 1.0 mm beside it. The emission from the axis is dominated by the molecular band spectrum of the WO molecule. The off-axis regions emit a broad structureless continuum, the incandescent radiation of tungsten clusters.

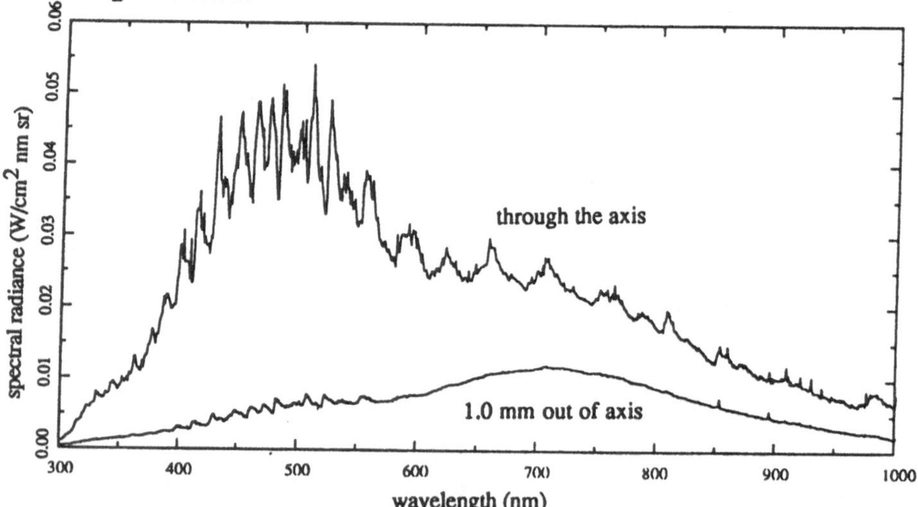

Fig.4 : Spectral radiance observed on lines of sight through the discharge axis and 1.0 mm beside it (microwave power 85 Watt, spectral resolution 1 nm).

The Theory of the Clusters' Emission Spectrum

To describe the clusters' emission spectrum theoretically we use the Rayleigh formula (7), an approximation of the Mie theory for objects much smaller than the wavelength of the emitted radiation.

$$\phi_\lambda \sim Im\left(3\frac{\varepsilon-1}{\varepsilon+2}\right)\frac{1}{\lambda}\cdot B_\lambda^{Planck}(T)$$

The above formula is valid for an ensemble of identical, non-interacting, homogeneous spheres of temperature T. Φ_λ is the spectral radiant flux of the cluster ensemble, $B_\lambda^{Planck}(T)$ the spectral radiant exitance of a black body of temperature T and ε the dielectric function for which we use bulk values at room temperature ((10), (11)).

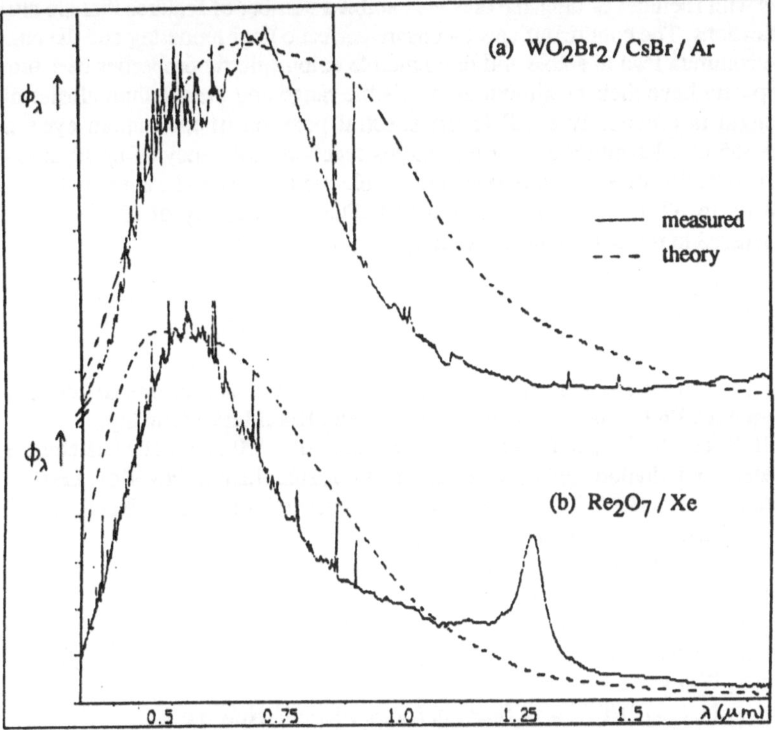

Fig.5 : Measured spectral radiant flux Φ_λ of a WO$_2$Br$_2$ / CsBr / Ar (a) and a Re$_2$O$_7$ /Xe (b) discharge. The dashed lines show calculated cluster spectra at 3500 K and 4500 K which are normalized to fit the observed continua.

We obtain a rough description of the clusters' emission spectrum (see fig.5a and 5b). From the position of the maximum of the emission we can deduce a mean temperature of the clusters: 3500 K for tungsten and 4500 K for rhenium clusters (we obtain rhenium clusters in a Re$_2$O$_7$ vapour discharge). These temperatures are consistent with the temperatures of the surrounding plasma (1).

1280

The details of the spectra, however, cannot be described by the model. The observed continua are distinctly narrower than the theoretical ones and there are experimental features like the peak of the rhenium cluster emission at 1275 nm that have no analogy in the model. Calculations including a more realistic cluster ensemble or bulk optical constants at higher temperatures do not improve the agreement between theory and observations distinctly.

Summary and Conclusions

We have developed a method to produce hot clusters by a regenerative chemical cycle inside a microwave excited high pressure discharge. The clusters have a temperature of about 3500 K (tungsten) or 4500 K (rhenium) and a size of about 2 nanometers. Due to their high temperature the clusters emit a thermal emission spectrum which we use for light generation.

Discharges with rhenium or tungsten clusters exhibit a number of features that are attractive for lighting applications. The continuous spectra ensure perfect colour rendering and the characteristic shapes of the continua lead to stable and reproducible colourimetric properties of a future cluster lamp. Both spectra have their maximum in the visible range and the rhenium cluster discharge's peak wavelength is practically equal to the spectral position of the human eye's maximum sensitivity at 555 nm. Rhenium cluster discharges reach an efficiency of up to 28% (tungsten 29%) for the conversion of microwave power to visible radiant flux and a corresponding luminous efficiency of up to 62 lm/W (tungsten 56 lm/W). The applicability of cluster discharges for lighting is discussed in more detail in reference (12).

References

(1)   Weber, B. (1990) 'Die mikrowellenangeregte Cluster-Entladung: eine neuartige Lichtquelle', Ph.D. thesis, Universität Karlsruhe, Karlsruhe, Germany.

(2)   Scholl, R. and Weber, B. (1991) 'The preparation of 4700 K hot clusters emitting incandescent radiation applicable to illumination', submitted to Phy. Rev. Lett.

(3)   Offermanns, S. (1990) 'Electrodeless high-pressure microwave discharges', J. Appl. Phys. 67, 115-123.

(4)   Offermanns, S. (1990) IEEE Trans. Microwave Theory Tech. 38, 904.

(5)   Schnedler, E. (1984) CALPHAD 8, 265.

(6)   Schnedler, E. (1985) Proc. Electrochem. Soc. 85-2, 95

(7)   Bohren, C.F. and Huffman, D.R. (1983) Absorption and Scattering of Light by Small Particles, Wiley, New York.

(8)   Zaidel, A.N. (1970) Tables of Spectral Lines, Plenum, New York.

(9)   Pearse, R.W.B. and Gaydon, A.G. (1965) The Identification of Molecular Spectra, Chapman and Hall, London.

(10)   Nomerovannaya, L.V., Kirillova, M.M., Noskov, M.M. (1971) 'Optical Properties of Tungsten Monocrystals', Sov. Phys. JETP 33, 405

(11)   Weaver, J.H. et al. (1981) 'Optical Properties of Metals', Physics Data 18-1, FIZ Karlsruhe, pp. 239-252.

(12)   Scholl, R. and Weber, B. (1991) 'A novel type of light source: Continuous radiation from small clusters in microwave excited discharges', submitted to J. Ill. Eng. Soc.

NANO-STRUCTURE FABRICATION BY ORGANIC MBE USING METALLOID PORPHYRINS AND PHTHALOCYANINES CONTAINING III, IV AND V ELEMENTS

K. TANIGAKI, T. W. EBBESEN AND S. KUROSHIMA
*Fundamental Research Laboratories, NEC Corporation, 34
Miyukigaoka, Tsukuba, Ibaraki 305, Japan*

ABSTRACT. A methodology for achieving quasi one-dimensional assemblies using metalloid phthalocyanines (Pc's) and porphyrins (Pr's) is presented, which contain III-V elements, $M_{III,IV,V}$'s, as a central atom. A recently developed organic MBE technique has been applied to make a high quality thin film crystal of gallium monochlorotetraphenyl porphyrin for this purpose. The possibility of obtaining molecular assemblies with one-dimensional properties is discussed with together some properties of $M_{III,IV,V}Pc$'s/Pr's.

1. Introduction

Phthalocyanines (Pc's) and porphyrins (Pr's) are of interest, since they can form ligands with various kinds of elements M at their center consisting of four nitrogen atoms [1], as shown in Fig.1. The parent skeleton of Pc's and Pr's is an insulator, but in some cases,

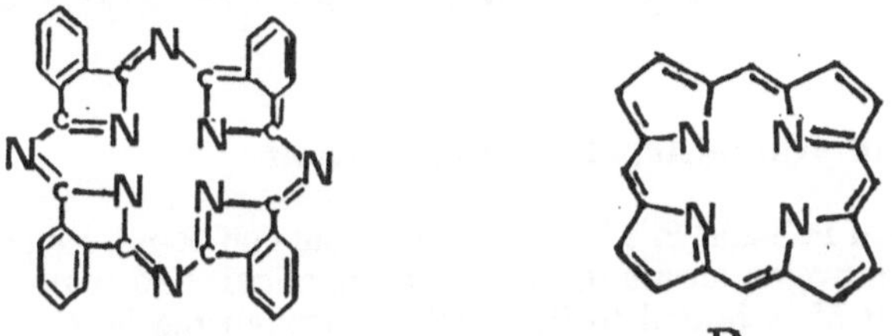

Pc   Pr

Fig.1 Molecular structures of Pc and Pr.

depending on the element M, the MPc's and MPr's may show semiconducting features [2]. In principle, structures with quasi one-dimensional properties could be attained by these molecular assemblies [3,4], which would be very difficult to fabricate with inorganic materials where extremely sophisticated lithographic techniques are needed.

For obtaining the quasi one-dimensional conduction paths of electrons and holes using MPc's or MPr's, $M_{III,IV,V}Pc$'s and $M_{III,IV,V}Pr$'s are promising, where $M_{III,IV,V}$ denotes the III-, IV- and V-elements. However, the knowledge of the properties of Pc's and Pr's containing $M_{III,IV,V}$'s is quite limited, and furthermore the technology for obtaining thin film crystals of MPc's and MPr's is not advanced compared with that of

1281

P. Jena et al. (eds.), Physics and Chemistry of Finite Systems: From Clusters to Crystals, Vol. II, 1281–1286.
© 1992 *Kluwer Academic Publishers.*

Si, GaAs, etc.. Here we report a methodology of nano-structure fabrication using the metalloid Pc's (H$_2$Pc: phthalocyanine) or TPPr's (H$_2$TPPr: mesc-tetraphenylporphyrin) containing III-V elements together with some properties of these molecules. We have also applied a recently-developed organic molecular beam epitaxy (organic MBE) technique to achieve high quality thin film crystals for gallium-containing TPPr.

2. Materials and Methods

M$_{III,IV,V}$Pc's and M$_{III,IV,V}$TPPr's used in this study were prepared from the reaction of H$_2$Pc or H$_2$TPPr and metal trichlorides or metal tetrachlorides according to standard procedures [5,6]. The molecules were purified by recrystallization or column purification prior to use. The samples were introduced into an organic MBE apparatus with a back pressure of around 5x10^{-10} Torr, the Knudsen cells of which were modified for organic molecules [7]. The molecules were evaporated on a quartz substrate in the thickness range of 100-10000 Å at a rate of less than 0.5 Å/min, after prebaking the substrates at 673K for 1 hour. UV-visible spectra were measured for the fabricated thin films and also after dissolving them in solvent in order to detect any structural change during the deposition.

The single crystal of GaTPPrCl was also epitaxially grown on a cleaved (001) surface of a 5x5mm^2 KBr substrate [8]. The obtained thin film was either transferred on to a grid for TEM observation after dissolving the KBr substrate in pure water, or it was directly subjected to 2θ X-ray analyses. The complete crystal structure was determined with the three-dimensional structure analyses using the bulk crystal grown in a vapor phase.

3. Methodology of Nano-Structure Fabrication

The special feature of Pc's and Pr's is that one element can be introduced into the center consisting of four nitrogen atoms through two chemical- and two coordinate-bonds. The central element M is bound to Pc's or Pr's through the octahedron type hybrid orbitals formed from mainly d$_2$sp^3 or sp^{3}d$_2$ hybridization. The two bonding orbitals of the element M, perpendicular to the molecular plane, could play a very important role in making a column structure. If the intermolecular interactions through these perpendicular orbitals are sufficiently strong so that conduction electrons or holes can move with a small barrier, a quasi one-dimensional conduction path is created.

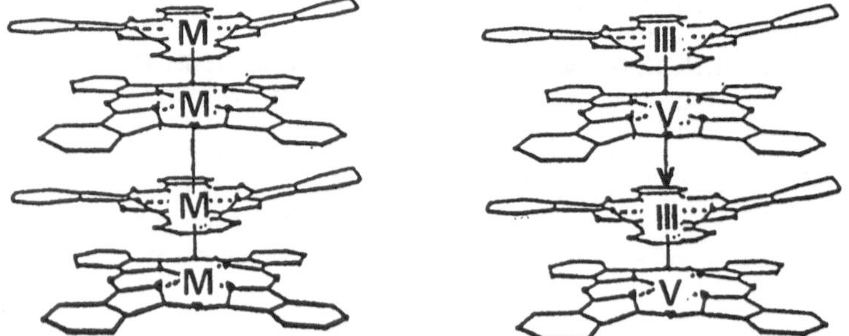

Fig.2 Quasi one-dimensional assemblies composed of phthalocyanine family molecules, where M is the VI-emement such as Si, Ge, III is the III-element such as Al, Ga, In, and V is the V-element such as P, As. The III←V bond is formed through charge-transfer from the V-element to the III-element.

From this point of view, attention should be paid to III, IV, V elements as the central element M, since the octahedral hybrid orbitals seem to be pure. In the case of IV-elements, quasi one-dimensional paths, such as -Si-Si- or -Ge-Ge-, are structurally constructed (Fig.2a). Furthermore, if III-elements and V-elements could be aligned in an alternative fashion like inorganic III/V crystals, such as -Ga-As- or -In-P-, charge-transfer conduction paths could be constructed (Fig.2b).

For this purpose, $M_{III,IV,V}PcX_n$ and $M_{III,IV,V}TPPrX_n$ (where X denotes the ligand and chlorine was presently used, and n=0,1,2,3 depends on the valence state of M) should be prepared. X could be removed during or after making their crystals and the conducting column structure could then be held together through the strong interaction of bonding orbitals perpendicular to the Pc/TPPr molecular plane.

4. Results and Discussion

Efforts were made to synthesize Pc's and TPPr's containing $M_{III,IV,V}$-elements, where M_{III}=Al, Ga, In; M_{IV}=Si, Ge; M_V=P, As were used. The results are shown in Table 1. Except for $AsPcX_n$, all derivatives were successfully obtained, although the yield of $SiTPPrCl_2$ and $GePcCl_2$ was quite low. $AsPcCl_n$ had been reported to be synthesized from the reaction of dilithium Phthalocyanine (Li_2Pc) with arsenic trichloride [9]. The procedure was followed for this experiment, but $AsPcCl_n$ was not obtained and only H_2Pc was recovered through the decomposition of Li_2Pc during the reaction.

Table 1. Syntheses of MIII,IV,V-Containing Metalloid TPPr and Pc.

	III		IV		V	
TPPr	Al	good	Si	poor	P	good
	Ga	good	Ge	good	As	poor
	In	good	Sn		Sb	
	Tl		Pb		Bi	
Pc	Al	good	Si	good	P	good
	Ga	good	Ge	poor	As	bad
	In	good	Sn	good	Sb	
	Tl		Pb		Bi	

Note: good; successfully carried out in a good yield. poor; can be carried out, but in a poor yield. bad; was not able to be synthesized.

The properties of $M_{III}PcCl$'s/$M_{III}TPPrCl$'s and $M_{IV}PCCl_2$'s/$M_{IV}TPPrCl_2$'s are very similar to those of typical metalloid Pc's/Pr's, such as CuPc and ZnTPPr, although M_{III}-containing PcCl's/TPPrCl's show very high resistance to acid. On the other hand, those of M_VPcCl_n's and $M_VTPPrCl_n$'s are very different from others, in that the metal valency changes, depending on the structure of the parent molecules. For example, it is reported that P(III)PcCl is the stable molecule in the Pc skeleton, while P(V)TPPrCl is the stable one in TPPr [10]. The situation of As-containing $PcCl_n$/$TPPrCl_n$ is much more complex. In our preliminary analyses of AsTPPrCln, the stable molecule seems to be As(III)TPPrCl. Furthermore, it should be noted that some structural changes tend to occur during the thin film fabrications of M_VTPPrCln's and M_VPcCln's mainly owing to the central M_V release from the parent skeletons.

Recently developed organic MBE technique [11] has been applied to make a high quality thin film crystals of GaTPPrCl [8], which is a one of the M_{III}-containing molecules, on a cleaved (001) surface of a KBr substrate. Fig.3 demonstrates a lattice image and an electron diffraction pattern observed for the GaTPPrCl thin film crystal grown under high vacuum with a organic MBE at a substrate temperature of 453 K. The observed black dots in the lattice image correspond to the molecules and show that the molecules stay at the right positions in the lattice. The electron diffraction pattern along the (001) direction of the substrate indicates that the crystal structure would be tetragonal with a plane spacing of 9.5 Å, since the electron diffraction spots shows 4-fold symmetry. The 2θX-ray analyses were also performed to estimate the d-spacing corresponding to the stacking distance and the value of 4.939 Å was obtained.

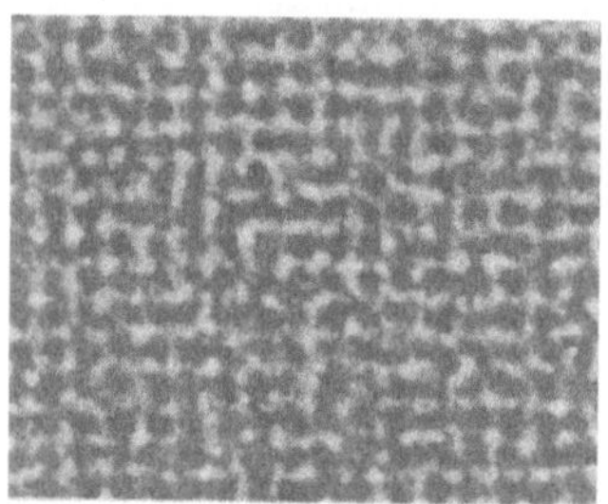

Lattice image of GaTPPrCl

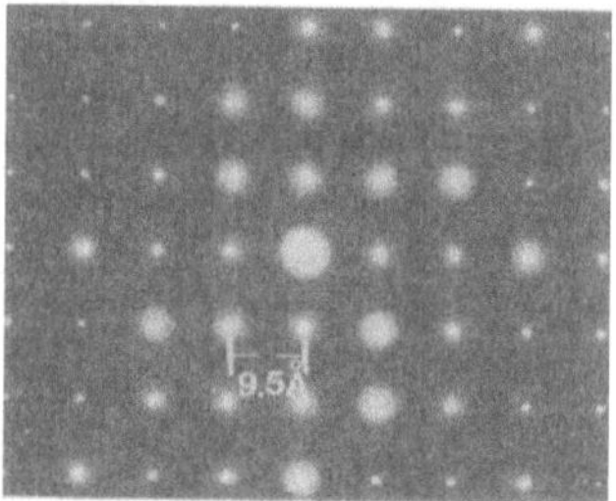

Electron diffraction pattern of GaTPPrCl

Fig.3 A lattice image (a) and an electron diffraction pattern (b) for the thin film single crystal of GaTPPrCl.

The determined value of 9.5 Å (plane spacing) and 4.939 Å (stacking d-spacing) are very small compared with the lattice constants of the ab- and c-directions reported for other Pc-family molecules in a tetragonal lattice [12-15]. This implies that the structure of the obtained thin film would be body-centered tetragonal (BCT), containing one molecule in the center of the lattice. Assuming this structure, the lattice parameters can be estimated to be a=b=9.5 Åx21/2=13.4 Å and c=4.939 Åx2=9.88 Å.

In order to confirm the crystal structure, a single crystal analysis was performed using a 0.2x0.2x0.2 mm³ GaTPPrCl bulk single crystal grown with a home-made vapor phase apparatus. Fig.4a shows a graphic model of GaTPPrCl determined by the X-ray analysis. The molecule has a mirror plane and the Ga element exists in the center of the molecular plane with 4/m symmetry. The phenyl rings are placed perpendicular to the mirror plane and the chlorine atom is placed on the 4-fold symmetry axis at a distance of 2.516 Å from the central Ga. The crystal structure was determined to be tetragonal with space group Im/4, the parameters of which are a=b=13.519 Å, c=9.850 Å and the Z=2. views from the c-axis and ab-axis (Figs.4b and 4c) show that the molecule is located with a rotation angle of 15º from the diagonal line of the square ab lattice plane. These data support the presumed BCT structure.

The 4-fold symmetry observed along the (001) direction of the substrate, in connection with the bulk single crystal analyses, indicates that the molecules are stacked on the KBr substrate with their molecular plane parallel to the substrate. The lattice parameter a=b=13.4 Å is extremely close to the value of 13.2 Å, which is just twice that of the lattice parameter of KBr. This implies that the GaTPPrCl single crystal epitaxially grows on the KBr substrate coincidentally in position with the (100) and (010) lattice directions of the substrate.

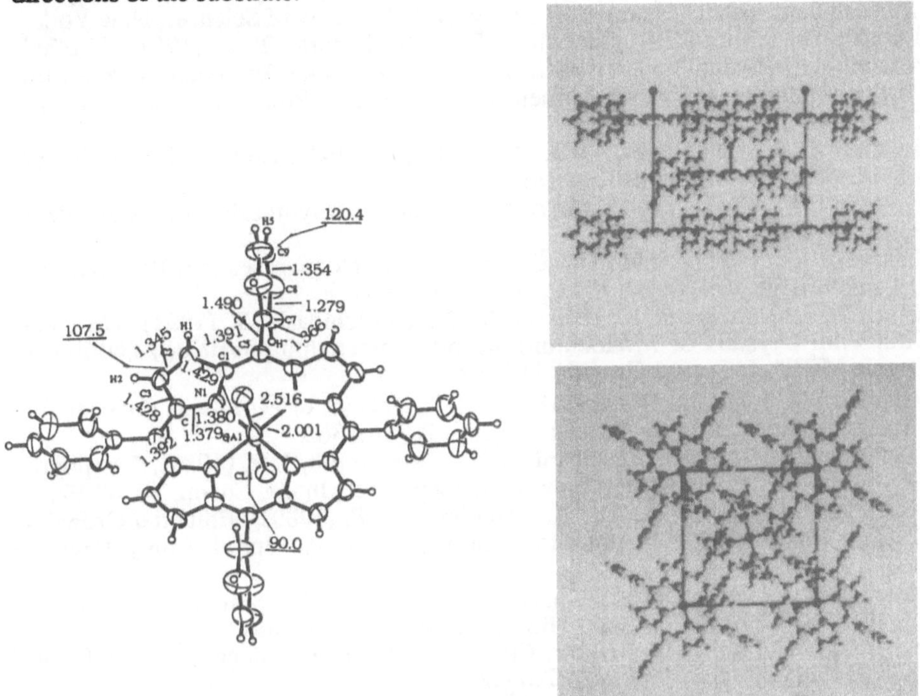

Fig.4 The molecular structure (a) and views of the crystal structure from a,b-axis (b) and c-axis. The view from the c-axis displays that the molecule is located with a rotation angle of 15° of from the diagonal line of the square ab lattice plane.

5. Conclusion

A possible approache for achieving molecular assemblies with quasi-one dimensional structures was presented. The present example of thin film fabrication shows that high quality single crystal of Pc-family molecules can be grown epitaxially on inorganic substrates with a sufficiently large area for optical and electrical measurements, using organic MBE techniques. $M_{III,IV,V}Pc's/Pr's$ thin film crystals could open a possibility of a new methodology to fabricate quasi one-dimensional molecular assemblies, and observing unique optical/electrical properties.

6. Acknowledgements

The authors are very grateful to Drs. P. M. Ajayan and S. Iijima for their discussions.

7. References

[1] Moser F. H., and Thomas A. L. (1983) "The Phthalocyanines" ,CRC Press, Florida, Vol.1 and Vol.2.

[2] Simon J. and Andre J.-J.(1985) "Molecular Semiconductors", Springer-Verlag, Berlin.

[3] for example, see Miller J. S. and Epstein A. J., (1978), "Synthesis and Properties of Low-Dimensional Materials" The New York Academy of Sciences, New York.

[4] Dirk C. W., Mintz E. A., Schoch K. F., Jr. and Marks T. J., (1981), "Cofacial Assembly of Partial Oxidized Metallomacrocycles as an Approach to Controlling Lattice Architecture in Low-Dimensional Molecular Solid", J. Macromol. Sci.-Chem., A16, 275-298.

[5] Leznoff C. C. and Lever A. B. P., (1989), "Phthalocyanines: Properties and Applications", VCH Publishers, Weinheim.

[6] Dolphin D., (1978), "The Porphyrins: Structure and Synthesis", PartA Academic Press, New York.

[7] Tanigaki, K. (1991) "Thin Film Crystals of Phthalocyanines and Porphyrins by Organic MBE", Mol. Cryst. & Liq.. Cryst., in press.

[8] Tanigaki, K, Kuroshima, S., Ebbesen, T. W., and Ichihashi, T. (1991) "Growth of Thin Film Crystals of Monochloro(meso-tetraphenylporphyrinato) Gallium by Orgnic MBE", J. Crystal Growth, 114, 3-6.

[9] Ruter, Jr., H. J. and McQeen, J. D. (1960) "Synthesis of ^{52}Mn and ^{74}As labelled phthalocyanines", J. Inorg. Nuclear Chem., 12, 361-363.

[10] Gouterman, M., Sayer, P., Shankland, E., and Smith, J. P. (1981), "Porphyrins 41. Phosphorus Mesoporphyrin and Phthalocyanine", Inorg. Chem., 20, 87-92.

[11] Hara M., Sasabe H. , Yamada A. and Garito A. F. (1989), "Epitaxial Growth of Organic Thin Films by Organic Molecular beam Epitaxy", Jpn. J. Apply. Phys., 28 L306-L308.

[12] Yamakado H., Yakushi K., Kosugi N., Kuroda H., Kawamoto A., Tanaka J., Sugano T. , Kinoshita M. and Hino S., (1989), "Preparation, Crystal Structure, and Solid State Properties of Highly Conductive (Phthalocyaninato)platinum Radical Salt", Bull. chem. Soc. Jpn, 62, 2267-2272.

[13] Darovskikh A. N., Frank-Kamenetskaya O. V., and Fundamenskii V. S. (1986), "Molecular and Crystal Structure of Tetragonal α-Phase of Neodymium Diphthalocyanine Radical", Kristallografiya, 31 901-905.

[14] Sugimoto H., Mori M., Masuda H. and Taga T. (1986), "Synthesis and Molecular Structure of a Lithium Complex of the Phthalocyanine Radical", J. Chem. Soc., Chem. Commun., 962-963.

[15] Schramm C. J., Scaringe R. P., Raymond P , Stojakovic D. R., Hoffman B. M., Ibers J. A., and Marks T. J. (1980), "Chemical Structural and Charge Transport Properties of The Molecular Metals Produced by Iodination of Nickel Phthalocyanine", J. Am. Chem. Soc., 102, 6702-6713.

X-RAY STUDY OF ALKOXIDE-DERIVED AMORPHOUS TiO$_2$ POWDER

Q. J. Wang, S. C. Moss, Physics Department, University of
Houston, Houston, TX 77204-5506
M. L. Shalz, A. M. Glaeser, Department of Materials Science
and Engineering, University of California, Berkeley, CA 94720
H. W. Zandbergen, National Centre for HREM, Delft University of
Technology, Rotterdamseweg137, 2628 AL Delft, The Netherlands
P. Zschack, NSLS X14, Building 725, Brookhaven National
Laboratory, Upton, NY 11973

ABSTRACT. Spherical TiO$_2$ particles (~0.35μm), prepared by controlled
hydrolysis of titanium tetraethoxide in ethanol, are porous agglomerates of an
approximately 6nm primary structure. We consider here the internal atomic
structure of these agglomerates, hitherto considered to be amorphous. X-ray
scans on a dried and pressed powder disk indicate a broad oscillatory diffuse
scattering pattern, S(k), with an approximate correlation range of 1nm.
Preliminary modelling of this diffuse pattern indicates a rough similarity to the
Debye scattering function for the brookite crystalline phase with approximately a
1nm particle size. The Fourier transform, G(r), also shows distances and
coordinations attributable to the (local) octahedrally coordinated crystalline
order in brookite. Initial high resolution electron microscope (HREM) studies, on
the other hand, reveal only occasionally ordered regions in an otherwise rather
amorphous structure and it therefore appears that this material may serve as an
interesting prototype for a discussion of nanocrystalline vs. amorphous
organization in which the brookite phase - the most distorted of the crystalline
polytypes - serves as a guide to the actual structure.

1. INTRODUCTION

In 1988 Edelson and Glaeser published an extensive study [1,2] of the sintering
and coarsening of nominally monodispersed ~0.35μm-diameter titania powders
produced by controlled hydrolysis of titanium tetraethoxide. The preparation
and purification (removal of organic residuals, etc.) of these essentially
monosized powders, described in Ref. [1,2] and in the M. S. thesis of Shalz [3],
results in mottled spherical particles which are composed of much finer
particulates on the order of 6-10nm. The atomic structure of these titania
particles was tentatively identified as "amorphous" by Edelson and Glaeser [1,2]
because the X-ray diffraction patterns [1] demonstrated only a nearly
featureless, slightly oscillating, diffuse scattering which increased strongly

P. Jena et al. (eds.), Physics and Chemistry of Finite Systems: From Clusters to Crystals, Vol. II, 1287–1294.
© 1992 Kluwer Academic Publishers.

toward low angles. On annealing for 1 h at 400°C the low angle scattering disappeared and the diffraction patterns could be clearly indexed as anatase with a particle size of ~10nm or close to the primary particulate size in the SEM photographs of the original material.

Our interest here is in the careful X-ray characterization of the original amorphous structure. From even a preliminary measurement, it is clear, as we shall see, that an effective X-ray "particle size," or correlation range, for this amorphous material is approximately 1nm. It is therefore intriguing to speculate on how TiO_2, in which the oxygen stoichiometry is nearly ideal, may be organized in a glassy or amorphous fashion. While a tetrahedral (SiO_4) random network model for vitreous SiO_2 is now well-established [4] an analogous network model linking TiO_6 octahedra has yet to be developed.

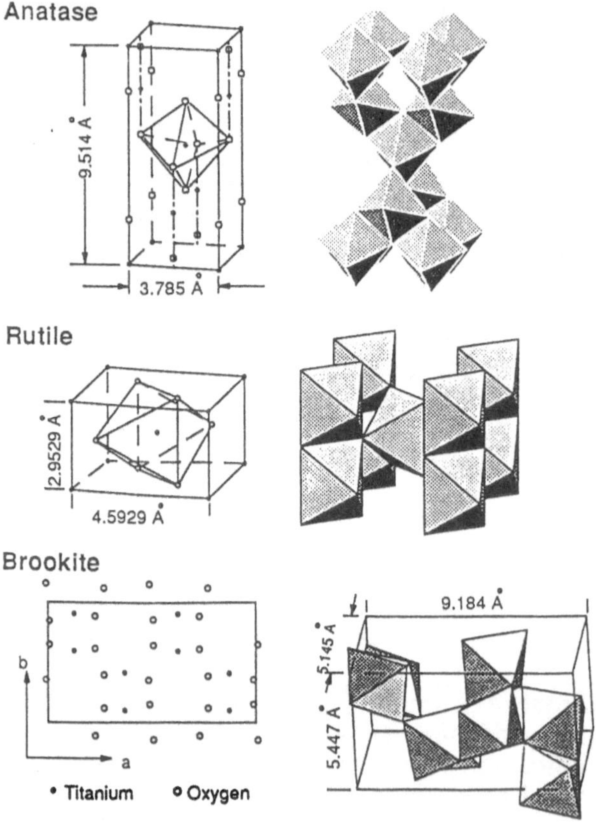

Figure 1   Illustration of the three crystalline polytypes of TiO_2. The brookite phase has both the most distorted TiO_6 octahedron and the most irregular octahedral attachments.

In contrast to SiO_2, and to many of the octahedrally coordinated metal oxides [3], TiO_2 has three crystalline structures in which the TiO_6 octahedra are assembled in different ways: <u>rutile</u>, where the octahedra alternate in sharing edges and corners in a layer with layers stacked in a AB/AB... sequence; <u>anatase</u> in which there is only edge sharing in a layer and the layers alternate ABCD/ABCD...; <u>brookite</u>, in which four rutile-linked units join along two edges (as in anatase) which meet at a corner. These three crystal structures are shown in Fig.1 and will be referred to later. It should also be noted that both rutile and anatase represent rather regular structures: rutile has two short and four long Ti-O distances (1.895Å and 1.961Å) while anatase has two long and four short distances (1.966Å and 1.937Å). All of the six distances in the brookite octahedron, however, are unique (1.863, 1.920, 1.945, 1.993, 1.995 and 2.040Å). This fact, together with its more varied organization of octahedra within a unit cell, will recommend brookite as a starting point for the modelling of the amorphous phase.

2. EXPERIMENT AND ANALYSIS

2.1 Data Collection

Our X-ray experiments were conducted on a dried compacted disk of TiO_2, 1.06mm in thickness and 1cm in diameter. Both in-house rotating anode and synchrotron X-ray studies were done. While we shall concentrate on the more accurate synchrotron radiation data, taken with 0.725Å X-rays, the Mok_α (λ=0.71Å) data collected in-house was generally in good agreement with it. The sample thickness of 1.06mm was a useful compromise which permitted symmetrical transmission and reflection data both to be collected with high accuracy. [The product of linear absorption and thickness for λ=0.725Å was measured to be μt=1.58 which is nearly ideal.] The data were taken on beamline X14 at the National Synchrotron Light Source where we used a thin evacuated Be dome over the sample to reduce the air scattering at low angles (evacuated incident and diffracted beam paths are always used). A novel analyzer/detector scheme, devised by Ice and Sparks [5] was also employed to remove the appreciable Compton scattering at modest to large angles. This scheme consists of a parafocussing graphite analyzer which disperses the scattered beam across a position sensitive detector and permits a careful spectrum fit to be made which essentially separates out the elastic component of interest.

Figure 2(a) shows a plot of the raw elastic scattering data, after Compton scattering has been removed and it has been corrected for absorption (the transmission and reflection runs overlap and can easily be matched). Figure2(b) is a plot of similar data taken on a Mo rotating anode unit (without correction) on material that had been stored for a year in air. The sharper peaks that are beginning to evolve out of the amorphous diffuse scattering are all indexable to anatase.

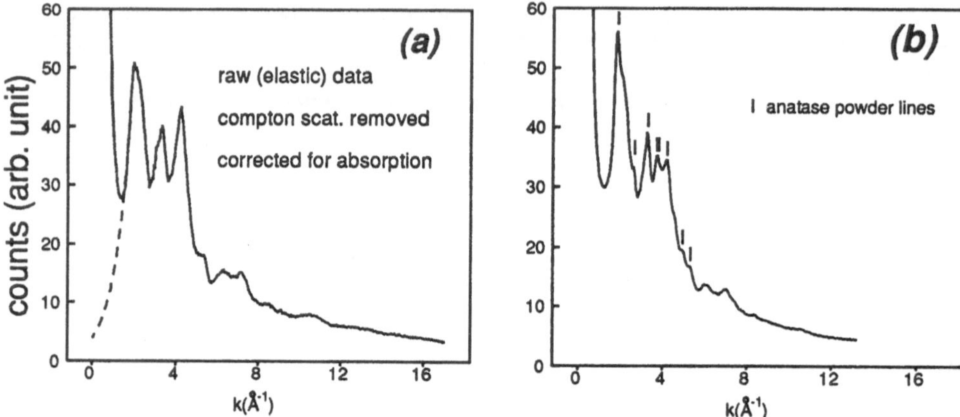

Figure 2   (a) The raw (elastic) synchrotron radiation X-ray data for a 1.06mm disk of amorphous TiO$_2$. The low angle scattering is pronounced but can be smoothly removed (dashed curve) for further data reduction; (b) raw rotating anode X-ray data for a similar sample left in air for ~ 1 year. The vertical bars indicate the Bragg positions for the emerging anatase phase.

2.2 X-ray Analysis

The X-ray pattern in Fig.2(a) was first roughly corrected for the appreciable small angle scattering by smoothly extrapolating, as indicated, the first peak at k~2.0Å to a small value at Q=0 (it is not so important, as it effects mainly large r features in the transform). The data was then normalized to the proper units and Fourier transformed to obtain the pair distribution function G(r). We let f_i be the scattering factor for the ith species (Ti or O) and C_i its concentration (C_{Ti}=1/3, C_O=2/3). The normalized X-ray intensity per atom can then be shown to be [6]:

$$I(k) = \langle f^2 \rangle + \sum_{i \neq j} \sum f_i f_j \int_0^\infty r G_{ij}(r) \frac{\sin kr}{kr}\, dr \quad , \quad G_{ij}(r) = 4\pi r [\rho_{ij}(r) - \rho_0]$$

$$\simeq \langle f^2 \rangle + \langle f \rangle^2 \int_0^\infty G(r) \frac{\sin kr}{k}\, dr$$

where $\langle f^2 \rangle = \sum_i C_i f_i$ and is used to normalize the data at large k and

$G(r) = \sum_i \sum_j W_{ij}(r)G_{ij}(r)$. In a simplified version of the weighting factors $W_{ij}(r)$,

$$W_{ij} \sim \overline{W_{ij}} = \frac{c_i c_j z_i z_j}{<f>^2} \quad ,z_{i,j} = \text{atomic numbers.}$$

The weighting factors are necessary for the interpretation and modelling of G(r); assuming them not to depend on k we can then extract a meaningful G(r) by Fourier transformation of the data. First we form the reduced expression ki(k):

$$ki(k) = \left[\frac{I(k) - <f^2>}{<f>^2}\right] k = \int_0^\infty G(r) \sin kr\, dr$$

where G(r) is related to the average <u>weighted</u> atomic density $\rho(r)$ by $G(r)=4\pi r[\rho(r)-\rho_0]$. Then by transformation between the experimental limits k_{min} and k_{max} we obtain

$$G(r) = 4\pi r[\rho(r) - \rho_0] = \int_{k_{min}}^{k_{max}} ki(k)\,(e^{-\alpha^2 k^2}) \sin kr\, dk$$

where $\exp(-\alpha^2 k^2)$ is an artificial damping factor often used to remove the oscillatory effects of truncation at k_{max}.

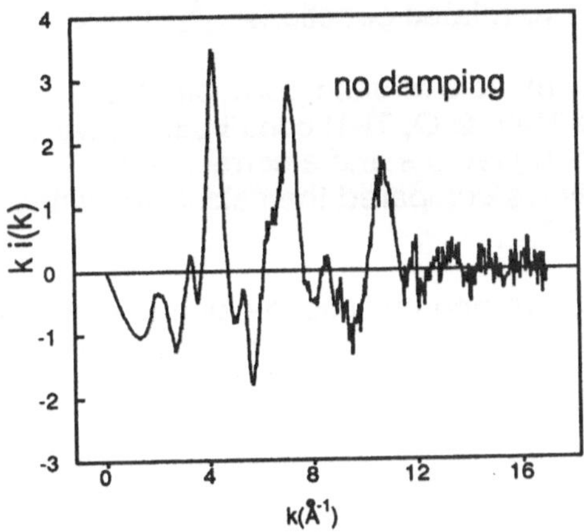

Figure 3 The final reduced intensity ki(k) obtained from the data in Fig.2(a)

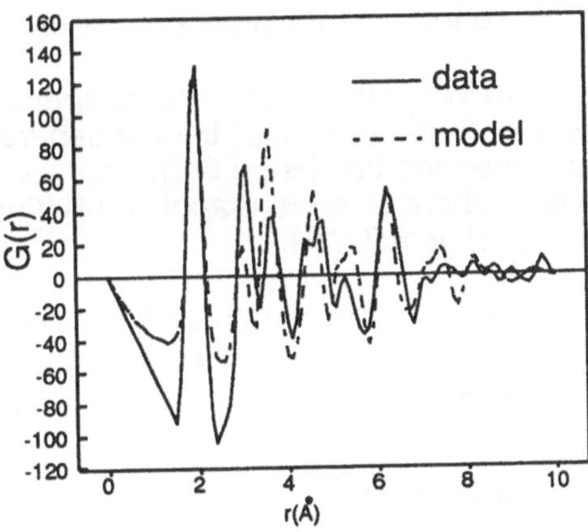

Figure 4 G(r) for the Fourier transform of ki(k), both for the data in Fig.3 and for a model of brookite, 1x2x2 unit cells in size, calculated using the Debye equation

Figure 3 presents our reduced intensity $ki(k)$ in which no damping or smoothing factors have been applied and there are thus occasional spikes or dips in the data which are artificial. The main features, however, are reproducible and are structurally significant. This data may then be transformed to $G(r)$ as above taking care to let $k_{max} i(k_{max})=0$ to minimize truncation oscillations at low r. Nonetheless these oscillations were still appreciable and an iterative procedure was employed to remove them, examining at each stage the back-transformed $ki(k)$ to insure that known data was not being altered in the process. Our final $G(r)$ is shown in Fig.4 in which no damping in k-space was employed and hence no artificial broadening in $G(r)$ occurs.

3. MODELLING

We have so far treated our X-ray intensity as we would any amorphous scattering pattern. We did, of course, first examine directly the diffraction patterns to determine if it could be indexed as crystalline. However, none of the known crystalline forms in Fig.1 have diffraction patterns which match our data although, as we shall see, brookite does a fair job. It should also be re-iterated that, if we treat the first peak at $k=2.0\text{Å}^{-1}$ as a Bragg peak, its half width yields a correlation range of $\sim$10-12Å which is a very small crystal, indeed. Nonetheless we must still look to the (three) crystalline forms for initial insight into the amorphous structure.

To this end we have proceeded in two related directions:

a) We initially constructed spherical models of each polytype and calculated separate $G_{ij}(r)$ to note separate Ti-O, O-O, Ti-Ti contributions; we then broadened the combined $G(r)$, after weighting factors and a correction for limited sphere size were applied [4]; finally we compared the calculation with the experimental $G(r)$.

b) While the above procedure is useful we nonetheless decided to use only the simpler procedure of calculating $I(k)$ from a limited size model using the Debye equation:

$$I(k) = \sum_{i,j} f_i f_j \frac{\sin kr_{ij}}{kr_{ij}} \quad ,$$

and then transformed the resulting $ki(k)$ to $G(r)$ <u>after damping to replicate thermal vibrations</u>. These comparisons are shown in Fig.5 for a sphere of 6.5Å (13Å diameter "particle"). The stoichiometries are wrong in each case because the surface atoms are not weighted properly. Nonetheless certain features are clear.

•   The first peak, due to Ti-O within an octahedron is similar in all (but weaker in rutile because of the surface atom problem).

•   In the region of r between 4-7Å only brookite has, at the least, peaks in the correct place.

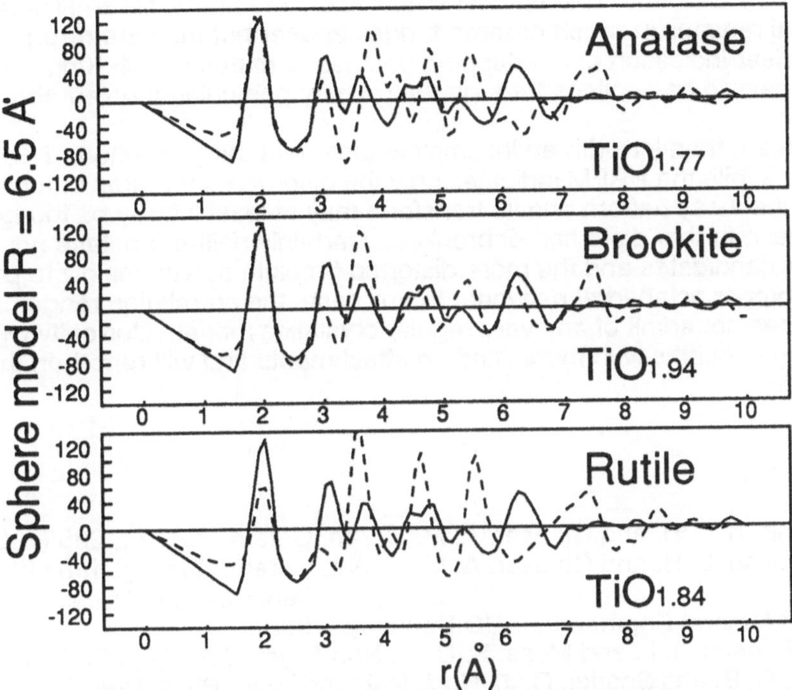

Figure 5   Comparison of the G(r) of small spherical models of anatase, brookite and rutile with the experimentally derived G(r). The stoichiometries are not exact because of errors introduced at the surface.

Because only the brookite calculation in Fig.5 appeared promising we confined our attention to this polytype. The results for a model of brookite of 1x2x2 unit cells along the a, b and c axes respectively, is also shown in Fig.4. Here the stoichiometry is exactly TiO_2 and the agreement with experiment is not too bad, although between r=2.7Å and r=4.0Å there is a reversal of the observed atom density. This region is mainly due to Ti-O and Ti-Ti distances between adjacent octahedra. It is thus very sensitive to the local disposition of units. Nonetheless all of the peaks come essentially at the experimental positions including the peak at r=6.5Å which is not reproduced by either anatase or rutile.

4. DISCUSSION

An important element in our initial interpretation has been the results of high resolution transmission electron microscopy HREM performed at 400 kV. While we do not present the images here (the effects are rather subtle and do not reproduce too well in this format) they indicate, in the thinnest sections of the material, only a wormy contrast characteristic of amorphous films. [There are occasional regimes in which ordered fringes appear but they are rare.] There is also the clear indication of overlapping patches of material of 4-10nm which would appear to be sections through the primary particulates noted above.

We are then left with an incomplete picture of the amorphous TiO_2 structure. While the HREM indicates only the disordered features of a glassy material, the X-ray pattern and its transform may at least initially be thought of perhaps as a further distortion of brookite. Certainly neither anatase nor rutile are useful candidates and the more distorted brookite serves mainly to point the way to a proper octahedral network. In any case, the correlation range of 10-12Å does not admit of any very regular crystallite model. Currently we are developing networks with more random attachments and will report on them at a later date.

REFERENCES

1. Edelson, L. H. and Glaeser, A. M., J. Am. Ceram. Soc. <u>71</u>, 225 (1988).
2. Edelson, L. H. and Glaeser, A. M., J. Am. Ceram. Soc. <u>71</u>, 198 (1988).
3. Shalz, M. L., thesis for the M. S. degree, Department of Materials Science and Mineral Engineering, UC Berkeley (1990).
4. Robertson, J. L. and Moss, S. C., J. Non-Cryst. Sol. <u>106</u>, 330 (1988).
5. Ice, G. E. and Sparks, C. J., Nucl. Inst. and Meth. Phys. Res. A <u>291</u>, 110 (1990).
6. Waseda, Y, "Novel Applications of Anomalous (Resonance) X-ray Scattering for Structural Characterization of Disordered Materials," Lecture Notes in Physics <u>204</u>, Springer-Verlag (New York, 1984).

ACKNOWLEDGEMENTS

This research was supported at the University of Houston by the U. S. Department of Energy, Office of Basic Energy Sciences on DE-FG05-87ER45325. At Berkeley support was by the Director, Office of Energy Research, Office of Basic Energy Sciences, Materials Science Division of the U. S. Department of Energy under contract no. DE-AC03-76SF00098. This research was performed in part at the Oak Ridge National Laboratory beam line X14 at the National Synchrotron Light Source, Brookhaven National Laboratory, sponsored by the Division of Materials Science and Division of Chemical Sciences, U. S. Department of Energy under contract no. DE-AC05-84OR21400 with the Martin Marietta Energy System, Inc.

ISOMERIZATION AND ICOSAHEDRAL FULLERENES

Brett I. Dunlap
Theoretical Chemistry Section, Code 6179
Naval Research Laboratory
Washington DC 20375-5000

ABSTRACT. A discussion of the evidence for isomerization in carbon clusters high-lights the special nature of Buckminsterfullerene. The predictions of Gaussian-type-orbital local-density-functional calculations on icosahedral fullerenes are reviewed. The larger icosahedral fullerenes have properties remarkably similar to C_{60} and should be easily made if special kinetic effects during vaporization and/or recondensation prevent or severely restrict isomerization as they appear to do.

1. Introduction

The most important technological problem in cluster science is to control the high energy processes, atom/ion bombardment, laser vaporization, electric arc, etc., used to make the new cluster materials in order to minimize the production of unwanted structural isomers. This isomer problem is easily seen in simple models of the cluster formation process [1-4]. In the recombination model of cluster creation there is enough energy available to completely atomize the starting material and thus to allow those atoms to reassemble in every fashion. In the bond-breaking picture of cluster formation different isomers correspond to different sets of broken bonds in the original material. As a function of cluster size, the number of isomers on a lattice grows exponentially [2,3]. Therefore, the larger the cluster the greater the chance of having to deal with unwanted isomers. Furthermore, this exponential growth means that the energy difference between isomers that are adjacent in energy must exponentially go to zero with increasing cluster size.

This last result does not imply that the energetic barrier to isomerization, or equivalently the melting temperature [5], vanishes exponentially with increasing cluster size. The difference is that several isomers can be almost isoenergetic but have large barriers separating each. In which case, the catchment volume in configuration space associated with each isomer would be rather large and assembling atoms at random would yield significant amounts of each of these isomers. If the barrier to isomerization is small, then for high enough temperatures or fast enough tunnelling rates a

P. Jena et al. (eds.), Physics and Chemistry of Finite Systems: From Clusters to Crystals, Vol. II, 1295–1303.
© 1992 *Kluwer Academic Publishers.*

cluster is characterized simply by listing the number of atoms that it contains. Clusters having high barriers to isomerization are likely to be technologically important because of a structural richness that could parallel the vast structural richness of organic chemistry, where writing a molecular formula, e.g. C_mH_n, is not sufficient to specify a substance. If high energy processes are to become a good source of a significant fraction of such rigid molecules containing tens or more of atoms, then the developmental keys will be kinetic rather than energetic.

In this regard, carbon clusters are interesting candidates for study because both bulk diamond and bulk graphite are refractory materials and stable isomers of finite carbon clusters have been found [6,7]. But carbon clusters are *truly fascinating* because, as created by high energy processes, small carbon clusters show isomerization and larger clusters so far do not! Multiple reaction rates for carbon clusters, C_7^+ to C_9^+, show that the laser vaporization process itself can create more than one isomer. Similar experiments with the large carbon clusters, C_n^+, n $\geq$ 40, find no reactions [8]. These larger, even-numbered carbon clusters are proposed to be fullerenes [9], hollow shells of carbon atoms arranged in an arbitrary number of six-membered rings and exactly twelve pentagons to ensure closure. The ^{13}C nuclear magnetic resonance spectra [10-12] of C_{60} and C_{70} fullerenes made by evaporating graphite in a helium atmosphere [13] are particularly simple when disolved in organic solvents, having one and five resonance positions respectively. Thus these fullerenes have the icosahedral and D_{5h} symmetries proposed in Refs. 14 and 15 respectively. Our linear combination of Gaussian-type orbitals (LCGTO) local density functional (LDF) calculations further support this structural assignment for C_{60}.

2. Buckminsterfullerene

The latest experimental [16,17] and theoretical [18-20] bond distances for C_{60} are compared in Table 1. Because C_{60} molecules spin in the pure solid at room temperature, the experimental X-ray diffraction bond distances listed in the table are averages of the appropriate bond distances in an osmylated compound [16]. The Hartree-Fock (HF) and second-order Möller-Plesset perturbation (MP2) corrected HF bond distances were computed using an all-electron triple-zeta plus polarization basis, 19 contracted Gaussian basis functions per carbon atom [18]. The LDF gas-phase calculations, like all the LDF calculations discussed herein, also used 19 contracted Gaussian basis fuctions per carbon atom, albeit obtained in a different fashion [20]. The LDF used is Perdew-Zunger [21]. These LDF (gas) bond distances are those of Ref. 20 expanded to four significant digits with a corresponding uncertainty added. In contrast to the case for *ab initio* methods, analytic basis sets cannot be used directly to evaluate the density functional total energy (because the exchange and correlation energy expressions invariably depend nonanalytically on the density). An additional three-dimensional numerical integration, or something equivalent to it such as finite Fourier expansion, is required. The LDF (gas) uncertainty in Table 1 is the standard deviation of ten optimizations of the 6-6 bond distance using ten different sets of sampling points [22]. Due to improvements in basis sets, these LDF (gas) bond distances are slightly different from 1.43 and 1.39 found in Ref. 4. Table 1 shows that the correlated *ab initio*, LDF and experimental bond distances are in quite close agreement. There is a remarkably close agreement between the LDF (gas) and (solid) values, the later using pseudopotentials and the constraint of icosahedral symmetry in a tetrahedral fcc environment. Table 1 does show, however, that the LDF single (5-6) bond length might be slightly too long compared to experiment and that the computed LDF (solid) double (6-6) bond length is slightly too short within the LDF model, unless this difference is a solid-state effect.

Table 1. The two symmetry-inequivalent sets of C_{60} equilibrium bond distances in Ångstroms. (The uncertainties are given in parenteses.) The pentagonal bond distance is labeled 5-6 and the bond distance shared only by hexagons is labelled 6-6.

Method	5-6	6-6	Ref.
HF	1.448	1.370	[18]
MP2	1.446	1.406	[18]
LDF(Solid)	1.444	1.382	[19]
LDF (Gas)	1.445(3)	1.387(3)	[this work]
NMR	1.450(15)	1.400(15)	[17]
X-ray	1.432(9)	1.388(5)	[16]

The electronic LDF structure of C_{60} is drawn at the bottom of the figure on the next page. At the top of the figure this electronic structure is used in a computation of cross-sections for photoexcitation in the ultraviolet (UPS) and X-ray (XPS) regions [23]. Comparison with experimental He II UPS data [24], so labelled leftmost in the top of the figure, gives one justification for the principal approximation of the calculation—the use of one-electron energy differences instead of the computationally more intensive differences between self-consistent (ΔSCF) final-state calculations. This agreement adds to the already quite conclusive evidence that C_{60} as made is purely icosahedral with no detectable degree of isomerization.

The special nature of C_{60} is rationalized as being a consequence of the fact that geometric constraints require that all energetically excited isomers of C_{60} and C_{70} as well as *all* isomers of C_{62}, C_{64}, C_{66}, and C_{68} have at least one energetically unfavorable pair of abutting pentagons [9,25,26]. It has been shown by semi-empirical and *ab initio* studies of isomers of C_{60} that shared bonds between two pentagons cost roughly one electron volt of energy each [27]. These isomerization energies are not significantly larger than the those predicted for C_7^+ to C_9^+ using the same semi-empirical method, and one finds the same relative abundances of the isomers of C_7^+ independent of source—graphite, diamond, or C_{10}^+ [28]. Furthermore, because of its unique nature one would expect C_{60} to have a higher barrier to isomerization than larger fullerenes. Conversely, one would expect larger fullerenes to be less resistant

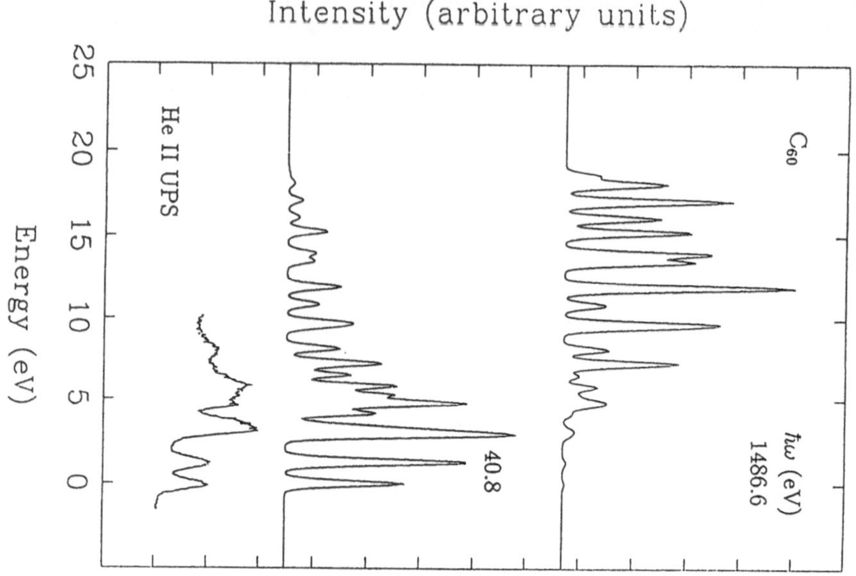
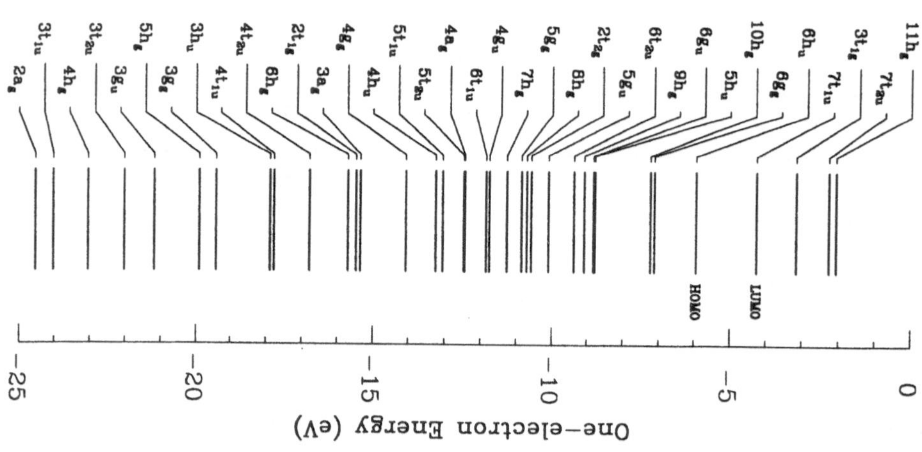

to isomerization. Stone and Wales [29] have proposed an elegant mechanism for interchanging positions two pentagons with two intervening hexagons that isolate the pentagons from each other. This mechanism was found to be Woodward-Hoffmann forbidden and thus fullerene isomerization is unlikely to occur spontaneously, at least by this mechanism. Unless an unexpected pathway to isomerization is found with a much smaller barrier to isomerization, one must assume that all fullerenes are rigid up to their decomposition temperature [30]. It is remarkable that no fullerene isomers have been detected to date, which prompts the following obvious yet astonishing question: What if isomerization does not occur for some reason in fullerenes made via vaporization?

3. Larger Fullerenes

Fullerenes larger than C_{70} can have more than one isomer in which all the pentagons are isolated. These isomers would likely comprise the ground state and any other isomers that might be created in the vaporization and recondensation process (the isolated pentagon rule [14]). Of these C_{76}, C_{84}, C_{90}, and C_{94} have been created in macroscopic amounts [31]. C_{76} has two fullerene structures with no adjoining petagons [32] and C_{84} has twenty-four [33]. The NMR spectrum of C_{76} has 19 lines [34], which is a relatively large number and thus could be associated with the presence of isomers, however 19 is also consistent with the number of sets of symmetry-inequivalent carbon atoms in a D_2 structure proposed for this molecule. This is more evidence against isomerization in the fullerenes.

One additional assumption leads directly to the large icosahedral fullerenes as a focal point for investigation. The isolated pentagon rule can be extended by postulating that the fullerenes made by vaporization have pentagons that are as far apart from each other as possible. (Such an assumption precludes the possibility that the double-helical motif [34] in the arrangement of the twelve pentagons in the D_2 noncentrosymmetric C_{76} and C_{84} structures will persist to significantly larger fullerenes and requires other methods for tubule preparation [35].) If the repulsive forces between pentagons is taken to be inversely proportional to the square of distance between pentagons, then, restricted to lie on a sphere but without further constraint, they will repel each other to the twelve five-fold vertices of an icosahedron [36].

We have studied icosahedral C_{180} and C_{240} using the LCGTO-LDF method [37]. It was shown that the difference between one-electron eigenvalues reproduced very accurately the results of ΔSCF calculation further justifying the use of the former in calculating the cross sections plotted in the figure. The results of the ΔSCF calculation of the two greatest electron affinities and the four smallest ionization potentials for each of these three molecules is given in Table 2. The calculated C_{60} ionization potential agrees with experiment [24,38]. The calculated C_{60} electron affinity is slightly higher than the experimental estimate of 2.65 ± 0.05 [39]. C_{180}

Table 2. The symmetry-restricted ΔSCF LDF electron afinitites for C_{60}, C_{180}, and C_{240} lie above the break in the table and ionization potentials below the horizontal space in the table. Under each molecule the orbital to which an electron is added or subtracted is followed by the affinity or ionization potential.

C_{60}		C_{180}		C_{240}	
$3\,t_{1g}$	1.71	$10\,t_{1g}$	3.21	$15\,t_{1g}$	3.31
$7\,t_{1u}$	2.82	$18\,t_{1u}$	3.56	$23\,t_{1u}$	3.61
$6\,h_u$	7.60	$19\,h_u$	6.91	$26\,h_u$	6.52
$6\,g_g$	8.78	$27\,h_g$	7.19	$34\,h_g$	6.93
$10\,h_g$	8.88	$18\,g_u$	7.24	$24\,g_g$	7.03
$6\,g_u$	10.47	$18\,g_g$	7.43	$14\,t_{2g}$	7.55

and C_{240} are 4.4 and 5.3 kcal/mole per carbon atom more stable energetically than C_{60}.

The remarkable coincidences of Table 2 are highlighted by considering the spherical π model of C_{60} [40]. In this model the valence electronic structure is dominated by a single π electron oriented radially on every carbon atom. To zeroth order—before the small perturbation of icosahedral symmetry—one must solve Schrödingers equation on the surface of a sphere. In this approximation each state is labeled by the conventional L and M quantum numbers, with L giving the nodal number of nodal surfaces and M giving the degeneracy. Thus increasing L corresponds to increasing energy. The number of π electrons needed to fill all levels through the L shell is $2(L+1)^2$. In the spherical model of a fullerene its radius is proportional to the square root of its surface area or, equivalently, its number of carbon atoms. Therefore one would expect the average radius of C_{180} and C_{240} to be given by $\sqrt{3}$ and 2 times the radius of C_{60}. Such an expectation is tested and met in Table 3. The appropriate $L_>$ values are odd for C_{60} and C_{180}, consistent with an ungarade highest occupied molecular orbital (HOMO) and lowest occupied molecular orbital (LUMO). On the other hand, $L_>$ is even for C_{240}, which is inconsistent with an ungarade HOMO and LUMO.

Table 3. Tests of the spherical model of the icosahedral fullerenes. The average radius, its standard deviation, and the average radius obtained by scaling the C_{60} radius by the square root of the number of carbon atoms are all in Ångstroms.

Molecule	Scaled Radius	Average Radius	Standard Deviation	$L_>$	$2(L_>+1)^2$
C_{60}	3.531	3.531	0.000	5	72
C_{180}	6.115	6.129	0.075	9	200
C_{240}	7.062	7.070	0.084	10	242

4. Conclusions

Structural isomers of carbon clusters can be and have been detected for smaller carbon clusters ions. The evidence is irrefutable that C_{60} is icosahedral and thus is a single isomer. C_{60} and other fullerenes have very special structures in which the carbons atoms are three-fold coordinated at the vertices of hexagons and twelve pentagons on a two-dimensional, roughly spherical, surface. This very restrictive set of constraints requires that fullerenes have more than 19 atoms, but does not limit the exponential growth of the number of isomers with the number of carbon atoms for larger clusters. These facts alone require that either C_{60} be extremely special or that the kinetics of the process making it be very special. The most compelling argument for C_{60} being special is that it is the only fullerene containing less than 68 carbon atoms that has no abutting pentagons on its surface. No similarly compelling arguments can be made for macroscopic samples of C_{70} and C_{76}, which are composed of at most a pair of stereoisomers. The answer to this dilemma must lie in the kinetics of the creation process.

If the fullerene creation process is so adiabatic in some sense that only ground-state structural isomers are created then one ought to be able to alter the process slightly in order to make the higher icosahedral fullerenes. C_{180} and C_{240} are more stable than C_{60} and have remarkably similar electronic structure, all possessing the same symmetry of their HOMO, LUMO and second LUMO. This coincidence of electronic structures might allow greater fine-tuning of the electronic properties fullerite materials, *e.g.* many models of the superconductivity in alkali-doped C_{60} would predict similar or larger transition temperatures in materials made by alkali-doping the larger icosahedral fullerenes. That the creation process cannot be alterred seems unlikely given the fact that C_{60} has been doped to yield $(K@C_{60})$ and $(La@C_{82})$ among other endohedral complexes [41]. The new '@' notation of Ref. 40 is interesting in that it both implies no important isomerization and encourages speculation about non-fullerene isomerization, e.g. a bound interior $(C_2@C_{58})$ or exterior $C_2(@C_{58})$ carbon dimer, as well as the question why not $(C@C_{44})$ [42]?

Acknowledgement

This work was supported by the Office of Naval Research (ONR) through the Naval Research Laboratory.

References

[1] J.E. Campana, T.M. Barlak, R.J. Colton, J.J. DeCorpo, J.R. Wyatt, and B.I. Dunlap, *Phys. Rev. Lett.* **47**, 1046 (1981).

[2] B.I. Dunlap, *Surf. Sci.* **121**, 260 (1982).

[3] J.E. Campana and B.I. Dunlap, *Int. J. Mass Spectrom. Ion Phys.* **46**, 523 (1984).

[4] B.I. Dunlap, *Int. J. Quantum Chem. Symp.* **22**, 257 (1988).

[5] R.S. Berry, *Sci. Am.*, October 1990, p. 68.

[6] S.W. McElvany, W.R. Creasy, and A. O'Keefe, *J. Chem. Phys.* **85**, 632 (1986).

[7] D.C. Parent and S.W. McElvany, *J. Am. Chem. Soc.* **111**, 2393 (1988).

[8] Q.L. Zhang, S.C. O'Brien, J.R. Heath, Y. Liu, R.F. Curl, H.W. Kroto, and R.E. Smalley, *J. Phys. Chem.* **90**, 525 (1986).

[9] H.W. Kroto, *Nature* **329**, 529 (1987).

[10] R. Taylor, J.P. Hare, A.K. Abdul-Dada, and H.W. Kroto, *J. Chem. Soc. Chem. Commun.*, 1423 (1990).

[11] R.D. Johnson, G. Meijer, and D.S. Bethune, *J. Am. Chem. Soc.* **112**, 8983 (1990).

[12] R.D. Johnson, G. Meijer, J.R. Salem, and D.S. Bethune, *J. Am. Chem. Soc.* **113**, 3619 (1991).

[13] W. Krätschmer, L.D. Lamb, K. Fostiropolous, and D.R. Huffman, *Nature* **347**, 354 (1990).

[14] H.W. Kroto, J.R. Heath, S.C. O'Brien, R.F. Curl, and R.E. Smalley, *Nature* **318**, 162 (1985).

[15] J.R. Heath, S.C. O'Brien, Q. Zhang, Y. Liu, R.F. Curl, H.W. Kroto, F.K. Tittel, and R.E. Smalley, *J. Am. Chem. Soc.* **107**, 7779 (1985).

[16] J.M. Hawkins, A. Meyer, T.A. Lewis, S. Loren, and F.J. Hollander, *Science* **252**, 312 (1991).

[17] C.S. Yannoni, P.P. Bernier, D.S. Bethune, G. Meijer, *J. Am. Chem. Soc.* **113**, 3190 (1991).

[18] M. Häser, J. Almlöf, G.E. Scuseria, *Chem. Phys. Lett.* **181**, 497 (1991).

[19] J.L. Martins, N. Troullier, and J.H. Weaver, *Chem. Phys. Lett.* **180**, 457 (1991).

[20] B.I. Dunlap, D.W. Brenner, J.W. Mintmire, R.C. Mowrey, and C.T. White, *J. Phys. Chem.* **95**, 5763 (1991).

[21] J.P. Perdew and A. Zunger, *Phys. Rev. B* **23**, 5048 (1981).

[22] R.S. Jones, J.W. Mintmire, and B.I. Dunlap, *Int. J. Quantum Chem. Symp.* **22**, 77 (1988).

[23] J.W. Mintmire, B.I. Dunlap, D.W. Brenner, R.C. Mowrey, and C.T. White, *Phys. Rev. B* **43**, 14281 (1991).

[24] D.L. Lichtenberger, K.W. Nebesny, C.E. Ray, D.R. Huffman, and L.E. Lamb, *Chem. Phys. Lett.* **176**, 203 (1991).

[25] T.G. Schmalz, W.A. Seitz, D.J. Klein, and G.E. Hite, *J. Am. Chem. Soc.* **110**, 1113 (1988).

[26] P.W. Fowler, J.E. Cremona, and J.I. Steer, *Theor. Chim. Acta* **73**, 1 (1988).

[27] K. Raghavachari, private communication.

[28] S.W. McElvany, B.I. Dunlap, S. O'Keefe, *J. Chem. Phys.* **86**, 715 (1987).

[29] A.J. Stone and D.J. Wales, *Chem. Phys. Lett.* **128**, 501 (1986).

[30] J. Milliken, T.M. Keller, A.P. Baronavski, S.W. McElvany, J.H. Callahan, and H.H. Nelson, *Chem. Mat.*, May/June, 386 (1991).

[31] F. Diederich, R. Ettl, Y. Rubin, R.L. Whetten, R. Beck, M. Alvarez, S. Anz, D. Sensharma, F. Wudl, K.C. Khemani, and A. Koch, *Science* **252**, 548 (1991).

[32] D.E. Manolopoulos, *J. Chem. Soc. Faraday Trans.* **87**, 2861 (1991).

[33] D.E. Manolopoulos and P.W. Fowler, *Chem. Phys. Lett.* **187**, 1 (1991).

[34] R. Ettl, I Chao, F. Diederich, and R.L. Whetten, *Nature* **353**, 149 (1991).

[35] S. Iijima, *Nature* **354**, 56 (1991).

[36] J.B. Weinrach, K.L. Carter, D.W. Bennett, and H.K. McDowell, *J. Chem. Educ.* **67**, 995 (1990).

[37] B.I. Dunlap, D.W. Brenner, J.W. Mintmire, R.C. Mowrey, and C.T. White, *J. Phys. Chem.* **95**, 8737 (1991).

[38] S.W. McElvany, M.M. Ross and J.H. Callahan, *Mat. Res. Soc. Symp. Proc.* **206**, 697-702 (1991).

[39] L.-S. Wang, J. Conceicao, C. Jin, and R.E. Smalley, *Chem. Phys. Lett.* **182**, 5 (1991).

[40] R.C. Haddon, L.E. Brus and K. Raghavachari, *Chem. Phys. Lett.* **125**, 459 (1986).

[41] Y. Chai, T. Guo, C. Jin, R.E. Haufler, L.P.F. Chibante, J. Fure, L. Wang, J.M. Alford, and R.E. Smalley *J. Phys. Chem.* **95**, 7574 (1991).

[42] D.W. Brenner, B.I. Dunlap, J.A. Harrison, J.W. Mintmire, R.C. Mowrey, D.H. Robertson, and C.T. White, *Phys. Rev. B* **44**, 3479 (1991).

FORMATION OF MUONIUM AND A MUONIC RADICAL IN FULLERENE

E. J. ANSALDO & J. J. BOYLE
TRIUMF and Dept. of Physics, University of Saskatchewan
Saskatoon, SK, Canada S7N 0W0

Ch. NIEDERMAYER
Universität Konstanz, Konstanz, Germany

G. D. MORRIS & J. H. BREWER
Dept. of Physics, U.B.C. and TRIUMF
Vancouver, BC, Canada V6T 2A3

C. E. STRONACH & R. S. CARY
Dept. of Physics, Virginia State University
Petersburg, VA 23806, USA

ABSTRACT. Three distinct electronic states were detected for positive muons (μ^+) after implantation into a C_{60} powder sample. About 60% of the μ^+ remained in the bare (diamagnetic") state, essentially an interstitial charged point particle. The rest of the muons were found to thermalize predominantly in two muonium ($Mu=\mu^+e^-$) atomic species. A "Vacuum" Mu state, with hyperfine coupling close to that of free Mu (perhaps at the molecular center), and a muonium substituted radical, i.e. a hydrogen-like Mu addition to double bonds on the carbon rings. This opens up a rich subfield of fullerene spectroscopy using muons. The diamagnetic μ^+ will allow the study of the properties of the superconducting (doped) cases. Muonium, as the muonated radical, will serve to map out the electron spin density in the rings, and as vacuum muonium on and near surfaces (molecules or grains), to study surface reactions with other adsorbed species.

1. Discussion

As a radioactive, light isotope of hydrogen, muonium ($Mu=\mu^+e^-$) has provided important and unique information on topics ranging from the electronic structure of hydrogenic centres in semiconductors (1-4) and of radicals in organic systems (5-9), to chemical reaction rates on surfaces (10,11) and other chemical, atomic and solid-state physics phenomena occurring within the time-scale of the 2.2 μs muon lifetime. For the particular case of the carbon allotropes, only the bare muon μ^+ was found in graphite, albeit with a high contact electron density (Knight shift), while two forms of muonium ("vacuum or normal" Mu and "anomalous" Mu^*) form in diamond (and other semiconductors).(3-5) In radical chemistry, addition of Mu

1305

P. Jena et al. (eds.), Physics and Chemistry of Finite Systems: From Clusters to Crystals, Vol. II, 1305–1309.
© *1992 Kluwer Academic Publishers.*

in unsaturated organic compounds has spawned a new spectroscopy.(9) Important information has been obtained on, for example electron spin densities and radical molecule reaction rate constants.(6-9) Finally, the quasifree Mu on granular surfaces (such as silica powder) permits the study of surface reactions.(10,11) We report here a first step in the application of μ-Mu spectroscopy to the fullerenes, namely the detection and characterization of two different states of Mu, in addition to the diamagnetic muon.

The measurements used the standard transverse field (muon spin rotation-relaxation) technique, which has been described in detail elsewhere.(4,12) Spin-polarized muons (from the TRIUMF cyclotron in Vancouver, BC, Canada) were implanted into a commercial sample of C_{60} (10-20% C_{70}),(14) located in a magnetic field oriented perpendicular to the initial muon spin direction. The formation of muonium was detected initially by the usual low field method (in 8 Oe applied field),(12,13) near room temperature. For the data shown in figure 1, the sample was held at a temperature of 210 K, and the applied field varied from 58 to 69 Oe. The quantity measured was the time-evolution of the muon spin polarization, given by the counting rate as a function of time for decay positrons along directions in the plane perpendicular to the applied field. The signal due to the diamagnetic μ^+, a spin ½ particle, is a simple Larmor precession at its characteristic frequency of 13.55 MHz/kOe. Its Fourier transform is a single sharp line, located below 1 MHz in figure 1. The lines for muonium can best be understood by reference to the Breit-Rabi diagram of figure 2 and previous work on solids and radicals.(1-13) Muon and electron are coupled by the hyperfine (Fermi contact) interaction, of strength v_0 (4463 MHz for free muonium).   At low fields (relative to the hyperfine interaction, Zeeman limit), four transitions are possible between the magnetic substates of the principal quantum number F. For the time resolution of the spectrometer, only the low frequency v_{12} and v_{23} are detectable. The central doublet in the C_{60} spectra of figure 1 could thus unambiguously be assigned to a "vacuum Mu" state, with hyperfine frequency v_0=4278 MHz, reduced from the full vacuum value by the medium (electron slightly pulled away from the muon), as expected for Mu located at cavities in the structure (likely inside the molecule). For higher fields (or lower hyperfine frequency, Paschen-Back régime) only a doublet is expected, corresponding to transitions in which only the spin of the muon flips. These transitions are labelled R_{12} and R_{34} in figures 1 and 2 because they are commonly observed in muonic radicals. For the fullerene sample studied, the doublet observed (of which only the lower frequency line, R_{12} is shown in figure 1) is due to a radical with hyperfine frequency of 325.7(5) MHz. The other member of the doublet, R_{34}, appears as expected around 260 MHz (field dependent). The detailed field dependence of the frequencies for both doublets was exactly as calculated for this assignment. No other significant lines were detected up to the 300 MHz spectral cutoff.

2. Results

In summary, the results have shown that, in addition to the diamagnetic state, a substantial fraction (30-40%) of muons form two distinct muonium species when implanted into fullerene. The "vacuum" Mu may reside inside the individual molecules, and/or be almost free at or near molecular surfaces in the solid. A muonic radical is formed by muonium addition to the double bonds, with hyperfine frequency typical of the beta position in the alkyl radical, i.e. the unpaired Mu electron density moves over to the radical center. Only line frequencies and relative amplitudes were needed for this identification. Important

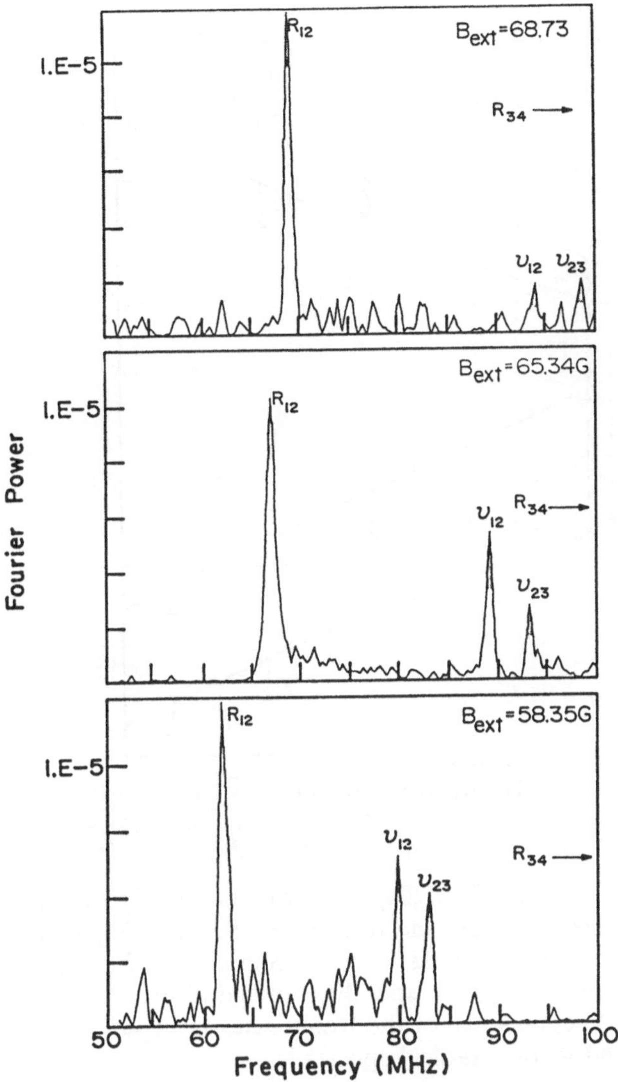

FIGURE 1. Fourier power spectra for the fullerene sample at 210K in the region of the main muonium lines for the radical (R_{12} line) and the vacuum Mu (v_{12} and v_{23} corresponding to a 4278 MHz hyperfine frequency). The μ^+ line is below 1 MHz in all cases. The only other significant feature of the spectra (up to the spectral limit of 300 MHz) is the R_{34} line, located at about 260 MHz, depending on applied field. The separation between R-lines gives the radical hyperfine coupling of 325.7 MHz mentioned in the text. Somewhat redundantly, data were taken at three fields to obtain an unambiguous assignment of lines into doublets.

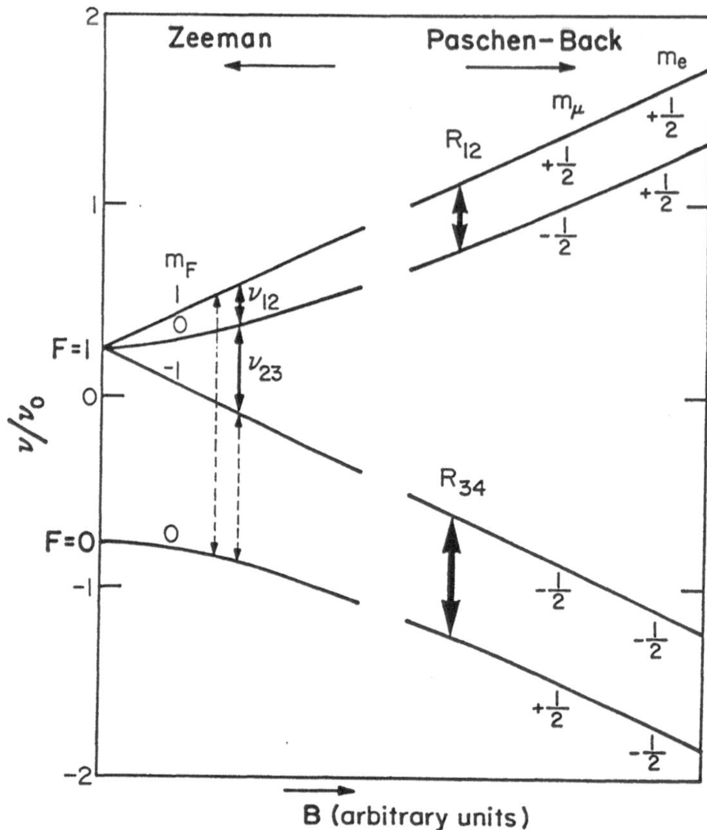

FIGURE 2. Schematic Breit-Rabi diagram for muonium eigenstates in an applied field. The solid arrows indicate the transitions shown in Figure 1 for "vacuum" Mu, which is well in the low field Zeeman region, and for the radical, in the Paschen Back regime. The corresponding analytical expressions were used to verify that the field dependence of the frequencies in Figure 1 were indeed due to the "vacuum" and muonic radical species.

information (not explored so far) is also contained in the phase of the signals (formation), the line widths (relaxation rates), and on further line splittings due to interactions with nuclei with spin. Thus muon spin spectroscopy may be used to explore the full range of behaviors that make the fullerenes such fascinating species. The diamagnetic muon may be used to study the superconductive state in particular,(15) and interstitial electron densities in general. The muonic radical(s) will permit further study of the electronic structure and chemical properties(16)(i.e. fullerene-hydrogen chemistry) of the molecular cluster, paralleling the extensive studies of free radicals already performed in other chemical systems. Finally, the "vacuum" Mu may be used as a probe of chemical reactions on both molecular and granular surfaces,(10,11) leading to an understanding of its catalytic properties.(17) It will also be interesting to test whether, similarly to silica, muonium atoms are ejected from the fullerene grains.(18)

3. Acknowledgements

Discussions with Professor Paul Percival were most helpful to the authors. Work at TRIUMF is supported by the NRC and NSERC of Canada. This work was also supported in part by US DOE grant DE-FG05-88ER45353 (CES) and US NASA grant NAG-1-416 (RSC).

4. References

1. J.H. Brewer and K. Crowe, *Ann. Rev. Nucl. Sci.* **28**, 239 (1978).
2. R.F. Kiefl *et al.*, *Phys. Rev. Lett.* **53**, 90 (1984).
3. B.D. Patterson, *Rev. Mod. Phys.* **60**, 69 (1988).
4. A. Schenck, *Muon Spin Rotation Spectroscopy. Principles and Applications in Solid State Physics* (Adam Hilger, Bristol, 1985).
5. J. H. Brewer *et al.*, *in Muon Physics*, edited by C.S. Wu and V.W. Hughes (Academic Press, New York, 1975), Vol. 3, p. 3.
6. P.W. Percival *et al.*, *Chem. Phys.* **32**, 353 (1978).
7. S.F.J. Cox and M.C.R. Symons, *Rad. Phys. Chem.* **27**, 53 (1986).
8. P.W. Percival *et al.*, *Chem. Phys.* **133**, 465 (1987).
9. E. Roduner, *Lecture Notes in Chemistry 49: The Positive Muon as a Probe in Free Radical Chemistry* (Springer-Verlag, Berlin, 1988).
10. R. Marzke *et al.*, *Ultramicroscopy* **20**, 161 (1986).
11. I.D. Reid *et al.*, *Nature* **345**, 303 (1990).
12. For details of muon and muonium spectroscopic applications, see the Proceedings of several International Conferences on Muon Spin Rotation, Relaxation and Resonanace, *Hyp. Int.*:**8** (1981), **17-19** (1984), **31-33** (1986), and **63-65** (1991).
13. D.C. Walker, *Muon and Muonium Chemistry* (Cambridge University Press, Cambridge, 1983).
14. Sample procured from MER Corp., Tucson, Arizona, USA.
15. R. Tycko *et al.*, *Phys. Rev. Lett.* **67**, 1886 (1991).
16. K. Holczer *et al.*, *Science* **252**, 1154-1157 (1991), and references therin.
17. P.J. Fagan *et al.*, *Science* **252**, 1160 (1991).
18. H.W. Kroto *et al.*, *Nature* **318**, 160 (1985).
19. G.M. Marshall *et al.*, *Phys. Lett.* **65A**, 351 (1978).

MAGNETIC SUSCEPTIBILITY OF PRISTINE C_{60} AND K-DOPED C_{60}, AND HEAT CAPACITY OF SOLID C_{60}

S.BANDOW,[1] H.OYA,[2] N.AKUZAWA,[3] H.SHINOHARA,[4] H.NAGASHIMA,[5] A.NAKAOKA,[5] M.OHKOHCHI,[6] Y.ANDO[6] AND Y.SAITO

Dept. of Electrical and Electronic Engin., Mie Univ., Tsu 514, Japan
[1]*Instrument Center, Inst. for Molecular Sci., Okazaki 444, Japan*
[2]*Suzuka Technical College, Suzuka 510-02, Japan*
[3]*Dept. of Industrial Chemistry, Tokyo National College of Tech., Hachioji 193, Japan*
[4]*Dept. of Chemistry for Materials, Mie Univ., Tsu 514, Japan*
[5]*School of Materials Science, Toyohashi Univ. of Tech., Toyohashi 441, Japan*
[6]*Dept. of Physics, Meijo Univ., Nagoya 468, Japan*

ABSTRACT. The diamagnetic susceptibility, χ_d, of crystalline C_{60} is investigated by using different kinds of crystalline powders, *hcp* and *fcc* structures. The obtained values of χ_d for *hcp* C_{60} are about twice larger than those of *fcc*. The typical value is -295 ppm emu/mol (C_{60} molar quantity) for *hcp* C_{60}, and -142 ppm emu/mol for *fcc* C_{60}. The superconducting upper critical field, $H_{c2}(0)$, of K-doped C_{60} is roughly estimated to be 100 ~ 130 kG. The heat capacity for solid (*fcc*) C_{60} is measured in the temperature range from 1.6 to 20 K and 300 to 400 K by an adiabatic and DSC methods, respectively.

1. Introduction

The C_{60} molecule is constructed with the 20 six-membered rings and 12 five-membered rings, and is considered to be an aromatic molecule due to sp^2 bond. The magnetic susceptibility of solid C_{60} and C_{70} has already been measured by Haddon *et al.* [1] and Ruoff *et al.* [2] to elucidate the π-elecron contribution to the magnetism. They showed very small diamagnetism contrary to the expectation from the ordinary aromatic molecules. On the other hand, the data on heat capacity of fullerenes are very few. In this paper, we report the magnetic and thermal properties of solid C_{60} and C_{70}.

P. Jena et al. (eds.), Physics and Chemistry of Finite Systems: From Clusters to Crystals, Vol. II, 1311–1316.
© 1992 *Kluwer Academic Publishers.*

1312

2. Experimental

C_{60} and C_{70} samples were prepared by the evaporation of carbon rod using arc discharge in He gas (20 ~ 100 Torr). The soluble extract from hexane/benzene or hexane/toluene eluents was chromatographically separated into pure fraction of C_{60} and of C_{70}. After the evaporation of solvent, the fullerenes were dried by heating at the temperature of 430 ~ 570 K under vacuum to remove the solvent.

The magnetic properties of fullerenes were measured by the high sensitivity Faraday balance [3] (main design from Oxford Instrument). This instrument has the sensitivity of 10^{-8} emu with the magnetic field higher than 10 kG and with the field gradient of 500 G/cm; which can be changed independently up to 50 kG for main field and to 1000 G/cm for field gradient. The temperature was measured by Pt thermometer ($T > 25$ K) and carbon resister ($T < 25$ K). The sample (ca. 30 mg for C_{60} and 9.10 mg for C_{70}) was put into quartz cell and fixed by paraffin liquid, whose correction was carefully conducted.

Heat capacity for higher temperature range (300 ~ 400 K) was measured by the differential scanning calorimetry (DSC, DAINI SEIKO-SHA SSC-560S). Before the measurement of heat capacity, the samples were heated up to 600 K with monitoring the differential thermal analysis (DTA) curve to confirm no solvent left in the sample. The measurements were repeated three times for each specimen in the temperature range from 300 to 400 K with the heating rate of 5 K/min. The Ar gas was purged to the specimen room at the flow rate of 60 ml/min during the measurement. The errors of the heat capacity are less than 3 % in the whole temperature range. The low temperature heat capacity (1.6 to 20 K) was measured only for C_{60} by an adiabatic method using the sample amount of 386 mg, which was put in the addenda and was sealed by woods alloy under 7 Torr of He ambient.

The crystal structure was also checked by a MAC Science powder X-ray diffractometer MXP^{3}V with a graphite monochromater equipped in front of the detector.

3. Results and Discussion

3.1. MAGNETIC PROPERTIES

3.1.1. *Fullerenes.* Typical temperature trace of the mass magnetic susceptibility χ_g of fullerene is shown in Fig. 1, which is for C_{60} from hexane/benzene eluent. Other samples show the similar temperature trace, *i.e.*, χ_g increases with decreasing temperature, indicating the existence of Curie term. This may come from the spurious paramagnetic impurities since both C_{60} and C_{70} have a large bandgap and the ground state is singlet. The spin concentration of each sample is estimated to be the order of 10^{-4} per a carbon atom. The values of molar diamagnetic susceptibility χ_d determined from the slope of $\chi_g T$ vs T plot and Curie constant from that of $(\chi_M - \chi_d)^{-1}$ vs T plot for the samples from different eluents are listed in Table 1, where χ_M (molar quantity of C_{60})

is represented by $720 \times \chi_g$. The χ_d of each sample seems to have the different values. However, referring to the X-ray diffraction data in the last column of the table, we can find that C_{60} with *fcc* structure has smaller χ_d compared with that of *hcp*; the averaged values of χ_d are -166 ppm emu/mol for *fcc* and -337 ppm emu/mol for *hcp*. Here it must be noticed that the samples assigned to *fcc* and *hcp* are not pure single phase but contain a small amount of the other phase, because the 111 reflection of *fcc* has a shoulder, indicating the existence of 002 reflection from *hcp* structure. Anyhow, we can say that the magnetic property is related to the crystal structure. The origin of this structure dependence is not yet clear, but it might be explained by the following way: The nearest intermolecular distance of solid C_{60} is 10.02 A and the molecular diameter is 7.1 A. Thus, the nearest C-C distance between molecules is considered to be 2.92 A. This value is a little shorter than that of the van der Waals diameter of carbon (3.4 A); the π-electron interaction between molecules can not be neglected in the solid C_{60}. Therefore, the π-orbital can be spread all over molecules. This was also found from the theoretical calculation of electronic structure of *fcc* C_{60} crystal by Saito and Oshiyama [4]: They found the considerable dispersion of π-bond level originating from the intermolecular overlap of π-bond. Hence, it is not so surprising that the electronic structure of solid C_{60} is affected by the crystal structure.

The χ_d determined in the present experiment does not coincide with the values obtained both by Haddon *et al.* [1] and Ruoff *et al* [2]. However, the χ_d obtained by averaging the values for *fcc* and *hcp*, -252 ppm emu/mol, is the same quantity obtained by them [1,2]. It seems that this difference comes from the crystal structure. Further, it must be noticed that the small values of χ_d for each C_{60} sample compared with the ordinary aromatic molecules may originate from the Van Vleck temperature-independent paramagnetism as Ruoff *et al.* [2] suggested. The χ_d for C_{70} is comparable with that obtained by Haddon *et al* [1].

3.1.2. *K-doped C_{60}.* The magnetization curves of K-doped C_{60} ($K_{3.8}C_{60}$, estimated from the mass of each element set into a quartz tube for doping) were measured by the Faraday balance with the magnetic field up to 50 kG and in the temperature range between 2 and 18 K (T_c = 18.4 K). This system behaves as Type-II superconductor with upper critical field H_{c2} higher than 50 kG at the temperature lower than 10 K. H_{c2} at 13.5 K can be roughly estimated to be 40 $\sim$ 50 kG, from which $H_{c2}(0)$ is evaluated to be 100 $\sim$ 130 kG by the relation, $H_{c2}(0) = 0.69(\partial H_{c2}/\partial T)_{T_c}$ [5], and the coherence length ξ at zero-temperature is estimated to be about 5 nm by $H_{c2}(0) = \Phi_0/2\pi\xi^2$, where Φ_0 is the fluxoid quantum. H_{c1} is estimated to be in the order of a hundred G at the temperatures lower than 6.4 K. These values are in the same order of magnitude with that reported by Holczer *et al.* [6]; H_{c2} = 490 kG and ξ = 2.6 nm.

1314

3.2. THERMAL PROPERTIES

Temperature dependence of heat capacity in high temperature range is shown in Fig. 2. Heat capacity increases gradually with increasing temperature. The heat capacity at 50 °C is 590 J/mol K for C_{60} and 751 J/mol K for C_{70}. The heat capacity per a carbon atom of fullerenes was calculated by the equation,

$$C_{p\ at} = (C_{p\ mol} - 6R)/N, \tag{1}$$

where $C_{p\ at}$ is the atomic heat capacity of carbon, $C_{p\ mol}$ the molar heat capacity of fullerene, R the gas constant and N the number of carbon atoms in the fullerene, *i.e.*, $N = 60$ for C_{60} and 70 for C_{70}. In the above formula, $6R$ corresponds to the heat capacity contributed from the molecular motions related to the translation and rotation. The translation mode is thought to be excited from low temperatures since the condensation force of molecules is very weak [7]. The rotation mode is also excited above 249 K where the order and disorder orientational transition occurs [8,9]. The heat capacity of carbon atom in fullerenes almost equal to that of graphite, as shown in Fig. 3.

The temperature dependence of low temperature heat capacity of C_{60} is shown in Fig. 4. The exact absolute value of heat capacity could not be determined due to the small amount of sample compared with addenda (386 mg of sample and *ca.* 14 g of addenda). The heat capacity at low temperatures shows the T^3 behavior below *ca.* 4 K. From the slope of C_{p}/T vs T^2 plot, the Debye temperature T_{Θ} for solid C_{60} is estimated to be 50 K. Since the intramolecular vibration is considered to be frozen at such low temperatures, the experimental value of 50 K should be the intermolecular T_{Θ}

Thanks are due to Prof. Keisaku Kimura for his valuable suggestions and discussions concerning the magnetic properties of aromatic molecules. The authors also express their thanks to Dr. Ping Wang for her critical reading of manuscript.

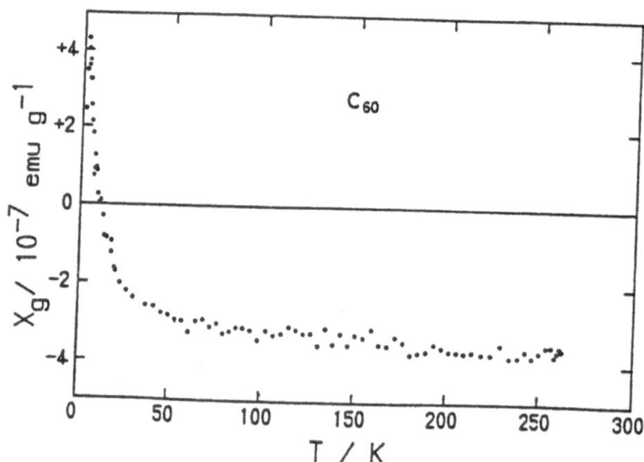

Fig. 1 Temperature dependence of magnetic susceptibility of C_{60}. The sample is eluted with hexane/benzene.

Table 1.   The values of diamagnetic susceptibility and Curie constant. The values are the C_{60} and C_{70} molar quantity.

C_{60}	χ_d $(10^{-6}$ emu/mol$)$	Curie constant $(10^{-5}$ emu K/mol$)$	Structure
Benzene	−295	419	hcp
HCl	−330	710	(hcp)
Toluene	−189	1480	fcc
	−142	1754	fcc
300	−387	1790	hcp + (fcc)
C_{70}			
Toluene	−544	1794	fcc

Benzene : from hexane/benzene solution.
HCl   : refluxed the sample "Benzene" by HCl solution to remove impurities.
Toluene : from hexane/toluene solution.
300   : heated the sample "Toluene" up to 300 °C.

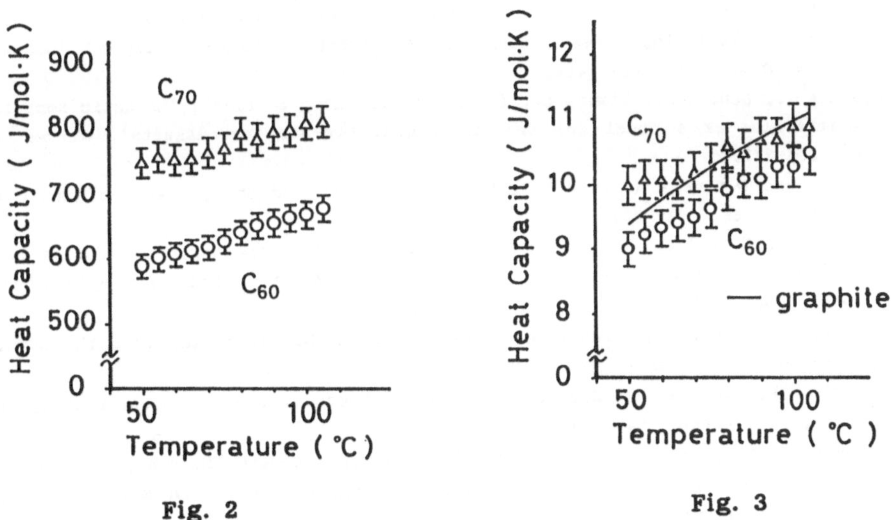

Fig. 2   High temperature heat capacity of fullerenes
Fig. 3   Heat capacity per a carbon atom in fullerenes. The values are from the equation (1) in the text. Solid line is for graphite.

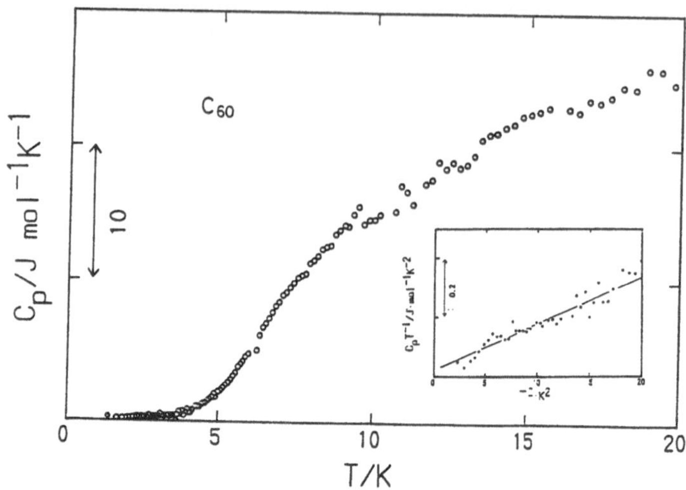

Fig. 4   Low temperature heat capacity of C_{60}.   Inserted figure is C_p/T vs T^2 plot to determine the Debye temperature.

References

[1]   Haddon R.C., Schneemeyer L.F., Waszczak L.V., Glarum S.H., Tycko R, Dabbagh G., Kortan A.R., Muler A.J., Mujsce A.M., Rosseeinsky M.J., Zahurak S.M., Makhija A.V, Thiel F.A., Raghavachari K., Cockayne E. and Elser V. (1991) 'Experimental and theoretical determination of the magnetic susceptibility of C_{60} and C_{70}', Nature, 350, 46-47.

[2] Ruoff R.S., Beach D., Cuomo J., McGuire T., Whetten R.L. and Diedrich F. (1991) 'Confirmation of Vanishingly Small Ring-Current Magnetic Susceptibility of Icosahedral C_{60}', J. Phys. Chem., 95, 3457-3459.

[3] Sugawara T., Bandow S., Kimura K., Iwamura H. and itoh K. (1986) 'Magnetic Behavior of Nonet Tetracarbene as a Model for One-Dimensional Organic Ferromagnets', J. Am. Chem. Soc., 108, 368-371.

[4] Saito S. and Oshiyama A. (1991) 'Cohesive Mechanism and Energy Bands of Solid C_{60}', Phys. Rev. Lett., 66, 2637-2640.

[5]   Werthamer N.R.,Helfand E. and Hohenberg P.C. (1966) 'Temperature and Purity Dependence of the Superconducting Critical Field, H_{c2}. III. Electron Spin and Spin-Orbit Effects', Phys. Rev., 147, 295-302.

[6] Holczer K., Klein O. and Gruner G. (1991) 'Critical Magnetic Fields in the Superconducting State of K_3C_{60}', Phys. Rev. Lett., 67, 271-274.

[7]   Guo Y., N.Karasawa N. and Goddard W.A. (1991) 'Prediction of fullerene packing in C_{60} and C_{70} crystals', Nature, 351, 464-467.

[8] Heiney P.A., Fischer J.E., McGhie A.R., Romanow W.J., Denenstein A.M., McCauley J.P., Smith A.B. and Cox D.E. (1991) 'Orientational Ordering Transition in Solid C_{60}', Phys. Rev. Lett., 66, 2911-2914.

[9]   Yannoni C.S., Johnson R.D., Meijer G., Bethune D.S. and Salam J.R. (1991) '^{13}C NMR Study of the C_{60} Cluster in the Solid State: Molecular motion and Carbon Chemical Shift Anisotropy', J. Phys. Chem., 95, 9-10.

Photoemission Spectra of C_{60} Clusters in Metal and Nonmetal Systems

S. J. Chase, R. Q. Yu, M. G. Mitch and J. S. Lannin
Dept. of Physics
Penn State University
University Park, PA 16802

Changes in the weighted electronic density of states of C_{60} molecules deposited on a variety of substrates have been studied by HeI ultraviolet photoemission (UPS). Spectra of C_{60} deposited on polycrystalline Ag exhibit a C_{60} LUMO-derived band which appears to originate in interactions at the interface. C_{60} was also exposed to Rb, developing a LUMO-derived band and similar phase characteristics to those observed in K_xC_{60} compounds. The effect of atomic H as an adsorbate was also investigated. While adsorbates generally show increased broadening of the π_p and σ_p occupied states of C_{60} with increasing coverage, exposure to atomic H resulted in spectra where the distinct molecular features were completely lost, bearing a distinct resemblence to graphite. However, Raman spectroscopy indicates the essential C_{60} structure has not been changed.

1. Introduction

Since the discovery of conductivity in alkali fullerides[1] there has been considerable interest in the effects of other elements on the electronic structure of C_{60}. In addition to several in-depth investigations of the phase behavior of the alkali-C_{60} compounds [2-6], Ohno et al. have studied the effects of metallic, semiconductor and semimetal substrates on the photoelectron spectra of deposited C_{60}[7]. Due to the relatively inert nature of C_{60}, the observed effects were largely attributed to Fermi level alignment and final state effects due to photoemission charge accumulation. This paper continues the investigation of Ag substrate interactions, and follows the line of inquiry with studies of H and Rb as adsorbates. We present spectra of Rb_xC_{60} which show similar characteristics to K_xC_{60}, exhibiting a LUMO-derived conduction band. We also present evidence of a conduction band formed from the interaction of a Ag substrate and deposited C_{60}. An interesting series of atomic hydrogen depositions illustrate the development of broad, graphite-like spectral features from the distinct molecular bands of C_{60}.

P. Jena et al. (eds.), Physics and Chemistry of Finite Systems: From Clusters to Crystals, Vol. II, 1317–1322.
© 1992 *Kluwer Academic Publishers.*

2. Experiment

Samples were made and measured at room temperature in an ultrahigh vacuum chamber with a base pressure of 3×10^{-10} torr. Ag was deposited by dc magnetron sputtering, while Rb and K were evaporated from well-degassed SAES getter alkali metal sources. Atomic hydrogen was created by passing a flow of hydrogen gas over a hot filament in the UHV chamber, the dosage being regulated by the gas pressure. C_{60} films were evaporated from a pure C_{60} source in high vacuum, samples then being moved directly to UHV without exposure to air. UPS measurements utilized a Microscience HA100 hemispherical analyzer and HeI photon energy of 21.2 eV. Raman spectra were taken using a Spex Triplemate spectrometer with a multichannel resistive anode detector. In-situ Raman measurements were made using multilayer substrates for interference enhanced Raman scattering, enabling the use of samples with as little as 0.5 monolayers of C_{60}. This permitted both UPS and Raman scattering to examine the same volume of sample for thicknesses up to 4 monolayers of C_{60}. No annealing of alkali samples was performed, as experiment showed no significant change in either UPS or Raman spectra after heating. Thicknesses of sputter deposited Ag and C_{60} were calibrated using a quartz crystal microbalance.

3. Results and Discussion

Figure 1 shows the UPS spectra for varying thicknesses of C_{60} on polycrystalline Ag. The spectrum of Ag indicates the dominant d band and a weaker structureless s band extending to the Fermi energy. The weak features ~1.8 eV above the main d peak are due to He ultraviolet satellite lines that have intensities ~4% of the main HeI intensity. With the addition of C_{60} new features are observed at 3.2 eV and 1.8 eV below E_F that are associated with the highest occupied molecular orbital (HOMO) π bands of C_{60}. As the Ag d band intensity decreases with C_{60} coverage it is also observed that changes occur in the form of the UPS spectra in the vicinity of E_F. Here the shape of the Fermi edge is

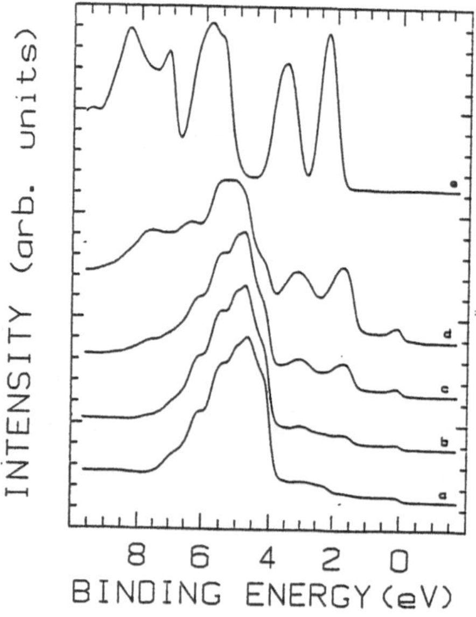

Fig.1 HeI spectra of C_{60} on Ag. a)Ag, b)0.2 monolayers C_{60}, c)0.6 ml, d)2 ml, e)5 ml.

modified, as shown in more detail in Fig. 2. A distinct peak at the Fermi edge is observed approximately 1.8 eV above the C_{60} HOMO band, which does not shift with increasing C_{60} deposition. This energy separation is significantly smaller than that observed for the alkali systems, although in other respects the Ag interfacial band and the alkali LUMO-derived bands are qualitatively similar. As the s band of pure Ag has an intensity of ~3% of the d band, an upper limit to the contribution of Ag s band states to Fermi edge spectral region can be obtained. This is shown by the dotted curves in Fig. 2, which demonstrate that the changes in this spectral region are associated with new occupied states of the C_{60}/Ag interface. These states, whose qual-

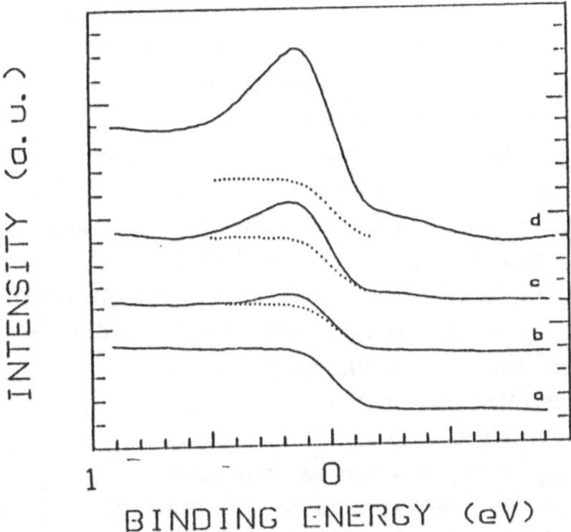

Fig.2 Fermi edge for HeI spectra of C_{60}/Ag. Dotted lines show approximate Ag s-band contribution. a)Ag, b)0.2 monolayers C_{60}, c)0.6 ml, d)2 ml

itative form is similar to that observed in K_3C_{60} [3,5], are attributed to population of a band derived from the lowest unoccupied molecular orbital (LUMO) π^* states of pure C_{60}.

As in the case of chemisorption of CO on Cu(100) [8], the new UPS band is associated with the formation of a metal-C_{60} resonance interaction. In contrast to CO/Cu, much larger changes due to a more intense UPS resonant tail response are observed here. This interaction results in a transfer of charge from the metallic Ag surface to the C_{60} cluster and partial occupancy of the resonant band. Surface enhanced Raman scattering measurements performed on rough Ag surfaces in UHV indicate a substantial shift of ~29 cm^{-1} of the $A_g(2)$ pentagonal pinch, high frequency optic mode due to the transfer of charge from the Ag surface to the C_{60} cluster. This hybrid Ag/C_{60} band appears to have its origin solely in the interface between the two materials. The new band becomes visible at C_{60} depositions of ~ 0.2 monolayers, and increases until ~ 2 monolayers coverage is obtained. Beyond this point increasing deposition does not increase the ratio of the hybrid peak intensity with respect to other C_{60} features. Possible explanations for the maximum peak intensity being reached at 2 monolayers instead of 1 are the well-known surface roughness of sputtered silver, which would provide greater surface area for interaction with the C_{60} clusters, or non-uniform C_{60} thickness. When UPS spectra of thick C_{60} are obtained the peak has disappeared completely.

Ohno et al. [7] recently reported UPS spectra for C_{60} on Ag, but did not observe any new features at the Fermi energy. We interpret this to be a consequence of the photon energy employed in measuring the spectrum. In the region near the Fermi edge the electron density is derived from silver s-electrons and carbon p electrons. The cross-section for the carbon p-electrons is much larger compared to that for silver s-electrons for hv=21.2 eV. In contrast, Ohno et al. used 110 eV, where the carbon p-electrons have a greatly reduced cross-section, effectively diminishing any C_{60} hybrid band contribution. Samples were also made where Ag was deposited on a thick (~ 4 monolayers) film of C_{60}. The hybrid LUMO band of Figs. 1 and 2 did not appear in this case, which confims the surface character of the hybrid band.

Figure 3 compares valence band spectra of Rb and K doped C_{60} for x=3. While the Rb_xC_{60} system shows qualitatively similar behavior to that of K_xC_{60}, in that it exhibits insulating or conducting behavior dependant on Rb concentration [5,6], Rb_xC_{60} does not show as sharp a Fermi edge as that seen with K. The separation of the HOMO and LUMO-derived band at x=3, of 2.1 eV, is ~0.2 eV less than that measured for the K system. The shift in the conduction band edge appear to be linked to changes in other spectral features, most notably the HOMO band, which first decreases, then increases in binding energy for both Rb and K with increasing deposition. Phase identification was further aided by Raman spectroscopy, done in tandem with the UPS. The Raman spectra [1,6] also indicate changes in the $A_g(2)$ mode due to charge transfer from the Rb to the C_{60} cluster.

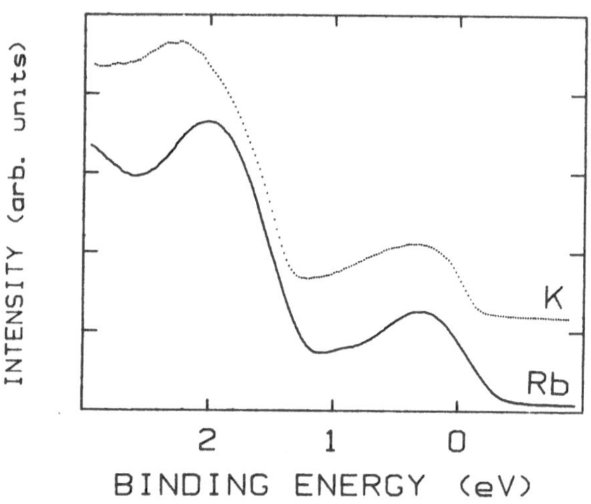

Fig.3 Fermi edge for HeI spectra of K_3C_{60} and Rb_3C_{60}.

By exposing C_{60} films to atomic hydrogen, the effects of terminating the dangling bonds on the surface of the fullerene structure were studied. Figure 4 shows the dramatic change in the UPS spectra of C_{60} after exposure. The distinctive features and sharp peaks are severely broadened, with the spectrum resembling graphite with its gently rounded π_p and σ_p bands. This might seem to indicate that the atomic structure giving rise to the spectral features is also absent. However, Raman spectroscopy exhibits only a slight broadening of the Raman active C_{60} vibrational modes, indicating that the fullerene cage is still intact. In addition, UPS spectra of the sample after heating show the return of C_{60} features. When C_{60} was evaporated in an atmosphere of atomic hydrogen of varying

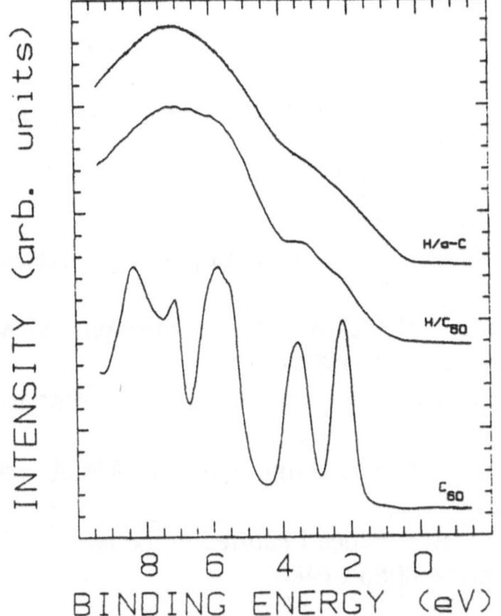

Fig.4 Comparison of UPS spectra of H on a-C and H on C_{60}.

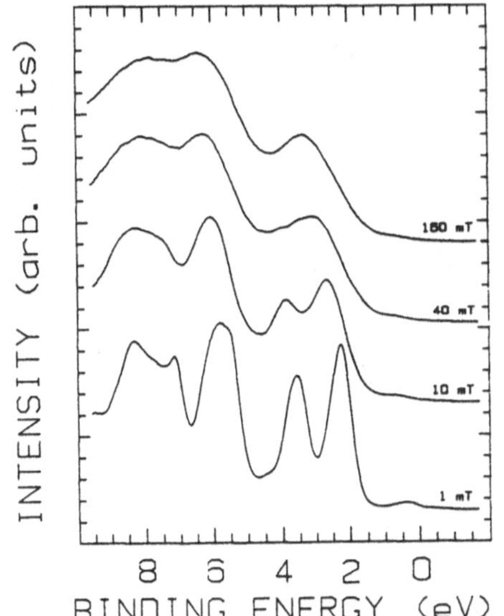

Fig.5 HeI spectra of C_{60} evaporated at various pressures of H.

pressure, UPS spectra show a gradual evolution with increasing pressure to the saturated spectrum. In addition, the spectra exhibit a shift to lower binding energy with increasing H concentration (Fig. 5). The loss of spectral features with the addition of H in Figs. 4 and 5 is most likely due to the broad distribution of H bonding sites on C_{60}.

4. Conclusions

We have studied the interaction of C_{60} clusters with a variety of systems, both adsorbate and substrate, utilizing UV photoemission. As of this writing, only two types of systems have been found that exhibit a conduction band in conjunction with C_{60}: alkali intercalates, and selected metal substrate surfaces. Rb_xC_{60} was found to have qualitatively similar spectra to K_xC_{60}, yet it had significant differences. This may possibly reflect the different electronic properties permitting superconductivity at higher temperatures than K-doped C_{60}. Also yielding insight into the electronic properties of C_{60} was H-saturated C_{60}, which illustrated the effects of saturating dangling bonds on the C_{60} surface.

This work was supported by US DOE Grant No. DOE-FG02-84ER45095.

REFERENCES
1. R. C. Haddon et al., 'Conducting Films of C_{60} and C_{70} by Alkali-Metal Doping', Nature (London) **350**, 354-358 (1990).
2. K. Holczer et al., 'Alkali-Fulleride Superconductors: Synthesis, Composition, and Diamagnetic Shielding', Science **252**, 1154-1157 (1991).
3. P. J. Benning et al., 'Electronic States of $K_x C_{60}$: Insulating, Metallic, and Superconducting Character', Science **252**, 1417-1421 (1991).
4. P. W. Stephens et al., 'Structure of Single-Phase Superconducting $K_3 C_{60}$', Nature (London) **351**, 632-634 (1991).
5. C. T. Chen et al., 'Electronic States and Phases of $K_x C_{60}$ from Photoemission and X-ray Absorbtion Spectroscopy', Nature (London) **352**, 603 (1991).
6. M. G. Mitch, S. J. Chase and J. S. Lannin, 'Raman Scattering and Electron-Phonon Coupling in $Rb_x C_{60}$', (to be published).
7. T. R. Ohno et al., 'C_{60}Bonding and Energy Level Alignment on Metal and Semiconductor Surfaces', Phys. Rev. B (in press).
8. B. Gumhalter, K. Wandelt and Ph. Avouris, '$2\pi^*$ Resonance Features in the Electronic Spectra of Chemisorbed CO', Phys. Rev. B **37**, 8048-8065 (1988).

ELECTRONIC STRUCTURE OF FULLERENES: ISOLATED MOLECULES AND METAL-DOPED CRYSTALS

M. R. PEDERSON, S. C. ERWIN,* W. E. PICKETT, K. A. JACKSON,+
and L. L. BOYER
Complex Systems Theory Branch, Naval Research Laboratory, Washington D.C. 20375
**Physics Dept., Univ. of Pennsylvania, Philadelphia, PA 19104-6272*
+Physics Dept., Central Michigan University, Mt. Pleasant, MI 48859

ABSTRACT. We analyze results from a variety of recent density-functional based calculations on isolated and crystalline fullerene molecules. From the generalized gradient-correction we predict an experimental cohesive energy of 7.25 eV per atom for the isolated molecule. Calculations of the band structures of alkali-doped fullerenes [K_6C_{60} and K_3C_{60}] are reported. We present cluster-based studies aimed at understanding the interactions governing the arrangement of the potassium atoms in the BCC K_6C_{60} structure. For the K_6C_{60} structure, the arrangement of potassium atoms is primarily governed by short-range carbon-potassium interactions. Calculations aimed at addressing the possibility of transition metal doped fullerene crystals are also presented.

1. Introduction

The primary goals of this paper are to review and further analyze some results from recent calculations on isolated fullerene molecules[1] and metal-doped crystals[1-3] and to present calculations aimed at explaining the crystalline charge states and some of the experimental findings related to the structures of the doped unit cells. In section 2, we start with the isolated fullerene molecule. In addition to calculating the molecular cohesive energy, to complement our calculations on BCC K_6C_{60} [2] we present energetics associated with highly charged fullerene molecules. In section 3, the bandstructure of the BCC K_6C_{60} is reviewed and some more recent calculations aimed at understanding the arrangement of the K atoms are presented. In addition, we compare the insulating BCC K_6 doped phase to the metallic FCC K_3 doped phase.[3] In section 4, we briefly review prospects for a different BCC metal doped structure of crystalline C_{60} that is stabilized by transition metal atoms.[1] Before continuing, we discuss the computational details.

All of the calculations presented here have utilized an approximation to the Hohenberg-Kohn-Sham density functional formalism.[4] Within this approximation, the total energy of a system of electrons and nuclei, is obtained by variationally adjusting the electronic charge density and the positions of the nuclei until the forces on each atom vanish and the electronic wavefunctions (from which the charge density is derived) self-consistently satisfy a Schroedinger equation. To solve the Schroedinger equation, the wavefunctions are expanded in terms of Gaussian-type orbitals.[5-6] Due to the large number of atoms associated with a fullerene molecule and the relatively large distance between individual components in the crystalline phase, we have found that a combination of calculations on periodic arrays of unit cells, isolated clusters, and fragments of the crystal are useful for identifying and studying the important interactions.

2. The Isolated Molecule: Electronic Structure and Charge States

To predict the cohesive energy of the isolated fullerene molecule, we have employed the

1323

P. Jena et al. (eds.), Physics and Chemistry of Finite Systems: From Clusters to Crystals, Vol. II, 1323–1328.
© 1992 *Kluwer Academic Publishers.*

generalized gradient approximation (GGA) of Perdew.[7] Several recent papers have shown that this approximation to the density-functional theory[4] yields significant improvements over the local density approximation (LDA) for calculating cohesive energies.[8-9] To calculate the cohesive energy, we have performed calculations on the isolated molecule at the experimentally determined pentagonal and hexagonal bondlengths of 2.764 and 2.651 Bohr respectively and compared the energy to that of the isolated ground state (spin polarized) atoms. Within the GGA, we obtain a cohesive energy of 7.38 eV per atom.[1] To estimate the actual experimental cohesive energy for C_{60}, we note that in a similar conjugated system (benzene) we find that the GGA overestimates the experimental cohesive energy by 0.1 eV per C:C bond.[8] Taking this and other small factors into account, we estimate that the actual experimental cohesive energy is approximately 7.25 +/-0.1 eV/atom. For details see Ref. [1]. The magnitude of the experimental heat of formation of the isolated molecule is expected to be slightly smaller due to zero point motion of the molecule. For diamond, which has an experimental heat of formation of 7.35 eV, the zero point motion has been calculated to be 0.195 eV per atom[9] suggesting that the experimental cohesive energy of diamond would be approximately 7.55 eV. Based on these considerations the energetic stability of C_{60} in comparison to diamond is quite evident.

In the right-hand panel of Fig. 1, the gaussian broadened density of states of the isolated molecule is presented. For the purpose of presentation, the energies are referenced with respect to the lowest unoccupied molecular orbitals [LUMO] of T_{1u} symmetry. The eigenvalue of the T_{1u} (LUMO) state for the isolated molecule is -4.49 eV. Fig. 1 shows a HOMO/LUMO gap of 1.74 eV

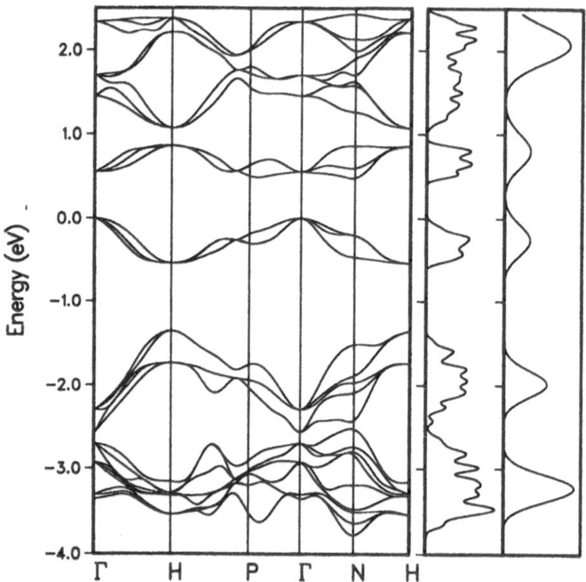

Figure 1. Electronic band structure of crystalline K_6C_{60},[2] referenced with respect to the T_{1u} band is shown on the left. The right panels compare the crystalline density of states to the gaussian broadened density of states for an isolated C_{60} molecule. In contrast to the crystalline case, the T_{1u} states are unoccupied for the isolated molecule.

between the five-fold H_u and the three-fold T_{1u} states. The isolated C_{60} molecule is known to bind as many as two extra electrons[10] and is expected to be stable in significantly higher charge states in the crystal. To address the energetics associated with highly charged negative C_{60}, we have performed calculations on the molecule as a function of the T_{1u} occupancy [Q = 0, 3, and 6]. First, using the self-consistent neutral C_{60} wavefunctions, we find that the energy and T_{1u} eigenvalues [eV] are adequately represented by $E(Bb) = -2.884 + 1.752(Q-1.283)^2$. The charge dependence of the T_{1u} eigenvalues may be obtained from the first derivative of this expression. We note that Q is the amount of charge added to the molecule $[0 < Q < 6]$. Repeating this calculation with self-consistent wavefunctions for all three charge states leads to energies that are represented by $E(Bb) = -3.21 + 1.52(Q-1.46)^2$. While the goal here is to predict the energetics of highly charged C_{60} molecules that are stabilized by crystal field effects, we note that these equations predict first electron affinities of 2.70 and 2.88 eV respectively in excellent agreement with the experimentally determined result of 2.74 eV.[10] We note that the results presented here neglect a variety of small effects such as spin polarization, multiplet effects and charge-induced geometrical relaxation. While such effects are insignificant for the purpose of this paper, they are expected to be important for an accurate determination of the second electron affinity. These details are presently being examined.

3. Alkali-Doped Fullerenes

In Fig. 2, the experimentally determined structure of the K_6C_{60} unit cell is shown. For this BCC structure with T_h symmetry, the unit cell is defined by placing a C_{60} at the origin and potassium atoms at the six sites that are equivalent to $(0,0.5,d)a$, with a the lattice constant and d the "dimerization coordinate". An "ideal" dimerization coordinate of 0.25 would lead to a square of potassium atoms on each face rather than the rhombus shaped ring shown in Fig. 2. Experimentally, the K atoms form a rhombus $(d = 0.28)$ and the lattice constant is found to be $a = 21.56$ Bohr. As shown in the figure it is useful to visualize a "cage" of 24 potassium atoms surrounding each

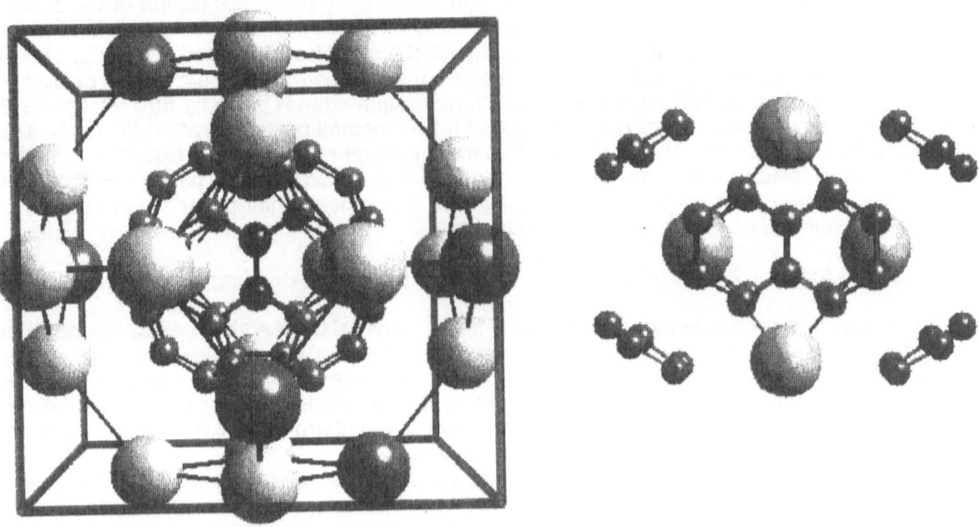

Figure 2. The unit cell for the K_6C_{60} BCC structure is composed of the fullerene molecule and the six dark potassium atoms (left). The forty carbon atoms that are closest to a potassium rhombus reside on six different fullerene molecules and are shown on the right.

fullerene molecule. However, for the experimental geometry, the nearest K-C distance [6.01 Bohr] is quite a bit smaller than the nearest K-K distance [7.68 Bohr]. To understand the interactions that determine the dimerization coordinate, it is first necessary to determine the charge states of the fullerene molecule and the potassiums. We now turn to this question.

In Fig 1, the bandstructure and density of states of the K_6C_{60} crystal is shown along with the density of states of the isolated cluster.[2] The bandstructure is found to be insulating with an indirect gap of 0.48 eV. The close correspondence between the broadened DOS of the isolated molecule and the DOS of the K_6 compound suggests that the occupied and unoccupied states near the Fermi level are primarily composed from carbon 2p states suggesting donation of K 4s electrons to the C band states. This point is confirmed by inspection of the Mulliken charges in Table 1.

Table 1. Mulliken charge associated with each of the four inequivalent atoms in the unit cell. The Mulliken charge suggests that the potassium atoms are ionized in the $K_6 C_{60}$ structure. Under these conditions, for an idealized case, each carbon atom would have a charge of 6.1 electrons. However, the twelve carbon atoms (Carbon 1) residing closest to the potassium atoms have a slight preference for an excess charge. The gray scale corresponds to how the C atoms are displayed in Fig. 2a.

Atom	Mulliken Charge	Gray Scale
Carbon 1	6.12	Dark
Carbon 2	6.08	Light
Carbon 3	6.07	Light
Potassium	18.17	

An alternative way of addressing the charge states is by direct calculation of the charge transfer reaction in the limit of a very large lattice constant. In this limit, a variational bound to the cohesive energy may be determined by accounting for the energies of isolated K and C_{60} charge states and the classical Madelung energy. In analogy to the C_{60} charge state calculations discussed in the previous section, we find that the energy of a potassium atom is given by $E(K) = -0.646 + 2.038(q+0.563)^2$ with q the amount of charge removed from a neutral potassium atom $[0 < q < 1]$. This formula predicts an ionization energy of 4.33 eV in excellent agreement with the experimental value of 4.34 eV. Combining this result with the equation for the C_{60} charge states, noting the constraint $q = Q/6$, and including the Madelung contributions for the experimental dimerization coordinate $(d = 0.28)$ the cohesive energy per cell is bounded by

$$E < -0.618 + 1.86(Q-0.576)^2 + 42.7Q^2/a \qquad [\, 0 < Q < 6, \text{ a large}]$$

with the lattice constant a in Bohr and the energy in eV. Analysis of the above equation indicates that complete transfer of six units of charge is energetically favorable for lattice constants smaller than 25.4 Bohr. It is easy to verify that for this lattice constant there is absolutely no overlap between the neighboring constituents so full charge transfer is nearly certain to occur at the experimental lattice constant (21.56) as well. With an understanding of the charge states, we can now identify the interactions that govern the dimerization coordinate.

While the Madelung energy is primarily responsible for overall stabilization of the crystal, it appears that short range interactions are primarily responsible for determining the dimerization coordinate. To demonstrate this point, it is first useful to discuss the Madelung energy as a function of the dimerization coordinate. From the results of Table 1 it is clear that the effects on the Madelung energy as a function of dimerization coordinate can be roughly understood by a simple model consisting of fully ionized potassium atoms ($q = 1$) and sixty carbon atoms with 6.1 units of charge. In Fig 3, the Madelung energy per unit cell is presented as a function of dimerization coordinate. As shown in the graph, one finds that the Madelung energy is minimized when the dimerization is chosen to be 0.18 rather than the experimentally observed value of 0.28. In other words a simple point charge model that neglects short range interactions between potassium atoms and carbon atoms leads to a _qualitatively incorrect_ dimerization coordinate. The primary reason for this is evident from examination of Fig. 2. A point charge model prefers to minimize the distance between oppositely charged nearest-neighbor ions. However, experimentally the potassium atoms arrange themselves to maximize, rather than minimize, the distance between themselves and the nearest carbon atoms.   To understand the short range repulsive C-K interactions we have performed calculations on the system shown in Fig. 2.   Pictured in Fig. 2 is a collection of forty carbon atoms and four potassium atoms that have been cut out of the bulk. The forty carbon atoms are taken from fragments of the six nearest fullerene molecules and correspond to successive nearest neighbors of the ring of potassium atoms. The forty nearest carbon atoms consist of two pairs of linked six-member carbon rings and four five-member carbon rings. We have deliberately chosen a model that includes

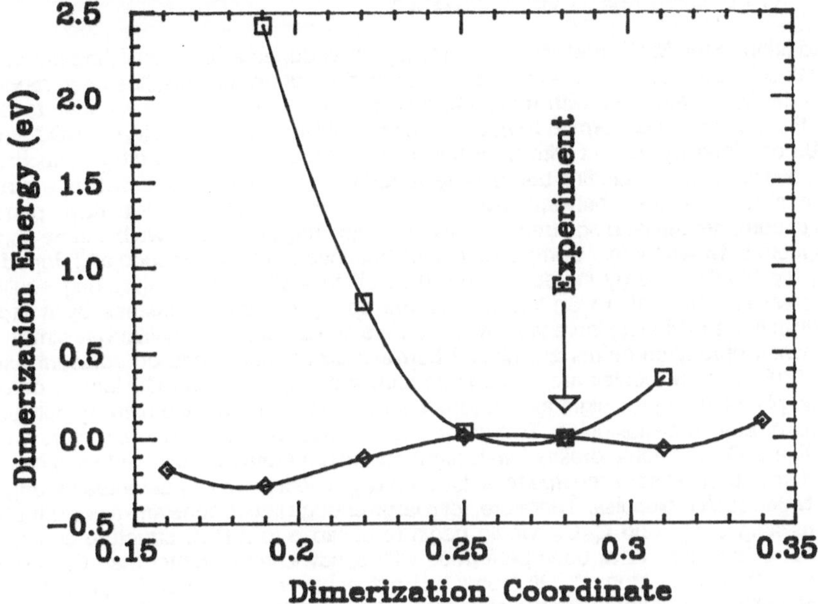

Figure 3. The results of two models for the dimerization energy per fullerene molecule is shown as a function of dimerization coordination. Considering Madelung contributions only (diamonds) leads to a double well with a qualitatively incorrect dimerization of 0.18. Using the right hand model in Fig. 2 leads to a dimerization coordinate of 0.27. For a structureless fullerene molecule, the Madelung energy exhibits a symmetric double well. As shown above, the real fullerene molecule breaks this symmetry.

the nearest carbon atoms and has the same stoichiometric ratio (1:10) as the insulating K_6C_{60} crystal.   At the experimental lattice constant of 21.56 Bohr, we have performed total energy calculations as a function of the dimerization coordinate and present the results in Fig. 3.[12] From this model we find a dimerization coordinate of 0.27 which is in good agreement with the experimental dimerization coordinate of 0.28. Given that the cluster-model is expected to correctly describe the short range interactions between carbon and potassium atoms but negelects a fair amount of the Madelung contribution, the results in Fig. 3 suggest that a model which includes both would tend to shift the dimerization coordinate even closer toward the experimental measurement. That is, the second shallower well obtained from the Madelung contribution will tend to favor a larger dimerization coordinate. While more detailed calculations are required to exactly determine the LDA prediction for the dimerization coordinate, the results of this paragraph suggest that in addition to strongly influencing the dimerization coordinate, the short-range potassium-carbon interactions are expected to play a role in the determination of the lattice constant.

Before continuing, we note that in addition to the BCC K_6C_{60} insulating phase, an FCC K_3C_{60} metallic phase exists as well. The unit cell for this structure is comprised of a fullerene molecule at the origin with the three potassium atoms occupying the (two) tetrahedral and (one) octahedral sites. While both theory and experiment show that the undoped[13] and fully doped[2] structures are insulators, the K_3C_{60} forms a metallic superconducting phase with a measured T_c of 18K.[14] Recently, the band structure, density of states and Fermi surface has been calculated for this prototypical superconductor.[3] The Fermi surface has an interesting shape with a free-electron-like sheet and a second multiply-connected sheet. More details can be found in Ref. [3].

4. Prospects for transition metal doped fullerenes

In addition to the M_6C_{60} and M_3C_{60} (M = Metal), alkali-doped arrays consisting of four metal atoms per unit cell have been reported as well. The structure of this unit cell has been reported to be similar to the M_6C_{60} structure with metal atom vacancies. In this section we wish to discuss calculations that have been performed to address the possibility of a different M_4C_{60} BCC structure that might be obtained by doping with transition metals. In analogy to the dibenzenechromium compounds, one expects that binding between neighboring fullerene molecules would be enhanced by placing a chromium atom between two six-member carbon atoms. We have performed calculations to compare the binding strength of two benzene rings to that of two benzene rings with a chromium atom between them. Comparing results obtained from the undoped and doped rings, we find that the binding energy increases from 0.06 eV to 4.37 eV, the ring-to-ring equilibrium distance decreases from 7.15 to 6.17 Bohr and the spring constant increases by a factor of twenty.[1] Given that the binding increases by two orders of magnitude, a qualitatively similar effect is expected when chromium atoms are placed between six-member rings on adjacent fullerene molecules. If fullerene molecules are oriented as shown in Fig. 2a on a BCC lattice, chromium atoms may be placed along the eight axes equivalent to <111> sandwiched halfway between six-member rings of adjacent molecules. To further address possibilities for this structure we have ascertained that a C_{60} molecule dressed with eight chromium atoms on the <111> faces of the molecule prefers to transfer some charge to an isolated C_{60} molecule which guarantees a long range attraction between such molecules. Therefore, for a large enough lattice constant, a cesium-chloride structure consisting of C_{60} and $C_{60}Cr_8$ would begin to collapse to a BCC structure consisting of $C_{60}Cr_4$. Similar calculations have been performed with sandwiched K, rather than Cr, atoms. In contrast to transition metal dopants, K atoms will not enhance binding when placed between benzene rings. This is in accord with the experimental results that indicate alkali atoms prefer to reside on the faces of the unit cell.

5. References

[1] M. R. Pederson, K. A. Jackson and L. L. Boyer, To appear in Phys. Rev. B.
[2] S. C. Erwin and M. R. Pederson, Phys. Rev. Lett. **67**, 1610, (1991).
[3] S. C. Erwin and W. E. Pickett, To appear in Science.
[4] P. Hohenberg and W. Kohn, Phys. Rev. **136**, B864 (1964); W. Kohn and L. J. Sham, Phys. Rev. **140**, A1133 (1965).
[5] M. R. Pederson and K. A. Jackson, Phys. Rev. B **41**, 7453, (1990).
[6] S. C. Erwin, M. R. Pederson and W. E. Pickett, Phys. Rev. B **41**, 10437 (1990).
[7] J. P. Perdew and Y. Wang, Phys. Rev. B **33**, 8800 (1986); J. P. Perdew, ibid. **33**, 8822 (1986).
[8] M. R. Pederson, K. A. Jackson and W. E. Pickett, Phys. Rev. B **44**, 3891 (1991).
[9] X. J. Kong, C. T. Chan, K. M. Ho, and Y. Y. Ye, Phys. Rev. B **42**, 9357 (1991).
[10] R. L Hettich, R. N. Compton and R. H. Ritchie, Phys. Rev. Lett. **67**, 1242, (1991).
[11] R. E. Smalley et al, Chem. Phys. Lett. **139**, 233 (1987).
[12] A problem that arises when attempting to model an infinite crystal by fragments is the treatment of the "dangling bond" states that occur due to improper termination. While improper in carbon systems, this is often dealt with by saturating carbon atoms with hydrogen atoms. See Ref. [5]. In this work, we have simply minimized the energy as a function of spin. This leads to system with a HOMO/LUMO gap similar to the band gap obtained in Ref. [2].
[13] S. Saito and A. Oshiyama, Phys. Rev. Lett. **66**, 2637 (1991).
[14] A. F. Hebard et al, a Nature 350, 500 (1991).

C_{60} ROTATIONAL DYNAMICS IN THE SOLID STATE

R. D. JOHNSON, C.S. YANNONI, H.C. DORN†, J.R. SALEM and
D.S. BETHUNE

*IBM Research Division,   Almaden Research Center
650 Harry Road, San Jose, CA 95120-6099*

†Dept. of Chemistry, Virginia Polytechnic U., Blacksburg, VA 24061

ABSTRACT.   The rotational dynamics of C_{60} in the solid state are probed using ^{13}C NMR. The unique dependence of the relaxation rate due to chemical shift anisotropy ($1/T_1^{CSA}$) on the square of the magnetic field enables us to separate this rate contribution from those due to other relaxation mechanisms and precisely determine its value. From $1/T_1^{CSA}$ and the measured values of the chemical shift tensor for C_{60}, the reorientational correlation time, τ, for the molecule can be calculated. At 283 K $\tau = 9.1 \pm 2\,ps$, corresponding to a rotational diffusion constant $D = 1.8 \pm 0.4 \times 10^{10}\,s^{-1}$. This value of τ is only 3 times longer than the value expected for completely unhindered rotation, and is shorter than the time found for C_{60} in solution (15.5 ps) at the same temperature. Below 260 K a second phase appears, characterized by significantly slower reorientation. The temperature dependence of τ for both the high temperature (rotator) and low temperature (ratchet) phases can be fit by Arrhenius expressions with activation energies 1.4 and 4.2 kcal/mol and pre-exponential factors $\tau_0 = 0.81\,and\,0.08\,ps$, respectively. The higher degree of hindrance of rotation indicated by the increased activation energy of the low temperature phase is consistent with recent reports of a phase transition involving orientational ordering in solid C_{60}, and with the arrest of the reorientation found by NMR below 100 K.

1. INTRODUCTION

1. INTRODUCTION   NMR provides a means to directly probe the dynamics of molecules in condensed phases. Normally NMR resonances in powder samples are greatly broadened by variation of the magnetic shielding of the nuclei with molecular orientation (an effect called chemical shift anisotropy or CSA). Earlier NMR studies[1,2] of solid C_{60} however, showed an ambient temperature spectrum consisting of a single, relatively narrow line (see **Figure 1**). This demonstrated that the molecules are rapidly reorienting in the solid state. Below 100 K the reorientation stops on the NMR timescale, enabling determination of the chemical shift tensor from the shape of the NMR line.[1] At the same time, the NMR longitudinal relaxation time, T_1, for solid C_{60} was measured as a function of temperature. A minimum in this time found at 233 K was interpreted as the classic T_1 minimum expected to occur for relaxation due to chemical shift anisotropy when the product of the molecular reorientation time τ and the NMR angular frequency is unity. A value $\tau = 2.1\,ns$ at 233 K was inferred.[1] CSA leads to relaxation because as the molecules reorient, the anisotropic shielding causes field fluctuations proportional to the applied magnetic field, B_0, with a concomitant relaxation rate proportional to B_0^2. This

P. Jena et al. (eds.), Physics and Chemistry of Finite Systems: From Clusters to Crystals, Vol. II, 1329–1334.
© 1992 *Kluwer Academic Publishers.*

mechanism is expected to dominate in both liquids and solids for purified samples at sufficiently high fields.

Recently Heiney et al.[3] found from X-ray studies and calorimetry that solid C_{60} exhibits a phase transition near 249 K, from a simple cubic structure at low temperatures to a FCC structure. This transition was examined by Tycko et al.[4] using NMR, who found a sharp break in the T_1 relaxation rate at the transition. These papers suggest that the phase above the transition temperature is characterized by free rotation or rotational diffusion and the phase below the transition is characterized by jump rotational diffusion between symmetry equivalent orientations. Here we report measurements of the rotational correlation time for solid C_{60} both above and below the phase transition.

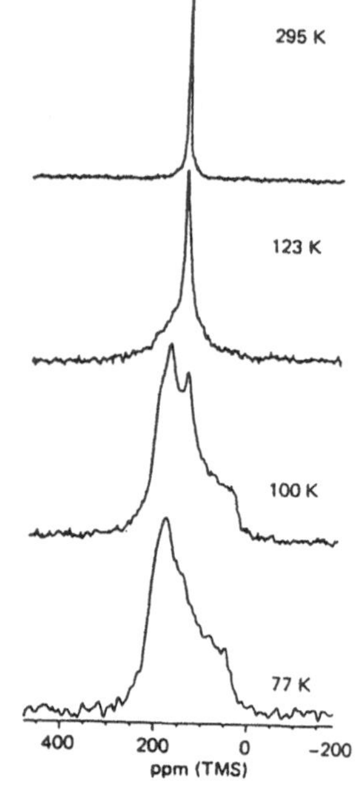

Figure 1. *^{13}C NMR spectra of solid C_{60} obtained at ambient, 123, 100, and 77 K.*

2. RESULTS The T_1 rates for the sample were measured at four magnetic field strengths (11.74, 7.06, 5.87, and 4.71 Tesla) using the inversion-recovery technique.[5] **Figure 2** shows the log of the inversion-recovery signal plotted against delay time t.

For each field the data show no deviation from linearity, indicating that within experimental error the recovery of inverted magnetization can be described by a simple exponential function with a single decay time T_1.

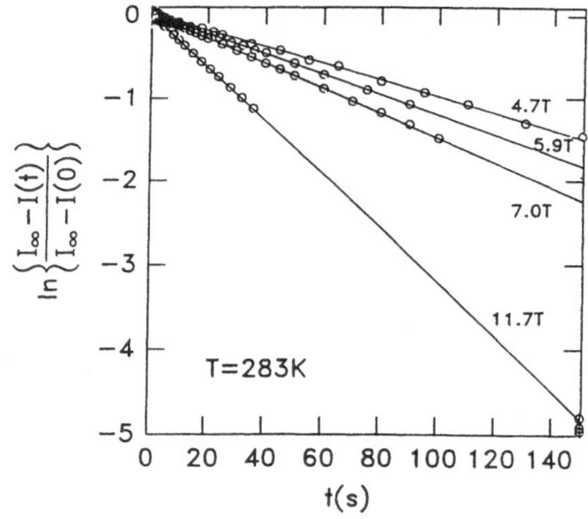

Figure 2. *Signal recovery versus delay time t (s) for the C_{60} sample at magnetic field strengths of 11.74, 7.06, 5.87, and 4.71T, at 283 K.*

In general the T_1 rate will include both CSA and non-CSA contributions: $1/T_1 = [1/T_1^{CSA} + 1/T_1^{NCSA}]$. Assuming that the molecular reorientation can be described as isotropic rotational diffusion, the CSA relaxation rate is[6, 7]

$$\frac{1}{T_1^{CSA}} = \gamma^2 B_0^2 \left(\frac{2}{3} A^2 \frac{\tau_A}{(1 + \omega^2 \tau_A^2)} + \frac{2}{15} S^2 \frac{\tau_S}{(1 + \omega^2 \tau_S^2)} \right) , \qquad (1)$$

where the ^{13}C magnetogyric ratio $\gamma = 67.31 \times 10^6 \, rad \, s^{-1} \, Tesla^{-1}$, and $\omega = \gamma B_0$ is the angular Larmor frequency. The factors A^2 and S^2 arise from the antisymmetric and symmetric parts of the shielding tensor σ, and τ_A and τ_S are the corresponding correlation times for first and second rank tensor functions of the molecular orientational angles. These times are simply related to the rotational diffusion constant D by $1/6D \equiv \tau = \tau_S = \tau_A/3$. For isotropic rotational diffusion, the mean squared angle of rotation in a time Δt about each axis is given by $<\theta_i^2> = 2D\Delta t$. In the limit of short correlation times ($\omega\tau << 1$), the expression on the right side of Eq. 1 shows the signature B_0^2 dependence of the CSA contribution to the T_1 rate noted above.

Figure 3 shows a plot of the T_1 rates found from fits to the data in Figure 2 vs B_0^2. The dependence is linear to an excellent approximation, with a best fit slope of 0.191 ± 0.003 mHz/Tesla^2 and an intercept $1/T_1^{NCSA} = 5.4 \pm 0.2$ mHz. While the CSA component of the relaxation rate dominates at these field strengths, the non-CSA contribution is significant, eg. 37% of the total rate at $B_0 = 7T$.

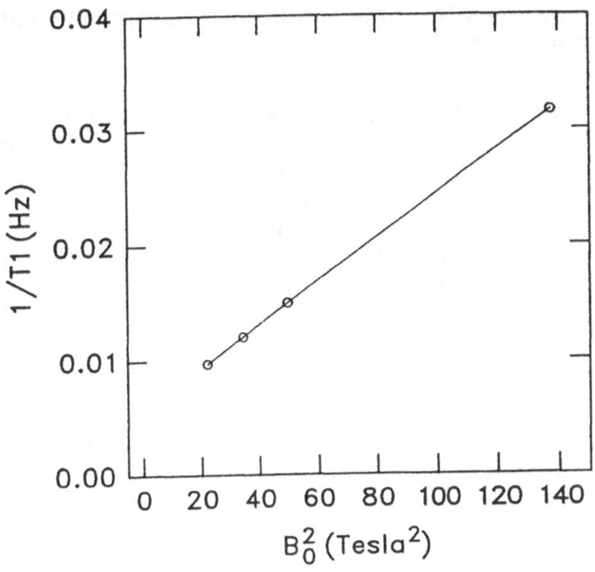

Figure 3. *Relaxation rates* $1/T_1$ *vs.* B_0^2, *for the experiments in Figure 2.*

The correlation time τ can be obtained from the slope of the $1/T_1$ vs B_0^2 plot using Eqn. 1, provided we know S^2 and A^2. S^2 is defined in terms of the three principal values σ_{11}, σ_{22} and σ_{33} of the symmetric part of σ : $S^2 = \Delta\sigma^2(1 + \eta^2/3)$, where $\Delta\sigma \equiv \sigma_{33} - (\sigma_{22} + \sigma_{11})/2$ and $\eta \equiv (\sigma_{22} - \sigma_{11})/(2\Delta\sigma/3)$. These principal values have been determined experimentally to be 180, 25, and 220 ppm for σ_{11}, σ_{22}, and σ_{33}, respectively,[1, 8] so that $S^2 = 3.17 \times 10^{-8}$. The quantity A^2 is defined in terms of the components of the antisymmetric part of σ. This quantity is extremely difficult to measure experimentally,[7] but it can be roughly estimated[9] that $A^2 \sim 1.4 \pm 1.4 \times 10^{-10}$, so that the anisotropic contribution to the relaxation rate should be relatively small ($\sim 6\%$). With these values for S^2 and A^2, from the relaxation rate $1/T_1^{CSA}$ at 283 K we calculate $\tau = 9.1 \pm 2 \, ps$, implying a rotational diffusion constant $D = 1.8 \pm .4 \times 10^{10} \, s^{-1}$.

1332

The phase transition has a dramatic effect on the relaxation behavior of solid C_{60}[4]. Below 260 K the inversion-recovery data on our sample shows a two-component decay. At 254 K for example, the measured decay curve at 7.04 Tesla can be fit using a biexponential with T_1 times of 1.6 s and 50 s. The large difference between these times allows them to be cleanly separated. The fraction of material in the rotator phase at 254 K is 74%, and falls to 25% at 241 K. The relative populations, the exact transition temperature, and the range of phase coexistence are all sample dependent, but for all samples examined the transition temperature indicated by NMR is in good agreement with that measured by DSC.

The temperature dependence of τ was studied for both phases. Although in the ratchet phase the motion cannot be described as isotropic diffusion,[3,4] for rapid jump motion between the 60 symmetry equivalent orientations Equation 1 will give an effective correlation time, τ, which retains its physical significance as the time required for molecular rotation through an angle of about one radian.

Figure 4 shows the dependence of the log of τ on 1/T for both phases. For the high temperature rotator phase (lower set of points), measurements were made in the temperature range 241 to 331 K, both above and below the phase transition (indicated on the plot by the vertical line). Over this temperature range τ varies from 14.9 to 6.8 ps.

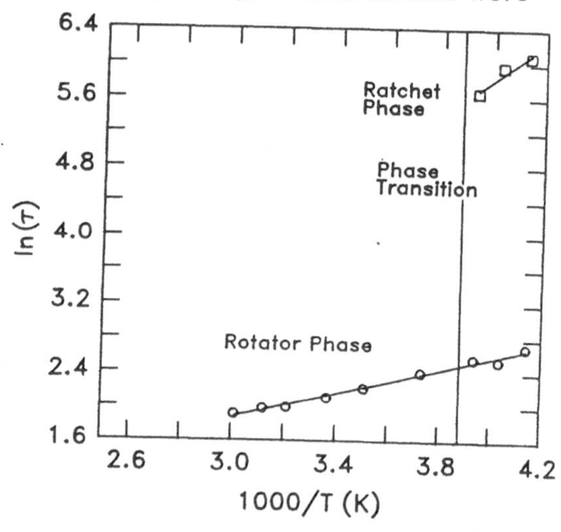

Figure 4. *Arrhenius plot of ln(τ) vs. 1000/T for the ratchet and rotator phases, with slopes giving activation energies of 4.2 and 1.4 kcal/mole-deg, respectively. τ is in ps.*

The NCSA rate $1/T_1^{NCSA}$ is approximately constant from 241-331 K, with a value $\sim 5 \pm 0.5 \times 10^{-3}\,s^{-1}$. The faster ratchet phase CSA relaxation rates correspond to much longer correlation times: τ varies from 0.44 to 0.29 ns as the temperature increases from 241 to 254 K. Despite the phase transition, the rotator phase data fall on a straight line over the entire temperature range. Fitting $\tau(T) = \tau_0 \exp(T_a/T)$ gives $\tau_0 = 8.1 \pm 1 \times 10^{-13}\,s$ and $T_a = 695 \pm 45\,K$ (or $E_a = 1.4 \pm 0.1\,kcal/mol$) for the rotator phase, while a similar fit to the ratchet phase data (upper part of Figure 4) yields $\tau_0 = 8 \times 10^{-14 \pm 1}\,s$ and $T_a = 2100 \pm 600\,K$ (or $E_a = 4.2 \pm 1.2\,kcal/mol$).

3. DISCUSSION AND CONCLUSIONS Above the phase transition, the rotational correlation times for solid C_{60} are remarkably short, and in fact are comparable to those for small molecules in solution. For tetrachlorocyclopropene in toluene-d_8, for example, $\tau = 7.2\,ps$ at room temperature, with an activation energy and pre-exponential factor very similar to those for rotator phase C_{60}.[7] Surprisingly, experiments on C_{60} dissolved in tetrachloroethane revealed $\tau = 15.5\,ps$ at 283K, showing that C_{60} reorients faster in the solid than in solution! How fast would the molecules reorient if they were completely unhindered? The correlation time for freely rotating molecules at temperature T is $\tau_{FR} \equiv (3/5)(I/kT)^{1/2}$.[10] From the measured bond lengths[11, 12] one derives a moment of inertia $I = 1.0 \times 10^{-43}\,kg\,m^2$ for C_{60}. Thus at 283 K, τ_{FR} is 3.1 ps, only three times shorter than the solid state value, and just at the border of the inertial regime.[13]

Below the phase transition the observed reorientational correlation time for the ratchet phase is about 30 times slower than the rotator phase. Extrapolating τ for the ratchet phase to lower temperatures using the fit Arrhenius parameters, one predicts a T_1 minimum at 206 K for a 7.06 T field, and a correlation time of 3 ms at 86 K. These temperatures are in reasonable accord with the previously reported T_1 minimum temperature of 233 K at 7 Tesla and the temperature at which a powder pattern due to CSA fully develops at a field of 1.4 T.[1]

The rotational diffusion constant we find at 260 K ($1.42 \times 10^{10} s^{-1}$) is in striking agreement with the value ($1.4 \times 10^{10} s^{-1}$) recently obtained using coherent quasielastic neutron scattering.[14] Tycko et al.,[2] obtained activation energies of ~ 1 and 5 kcal/mol for the rotator and ratchet phases from the temperature dependence of T_1 measured at a single field, also in reasonable agreement with our results. Somewhat longer correlation times were found in that work than reported here; some of this discrepancy may result from our accounting for non-CSA relaxation. A recent molecular dynamics calculation[15] yielded, at 283 K, $\tau = 0.18$ ps, and a rotational barrier of 300 K, values significantly lower than given by experiments, but in qualitative agreement.

We have shown that by measuring ^{13}C relaxation rates of C_{60} as a function of magnetic field, the contribution due to chemical shift anisotropy can be obtained and used to give precise values of the reorientational correlation time. The fact that the rate of reorientation of C_{60} in the rotator phase approaches the value it would have in the gas phase and exceeds the value we found in solution reflects the extraordinary smoothness of the rotational potential for C_{60} in the solid. The dynamical behavior of C_{60} is in many respects quite parallel to that of the classic solid state rotator adamantane.[16] Adamantane also undergoes a phase transition, from tetragonal to orientationally disordered face-centered cubic at 213 K,[17] and shows a drop in activation energy across the transition. For both of these molecules the difference in activation energies observed for the two phases suggests a picture where (1) the orientationally ordered phase has a rotational potential with fairly deep minima, and (2) above the phase transition, random orientations of neighboring molecules lead to a much smoother potential, with numerous shallow potential minima.

We would like to acknowledge the efforts of Professor L. Taylor in preparing the sample, the technical assistance of Gregory May, and stimulating conversations with Robert Tycko.

REFERENCES

1. C. S. Yannoni, R. D. Johnson, G. Meijer, D. S. Bethune, and J. R. Salem, *J. Phys. Chem.* **95**, (1991) 9-10.
2. R. Tycko, R. C. Haddon, G. Dabbagh, S. H. Glarum, D. C. Douglass, and A. M. Mujsce, *J. Phys. Chem.* **95**, (1991) 518.
3. P.A. Heiney, J.E. Fischer, A.R. McGhie, W.J. Romanow, and A.M. Denenstein, J.P. McCauley Jr. and A.B. Smith III, *Phys. Rev. Lett.* (1991) 2911-14.
4. R. Tycko, G. Dabbagh, R.M. Fleming, R.C. Haddon, and A.V. Makhija and S.M. Zahurak. preprint, submitted to PRL.
5. R. Freeman, **Dynamic Nuclear Magnetic Resonance Spectroscopy**, eds. L.M. Jackson and F.A. Cotton, (New York, 1975).
6. H. W. Spiess, *NMR: Basic Principles and Progress* **15**, eds. P. Diehl, E. Fluck, and R. Kosfeld (Springer Verlag, Berlin, 1978) 55.
7. F.A.L. Anet, D.J. O'Leary, C.G. Wade, and R.D. Johnson, *Chem. Phys. Lett.* **171**, (1990) 401.
8. P.P. Bernier, D.S. Bethune, R.D. Johnson, R.D. Kendrick, G. Meijer, J.R. Salem, and C.S. Yannoni. Experimental NMR Conference, abstracts, St. Louis, Mo., 1991
9. J.C. Facelli, A.M. Orendt, D.M. Grant, and J. Michl, *Chem. Phys. Lett.* **112**, (1984) 147.
10. R.T. Boere and R.G. Kidd, *Annual Reports on NMR Spectroscopy* **13**, ed. G. Webb (Academic Press, London, 1982) 319.
11. C.S. Yannoni, P.P. Bernier, D.S. Bethune, G. Meijer, and J.R. Salem, *J. Am. Chem. Soc.* **113**, (1991) 3190-3192.
12. K. Hedberg, L. Hedberg, D.S. Bethune, C.A. Brown, M. de Vries, H.C. Dorn, and R.D. Johnson, *Science* **(in press)**, (1991) xx-xx.
13. W.B. Moniz, W.A. Steele, and J.A. Dixon, *J. Chem. Phys.* **38**, (1963) 2418-26.
14. D.A. Neumann, J.R.D. Copley, R.L. Cappelletti, W.A. Kamitakahara, R.M. Lindstrom, K.M. Creegan, D.M. Cox, W.J. Romanow, N. Coustel, J.P. McCauley Jr., N.C. Maliszewskyj, J.E. Fischer, and A.B. Smith III, *Phys. Rev. Lett.* **submitted**, (1991).
15. A. Cheng, M. L. Klein J. Chem. Phys., submitted for publication.
16. H. A. Reiley, *Mol. Cryst. Liq. Cryst.* **101**, (1969) 9.
17. C. E. Nordmann and D. L. Schmitkons, *Acta Cryst.* **18**, (1965) 764.

AB-INITIO MOLECULAR DYNAMICS SIMULATION OF C_{60}

H. KAMIYAMA, K. OHNO, Y. MARUYAMA, and Y. KAWAZOE
Institute for Materials Research, Tohoku University
2-1-1 Katahira, Aoba-ku, Sendai, 980 JAPAN

ABSTRACT.   The electronic structures and atomic dynamics in the ground state of C_{60} are simulated with full potentials by means of *ab-initio* molecular dynamics using the mixed bases which consist of both plane waves and 1s atomic orbitals for electron wave functions.

1. Introduction

Recently, the C_{60} microcluster has attracted considerable interest, since it is unbelievably stable and has a peculiar succor-ball shape. It was named "Fullerene" on behalf of Buckminster Fuller who proposed this structure. The crystalline states of this cluster doped with K or Rb exhibit superconductivity at unexpectedly high temperatures. Both theoretical and experimental studies on this microcluster have been performed very rapidly[1]–[5]; theoretical treatments have been performed on the electronic structure under assumed atomic arrangements [3] except for few recent works.[4],[5]

Ab-initio molecular dynamics simulations and related simulated annealing methods have successfully been used combined with the density functional theory and the pseudopotentials to calculate electronic and atomic structures of microclusters, solid surfaces, and point defects.[6] It has been demonstrated that such molecular-dynamics simulations, originally introduced by Car and Parrinello (CP) [6], which do not need any conventional matrix-diagonalization techniques, are efficient for computations with relatively small computer memory size. There have only a few papers been published, which apply the method to C_{60} quite recently, based on the plane wave expansion.[4],[5] Although these calculations are called *first-principle*, they assumed pseudopotentials. If one wants to treat light atoms such as Li, Be, B and C, construction of good pseudopotentials becomes a rather difficult task and one generally needs a large number of plane waves in order to treat the valence electrons correctly, since the screening region due to the 1s core electrons is so small that the strong Coulomb potential around the nucleus is hardly to be removed from the pseudopotential. Hence for these light atoms the full potential calculation in dealing with all core electrons as well as all valence electrons is required to make the analysis better.

In this paper, an ab-initio full-potential calculation of C_{60} are carried out tentatively to analyze the vibrational modes of this microcluster. Here the electronic states are calculated by using the local-density approximation (LDA) in density functional theory. The atomic dynamics are assumed to be governed by classical mechanics with potentials determined by the electronic total energy. In our calculation, wave functions are basically treated

P. Jena et al. (eds.), Physics and Chemistry of Finite Systems: From Clusters to Crystals, Vol. II, 1335–1340.

in real space which is divided into 64x64x64 meshes. For the sake of saving memories, the wave function is expanded by the *mixed bases* which consist of not only plane waves, adopted by the original CP, but also 1s atomic orbitals. This resembles in part to molecular orbital approach. The present calculation treats core electrons as well as valence electrons "exactly" in the sense that we do not employ any pseudopotential.

2. Mixed-basis approach in the Car-Parrinello formalism

In this paper, we propose a new method allowing an *ab-initio* molecular dynamics simulation for the full-potential and all-electron calculation in the *mixed-basis approach* which uses both plane waves (PW) and the 1s core atomic orbitals (AO) all mixed.[7] Historically the mixed-basis was first introduced by Louie, Ho and Cohen [8] in order to treat the spatial locality and asymmetry of the *d*-orbitals in the transition metals, and the present method may also applicable for such purposes. The present work is the first attempt to incorporate this method to the *ab-initio* molecular dynamics simulation of the light atoms like carbon.

For the dynamics of electrons, we use the usual steepest decent (SD) method having the first derivative with respect to t so as to keep the electronic states near the Born-Oppenheimer (BO) surface at each time step. In order to orthogonalize different electronic levels, here we adopt the Gram-Schmidt orthogonalization and the Payne algorithm [9] for the choice of the Lagrange multiplier associated with the orthogonal condition. On the other hand, we treat the atomic motion by means of the classical Newton equation. However such choices should not be regarded as any specific feature of the mixed-basis approach; the reader should note that the mixed-basis approach is basically applicable to other algorithms of *ab-initio* simulation mentioned above.

Because the bases are not mutually diagonal in the mixed-basis approach, we start from the modified equation which guarantees the orthogonality;

$$\mu S \dot{\Psi}_i = -(H - \Psi_i^\dagger H \Psi_i) \Psi_i, \tag{1}$$

$$M_n \ddot{\mathbf{R}}_n = -\nabla_n E, \tag{2}$$

where μ denotes the fictitious "mass" for electron wave functions Ψ_i, while M_n represents the real nucleus mass of the nth atom; $H \ (= \langle k|H|l \rangle)$ and E denote the Hamiltonian of the electrons and the total energy of the system, respectively. Hereafter we use subscripts ij for electronic levels and subscripts kl for basis electron wave functions of atoms.

The distinction of the present equations of the mixed-basis approach, (1) and (2), from those of original PW approach is the presence of the overlap matrix $S \ (= \langle k|S|l \rangle)$ in Eq.(1), which is due to the fact that the bases are not mutually orthogonal. Introducing the lower half triangular matrix U which satisfies $S = UU^\dagger$, and writing $U^\dagger \Psi_i = \Phi_i$ and $H' = U^{-1} H U^{\dagger-1}$, we have

$$\mu \dot{\Phi}_i = -(H' - \Phi_i^\dagger H' \Phi_i) \Phi_i, \tag{3}$$

Once we adopt this representation, the main algorithm of updating the wave function

Φ_i is the same as the earlier PW approach. In evaluating the charge density one needs to trace the wave functions to those in the original nondiagonal frame, $\Psi_i = U^{\dagger-1}\Phi_i$. In our preliminary calculation, exponential damping rates α_n for 1s atomic wave function are regarded as time independent parameters to simplify the calculation. Nevertheless one should note that these parameters can also be treated as time dependent variables, and can be calculated with the following equation.

$$\nu_n \dot{\alpha}_n = -dE/d\alpha_n. \tag{4}$$

The effective one-electron Hamiltonian leads

$$H = T + V, \quad T = -\frac{1}{2}\nabla^2, \tag{5a}$$

$$V(\mathbf{r}) = -\sum_n \frac{Z_n}{|\mathbf{r} - \mathbf{R}_n|} + \int d\mathbf{r}' \frac{\rho(\mathbf{r}')}{|\mathbf{r} - \mathbf{r}'|} + V^{ec}(\mathbf{r}), \tag{5b}$$

where $V^{ec}(\mathbf{r})$ denotes the exchange-correlation potential which is evaluated in real space under LDA.

Once the potential function is determined, we then proceed to evaluate the Hamiltonian matrix element $\langle k|H|l\rangle = \langle k|T|l\rangle + \langle k|V|l\rangle$. Here the basis $|l\rangle$ denotes PW for $l = 1 \sim N_{PW}$ and AO for $l = (N_{PW}+1) \sim (N_{PW}+N_A)$, where N_{PW} and N_A are the number of plane waves and the number of atomic orbitals, respectively. In the calculation of potential matrix element $\langle k|V|l\rangle$ (and also in the calculation of the charge density $\rho(r)$), there are three types of combinations: (1) PW-PW, (2) PW-AO, and (3) AO-AO. Because the 1s (core) atomic orbitals (AO)

$$\varphi_n(\mathbf{r}) = \langle \mathbf{r}|n\rangle = \sqrt{\frac{\alpha_n^3}{\pi}} e^{-\alpha_n|r-\mathbf{R}_n|} \tag{6}$$

are well localized around each nucleus site, the combinations, PW-AO and AO-AO, are rather easily calculated straightforwardly in the real space (in the small area around each atomic nucleus), while for PW-PW the standard calculation in the Fourier space functions. Moreover, for AO-AO one may drop all cross terms between different atomic orbitals at different nucleus sites as a good approximation.

In the present algorithm of the atomic dynamics, the spatial derivatives of the total energy is expressed as

$$\frac{\partial E}{\partial R_n} = \frac{\partial \sum_i \Psi_i^\dagger T \Psi_i}{\partial R_n} + \int \frac{\partial \rho}{\partial R_n} V dr - \frac{\partial}{\partial R_n} \sum_{m \neq n} \frac{Z_n Z_m}{|R_n - R_m|} + Z_n \int \rho \frac{\partial}{\partial R_n} \frac{dr}{|R_n - r|}. \tag{7}$$

The first two terms represent all the terms involving the derivatives of the atomic orbitals (AO) which depend explicitly on R_n via Eq.(6), while the last two terms represent the Hellmann-Feynman force which already exists in the PW approach.

3. Condition and parameters for Numerical Simulation

For C_{60} sixty 1s atomic orbitals, each being located at the atomic sites, and $11 \times 11 \times 11 = 1331$ plane waves are adopted as a basis set to derive the full potential and full charge density without using any pseudopotential. Furthermore, the real space is divided into $64 \times 64 \times 64$ meshes, where 2 meshes in length correspond approximately to 1a.u.=0.529A.

For the computation, we have used NEC SX-2N supercomputer, with which we have achieved 98.5% vector operation ratio.

The basic time step Δt adopted here is 8a.u., which corresponds to 1.2×10^{-15} sec. The exponential damping rates α_n for 1s atomic wave function are fixed to $\alpha = 1/0.3 \text{a.u.}^{-1}$.

N_{pw}	1331
real space mesh size	$64^3 \{=(64/1.9)^3 \text{a.u.}^3\}$
μ	100 a.u.
α	$1/0.3 \text{ a.u.}^{-1}$
dt	8 a.u. ($= 1.2 \times 10^{-15} \text{sec}^{-1}$)
initial C-C length	2.7 a.u. ($= 1.43$ Å)
convergence of initial electron system	Payne algorithm 90 steps

Table I Parameters used for present computation. N_{pw} denotes number of plane waves, μ denotes fictitious mass for electron wave functions Ψ_i, while α is exponential damping rate for 1s atomic wave function.

Initial C-C length is assumed to be 2.7 a.u. between all atoms. Parameters used for computation in this paper are summarized in Table I.

4. Results and Discussions

First, we fix the positions of atoms until the electronic states well converge to the BO surface. This process typically needs 70 steps; each step takes CPU time of approximately 100 seconds by SX-2N. The resultant time sweep of the electronic energy levels and the total energy of the electron system derived by SD algorithm in Eq.(1) are shown in Fig.1. From this figure, the electronic states converge comparably fast within 10 steps. Figure 2 shows the resulting charge densities after 70 iteration where the electronic states well converged to the energy minimum. As expected, there are two kinds of bonding between carbon atoms corresponding to single and double bonds.

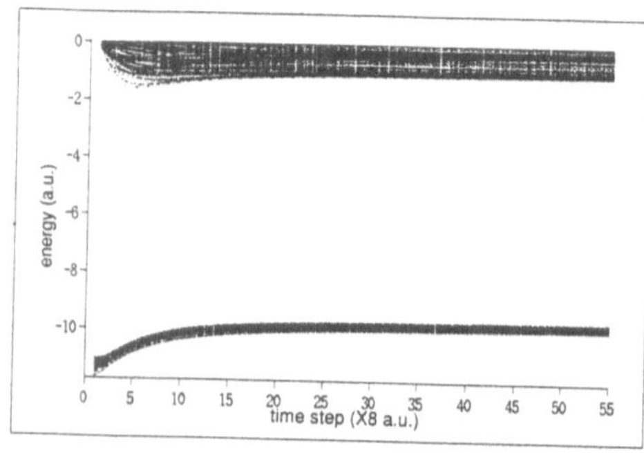
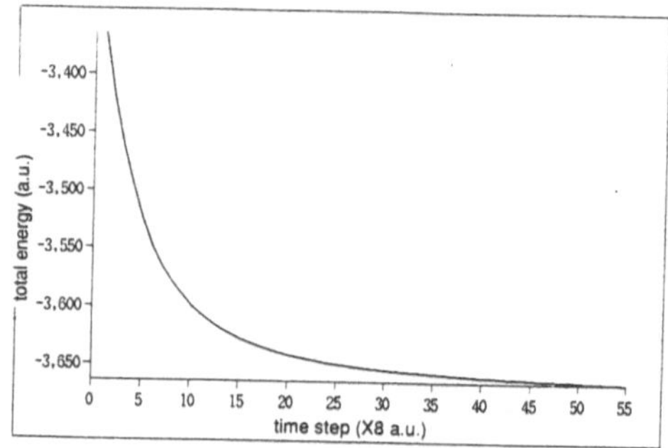

Fig.1 (a) The time sweep of the electronic energy levels and (b) the total energy of the electron system for fixed atoms, derived by the steepest descent algorithm.

In the hexagonal ring, there are equal number of single and double bonds alternating to each other. From Fig.2, the bond alternation can clearly be seen from the surface which cuts the bond centers of one hexagonal ring vertically.

Two solid circles in the figure indicate the equal electron density and these circles have obviously have different radius indicating that charge density at the double bond site is higher than that at the single bond site. Starting from these converged electronic states, all the constraints on the atomic sites are released, and the molecular dynamics simulation is performed according to Eqs.(1) and (2). In this algorighm, one basic loop for iteration takes approximately 130 seconds with SX-2N.

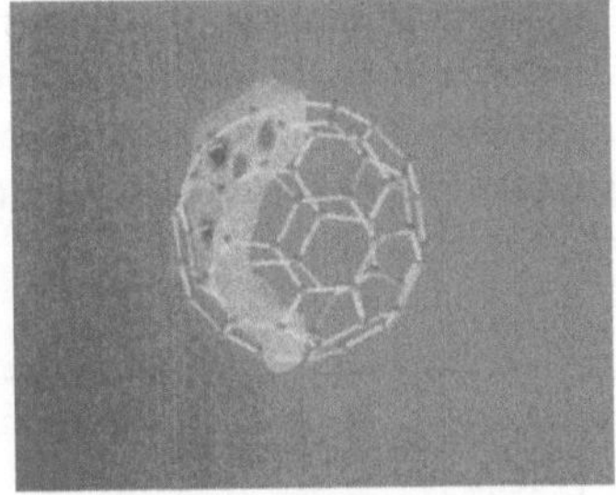

Fig.2 Resulting charge densities after the electron system converged.

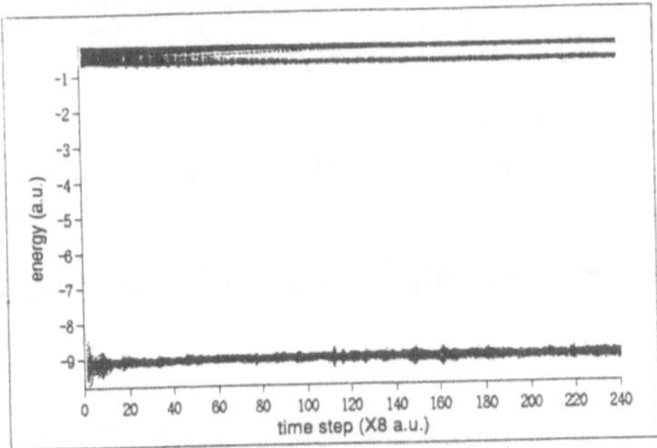
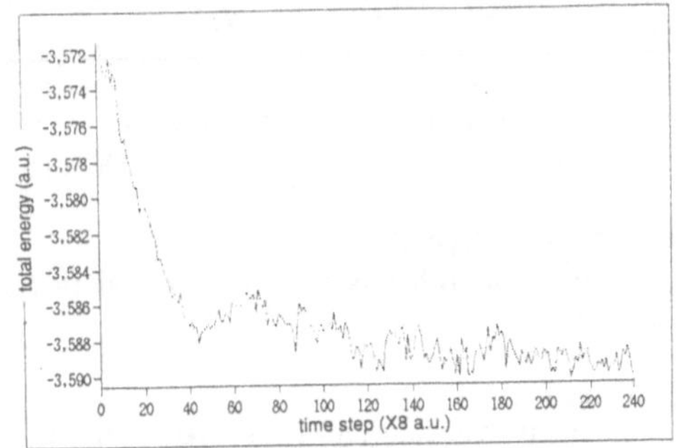

Fig.3 (a) The time sweep of electronic states and (b) electron total energy after the constraints on the atomic sites are released.

Variation of electronic energy levels and the total energy of electron system are shown in Fig.3. Since we adopted SD algorithm in the calculation of the electronic state, fictitious kinetic energy decreases step by step and variation in total energy decreases finally converging to the energy minimum. This specific tendency is seen in Fig.3(a). Total electron energy in Fig.3(a) seems to decrease toward the energy minimum, when vibration amplitude becomes small.

This affects also the atomic vibration motion. Figure 4 shows the motion of specific atoms which have maximum z-coordinate in C_{60} cluster. At the initial stage, all the atoms are fixed and the numerical simulation starts with zero velocity. After several time steps of the calculation, however, atoms begin to move and their kinetic energy increases, and the molecule goes into a finite temperature. This may cause several vibration modes of C_{60} to be excited. In Fig.4, the atoms move from their initial position and start vibrational motion: several vibration modes seem to be mixed all together. From Fig.4(a), the vibrational period along y-axis is approxmately 30 steps while period of vibration along z-axis is approxmately 70 steps. This indicates that the vibrational frequency parallel to the surface of C_{60} (mode I) is very high and its characteristic energy is given by 1340K, while the frequency of the vibration normal to the surface (mode II) is rather low compared to mode I corresponding to a temperature of 600K. Rosseinsky *et al* [10),11)] reported that the intramolecular phonon modes possess characteristic energies of 1000-2000K, which are extraordinarily high compared to intermolecular phonon modes. This value of the energy is consistent with the present result of mode I in our calculation with vibrational energy of 1340K.

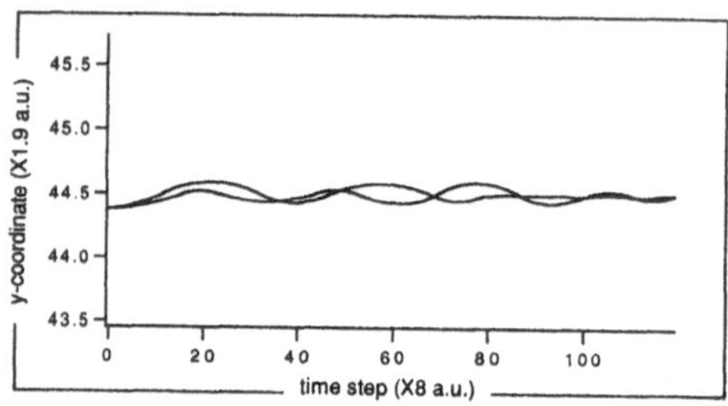
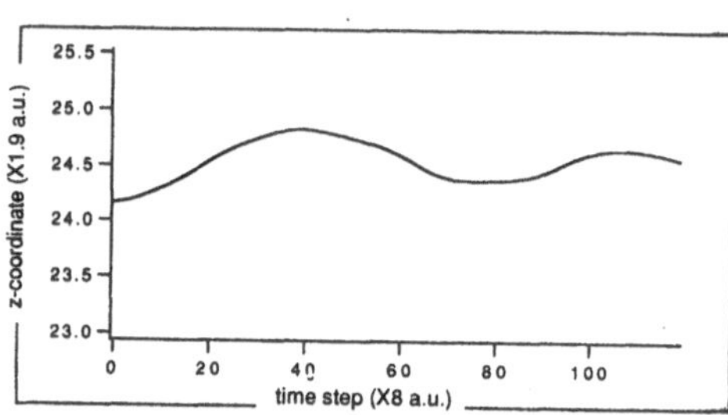

Fig.4 Variation of atomic coordinates for several specific atoms. At this atomic site, (a) y-axis is approxmately parallel to C_{60} surface, while (b) z-axis is vertical to the surface.

Acknowledgement

This research has been supported in part by "Partnership Program' with IBM Japan Co., Ltd.

References

1) J. H. Weaver, J. L. Martins, T. Komeda, Y. Chen, T. R. Ohno, G. H. Kroll, N. Troullier, R. E. Haufler, and R. E. Smalley, Phys. Rev. Lett. **66** (1991) 1741.

2) Y. Kajii, T. Nakagawa, S. Suzuki, Y. Achiba, K. Obi, and K. Shibuya, Chem. Phys. Lett., in press.

3) S. Saito and A. Oshiyama, Phys. Rev. Lett. **66** (1991) 2637.

4) Q. M. Zhang, J. Y. Yi, and J. Bernhole, Phys. Rev. Lett. **66** (1991) 2633.

5) B. P. Feuston, W. Andreoni, M. Parrinello, and E. Clementi, Phys. Rev. **B44** (1991) 4056.

6) R. Car and M. Parrinello, Phys. Rev. Lett. **55** (1985) 2471.

7) K. Kawazoe, H. Kamiyama, Y. Maruyama, and K. Ohno, to be published.

8) S. G. Louie, K. M. Ho, and M. L. Cohen, Phys. Rev **B19** (1979) 1774.

9) M. C. Payne, M. Needels, and J. D. Joannopoulos, Phys. Rev. **B37** (1988) 8138.

10) M. J. Rosseinsky, A. P. Ramirez, S. H. Glarum, D. W. Murphy, R. C. Haddon, A. F. Hebard, T. T. M. Palstra, A. R. Kortan, S. M. Zahurak, and A. V. Makhija, Phys. Rev. Lett. **66** (1991) 2830.

11) R. C. Haddon, A. F. Hebard, M. J. Rosseinsky, D. W. Murphy, S. J. Duclos, K. B. Lyons, B. Miller, J. M. Rosamilia, R. M. Fleming, A. R. Kortan, S. H. Glarum, A. V. Makhija, A. J. Muller, R. H. Eick, S. M. Zahurak, R. Tycko, G. Dabbagh, and F. A. Thiel, Nature (London) **350** (1991) 321.

RADIAL DISTRIBUTION FUNCTION STUDIES OF THE STRUCTURE OF C_{60}

F. LI and J. S. LANNIN
Department of Physics
Penn State University
University Park, PA 16802

ABSTRACT.   Neutron diffraction measurements extending to high Q values of 45^{-1} have provided direct information on the structure of C_{60} at both 300K and 10K.   Radial distribution functions (rdf) indicate relatively undistorted fullerenes whose bond distances are found to be independent of temperature.   Improved double and single bond distances of 1.39A and 1.46A are obtained from the rdfs.   The rdfs exhibit subtantial narrowing of longer range correlations for r >8A, due to hindered rotations at low temperature.

A high degree of interest exists in the structure and physical properties of the C_{60} cluster based solids known as Buckminster–fullerene (BF)[1–4].   This includes both pure C_{60} as well as systems with covalent or ionic bonding to C_{60}, from either larger chemical groups or individual alkali atoms, respectively.   With the addition of alkali dopants or covalently bonded molecules, the rotational degrees of freedom of C_{60} clusters tend to be frozen out at 300K, while the structure of individual C_{60} clusters is modified to an unknown degree. Generally, it has been assumed that small distortions of the C_{60} cluster occur with chemical bonding.

While diffraction measurements have accurately determined crystal symmetries, the precise nature of possible disorder within individual C_{60} clusters or intercluster disorder has not been directly studied. Specifically, Rietveld analysis of powder diffraction data have modeled the C_{60} structure by assuming a truncated icoshedron with two or more parameters corresponding to single and double bond distances.   Different sets of bond distances have been estimated to date[5–9] that vary, as do a number of quantum calculations[10–14].

With decreasing temperature x–ray diffraction measurements on pure C_{60} have shown that a transition from an FCC to SC phase occurs at ~250K[15].   NMR has also shown that C_{60} rotation disorder does not occur below ~100K[16].   Additional NMR studies, using the Carr–Purcell splitting have suggested bond distances of 1.40A ± 0.15A and 1.45A ± 0.015A at 77K[16].   These values differ, for example, from an osymated chemical group bonded to C_{60}, where distances of 1.38A and 1.42A were obtained for single crystals[7].

P. Jena et al. (eds.), Physics and Chemistry of Finite Systems: From Clusters to Crystals, Vol. II, 1341–1346.
© 1992 Kluwer Academic Publishers.

In the present study, an alternative means of studying C_{60} both at 300K and 10K is presented. Radial distribution function (rdf) analysis of the Fourier transform of a wide range of Q space data directly yields weighted averages of both nearest neighbor distances as well as longer intracluster distances. The results provide accurate estimates of the averaged first, second and third neighbor bond distances which may be compared to theoretical calculations and to other experiments. In addition, the higher neighbor rdf peaks yield direct information on intracluster distances, confirming that a relatively undistorted truncated icosohedron is consistent with experimental data. The low temperature rdf spectra also clearly indicate the freezing of rotational disorder.

The radial distribution functions in this study were obtained from pulsed neutron diffraction measurements with quite large maximum wavevectors, Q_{max}, of $45A^{-1}$. The intrafullerene features of the rdf directly reveal the bonding and distortion of C_{60} molecules, while the atomic correlations of between fullerenes exhibit the orientational order at low temperature.

A 580mg sample of C_{60} was prepared by the standard chromatographic separation procedure[17]. Fourier transform infrared analysis confirmed the high purity of the C_{60} phase, while NMR and neutron absorption prompt gamma analysis indicated ~1at%H. Neutron diffraction measurements were performed on the SEPD instrument of the Intense Pulsed Neutron Source (IPNS) of the Argonne National Laboratory. For the low temperature measurement, the sample was sealed in a vanadium container with He gas and cooled by a Displex closed—cycle refrigerator. Data obtained for wavevectors, Q extending to $45A^{-1}$ was employed after calibrations and background corrections to obtain rdf's.

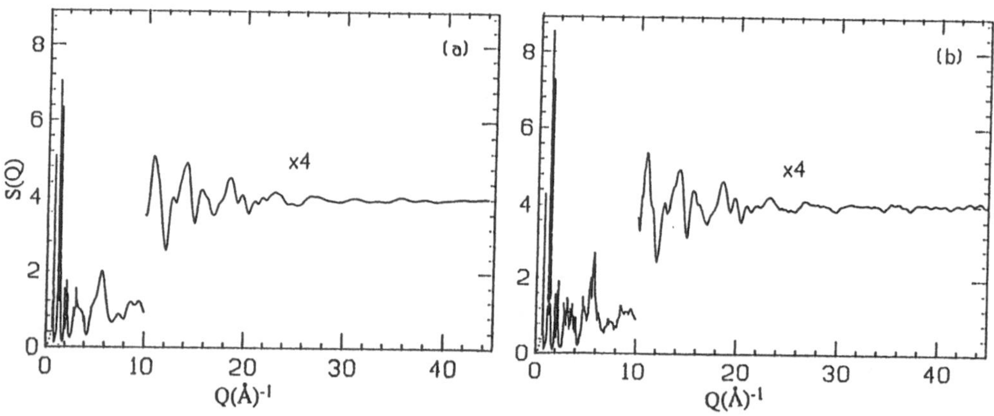

Fig. 1.   Comparison of the structure factor of C_{60} at 300K (a)
          and 10K (b).

In Figure 1, the structure factors of solid C_{60} are shown for Q between $0.7A^{-1}$ to $45A^{-1}$ for temperatures of 300K and 10K. A quadratic extrapolation to Q=0 is indicated by a dotted curve. The sharp features in the two S(Q) curves extending to Q≈4–8A^{-1} are similar to those obtained in the other diffraction experiments. These peaks are associated with the crystalline FCC lattice at 300K and SC lattice at 10K[5,15]. The present Bragg peaks are broader compared to the other diffraction results because of the shorter coherence lengths in this polycrystalline C_{60} sample. The broad, damped features at higher Q reflect short range interference effects from atoms within individual C_{60} lusters or on neighboring clusters.

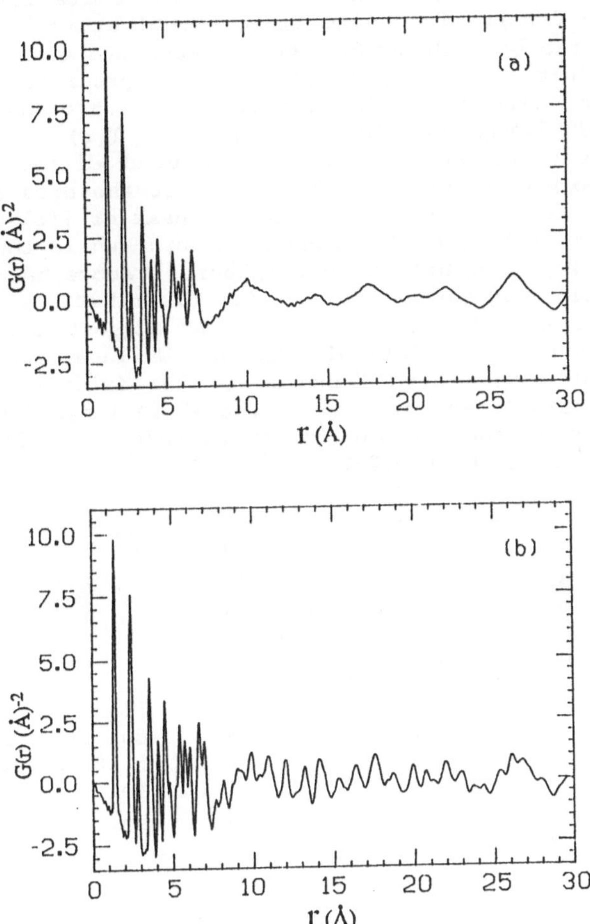

Fig. 2. Comparison of pair correlation functions of C_{60} at 300K (a) and 10K (b)

Fourier transformation of F(Q)=Q[S(Q)-1] yields the pair distribution function, G(r). A Lorch modification function in the form of (sinx)/x, where x=πQ/Q_{max}, has been used in the present analysis to reduce the noise caused by the Q_{max} termination effect. The resultant

G(r) functions for 300K and 10K are shown in Fig. 2(a) and (b), respectively. The G(r) spectra are qualitatively separated into two regions: a) intense peaks below ~7.5A which contain both intrafullerene and weak interfullerene radial correlations, and b) features above ~7.5A which are associated with only interfullerene correlations.

For a discussion of the peaks in G(r) below 7.5A, it is useful to employ the modified radial distribution function $T(r) = G(r) + 4\pi r\rho = J(r)/r$, where J(r) is the radial distribution function (rdf). Figure 3 shows the T(r) functions of C_{60} at 300K and 10K from 0—8A. The constraint that T(r) oscillates about zero intensity below its first peak yields the microscopic density for solid C_{60}, $\rho = 0.081$ atoms/A^3. The thirteen features presented by each curve, twelve peaks and a shoulder at ~5.2A, are consistent with the distance distribution predicted by a truncated icosohedral, BF structure. The truncated icosohedron consists of 12 pentagons and 20 hexagons. The first peak of T(r) is associated with three first neighbor atoms, two at the bond distance shared by a pentagon and hexagon (single bond) and one at the bond distance shared by two hexagons (double bond). The second peak of T(r) consists of two pentagonal and four hexagonal second neighbors, while the third peak of T(r) contains two hexagonal third neighbors. Above 3A, interfullerene correlations will contribute to the T(r) distributions. For the 300K result, the form of G(r) at larger r values suggest that the influence of these correlations is relatively smooth. While at 10K interfullerene correlations may cause extra features in the interval between ~3–7A. Despite the interfullerene overlap in the ~3–7A range, the positions of the twelve peaks at 10K are very close to those at 300K, with a small averaged deviation of about 0.22%.

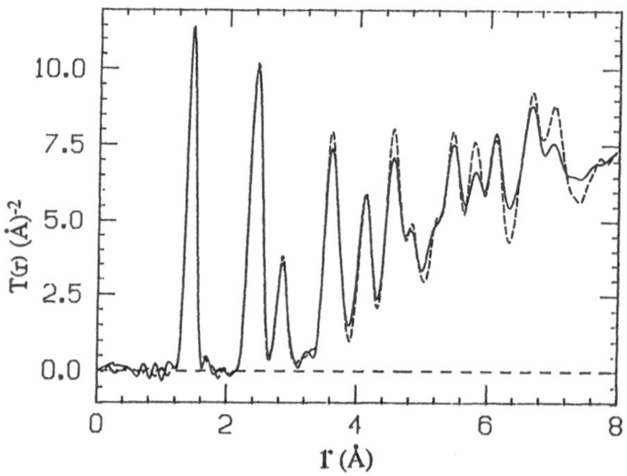

Fig. 3. Comparison of modified rdf of C_{60} for r <8A at 300K (solid) and 10K (dashed)

The first three peaks in T(r) provide important information on double and single bond distances in individual C_{60} molecules. These distance values are evaluated by comparing the weighted averages of distances from the rdf, J(r), and from a BF structure. For the nth peak of the rdf, the weighted average of atomic distances are obtained by first moments of the J(r) peaks. As indicated by Figure 3, the 300K and

10K results have the same values of the first three distances, 1.440A, 2.432A and 2.850A. For the BF structure with r_d as the double bond length and r_s as the single bond length, the weighted averages of atomic distances are estimated from the three equations: $<r_1>=(r_d+2r_s)/3$, $<r_2>=[2\sin 54° \cdot r_s + 2(r_d^2 + r_s^2 + r_d r_s)^{\frac{1}{2}}]/3$, $<r_3>=r_d+r_s$. This comparison suggests that the C_{60} molecule is most likely to have double bonds of 1.39A and single bonds of 1.46A. It is worth noticing that there is no noticeable change of bond lengths between 300K and 10K.

The estimated bond lengths of 1.39A and 1.46A are within the range of $1.40\pm0.015A$ and $1.45\pm0.015A$ obtained by modeling studies of low temperature NMR measurements[16]. The present bond distances also agree with the averaged values, 1.391A and 1.455A, obtained in a recent diffraction study of C_{60} at low temperature[9]. However, that study indicates that the bond lengths of ordered C_{60} molecules fluctuate depending upon the orientational positions of C_{60}. If such variations do occur, the present study indicates from the rdf peak widths that the interatomic distances at 10K fluctuate less than at 300K. Other structural studies on C_{60} compounds or derivatives suggest various C–C bond values, e.g., 1.44A in K_6C_{60}[5], 1.32A and 1.49A in K_3C_{60}[9], 1.388A and 1.432A in the Os derivative[7], and 1.388A and 1.445A in the Pt derivative[8]. All these results yield the first or higher neighbor distances significantly deviated from the current rdf values. With values of 1.39A and 1.46 bond lengths, the additional peaks in T(r) in Fig. 3, for r >3A, are all found to agree within 0.5% of the values for a truncated icosohedron. This directly confirms the BF structure.

As shown in Figure 2, there are significant differences in the high r region between the 300K and 10K results. The G(r) function of 300K shows broad features centered at 10.1A, 14.3A and 17.6A, which are very near to the ratio of $1:\sqrt{2}:\sqrt{3}$, indicating that they are associated with the average interfullerene distances in an FCC like environment. The broadness of these peaks is a consequence of the rotational disorder in C_{60}. At low temperature, the orientational order of C_{60} molecules results in sharp interfullerene correlation peaks, as shown in Fig. 2(b).

At low temperature, the broad contours of the 300K rdf have transformed into many sharp correlation peaks, as shown in Fig. 2(b). The presence of these sharp features indicate that interatomic order exists between clusters at this temperature. This confirms the crystallographic results that neighbor C_{60} units are ordered by a rotation angle. Modeling of the interfullerene features to reveal the rotation angle and possible disorder is in progress.

In summary, the radial distribution function study on C_{60} solid has observed the molecular structure and interfullerene correlation of this large cluster at above and well below the transition temperature. It has been shown that the truncated icosohedron, BF has very small distortions and consistent molecular structure at both 300K and 10K. The estimated values of double and single bond distances of C_{60} are in agreement with some other low temperature measurements, but differs from the results obtained in C_{60} compounds and derivatives. The rdf at 10K, with sharp interfullerene correlation peaks, directly confirms the orientational ordering of C_{60} at low temperature associated with hindered rotations.

ACKNOWLEDGEMENTS. This work was supported by USDOE grant DE-FG02-84ER45095 and NSF grant DMR-8922305-1 and USDOE Basic Energy Sciences support to the IPNS program at Argonne National Laboratory. We wish to thank J. R. D. Copley for a C_{60} coordinates program, J. Berholc for communicating an unpublished rdf, and P. Sokol, R. Q. Yu, and D. Montague for useful discussions.

BIBLIOGRAPHY.
[1] H.W. Kroto, J.R. Heath, S.C. O'Brien, R.F. Curl and R.E. Smalley, Nature, 318, 162 (1985).
[2] E.A. Rohlfing, D.M. Cox, and A. Kaldor, J. Chem. Phys. 81, 3322 (1984)
[3] W. Kratschmer, K. Fostiropoulos and D.R. Huffman, Chem. Phys. Lett. 170, 167 (1990).
[4] Materials Research Society Proc. 206: Clusters and Cluster Assembled Materials, ed. R. S. Averbach, J. Bernholc and D. L. Nelson, (Materials Research Society, Pittsburgh, 1991).
[5] O. Zhou, et al., Nature 351, 462 (1991).
[6] P. W. Stephens, L. Mihaly, P. L. Lee, R. L. Whetten, S. Huang, R. Kaner, F. Deiderich, and K. Holczer, Nature 351, 632 (1991).
[7] J.M. Hawkins, A. Meyer, T.A. Lewis, S. Loren, and F.J. Hollander, Science, 252, 312, (1991).
[8] P. J. Fagan, J. C. Calabrese, and B. Malone, Science 252, 1160 (1991).
[9] W. I. F. David, et al., Nature 353, 147 (1991).
[10] G.E. Scuseria, Chem. Phys. Lett. 176, 423 (1991).
[11] Q.-M. Zhang, J.-Y. Yi and J. Bernholc, Phys. Rev. Lett. 66, 2633 (1991).
[12] B.P. Feuston, W. Andreoni, M. Parrinello and E. Clementi (to be published).
[13] R. L. Disch and J. M. Schulman, Chem. Phys. Lett. 125, 465 (1986).
[14] H. P. Lüthi and J. Almlöf, Chem. Phys. Lett. 135, 357 (1987).
[15] P. A. Heiney, J. E. Fisher, A. R. McGhie, W. J. Romanow, A. M. Denenstein, J. P. McCauley, A. B. Smith and D. E. Cox, Phys. Rev. Lett. 66, 2911, (1991).
[16] C.S. Yannoni, R.D. Johnson, G. Meijer, D.S. Bethune and J.R. Salem, J. Phys. Chem. 95, 9 (1991).
[17] R. L. Cappelletti, J. R. D. Copley, W. A. Kamitakahara, F. Li, J. S. Lannin and D. Ramage, Phys. Rev. Lett. 66, 3261 (1991).

RELATIVE ENERGETICS OF C$_{44}$ FULLERENE ISOMERS[†]

M. LYONS,[‡] B.I. DUNLAP, D.W. BRENNER, D.H. ROBERTSON
R.C. MOWREY, J.W. MINTMIRE and C.T. WHITE
Theoretical Chemistry Section, Code 6179
Naval Research Laboratory
Washington DC 20375-5000

ABSTRACT. We have carried out an exhaustive search of the energetics of all 87 isomers of the fullerene C$_{44}$ using an empirical potential function. We find no single structural isomer with exceptional stability, suggesting that the enhanced abundance of this cluster observed in some studies may be due to a mixture of different molecules. We find in general that smaller, more spherical clusters and clusters whose pentagons are more isolated tend to be more stable, although the energy differences are not sufficiently large with our classical potential function to identify a particular criterion that distinguishes the stability of various similar isomers.

1. Introduction

From their first detection in molecular beams[1] through their isolation in macroscopic quantities[2], carbon fullerene clusters have generated interest both for their unusual properties and inherent beauty. The C$_{60}$ cluster 'Buckminsterfullerene' has been especially savored, and its exceptional stability has been attributed to its perfectly spherical shape and high symmetry. This property allows both a uniform distribution of stress around the molecule and resonance stability of the π electrons. It is also the smallest fullerene to allow complete isolation of the 12 pentagons, another criterion that has been proposed to contribute to its exceptional stability.[3]

There are other fullerene clusters, notably C$_{44}$, C$_{50}$, and C$_{70}$, whose exceptional abundance suggests that they may also have properties that make them especially stable.[1,2,4,5] The large number of isomers possible for each of these clusters (which grows exponentially with the number of atoms in the fullerene) makes it very difficult to study every isomer to determine which may be most responsible for their enhanced abundance. Furthermore, the relative contribution of the various criteria for stability[3] (namely size, shape, isolation of pentagons, and resonance stability of the π electrons) has not yet been fully quantified so that choosing among the various criteria for a given isomer often requires a full calculation of relative stabilities. As larger and more diverse fullerene structures are explored, the relative contribution of each of the stability criterion to structural energetics will become crucial in identifying candidate isomeric structures.

---

† Supported by the U.S. Office of Naval Research.
‡ Present address: University of Virginia, Charlottesville, VA.

P. Jena et al. (eds.), Physics and Chemistry of Finite Systems: From Clusters to Crystals, Vol. II, 1347–1351.
© 1992 *Kluwer Academic Publishers.*

To begin to explore the stability of various fullerene isomers and to evaluate the relative contributions of three stability criteria (size, shape and isolation of pentagons), we have carried out an exhaustive study of the energetics of all of the isomers for the fullerene cluster C_{44} using an empirical many-body potential. This cluster was chosen both because it has a tractable number (87) of possible isomers[3], and because it has been observed in enhanced abundances when produced by the laser vaporization of graphite.[5] Although we find that binding energies scale approximately with each criterion, no one isomer is found that is exceptionally stable. This suggests that either the exceptional stability of a single isomer is a result of aromatic contributions to the binding energy, or that in contrast to C_{50}, C_{60}, and C_{70} where one isomer is thought to predominate, the enhanced signal is composed of a mixture of isomers. If neither is the case, then a kinetic mechanism may be indicated to explain the lack of a mixture.

II. Details of the Calculation

We found that all but four of the 87 C_{44} fullerenes could be generated starting from a single structure by applying the Stone-Wales mechanism for generating isomers.[6] In this mechanism pentagons are moved around the cluster by rotating bonds that join separated pentagons by 90°. The remaining structures were generated by doing two successive rotations, a rotation of a bond connecting hexagons which yields a seven-membered ring followed by a rotation to reform five- and six-membered rings. Specific details of the rotation mechanisms and structures for all of the 87 clusters will be given in a more complete report of this work. The various isomers were characterized according to their radius (calculated as the average distance from the center of mass of the optimized cluster geometry to each of the atoms), their lack of roundness (given as the mean squared deviation from their radius), the number of pentagons sharing common sides, and the number of sites where three pentagons meet. The latter two quantities have been suggested by Liu *et al.* as quantitative measures of stability.[7]

Once generated via the Stone-Wales mechanism, structures and binding energies were obtained for each of the clusters by minimizing the energy as given by an empirical potential.[8] Full minimization with respect to all coordinates was performed with no constraints on the symmetry of the clusters. We have used this empirical potential in previous work both to obtain minimum energy structures for subsequent electronic structure calculations[9-11] and to model reactive dynamics of fullerene clusters.[12] Comparisons between the structure and energetics given by the empirical potential and a first principles method for C_{60}, and one isomer each of C_{44} and C_{45} are given in references 9 and 11. For the structures we have examined previously, the empirical potential yields relative stabilities that agree with first principles local-density functional calculations. As mentioned above, however, electronic effects are not included in the empirical potential so special stability due to resonance is not accounted for in the present work.

III. Results

Shown in Fig. 1 is the energy per atom versus the average radius for each of the 87 clusters. We find that no one cluster appears exceptionally stable, and that, in general, the energy per atom increases (i.e. the cluster becomes less stable) as the average radius increases. This trend is most apparent for radii greater than ≈ 3.07 Å with considerable scatter throughout the data. This suggests that the larger clusters are noticably less stable, but that in general size is not necessarily a strong indication of energetic stability for clusters with similar sizes.

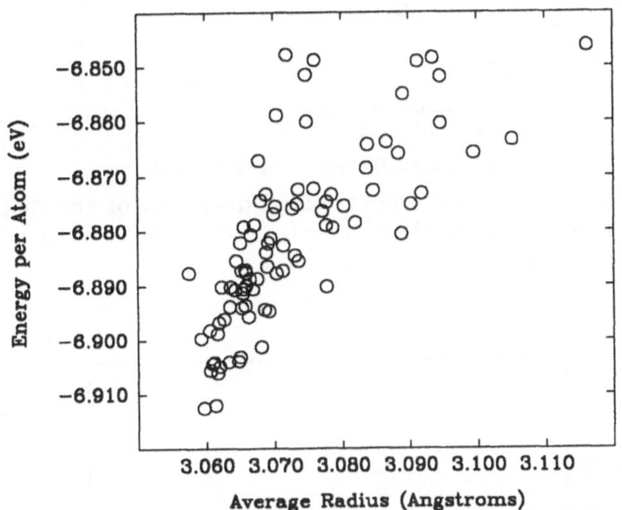

Figure 1. Energy per atom versus average radius of all 87 isomers of C_{44}.

Shown in Fig. 2 is the energy per atom versus the standard deviation of the distance of each atom from the center of the cluster for each of the isomers. We find that in general the smaller the standard deviation the more stable the cluster, but that there is again considerable scatter in the data. This suggests that highly distorted clusters are less stable, but again this is not necessarily a strong indication of energetic stability among similar structures.

The third criterion for stability we have examined is the isolation of pentagons. Shown in Fig. 3 is the energy per atom versus the number of bonds shared between pentagons for the various isomers. In this case several isomers have the same number of shared pentagons, and so we have averaged the energy for these sets of isomers. This removes some scatter from the data as compared to Figs. 1 and 2. The isolation of pentagons criterion for stability is apparent, where the sets of isomers having fewer adjacent pentagons are more stable. The three stability criteria tested are not independent and correlations among them will be discussed in detail in a more complete report of this work.

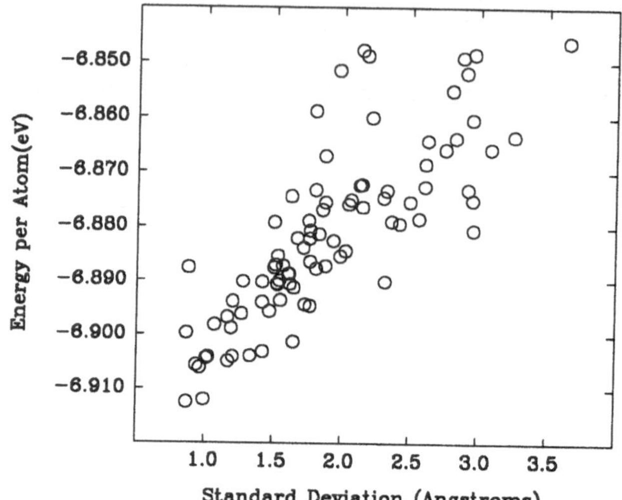

Figure 2. Energy per atom versus the standard deviation of the distance of each atom from the center of the cluster for each isomer of C_{44}.

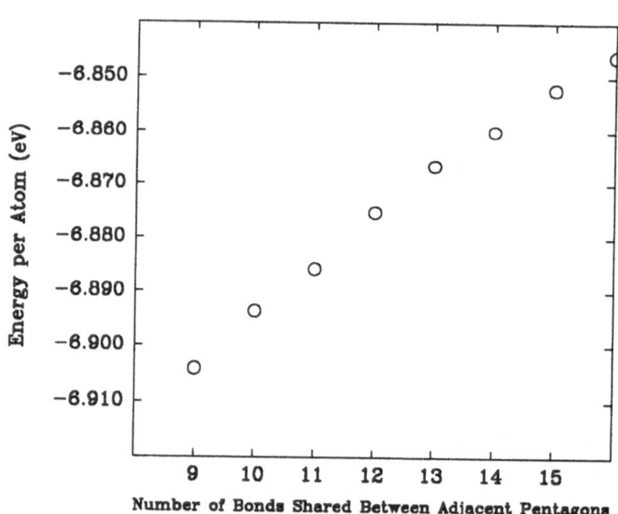

Figure 3. Average energy per atom for clusters with the same number of adjacent pentagons versus the number of bonds shared between adjacent pentagons for isomers of C_{44}.

The final criterion for stability studied was the number of atoms in each cluster shared among three pentagons. No trends relating this criterion to relative stability was observed in the data, suggesting that this is not a strong criterion for assessing stability.

IV. Summary

We have examined the total energy given by an empirical potential for all 87 possible isomers of the fullerene C_{44}. We find in general that smaller clusters that are more spherical tend to be more stable, as are clusters where the pentagons are better isolated. We find considerable scatter in our results, however, suggesting that these criteria are only qualitative guides to stability. No single isomer was found to be exceptionally stable with respect to the others. This suggests that the enhanced abundance of C_{44} observed in some studies[5] is due to a mixture of isomers, or that stability via resonance stabilization of the π electrons determines a single exceptionally stable cluster. If neither is the case, then a kinetic mechanism of formation may be indicated to explain the lack of a mixture.

We thank Professor D.J. Klein for discussions involving generating the various isomeric structures. This work was supported by the U.S. Office of Naval Research through the Naval Research Laboratory.

V. References

[1] R.F. Curl and R.E. Smalley, Science *242*, 1017 (1988); and references therein.

[2] W. Krätschmer, K. Fostiropoulos, and D.R. Huffman, Chem. Phys. Lett. *170*, 167 (1990); W. Krätschmer, L.D. Lamb, K. Fostiropoulos, D.R. Huffman, Nature *347*, 354 (1990).

[3] T.G. Schmalz, W.A. Seitz, D.J. Klein and G.E. Hite, J. Am. Chem. Soc. *110*, 1113, (1988), and references therein.

[4] E.A. Rohlfing, D.M. Cox and A. Kaldor , J. Chem. Phys. *81*, 3322(1984).

[5] A. O'Keefe, M.M. Ross, and A.P. Baronavski, Chem. Phys. Lett. *130*, 17(1986).

[6] A.J. Stone and D.J. Wales, Chem. Phys. Lett. *128*, 501, (1986).

[7] X. Liu, D.J. Klein, T.G. Schmalz and W.A. Seitz, preprint.

[8] D.W. Brenner, Phys. Rev. B *42*, 9458 (1990).

[9] B.I. Dunlap, D.W. Brenner, J.W. Mintmire, R.C. Mowrey, and C.T. White, J. Phys. Chem. *95*, 5763 (1991).

[10] B.I. Dunlap, D.W. Brenner, J.W. Mintmire, R.C. Mowrey and C.T. White, J. Phys. Chem. *95*, in press (1991).

[11] D.W. Brenner, B.I. Dunlap, J.A. Harrison, J.W. Mintmire, R.C. Mowrey, D.H. Robertson and C.T. White , Phys. Rev. B *44*, 3479 (1991).

[12] R.C. Mowrey, D.W. Brenner, B.I. Dunlap, J.W. Mintmire, and C.T. White, J. Phys. Chem. *95*, 7138 (1991).

MOLECULAR-DYNAMICS SIMULATIONS OF C_{60}/He COLLISIONS

R.C. MOWREY, D.W. BRENNER, B.I. DUNLAP, J.W. MINTMIRE
and C.T. WHITE
Theoretical Chemistry Section, Code 6179
Naval Research Laboratory
Washington DC 20375-5000

ABSTRACT. It has been reported[1] that collisions between C_{60}^+ and He with energies in the tens of electron volts in the center-of-mass reference frame produce not only the expected smaller fullerene fragments but also many of their He adducts. More recently, $C_{60}He^+$ has been observed as a collision product[2] and some evidence indicates that the He atom resides within the fullerene cage. We have performed molecular dynamics simulations for collisions between C_{60} and He with the goal of studying the details of the trapping process. Direct scattering from the outside of the cluster, transmission through the cluster, and trapping of the He atom within the cluster were observed. Relative abundances of $C_{60}He^+$ are calculated as a function of the collision energy and are compared with experimental measurements.

1. Introduction

In recent experiments studying the collision-induced dissociation of the fullerenes C_{60}^+ and C_{70}^+ with helium target atoms using an acceleration potential of 8 keV, smaller fullerene ions arising from the loss of C_{2n} and their helium adducts were detected[1]. Repetition of the experiment using ^{3}He supported the conclusion that the detected fragments represented fullerene-helium adducts. Other experiments confirmed these results and demonstrated the formation of $C_{60}He^+$ during the collision[2]. An analysis of the fragments formed by the buckminsterfullerene-helium complex when it was bombarded with Xe indicated that the helium atom was located in the interior of the cluster.

We have performed a series of molecular dynamics simulations of the collision of helium atoms with buckminsterfullerene to study the trapping of the rare gas atom within the cluster. In these calculations the positions and velocities of the helium atom and the individual carbon atoms of the buckminsterfullerene cluster were determined as a function of time. The helium-carbon interaction was described by a repulsive potential and the carbon-carbon potential permitted bond breaking and formation. The probability for trapping the helium atom depended on the collision energy and on the impact point of the atom on the cluster. The simulations predict larger trapping probabilities at each collision energy than are observed experimentally. The maximum trapping probability occurs at a higher collision energy in the calculations than in the experimental systems. The differences in the experimental

1353

P. Jena et al. (eds.), Physics and Chemistry of Finite Systems: From Clusters to Crystals, Vol. II, 1353–1358.
© *1992 Kluwer Academic Publishers.*

and theoretical results are attributed to unimolecular dissociation of $C_{60}He^+$. This process occurs in the experiments but is not observed in the simulations because of the short times over which the dynamics are calculated.

2. Outline of the Calculation

The potential energy surface that governs the dynamics of the nuclei of the carbon atoms is defined by an empirically derived, many-body expression that was originally constructed to describe bonding in hydrocarbon systems[3]. This potential reproduces the structures and energetics of these systems and allows bond breaking and formation, in contrast to traditional force-field potentials. The potential is based upon the Tersoff bond-order expression[4] but is extended to include terms to account for nonlocal effects and correct for overbinding of radicals. The helium-carbon interactions are described by a pair-wise screened Coulomb interatomic potential. We used the Moliere[5] screening function and the screening radius given by O'Connor and MacDonald[6]. With this potential the barrier heights for the penetration of He into the interior of the cluster through the center of five- and six-membered rings are 13.1 and 9.35 eV, respectively. These values are in good agreement with the estimate of 10 eV from *ab initio* calculations in which helium is forced through a C_6H_6 plane[1].

Although the atoms in the clusters in experimental systems began with a variety of different values of positions and velocities we chose to begin each simulation with the cluster atoms in their equilibrium positions and with zero velocity because of constraints imposed by computational resources. However, trajectories were calculated for collisions with different impact points of the He atom on the cluster and for different orientations of the cluster. The impact points were determined by placing an evenly spaced grid on the cluster. The cluster was oriented with either a C_{5v} or C_{3v} axis oriented along the Z axis which corresponds to this axis being perpendicular to either a five- or six-membered ring of the cluster, respectively. The He atom was initially placed a large distance from the cluster in the positive Z direction and with its momentum parallel to the Z axis. The classical equations of motion were integrated using a third-order Nordsieck predictor-corrector algorithm with variable time steps[7]. Total integration times were between 350 and 500 fs.

At the conclusion of each trajectory calculation the position and energy of the helium atom were examined to determine whether trapping occurred. A helium atom was considered to be trapped if at the end of the calculation it was located within the cluster lattice. Examination of the final energy of the helium atom for all trajectories that satisfied this criterion showed that the energy was reduced to values on the order of a few electron volts, far less than the amount needed to escape from the cluster. Trapping probabilities were calculated by taking the ratio of the number of trajectories that led to trapping and the total number of trajectories.

3. Results

Preliminary experimental results suggest that the greatest abundance of $C_{60}He^+$ is observed at a center-of-mass collision energy near 33 eV. More recently, mass-analyzed ion kinetic energy spectrometry (MIKES) quadrupole scans have been used to measure the relative abundance of $C_{60}He^+$ and C_{60}^+ as a function of collision energy[8]. A comparison of the measured and predicted relative abundances of $C_{60}He^+$ and C_{60}^+ as a function of collision energy is shown in Figure 1. Note that the experimental values in this figure have been multiplied by a factor of ten. The theoretical and experimental results predict a maximum abundance at a collision energy of 44 and 27.6 eV, respectively. Also, the relative abundance of $C_{60}He^+$ that is observed in the experiments is much less than the theoretical predictions throughout the energy range shown. It is important to emphasize that it is the raw data from the simulations and the experiments that are shown in Figure 1. As is discussed below, the agreement between the two sets of results improves significantly when the data are treated to remove important differences in the experimental system and the model that is treated in the simulations.

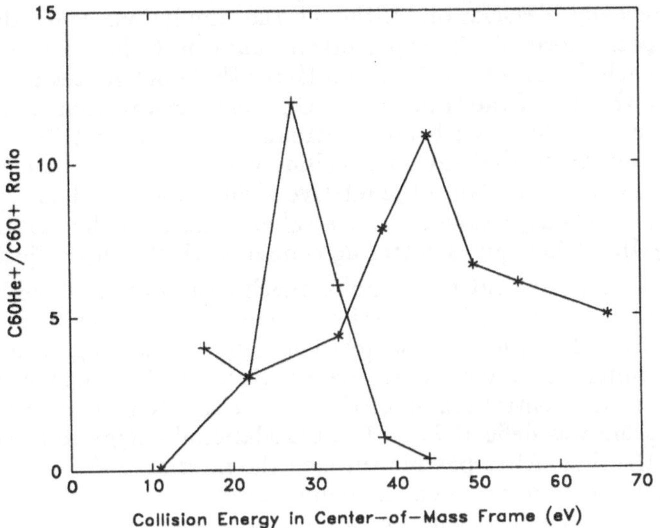

Figure 1. Ratio of $C_{60}He^+/C_{60}^+$ as a function of the center-of-mass collision energy. + experiment ($\times$ 10) and $*$ simulation. The data points are connected by lines to guide the eye.

The difference in the measured and predicted energy at which the maximum abundance of $C_{60}He^+$ occurs probably results from unimolecular dissociation of the ion. During the formation of $C_{60}He^+$ the center-of-mass collision energy is transferred to internal energy of the complex. Using an estimated value of 10 eV as the activation energy for the formation of $C_{60}He^+$ ample energy is available for formation of the complex and for unimolecular dissociation at collision energies of 44 and 27.6 eV. However, the complex formed at the higher collision energy will have an additional 16.4 eV of internal energy and, based upon an analysis using RRKM theory, would be expected to have a larger fragmentation rate. This analysis is confirmed by experimental measurements of the abundances of $C_{60}He^+$ and $C_{58}He^+$ as a function of energy[8]. The ratio of $C_{58}He^+/C_{60}He^+$ was relatively constant with values near 1% up to a center-of-mass collision energy of 33 ev. It rapidly increased to a value of 67% at an energy of 44 eV indicating that the fragmentation rate was strongly dependent on the internal energy. This suggests that the measured abundance of $C_{60}He^+$ reflects the competing effects of an increasing rate of capture of He by C_{60}^+ and an increasing fragmentation rate of the collision complex with increasing collision energy. In comparison, no fragmentation is observed in the theoretical simulations because the time scales over which the dynamics are followed (less than one picosecond) are much too short for this process to occur. Presumably, the trapping probabilities computed from the simulations would decrease and the maximum probability would shift to a lower energy if the trajectories were computed for timescales comparable to those occurring in the experimental systems.

The low abundance of $C_{60}He^+$ observed experimentally is probably also partially caused by unimolecular dissociation. Although the simulations count the total number of $C_{60}He^+$ species formed, the experiments count only the fraction that survive long enough to reach the detector. A second factor that contributes to the low abundance arises from the use of the transmitted C_{60}^+ beam as a reference. While passing through the target gas the main beam is attenuated by about 20%. Because it is the 80% of the main beam that does not collide with the target that is used as the reference for the $C_{60}He^+$ abundance the relative yield of $C_{60}He^+$ should be multiplied by a factor of four. This adjustment yields a value of 4.8% for the maximum relative abundance of $C_{60}He^+$ which gives better agreement with the theoretical prediction.

Qualitative information about the trapping mechanism was obtained by examining the atomic positions and velocities at various times during typical collisions. At low collision energies (~ 22 eV) most trajectories leading to trapping occurred for collisions with small impact parameters. In these trajectories the incident helium atom penetrated through the central region of the five- or six-membered ring surrounding the Z axis. The atom was deflected and lost translational energy to the cluster while passing through the ring. After passing through the interior of the cluster the helium atom collided with a carbon atom on the opposite side and lost additional energy as its direction of motion was reversed. The atom continued to move across the interior of the cluster but had insufficient translational energy to escape through the cluster. Most of the kinetic energy of the helium atom was transferred to vibrational energy of the cluster during this process. For collisions with large impact parameters the barrier for penetrating into the interior of the cluster is higher because of the orientation of the rings of carbon atoms. Therefore, collisions with low energy and large impact parameters tend to scatter from the exterior of the cluster. As the collision energy increases to 33 eV and higher, trajectories with small impact parameters tend to be transmitted through the cluster. The helium atom is deflected only slightly as it penetrates through the front side of the cluster and passes through an opening in the ring on the opposite side instead of striking a carbon atom and becoming trapped. Collisions with large impact parameters become the dominant contributors to the trapping probability because the helium atom has sufficient energy to penetrate into the interior of the cluster instead of being deflected. The trapping probability decreases at high collision energies because even though the incident helium atom loses energy as it penetrates into the interior of the cluster it retains sufficient translational energy to escape through the cluster lattice.

4. Summary

The molecular dynamics simulations of the collision of helium with buckminster-fullerene predict trapping probabilities that are larger than the experimental values throughout the range of collision energies examined. This discrepancy and the result that the maximum trapping probability occurs at a higher collision in the simulations probably result from the different time scales treated in the simulations and the experiments. During the short times treated in the simulations there is no evidence of unimolecular dissociation of the $C_{60}He^+$ cluster. However, the experimental results provide clear evidence of the importance of this process at high collision energies.

Acknowledgements

This work was supported by the Office of Naval Research through the Naval Research Laboratory. We thank the NRL Research Advisory Committee for a grant of computer time that was used in performing a portion of these calculations.

References

1. Weiske, T., Böhme, D.K., Hrušák, J., Krätschmer, W. and Schwarz, H. (1991) 'Endohedral cluster compounds: inclusion of helium within C_{60}^+ and C_{70}^+ through collision experiments', Angewandte Chemie International Edition in English 30, 884-886.

2. Ross, M.M. and Callahan, J.H. (1991) 'Formation and Characterization of $C_{60}He^+$', Journal of Physical Chemistry 95, 5720-5723.

3. Brenner, D.W. (1990) 'Empirical potential for hydrocarbons for use in simulating the chemical vapor deposition of diamond films', Physical Review. B 42, 9458-9471.

4. Tersoff, J. (1988) 'New empirical model for the structural properties of silicon', Physical Review Letters 56, 632-635; (1988) 'New empirical approach for the structure and energy of covalent systems', Physical Review B 37, 6991-7000.

5. Moliere, G. (1947) 'Theorie der Streuung schneller geladener Teilchen I', Zeitschrift für Naturforschung 2a, 133-145.

6. O'Connor, D.J. and MacDonald, R.J. (1977) 'A correction factor to the inter-atomic potential screening function for use in computer simulations', Radiation Effects 34, 247-250.

7. C.W. Gear, C.W. (1971) *Numerical Initial Value Problems in Ordinary Differential Equations*, Prentice-Hall. Englewood Cliffs, New Jersey.

8. Mowrey, R.C., Ross, M.M. and Callahan, J.H., 'Molecular dynamics simulations and experimental studies of the formation of endohedral complexes of buckminsterfullerene', Journal of Chemical Physics, submitted.

FORMATION AND CHARACTERIZATION OF $C_{60}He^+$

Mark M. Ross and John H. Callahan
Chemistry Division/Code 6113
Naval Research Laboratory
Washington, D.C. 20375-5000

ABSTRACT. It is shown that kilovolt collisions of C_{60}^+ with helium result in not only the expected dissociation to yield smaller fragment carbon cluster ions but also the formation of fullerene-helium adduct ions corresponding to C_xHe^+, where x is even and varies from 48 to 58, in agreement with Schwarz et al. This report provides the first unambiguous identification of the $C_{60}He^+$ product, indicating that the helium adduct formation does not require fragmentation. Using hybrid tandem mass spectrometric techniques the mechanism of the formation of these unusual adduct species was investigated. The nature of the collisions, the energetics of the process, and the products of dissociation of $C_{60}He^+$, provide strong evidence that the helium atom is inside the fullerene cage structure.

1. INTRODUCTION

While large, gas-phase carbon clusters have been studied for more than six years[1], there has been a recent, dramatic increase in interest due to the discovery of a process whereby large quantities of the all-carbon molecules, fullerenes, can be produced[2]. This development has allowed detailed characterizations of the structures and properties of fullerenes (C_x, where x is even and greater than 32), and, in particular, of the sixty-atom, soccer ball-shaped carbon molecule, C_{60}, buckminsterfullerene. Recent research efforts have focused on reacting, derivatizing, or doping these molecules or films of these molecules and have yielded new fullerene-based compounds and materials, which may provide unique or advantageous properties. One modification that has proven to be difficult is the incorporation of an element inside of the fullerene structure.

Some earlier experiments at Rice University were directed at putting a variety of metal atoms inside of C_{60} as this and other fullerenes were produced by laser vaporization of a metal-impregnated graphite rod in a molecular beam source[3]. These investigations of the photodissociation of the $C_{60}M^+$ adducts provided evidence for trapping of the metal atom inside of the C_{60} soccer-ball structure. However, other results were interpreted to indicate that the metal did not reside inside C_{60} [4], and until recently, there have been no other experimental investigations of this type.

P. Jena et al. (eds.), Physics and Chemistry of Finite Systems: From Clusters to Crystals, Vol. II, 1359–1364.
© 1992 *Kluwer Academic Publishers.*

1360

With the large quantities of fullerenes numerous mass spectrometric studies have been performed[5-8], including collision-induced dissociation by tandem mass spectrometry[6,8]. These investigations have shown that fullerene ions dissociate to smaller fullerene fragment ions, corresponding to C_2 losses, as well as to smaller carbon cluster ions. In a recent study of the high-energy (8-keV) collision-induced dissociations of fullerene ions with helium target gas, Schwarz et al.[9] observed fragment ions corresponding to $[C_x+4]^+$ in addition to the expected all-carbon fullerene fragment ions, C_x^+, with x even and between 48 and 58. The postulation that the unexpected ions were fullerene-helium adducts, C_xHe^+, was supported further by observation of $[C_x+3]^+$ product ions when ^3He was used as the target gas.

The purpose of this study was to confirm this unexpected and unprecedented result and to elucidate some the details of the energetics and mechanism of this phenomenon.[10] We report the formation of not only the previously-observed C_xHe^+ product ions but also unambiguous detection of $C_{60}He^+$. Using hybrid tandem mass spectrometric techniques some insights were gained into the mechanism of this reaction and on the location of the helium atom.

2. EXPERIMENTAL

The fullerene samples were prepared according to published methods[11]. Thermal desorption of the C_{60}/C_{70} mixture followed by electron ionization was performed in a VG ZAB-2FQ mass spectrometer (VG Analytical, Ltd.) as described previously[8]. This mass spectrometer is a three-sector instrument with a hybrid configuration (BEqQ), consisting of a magnetic sector (B), electrostatic analyzer (E), rf-only collision quadrupole (q), and a mass analyzer quadrupole (Q). Three scan modes were used in these experiments. The first scan mode is a conventional collision-induced dissociation/mass-analyzed ion kinetic energy spectrometry (CID/MIKES) experiment in which the magnetic sector transmits one mass-to-charge ratio (m/z) ion, which collides with a target gas at kilovolt energy (3 to 8 keV) in the subsequent field-free region, and the electrostatic sector voltage is scanned to provide a kinetic energy spectrum of the collision-induced fragment ions. The kinetic energy at which a fragment ion occurs is determined by the ratio of the fragment ion mass to the parent ion mass. However, because fragmentation processes have significant activation energies (particularly as ion mass increases), the kinetic energies at which fragment ions appear are shifted to somewhat lower values (by an amount corresponding to the energy uptake). The second scan mode, similar to the first, consists of setting the magnetic sector to transmit one m/z ion (A^+), which undergoes keV collisions. The mass analyzing quadrupole is then set to transmit an ion of one m/z (B^+) and the voltage of the electrostatic analyzer is scanned. This provides an energy spectrum of those ions corresponding to the process of $A^+ \longrightarrow B^+$. The third scan mode involves consecutive CID. The magnetic sector transmits one m/z ion, followed by keV collisions, and the electrostatic analyzer is set at a fixed voltage. Ions of one energy are transmitted to the rf-only quadrupole, and these ions may correspond to a range of m/z depending upon the energy and process studied. Subsequent 0 to 500-eV collisions of these ions with a target gas produce further fragment ions that are mass analyzed by the second quadrupole.

3. RESULTS AND DISCUSSION

Figure 1 shows the CID/MIKE spectrum that results from 8-keV collisions of C_{60}^+ with helium. The onset of the parent ion, C_{60}^+, appears at a kinetic energy of approximately 8000 eV, and the fragment ions, which are transmitted at lower kinetic energies through the electrostatic analyzer, are broad peaks due to the nature of high-energy CID. In this experiment the center-of-mass energy is approximately 44 eV and the parent ion was attenuated by approximately 40% (+/- 5%) by the target gas. At this attenuation, approximately 75% of the collision events involve only single collisions. The peak corresponding to the C_{60}^+ parent species is off scale so that some of the fullerene fragment ions, C_x^+ with $x \leq 58$, can be observed. A magnification of the spectrum at energies lower than that of the parent ion shows that some of the fragment ion energy regions consist of multiple peaks. In agreement with previous work[8], the kinetic energies of some of the fragment ions are shifted to values that are significantly lower than those calculated from the fragment/parent mass ratio and the full accelerating potential. In order to determine if C_xHe^+ is produced, a more careful examination of this spectrum is required.

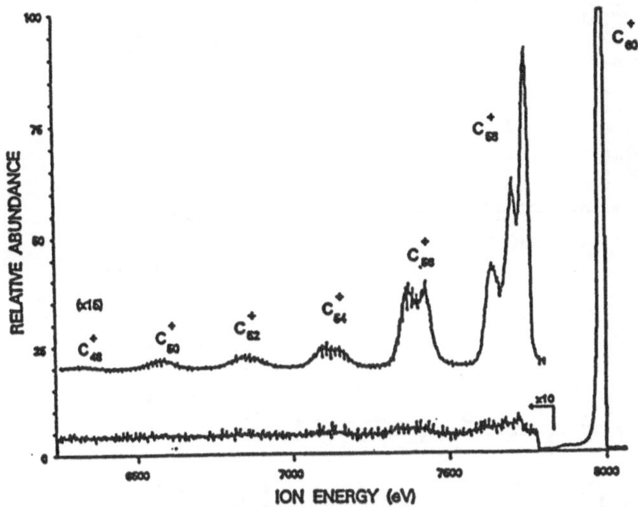

Figure 1. 8-keV collision-induced dissociation/mass-analyzed ion kinetic energy (CID/MIKE) spectrum of C_{60}^+ with helium (approximately 45% attenuation), showing the kinetic energy range that includes the parent C_{60}^+ and fragment ions from C_{48}^+ to C_{58}^+.

Expansion of Figure 1 (the upper trace is the same scan obtained at higher sensitivity) around the region where the C_{58}^+ fragment ion appears reveals three peaks at approximately the kinetic energy where this ion is expected ((696/720)*8000 eV=7733 eV). The peak at highest energy represents those C_{58}^+ species that result from unimolecular dissociation of C_{60}^+, because this peak is observed in the absence of collision gas and does not shift in kinetic energy due to collision processes. The other two peaks were identified by using the second scan mode. When the mass analyzer quadrupole was set to transmit m/z 696 (C_{58}^+) and the electrostatic sector voltage was scanned over a narrow range, the resultant spectrum shows the kinetic energies of those ions that result from the 8-keV collisions of C_{60}^+ with helium and yield an m/z 696 (C_{58}^+) ion that is transmitted by the quadrupole. Two peaks are observed, corresponding to the high and low energy peaks in the C_{58}^+ region, and represent species that arise from unimolecular dissociation (the higher energy peak) and those that result from CID of C_{60}^+ (shifted to lower energy, ca. 7630 eV). The spectrum that is obtained when the quadrupole is set to transmit m/z 700 ($C_{58}He^+$) shows that the middle peak in the C_{58}^+ region corresponds to $C_{58}He^+$. These product ions have a range of kinetic energies that reaches a maximum at approximately 7700 eV,

1362

which is between the energies of the C_{58}^+ species that arise from the two C_{60}^+ dissociation processes (unimolecular and collision-induced dissociation). Similar results were obtained for other fullerene fragment ions, C_x^+, where x is even and extends from 56 to 48, which were broad peaks or visible doublets in which the lower energy peak corresponded to C_x^+ and the higher energy peak was the helium adduct, C_xHe^+. The relative kinetic energies of C_x^+ and C_xHe^+ indicate that the helium is taken up by the ion <u>before</u> fragmentation. If the fullerene-helium adducts were formed after fragmentation (in a second collision) it is reasonable to expect that this ion would appear at lower kinetic energy than the all-carbon collision-induced fragment ions (at ca. 7630 eV) because of the additional loss of kinetic energy due to addition of the helium atom in a subsequent collision.

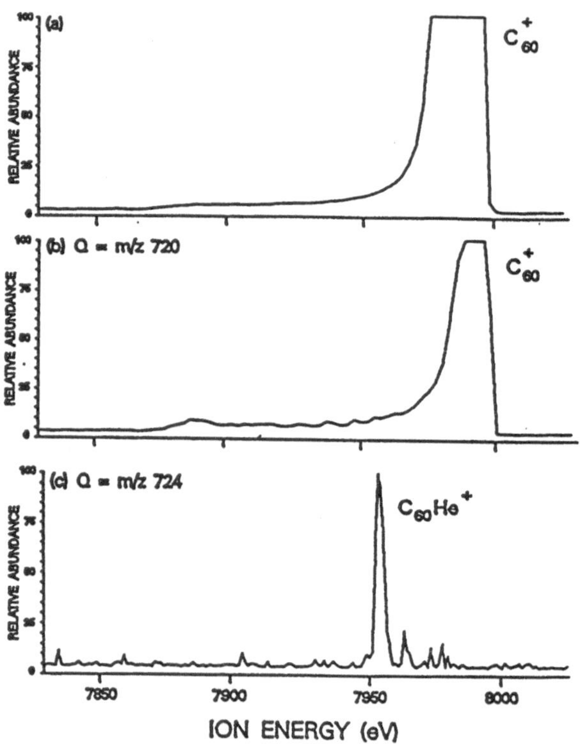

Figure 2. 8-keV C_{60}^+/He CID/MIKE spectrum (a) showing the ions that are transmitted through the electrostatic analyzer over a narrow energy range that includes C_{60}^+, (b) the same energy range but with additional mass selection of m/z 720 ions (C_{60}^+) by the quadrupole, and (c) over the same energy range with mass selection of m/z 724 ions ($C_{60}He^+$) by the quadrupole.

Following confirmation of the formation of the helium adducts of fullerene fragment ions, the $C_{60}He^+$ product ion was sought. The observation of this ion would demonstrate that fragmentation of C_{60}^+ is not a prerequisite for or concomitant with helium attachment. Figure 2(a) shows an expansion of the energy range in Figure 1 in which the parent C_{60}^+ species appears. The low-energy tail is prominent and the unusual shape of the kinetic energy distribution is distinctive (careful inspection reveals some structure, which will discussed in a subsequent report) and was not observed using other collision gases (H_2, Ne, Ar, O_2). Figure 2(b) shows the spectrum that results from the second scan mode in which the electrostatic sector voltage is scanned over the same energy range as in 2(a) while m/z 720 ions are transmitted by both the magnetic sector and the quadrupole. This energy spectrum has the same general shape as that in Figure 2(a). However, Figure 2(c) shows the spectrum that results when the quadrupole is set to transmit m/z 724 ($C_{60}He^+$) ions. The C_{60}^+ ions that collide with helium at 8 keV and pick up a helium atom yield m/z 724 product ions that have a very narrow range of kinetic energy (approximately 10 eV), while subsequent dissociation results in fragment ions with a wide range of kinetic energy. The $C_{60}He^+$ ions appear at lower energy than the full acceleration energy (8 keV) because of the kinetic energy lost upon uptake of the helium atom. The position in energy of this adduct ion is lower than the onset energy of the parent C_{60}^+ species by about 50 eV, which is approximately the C_{60}/He center-of-mass collision energy of 44 eV. This uptake of the full collision energy may indicate a head-on collision, which might be necessary for the helium atom to penetrate the C_{60} cage.

The kinetic energy of $C_{58}He^+$ is consistent with the unimolecular dissociation of $C_{60}He^+$ (($KE(C_{60}He^+)*(700/724) \approx KE(C_{58}He^+)$). The formation of $C_{60}He^+$ was observed from C_{60}^+/He collisions at laboratory collision energies from 8 to 3 keV, and the shift in the $C_{60}He^+$ kinetic energy relative to that of C_{60}^+ decreases roughly according to the decreasing center-of-mass collision energy. For example, when the center-of-mass collision energy was varied from 44 to 17 eV (corresponding to 8 to 3-keV collisions with He, respectively), the shift in the energy of $C_{60}He^+$ from that of C_{60}^+ changed from approximately 50 to 20 eV. The most abundant formation of $C_{60}He^+$ was observed with 5-keV collisions of C_{60}^+ with helium. Recent theoretical calculations suggest that the highest trapping efficiencies should occur near 8 keV, but experimental observations are hindered by the fact that some ions undergo unimolecular decomposition.

While the second type of scan provided definitive detection of $C_{60}He^+$, the location of the helium atom (on the outside or inside of the C_{60}^+ sphere) was unclear. The third scan mode provided insights into this question. Figure 3(a) shows the mass spectrum that results when the magnetic sector transmits C_{60}^+, the voltage of the electrostatic sector is fixed at the value corresponding to maximum transmission of $C_{60}He^+$, and the quadrupole is scanned. As is seen in the figure, both m/z 724 and 720 are transmitted through the electric sector at this energy, and, in addition, m/z 700 is detected, which is likely the result of unimolecular loss of C_2 from $C_{60}He^+$. Figure 3(b) shows the same scan when m/z 724 and 720 undergo collisions with xenon at 200 eV in the rf-only, collision quadrupole. In addition to the formation of $C_{58}He^+$, fragment ions corresponding to C_{58}^+, $C_{56}He^+$, C_{56}^+, $C_{54}He^+$, and C_{54}^+ are observed. This series of doublets, separated by 4 mass units, is similar to that observed by Schwarz et al.[9]. It cannot be ruled out that some $C_{60}He^+$ dissociates to yield C_{60}^+.

The dissociations of $C_{60}He^+$ caused by multiple 200-eV collisions with xenon yield product ions that retain the helium atom and correspond to losses of C_2 units, which is analogous to dissociation of C_{60}^+. Recent CID studies of adducts of C_{60}^+ formed in the gas-phase, including $C_{60}H^+$, $C_{60}C_3H_7^+$, and $C_{60}C_4H_9^+$, revealed that, unlike $C_{60}He^+$, the major fragmentation pathway is loss of the substituent atom or group[7]. In this experiment, the C_xHe^+ (x < 60) products indicate a significant stability of the fullerene-helium adducts and that the helium atom is not bound to the outside of the ball, unlike the location of the substituent atoms in the other adducts mentioned. The formation of $C_{60}He^+$ at high energy with a low number of collisions and the retention of the helium atom by fullerene fragment ions produced by CID constitute strong evidence that the helium is inside of the C_{60} spherical cage structure.

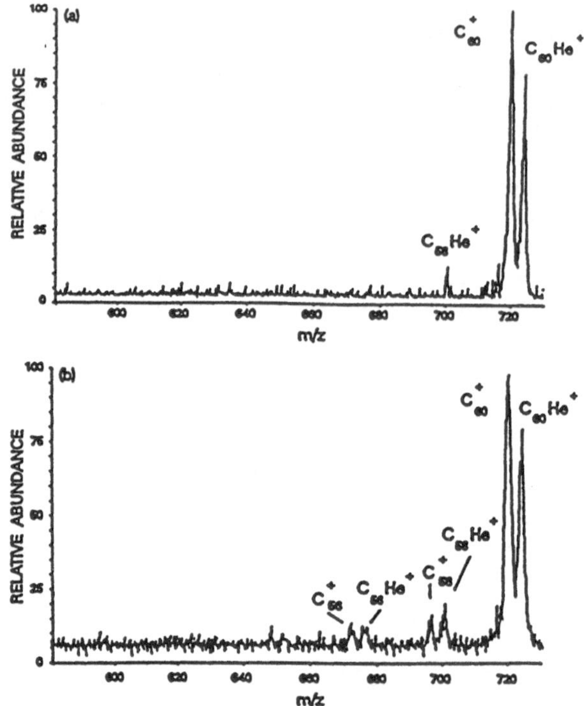

Figure 3. 8-keV C_{60}^+/He CID with the electrostatic analyzer voltage set to maximize the transmission of m/z 724 ($C_{60}He^+$) followed by scanning the quadrupole for fragment ions resulting from (a) unimolecular dissociation (no collision gas) and (b) 200-eV xenon CID (in the rf-only quadrupole) of the transmitted C_{60}^+ (m/z 720) and $C_{60}He^+$ (m/z 724).

4. REFERENCES

(1) Kroto, H.W.; Heath, J.R.; O'Brien, S.C.; Curl, R.F.; Smalley, R.E. *Nature* 1985, *318*, 162.

(2) Krätschmer, W.; Lamb, L.D.; Fostiropoulos, K.; Huffman, D.R. *Nature* 1990, *347*, 354.

(3) Weiss, F.D.; Elkind, J.L.; O'Brien, S.C.; Curl, R.F.; Smalley, R.E. *J. Am. Chem. Soc.* 1988, *110*, 4464.

(4) Cox, D.M.; Trevor, D.J.; Reichman, K.C.; Kaldor, A. *J. Am. Chem. Soc.* 1986, *108*, 2457.

(5) Luffer, D.R.; Schram, K.H. *Rapid Commun. Mass Spectrom.* 1990, *4*, 552.

(6) Young, A.B.; Cousins, L.; Harrison, A.G. *Rapid Commun. Mass Spectrom.* 1991, *5*, 226.

(7) McElvany, S.W.; Callahan, J.H. *J. Phys. Chem.* 1991, *95*, 6186.

(8) Doyle, R.J.; Ross, M.M. *J. Phys. Chem.* 1991, *95*, 4945.

(9) Weiske, T.; Böhme, D.K., Hrusák, J.; Krätschmer W.; Schwarz, H. *Angew. Chem. Int. Ed. Engl.* 1991, *30*, 884.

(10) Ross, M.M.; Callahan, J.H. *J. Phys. Chem.* 1991, *95*, 5720.

(11) Aije, H.; Alvarez, M.M.; Anz, S.J.; Beck, R.D., Diederich, F.; Fostiropoulos, K.; Huffman, D.R.; Krätschmer, W.; Rubin, Y.; Schriver, K.E.; Sensharma, D.; Whetten, R.L. *J. Phys. Chem.* 1990, *94*, 8630.

ELECTRON MICROSCOPY, ELECTRON ENERGY LOSS AND X-RAY EMISSION SPECTROSCOPY OF SOLID C60 AND C70

Y. SAITO, N. SUZUKI[1], M. TERAUCHI[2], R. KUZUO[2], M. TANAKA[2],
H. SHINOHARA[3], A. OHSHITA, M. OHKOHCHI[4], and Y. ANDO[4]

Department of Electrical and Electronic Engineering, Mie
University, Tsu 514, Japan
[1] Toyota Central Research and Development Laboratories,
Nagakute, Aichi 480-11, Japan
[2] Research Institute for Scientific Measurement, Tohoku
University, Sendai 980, Japan
[3] Department of Chemistry for Materials, Mie University, Tsu
514, Japan
[4] Department of Physics, Meijo University, Nagoya 468, Japan

ABSTRACT. Crystal structures and electronic structures of C_{60} and C_{70}
in a solid phase have been examined. C_{60} platelets grown from benzene
solution have the hcp structure, whereas vacuum deposited C_{60} films
have the fcc structure. On the other hand, C_{70} films formed by vacuum
deposition have the hcp structure. For solution growth of C_{70} crystals, not only hexagaonal plates with the hcp structure but also cube
or rectangular-shaped crystals with unknown structure segregated. The
electron energy loss spectrum in a low-loss region of solid C_{60} shows
eleven single-electron excitations as well as two collective excitations. The carbon K-edge excitation spectrum of C_{60} provided information on unoccupied π^* and σ^* bands. X-ray emission spectra, which
reflect the density of states of occupied valence bands derived from
p-orbitals, are also presented for both the C_{60} and C_{70} fullerites.

1. Introduction

Crystal structures of solid C_{60} and C_{70} (fullerites) have been studied
so far by X-ray diffraction[1,2], electron diffraction[3,4] and STM
[5]. Vacuum deposited C_{60} films have the fcc structure (a=14.198 Å)
[1]. On the other hand, solid C_{60} grown from solutions have a variety
of crystal structures depending on solvents used. For example, C_{60}
fullerites grown from benzene solution have the hcp structure (a=
10.018 Å, c=16.374 Å) with stacking disorders[6], but that grown from
toluene solution, the fcc structure (a=14.171 Å)[1,2]. The electronic
structure of the molecular C_{60} has been studied by optical spectroscopy in hexane solutions[7] and photoemission spectroscopy in the gas
phase of C_{60}^- ions[8]. In the crystalline form, there had been

P. Jena et al. (eds.), Physics and Chemistry of Finite Systems: From Clusters to Crystals, Vol. II, 1365–1370.
© 1992 *Kluwer Academic Publishers.*

studies with nearly every spectroscopic technique available, e.g., photoelectron[9], inverse photoelectron[10], X-ray absorption[11,12], X-ray emission[13] and electron energy loss spectroscopy[14-16].

In this paper we report on TEM (transmission electron microscopy), EELS (electron energy loss spectroscopy), and XES (X-ray emission spectroscopy) studies of C_{60} and C_{70} fullerites grown from benzene solution as well as those formed by vacuum deposition.

2. Experimental

2.1. Preparation of fullerites

C_{60} and C_{70} were extracted from fullerene-rich carbon soot with benzene. Purification and separation of C_{60} and C_{70} was accomplished by chromatography using a column of neutral alumina eluted with hexane/benzene or hexane/toluene mixtures. Since C_{70} obtained through only chromatographic separation was contaminated by hydrocarbons, chromatographed C_{70} was washed with ether to remove hydrocarbon impurities. Purity of C_{60} (99 %) and C_{70} (98 %) was checked by ^{1}H, ^{13}C NMR and UV absorption spectroscopy as well as HPLC.

Specimens for TEM and EELS were obtained by dropping benzene solutions of the chromatographed C_{60} (or C_{70}) on copper grids covered with a perforated (holey) carbon film. After evaporation of the benzene solvent under ambient temperature and atmosphere, crystallites of C_{60} (or C_{70}) were left. We examined plate crystallites for TEM and EELS because this morphology provides thickness of 50 to 500 nm suitable to TEM and EELS.

Thin films of C_{60} and C_{70} with thickness of 40-100 nm were formed by evaporating purified C_{60} and C_{70}, respectively, on a NaCl(100) surface at room temperature. The evaporation was carried out under high vacuum (ca. 10^{-4} Pa) from a resistively-heated Mo boat which was heated up to ca. 500°C. The deposited films were removed from the substrate by dissolving the substrate in distilled water and then supported on Cu grids for TEM.

For XES study, fullerites as large as a few mm in width and more than 1 μm in thickness, which segregated from benzene solution[13], were used.

2.2. TEM, EELS and XES apparatus

High-resolution TEM (HRTEM) images were taken with a JEM-2000EX (C_s = 0.7 mm) microscope operated at 200 kV. Some of high-resolution (HR) images were processed to filter out noise in the images with an LUZEX III image processor.

The high-resolution energy loss spectra in a low-loss region were obtained with a specially designed high energy EELS aparatus which was equipped with a field emission (FE) gun as the electron source and two Wien filters as the monochrometer and the analyzer. Another electron microscope (HF2000) equipped with a FE gun and a parallel EELS analyzer (Gatan Model 666) was also employed. The acceleration voltage was 60 kV for the former EELS machine, and 200 kV for the latter.

For measurement of soft X-ray of the carbon K-emission band, an electron probe micro-analyzer (EPMA) was used[13].

3. Transmission electron microscopy (TEM)

3.1. C_{60} fullerites

C_{60} fullerites grown from benzene solution have morphology of hexagonal (or rhombic) plates and rods. The crystal structure of the fullerites with the former morphology is the hcp (a=10.1$\pm$0.1 Å, c=16.5 $\pm$0.2 Å), and that of rod fullerites is unknown, as has been reported previously[4].

Figure 1 shows a HR image of an edge region of a C_{60} platelet and an electron diffraction pattern (EDP) indicating that the incident beam is parallel to the c-axis of the hcp structure. A rhombus superimposed in the picture is a unit cell of the hcp lattice projected along the c-axis. Circles corresponding to the size of a C_{60} ball (7.1 Å in diam.) arranged on a closed packed layer are also shown. It is found that the bright regions correspond with the shell of the C_{60}, and each dark spots, cavities inside the molecules.

Figure 2 shows a HR image of a C_{60} film formed by vacuum deposition. The image and an EDP reveal that the film is polycrystalline,

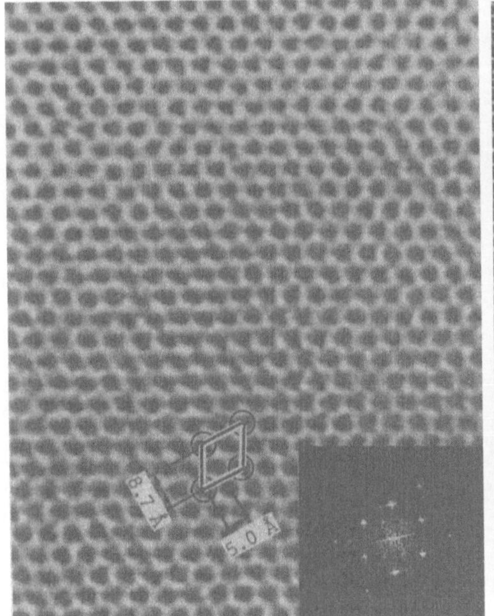

Figure 1. HRTEM image and EDP of a C_{60} plate.

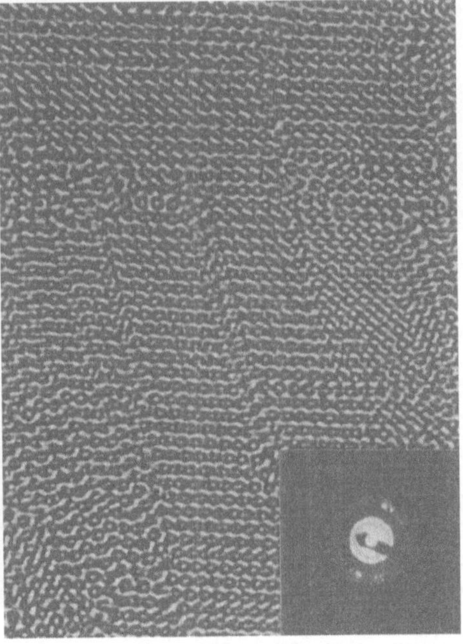

Figure 2. HRTEM image and EDP of a vacuum deposited C_{60} film.

consisting of small grains (several tens to a few hundred Å) with the fcc structure (a=14.3±0.2 Å).

3.2. C_{70} fullerites

Solution-grown C_{70} fullerites have morphology of hexagonal plates and cubes (or rectangular shapes). The hexagonal plates of C_{70}, a HRTEM image of which is shown in Fig. 3, have also the hcp structure (a=10.6 ±0.2 Å, c=17.4±0.3 Å). The crystal structure of the fullerites showing rectangular shapes is unknown as yet. Figure 4 shows a HRTEM image and an EDP of a rectangular C_{70} fullerite. Streaks in the EDP are due to a lamella texture observed in the image.

Vacuum deposited C_{70} films also have the hcp structure and prefer to orient its c-axis to the surface normal of the NaCl substrate, details of which will appear elsewhere.

Figure 3. HRTEM image and EDP of a C_{70} hexagonal plate.

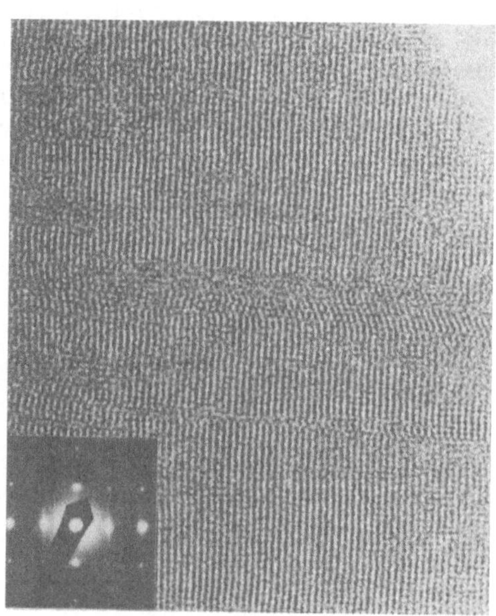

Figure 4. HRTEM image and EDP of a C_{70} rectangular crystal.

4. Electron energy-loss spectroscopy (EELS) of C_{60} fullerites

4.1. Low-loss region

Figure 5(a) shows a high-resolution energy-loss spectrum (energy resolution of 0.16 eV, FWHM of a zero-loss peak) in a region of less than 12 eV of a C_{60} fullerite grown from benzene solution. Eight shoulders and/or peaks (indicated by vertical lines) other than the plasmon peak at 6.4 eV (arrow) are observed at 2.1, 2.8, 3.7, 5.1, 6.1, 7.5, 9.5 and 10.7 eV. On the low-energy side of the so-called σ plasmon peak at 25.5 eV, which is not shown here, there exist three other shoulders at 15, 18, and 20 eV[16].

The strong peak at 3.7 eV clearly corresponds to the 329 nm peak observed in the UV-visible absorption spectrum (AS) of C_{60} in hexane solution[7] and the 339 nm peak for thick C_{60} samples[3] as well as X-ray photoelectron spectroscopy (XPS)[9]. This peak is assigned to the first optically allowed transition (h_u-t_{1g}). The hump at 2.1 eV in this study have no counterpart in either AS or carbon 1s XPS. The hump at 2.8 eV possibly correspond to the weak peak at 404 nm in AS, but absent in XPS. The 5.1 eV peak, which correlates with the strong UV absorption band at 260 nm and with the strong XPS peak at 4.8 eV, is due to single-electron transitions from π to π^* orbitals. The other features also originate presumably from interband transitions.

4.2. K-shell excitation

A K-edge excitation spectrum of a solution-grown C_{60} fullerite with energy resolution of 1.0 eV is shown in Fig. 5(b), in which the density of unoccupied states calculated by Saito and Oshiyama[17] is superimposed. The experimental spectrum exhibits fine features, which correlate well with the calculated density of states (DOS).

Peaks at 284, 286 and 288 eV correlate with transitions into π^* bands derived from t_{1u}, (t_{1g}, t_{2u}, h_g), and (h_u, g_g, g_u, t_g) molecular

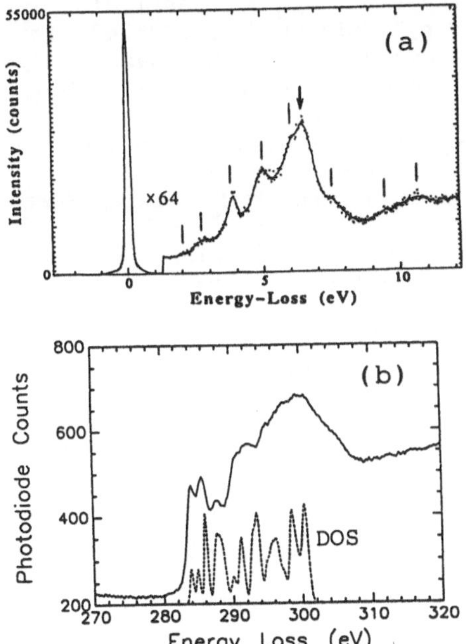

Figure 5. EELS of solid C_{60} in (a) a low-loss region and (b) a K-edge region.

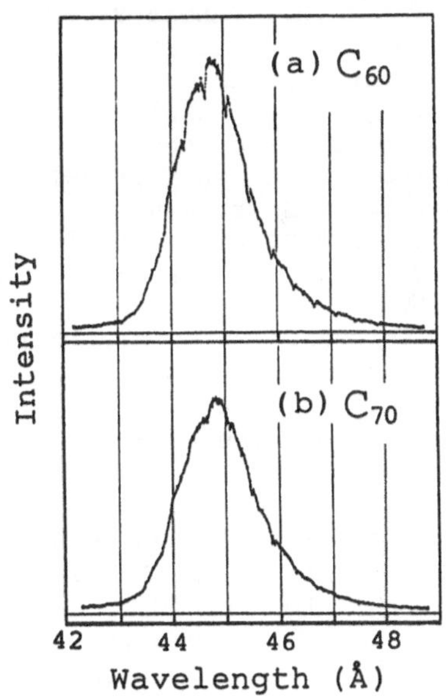

Figure 6. XES of (a) solid C_{60} and (b) solid C_{70}.

orbitals, respectively. There may be some contribution from core-exciton formation to the intensity of the first peak (284 eV) because the peak assgined to the t_{1u} transition is too strong for its low DOS.

5. X-ray emission spectroscopy (XES)

Figure 6 shows carbon K-emission spectra of C_{60} and C_{70} fullerites. The spectrum of C_{60} shows nine features over the K-emission band from 284 to ~260 eV. The experimental features agree excellently with the calculated DOS of valence bands[13]. The spectrum of C_{70} is less resolved, and farily resembles that of graphite.

References
[1] Fleming,R.M. et al.(1991) "Diffraction symmetry in crystalline, closed-packed C_{60}" Mater. Res. Soc. Symp. Proc. 206, 691-695.
[2] Heiney,P.A. et al.(1991) "Orientational ordering transition in solid C_{60}" Phys. Rev. Lett. 66, 2911-1914.
[3] Krätschmer,W. et al.(1990) "Solid C_{60}: a new form of carbon" Nature 347, 354-358.
[4] Saito,Y. et al.(1991) "Crystal structure and morphology of solid C_{60}/C_{70} and C_{60} grown from benzene solution" Jpn. J. Appl. Phys. 30, in press.
[5] Li,Y.Z. et al.(1991) "Ordered overlayers of C_{60} on GaAs(110) studied with scanning tunneling microscopy" Science 252, 547-548.
[6] Takata,M., Kubota,Y., Sakata,M. and Harada, J. unpublished.
[7] Ajie,H. et al.(1990) "Characterization of the soluble all-carbon molecules C_{60} and C_{70}" J. Phys. Chem. 94, 8630-8633.
[8] Yang,S.H. et al.(1987) "UPS of buckminsterfullerene and other large clusters of carbon" Chem. Phys. Lett. 139, 233-238.
[9] Weaver,J.H. et al.(1991) "Electronic structure of C_{60}: Experiment and theory" Phys. Rev. Lett. 66, 1741-1744.
[10] Jost,M.B. et al.(1991) "Band dispersion and empty electronic states in solid C_{60}: Inverse photoemission and theory" Phys. Rev. B 44, 1966-1969.
[11] Shinohara,H. et al.(1991) "Carbon K-shell X-ray absorption near-edge structure of solid C_{60}"Jpn. J. Appl. Phys. 30, L848-L850.
[12] Terminello,L.J. et al.(1991) "Unfilled orbitals of C_{60} and C_{70} from carbon K-shell X-ray absorption fine structure" Chem. Phys. Lett. 182, 491-496.
[13] Saito,Y. et al.(1991) "X-ray emission spectrum of solid C_{60}" J. Phys. Soc. Jpn. 60, 2518-2521.
[14] Saito,Y. et al.(1991) "Bulk plasmons in solid C_{60}" Jpn. J. Appl. Phys. 30, L1068-L1070.
[15] Hansen,P.L, et al.(1991) "An EELS study of fullerite - C_{60}/C_{70}" Chem. Phys. Lett. 181, 367-372.
[16] Kuzuo,R. et al. (1991) "High-resolution electron energy-loss spectra of solid C_{60}" Jpn. J. Appl. Phys. 30, in press.
[17] Saito,S. and Oshiyama,A.(1991) "Cohesive mechanism and energy band of solid C_{60}" Phys. Rev. Lett. 66, 2637-2640.

SUPERCONDUCTIVITY IN ALKALI INTERCALATED C_{60}

M. SCHLUTER, M. LANNOO*, M. NEEDELS, G. A. BARAFF
AT&T Bell Laboratories
Murray Hill, NJ 07974

D. TOMANEK
Dept. of Physics and Astronomy
and Center for Fundamental Materials Research
Michigan State University
East Lansing, MI 48824-1116

ABSTRACT. A model for superconductivity in alkali intercalated C_{60} is proposed. The key to the high observed transition temperatures is the molecular nature of fullerite. Two different energy scales $t_{intra}/t_{inter} \geq 5-10$ for electron hopping on the balls and between balls respectively allow for a large electron phonon coupling constant $\lambda = 2\,N \cdot V$. Electron scattering occurs mainly by vibrations on the balls and is dominated by t_{intra}, while the density of states N is given by inter-ball hopping and $-t_{inter}^{-1}$. Combined with a high Debye frequency this results in high transition temperatures. First-principles and semi-empirical methods are used to calculate the relevant parameters. A comparison with intercalated graphite shows crucial differences resulting in much smaller T_c values.

Superconductivity with unusually high transition temperatures ($T_c \geq 30K$) has recently been discovered [1] and has raised intense theoretical speculations.[2-7] We have investigated the electronic structure of these materials and calculated the conventional electron-phonon coupling properties. We have found a very unusual situation in which the existence of two largely different energy scales, caused by the molecular nature of C_{60} fullerite, can lead to an optimum electron-phonon coupling constant $\lambda = 2\,N \cdot V$.

The strong bonding within the C_{60} molecules dominates the electron scattering V which is given by the characteristic energy scale $t_{intra-ball}$ of (predominantly) π-electrons moving on an isolate C_{60} molecule. Maximum scattering occurs if the molecular electron-state is degenerate. For conduction to occur, inter-ball hopping takes place on a much smaller energy scale $t_{inter-ball}$ which dominates the density of states at the Fermi-level $N(\varepsilon_F)$ via $t_{inter-ball}^{-1}$. This leads to a large optimal λ. This is in strong contrast to the standard single band picture in which a single energy scale t determines both N and V

---

*   Permanent Address: ISEN, Lille, France

P. Jena et al. (eds.), Physics and Chemistry of Finite Systems: From Clusters to Crystals, Vol. II, 1371–1377.
© 1992 *Kluwer Academic Publishers.*

and essentially cancels out. The optimal value of λ is combined in C_{60} with a high Debye frequency, the prefactor determining T_c. This again is due to the relatively stiff on-ball modes given by $t_{intra-ball}$ and to the light mass of carbon atoms. This situation is a direct consequence of the molecular nature of fullerite. It is important to realize that this picture requires relatively large molecular units (C_{60}) which are (isotropically) weakly coupled. For instance, sheets or chain structures will not fall into this category since here the density of states for conduction along the preferred directions is small and also determined by t_{intra}^{-1} as opposed to t_{inter}^{-1}. As to the optimal size of the molecular units, one may consider the following: Because of the particular nature of the conduction states of carbon fullerenes, the electron scattering matrix element V is enhanced for small structures. The conduction states are of predominantly π-character with some σ-admixture, increasing with increasing curvature of the molecular unit.[3] Since $t_\sigma > t_\pi$, V increases with decreasing fullerene size $\sim 1/R^2$. We stress, however, that we do not propose electron pairing localized on individual molecules. The measured coherence length [8] of $\xi \approx 25 \,\text{Å}$ delocalizes the Cooper pair over 10-20 molecules and the Coulomb interaction μ^* is not expected to be significantly enhanced over standard values. A detailed analysis of μ^* for C_{60} has, however, still to be carried out.

To emphasize the unique nature of our picture we contrast C_{60} alkali doped fullerite with alkali intercalated graphite. Both factors determining $\lambda = 2\,N \cdot V$ are not enhanced in graphite. First N, being largely given by the intra-sheet hopping is smaller [9] than for C_{60}, although it is somewhat re-enhanced by the higher doping density of e.g. KC_8 as compared to e.g. K_3C_{60}. Secondly, V is smaller since the important vibrations couple less effectively to the π electrons in flat graphite sheets than on curved C_{60} balls. In particular, buckling-type vibrations do not couple at all to first order in graphite.[6] Our calculations, to be described below yield a λ value about a factor of five smaller for intercalated graphite as compared to C_{60}, a qualitative difference. This leaves us with a rather unique situation in C_{60} where N is given by intermolecular compound properties, largely dependent on the specific intercalation compound, to be multiplied by V which is a rather constant intra-molecular quantity unaffected by intercalation.

This simple picture is beautifully confirmed by several experimental observations. First, for a given compound, T_c decreases drastically with increasing pressure [10] which can be explained by the soft compressibility of fullerite resulting in decreased density of states values N with decreasing intermolecular distances. Secondly, the observed [11] increase in T_c with increasing alkali intercalant size again supports the same density of states argument. In fact, these observations can be explained quantitatively assuming simply $T_c \approx \hbar\omega_D \exp[-\dfrac{1}{\lambda - \mu^*}]$ and $\lambda = 2\,NV$ with $N \sim t_{inter}^{-1} \sim d^2$ as is commonly done for p-electron overlaps. Then, the same value of $(\lambda - \mu^*)$ needed to explain the absolute value of T_c also quantitatively describes the variation of T_c between K_3C_{60} and $Rb_2Cs C_{60}$, solely on the basis of interball distance (d) variation.[11] A further confirmation of the picture can be found in the apparent disappearnce of on-ball Raman phonon-lines with metallic intercalation.[12] The strong on-ball electron-phonon coupling V yields an increased phonon line-width, calculated by us to be of order 5%

which should wash out most of their spectral features.[7]

We now describe some of the detailed calculations done to support this picture. The electronic structure of C_{60} has been studied by many groups.[13] A simple picture emerges. C_{60} is insulating with a gap of order 1.5 - 2eV separating a five-fold degenerate h_u level (HOMO) from a three-fold degenerate t_{1u} level (LUMO). The levels near the gap derive from predominantly π-like states centered at the individual carbon atoms with some small s,p_σ admixtures due to the finite curvature of C_{60}. The C_{60} envelope of these orbitals has predominantly $\ell = 5$ character for states near the gap. When placed in an fcc lattice the C_{60} molecular levels weakly interact via t_{inter} resulting in a bandstructure with a ~0.5eV wide conduction band, reasonably well separated from higher bands. Details of this band depend on the relative orientation of the C_{60} molecules which are so far unknown. Upon alkali intercalation, well ordered compounds like e.g. K_3C_{60} result with the alkali atoms occupying octhedral and tetrahedral fcc interstitial sites. Detailed calculations [3] show that the rigid band picture is a good first approximation and that the alkali (donor) electrons can be viewed as partially filling the C_{60} conduction band. This is also in accord with the observation that K_6C_{60} becomes an insulator, as expected from a complete filling of the t_{1u} derived conduction bands. We have carried out a series of Density Functional (LDA) calculations [14] for fcc C_{60} serving as anchor points for simpler empirical tight binding (ETP) calculations which we use to evaluate the electron-phonon coupling matrix elements. The ETB Hamiltonian is based on four orbitals (sp^3) per atom with its parameters fitted to a large database of carbon molecules and solid structures.[15] For C_{60} its predictions agree well with the LDA results. The fit of the empirical parameters contains also a d^{-2} distance dependence of all hopping matrix-elements which is an essential ingredient for the evaluation of the electron-phonon coupling. Interball hopping is only approximately described by this procedure. Quantities like the conduction bandwidth, the Fermi surface and the density of states $N(\varepsilon_F)$, however, depend on details of these inter-ball interactions. We feel that these quantities cannot be determined reliably at present and we will, therefore, consider reasonable ranges of values when needed. To determine the phonon eigenmodes of C_{60} we use a simple Keating model [16] with two nearest-neighbor on-ball force constants with a variable ratio α/β of bond stretching to bond bending. We fix the overall scale to reproduce the highest modes of C_{60} observed near $1600cm^{-1}$ and we vary the ratio to best reproduce the spectrum of more elaborate vibrational calculations.[17] Final values for the electron-phonon coupling V are rather independent on the β/α ratio while details of the spectral distribution of mode coupling depend on it. The on-ball modes range from ~$300cm^{-1}$ up to ~$1600cm^{-1}$. Modes with predominantly radial displacements are at the lower end of the spectrum (the quadropolar deformation of C_{60} is the lowest, buckling modes are somewhat higher) while the high energy modes have mostly tangential displacements. There is a strong similarity to graphite where the optical layer stretching modes are near $1600cm^{-1}$, while the layer buckling modes occupy the lower end of the spectrum.[17]

It is instructive and useful to first consider the static Jahn-Teller effect of C_{60}, which occurs when electrons are placed into the 3-fold orbital degenerate t_{1u} conduction state.

Group theory shows that only two symmetries of modes, the totally symmetric non-degenerate A_g modes and the five-fold degenerate H_g modes couple to t_{1u}. For C_{60} there are two A_g modes, the overall radial breathing mode (A_g^1) near $500\,cm^{-1}$ and a pentagon stretching mode (A_g^2) near $1500\,cm^{-1}$. These A_g modes couple diagonally to the t_{1u} states and do not lift any degeneracies. The Jahn-Teller coupling problem of a 5-fold vibrational mode (d-like) to a 3-fold electronic state (p-like) is interesting and has been studied in detail.[18-19] The five distortion coordinates of H_g can be grouped into two sets, the trigonal and tetragonal distortions. For icosahedral C_{60} the coupling constants for the two sets are degenerate while for point-defects in tetrahedral semiconductors (e.g. the Si vacancy) this degeneracy is lifted. The analogy of t_{1u} of C_{60} to the T_2 state of the silicon vacancy is particularity instructive and has been discussed elsewhere.[19] For C_{60} the Jahn-Teller energy surface can be evaluated analytically [19] knowing the individual coupling matrix elements V_{ij}^m between states i,j. of t_{1u} and the mode frequencies ω_m. Because of the high symmetry of C_{60} this energy surface is two dimensionally orientationally degernate ("mexican hat") i.e. E_{JT} dependes only on three of the five distortion coordinates.

For $n = 1, 2, 3$ electrons the Jahn-Teller energy lowering is $1, 4, 3 \times E_{JT}$ respectively, with $E_{JT} = \dfrac{2}{15} \sum_m \sum_{ij}^3 |V_{ij}^m|^2 / M\omega_m^2$. For more than half filling the problem is electron-hole symmetric. Among the eight H_g modes of C_{60} we calculate significant coupling to mainly two modes, a low energy buckling mode H_g^1 and a high energy optic mode H_g^2. This strong "selection rule" is successively more relaxed as the strength of the bond-bending force constant β is increased. However, the sum over all modes remains remarkably invariant within a few percent for α/β ranging from 0.1 to 0.3. For $\alpha/\beta = 0.1$ we find dominant contributions of $\sim 12\,meV$ from H_g^1, of $\sim 7\,meV$ from H_g^2 which together with a $\sim 2\,meV$ diagonal contribution from A_g^2 and some other minor contributions yield an overall energy lowering $\Delta E \approx 28\,meV$.

To test the numerical accuracy of these values we performed frozen phonon LDA calculations for the A_g^2 mode and obtained virtually identical results. Using the calculated value of E_{JT} we can evaluate the "negative U" contribution which measures the tendency towards charge disproportionation or electron binding on individual C_{60} molecules. For half-filing of t_{1u} we find $U \approx -60\,meV$ which is clearly too small to overcome classical Coulomb repulsion effects which can be estimated to be $\geq 0.5\,eV$ in agreement with experimental charging studies of C_{60} in electrolyte solutions.[20] While the static value of E_{JT} seems small, dynamically the same type of coupling can be significant for electrons travelling in Bloch states made from the t_{1u} orbitals.

The standard BCS expression for electron-phonon scattering is [21] $\lambda = 2N(\varepsilon_F) \cdot V = 2N(\varepsilon_F) \cdot \sum_m \dfrac{\ll V_m^2 \gg}{M\omega_m^2}$ where the sum runs over all vibrational modes m of the extended system and where the double bracket denotes a double Fermi-surface average for electrons scattered from Bloch states k to k'. This average can be carried out analytically in the limit of vanishing dispersion or vanishing t_{inter} as

$V = \frac{5}{6} E_{JT}$. Details of the derivation of this remarkable equality are given elsewhere.[19] The $q = k - k'$ dependence of the scattering is generally not important since the strength is given by the relatively dispersion-less on-ball vibrations. For the H_g Jahn-Teller modes the scattering is dominated by the inter-band terms (within t_{1u}) including $q = 0$. For the A_g symmetric mode $q = 0$ scattering is zero since it corresponds to a coherent overall shift of all electronic levels. For finite wave vector $q \neq 0$, A_g mode scattering is strong, again given by the on-ball coupling. This situation is not unlike the electron scattering from diagonal disorder in random alloys. As a consequence of these results phonon linewidths should be broadened upon intercalation. For $q = 0$, only H_g modes should be affected, as clearly seen in Raman scattering, [12] and as predicted by Varma et al. [7] However, for finite q, A_g modes should also be affected.

The question of the size of $N(\varepsilon_F)$ entering λ is largely unsettled at this point. Estimates range from values of ~2 states/eV-spin-C_{60} derived from photoemission data, [22] to values of ~$6-20$ derived from bandstructure estimates, [13] up to values of 10-15 and >20 inferred from susceptibility [23] and NMR data [24] respectively. An average value of $N \approx 15$ yields a $\lambda \approx 0.6$ which with $\mu^* \approx 0.1-0.2$ yields transition temperatures in the 20 - 30K range. For this the average cutoff Debye frequency $\hbar\omega_{log}$ has been estimated to be of order ~1400K. The simple mode frequency averaging procedure [21] used here can be simply improved upon as described in the literature, [25] but we do not expect drastic differences.

We conclude from these calculations that the observed superconductivity in alkali intercalated C_{60} compounds can be understood in terms of conventional electron coupling, albeit strongly modified by the molecular nature of the material. While the electron-phonon coupling λ has been calculated in some detail, the opposing Coulomb repulsion μ^* has not yet been investigated on a comparable level. This remains to be done for systems with a several largely different energy scales and strong spatial inhomogemities, to establish the validity of our model. Considerations for conventional materials are known lead to limiting cancelling effects between λ and μ^*.[26] A further caveat is in place because of the rather close proximity of the scales of t_{inter} and $\hbar\omega_{log}$ as well as the electrostatic on-ball Coulomb repulsion for C_{60}. This situation appraoches the limits of validity for retarded electron-phonon scattering on one hand and the band picture on the other hand.

We thank W. Zhong and Y. Wang for assistance with numerical calulations, S. Duclos, R. Tycko, R. C. Haddon and A. P. Ramirez for the discussion of their measurements, P. B. Littlewood, J. C. Tully, S. Shastri, J. Zaanen, C. M. Varma, K. Raghavachari, and A. J. Millis for useful discussions.

References

1. Rosseinsky, M. J., Ramirez, A. P., Glarum, S. H., Murphy, D. W., Haddon, R. C., Hebard, A. F., Palstra, T. T. M., Kortan, A. R., Zahurak, S. M., Makhija, A. V., Phys. Rev. Lett., **66**, 2830 (1991).

1376

    Hebard, A. F., Rosseinsky, M. J., Haddon, R. C., Murphy, D. W., Glarum, S. H., Palstra, T. T. M., Ramirez, A. P., Kortan, A. R., Nature, **352**, 222 (1991).

2. Chakravarty, S., Gelfand, M. P., Kivelson, S., to be published.

3. Martins, J. L., Troullier, N., Schabel, M., to be published.

4. Baskaran, G., Tosatti, E., to be published.

5. Zhang, T. C., Ogato, M., Rice, T. M., to be published.

6. Schluter, M., Lannoo, M., Needels, M., Baraff, G. A., Tománek, D., to be published.

7. Varma, C. M., Zaanen, J., Raghavachari, K., to be published.

8. Holczer, K., Klein, O., Gruner, G., Thompson, J. D., Diedetich, F., Whetten, R. L., Phys. Rev. Lett. **67**, 271 (1991).

9. Fisher, J. E., in "Intercalated Layer Materials", ed. Levy, F. A., Reidel Publ., Dordrecht (1979), p. 481.

10. Sporn, G., Thompson, J. D., Huang, S. M., Kaner, R. B., Diederich, F., Whetten, R. L., Gruner, G., Holczer, K., Science **252**, 1829 (1991).
Schirber, J. E., Overmyer, D. L., Wang, H. H., Williams, J. M., Carlson, K. D., Kini, A. M., Pellin, M. J., Welp, N., Kwok, W. -K., Physica C **178**, 137 (1991).

11. Fleming, R. M., Ramirez, A. P., Rosseinksy, M. J., Murphy, D. W., Haddon, R. C., Zahurak, S. M., Makhija, A. V., Nature **352**, 787, (1991).

12. Duclos, S., Haddon, R. C., Glarum, S., Hebard, A. F., Lyons, K. B., Science, to be published.

13. Haddon, R. C., Brus, L. E., Raghavachari, K., Chem. Phys. Lett., **125**, 459 (1986).
Weaver, J. H., Martins, J. L., Komeda, T., Chen, Y., Ohno, T. R., Kroll, G. H., Troullier, T., Phys. Rev. Lett., **66**, 1741 (1991).
Zhang, Q., Yi, J. Y., Berhnolc, J., Phys. Rev. Lett., **66**, 2633 (1991).
Saito, S., Oshiyama, A., Phys. Rev. Lett. **66**, 2637 (1991).
Feuston, B. P., Andreoni, W., Parrinello, M., Clementi, E., to be published.

14. Needels, M., Schluter, M., to be published.

15. Tománek, D., Schluter, M., Phys. Rev. Lett., to be published.

16. Martin, R. M., Phys. Rev. **B4**, 4005 (1970).

17. Onida, G., Benedeck, G., to be published.

18. O'Brien, M. C. M., J. Phys. C. **4**, 2524 (1971).

19. Lannoo, M., Baraff, G. A., Schluter, M., Tománek, D., to be published.

20. Miller, B., private communications.

21. McMillan, W. L., Phys. Rev. **167**, 331 (1986).

22. Chen, C. T., Tjeng, L. H., Rudolf, P., Meigs, G., Rowe, J. E., Chen, J., McCauley, J. P., Smith, A. B., McGhie, A. R., Romanow, W. J., Plummer, E. W., Nature **352**, 603 (1991).

23. Ramirez, A. P., Rosseinsky, M. J., Murphy, D. W., Haddon, R. C., to be published.

24. Tycko, R., Dabbagh, G., Rosseinsky, M. J., Murphy, D. W., Fleming, R. M., Ramirez, A. P., Tully, J. C., Science **253**, 884 (1991).

25. Bergman, G., Rainer, D., Z. Physik **263**, 59 (1973).

26. Cohen, M. L., Anderson, P. W., in Superconductivity in D and F Band metals (D. H. Douglass ed.) p. 17, AIP Conf. Proc. #4 (1972).
Anderson, P. W., to be published.

Extraction and Characterization of Large All-Carbon Fullerenes

H. Shinohara, H. Sato, Y. Saito[1], A. Izuoka[2],
T. Sugawara[2], H. Ito[3], T. Sakurai[3], and T. Matsuo[3]

Department of Chemistry for Materials
Mi'e University, Tsu 514, Japan
[1]Department of Electrical and Electronic Engineering,
Mi'e University, Tsu 514, Japan
[2]Department of Pure and Applied Science
The University of Tokyo, Tokyo 153, Japan
[3]Institute of Physics, College of General Education,
Osaka University, Osaka 560, Japan

Abstract. Larger fullerenes than C60 and C70 are successfully extracted via several organic solvents from fullerene-rich carbon soot produced by a carbon arc generator. The present extracts are not soluble in benzene or toluene which are normally used as extraction solvents for C60 and C70. The secondary ion mass spectrometry (SIMS) of the obtained extracts reveals, for the first time, fullerenes C70+2n ($1 \geq n$) up to at least C500. The SIMS spectra exhibit a remarkably wide distribution of larger fullerenes which has a broad maximum around C130 - C150. The position of the maximum is a sensitive function of extraction solvents which are used in the sample preparation. The possibility for the isolation of such large fullerenes is also presented.

1. Introduction.

The very stable, nonreactive all-carbon fullerenes C60 and C70 were first identified in the condensing carbon vapor produced by laser vaporization of graphite.[1] Recently, they have been produced in macroscopic quantities by vaporization of graphite rods with resistive heating[2] or contact arc.[3] The fullerene-rich carbon soot contains larger fullerenes such as C76, C78, C84, C90 and C94 which are recently isolated by column chromatographic separation of benzene(toluene) soluble extracts.[4,5]

In a recent laser vaporization experiment on graphite,[6] carbon clusters up to C200 have been observed by a reflectron time-of-flight mass spectrometer. An excimer laser ablation of polyimide has produced carbon clusters as large as C400.[7] Moreover, a CO_2 laser desorption Fourier transform mass spectrometry

P. Jena et al. (eds.), Physics and Chemistry of Finite Systems: From Clusters to Crystals, Vol. II, 1379–1384.

of benzene soot has also produced a series of large carbon clusters extending up to C600.[8] All of these studies jointly imply that these large-sized carbon molecules are also present in laboratory-produced carbon soot. Specifically, it is extremely interesting and important to know whether or not such large-sized carbon molecules also have polyhedral, spheroidal fullerene structures composed of pentagons and hexagons like C60 and C70.

The present study reports extraction and mass spectroscopic characterization of very large fullerenes up to C500 obtained from fullerene-rich carbon soot. In our previous studies,[9,10] we have found that quinoline is a far better extraction solvent than benzene(toluene) for large fullerenes. In this study, quinoline/ pyridine extracts from carbon soot are subjected to mass spectroscopic analyses.

2. Experimental

Details of the production of fullerene-rich carbon soot by the contact arc method[3] have been described previously.[9-13] Briefly, an arc between two graphite rods (13.5 mm in diameter) is sparked at 200 - 250 A in the direct current (DC) mode in a He (20 - 100 Torr) atmosphere. The DC mode was much more effective than the AC mode to evaporate carbon where a positive graphite electrode was preferentially consumed. The optimum yield (12 % in weight) of various fullerenes was observed at a He pressure of ca. 20 Torr in the present experimental configuration.

The fullerene-rich carbon soot (3.44 g) was first rinsed for 24 hours with benzene (500 ml). The soot was then refluxed for 4 hours with quinoline (300 ml) and filtered. The obtained solution was concentrated and the resultant residue was rinsed with hexane, methanol, and benzene. With this procedure, large fullerene-rich black powder (extract I 380 mg) was obtained. Extract I was further refluxed for 1 hour with pyridine (70 mg) and filtered. From the pyridine solution we have obtained another fullerene-rich black extract (extract II 170 mg) and the residual powder (210 mg). These procedures are illustrated in Fig.1.

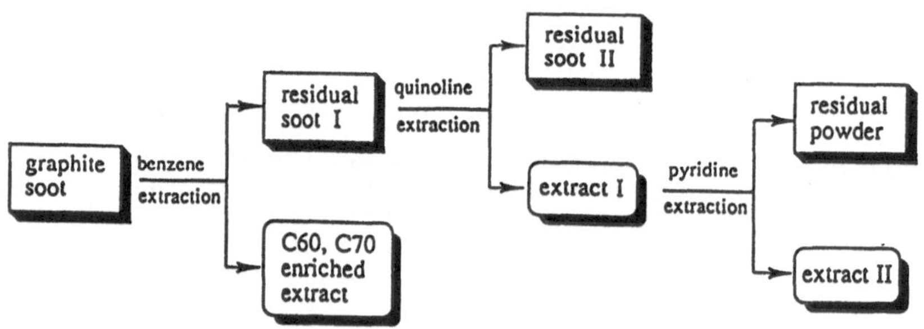

Fig.1 Schematic diagram of the preparation of the large fullerene rich carbon powder via quinoline/pyridine extraction.

Mass analyses of the present extracts were performed by second-ary-ion-mass spectrometry (SIMS) with the grand scale mass spec-trometer at Osaka University (GEMMY).[14] The sample were suspend-ed in m-nitrobenzyl alcohol and ionized by Cs^+ ion bombardment. Both primary and secondary ions were accelerated to 15 keV.

3. Results and Discussion

Figure 2 exhibits a typical Cs^+ SIMS mass spectrum of ex-tract II in the mass spectral region of m/z = 1200 - 1800 (C130 - C266). One of the salient features of the spectrum is the presence of a remarkably wide distribution of even-numbered carbon molecules C70+2n.[9,10] The high resolution mass spectra of the same sample (not shown) indicate that the observed series of peaks are composed of all-carbon molecules and not of hydro-fullerenes such as (C70+2n)H and (C70+2n)H_2 or any other reaction products by impurities. The observed intensity distribution exhibits a broad maximum around m/z = 1600 - 1800 (C130 - C150) and then almost smoothly decreasing trend at higher masses. Con-trary to the previous expectation,[15,16] the intensity of the fullerene C240 is not particularly enhanced in the mass spectrum.

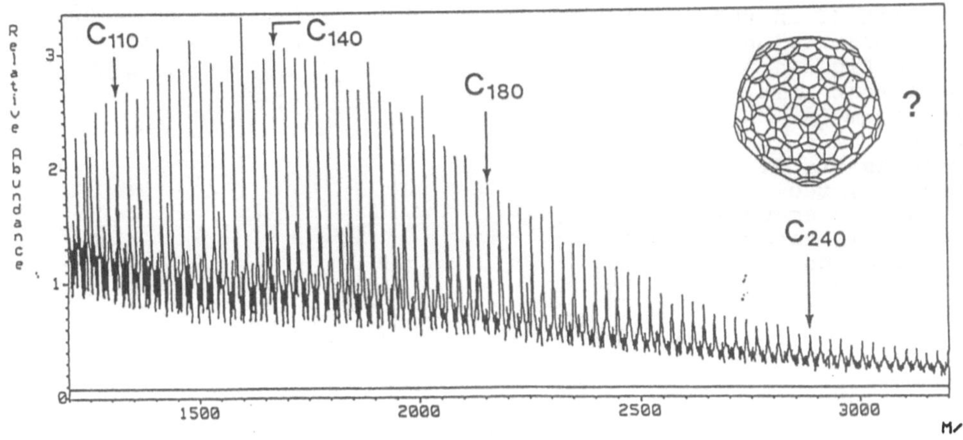

Fig.2 A Cs^+ SIMS spectrum of a large fullerene-rich black powder (extract II) via quinoline and pyridine extraction. A part of the mass spectrum in the range m/z = 1200 - 3200 (C100 - C266) is presented. The C240 is not particularly enhanced in the spectrum.

The appearance of the broad maximum in the distribution is characteristic to the present sample preparation. A SIMS spectrum of the residual powder (cf. Fig.1) has a similar maximum (but shifting to higher m/z values) around m/z = 1900 -2200 (C160 -

C180), which is shown in Fig.3-a. Fig. 3-b is a SIMS spectrum of extract II for a comparison, which is essentially similar to Fig.2. As shown in Fig.1, extract II is a pyridine extract from extract I after quinoline extraction, whereas the residual powder is a portion that is not extracted with pyridine. Based on this experimental procedure it is understandable that the residual powder contains larger fullerenes than extract II involves. Moreover, the above results present a possibility that if extraction solvents are carefully chosen on the basis of specific solubility for the different sizes of fullerenes in the whole sample preparation scheme, large fullerene-rich extracts having much narrower size distribution of fullerenes can be obtained. Using an HPLC separation [4,5] of such an extract, it will not be difficult to isolate very large fullerenes.

Fig.3 a. A SIMS spectrum of the residual powder (cf.Fig.1) in a mass spectral range of m/z= 800 - 3000 (C68-C250); b. A SIMS spectrum of extract II in the same m/z region as in Fig.3-a. Note that this spectrum is essentially the same as Fig.2. However, the number of accumulation in recording the spectrum is different.

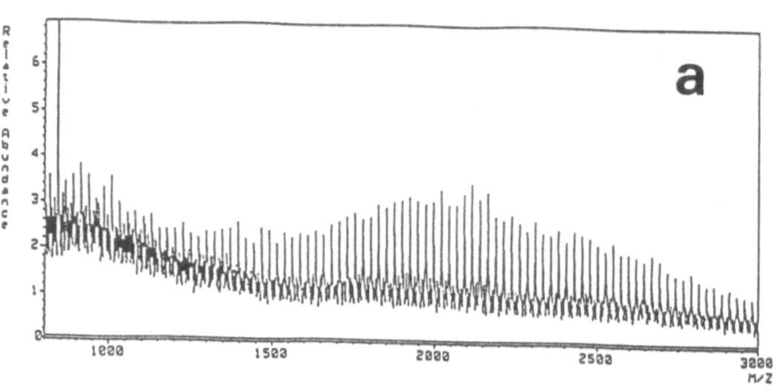

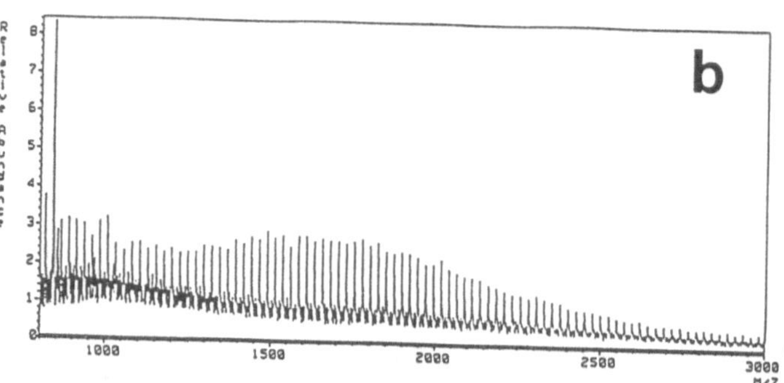

Fig.4 exhibits a SIMS spectrum of extract II in a higher mass range m/z = 2900 - 5100 (C242 - C424). As is seen, the series of even-numbered peaks C70+2n are still clearly observable in this high mass region. Giant fullerenes extending up to at least C500 are also seen in a much higher mass spectral range. the present mass spectroscopic observation implies that not only certain sizes of large fullerenes such as C240 and C540[15,16] but other even-numbered fullerenes also exist in laboratory-produced carbon soot.

In analogy to C60 and C70, all even-numbered clusters C70+2n (n≥1) are believed to correspond to polyhedral, spheroidal

carbon cages.[4,5,17,18] The fullerene hexagon + pentagon structures are especially suitable for large C70+2n clusters because
they have minimal 12 pentagons needed for geometric closure but
retain most of the graphite stabilization energy because of their
hexagonal rings.[19] Very recent theoretical studies[18,19] on the
stability of fullerenes indicate that at least one fullerene-type
(12 pentagons and n-1 hexagons cage) structure exists for all
even C2n (2n$\geq$1).

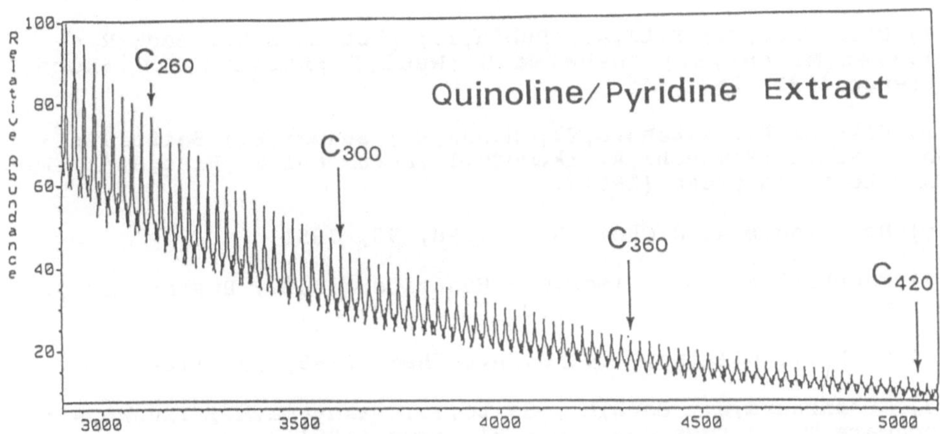

Fig.4 A Cs[+] SIMS spectrum of a large fullerene-rich black powder
(extract II in Fig.1) via quinoline/pyridine extraction in the
mass spectral region m/z = 2900 - 5100 (C242 - C424). The C2n
series is observed up to C500.

The present observation of even-numbered large carbon molecules strongly suggests the presence of a series of such fullerenes in laboratory-produced carbon soot. A similar series of
even-numbered large fullerenes is also observed in the negative
ion detection SIMS of the present sample.[20] It is extremely
interesting and important to know whether or not such a gigantic
carbon molecule has also a spheroidal fullerene structure.

Acknowledgment

H.S. thanks the Japanese Ministry of Education, Science and
Culture (Grant-in-Aid General Scientific Research and for Scientific Research on Priority Areas) for the support of the present
study.

References

(1) Kroto,H.W.; Heath,J.R.; O'Brien,S.C.; Curl,R.F.; Smalley,R.E. Nature 1985, 318, 162.

(2) Kraetschmer,W; Lamb,L.D.; Fostiropoulos,K; Huffman,D.R. Nature 1990, 347, 354.

(3) Haufler,R.E; Chai,Y.; Chibante,L.P.F.; Conceicao,J.; Changming Jin; Lai-Sheng Wang; Maruyama,S.; Smalley,R.E. Mat. Res. Soc. Proc. in press (1991).

(4) Diederich,F.; Ettl,R.; Rubin,Y.; Whetten,R.L.; Beck,R.; Alvarez,M.;Anz,S.;Sensharma,D.;Wudl,F.;Khemani,K.C.;Koch,A. Science 1991, 252, 548.

(5) Kikuchi,K.; Nakahara,Y.; Honda,M.; Suzuki,S.; Saito,K.; Shiromaru,H.;Yamauchi,K.;Ikemoto,I.;Kuramochi,T.;Hino,S.;Achiba,Y. Chem.Lett. in press (1991).

(6) Rohlfing,E.A. J.Chem.Phys. 1990, 93, 7851.

(7) Campbell,E.E.B.; Ulmer,G.; Hasselberger,B.; Busmann,H.-G.; Hertel,I.V. J.Chem.Phys. 1990, 93, 6900.

(8) So,H.Y.; Wilkins,C.L. J.Phys.Chem. 1989, 93, 1184.

(9) Shinohara,H.; Sato,H.; Saito,Y.; Takayama,M.; Izuoka,A.; Sugawara,T. J.Phys.Chem. Oct.31 issue (1991).

(10) Shinohara,H.; Sato,H.; Saito,Y.; Izuoka, A.; Sugawara,T.; Ito,H.; Sakurai, T.; Matsuo,T. J.Am.Chem.Soc. submitted.

(11) Shinohara,H.; Sato,H.; Saito,Y.; Tohji,K.; Udagawa,Y. Jpn.J.Appl.Phys. 1991, A30, L848.

(12) Shinohara,H.; Sato,H.; Saito,Y.; Tohji,K.; Matsuoka,I.; Udagawa,Y. Chem.Phys.Lett. 1991, 183, 145.

(13) Saito,Y.; Shinohara,H.; Ohshita,A. Jpn.J.Appl.Phys. 1991, A30, L1068.

(14) Matsuda,H.; Matsuo,T.; Fujita,Y.; Sakurai,T.; Katakuse,I. Int.J.Mass Spectrom.Ion Processes 1989, 91, 1.

(15) Kroto,H.W.; Mckay,K. Nature 1988, 331, 328.

(16) Kroto,H.W. Science 1988, 242, 1139.

(17) Fowler,P.W. Chem.Phys.Lett. 1986, 131, 444.

(18) Fowler,P.W. J.Chem.Soc.Faraday Trans. 1990, 86, 2073.

(19) Bakowies,D.; Thiel,W. J.Am.Chem.Soc. 1991, 113, 3704.

(20) Shinohara,H; Saito,Y.; Ito,H.; Matsuo,T. to be published.

Carbon K-Edge XANES and EXAFS of C60, C70, and K_3C60

H. Shinohara, H. Sato, Y. Saito[1], M. Kobayashi[2],
Y. Akahama[2], H. Kawamura[2], and K. Tohji[3]

Department of Chemistry for Materials
Mi'e University, Tsu 514, Japan
[1]Department of Electrical and Electronic Engineering,
Mi'e University, Tsu 514, Japan
[2]Department of Material Science, Faculty of Science
Himeji Institute of Technology
Hyogo 678-12, Japan
[3]Department of Resources Engineering
Tohoku University
Sendai 980, Japan

Abstract. The carbon K-edge absorption spectra of microcrystalline C60, C70 and potassium-doped K_3C60 have been measured using 250 - 500 eV synchrotron radiation. The C60 and C70 fullerenes exhibit π^* and σ^* resonances as has been observed in the carbon K-edge XANES (X-ray absorption Near-Edge Structure) spectra of graphite. However, the XANES spectra of C60 and C70 show fine structures in both of the resonances, and most of the spectral positions of the observed peaks are in good agreement with a reported density-of-state (DOS) calculation. The XANES feature of K_3C60 consists of the carbon K-edge and the potassium L-edge absorption which is overlapped with the σ^* transition of the carbon K-edge absorption. While EXAFS features of pure C60 and C70 are very weak, those of the potassium-doped K_3C60 are clearly observable. A Fourier transformation of the extracted EXAFS oscillation with an assumed phase factor gives the nearest-neighbour K-C distance of 3.1 A for $K_3 C60$.

1. Introduction

The very stable, nonreactive all-carbon fullerenes C60 and C70 were first identified in the condensing carbon vapor produced by laser vaporization of graphite.[1] Recently, they have been produced in macroscopic quantities by vaporization of graphite rods with resistive heating[2] or contact arc.[3] It is presumed that solid C60 and C70 exhibit unique and completely different electronic properties as compared, for example, with graphite and diamond. Perhaps most intriguing is the spherical shape of the non-planar conjugated pi-bond network of those fullerenes which is (formally) composed of sp^2 hybridized carbon atoms. Therefore,

P. Jena et al. (eds.), Physics and Chemistry of Finite Systems: From Clusters to Crystals, Vol. II, 1385–1390.
© 1992 Kluwer Academic Publishers.

it is particularly important to shed light, experimentally, on the π and σ electronic structure of C60 and C70.

In general, XANES (X-ray absorption near-edge structures) features are sensitive to the local symmetry and the type of chemical bonding of the exciting atom and can be useful as fingerprints to distinguish structures. As expected, the carbon K-shell XANES features of the microcrystalline C60 and C70 are quite different from those of graphite and diamond. The characteristic features observed in the π^* band in the C60 and C70 spectra are attributed to a cage-strained pi-bond network constrained in the spherical cage structures. The observed band structures are in good agreement with a theoretical prediction recently reported by Saito and Oshiyama.[4]

In this report EXAFS (Extended X-ray Absorption Fine Structure) analyses are also performed for potassium-doped K_3C60 microcrystals in order to obtain structural information on this superconductor. It is particularly useful for the structural study of the materials which lack long range order. The amplitudes of EXAFS oscillations are very small for pure C60 and C70, whereas those of K_3C60 are clearly observable. The obtained nearest-neighbour K-C distance is 3.1 A, which is shorter than the corresponding value determined by X-ray powder pattern analyses.[5]

2. Experimental

The XANES and EXAFS experiments have been performed on the BL-2B1 Beam Line of UVSOR (Ultra-Violet Synchrotron Radiation) at the Institute for Molecular Science. The beam line is equipped with a grasshopper monochromator (600 lines/mm mechanically ruled grating). The energy resolution of the monochromator with grating is 0.8 eV around the carbon K-edge. Background structures exist at the K absorption edge of carbon, which are due to carbon contaminants deposited on the optical surfaces. In order to correct for these structures and to eliminate intensity fluctuations arising from possible instabilities of the storage ring, a normalization-subtraction procedure is employed.[5] The background features are eliminated by subtracting the signal stemming from the Au-coated grid reference from that of C60 deposited on a substrate. This ensures optimum cancellation of all background structures. Energy calibration is performed by use of two transmission minima at 284.7 and 291.0 eV, which come from the carbon contamination.

Details of the production of fullerene-rich carbon soot by the contact arc method[3] have been described previously. Briefly, an arc between two graphite rods (13.5 mm in diameter) is sparked at 200 - 250 A in the direct current (DC) mode in a He (20 - 100 Torr) atmosphere. The optimum yield (12 % in weight) of various fullerenes was observed at a He pressure of ca. 20 Torr in the present experimental configuration.

Separation and purification of C60 and C70 are performed by column chromatography on neutral alumina (ICN Biomedicals Akt.I) with hexane/toluene solvent. Evaporation of the solvent is done at 250 °C.

3. Results and Discussion

The carbon K-edge photoabsorption spectra of microcrystalline C60 and C70 at room temperature are displayed in Fig.1. Several band peaks and shoulders are clearly seen. The resonance labeled π^* corresponds to a transition to π^* antibonding orbitals (or, more precisely, π^* bands in the case of the solids). Between 290 and 310 eV, transitions to σ^* antibonding orbitals are observed. The XANES spectra of C60 and C70 resemble that of graphite in terms of the presence of an $1s \longrightarrow \pi^*$ transition appearing at 284.8 eV, though the overall spectral features are entirely different (vide infra).

The presence of the π^* resonance at 284.8 eV is characteristic of unsaturated (sp^2 or sp) carbon bonds. Its prominent intensity in the C60 and C70 cluster therefore indicates the presence of unsaturated C-C bonds as in graphite and amorphous carbon (but in contrast to diamond). However, another prominent π^* peak is observed at 287.4 eV in the C60 spectrum. This is in marked contrast to the XANES spectra of the other forms of solid carbon. A similar fine structure is also observed in the C70 spectrum. In order to examine the angular dependence of the XANES features, the microcrystalline samples were rotated 20 - 80 degrees relative to the incident X-ray beam. No spectral changes were observed. This guarantees that the present absorption spectra reflect the density-of-states of the unoccupied states of C60 and C70.

Fig.1 Carbon K-shell XANES spectra of microcrystalline (a) C60 and (b) C70. Between 285 and 292 eV, transition to π^* antibonding orbitals are observed. The intensities of the two spectra are normalized at 325.8 eV, so that a direct comparison of the absorption intensity is possible.

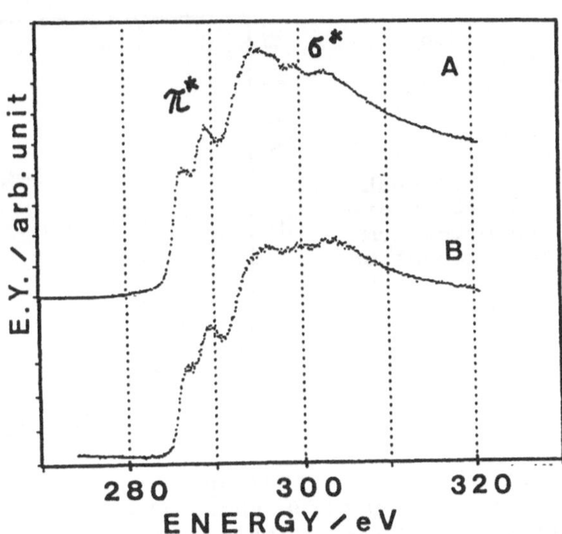

These spectral features agree well with the recent theoretical studies done by Saito and Oshiyama.[6,7] In the light of their calculation, for example, the origin of the π^* band peaks at 285.5 and 287.4 eV can be assigned as the $g_g \longrightarrow t_{2u}$ and $g_g \longrightarrow h_u$ transitions, respectively. They found that most levels between -6 eV and +7 eV (HOMO = 0 eV) had considerable dispersions in the fcc crystal because of large intercluster overlaps of the bond spreads outside the clusters.[7] Very recently, Terminello et al.[8] have reported a high resolution XANES study on thin film C60 and C70. Their XANES spectra show very strong peaks due to the 1s ⟶ LUMO transitions, which is weak in our spectra.

The relative intensity and the band width of the π^* transition (with respect to the σ^* transition) are increased on going from C60 (5.6 eV, fwhm) to C70 (6.3 eV). These results are indicative of the fact that in the C70 solid the level density of both the π^* and the σ^* transitions are increased in reference to C60 and that those of the π^* resonance are particularly increased. A very recent density-of-state (DOS) calculation of C70 by Saito and Oshiyama[6,7] using the local-density approximation in the density-functional theory also indicates that the DOS of the state of C70 is larger than that of C60. Furthermore, most of the spectral positions of the observed peaks and shoulders (both π^* and σ^* states) are in good agreement with the DOS calculation.

A part of the C70 surface thus becomes a smoothly curving net with more or less flat ellipsoidal surface. For example, there exist curved "tetracene-" and "perene-like" pi-bond networks of hexagonal rings on the C70 surface, which are not existent on the C60 cage. In this context, C70 is much more "graphite-like" than C60. The increasing graphite-like character in C70 is consistent with the XANES observation that the level density of the π^* transition of C70 is particularly increased.

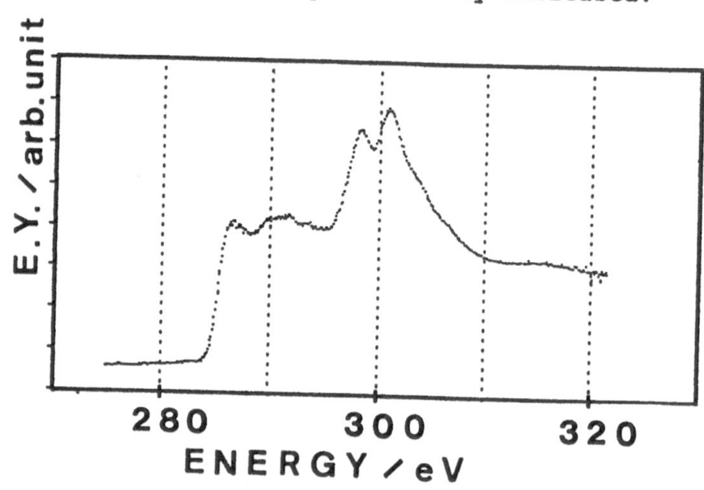

Fig.2 XANES spectrum of potassium-doped K3C60 in the same energy range as Fig.1. The intense peaks around 300 eV are due to potassium L-edge absorption. The K3C60 sample is prepared ex-situ and shows a superconducting transition at 19 K.

Fig.3 shows a XANES spectrum of potassium-doped K3C60 microcrystals. The spectrum shows the carbon K-edge π^* resonance and σ^* resonance which is partly overlapped with the potassium L-edge absorption. The intensity of the K L-edge absorption increases as the number of potassium atoms (n) in KnC60 increases. The results are consistent with a recent report[9] on the XANES spectra on the same system with much higher spectral resolution.

Fig.3 exhibits EXAFS features of C60, C70, and K3C60 which are present at much higher energy regions. As is clear, the EXAFS oscillations are very weak in pure C60 and C70, so that at present the analyses of the oscillations are not performed.

However, the EXAFS feature of K_3C60 is clearly observable.

Fig.3 EXAFS features of micro-crystalline (a) C60, (b) C70, and (c) K_3C60 at room temperature. Notice the difference in the background absorption.

Fig.4 shows the extracted EXAFS oscillation $\chi(k)$ which is obtained after subtraction of the smooth X-ray absorption background and normalization.[10,11] Here, the wave vector k is defined by the following equation:

$$k = \sqrt{2m(E-E_0) / \hbar^2} \qquad (1)$$

,where E is the energy of the incident X-ray beam and E_0 is the threshold energy of a L_{III} electron of K atoms. The E_0 value was set to be 293 eV in the present analysis. The function $k^3\chi(k)$ was Fourier transformed over the range of $3.0 \leq k \leq 7.4$.
The obtained magnitude, FT, is shown in the right side of Fig.4.

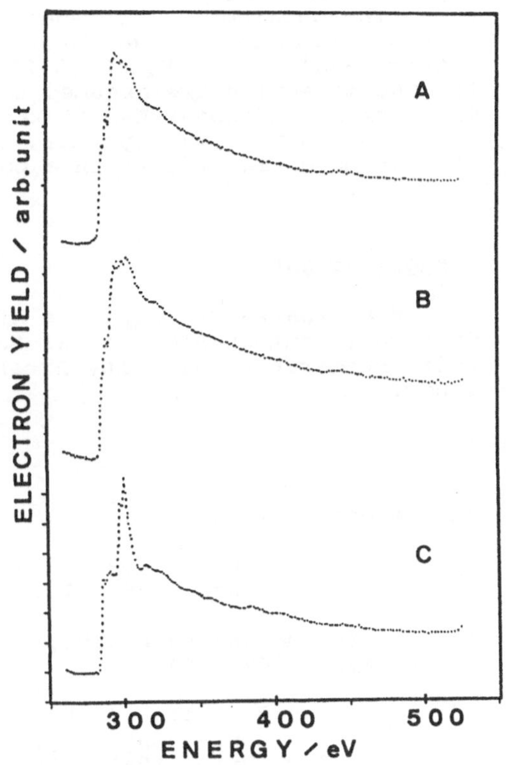

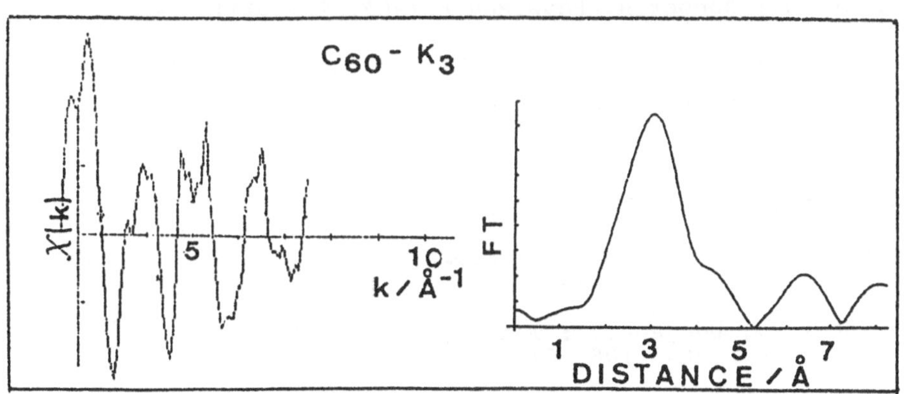

Fig.4 Extracted oscillation of Fig.3-c and its Fourier transform.

In the present calculations the values for the back scattering amplitude and the phase shift reported by Teo and Lee[12] were used. The obtained nearest-neighbour K-C distance is 3.06 Å. The present value is shorter than that obtained by an X-ray powder pattern analysis of K_3C60 (3.27 Å for tetrahedral site),[13] where the C60 molecules are assumed to be locked into a fixed position. The nearest-neighbour K-C distance decreases if the C60 molecules rotate in the K_3C60 crystals. The present results implies that the C60 molecules rotate in a cooperative manner (jump motion).

Acknowledgment

H.S. thanks the Japanese Ministry of Education, Science and Culture (Grant-in-Aid General Scientific Research and for Scientific Research on Priority Areas) for the support of the present study.

References

(1) Kroto,H.W.; Heath,J.R.; O'Brien,S.C.; Curl,R.F.; Smalley,R.E. Nature 1985, 318, 162.

(2) Kraetschmer,W; Lamb,L.D.; Fostiropoulos,K; Huffman,D.R. Nature 1990, 347, 354.

(3) Haufler,R.E; Chai,Y.; Chibante,L.P.F.; Conceicao,J.; Changming Jin; Lai-Sheng Wang; Maruyama,S.; Smalley,R.E. Mat. Res. Soc. Proc. 1991, 206, 627.

(4) Saito,S.; Oshiyama,A. Phys.Rev.B submitted.

(5) Stöhr,J.; Jaeger,R. Phys.Rev.B 1982, 26, 4111.

(6) Saito,S.; Oshiyama,A. Phys.Rev.Lett. 1991, 66, 2637.

(7) Saito,S.; Oshiyama,A. to be published.

(8) Terminello L.J. et al. Chem.Phys.Lett. 1991, 182, 491.

(9) Chen,C.T. et al. Nature 1991, 352, 603.

(10) Lytle,F.W.; Sayers,D.E.; Stern,E.A. Phys.Rev.B, 1975, 11, 4825.

(11) Tohji,K.; Udagawa,Y.; Tanabe,S.; Ueno,A. J.Am.Chem.Soc. 1984, 106, 612.

(12) Teo,B.K.; Lee,P.A. J.Am.Chem.Soc. 1979, 101, 2815.

(13) Stephens,P.W. et al. Nature 1991, 351, 632.

TIGHT-BINDING MOLECULAR DYNAMICS STUDY OF C$_{60}$ AND OTHER CARBON CLUSTERS

C. Z. Wang, C. H. Xu, B. L. Zhang, C. T. Chan, and K. M. Ho

Ames Laboratory-USDOE,
and Microelectronics Research Center and Dept. of Physics
Iowa State University
Ames, Iowa 50011, USA

ABSTRACT

The structure and dynamics of C$_{60}$ buckyball and carbon clusters C$_n$ (n=2–90) have been studied with molecular dynamics simulations using a tight-binding potential model. The studies show that it is possible to nucleate a 'buckyball-like' cluster by cooling and compressing carbon atoms from the gas phase. The studies also show that there is a transition from one-dimensional linear and cyclic structures to two-dimensional cage structures as the number of carbon atoms reaches n=20. Magic numbers for fullerene formation energy are observed at n=50,60,70 and 84.

1. INTRODUCTION

The breakthrough in synthesis[1] of C$_{60}$ buckyballs and other fullerenes creates a very active research area in physics and chemistry. It is now well established that the ground state geometry of the C$_{60}$ molecule is a truncated icosahedron consisting of 20 six-membered rings and 12 five-membered rings as proposed several years ago by Kroto et al.[2] Other fullerenes such as C$_{70}$ and C$_{76}$ have also been identified by experiments.[3–5] Fullerenes emerged as a new phase of carbon with a range of structural and chemical properties intermediate between small molecules and bulk phases and have attracted considerable scientific as well as technological interests. Nevertheless, the microscopic mechanisms of formation, fragmentation, as well as stability of the fullerenes are still not well understood. Recently, we have addressed above issues through extensive molecular-dynamics studies of the structure and dynamics of the fullerenes using a tight-binding potential model. In this paper, we report on the studies of vibration, disintegration and nucleation of the C$_{60}$. We also report on our systematic study of the structure of carbon clusters for cluster sizes ranging from n=2 to n=90.

P. Jena et al. (eds.), Physics and Chemistry of Finite Systems: From Clusters to Crystals, Vol. II, 1391–1396.
© 1992 *Kluwer Academic Publishers.*

2. TIGHT-BINDING POTENTIAL MODEL FOR CARBON

In our tight-binding (TB) approach, the total potential energy of the system is expressed as

$$E(\{\vec{r}_i\}) = \sum_{n}^{occupied} < \psi_n | H_{TB}(\{\vec{r}_i\}) | \psi_n > + E_{rep}(\{\vec{r}_i\}), \tag{1}$$

where $\{\vec{r}_i\}$ denotes the positions of the atoms (i=1,2,...N). The first term in (1) is the electronic band-structure energy calculated by a parametrized tight-binding Hamiltonian $H_{TB}(\{\vec{r}_i\})$, and the second term is a short-ranged repulsive energy. The orthogonal sp^3 basis tight-binding Hamiltonian $H_{TB}(\{\vec{r}_i\})$ is described by the following parameters: ϵ_s and ϵ_p are the on-site atomic energies, $v_{ss\sigma}(r_{i,j})$, $v_{sp\sigma}(r_{i,j})$, $v_{pp\sigma}(r_{i,j})$ and $v_{pp\pi}(r_{i,j})$ are the hopping TB matrix elements as a function of interatomic distance $r_{i,j}$. The repulsive energy is expressed as $E_{rep} = \sum_i f[\sum_j \phi(r_{i,j})]$. The above parameters and functions are determined by fitting to first-principles calculation results of the electronic band structure and volume-dependent binding energies of various crystalline carbon phases with emphasis on the graphite, diamond and the linear chain structures. The resultant TB potential reproduces well the binding energies and bond lengths of crystalline carbon with different coordination numbers. It also describes well the properties of carbon in the liquid and amorphous states.[6] The reliability of the potential for applications to carbon clusters is tested by comparing the ground state geometries obtained by the present tight-binding model with the results of accurate quantum mechanical calculations. In the range $5 \leq n \leq 11$, we found that odd-numbered clusters prefer a linear structure, while even-numbered clusters prefer a ring structure. This result agrees well with the accurate ab-initio calculation results of Raghavachari et al.[7] Since the potential parameters are determined by fitting to the bulk properties, we expect that the calculation will become even more reliable as the cluster size gets bigger.

3. STRUCTURE AND DYNAMICS OF C$_{60}$

Molecular-dynamics simulation using the above potential model showed that the ground state geometry of C$_{60}$ is an ideal buckyball with 1.40 Å for the 'double bonds' (the shared hexagon edges) and 1.46 Å for the 'single bonds' (the pentagon edges). These values are in good agreement with the experimental(NMR) data[8] and first-principles density functional calculation results[9] of 1.40 Å and 1.45 Å. Finite temperature simulations show that the buckyball is very stable against thermal disintegration. The carbon-carbon bonds of the buckyball did not break until the temperature reaches 5500K. At 6000K, the buckyball disintegrates rapidly into chain-like segments. Fragmentation of the C$_2$ dimers from the cluster is also observed in the disintegration process.

The vibrational spectrum of the C$_{60}$ buckyball is obtained from the velocity-velocity correlation function in the simulation. The results at T=11K (Fig. 1) show 46 distinct vibrational modes divided equally into even and odd parities. Four odd-parity modes are infrared-active and ten of the even-parity modes are observable by Raman spectroscopy. The present theoretical results are in good agreement with the

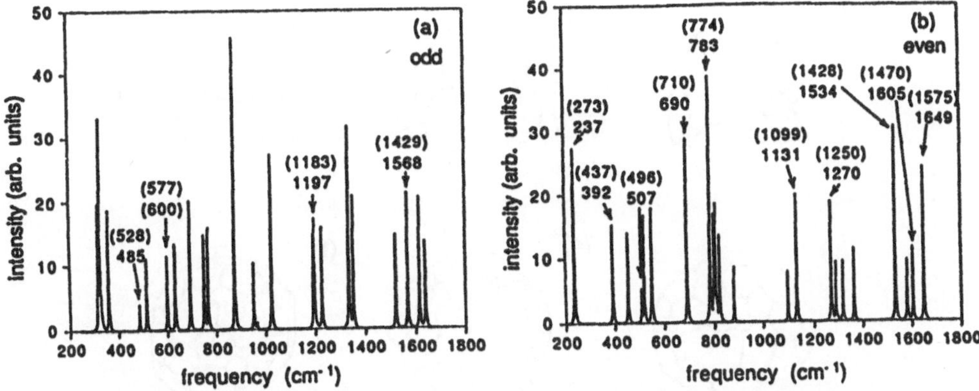

Fig.1. Vibrational spectrum of C_{60} from molecular dynamics simulation. The modes with odd parity and even parity are shown in (a) and (b) respectively. The arrows in (a) and (b) indicate the infrared active and Raman active modes. Calculated frequencies of these mode are shown in comparison with experimental data(the numbers in the parentheses).

available infrared and Raman data,[10,11] considering the fact that the properties of C_{60} have not been used in the fitting bases to determine the potential parameters.

In a first step towards understanding the formation and growth of the C_{60} molecules, we enclosed 60 carbon atoms in a hollow sphere of radius R. Perfect specular reflection occurs when the atoms hit the inner surface of the sphere. We start the simulation by heating the carbon atoms to very high temperatures (10000 K) in a large sphere (R=9.22 Å) with the carbon atoms in the gas phase. The temperature and the radius of the sphere are gradually reduced. At 6000K nucleation of the cluster occurs, proceeding from chain-like structure to cap-like fragments (Fig. 2). Further reduction of the radius to 3.832 Å accelerates the formation of a closed cage structure. We note that the structure of the cage is very similar to that of the buckyball apart from some defects which are probably due to the rapidity of the compression and cooling in the simulation. Further cooling to T=0 K results in a metastable C_{60} whose cohesive energy is higher than that of the optimized buckyball by 0.12 eV/atom.

4. STRUCTURAL TREND OF CARBON CLUSTERS

In order to understand the general structural trend and relative stability of the fullerenes, we have performed a systematic study of the structure and formation energy of the carbon clusters as a function of cluster size. For small clusters C_n ($n \leq$ 11), the ground state geometries are easily obtained by simulated annealing. For $n \geq$ 12, we use the 'pressure cooker' scheme as described in Section 3. We found that monocyclic rings are energetically favorable for the clusters with $12 \leq n \leq 19$, while for clusters with $20 \leq n \leq 60$, the most stable structures are cages made up of 5- and 6-membered rings. In Fig. 3, we plot the formation energy as a function of cluster size ranging from n=2 to n=70 for the linear chain, monocyclic ring, graphitic fragment

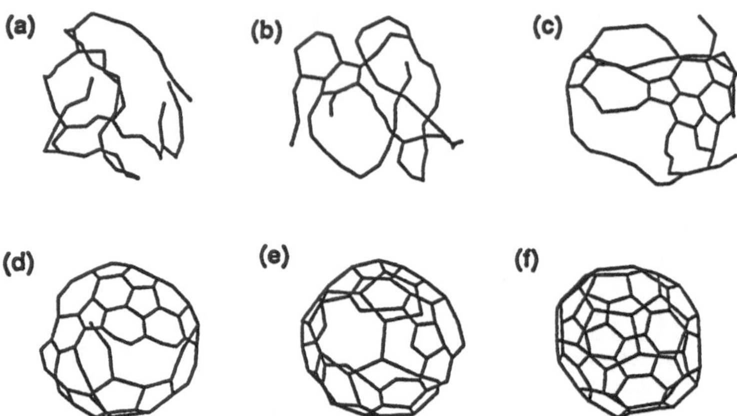

Fig. 2. Perspective view of C_{60} cage formation process at T=6000 K. The carbon atoms are connected by straight lines when the interatomic distances are less than 1.8 Å. (a), (b), and (c) are snapshots at 2.8ps, 4.2ps, and 5.6ps respectively and with the sphere radius R=5.32 Å. (d), (e), and (f) are typical snapshots when the sphere radius is reduced to 4.61 Å, 3.90 Å, 3.832 Å respectively. Note a closed cage forms when R=3.832 Å.

and cage structures respectively. The transition from one-dimensional ring structures to two-dimensional cage structures when n reaches 20 is clearly shown. C_{20} is the smallest closed cage consisting of 12 pentagons. If we further compress C_{18} and C_{19} from the monocyclic rings we will obtain less stable multiple ring structures. Even-numbered cages are energetically more favorable than the odd-numbered ones. This is because, geometrically, closed cage structures cannot be formed with all the atoms three-fold coordinated when n is odd.[12]

For $n \geq 30$, the cluster structures obtained from the 'pressure cooker' contain defects such as 4-membered and 7-membered rings, or topological faults in the arrangement of the 5-membered and 6-membered rings. The presence of such defects is due to the rapidity of the compression and cooling in the simulation. It is very difficult to remove these defects by simulation annealing since the energy needed to break and to rearrange the network is very large. In order to overcome this difficulty, we developed a new scheme in which the fullerene networks with the correct topological connections are first generated by the computer. The geometry are then optimized by tight-binding molecular dynamics to get the nearest local minimum. Fig. 4 shows the formation energy of the optimized even-number fullerenes ranging from n=20 to 90 obtained by the new scheme. The results of the new method are identical to those obtained by the 'pressure cooker' method for $n \leq 30$. However, for larger fullerenes, the new method is found to be much more efficient. In particular, the ground state structures of $C_{60}(I_h)$, $C_{70}(D_{5h})$ and $C_{76}(D_2)$ predicted by this scheme are in excellent agreement with experiment. An interesting result is that the ground state structure of C_{84} predicted by the present scheme is found to be energetically more stable than the chiral structure (similar to that of C_{76}) proposed by Fowler,[13] although they both have the D_2 point group symmetry.

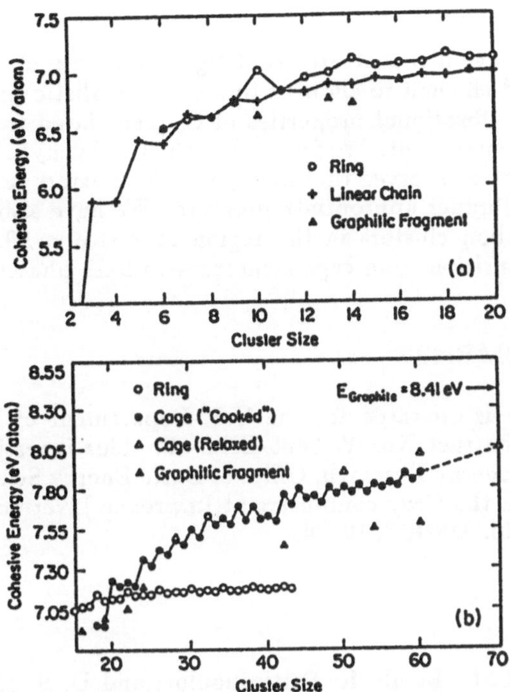

Fig. 3. Formation energy as a function of carbon cluster size for (a) chains, rings, and graphitic fragments with $2 \leq n \leq 20$, and (b) rings, cages, and graphitic fragments with $15 \leq n \leq 60$. Cages indicated by ('Cooked') are obtained from the 'pressure cooker'. Cages indicated by (Relaxed) are obtained by relaxing the structures proposed by Kroto (Ref. 14) based on geometry considerations.

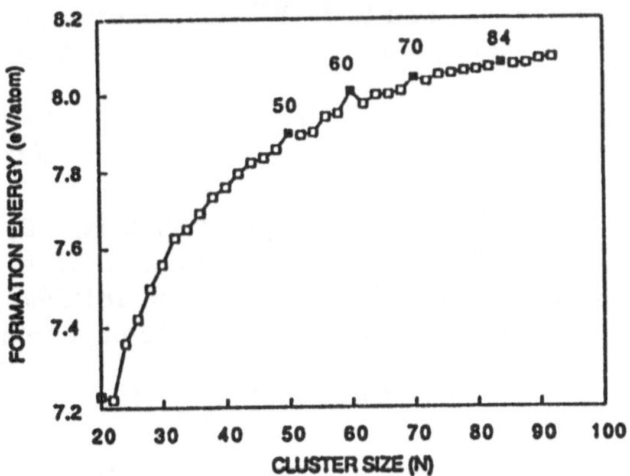

Fig. 4. Formation energy as a function of carbon cluster size with $20 \leq n \leq 90$ obtained by the new scheme.

5. SUMMARY

In summary, we have demonstrated that tight-binding molecular-dynamics is sufficiently accurate and efficient to allow us to perform realistic simulations of fullerenes. The structural and vibrational properties of C_{60} calculated from the present scheme compare well with experiment. We found that the C_{60} buckyball is very stable against disintegration. We also observed that buckyball-like clusters can grow at high temperatures ($T \approx 6000$ K) under appropriate pressure. We have also studied the structure and stability of carbon clusters in the region of $2 \leq n \leq 90$. The study of larger clusters and the transition from cage structures to bulk phases are in progress.

ACKNOWLEDGMENTS

Ames Laboratory is operated for the U.S. Department of Energy by Iowa State University under Contract No. W-7405-ENG-82. This investigation was supported by the Director for Energy Research, Office of Basic Energy Sciences including a grant of computer time on the Cray computers at Lawrence Livermore Laboratory and by NSF under Grant No. DMR-8819379.

REFERENCES

1. W. Krätschmer, L. D. Lamb, K. Fostiropoulos, and D. R. Huffman, Nature, **347**, 354(1990).
2. H. W. Kroto, J. R. Heath, S. C. O'Brien, R. F. Curl, and R. E. Smalley, Nature, **318**, 162(1985).
3. R. Taylor, J. P. Hare, A. K. Abdule-Sada, and H. J. Kroto, J. Chem. Soc., Chem. Commun., **20**, 1423(1990).
4. R. D. Johnson, G. Meijer, J. R. Salem, and D. S. Bethune, J. Am. Chem. Soc., **113**, 3619(1991).
5. R. Ettl, I. Chao, F. Fiederich, and R. L. Whetten, Nuture, **353**, 149(1991).
6. For details of the tight-binding model, see C. H. Xu, C. Z. Wang, C. T. Chan and K. M. Ho, to be published.
7. K. Raghavachari and J. S. Binkley, J. Chem. Phys., **87**, 2191(1987).
8. C. S. Yannoni, P. P. Bernier, D. S. Bethune, G. Meijer, and J. R. Salem, J. Am. Chem. Soc., **113**, 3190(1991).
9. Q. Zhang, Jae-Yel Yi, and J. Bernholc, Phys. Rev. Lett., **66**, 2633(1991).
10. D. S. Bethune, G. Meijer, W. C. Tang, H. J. Rosen, W. G. Golden, H. Seki, C. A. Brown and M. S. de Vries, Chem. Phys. Lett., **179**, 181(1991).
11. W. Krätschmer, K. Fostiropoulos, and D. R. Huffman, Chem. Phys. Lett., **170**, 167(1990).
12. T. G. Schmalz, W. A. Seitz, D. J. Klein, and G. E. Hite, J. Am. Chem. Soc., **110**, 1113 (1988).
13. P. W. Fowler, J. Chem. Soc,. Faraday Commun., **87** 1945(1991).
14. H. W. Kroto, Nature, **329**, 529 (1987).

VIRTUAL SYMMETRIC CHARGE TRANSFER SUPERCONDUCTING PAIRING EXCITATIONS IN C_{60}

C. T. WHITE, M. R. COOK, B. I. DUNLAP, R. C. MOWREY,
D. W. BRENNER, P. P. SCHMIDT, and J.W. MINTMIRE
Naval Research Laboratory, Code 6179
Washington DC 20375-5000

ABSTRACT. The recent discovery of superconductivity in the alkali-fullerides has focused intense interest on the origin of the superconductivity in these new cluster materials. These materials' narrow conduction bands, short coherence lengths, and unusually high T_c's suggest that electron-electron interactions are potentially important in explaining their superconductivity. We have examined the character of the lowest energy, most retarded, virtual charge transfer excitations of a single C_{60} molecule. We find that these excitations produce identical intermediate polarizations on opposite sides of the C_{60} molecule leading to a pairing interaction that operates across the equatorial plane of the C_{60} molecule. These excitations arise from local field effects and, in contrast to more usual dynamic polarization mechanisms, are not dipole active. To the extent that this mechanism proves important, it begins to provide a cogent framework for the design of other cluster- and molecular-based superconductors.

1. Introduction

The production of uniquely abundant, preeminently stable C_{60} clusters in laser vaporization supersonic cluster beam experiments immediately led to the suggestion of a beautiful truncated icosahedral structure for this special cluster [1]. Although chemically and aesthetically appealing, this proposed structure had to wait five years to be definitively verified by NMR experiments, made possible by breakthroughs yielding macroscopic amounts of this single species [2]. With the production and purification of large quantities of C_{60}, progress in using this new allotrope of carbon to synthesize cluster materials with remarkable properties has proceeded at a breathtaking pace from crystalline molecular solids [3], to metals [4], to superconductors [5].

Hebard *et al.'s* discovery [5] of superconductivity at 18 K in potassium doped C_{60} followed by the observation of superconductivity in Rb_xC_{60} [6,7], Cs_xC_{60} [8], and $Cs_xRb_yC_{60}$ [9] alkali-fullerides, with superconducting transition temperatures, T_c's, close to and in excess of 30 K, has focused intense interest in the origin of the superconducting pairing interactions in these materials. These stunning experimental results—so far conservatively yielding T_c's within striking distance of the high T_c copper-oxides—establish the alkali-fullerides as by far the best organic based superconductors yet synthesized. In spite of the intense interest, the origin of the pairing interaction in these materials remains unknown with many crucial experiments such as isotope studies yet to be performed.

Although phonon-mediated pairing alone may eventually account for the superconductivity in these cluster materials [10,11], their narrow conduction bands [12], fairly short coherence lengths [13], and unusually high T_c's [5-9] suggest that a more ex-

1397

P. Jena et al. (eds.), Physics and Chemistry of Finite Systems: From Clusters to Crystals, Vol. II, 1397–1402.
© 1992 *Kluwer Academic Publishers.*

otic pairing mechanism may also be important. Possibilities include the RVB model [14], spin bags [15], negative U models [16-19], and polarization (charge-transfer) mechanisms [20-22] to name but a few. In the next Section we discuss two of these interrelated possibilities: the negative U and dynamic charge-transfer polarization-mediated pairing models. This discussion leads up to our introduction in Section 3 of a novel pairing mechanism acting across the C_{60} molecule which is mediated by virtual, symmetric charge-transfer excitations. This type of pairing interaction appears to enjoy many of the advantages of the negative U and classic exciton models while perhaps avoiding some of their drawbacks. The idea of a pairing interaction mediated by somewhat retarded, virtual, symmetric charge-transfer excitations is not restricted to the alkali-fullerides [23]. Hence—to the extent this mechanism proves important—it begins to provide a cogent framework for the design of other molecular and cluster based superconductors.

2. Background

Because of its large number of internal degrees of freedom and comparatively hefty size it is possible that each C_{60} molecule in the alkali-fullerides acts as an Anderson negative U center [24]. Indeed just such a suggestion has recently been made by Chakravarty and Kivelson [25] who rely on the internal electronic degrees of freedom of the C_{60} molecule to mediate a local (on-ball) attractive effective electron-electron interaction in the *static limit*. The possibility of negative U mediated and enhanced superconductivity arising from complex internal molecular degrees of freedom was pointed out over a decade ago [16,17] and current developments in this area have been recently reviewed [19].

If each C_{60} acts as a negative U center, then in contrast to phonon mediated pairing which is cutoff within the Debye frequency, ω_D, of the Fermi level, the pairing interaction acts over the entire Brillouin zone. As a result, the effective half-bandwidth, $B/2$, and not ω_D acts as the cutoff. Because of the large number of electrons that can then participate in the pairing, this large cutoff can lead to relatively short coherence lengths and fairly high T_c's even for moderate values of U. In addition, because this large cutoff will be modulated by the degree of band filling, this model suggests a maximum T_c in the vicinity of a half-filled band. Such a maximum has been observed for K_3C_{60} [7]. These properties of the negative U model can be readily seen from the weak-coupling treatment of a single band negative U model, which assuming a constant density of states per spin of $\rho = 1/B$, yields [19], $k_B T_c = 1.134(B/2)\sqrt{4n(1-n)}e^{-B/|U|}$, in terms of the degree of band-filling n ($0 < n < 1$). This result shows that T_c can decrease with increasing B and hence this model is also consistent with the observed pressure dependence of T_c in K_3C_{60} [26]. Thus, the postulate that each C_{60} acts as a negative U center in the alkali-fullerides is enticing.

However, defeating the long-range direct Coulomb repulsion to achieve a negative U center is not easy. In general for this to occur requires that in the static limit that the effective local electron-electron interaction given by [27]:

$$V_{\text{eff}} = E_0 + E_2 - 2E_1, \tag{1}$$

is less than zero, where E_0, E_1, and E_2 are the total energies of the empty, and singly and doubly occupied center respectively. For C_{60} this implies that $2C_{60}^{(-1)}$ will disproportionate in the alkali-fullerides: $2C_{60}^{(-1)} \rightarrow C_{60}^{(-2)} + C_{60}$. Complex disproportionation reactions do occur and examples of negative U centers in solids are known but hardly common. A well-known example is the single vacancy in crystalline Si [28,29]. In this instance the local attractive effective electron-electron interaction is mediated by a self-consistent relaxation in the neighborhood of the vacancy caused by changing occupancy [29]. Because the vacancy in Si is smaller than C_{60}, it may seem that if the direct Coulomb repulsion is defeated there, then it can be easily defeated in C_{60}. But this vacancy is embedded in a highly-covalent solid while C_{60} is embedded in a poor metal and this makes avoiding the direct Coulomb repulsion difficult. Electrochemical doping experiments should be able test if C_{60} is a negative U center.

Another enticing idea for exotic pairing in the alkali-fullerides is the polarization based models of Refs. 20-22 with C_{60} adopting the role of the polarizable complex. Within these models one electron causes a particle-hole excitation in the complex with energy Γ and matrix element M_1 and a second electron returns this excited species to its ground state with matrix element M_2. Because both electrons are assumed to act at the same location, $M_2 = M_1^*$ and this exchange of virtual excitons between these two electrons results in an effective attractive component to their overall interaction of the form, $V_{\text{eff}} = -4|M_1|^2/\Gamma$, which is cutoff by Γ due to retardation [30]. These models rely on a strong dipolar response in the material to produce an attractive local component to the overall electron-electron interaction. A hallmark of such a strong dipolar response is excitonic structure in the frequency-dependent optical dielectric function of the material [21,22]. However, the intermediate dipole potential is short-ranged and hence operates where the direct Coulomb repulsion between the electrons is large. Thus, the negative U and exciton models both have to contend with the direct Coulomb repulsion. Indeed, for a two-level system it is easy to show that the attractive interaction due to the exchange of excitons is just the lowest-order term of Eq. 1. One way around the direct Coulomb repulsion is to assume that the effective interaction, which is actually frequency dependent, is strongly retarded, $i.e.$ Γ is small, so that this replusion can be avoided. But if the coherence length is short and the T_c's are high, then many electrons would have to participate in the pairing so that interaction cannot be too retarded. To avoid the direct Coulomb repulsion while maintaining the short coherence lengths then becomes a difficult although perhaps not impossible balancing act. In the next Section, building upon the negative U and exciton models, we suggest a way around these difficulties that may apply to the alkali-fullerides.

3. Symmetric Charge Transfer Mediated Pairing

Our approach is to view the C_{60} molecule as a polarizable group just as in the exciton models. Our principal result is depicted in these terms in Fig. 1, where an electron at one side but exterior to the C_{60} molecule is pictured as inducing a symmetric response across the C_{60} molecule, yielding an intermediate attractive potential for an electron on the opposite side but also exterior to the molecule [31]. We do not assert that this is the static response of the molecule, which could lead to a negative U center, but rather that component of the dynamic response resulting from the lowest-energy, most-retarded, virtual excitations across the HOMO-LUMO gap. Although this type of response is by no means typical, we find that for C_{60} it is required, because both the h_u HOMO and t_{1u} LUMO states are antisymmetric under inversion. This requires, to second order in perturbation theory, that these virtual HOMO-LUMO excitations mediate an identical attractive electron-electron interaction for electrons on the same and opposite sites across the molecule. However, if these electrons are at different sides of the molecule then their direct Coulomb repulsion—reduced by screening and retardation—can be better avoided. Therefore, the exchange of these virtual excitations should favor pairing across the molecule.

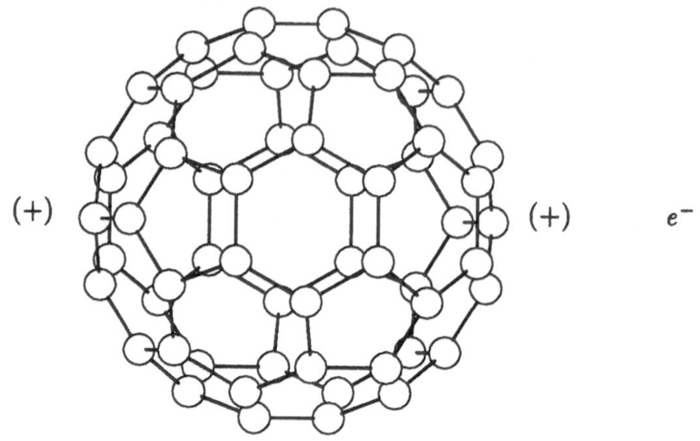

Fig. 1 Structure of C_{60} molecule showing the intermediate HOMO-LUMO polarization resulting from a minus charge at the right.

To explore if this is an appreciable effect, we have carried out a preliminary calculation using a Hückel model for the π orbitals normal to the surface of the molecule. These calculations were implemented by assuming off-diagonal matrix elements of the form $V_o \exp[-2(R/R_o - 1)]$. The parameters R_o and V_o, the "*equilibrium*" separation and hopping matrix element, were chosen to be 1.4 Å and -2.4 eV respectively. Eq. 1 for V_{eff} was then evaluated by assuming that each exterior electron only affects a single pentagon in the cage by shifting the corresponding site diagonal levels by W. V_{eff} was then calculated from Eq. 1 by obtaining the Hückel energy of the molecule without any perturbation, with only one pentagon affected, and with two pentagons on opposite sides of the molecule affected. Within this simplified approach these three quantities correspond to E_0, E_1, and E_2 respectively in Eq. 1. Two different V_{eff}'s were defined: one by summing over all the occupied states V_{eff}^0, and the other by summing over only the original h_u states, V_{eff}^1. The results are depicted in Fig. 2. To second order in W, V_{eff}^1 will correspond to the component of the effective interaction across the molecule resulting from virtual excitations from the HOMO states to intermediate higher unoccupied states of the molecule only. As can be seen, although V_{eff}^0 is small, V_{eff}^1 is large leading us to expect that this component of the interaction may prove important to understanding the superconductivity properties of the alkali-fullerides. This response is not the usual response caused by bonding to antibonding excitations across the gap. Such an intermediate dipolar response will lead to a repulsive contribution to V_{eff} across the complex. The magnitude of this response is also unusually large because of the large degeneracy of the HOMO states. These results show that this pairing interaction depends crucially on valence and symmetry. The fact that we find such a large intermediate response suggests that the C_{60} impurities embedded in normal superconductors might even enhance

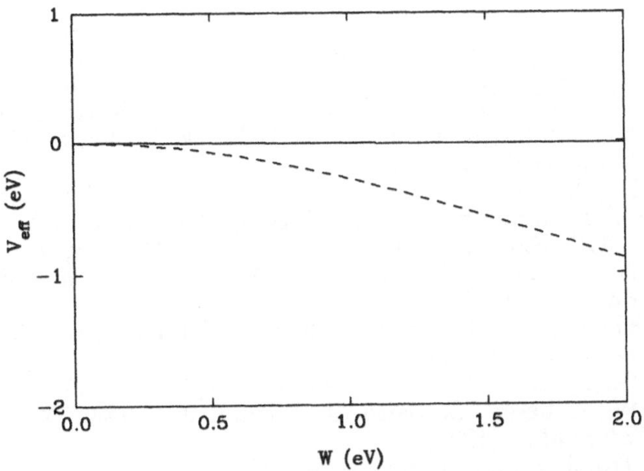

Fig. 2 V_{eff} as a function of W arising from all the occupied states (solid line) and from the HOMO h_u states (dashed line) only.

their T_c's.

This symmetric charge-transfer mediated pairing draws a compromise between static (negative U) and dynamic (exciton) charge-transfer mediated pairing. The overall polarization need not lead to a negative U but the interaction acts over a large enough distance that the cutoff need not be that low either.

4. Summary

Building on the negative U and exciton models we have pointed out that pairing interactions across the C_{60} molecule mediated by the exchange of virtual symmetric charge transfer excitations may prove important in understanding the occurrence of superconductivity in the alkali-fullerides. To the extent such excitations prove important they represent a paradigm for designing other molecular based superconductors.

Acknowledgements

This work was supported by the Office of Naval Research. One of us (CTW) thanks S. Chakravarty for sending him a preprint of Ref. 25 prior to publication.

References

[1] H.W. Kroto, J.R. Heath, S.C. O'Brien, R.F. Curl, and R.E. Smalley, *Nature* **318**, 162 (1985).
[2] W. Krätschmer, L.D. Lamb, K. Fostiropoulos, and D.R. Huffman, *Nature* **347**, 354 (1990).

[3] R.M. Fleming, T. Siegrist, P.M. Marsh, B. Hessen, A.R. Kortan, D.W. Murphy, R.C. Haddon, R. Tycko, G. Dabbagh, A. M. Mujsce, M.L. Kaplan, and S. M. Zahurak, *Mat. Res. Soc. Symp. Proc.* **206**, 691 (1991).

[4] R.C. Haddon, A.F. Hebard, M.J. Rosseinsky, D.W. Murphy, S.J. Duclos, K. B. Lyons, B. Miller, J.M. Rosamilia, R.M. Fleming, A.R. Kortan, S.H. Glarum, A. V. Makhija, A.J. Muller, R.H. Eick, S.M. Zahurak, R. Tycko, G. Dabbagh, and F.A. Thiel, *Nature* **350**, 320 (1991).

[5] A.F. Hebard, M.J. Rosseinsky, R.C. Haddon, D.W. Murphy, S.H. Glarum, T. T. M. Palstra, A.P. Ramirez, and A.R. Kortan, *Nature* **350**, 601 (1990).

[6] M.J. Rosseinsky, A.P. Ramirez, S.H. Glarum, D.W. Murphy, R.C. Haddon, A. F. Hebard, T.T.M. Palstra, A.R. Kortan, S.M. Zahurak, and A.V. Makhija, *Phys. Rev. Lett.* **66**, 2830 (1991).

[7] K. Holczer, O. Klein, S-M. Huang, R.B. Kaner, K-J Fu, R.L. Whetten, and F. Diederich, *Science* **252**, 1154 (1991).

[8] S.P. Kelty, C-C. Chen, and C.M. Lieber, *Nature* **352**, 223 (1991).

[9] K. Tanigaki, T.W. Ebbesen, S. Saito, J. Mizuki, J.S. Tsai, Y. Kubo, and K. Kuroshima, *Nature* **352**, (1991).

[10] J.L. Martins, N. Troullier, and M. Schabe, preprint.

[11] M. Schlüter, M. Lannoo, M. Needels, G.A. Baraff, and D. Tománek, preprint.

[12] S. Saito and A. Oshiyama, preprint.

[13] K. Holczer, O. Klein, G. Gruner, J.D. Thompson, F. Diederich, and R.L. Whetten, *Phys. Rev. Lett.* **67**, 271 (1991).

[14] P.W. Anderson, *Science* **235**, 1196 (1987).

[15] J.R. Schrieffer, X.G. Wen, and S.C. Zheng, *Phys. Rev. Lett.* **60**, 944 (1988).

[16] E. Simanek, *Solid State Commun.* **32**, 731 (1979).

[17] C.S. Ting, K.L. Ngai, and C.T. White, *Phys. Rev. B* **22**, 2318 (1980).

[18] S. Robaszkiewicz, R. Micnas, and K.A. Chao, *Phys. Rev. B* **24**, 4018 (1981); *ibid.* **26**, 3915 (1982).

[19] R. Micnas, J. Ranninger, and S. Robaszkiewicz, *Rev. Mod. Phys.* **62**, 113 (1990).

[20] W.A. Little, *Phys. Rev.* **134**, A1416 (1964).

[21] V.L. Ginzburg, *Contemp. Phys.* **9**, 355 (1968).

[22] D. Allender, J. Bray, and J. Bardeen, *Phys. Rev. B* **7**, 1020 (1972).

[23] C.T. White, preprint.

[24] P.W. Anderson, *Phys. Rev. Lett.* **34**, 953 (1975).

[25] S. Chakravarty and S. Kivelson, *Europhys. Lett.*, in press.

[26] G. Sparn, J.D. Thompson, S.-M. Huang, R.B. Kaner, F. Diederich, R.L. Whetten, G. Grüner, and K. Holczer, *Science* **252**, 1829 (1991).

[27] C.T. White, and K.L. Ngai, *J. Vac. Sci. Technol.* **16**, 1412 (1979); K.L. Ngai, and C.T. White *J. Appl. Phys.* **52**, 320 (1981).

[28] G.D. Watkins and J.R. Troxell, *Phys. Rev. Lett.* **44**, 593 (1980).

[29] G.A. Baraff, E.O. Kane, and M. Schlüter, *Phys. Rev. Lett.* **43**, 956 (1979).

[30] This result from 2nd order perturbation theory assumes a two-level model with the lower state occupied.

[31] These two additional electrons are taken exterior to the complex for simplicity only—they could also be on the same molecule.

INTERNATIONAL ADVISORY BOARD

Honorary Chairman: L. Pauling (U.S.A.)
Chairman: P. Jena (U.S.A.)
Co-Chairman: C. Janot (France)
Secretary: S. N. Khanna (U.S.A.)
Treasurer: B. K. Rao (U.S.A.)

A. S. Arrott (Simon Fraser University, Canada)
R. S. Berry (University of Chicago, U.S.A.)
J. Buttet (Institut de Physique Experimentale, Switzerland)
H. S. Chen (AT&T Bell Laboratories, U.S.A.)
M. Ciftan (Army Research Office, U.S.A.)
J. Friedel (Université Paris-Sud, France)
G. C. Hadjipanayis (University of Delaware, U.S.A.)
J. Jortner (Tel Aviv University, Israel)
A. Kaldor (Exxon Research and Engineering Co., U.S.A.)
E. Recknagel (Universität Konstanz, Germany)
S. J. Riley (Argonne National Laboratory, U.S.A.)
D. Shechtman (Technion University, Israel)
R. E. Smalley (Rice University, U.S.A.)
P. J. Steinhardt (University of Pennsylvania, U.S.A.)
S. Sugano (University of Tokyo, Japan)

LOCAL ORGANIZING COMMITTEE

L. Bloomfield (University of Virginia)
S. El-Shall (Virginia Commonwealth University)
D. Hartman (Virginia Commonwealth University)
P. Jena (Virginia Commonwealth University)
W. Jesser (University of Virginia)
S. N. Khanna (Virginia Commonwealth University)
A. C. Lilly (Philip Morris Research Center)
G. Melson (Virginia Commonwealth University)
D. Phillips (NASA Langley Research Center)
J. Poon (University of Virginia)
B. K. Rao (Virginia Commonwealth University)
H. Schone (College of William & Mary)
J. Schug (Virginia Polytechnic Institute & State University)
C. Spencer (Virginia Polytechnic Institute & State University)
C. E. Stronach (Virginia State University)

Dr. Yohji Achiba
Department of Chemistry
Tokyo Metropolitan University
Hachiohji 192-03 Tokyo
Japan

Dr. Ashraf Ali
Goddard Space Center
NASA
Greenbelt, MD 20771

Dr. Francois Amar
Dept. of Chemistry
University of Maine
Orono ME 04469

Dr. R. Anderson
Physics Department
University of Maryland
College Park, Maryland 20742

Dr. Wanda Andreoni
IBM Research Division-Zurich
Ruschlikon, CH-8803
Switzerland

Dr. Ronald P. Andres
School of Chemical Eng.
Purdue University
W. Lafayette, IN 47907

Dr. Alex Antonelli
Physics Dept.
Box 2000
Virginia Commonwealth University
Richmond, VA 23284-2000

Dr. Anthony S. Arrott
Physics Dept.
Simon Fraser University
Burnaby, B.C.
Canada V5A 1S6

Dr. Jaap Baak
Kamerlingh Onnes Lab.
PO Box 9506
Leiden 2300 RA
The Netherlands

Dr. Wolfgang Bacsa
Physics Dept./Davey Lab.
PA State Univ.
University Park, PA 86802

Dr. C. Baldwin

Dr. Philip C. Ball
Nature
4 Little Essex St.
London WC2R 3LF, UK

Dr. Patrick G. Barber
Dept. of Natural Sciences
Longwood College
Farmville, VA 23901

Dr. Mustansir Barma
Tata Inst. of Fundamental Research
Homi Bhabha Road, Colaba
Bombay 400 005, India

Dr. R. A. Barrio
Instituto de Fisica
UNAM, APDO Postal 20-364
Mexico D.F. 01000
Mexico

Dr. Lawrence S. Bartell
Dept. of Chemistry
University of Michigan
Ann Arbor, MI 48109-4865

Dr. Thomas L. Beck
Chemistry Department
Univ. of Cincinnati
Cincinnati, OH 45221

Dr. Paul S. Bechthold
Inst. fur Festkorperforschung
Postfach 1913
D5170 Julich
Germany

Prof. Peter A. Beckmann
Physics Department
Bryn Mawr College
Bryn Mawr, PA 19010

Dr. Esther Belin
Laboratoire de Chimie Physique
UA 176, 11 Rue Pierre et Marie Curie
75231 Paris Cedex 05
France

Mr. Franck M. Beniere
C.P.M.A. Bat 690
Universite Paris Sud Orsay
Orsay 91405
France

Prof. K. Bennemann
Freie Universitat Berlin
Inst. fur Theoretische Physik-WE2
Arnimallee 14, D-1000 Berlin 33
Germany

Dr. Jerry Bernholc
Dept. of Physics
NC State University
Raleigh, NC 27695-8202

Dr. R. Stephen Berry
Chemistry Department
University of Chicago
Chicago, IL 60637

Dr. Donald Bethune
IBM
650 Harry Road
K34/802
San Jose, CA 95193

Dr. N. D. Bhaskar
Aeorspace Corporation
PO Box 92957
Los Angeles, CA 90009

Dr. Christopher Binns
Department of Physics
University of Leicester
Leicester LE1 7RH
England

Dr. L. M. Bloomfield
Dept. of Physics
University of Virginia
Charlottesville, VA 22901

Dr. V. Bonacic-Koutecky
Freie Universitat Berlin
Inst. fur Physikalische und
   Theoretische Chemie
Takustrasse 3, D-1000 Berlin 33
Germany

Dr. Kit Bowen
Department of Chemistry
Remsen Hall-Dunning Hall
Baltimore, Maryland 21218

Dr. Catherine Brechignac
Laboratoire Aime Cotton
CNRS II, Bat. 505
Campus d'Orsay , 91405 Orsay Cedex
France

Dr. Michel Broyer
Bat 205
University Lyon I
43 Bd due 11 Novembre 69622
Villeurbanne, France

Dr. Jean-Pierre Bucher
IPE, EPFL
PHB-Ecublens
Ch-1015 Lausanne
Switzerland

Dr. Sergei Burkov
McMaster University, IMR
Hamilton, Ontario L8S 4L8
Canada

Dr. Jean Buttet
Inst. Phys. Experimental
PHB Ecublens
Lausanne 1015
Switzerland

Dr. Philippe Cahuzac
Lab. AIME-Cotton CNRS II, bat 505
Orsay Cedex 91405
France

Dr. John H. Callahan
Code 6113, Chemistry Div.
Naval Research Lab
Washington, D.C. 20375-5000

Dr. Alfredo Caro
Paul Scherrer Inst.
Villigen 3232
Switzerland

Mr. Randolph S. Cary
Box 358
Virginia State University
Petersburg, VA 23803

Dr. Sabrina Chase
Dept. of Physics
Penn State University
University Park, PA 16802

Dr. Soumitra Chattopadhyay
223 Physics Building
University of Missouri
Columbia, MO 65211

Dr. S. Chekmarev
Inst. of Thermophysics
630090 Novosibirsk
USSR

Dr. Ho S. Chen
3 Country Place
Lebanon, NJ 08833

Dr. Jian Chen
Dept. of Chemistry
University of MI
Ann Arbor, MI 48109

Dr. J. L. Chen
Dept. of Physics & Astronomy
Univ. of Maryland
College Park, MD 20742

Dr. Hai-Ping Cheng
Chemistry Department
University of Chicago
Chicago, IL 60615

Dr. Pierre Cheyssac
Lab. de Physique de la Matiere Condensee
Universite de Nice-Sophia Antipolis
Parc Valrose, 06034 Nice Cedex, France

Dr. C.L. Chien
Physics Dept.
Johns Hopkins Univ.
Baltimore, MD 21218

Dr. M. Ciftan
Physics Division
Army Research Office
Research Triangle Park, N.C. 27709

Dr. Noel Clark
University of Colorado
Dept. of Physics, Box 390
Boulder, CO 80309

Mr. Brian Constance
Physics Department
Box 2000
Richmond, VA 23284-2000

Dr. Donald M. Cox
Exxon Research & Eng. Co.
Annandale, NJ 08801

Dr. Francoise Cyrot-Lackmann
C.N.R.S.-L.E.P.E.S.
BP 166-25, Avenue des Martyrs
Grenoble 38042 Cedex, France

Mr. George Daly
Chemistry Department
Virginia Commonwealth University
Richmond, VA 23284-2000

1408

Dr. T.P. Das
Dept. of Physics
SUNY at Albany
Albany, NY 12222

Mr. Michael R. Davis
Box 358
Virginia State Univ.
Petersburg, VA 23803

Dr. Walt A. DeHeer
IPE
Ecole Polytechnique Federale de Lausanne
Lausanne, Switzerland

Dr. L. J. de Jongh
Kamerlingh Onnes Laboratory
PO Box 9506
Leiden 2300 RA
The Netherlands

Dr. Theodore S. Dibble
Dept. of Chemistry
University of Michigan
Ann Arbor, MI 48109

Dr. A. Ding
Optisches Institut
Technische Universitat
Berlin
Germany

Dr. Calvin J. Doss
Physics Dept.
VPI & SU
Blacksburg, VA 24061-0435

Dr. David C. Douglass
Dept. of Physics
University of Virginia
Charlottesville, VA 22901

Dr. Kenneth Douglas
Condensed Matter Lab.
Dept. of Physics
University of Colorado
Campus Box 390
Boulder, CO 80309-0390

Dr. Hughes Dreysse
Lab. Physique du Solide
BP 239, Vandoeuvre Les Nancy
France

Dr. Manfred Dubiel
Martin Luther Universitat
Halle-Wittenberg
Fachbereich Physik, PF
O-4010 Halle, Germany

Dr. Frederic J. Dulles
320 S. Huron St. #1
Ypsilanti, MI 48197

Dr. B. I. Dunlap
Chemistry Division
Code 6119
Naval Research Laboratory
Washington, DC 20375-5000

Dr. Richard A. Dunlap
Physics Department
Dalhousie University
Halifax, Nova Scotia B3H 3J5
Canada

Dr. Olof E. Echt
Physics Department
University of New Hampshire
Durham, NH 03824

Dr. A. S. Edelstein
Naval Research Laboratory
Washington, DC 20375

Dr. Takeshi Egami
Dept. MSE
Univ. of PA
Philadelphia, PA 19104-6272

Dr. S. El Shall
Chemistry Dept.
Box 2019
Virginia Commonwealth University
Richmond, VA 23284-2006

Dr. David W. Ewing
Department of Chemistry
John Carroll Univ.
Cleveland, OH 44118

Dr. Pierre Fayet
Inst. of Exp. Physics, BSP
University of Lausanne
CH-1015 Lausanne
Switzerland

Dr. Clarence Finley
3550 7th St. Road
New Kensington, PA 15068

Dr. Martin Foltin
Inst. für Ionenphysik
Univ. Innsbruck
Technikerstrasse 25
Innsbruck A-6020
Austria

Dr. F. W. Froben
Free University
Arnimallee 14
D-1000 Berlin 33
Germany

Dr. Nobuhisa Fujima
Faculty of Engineering
Shizuoka University
Hamamatsu, Japan

Dr. Takeo Fujiwara
Dept. of Applied Physics
University of Tokyo
Hongo-Bunkyo-ku
Tokyo 113, Japan

Dr. Kiyokazu Fuke
Inst. for Molecular Science
Myodaiji, Okazaki 444
Japan

Dr. R. A. Fusina
Code 4694
Naval Research Lab
Washington, DC 20375-5000

Ms. Sunita Gangopadhyay
Physics Department
Univ. of Delaware
Newark, Delaware 19716

Dr. Sergei V. Gaponenko
Stepanov Inst. of Physics
Leninskii pr. 70
Minsk 220602
USSR

Dr. James Garvey
Dept. of Chemistry, SUNY
Buffalo, NY 14214

Dr. Ignacio L. Garzon
Inst. de Fisica, UNAM
PO Box 439027
San Diego, CA 92143-9027

Mr. Horst J. Göhlich
Heisenbergstr 7
Max Planck Inst. fur Festkorperforschung
7000 Stuttgart 80
Germany

Dr. Alan I. Goldman
3213 Oakland St.
Ames, IA 50010

Dr. James L. Gole
Dept. of Physics
GA Inst. of Technology
Atlanta, GA 30332

Dr. Margaret Gorska
Physics Dept.
University of Maryland
College Park, MD 20742

Dr. R. H. Gowdy
Physics Department
Virginia Commonwealth University
Richmond, VA 23284-2000

1410

Mr. V.P. Gregory
Chemistry Dept.
VPI&SU
Blacksburg, VA 24061

Dr. J. Gspann
Univ. Karlruhe
Inst. of Mikroshr.
W-7500 Karlsruhe
Postfach 3640
Germany

Dr. Claude Guet
Dept. de Recherche Fondamentale
   sur les Ions
Centre d'Etudes Nucleaires de Grenoble
85X 38041 Grenoble Cedex, France

Dr. Bernd Gunther
FH-IFAM
Lesumer Heerstr 36
D-2820 Bremen 77
Germany

Dr. Z. B. Guvenc
Chemistry Div.
Argonne National Lab
9700 So. Cass Avenue
Argonne, IL 60439

Dr. Pierre A. Guyot
LTPCM/ENSEEG BP75
St. Martin d'Heres 38402
France

Dr. Tom Haas
Graduate Eng. Program
Virginia Commonwealth Univ.
Richmond, VA 23284-2019

Dr. Hellmut Haberland
Fakultat fur Physik
H. Herderstr. 3
7800 Freiburg, Germany

Dr. Peter A. Hackett
Steacie Inst. for Molecular Science
NRC 100 Sussex Drive
Ottawa, Ontario KIA OR6
Canada

Dr. G. Hadjipanayis
Department of Physics & Astronomy
University of Delaware
Newark, Delaware 19716

Dr. Jurgen Hafner
Inst. f. Theor.Physik
Techn. Universitat
Wiedner Hauptstr. 8/10
Wien A1040
Austria

Dr. Frank Hagelburg
30 Mercer Street
Albany, NY 12203

Dr. Otto F. Hagena
Inst. f. Mikrostruktechnik
Postfach 3640
D-7500 Karlsruhe 1
Germany

Dr. Wolfgang Harbich
IPE, EPFL
1015 Ch Lausanne
Switzerland

Dr. Andreas Hartmann
Dept. of Chemistry
PA State University
University Park, PA 16802

Dr. Y. He
Dept. of Physics, Nanjing University
Nanjing 210008
People's Republic of China

Dr. Arthur F. Hebard
AT&T Bell Laboratories
Room 1D-460
Murray Hill, NJ 07974

Professor Shinohara Hisanori
Department of Chemistry
Mie University
Mie, Tsu 514
Japan

Dr. Kai Ming Ho
Physics Dept.
Iowa State Univ.
Ames, IA 50011

Dr. Herbert Hofmeister
Inst. of Solid State Phy & Elec. Microscopy
Halle O-4050
Germany

Dr. James W. Hovick
2404 Stone Drive
Ann Arbor, MI 48105

Dr. Carmen A. Huber
Naval Surface Warfare Center-R41
10901 New Hampshire Avenue
Silver Spring, MD 20903-5000

Dr. S. Iijima
34 Miyukogaoka
Tsukuba Ibarki 305
Japan

Dr. Tsutomu Ikegami
Dept. of Chemistry
University of Tokyo
Bunkyo-ku
Tokyo 113
Japan

Dr. Z. A. Insepov
Kazakh Polytechnical Inst.
Satpaev Str. 22
Alma-Ata 480013
Kazakhstan

Dr. Neil Isenor
Dept. of Physics
University of Waterloo
Waterloo, Ontario
N2L 3G1 Canada

Dr. Y. Ishii
Dept. of Material Science
Himeji Inst. of Tech.
Kamigouri-Cho, Akou-gun
Hyogo 678-12
Japan

Dr. S. Itoh
Central Research Laboratory
Hitachi Ltd, Kokuburji
Tokyo 185, Japan

Dr. Julius Jellinek
Chemistry Division
Argonne National Lab.
Argonne, IL 60439

Dr. Puru Jena
Physics Department
Virginia Commonwealth University
Richmond, VA 23284-2000

Dr. W. A. Jesser
Materials Science
University of Virginia
Charlottesville, VA 22901

Professor Ping Jiang
Physics Department
Fudan University, Shanghai 200433
Peoples' Republic of China

Dr. Malte Joppien
Laruper Chausee 149
200 Hamburg 53
Germany

Prof. J. Jortner
School of Chemistry
Tel Aviv University
69978 Tel-Aviv
Israel

Dr. F. Kaatz
Naval Research Laboratory
Washington, DC 20375

Dr. Hiroshi Kamiyama
Inst. for Materials Research
Tohoku University
2-1-1 Katahira
Aoba-ku, Sendai 980
Japan

Dr. Manfred M. Kappes
Chemistry Department
Northwestern University
Evanston, IL 60202

Dr. Albert A. Katsnelson
Dep. Solid State Physics
Moscow State Univ.
Moscow 119899
USSR

Professor Ryoichi Kawai
Department of Physics
University of Alabama at Birmingham
Birmingham, AL 35294

Dr. Yoshiyuki Kawazoe
2-1-1 Katahira Aoba-ku
Sendai 980
Japan

Dr. Eric Kay
IBM Almaden Research Center
San Jose, CA 95120-6099

Dr. Koji Kaya
Dept. of Chemistry
Keio University
Hiyoshi, Yokohama 223
Japan

Dr. Shiv N. Khanna
Physics Department
Virginia Commonwealth University
Richmond, VA 23284-2000

Dr. M. Kleman
Laboratoire Physique des Solides
Université Paris-Sud
91405 Orsay Cédex
France

Dr. Walter D. Knight
Physics Department
Univ. of California
Berkeley, CA 94720

Dr. Tamotsu Kondow
Dept. of Chemistry
University of Tokyo
Bunkyo-ku, Tokyo 113
Japan

Dr. Refik Kortan
AT&T Bell Laboratories
1B-314
600 Mountain Avenue
Murray Hill, NJ 07974

Dr. John A. Kovacich
4201 N. 27th St., Dept. H604
Milwaukee, Wisconsin 53216

Dr. Uwe Kreibig
I. Physical Institute
 of the RWTH
Sommerfeld STR. 28
D-5100 Aachen, Germany

Dr. D. Kreisle
Universitat Konstanz
Fakultat fur Physik
PO Box 5560
Konstanz 7750, Germany

Mr. Vitaly V. Kresin
Physics Department
University of California
Berkeley, CA 94720

Dr. Hiroki Kumahora
Energy Research Lab, Hitachi Ltd.
1168 Moriyama-cho, Hitachi-shi
Ibaraki-ken 316, Japan

Dr. Jitendra Kumar
Indian Inst. of Technology Kanpur
Materials Science Program
Kanpur, U.P. 208017
India

Dr. Holger C. Kunstle
Witgreuksweg 383-312
Enschede 7522ZA
The Netherlands

Dr. Pavel B. Kurasov
Manne Siegbahn Inst.
Freskativagen 24
10405 Stockholm
Sweden

Dr. G. Lacueva
Physics Department
John Carroll University
University Heights, OH 44118

Dr. Uzi Landman
GA Inst. of Tech.
Physics Dept.
Atlanta, GA 30332-0430

Dr. J. Lannin
Dept. of Physics
PA State University
104 Davey Lab
University Park, PA 16802

Dr. Thomas Leisner
Physics Dept.
Univ. of Konstanz
W-7750 Konstanz
Germany

Dr. Derek M. Lindsay
Chemistry Dept.
City College
Convent Avenue at 138th St.
New York, New York 10031

Dr. Feng Liu
Ceramic Engineering
Rutgers Univ.
Piscataway, NJ 08855-0909

Dr. Jeffrey Z. Liu
NEC Research Inst.
4 Independence Way
Princeton, NJ 08540

Dr. Maria J. Lopez
Depto. Fisica Teorica
Facultad de Ciencias
Valladolid 47011
Spain

Dr. Bruce Lossee
Philip Morris Research
  Center
Richmond, VA

Dr. Hans O. Lutz
Fakultat fur Physik
Univ. Bielefeld
4800 Bielefeld 1
Germany

Ms. Michelle Lyons
457-5 Lambeth
Charlottesville, VA 22904-0033

Dr. S. D. Mahanti
Physics Dept.
Michigan St. Univ.
E. Lansing, MI 48824

Dr. Mary Mandich
AT&T Bell Labs
Materials Science Div.
ID251
600 Mountain Avenue
Murray Hill, NJ 07974

Dr. Matti J. Manninen
Physics Department
University of Jyvaskyla, PO Box 35
SF-40351 Jyvaskyla, Finland

Dr. Nils Mårtensson
Physics Dept.
Box 530
S-751 21 Uppsala
Sweden

Dr. T. Patrick Martin
Max Planck Inst., FKF
7000 Stuttgart 80
Germany

Dr. Jose L. Martins
Dept. of Materials Science
Univ. of Minnesota
Minneapolis MN 55455

Dr. R. W. McCallum
106 Wilhelm
Ames Lab, ISU
Ames, IA 50011

Dr. Tom McMullen
Physics Department
Virginia Commonwealth University
Richmond, VA 23284-2000

Dr. K.H. Meiwes-Broer
Physics Dept.
University of Bielefeld
4800 Bielefeld 1
Germany

Dr. Hans Micklitz
II Phys. Inst.
Universitat du Koln
Zulpicher Str. 77
Koln D-5000, Germany

Dr. Michael Mitch
Department of Physics
Penn State University
University Park, PA 16802

Dr. D. Mitchell
Physics Dept.
SUNY-Albany
Albany, NY 12222

Dr. Taro Moriwaki
Tokyo Metropolitan University
Minami-Osawa
Hachioji Tokyo
Japan

Dr. Simon C. Moss
Physics Dept.
University of Houston
Houston, TX 77204-5504

Dr. Wolfgang Mueller
29904 Sycamore Oval
Westlake, Ohio 44145

Dr. Mamoun Muhammed
Royal Inst. of Tech.
Inorganic Chemistry
S-10044 Stockholm
Sweden

Dr. Roland Muller
Laruper Chaussee 149
200 Hamburg 50
Germany

Dr. Manu Multani
Tata Inst. of Fundamental Research
Homi Bhabha Road, Colaba
Bombay 400 005
India

Dr. M. L. Myrick
Lawrence Livermore Nat'l Lab.
P.O.B. 808, L-524
Livermore, CA 94550

Dr. Takashi Nagata
Deptl of Chemistry
University of Tokyo
Bunkyo-ku, Tokyo 113, Japan

Dr. Atsushi Nakajima
Dept. of Chemistry
Keio University
3-14-1 Hiyoshi, Kohoku-ku
Yokohama 223, Japan

Dr. Yuichiro Nishina
Inst. for Materials Research
Tohoku University
Sendai 980-91
Japan

Dr. David Noakes
Physics Department
Virginia State University
Petersburg, Virginia

Dr. Fujima Nobuhisa
Shizuoka University
Faculty Engineering
3-5-1 Johoku, Hamamatsu 432
Japan

Dr. Shinji Nonose
Dept. of Chemistry
University of Tokyo
Bunkyo-ku, Tokyo 113, Japan

Dr. Jan A. Northby
Physics Department
University of Rhode Island
Kingston, RI 02881

Dr. Stacie S. Nunes
Electrical Engineering Dept.
SUNY New Paltz
New Paltz, NY 12561

Dr. M. Oda
Vacuum Metallurgical Co.
516 Yokota, Sanbu-cho
Sanbu-gun, Chiba-pref. 289-12
Japan

Dr. Jorge M. Pacheco
Fritz Haber Inst. der MPG
Faradayweg, 4-6
1000 Berlin
Germany

Dr. Fernando Palacio
Inst. Ciencia de Materiales de Aragon
CSIC Universidad de Zaragoza
Pza. de San Francisco s/n
50009 Zaragoza
Spain

Dr. D. Papaconstantopoulos
Naval Research Laboratory
Code 4693
Washington, D.C. 20375-5000

Dr. Denise C. Parent
Code 6113
Naval Research Laboratory
Washington, DC 20375-5000

Dr. Eric K. Parks
Bldg. 200
Argonne National Lab.
Argonne, IL 60439

Dr. Mark R. Pederson
Naval Research Laboratory-4692
Washington, DC 3\20375-5000

Dr. Donald H. Phillips
Mail Stop 234
NASA, Langely Research Center
Hampton, VA 23665

Dr. Kim W. Pierson
8336 Claremont Woods Dr.
Alexandria, Virginia 22309

Dr. S. J. Poon
Physics Department
University of Virginia
Charlottesville, VA 22901

Dr. John Prater

Dr. Mehernosh Press
Tata Inst. of Fundamental Research
Homi Bhabha Road, Colaba
Bombay 400 005
India

Dr. Kathleen A. Puech
Dept. of Pure and Applied Physics
Trinity College
Dublin 2, Ireland

Dr. K. Raghavachari
AT&T Bell Labs
Murray Hill, NJ 07974

Dr. Bijan Rao
Physics Department
Virginia Commonwealth University
Richmond, VA 23284-2000

1416

Dr. K.V. Rao
Dept. of Condensed Matter Physics
Royal Inst. of Tech.
Stockholm, Sweden S10044
Sweden

Mr. Budda Reddy
Physics Department
Virginia Commonwealth University
Richmond, VA 23284-2000

Dr. Ronald Reifenberger
Purdue Univ.
Dept. of Physics
W. Lafayette, IN 47907

Dr. Francois A. Reuse
E.P.F.L.
1015 Ch Ecublens
Lausanne
Switzerland

Dr. Stephen J. Riley
Chemistry Divison
Argonne National Lab
Argonne, IL 60439

Dr. N. Y. Rivier
Blackett Laboratory
Imperial College
London SW7 2BZ
England

Dr. Jack L. Robertson
NIST
Bldg. 235, Room E-151
Gaithersburg, MD 20899

Dr. J. Robles
Dept. de Quimica
Univ. Autonoma Metropolitana, APDO
Postal 55-534, Iztapalapa 09340, DF
Mexico

Dr. Arne Rosen
Dept. of Physics
Chalmers Univ.
S-41296 Göteborg
Sweden

Dr. Johannes W. Roth
Inst. fur Theoretische und Angervundte Physik
    der Univ. Stuttgart
Pfaffenwaldring 657
D-7000 Stuttgart 80, Germany

Dr. Jean-Philippe Roux
Laboratoire Aime Cotton
Bat 505
Orsay Cedex 91405, France

Dr. Vincent Russier
University P. et M. Curie
4 Place Jussieu
75252 Paris Cedex 05, France

Dr. Yahachi Saito
Mie University
Faculty of Engineering
1515 Kamihamacho, Tsu 514
Japan

Dr. D. R. Salahub
Dept. of Chemistry
University of Montreal
Montreal, Quebec
Canada H3C 3V1

Dr. Harry Sarkas
c/o Kit Bowen
582 W. University Parkway
Baltimore, MD 21210

Dr. I. Satoshi

Dr. Klaus Sattler
Physics Program
University of Hawaii at Manoa
2505 Correa Road
Honolulu, Hawaii 96822

Dr. Winston A. Saunders
Cal Tech, Mail Stop 128-95
1201 E. California Blvd.
Pasadena, CA 91125

Dr. Shin-ichi Sawada
Fundamental Res. Labs.
NEC Corporation
34 Miyuki-ga-oka
Tsukuba Ibaraki 305, Japan

Dr. Michael Schluter
6700 Mountain Avenue
Murray Hill, NJ 07974

Dr. Rudiger Schmidt
Fakultat fur Physik
Universitat Bielefeld
4800 Bielefeld 1
Germany

Dr. Robert Scholl
Philips GmbH
Forschungslab Aachen
Postfach 1980
D-5100 Aachen
Germany

Dr. Harlan Schone
Physics Dept.
College of William & Mary
Williamsburg, Virginia

Dr. John C. Schug
Chemistry Dept.
VPI & SU
Blacksburg, VA 24061

Dr. David L. Schutt
Frick Chemical Laboratory
Princeton University
Princeton, NJ 08544

Dr. Adrian Selinger
Chemistry Department
Penn State University
152 Davey Laboratory
University Park, PA 16802

Dr. Martine Sence
IRSAMC-118
Route de Narbonne
Toulouse 31062
France

Dr. Frederick A. Senese
MS 234 NASA
Langley Research Center
Hampton, VA 23665

Dr. Jeffrey E. Shield
47 Wilhelm Hall
Iowa State University
Ames, IA 50011

Dr. Donald Shillady
Chemistry Department
Virginia Commonwealth University
Richmond, VA 23284-2006

Dr. R. W. Siegel
Materials Science Division
Argonne National Lab.
Argonne, IL 60439

Dr. Thomas J. Silva
Physics Dept.
University of Calif-San Diego
9500 Gillman Drive, Box 319
La Jolla, CA 92093-0319

Dr. Richard E. Smalley
Rice University
Chemistry Dept., PO Box 1892
Houston, Texas 77251

Dr. Daniel Solgadi
Laboratoire de Photophysique Moleculaire
Universite de Paris-Sud
Batiment 213, 91405 Orsay, France

Dr. Fernand Spiegelman
11 Rue Bel-Air
Toulouse 31400
France

Dr. P. Steinhardt
Physics Department
University of PA
209 Wo. 23rd St.
Philadelphia, PA 19104-6396

Dr. Valerij S. Stepanuyk
Dept. Solid State Physics
Moscow State University
Moscow 119899
USSR

Dr. Carey E. Stronach
Box 358
Virginia State University
Petersburg, VA 23803

Dr. Satoru Sugano
Faculty of Science
Himesji Inst. of Tech.
Kamigori Ako-gunn 678-12
Japan

Mr. S. B. Sulaiman
Physics Department
SUNY at Albany
Albany, NY 12222

Dr. Alfred C. Switendick
Division 1151
Sandia National Labs.
PO Box 5800
Albuquerque, NM 87185

Dr. G. Bruce Taggart
National Science Fondation
Room 408
1800 G. Street, NW
Washington, DC 20550

Dr. Ikeda Takashi
NEC 34 Miyukogaoka
Tsukuba Ibaraki 305
Japan

Dr. Katsumi Tanigaki
NEC Corporation
34 Miyukigaoka, Ibaraki 305
Tsukuba, Ibaraki 305
Japan

Dr. Rene Tarento
Labo de Physique des Materieuse
CNRS Bellevue, 1 Place Austride Briand
Meudon 92295
France

Dr. Roberto Teghil
Universita Di Basilicata
Dipart. di Chimica
Potenza 85100
Italy

Dr. Jussi T. Timonen
Physics Dept.
University of Jyvaskyla
SF-40351 Jyvaskyla, Finland

Dr. Gerard J. Torchet
Physique des Solides
Bat 510
Universite de Paris-Sud
Orsay 91405, France

Dr. Frank Träger
Fachbereich Physik
Universität Kassel
Heinrich-Plett-Strasse 40
D-3500 Kassel, Germany

Dr. A.P. Tsai
Katahira 2-1-1
Aoba-ku, IMR
Tohoku University
Sendai 980, Japan

Dr. T. T. Tsong
Inst. of Physics
Academia Sinia
Nankang, Taipei, Taiwan 11529

Ms. Tonia Tsoukatos
Department of Physics
University of Delaware
Newark, Delaware 19716

Dr. K. M. Unruh
Physics Dept.
University of Delaware
Newark, DE 190716

Dr. John M. Vail
Dept. of Physics
University of Manitoba
Winnipeg, MB
Canada R3T 2N2

Dr. J. J. van der Klink
IPE EPFL
Lausanne CH-1015
Switzerland

Dr. Benjamin W. Van de Waal
Physics Dept., CT1324
University Twente, POB 217
7500 AE Enschede
The Netherlands

Dr. Daniel Van Der Putten
Kamerlingh Onnes Laboratory
PO Box 9506
Leiden 2300 RA
The Netherlands

Dr. Jan M. Van Ruitenbeek
Kamerlingh Onnes Lab.
PO 9506
Leiden 2300 RA
The Netherlands

Dr. J.L. Verger-Gaugry
LTPCM/INPG, BP 75
Domaine Universitaire
St. Martin d'Heres 38402
France

Dr. Eugenio E. Vogel
Dept. Fisica
Universidad de la Frontera
Temuco
Chile

Dr. A. A. Vostrikov
Inst. of Thermophysics
Siberian Branch, Academy of Sciences
Novosibirsk 630090
USSR

Dr. Cai-Zhuang Wang
Physics Dept.
Iowa State Univ.
Ames, IA 50011

Dr. Chumin Wang
Instituto de Investigaciones
 en Materiales, UNAM
Mexico City, C.P. 04510
Mexico

Dr. Ping S. Wang
Physics Dept.
University of Maryland
College Park, MD 20742

Dr. Q. J. Wang

Mr. Scott Weber
Physics Department
Virginia Commonwealth University
Richmond, Virginia 23284-2000

Dr. Robert L. Whetten
Chemistry Department
University of California
Los Angeles, CA 90024-1569

Mr. Douglas Wright
Chemistry Department
Virginia Commonwealth University
Richmond, VA 23284-2006

Ms. Jie Xhie
Univ. of Hawaii
Dept. of Physics
2505 Correa Road
Honolulu, Hawaii 96822

Ms. Ping Xia
Dept. of Physics
University of Virginia
Charlottesville, VA 22901

Dr. Shimin Xu
Department of Chemistry
University of Michigan
Ann Arbor, MI 48109

Dr. Miguel J. Yacaman
Inst. de Física, UNAM
Apartado Postal 20-364
Mexico D.F. 01000
Mexico

Dr. Isao Yamada
Ion Beam Eng. Experimental Lab.
Kyoto University
Sakyo, Kyoto 606
Japan

Dr. Y. Yamada
Penn State University
Chemistry Department
152 Davey Laboratory
University Park, PA 16802

Dr. Constantine Yannouleas
Physics Department
Virginia Commonwealth University
Richmond, Virginia 23284-2000

Dr. Richard Zallen
Department of Physics
VPI & SU
Blacksburg, VA 24061

Dr. Nurbosyn U. Zhanpeisov
Inst. of Catalysis
Novosibirsk 630090
USSR

Yung K.C. 1171